Series on Directions in Condensed Matter Physics – Vol. 17

INSULATING AND SEMICONDUCTING GLASSES

SERIES ON DIRECTIONS IN CONDENSED MATTER PHYSICS

Published

Series on Directions in Condensed Matter Physics – Vol. 17

INSULATING AND SEMICONDUCTING GLASSES

Editor

P Boolchand

University of Cincinnati

World Scientific
Singapore • New Jersey • London • Hong Kong

Published by

World Scientific Publishing Co. Pte. Ltd.

P O Box 128, Farrer Road, Singapore 912805

USA office: Suite 1B, 1060 Main Street, River Edge, NJ 07661

UK office: 57 Shelton Street, Covent Garden, London WC2H 9HE

British Library Cataloguing-in-Publication Data
A catalogue record for this book is available from the British Library.

Cover Design:
The picture shows a model of an amorphous-Si film grown on a crystalline-Si substrate. It was published in *Journal of Non-Crystalline Solids* **114**, 681 (1989) by Dr. F. Wooten and is reproduced by permission of the author.

INSULATING AND SEMICONDUCTING GLASSES
Series on Directions in Condensed Matter Physics — Volume 17

ISBN 981-02-3673-5

This book is printed on acid and chlorine free paper.

Printed in Singapore by World Scientific Printers

Preface

Most inorganic materials can be rendered noncrystalline by vapor condensation onto cooled substrates, but only a select few can be cooled from the molten state to yield a bulk glass that displays a softening at the glass transition temperature (T_g). A glass is formed by cooling (or supercooling) a melt and is widely regarded as a frozen solution.

The origin of the glass forming tendency has been the subject of discussions and debate in the scientific community for over 60 years. In recent years, the development and maturing of new ideas based on interatomic forces acting as mechanical constraints, the advent of new experimental probes of glass structure and dynamics, and large scale molecular dynamic simulations of covalently bonded networks, is providing, new insights into the morphological structure of network glasses. In Condensed Matter Science, we are used to the notion that atomic scale structure forms the starting point to determine the physical behavior of crystalline solids. One expects that in glassy solids as well, the atomic scale structure will be intimately tied to the physical behavior of glasses, including the glass forming tendency. And while there are important kinetic effects that control the glass formation process, these effects are not decoupled from the structural organization of matter as atomic motion is arrested at T_g. The nature of the glass transition continues to be a subject of profound discussions in basis science.

The scope of this volume is on insulating and semiconducting glasses focussing on inorganic materials that encompass a variety of optical and electronic materials. Glasses based on metallic systems and organic polymers are not covered in this volume, although several reviews on these materials have appeared in the literature.

The book is fortunate to have *Austen Angell* (Chapter 1) and *James Phillips* (Chapter 2) introducing ideas on the nature of the glass transition with the former providing an overview, and the latter commenting specifically on molecular glasses. A more traditional view of glass phenomenology is provided by *Ivan Gutzow* in Chapter 3. Interatomic forces serve as mechanical constraints and play a central role in determining the mechanical and chemical behavior of glassy networks. When a network structure becomes barely rigid, the glass forming tendency is apparently also optimized. In Chapter 4, *Michael Thorpe* reviews ideas on the onset of rigidity in atypical and generic networks. In the real world, structure of glasses are probed by elastic scattering methods [Chapter 5(A)] as reviewed by *Adrian Wright* and by an array of chemically specific nuclear resonance methods that have proved to be rather powerful in decoding local structures. These include the Mössbauer effect reviewed by *Punit Boolchand* [Chapter 5(B)], Nuclear Quadrupole resonance reviewed by *Phillip Bray* [Chapter 5(C)], and Solid State Nuclear Magnetic resonance reviewed by *Hellmut Eckert* [Chapter 5(D)]. New insights into atomic and molecular dynamics in glasses have emerged from an impressive array of vibrational spectroscopies; including inelastic neutron scattering as reviewed by *Ron Cappelletti* [Chapter 6(A)], Lamb-Mössbauer factors as a probe of low frequency vibrational excitations as reviewed by *Punit Boolchand* [Chapter 6(B)], Raman scattering as reviewed by *Kazuo Murase* [Chapter 6(C)] and very low frequency acoustic excitations probed by Brillouin scattering as reviewed by *Claire Levelut* [Chapter 6(D)].

In the early sixties, one discovered that an excess of very low frequency vibrational excitations over Debye-like ones represents a generic feature of the glassy state of matter. Thirty years later these excitations continue to be identified with tunneling states although their microscopic origin continues to pose a challenge. Model systems for such tunneling states are identified in crystalline solids as well. The subject is reviewed by *Siegfried Hunklinger* and *Christian Enss* in Chapter 7. The physics of electrical transport in disordered semiconductors is reviewed by *Harold Overhof and Peter Thomas* in Chapter 8.

Since the early 1990's, the rapid growth of computing power accessible even on small machines, and the development of efficient computing algorithms has made feasible molecular dynamic simulations of covalently bonded random networks, as elegantly illustrated by *David Drabold* in Chapter 9. He develops the interaction from first principles and is able to handle ensembles of typically hundreds of atoms.

Hellmut Fritzsche reviews light induced effects in chalcogenide glasses (Chapter 10), addressing both the scalar and vector effects, an area in which our understanding of the mechanistic details is currently evolving. Chalco-halide glasses, based on alloys of the chalcogens (S, Se and Te) and halogens (Cl, Br and I) with the tathogens (C, Si, Ge, Sn) and the pnictides (P, As, Sb), have gained popularity as low loss infrared optical fiber materials. *Jacques Lucas* has reviewed the subject in Chapter 11.

The economic well-being of a field of scientific endeavor is ultimately tied to the evolution of commercial products. Glass science has been particularly fertile in this regard as illustrated in the last chapter of the book. Apart from xerography and silica fibers, one of the other large applications of glasses resides in rewrittable optical memory discs, now in use in several large sized personal computers, as discussed by *Stan Ovshinsky* in Chapter 12(A). Amorphous semiconductors have also found applications in X-ray imaging with important medical benefits as illustrated by *Safa Kasap* and *John Rowland* in Chapter 12(B). Light induced effects in glasses have also served as the base for digital and holographic storage of information, and its reading in real time as reviewed by *Maria Mitkova* in Chapter 12(C). The antiquated cathode ray tube as a display device may be replaced by several new emerging technologies. One such, based on diamond-like carbon films, utilizes their negative electron affinity, and are finding use as active elements in flat-panel displays as reviewed by *James Jaskie* in the final Chapter 12(D) of the book.

I hope the book will be useful to entry level graduate students, and also to established researchers, interested in glass science and applications of these fascinating materials, in a variety of disciplines in the physical sciences (physics, chemistry and geology) and engineering (materials science and engineering, electrical engineering).

Punit Boolchand
Editor
Cincinnati, Ohio

Editorial Consultants

1. Marty Abkowitz
2. Austen Angell
3. Tom Beck
4. Wayne Bresser
5. Marc Cahay
6. Ronald Cappelletti
7. John deNeufville
8. Hellmut Eckert
9. Hellmut Fritzsche
10. Bernard Goodman
11. Phillip Gradinetti
12. Prabhat Gupta
13. Darl McDaniel
14. John Mackenzie
15. Maria Mitkova
16. Kazuo Murase
17. Amar Nath
18. Robert Pohl
19. James Phillips
20. Kurt Schwartz
21. Craig Taylor
22. Michael Thorpe
23. Adrian Wright
24. Richard Zallen

Contents

1

Glass Formation and the Nature of the Glass Transition

C. A. ANGELL

Department of Chemistry, Arizona State University Tempe,
AZ 85287-1604
caa@asu.edu

Contents

1. Introduction: Questions, Concepts, and Terminology

Insulating and semiconducting glasses constitute a technologically important subset of a much wider field of glassy substances. It is now widely appreciated that, irrespective of the class of liquid considered (molecular, ionic, metallic, or polymeric), there are always some examples in which the familiar process of crystallization does not occur during cooling, and instead brittle glassy solids form by continuous increase of viscosity. This is a circumstance of enormous importance to our technological society, which depends heavily on non-crystalline solid materials for structural, optical, and communications applications. In consequence, the quest for understanding of supercooling and glass formation has commanded an increasing level of increasingly sophisticated attention.

An account of the phenomenology of glass formation from cooling liquids and solutions requires consideration of two major problem areas, which can be summarized by the two questions:

(1) Why do certain metastable liquids fail to generate crystals, the thermodynamically stable state, during cooling at reasonable rates?

(2) Why do liquids, in which crystals fail to form during cooling, quite abruptly become non-diffusive, hence brittle, at temperatures of roughly 2/3 of the expected (or observed) melting points?

In this chapter, the separate intellectual challenges presented by each of these questions are highlighted. Answering the first question requires understanding of nucleation kinetics and crystal growth phenomena, particularly the former, but even a full theoretical account of the kinetics of nucleation and growth will not satisfactorily answer the basic question of "why some substances but not others?" This is a deeper question related to the mechanical stability of crystals and all the factors which finally decide melting points, because it is the viscosity at the melting point, more than any other single factor, which determines crystallization kinetics, hence glassforming ability [1]. Most practical glassforming systems involve more than one component, so a further component of the first question must involve an understanding of stabilization of liquids viz-à-viz crystals by non-ideal mixing in the liquid state.

Answering the second question requires understanding of the diffusion mechanism in liquids and the relation of diffusion to all the relaxation functions and their thermodynamic and mechanical counterparts.

Most of the interest in the "glassy state problem" is at this time focused on the second problem area, so we will give the initial section of the article to presenting some of the key concepts, definitions, and vocabulary encountered in this phenomenology. In the second section, we will return to the first problem area and discuss the origin of slow crystallization for some pure substances and also for mixtures of pure substances, which individually crystallize easily. The next two sections will then deal in depth with the viscous liquid and glass transition phenomenology, thermodynamic and kinetic aspects. We will give some emphasis in the second of these sections to the case of model systems which would seem especially amenable to theoretical treatment because of the simplicity of constitution — atoms joined by single, breakable covalent bonds — and which, despite this simplicity, exhibit all the richness of the viscous liquid phenomenology. A final section readdresses the glass transition as a problem in solid state, rather than liquid state, physics.

1.1. *What is a glass?*

Nowadays, many different sorts of "glasses" are under discussion. The general concept of "glass" relates to systems, which have some degree of freedom that (1) fluctuates at a rate, which depends strongly on temperature or pressure, and that (2) becomes so slow at low T or high P that the fluctuations become frozen. At this point, properties determined by the slow degree of freedom change value more or less abruptly, giving the "glass transition." For instance, in spin glasses, it is the magnetic susceptibility which decreases suddenly as the fluctuations in magnetization freeze in, while with glassforming liquids, it is the heat capacity, compressibility, expansivity, and dielectric susceptibility. Thus the most general definition we can give for a "glass" is as follows:

> "A glass is a condensed state of matter which has become non-ergodic by virtue of the continuous slow-down of one or more of its degrees of freedom."

Satisfying this definition, there are spin glasses and orientational glasses (dipole, quadrupole, and octapole) and vortex glasses, as well as the classical glasses which themselves have now become known as "structural" glasses. Even ordinary crystals can be considered "glasses" by this definition, since, when defect concentrations become frozen-in during cooling (because defect population change require migration of defects to or from the crystal surface — which can be slow), there is a quite sudden change in properties, e.g. the temperature dependence of electrical conductivity in the case of ionic crystals. While this is a simple case as far as the freezing in of the degree of freedom (the defect population) is concerned, it is in fact atypical because the "freezing-in temperature" depends on the crystal size. In the glasses which are getting the most attention today, the time scale for establishing equilibrium is intrinsic to the substance provided the samples are large enough that surface layer effects are negligible: the "defects" or "structural states" are generated internally.

"Structural" glasses are distinguished from most others by the large change in heat capacity (thermal susceptibility) which accompanies the freezing-in of a particular structural state, or defect population. Consequently, the Kauzmann paradox to which so much attention has been given in structural glass studies (as detailed below) is not discussed in other glass physics circles. The only cases of structural glasses which lack the heat capacity jump are those at the "strong" extreme of the overall strong/fragile behavior pattern (see below) and these have been considered the least interesting (at least until recently when "polyamorphism" in strong liquids was recognized [2, 3]). The importance of the heat capacity jump in the phenomenology of "structural" glasses, and the importance of the equilibrated state being kinetically stable enough to study and characterize, are both included in the definition of a "structural" glass preferred by this author:

> "A (structural) glass is an amorphous solid which is capable of passing continuously into the viscous liquid state, usually, but not necessarily, accompanied by an abrupt increase in heat capacity."

Note that this definition relegates most of the "metallic glass" materials to the grey world of "amorphous solids" because, although formed from a liquid (by ultra fast quenching), they crystallize before ever achieving the supercooled liquid state (exceptions are now becoming known). On the other hand, the definition admits many substances produced initially by routes which never involve a liquid state [3].

The various possible routes to the glassy state are summarized in Fig. 1. Note that the non-liquid routes all involve some more or less drastic departure from the initial state, and this complicates systematic analysis of the process.

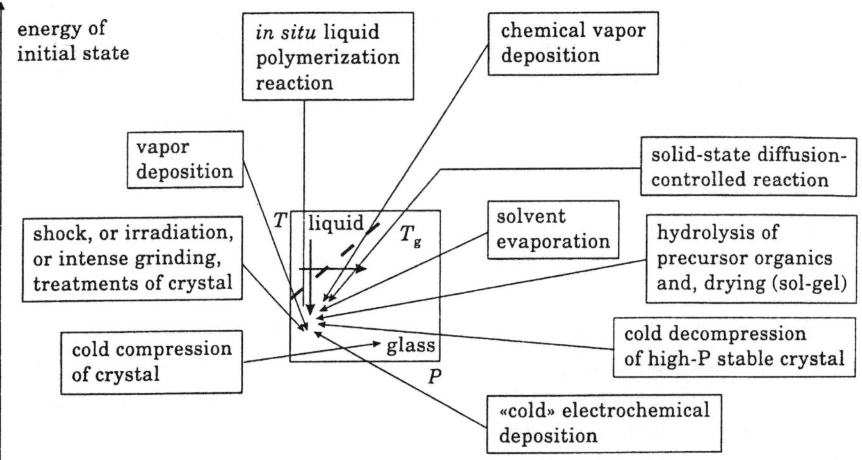

Fig. 1. Various routes to the glassy state, roughly indicating the energies of the initial states relative to the final glassy states. The route of crystal compression below the glass transition temperature (T_g) may yield glasses that are thermodynamically distinct from those obtained by the other route but that may transform to them via non-equilibrium first-order transitions. [Reproduced with permission from Ref. [2] (copyright American Association for the Advancement of Science).]

Thus for the purposes of these lectures, we will restrict attention to structural glasses formed from liquids and furthermore, from liquids in which no changes of composition occur during vitrification.

1.2. *Ergodicity-breaking and the glass transition*

The freezing-in of a structural state during cooling of the liquid, means that the state of internal equilibrium possessed by the initial liquid is lost. To modernize the classical description, "vitrification," this process has recently [4] been called "ergodicity breaking." This is because states at temperatures above the glass transition satisfy the ergodic hypothesis of statistical mechanics [5] (i.e. systems in equilibrium in the course of fluctuations revisit, or may revisit, the same state within the observation period), hence are called "ergodic states." Glasses are "non-ergodic."

The definitions given above are both consistent with the latter statement. However, it should be recognized that common usage among physicists is obscuring this distinction. There is a strong tendency to use the term "glassy" for any aspect of a system capable of generating a glass on sufficient cooling.

Thus one sees frequent reference to (a) "glassy dynamics" — by which is usually meant the dynamics of viscous *liquids* approaching the glass transition temperature from above, and (b) "strong, or fragile glasses" in reference to phenomenology of supercooled liquids above T_g. While such a blurring of the distinction between liquids and glasses is regrettable, it may be inevitable.

The crossover from liquid to glassy (ergodic to non-ergodic) behavior occurs over a range of temperature called the "glass transformation range." This is inevitable for a kinetically controlled phenomenon, and it is probably inappropriate that the term glass "transition" is used to describe the ergodicity-breaking phenomenon. The term is, however, firmly entrenched, and there is little question of changing it. Because of the range of temperature involved and also because of its dependence on cooling rate, there is a problem in attaching a characteristic "glass transition temperature" to the phenomenon for any particular material. This is compounded when, as is commonly the case, the "glass temperature" T_g is determined during heating because there is then a further (poorly appreciated) dependence on annealing history as well as on heating rate. Nevertheless, when appropriately measured, T_g is very reproducible and has become recognized as an important material parameter. We now discuss its assignment.

T_g has traditionally been defined (unambiguously, except for measurement difficulties) as the temperature at which the viscosity reaches 10^{12} Pa.s. In inorganic glasses of technological interest, this temperature was observed to correspond more or less with the temperature at which a break in the length-temperature relation for glass rods was observed (due to the higher expansion coefficient in the ergodic state). Partly because this correspondence was totally lost in the case of high polymers, and partly because of the simplicity of detection of the "transition" by thermal measurements sensitive to changes in heat capacity, interest then passed to definitions based on differential thermal analysis DTA or differential scanning calorimetry DSC scans [6, 7]. A case is illustrated in Fig. 2 for the popular model system CKN [8, 9, 10] (which consists of simple ions in the proportions 40% $Ca(NO_3)_2 \cdot 60\%$ KNO_3), when scanned at the standard rate of 10 K min^{-1} during cooling and subsequent heating *without* any "annealing" ("waiting" in spin glass terminology, "aging" in polymer terminology) below T_g.

The most commonly used definition of T_g is the point marked "$T_{g,\mathrm{EM}}$" (EM for ergodicity-making) which is called the "$C_{p,\mathrm{onset}}$" definition [11, 12]. This

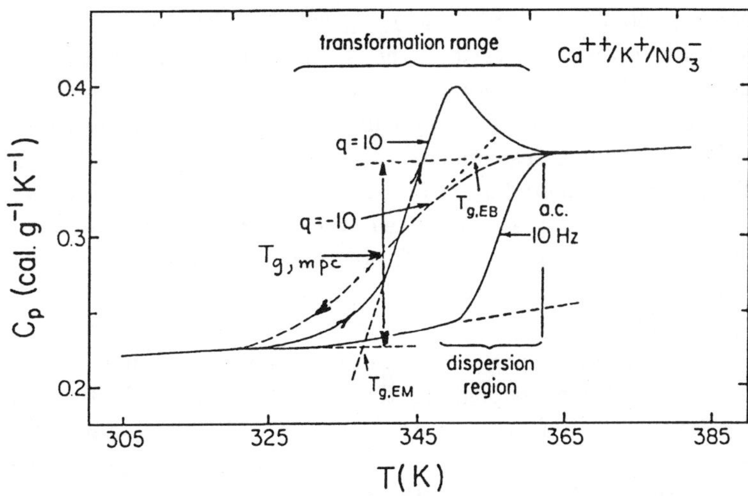

Fig. 2. Heat capacity vs. temperature relations for CKN showing how $T_{g,\mathrm{EM}}$, $T_{g,\mathrm{EB}}$, midpoint cooling glass temperature and $T_{g,\mathrm{mpc}}$ are defined from heat capacity upscan and downscans respectively. (Based on original data from Ref. [8] and including estimated a.c. heat capacity dispersion curve at constant frequency of 10 Hz from C.A. Angell, L.M. Torell, *J. Chem. Phys.* **78** (1983) 15 which has since been closely confirmed by experiment). Note that the "dispersion range" in an a.c. experiment is not usually the same as the "transformation range" observed in a cooling experiment in which the final state is a glass, i.e. a non-ergodic state.

corresponds to the temperature at which molecular liquids have viscosities of $\sim 10^{10}$ Pa.s (at this temperature, polymers have higher and often unmeasurable viscosities depending strongly on polymer chain length). Another commonly used definition is the "C_p, midpoint" determined during heating [12], where the viscosity is $\sim 10^9$ Pa.s. All of these temperatures depend on the precise manner in which the system under study was prepared. This is due to the sensitivity of the heat capacity evolution during heating to the initial enthalpy of the non-ergodic state being heated [13].

The only unambiguous definitions of the glass transition temperature are those which depend only on the cooling rate e.g. the volume crossover on cooling advocated by Plazek [14], the Hodge cooling fictive temperature [13], and the ergodicity breaking (EB) point defined in Fig. 2 from heat capacity data during steady cooling. The "fictive" temperature [15] as defined by Moynihan and co-workers from heating scans [9] does not depend on heating rate but does depend on annealing history after an initial EB on cooling.

For a cooling rate of 10 K/min, the $T_{g,\text{EB}}$ of Fig. 2 falls at a relaxation time of $\sim 10^{-1}$ sec, rather shorter than the $\sim 10^2$ s characteristic of the normal 10 K/min C_p onset (or ergodicity-making EM) temperature. The fictive temperature from equal 10 K/min cooling and heating rates lies close to the latter. Since scanning instruments can now be calibrated during cooling, using liquid crystal mesophase transitions which have negligible hysteresis [V Velikov, Q Lu, CA Angell, submitted], the only disadvantage of defining T_g during cooling is that, for liquids with weak glass transitions, it is difficult to detect. Since most of the interest in the glass transition problem currently is with "fragile" glassformers which tend to have strong C_p manifestations [16], this problem is not too important. However, because the temperature interval $T_{\text{EB}} - T_{\text{EM}}$ will depend strongly on the liquid character [11] (strong or fragile — see below and Ref. [16]), it would be better if the downscan T_g could be defined so as to coincide more closely with the common upscan definition. Thus a preferred choice might be the C_p midpoint temperature on cooling, illustrated in Fig. 2, a choice which would correspond closely with Plazek's volume crossover T_g [14] and Hodge's cooling fictive temperature [13]. The relation of these arbitrarily defined temperatures to a more fundamental glass transition temperature underlying the kinetically determined phenomena, will be considered later.

Before leaving this introductory section, the "fictive temperature" mentioned above should be elucidated because it is an important relative of the glass transition temperature and plays a role in phenomenological descriptions of the glass transition. It is a way of characterizing the "state" of the system, which is preserved when ergodicity is broken. Defined by enthalpy measurements, the fictive temperature of the glass is the temperature at which the ergodic system has the same enthalpy as the glass. The fictive temperature of a glass may be lowered by annealing.

Unfortunately, because different degrees of freedom in supercooled liquids can have different relaxation times, and hence annealing rates, the fictive temperature defined by measurements on one variable, e.g. enthalpy, may not correspond exactly with that defined by measurements on another, e.g. volume, though they may be close [17]. This reflects simply that no single order parameter description of the glass transition phenomenon is applicable.

Finally, in this section, it should be pointed out that, while most of the non-vibrational heat capacity is lost at the glass transition, a part (which is important to developing micro-heterogeneous models of the supercooled liquid state) remains, and is only lost at much lower temperatures. This is a weak component associated with a fast process (or processes) which is less cooperative than the principal relaxation and has different dynamic character. These are called secondary relaxations (β-, γ-relaxations) to distinguish them from the primary (α-) relaxation which carries most of the thermodynamic strength. They have been long known in polymer science where they are usually associated with side chain motion, but their general occurrences in simple glasses was not expected until Goldstein [18] predicted them as a necessary consequence of his landscape picture of glassforming systems. Their calorimetric characterization is a recent development [19, 20, 21]. In many cases, they appear to be the continuation of the high temperature relaxation process before any cooperative processes set in [21], but this matter can only be dealt with properly after the subject has been further developed.

2. Origin of Glassforming Ability

Here we look into the question of why glasses exist and how they can be formed when needed. After all, they shouldn't exist. In almost every case, they are *metastable* with respect to the crystal state of the same composition (if it is a one-component system) or a combination of crystals (if it is a multi-component system). This means they only exist because the system did not have time to reach the lowest available free energy surface. "Not having time" means that crystals do not have time to nucleate and grow during the process of cooling the liquid to below the glass transition temperature. For crystals to nucleate and grow, fluctuations in order must occur, and these will occur more slowly the more viscous the liquid. As noted earlier, liquids, which vitrify easily, are almost always rather viscous at their freezing temperatures.

Let us try to understand why this happens, so we can know how to *make* it happen if we want to.

Figure 3 shows the free energy vs. temperature relations for a hypothetical system which has a variety of possible crystal states of different lattice energy (the lattice energy is G at $T = OK$ when the reference state is the dilute gas state). For simplicity, we assume the same heat capacity-temperature relations, hence the same curvature of G. The liquid phase, which has higher entropy, is drawn to cut across each of the crystal curves. The intersection of G(liquid) with each G(crystal) determines the melting point each would have if transitions between crystal polymorphs were excluded. Clearly the polymorph with the poorest crystal packing (smallest lattice energy) has the lowest melting point, and would accordingly melt to the liquid of highest viscosity.

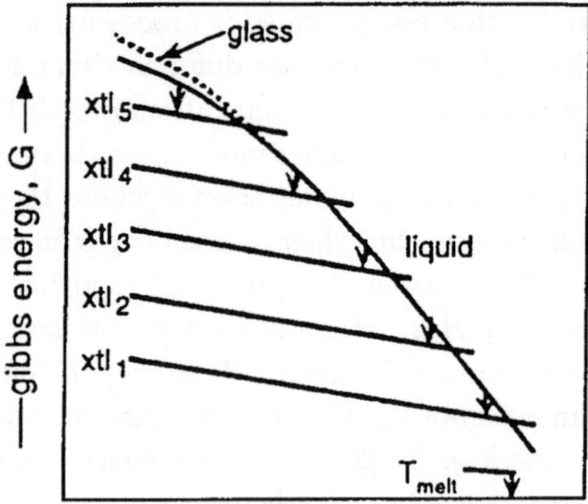

Fig. 3. Schematic variation of Gibbs free energy (G) with temperature, showing access to different metastable crystals from the supercooled liquid. Slower cooling rates move the glass transition to lower temperature, permitting access to more metastable crystalline forms.

From this "thought" example, we would conclude that systems which vitrify easily should be those which lack any large lattice energy crystal packing arrangements. We see, by this argument that the ability of a substance to vitrify is determined in the crystalline state, not in the liquid state as is often supposed.

To support this line of thought, we consider the case of the three isomers of the disubstituted benzene, xylene. The liquids all boil within a few degrees of each other, hence we know that the attractive forces between the molecules are roughly the same, and consistent with this, the viscosities of the three isomers measured at the same temperature are very similar. However, only the meta-isomer is observed to form a glass on fast cooling [22, 23]. At a simple level, the explanation follows immediately on examination of the melting points of the three isomers — the meta-isomer has quite the lowest fusion point (185 K compared with 211 K for the *o*-isomer and 216 K for the *p*-isomer). This means that, consistent with our earlier generalization, the meta-isomer has the largest viscosity at the temperature at which it becomes thermodynamically metastable. Our problem, therefore, becomes one of explaining why the meta-isomer should have the lowest melting point of the three.

The explanation according to our argument above should lie in the difficulty of packing the meta-isomer molecules in long range three-dimensional order, and in this case, we can prove it by calculating that there is a smaller crystal lattice energy for the meta-isomer than for either of the others. This is done in Fig. 4 using precise thermochemical data available for these substances [24]. The difference in lattice energy responsible for the ~ 70 K difference in melting points of m- and p-isomers is only ~ 1 kcalmol^{-1}.

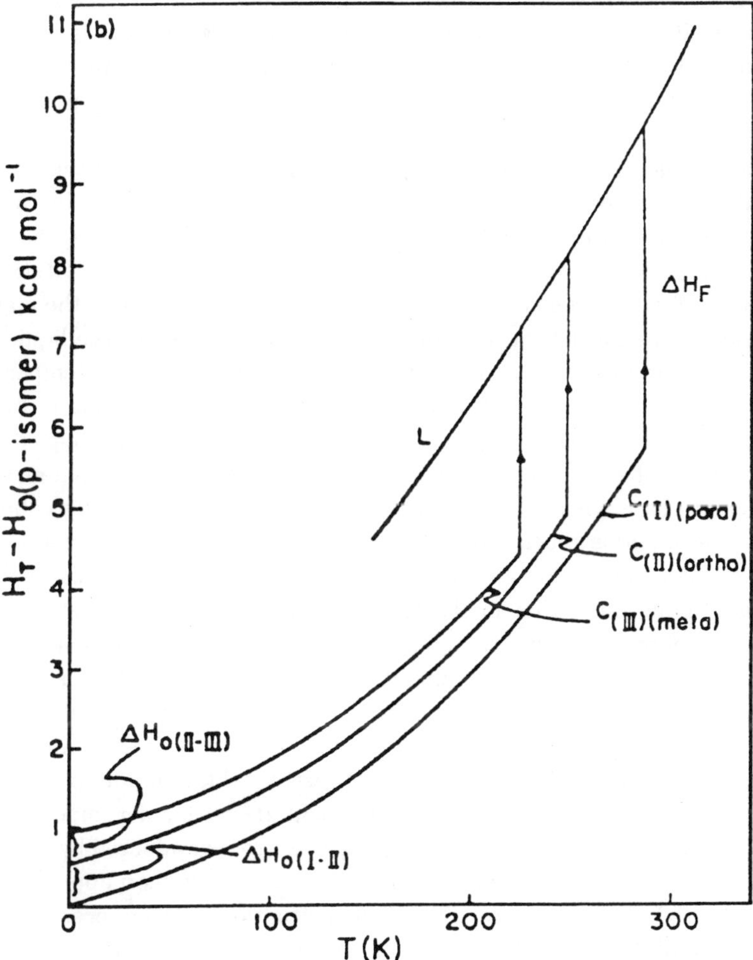

Fig. 4. Enthalpy vs. temperature relations relative to enthalpy of p· xylene at *OK* for crystal and liquid isomers of xylene, showing how smaller lattice energy (less negative H_o) due to packing problems in case of m-isomer leads to low fusion temperature, hence greater glassforming ability for this isomer.

In summary, therefore, we attribute the stability of certain materials in the amorphous state to the failure of nature to find suitable solutions to the three-dimensional long-range order-packing problem in the case of these substances.

The probability of glass formation in such cases will clearly be enhanced by the formation of liquid mixtures, in which the free energy of the liquid phase is decreased, while that of the phase which is to crystallize remains unchanged in the general case in which the crystallizing solid is pure. The more strongly the components in the mixture interact, the lower the activity coefficient of the solvent, hence the more rapidly the freezing point of the solvent will be depressed. The more the freezing point is depressed, the more viscous becomes the liquid at the point (the liquidus) where thermodynamic instability is reached, hence the slower the growth of nucleogenic fluctuations, and the less likely the crystallization.

However, if attractive interactions become too great, a new crystal structure will become favored. The more stable the new compound formed, the higher in temperature one must heat it before the increased entropy of the liquid state leads to free energy crossover, hence fusion. The higher the fusion temperature, the less viscous the liquid at the liquidus temperature and the more probable the crystallization of the *compound* on cooling. The most favorable situation for formation of glasses on cooling of binary solutions will therefore be the intermediate case where the solvent-solute interactions are strong enough to depress the liquidus much more than calculated for ideal solutions, but not strong enough to generate a new and competitively crystallizing compound.

It is usually found that in the glassforming region of such systems, there are one or more crystalline compounds with very low melting points, implying poor structures. The optimum situation for glass formation is when the exothermic mixing is strong enough to produce a compound, but only just. This maximizes the melting point lowering — thus we name this approach to finding glassforming compositions as the "barely stable compound" approach. This implies that, rather than multi-component glasses having some especially stable amorphous state arrangement of particles (to be evaluated by scattering techniques, and appreciated for its stability, as is the common approach to interpretation of glass structure). The actual situation is the opposite. Glasses form as a weak, disordered reflection of a three-dimensional crystal structure, which is so energetically incompetent that it can barely compete with the disordered form at 0 K.

It is quite consistent with the above line of thought that, in metallic systems, those which vitrify most easily are those which do not show the existence of any stable binary compounds, but which have strong solvent-solute interactions, hence deep eutectic temperatures. A good example is the system Au–Si [25]. It is furthermore consistent that, during warm-up of glasses formed during rapid quenching of the binary liquids, crystalline *compounds* can be generated which are metastable, have no thermodynamically stable range, and later (at higher T) decompose to a combination of the components (or of one component plus a binary solution).

We illustrate the above argument by showing in Fig. 5 a representative series of possible phase diagrams for the system $A + B$ in which the strength of the $A - B$ interaction increases systematically in the sequence $1 - 8$. In the unlikely event that the viscosity were to be independent of composition in

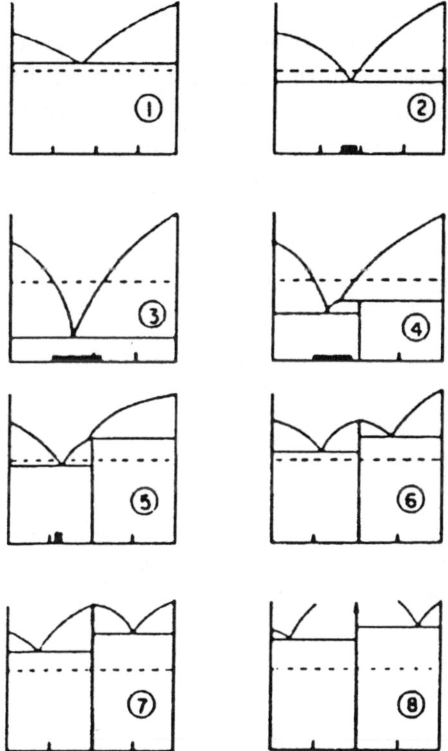

Fig. 5. Illustration of how effects of increases in solute-solvent interaction may influence liquidus temperatures in binary systems, hence glassforming composition ranges (hatched bars) and glass stability within that range.

all these cases, hence a function only of temperature, then the glassforming composition ranges in the above systems could be related straightforwardly to the position of the liquidus curves, and for a fixed cooling rate would be those indicated in the diagram. Although the glassforming range would be similar in cases 3 and 4, system 3 would have the most stable glass composition (near the deep eutectic). A dependence of viscosity on composition would modify these considerations quantitatively. Nevertheless the basic ideas suffice to qualitatively explain the glassforming ranges observed in many binary system families of limited glassforming ability, e.g. the $BiCl_3$-alkali chloride glasses described by Topol *et al* [26, 27]; nitrate glasses [28], and calcium-aluminate glasses [29]; in the latter case, melting point lowerings of ~ 1000 K are involved.

To the extent that the above line of reasoning is valid, theories for stable glassforming ranges in multi-component systems can be regarded as theories for free-energy lowering of given solid phases on dissolution of solutes, as a function of solute character. Even the famous Zachariasen rules for glassforming substances may be regarded as rules for predicting structures which have lattice energies which are weakly competitive with their amorphous equivalents, hence which have low melting points relative to the forces acting between the particles. There are now known a great many inorganic glassforming systems, usually mixtures satisfying the above "barely stable compound" principle, in which the "glassformer," e.g. ZrF_4 [30], does not conform to the Zachariasen rules.

The understanding of glassforming ability, which we are promoting here, can be summed up by the banality, "glasses form when crystals do not."

Now it is time to return to the central thermodynamic problem presented by the behavior of many of the liquids, which fail to crystallize, as they approach the glass transition temperature. It is primarily a problem of the so called "fragile" liquids. These are also the liquids which exhibit viscosities and relaxation times which tend to diverge at finite temperatures, thereby providing the theoretician with a challenge which has withstood decades of enquiry.

3. The Kauzmann Paradox and the Potential Energy Hypersurface

In Fig. 6, we reproduce Kauzmann's original presentation of the entropy problem in supercooling liquids; Kauzmann used the difference in entropy between the crystal and liquid at the melting point, ΔS_f, as a scaling parameter to permit a comparative display of the manner in which the difference in

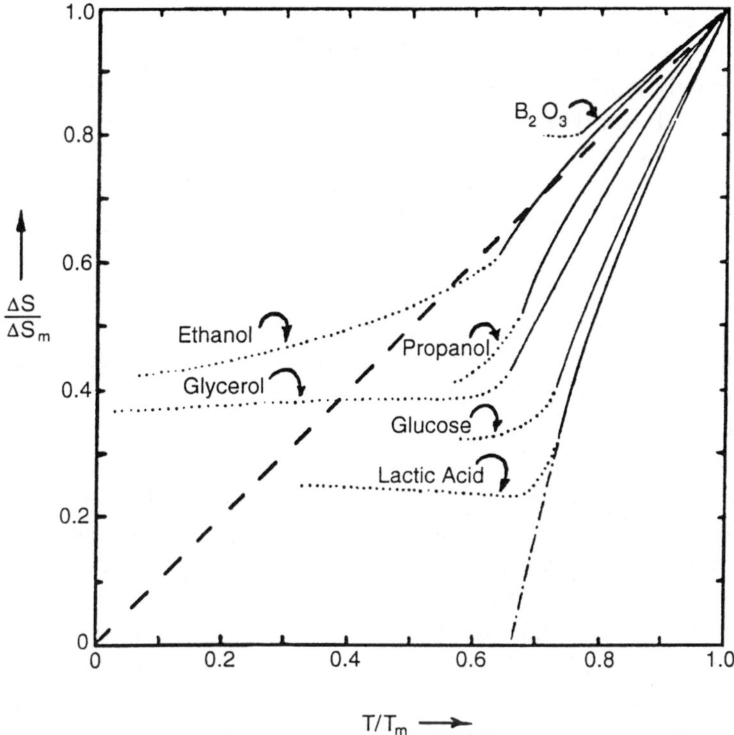

Fig. 6. Kauzmann's presentation of the entropy crisis which bears his name. The figure shows the rate at which the difference in entropy between liquid and crystal, normalized at the fusion point, disappears as T is lowered towards absolute zero. For B_2O_3, now known as a "strong" liquid, the liquid would always be of higher entropy than the crystal, even if the glass transition did not intervene, at high T/T_m, to change the heat capacity. At the other extreme, lactic acid loses its excess entropy so rapidly on cooling that if T_g did not intervene to arrest the loss, liquid would arrive at the same entropy as the crystal at 2/3 of the melting point. [This is the temperature usually associated with the temperature of the glass transition itself (the 2/3 rule which this set of data only weakly support).] Lactic acid is an example of a "fragile" liquid. Other examples of these plots for fragile liquids are given in Refs. [3, 31].

entropy between the liquid and crystal states, for six different substances, varies during their supercooling. All cases show positive slopes, reflecting the fact that liquids have higher heat capacities than the corresponding crystals. What is interesting is the magnitude of the slope and the related extrapolation to zero excess entropy.

In the case of boron trioxide, now much spoken of as a "strong" liquid [16, 31, 32, 33], the slope is such that the excess entropy of the liquid over crystal is only tending to disappear in the vicinity of 0 K — which raises no concern at all. On the other hand, to various degrees, the other liquids in the figure show provocative behavior. In the case of lactic acid, which we would now call the most "fragile" of the six, the excess heat capacity of the liquid over crystal is causing the excess entropy to decrease so quickly that a simple extension of its behavior to lower temperatures would lead that excess to vanish at a temperature which is only $\sim 2/3$ of the fusion temperature. As far as can be told from the data, all that prevents this at-first-sight-mind-boggling *thermodynamic* inversion from occurring, is the occurrence of a *kinetic* phenomenon, the glass transition (i.e. the trapping) at the temperature T_g — hence the term "paradox."

Two quite profound theoretical problems are presented by the data of Fig. 6. The first [34, 35] is the problem of constructing an equilibrium theory for the liquid state which contains an explanation of how, on infinite time scales, the system evolves so as to undergo a rather abrupt, if not singular, change in heat capacity at some temperature between the glass transition temperature and absolute zero. Part of this problem involves the interpretation of fragility of liquids and the coupling of vibrational to configurational degrees of freedom. The second [34, 36] is the problem of constructing a theory which explains in a satisfying manner the reason why, in every case known, the *kinetic* characteristics of the liquid (which can to first approximation be represented by its diffusivity) evolve with temperature in such a way as always to generate equilibration times of the order of experimental time scales *before* the thermodynamic crisis arrives.

Before considering these further, let us consider what is implied about the topology of the potential energy hypersurface, which must be representative of phases exhibiting this type of behavior. Here we merely rephrase much of what was written by Gibbs [34] and his contemporary Goldstein [18] in articles written some 25 years ago.

The fact that glasses are brittle solids at temperatures below their glass transition temperatures implies that the arrangement of particles taken up as a liquid cools below T_g can be described by a point in configuration space near the bottom of a potential energy minimum in this space [18, 37]. If this were not so, the system would move in the direction dictated by the collective unbalanced force acting on it, and some sort of flow would occur. Notwithstanding the legend about medieval cathedral windows [38], this does not occur in glassy systems held at temperatures less than half their glass transition temperatures, even on geological time scales. On the other hand, the existence of the annealing phenomenon, in which the density and energy of a glass formed during steady cooling can change with time on holding at a temperature below but close to the "glass transition temperature" means that there is more than one such mechanically stable minimum available to the system. Indeed, there would appear to be an almost infinite number, of order e^N, where N is the number of particles in the system [39, 40, 41]. The minima, or "basins," obviously are distributed over a wide range of energies, usually scaling with density. However, there are also many ways of organizing the same collection of particles into minima which differ negligibly in energy from one another. Each minimum is a configurational microstate, or "configuron" of the system. Each laboratory "glass" is thus a palpable configuron, though there is evidence that the distinguishing properties of a glass, viz. its structure, heat capacity, and diffusivity are determined within very small quasi-independent regions, so that the properties of a glassformer can be evaluated by consideration of only a tiny subset of its particles (provided no free surfaces are permitted).

The fact that annealing proceeds more slowly the lower the temperature at which the annealing is carried out suggests that the process of finding deeper minima becomes more difficult statistically as the temperature is decreased. One arrives at the notion of an interconnected series of minima on a landscape of inconceivable complexity, in which increasing depth is associated with decreasing configuron population, see Fig. 7 {from Ref. [16(b)]}. The important implication of Kauzmann's presentation in Fig. 6 is that for each system, at least for each "fragile" system, there must exist a statistically small number of minima at energies still well above that representing the crystal and that these must set an absolute limit on the energy lowering achievable by annealing the amorphous system. It is into one of these last few minima that the system is tending to settle at the temperature where the excess entropy is tending to vanish.

The temperature characteristic of this ground state for amorphous packing has become known as the Kauzmann temperature for that system, although the issue of the absolute value of the entropy relative to that of the crystal, which would be appropriate for a ground state system, remains unresolved. The issue is addressed in an important new paper by Speedy and Debenedetti [41] who succeed in evaluating quantitatively the number $\exp[(1.2N)]$ and distribution of minima (each of which is aptly termed a "glass" by these authors) for a model tetravalent system [42]. This is an "inherent structure" [39] analysis which

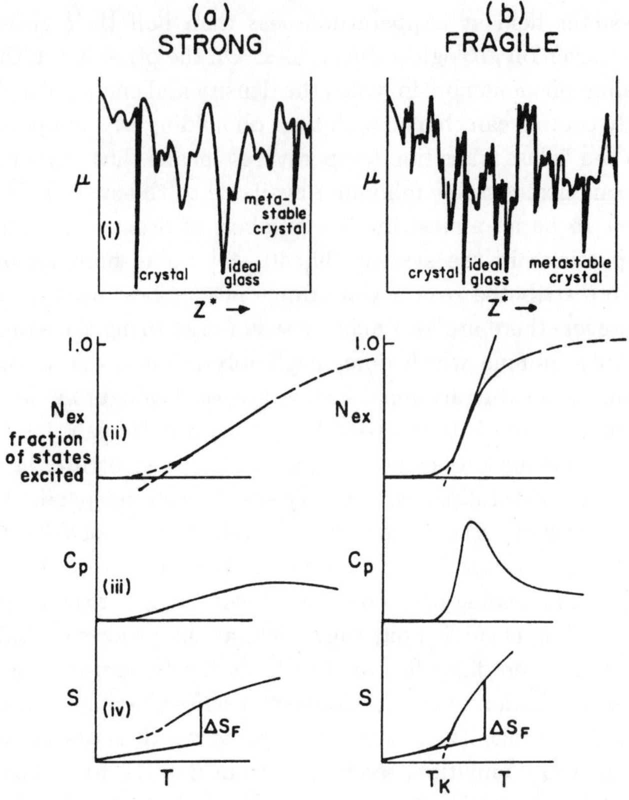

Fig. 7. Sections through the $3N + 1$ dimensional energy hypersurfaces of (a) strong and (b) fragile liquids. Differences in the "density of minima" can be understood at a very elementary level in terms of two-state models in which there are small increases (strong) or large increases (fragile) in the number of distinct packings per elementary bond-breaking event [16(b), 43(a)]. The consequences are represented in parts (ii), (iii), and (iv) of the figure for (ii) level of excitation, (iii) configurational heat capacity, and (iv) immediacy of the Kauzmann crisis, respectively {from Ref. [16(b)] by permission}. The difference between the two types of behavior is currently being associated with the difference in the entropy excited in the elementary excitation event which raises the system point on the energy landscape by an incremental amount, [see Refs. 43(b)].

succeeds in obtaining a quantitative description of the liquid thermodynamics in terms of the inherent structure, and which appears to identify the density of the "ideal glass" for the model considered. The authors argue that this ideal state would be reached by a second order transition during slow (but not necessarily infinitely slow [41]) densification, and conclude it would have an entropy in excess of the crystal at the same PV_o/NkT. Simple models of the two-state variety [43, 44] would suggest the ideal glass would only be approached at 0 K despite linear extrapolations which would indicate otherwise [see Fig. 7(ii)]. Stillinger has however, disputed the existence, in principle, of an ideal glass state [45].

The more minima per unit of energy, the larger the configurational component of the total heat capacity should be, hence the larger the drop in C_p observed at the glass transition when ergodicity is broken during cooldown. Of course, the change in C_p reflects the density of minima at the level of the landscape at which the system gets trapped during the cooling process, and this level will depend on the height of energy barriers separating the minima as well as the total degeneracy, as will be described in the next section. Both features of the landscape derive from the nature of the interaction potential for the particles, but our knowledge of exactly how is in an elementary stage [39, 40, 41]. It is commonly found that fragile liquids, like the lactic acid of Fig. 6, have large changes in heat capacity at their T_g, implying highly degenerate landscapes even quite close to T_K.

It is simpler to discuss the potential energy hypersurface as opposed to the chemical potential (free energy) hypersurface, but it may also be less fruitful. For each interaction potential, there must exist a single immutable potential energy hypersurface for an N particle system; however, the free energy hypersurface will depend on temperature. While a hard sphere system will have a potential energy hypersurface, which is totally degenerate — all configurations have zero energy — all configurations do not have the same *free* energy except at absolute zero. The different configurations have different amounts of spare volume, and the free flight motion of spheres adjacent to such "loose spots" provides an entropy-generating mechanism. Thus different packings yield different entropies and hence different *free* energy minima. These have been evaluated by Dasgupta using a density functional approach [46]. Since we are most commonly interested in the behavior of systems at different temperatures, it seems that the most relevant hypersurface is the chemical potential hypersurface. In any case, both of these hypersurfaces are quite impossible to conceptualize given that they exist in a space of dimensionality of the order of the number of particles. The attempt to represent them by two-dimensional slices, such as illustrated in Fig. 7, is a grotesque, but frequently practiced, oversimplification. The community only permits it because it provides a way of thinking about such problems as the annealing of glasses and the configurational entropy of disordered systems.

The "relative height of the landscape," T_{top}/T_K, can be estimated, as described in Ref. [47], by assuming that the *increase* in heat capacity observed at the glass transition ΔC_p is all due to access, above T_g, to the configurational microstates represented by the minima of Fig. 7, and integrating from the temperature T_K (where the system at equilibrium would settle into the lowest state on the hypersurface) to the temperature at which all the states are populated and the full entropy of the landscape ($k_B \ln W$) is excited. This requires knowledge of the total number of states for the N-particle system and the form of ΔC_p with T. Since the number of states is believed to exponentiate with the number of heavy atoms, $W \approx e^N$, the entropy of the landscape is approximately R entropy units per mole of heavy atoms. For each of two reasonable alternative $\Delta C_p(T)$ functions, T_{top}/T_K is found [47] to be about 1.6 for fragile systems [43(b), 43(c)]. This value is close to the T_K-reduced temperature at which a number of lines of investigation indicate a dynamical crossover of some type occurs in the liquid.

To be in equilibrium (except with respect to crystallization), a system must be able to visit, move between, a representative subset of the minima characterizing its chemical potential hypersurface. In this case, it is not correct to speak, as is often done, of a system well below its glass transition temperature as existing in any one of a manifold of "metastable states." These are "mechanically stable" but not "metastable" states. When held at a temperature near but below the glass transition temperature established during normal cooling, the system "anneals" by exploring the lower energy minima which were inaccessible time-wise during the initial cooling. Since, for most potentials, these are minima in which the particles are more densely packed, the glass volume will usually decrease during annealing. Even at constant density, however, annealing can occur since, in a complex system, configurations of different energy may have the same volume. Equilibration is made simpler by the fact that a macroscopic system consists of statistically independent nanoscopic regions such that the mean distance moved, by a single particle during equilibration, is only a fraction of a molecular diameter [48].

Annealing occurs more slowly at lower temperatures for two reasons. Firstly there are energy barriers to be crossed in the process of passing from minimum to minimum — the system collectively vibrates for increasingly long periods before some chance fluctuation (associated with exceptionally anharmonic excursions on the parts of some particles) permits it to rearrange, hence to enter a new minimum. Secondly, there are entropic barriers because the lower energy minima are more distantly spaced, hence the rearranging group must be larger or the number of successive rearrangements which must be made in order to arrive at one of lower energy states must increase as temperature decreases.

This line of thought was made more quantitative by the development of the "entropy theory" of Adam and Gibbs [36], and the related entropic relaxation argument by Mohanty *et al* [49]. We now discuss their usefulness in more detail.

4. Relaxation and Entropy

The Adam–Gibbs theory, which was based on a modification of conventional transition state theory to accommodate the notion that, in viscous liquids, the rearrangements over energy barriers must be cooperative, led to an expression for the relaxation time which contains the excess (configurational) entropy, S_c of the Kauzmann paradox, in the exponent denominator

$$\tau = \tau_o \exp\left(\frac{C'\Delta\mu}{TS_c}\right) \tag{1}$$

where $\Delta\mu$ is the conventional free energy barrier to rearrangements, and C' is a constant. The familiar departure from Arrhenius behavior comes from the temperature dependence of S_c which itself depends on the value of the configurational heat capacity. This is manifested at the glass transition by the change in heat capacity ΔC_p. Figure 8 depicts the connection between hypersurface topology and non-Arrhenius relaxation via heat capacity according to Eq. (1).

4.1. *Relaxation in the non-ergodic state*

For a fixed value of the configurational entropy, Eq. (1) predicts that such relaxations as may be observed, will have Arrhenius character, the slope of the Arrhenius plot being inversely proportional to the value of the entropy S_c held constant.

Evidence for the essential correctness of this prediction can be obtained from different sources. The first example is the Arrhenius variation of the electrical conductivity of ionic glasses. Well above T_g, the ionic conductivity

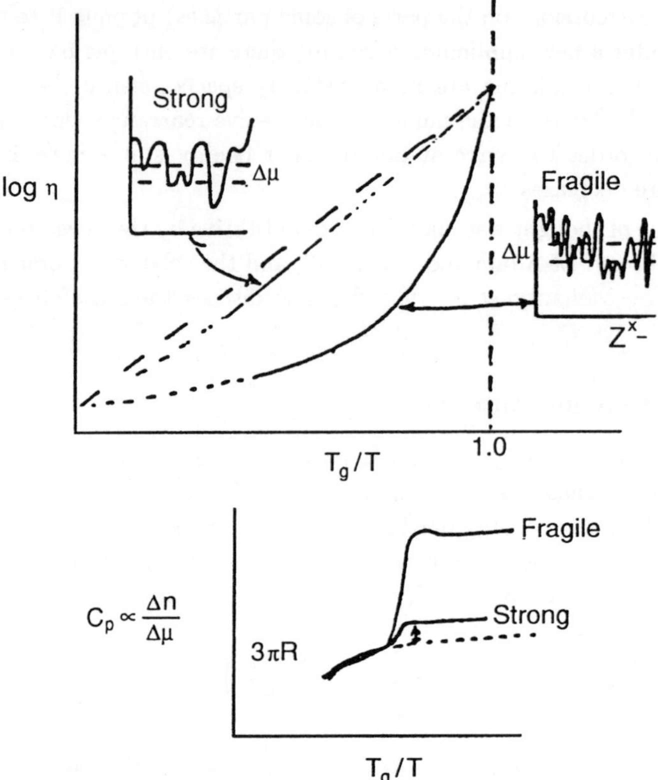

Fig. 8. Relation between heat capacity change at T_g, hypersurface topology, and departure from Arrhenius relaxation time-temperature dependence for strong and fragile liquids.

follows the inverse viscosity in its tendency to vanish (viscosity to diverge) near the Kauzmann temperature [50, 51]. However close to T_g it tends to decouple, assuming a smaller temperature dependence, and then finally changes slope again at T_g to assume its glassy state value [52, 53]. As Eq. (1) predicts, the Arrhenius activation energy is proportional to the amount of annealing which has been imposed on the glass since this lowers S_c. The smaller S_c becomes, the larger the glassy state activation energy. This correlation is more pronounced the more closely the conductivity is coupled to the viscosity [54].

A more direct example is offered by studies of vapor-quenched glasses of high excess entropy [55]. These can be studied at different entropy levels, fixed by annealing the deposits for different annealing times at higher temperatures. Equation (1) then predicts that, when the temperature is chosen to bring the relaxation into the experimental time window, the relaxation time (which can be obtained from sensitive calorimetric studies at the lower temperatures [55]) will be a linear function of the inverse product TS_c. The validity is demonstrated in Fig. 9 [55]. As predicted, the slope is larger (and τ is longer) the smaller the average value of S_c.

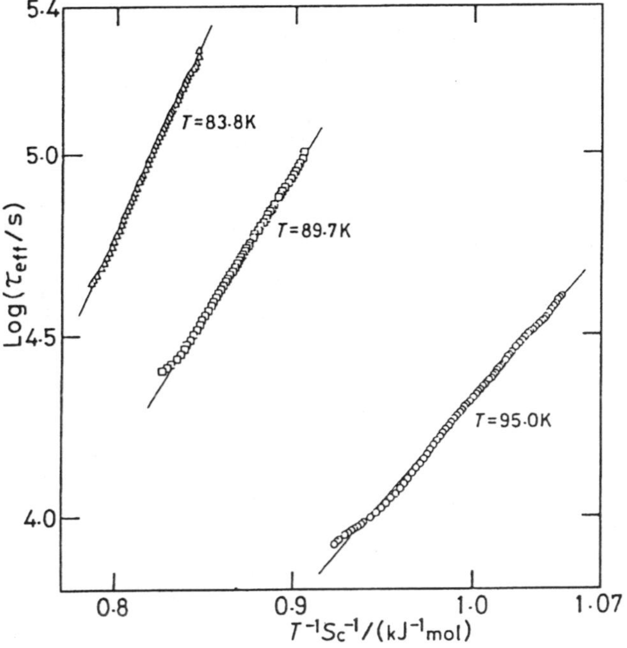

Fig. 9. Test of the Adam–Gibbs equation for the relaxation time of butyronitrile vapor-deposited samples (from Ref. [54] by permission).

4.2. *Relaxation in the ergodic domain*

Most of the tests of Eq. (1) have been carried out in the region of temperature above T_g where the value of S_c is an equilibrium quantity and one, which changes systematically with the temperature. Equation (1) predicts that under these circumstances, there will only be a single relaxation time-vs.-temperature relation and that it will be a linear one for $\log \tau$ vs. $(TS_c)^{-1}$. An example is given in Fig. 10 for the case of trinaphthyl benzene [56], for which the configurational entropy was obtained from differential scanning calorimetry. Log (viscosity) is seen to be linear in $(TS_c)^{-1}$, at low temperatures, and the curvature at high temperatures is as likely to be due to uncertainties in the assessment of S_c {taken as either the difference between liquid and crystal entropies, ignoring differences in vibrational entropy at high T, (upper curve) or between liquid and glass at T_g (lower curve)} as it is to failure of the theory.

A more frequently used, though not as direct, test of the Adam–Gibbs equation is to develop the equation into a Vogel–Fulcher-like form and then demonstrate that the T_o parameter has a value close to that of the Kauzmann

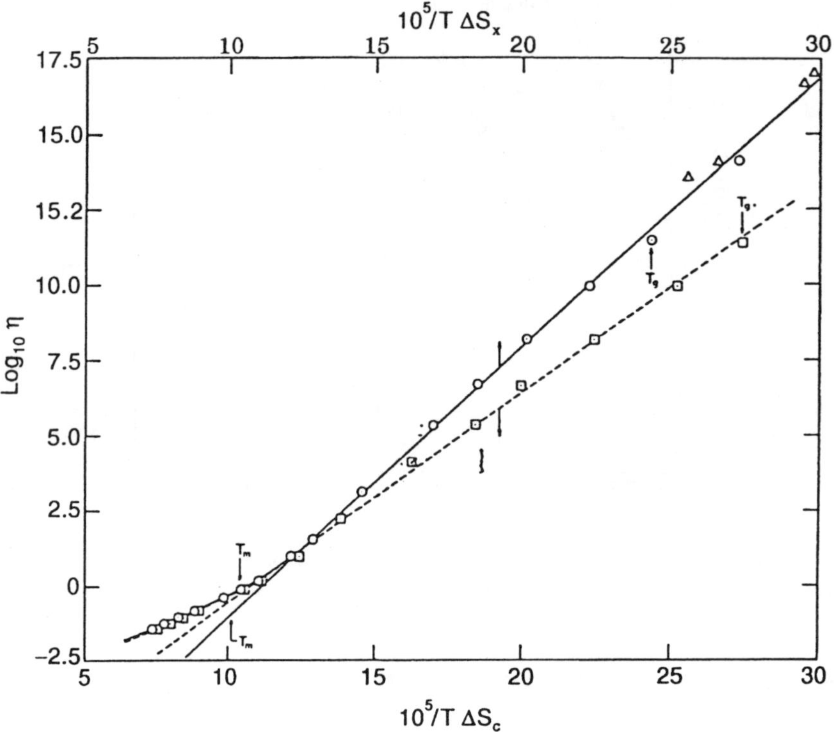

Fig. 10. Test of the Adam–Gibbs equation for viscosity of tri-naphthyl benzene at temperatures above T_g. S_c has been assessed in two different ways leading to two different plots (see text), each of which is seen to be linear over a wide range of the variable $(TS_c)^{-1}$ (from Ref. [56] by permission).

temperature obtained by the purely thermodynamic route. In their original treatment, Adam and Gibbs made the simplest assumption for the excess heat capacity, which determines the configurational entropy temperature dependence, viz. that it is a constant. This yields

$$S_c = \Delta C_p \ln T/T_K \tag{2}$$

Substitution into Eq. (1) then yielded the Vogel–Fulcher equation as an approximation, valid near T_K. However $\Delta C_p = $ constant does not describe many molecular systems. More accurate [57, 58] is $\Delta C_p = K/T$, from which

$$\Delta S(T) = \frac{K(T - T_K)}{T_K T} \tag{3}$$

which leads to the Vogel–Fulcher equation as an identity

$$\tau = \tau_o \exp[DT_o/(T - T_o)] = \tau_o \exp -(F\varepsilon) \tag{4}$$

where F is a fragility parameter, $0 < F < 1$, and $\varepsilon = (T/T_o - 1)$.

There has always been dispute concerning the validity of Eq. (4), and this has been revived with vigor recently in the light of a temperature derivative analysis by Stickel *et al* [59, 60, 61]. Stickel *et al.* show that, particularly for fragile liquids, there is a region at relatively high temperatures where the V–F equation fits quite well but yields a T_o that is considerably higher than T_K and often is also larger than T_g (a result which is unphysical). This analysis emphasizes the high temperature data whereas an analysis focusing on the last five to six decades in relaxation time (covering a small range of ordinate values on the Stickel plot) yields a lower value of T_o, one which usually agrees rather well with the Kauzmann temperature, while also yielding a physical (phonon-like) pre-exponential τ_o. We have documented this elsewhere for a large number of different glassformers. For intermediate liquids such as glycerol, even the Stickel analysis yields a T_o in good accord with T_K [58], probably because the landscape remains influential too much higher T/T_g for such liquids.

The table in Ref. [47] shows that T_K/T_o values close to unity are obtained for some 40 glassformers with T_g varying from 50 to 1,000 K, utilizing the low temperature T_o fitted on data nearest T_g and giving pre-exponents with the physical (phonon) values in all cases. Since the liquids represented in these tables range from molecular through covalent (Se, As_2Se_3) to complex ionic oxides, we judge the case for the Adam–Gibbs approach to relaxation in glassforming liquids to be quite a strong one.

The fragility of the different liquids, which is discussed below, is expressed in two ways in these tables, the first by the parameter F in Eq. (5) [62], and the second by the slope of the Arrhenius plot for each liquid at its T_g normalized by T_g (dlog $\tau/dT_g/T$), which is designated m [63, 64, 65]. A third definition which, like F, has the advantage of varying between 0 and 1, is T_K/T_g which is usually the same as T_o/T_g, especially if data fits are constrained by fixing τ_o at the physical value of 10^{-14} s. For polymers in which T_K is usually not available, an equivalent fragility index F' is obtainable from the common Williams Landel Ferry (WLF) equation parameter C_2 via the relation

$$F' = 1 - C_2/T_g = T_o/T_g \qquad (5)$$

This is obtained from the well-known equivalence of the WLF and VTF equations. For this relation to be a reliable index of fragility, C_1 must be fixed at 16 for relaxation data which, as explained elsewhere [66], is the equivalent of fixing $\tau_o = 10^{-14}$ s, the quasi-lattice vibration period.

4.3. *Relaxation in the non-ergodic state near T_g*

Near T_g, relaxation is complex because the quantity S_c of Eq. (1) is time-dependent. Thus a measurement performed at a constant temperature will yield a relaxation time for recovery of the equilibrium state which is never a linear function of the displacement because S_c itself relaxes according to an Adam–Gibbs equation. When S_c finally reaches its equilibrium value, the relaxation time, of course, becomes time-independent. Behavior in this region has been treated in detail in the review by Hodge [13]. Here we note only that if the temperature of the isotherm falls below T_K, then the relaxation time will necessarily diverge as time increases. A diagram summarizing the relation between the accessible microstates, relaxation time and configurational entropy is given in Fig. 11.

4.4. *Entropy at the glass transition and the validity of Ehrenfest-like thermodynamic relations*

The similarity between the changes in C_p, expansivity α, and isothermal compressibility K_T, seen at the glass transition, and those expected in Ehrenfest second order transitions has naturally prompted enquiry about the validity of the two Ehrenfest equations for the pressure dependence of the transition, notwithstanding the fact that what is observed is not a *thermodynamic* transition. In particular, Davies and Jones [67, 68] demonstrated that a sufficient

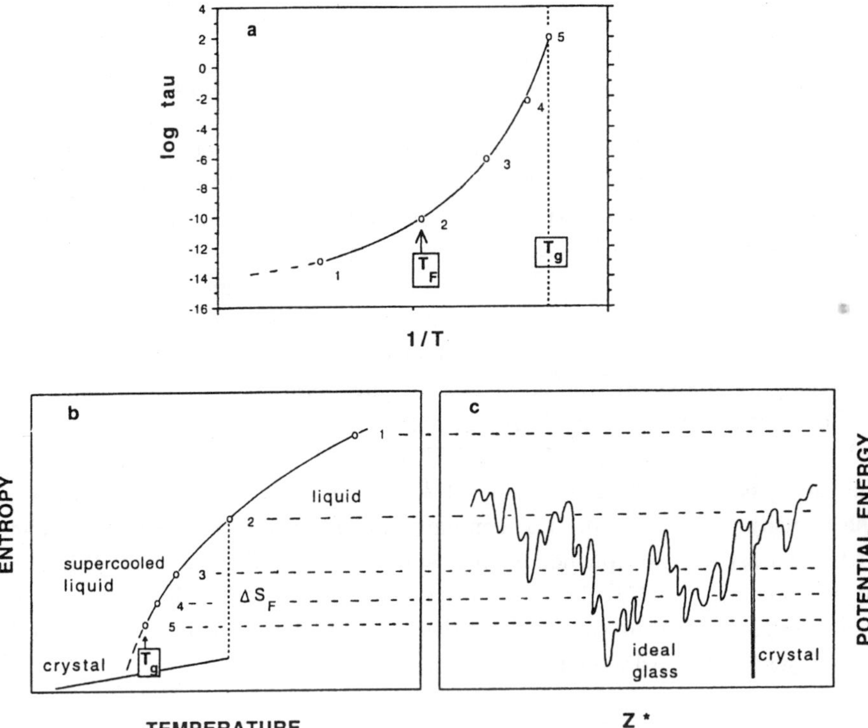

Fig. 11. Illustration of the relation between relaxation time, entropy and excitation level on the potential energy hypersurface for a fragile glassforming system. Point 1 is in the free diffusion regime, unperturbed by any barriers to cross or traps to escape from. This is the regime of mode coupling theory validity. Around point 2, the melting point of this model glassformer, the system begins to "sense" the landscape, and it becomes increasingly enmeshed as T approaches T_g at point 5. With relatively little excess entropy (the "lifeblood of the liquid state") remaining, the system falls out of equilibrium, becoming trapped in a single minimum (becomes a "glass") as its relaxation time rapidly increases beyond the normal experiment measurement timescales.

condition for validity of these equations would be that the excess entropy stay the same irrespective of the pressure at which the system goes through the transition. Following Davies and Jones, the Ehrenfest equations take the forms:

(From the condition $\Delta S = $ constant at T_g) $dT_g/dP = VT\Delta\alpha/\Delta C_p$ (6)

and

$$\text{(From the condition } \Delta V = \text{constant at } T_g) \qquad dT_g/dP = \Delta_K/\Delta\alpha \qquad (7)$$

The validity of Eqs. (6) and/or (7), when T_g is measured at the same rate at each constant pressure, may be seen as a test of the validity of theories which propose that the glass transition is a consequence of diminishing "excess" or "free" volume on the one hand or "excess entropy" on the other. If these excess quantities were both determined by the same single order parameter, then both equations would be valid, and the Prigogine Defay ratio obtained by combining the Eqs. (6) and (7) would be unity. The relations have been tested by a number of workers over the years, and Eq. (6) has been found to hold within experimental error in almost all cases. Equation (7) on the other hand, fails regularly [17, 69, 70]. Thus the Adam–Gibbs equation passes another test [71], and further support for a thermodynamic underpinning of the transition is provided. The complex multiple order parameter character of the glass transition phenomenon is substantiated. The work of Tahahara *et al.* cited in Ref. [71] demonstrates the important point that the product TSc retains a constant value at the glass transition when the temperature is varied, just as the Adam–Gibbs equation implies.

5. Kinetic Aspects of Vitrification: Strong and Fragile Liquids

There has been much discussion in recent years of the behaviour of glassforming liquids and polymers in terms of the property known as fragility, referred to in earlier sections. To reiterate, fragility is a qualitative concept related to the rapidity with which a liquid structure, which was arrested at the glass transition during cooling, becomes disrupted on reheating. The disruption is monitored by the structural relaxation time of the liquid which decreases, concomitantly, from the values of T_g (10^2 s) toward the picosecond values characteristic of highly fluid systems. The relaxation time for shear stress perturbations is simply related to the viscosity by a Maxwell relation

$$\tau_\eta = \eta s/G_{s,\infty} \qquad (8)$$

The behaviour of a selection of liquids is shown in Fig. 12. The data are presented in a scaled Arrhenius form [16] using, as scaling parameter, the temperature at which the glass transition is manifested during heating at 10 K/min. This corresponds to the temperature at which the enthalpy

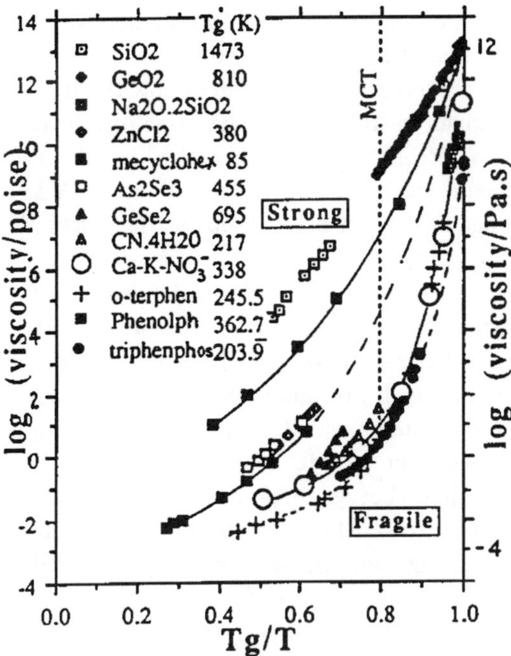

Fig. 12. T_g-scaled Arrhenius plot for viscosity of various liquids including highly fragile case, tri-phenyl phosphite and convenient (T_g > ambient) molecular system, phenolphthalein.

relaxation time is $\sim$ 100 sec. The diagram [61] shows that the more fragile liquids have viscosities which are well below the traditional 10^{12} Pa.s at the T_g defined from scanning calorimetry as above, but would be well above 10^{12} Pa.s if T_g were defined to be $0.9\,T(\tau_H = 100\,\text{s})$. An intermediate choice $T_\tau H \approx 10^3$ s would consolidate the viscosities at $\sim 10^{13.5}$ p. While this would serve to maintain, roughly, the traditional view of the glass transition as an isoviscous phenomenon, it would be achieved at some expense. Variations of G_∞ between different liquids would then mean the shear relaxation times, which are of more fundamental interest (but are usually not available), would be different at T_g.

The liquids selected in Fig. 12 include the chalcogenides GeSe$_2$ and As$_2$Se$_3$ which are compounds occurring in the model covalent glassformer system Ge–As–Se on which we focus some attention in the next section. Se, studied very near T_g, appears more fragile than either of these compounds due to its 1-dimensional chain structure and the additional influence of a ring-chain equilibrium [65] which sets in near T_g. The position of Se in the large scale pattern is like that of other inorganic chain polymers such as $NaPO_3$ which is found to be intermediate in the Fig. 12 pattern.

The latter case emphasizes the problem of how the fragility is best defined. Two metrics were given in Table 1, m, the slope of the scaled Arrhenius plot (Fig. 12) and $D^{-1} = F$ from the curvature of the Fig. 12 curves via the modified VTF equation. These would be interchangeable if the VTF equation applied over the whole range of data in Fig. 12, but as discussed above, this is rarely the case [59].

Thus some confusion between fragilities defined in different temperature domains can be expected. Since much of the usefulness of the fragility concept lies in the way its value seems correlated with other properties of materials interest [72], some of which are manifested below T_g (e.g. physical aging of polymers), a definition made near T_g might seem most appropriate. On the other hand, for those interested in liquid state behaviour, for instance, the testing of mode coupling theory, a definition based on overall behaviour such as a best fit value of F could be more useful. In this case, some ambiguity can be removed by fixing the parameter τ_o at its physical value 10^{-14} s and best fitting the remaining data to Eq. (4) to obtain D (or F) and then comparing the slope at T_g with that directly measured. This issue can only be resolved by finding which definition leads to the most useful correlations with other properties, e.g. non-exponentiality and non-linearity of relaxation [3, 13, 64, 73]. Since this chapter went to print, this issue has been addressed, in a manner which the author feels is very helpful, by defining the $F_{1/2}$ fragility. This is defined by the experimental data alone at a point midway (on log scale) between the relaxation time at the standard T_g (100 s) and the infinite temperature (or phonon frequency) value of 10^{-14} s. This has now been tabulated and compared with other measures in Refs. [116] and [117] and a simple one scan means of obtaining it is described in Ref. [116]. Sometimes interesting differences appear between this new measure and the m fragility (steepness index) defined at T_g itself.

5.1. *The glass transition in covalent systems*

From this point forward, we concentrate attention on the behaviour of glasses and liquids in the chalcogenide system Ge–As–Se because their simple constitution makes them good model systems for analytical treatment. Ge–As–Se is representative of the general class of covalent bonded glasses in which individual elements express different capacities for bonding to neighbors according to the $8 - n$ rule, where n is the periodic table group number.

Chalcogenide systems have been recognized for a long time [74] as remarkable for their strongly glassforming characteristics in certain composition domains. Interest in these systems ran very high in the 1960's because of their potential as IR transmitting media and reversible switching materials [75, 76]. The interest content of chalcogenide systems was raised to a new level by the work of Phillips [77, 78] and Thorpe [78, 79] who saw that their covalent bonding character made them model systems for testing of constraint theory predictions [78, 79]. Thorpe in particular predicted that a rapid increase of mechanical rigidity would set in when the bond density passed a critical value [79]. The bond density, also called the average coordination number (meaning coordination of covalently bonded, as opposed to merely spacefilling, neighbors), was defined as

$$\langle r \rangle = \Sigma n_i X_i \tag{9}$$

where n is the number of covalent bonds made by species i according to the $(8 - n)$ valence electron rule and X is its mole fraction in the multi-component solution. The argument [79] counts off the total degrees of freedom for a mole of atoms against the total of bond length and bond angle constraints imposed by the intact covalent bonds and predicts that rigidity should percolate through the structure for any value of $\langle r \rangle$ greater than 2.4.

Ge–As–Se is an ideal system for testing these ideas because all components are of similar mass (neighbors on the periodic table), and the various pair bonds which can form are of comparable strength. Because of the different valences, 4, 3, and 2 respectively, the condition $\langle r \rangle = 2.4$ can be satisfied

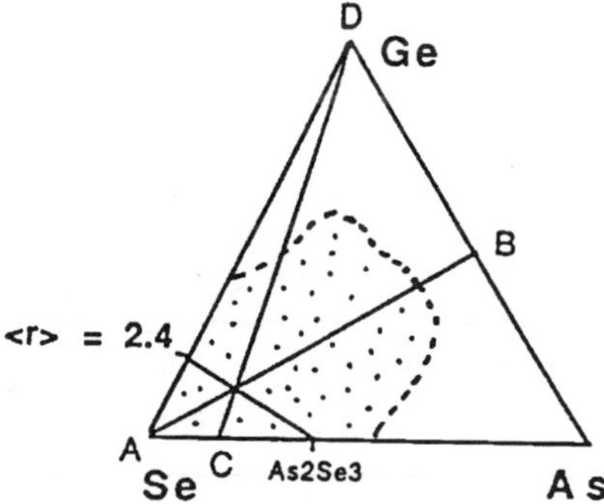

Fig. 13. The model covalent glass system Ge–As–Se, showing the domain of easy glass formation and the condition for rigidity percolation, $\langle r \rangle = 2.4$, within this domain. The ternary composition line on which interest is focused in the following discussion is AB.

for a range of different ternary mixtures [80] as shown by the line marked $\langle r \rangle = 2.4$ in Fig. 13. Figure 13 also shows the glassforming region for moderate rate cooling of the liquid alloys. Although the predictions of breaks in the mechanical properties at $\langle r \rangle = 2.4$ has met with only moderate success [81] because of neglect of non-bonding interactions and other factors, it was found by Tatsumisago *et al.* [82] that properties characteristic of the viscous liquid and the glass transition showed special behaviour near this $\langle r \rangle$ value.

We will give the next part of this paper to a review of the latter findings for these systems, which suggest that they may provide an excellent testing ground

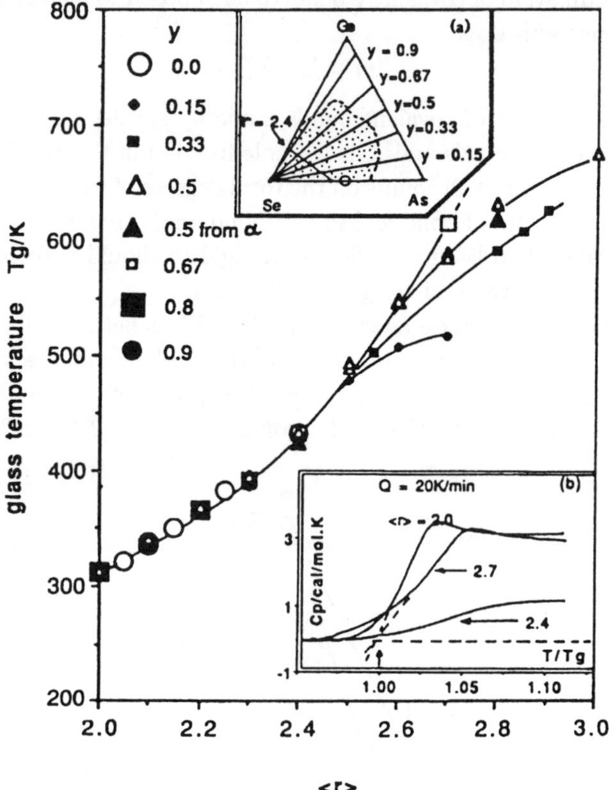

Fig. 14. $\langle r \rangle$ dependence of T_g from DSC and expansivity, for Ge–As–Se glasses along various pseudobinary cuts $\mathrm{Se}_{(1-x)}(\mathrm{Ge}_y\mathrm{As}_{1-y})_x$ as indicated in inset (a). Inset (b) shows behaviour of the heat capacity C_p through the glass transition for $\langle r \rangle$ values at, and on either side of, the rigidity-percolation threshold along the $y = 0.5$.

for general theories of ergodicity breaking and the overall problem of the glass transition. Elsewhere [83], it is argued that a phenomenological relationship exists between silicon- (or germanium-) based covalent bonded alloy solutions and two other major classes of solutions which together control much of what happens in the liquid state of the natural world. These are the aqueous solutions, typified by the well studied system H_2O + LiCl [84], and the siliceous solutions typified by the even more thoroughly studied system SiO_2 + Na_2O [85].

The first finding of interest in the Ge–As–Se system is that for glasses in the composition range below and a little above $\langle r \rangle = 2.4$ have T_g values which depend only on $\langle r \rangle$. This is shown in Fig. 14. It was noted that the glass transition near $\langle r \rangle = 2.4$ was very "smeared out" (see Fig. 14 insert). To investigate this further, we study behaviour along a single line, AB, in Fig. 13 along which Ge:As = 1. Figure 15 shows how both the activation energy for viscosity and the change of heat capacity at T_g go through minimum values close to $\langle r \rangle = 2.4$. A recent paper by Senapati and Varshneya [86] shows similar behavior for the related system Ge–Sb–Se and shows that a minimum also occurs in the change in expansion coefficient.

The correlation seen in Fig. 15 is predictable from the C_p minimum via the Adam–Gibbs equation for relaxation processes [36]. For instance, the shear relaxation time, τ_s, whose temperature dependence dominates that of the viscosity, is written according to Adam and Gibbs as

$$\tau_s = \eta/G_\infty = \tau_o \exp(C/TS_c) \tag{10}$$

where η is the viscosity, G_∞ is the shear modulus at high frequencies, $\omega\tau \gg 1$. Because ΔC_p controls the temperature dependence of S_c in Eq. (4), it also controls the magnitude of deviations from the Arrhenius equation, hence also the slope of the Arrhenius plot of η measured at T_g.

The reason for ΔC_p having a minimum at $\langle r \rangle = 2.4$ is less obvious but must have to do with the bond distribution being optimal at this composition hence the drive to thermal disruption via increased degeneracy, a minimum. In terms of the potential energy hypersurfaces discussed earlier, $\langle r \rangle = 2.4$ must correspond to a hypersurface with a relatively small number of minima, presumably because of a more or less unique bond distribution. This would correspond to a small entropy and a small increase in entropy per unit increase in energy. Since this work went to print, a major advance has been made in the understanding of this problem by recognizing the role of the fragility

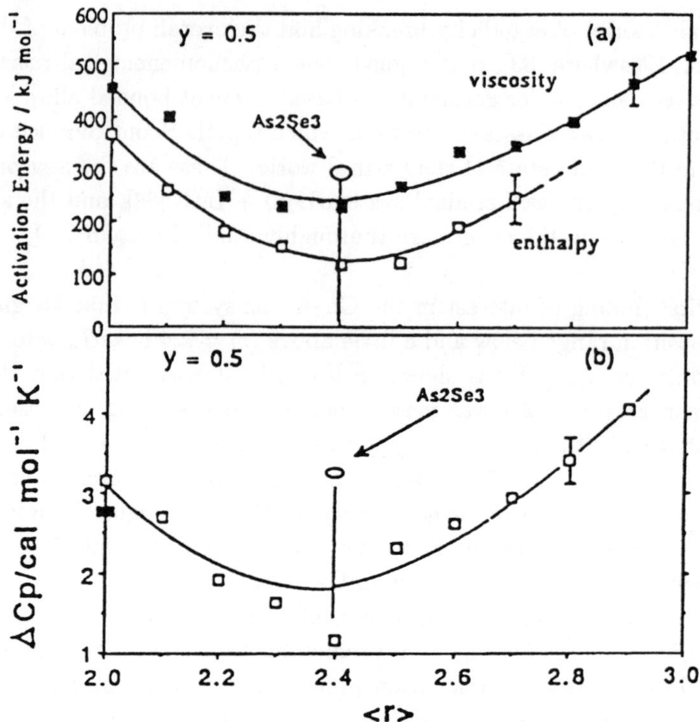

Fig. 15. Excess molar heat capacities, and activation energies for viscosity and enthalpy at T_g, along the line AB of Fig. 5 (adapted from Ref. [82]).

in arresting the system during cooling in a state in which the constraints are not the values that would be expected ideally (hence theoretically) from the composition. A number of constraints, which depends on the cooling rate, are retained in the broken state in the glass and distorts the behavior from that predicted by the usual constraint-counting. When this is recognized and the "ideal glass transition temperature is calculated using the Vogel Fulcher law for data obtained above T_g, it is found that the ideal glass temperature goes through a sharp minimum at $\langle r \rangle = 2.4$ in the Ge–As–Se system, quite different from the behavior of the ordinary T_g seen in Fig. 14. This development is described in Refs. [43(b)] and [43(c)].

It has been shown elsewhere [82] how the viscosity behavior at $\langle r \rangle = 2.4$ is that of a "strong" liquid — even stronger than B_2O_3 and certainly stronger than the tetrahedral network compound $GeSe_2$, seen in Fig. 4. The system

Se-(Ge:As) thus exhibits the full range of strong to fragile behavior previously only seen by reference to a variety of other liquids, hence this system is rich in the phenomenology of the liquid-to-glass transition. Being simply constituted (no internal degrees of freedom) it is therefore to be regarded as a model system for experimental and theoretical studies. Addition of iodine, which would break up chain structures, might be expected to lead to even more fragile behavior, but this is not strongly supported by the available data [87].

Consistent with strong liquid behavior at $\langle r \rangle = 2.4$ is the relation between T_g and the Kauzmann temperature T_K at which extrapolations of the excess of liquid entropy over crystal entropy (which is ΔS_f at the fusion temperature) show the excess vanishing. The large value of T_g/T_K for As_2Se_3, 1.90 is contrasted with the small value, 1.23, for the relatively fragile Se.

Two further correlations of liquid behaviour with fragility may be made using data from the Ge–As–Se systems.

Firstly it is found that the manner in which the liquids approach equilibrium after a perturbation, i.e. the relaxation function, is controlled by the fragility. An example of linear relaxation in the time domain is shown in Fig. 16 for a chalcogenide alloy [63]. If the relaxation function is represented by the popular "stretched exponential" function

$$\theta(t) = \exp-([t/\tau_o]^\beta), \quad 0 < \beta < 1.0 \tag{11}$$

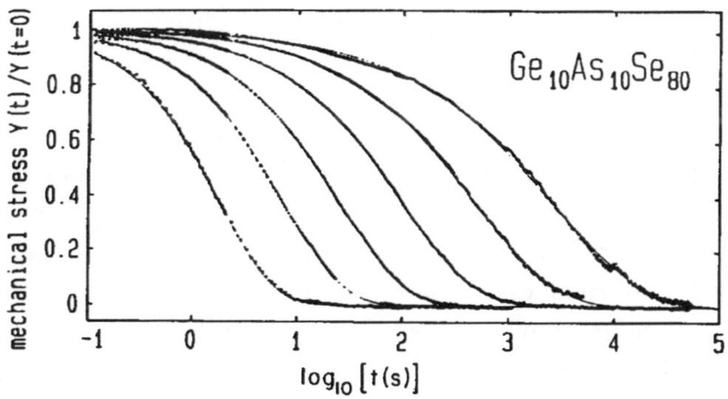

Fig. 16. Normalized stress relaxation of a Ge–As–Se supercooled liquid as measured at temperatures (from left to right) 433.4, 422.6, 413.3, 403.7, 394.5, and 385.3 K. The *solid lines* are fits with Eq. (11). The stretching exponent *b* is temperature dependent. The *dashed line* represents mono-exponential relaxation (adapted from Ref. [63]).

which fits the data very well except at very short times, then β (the "non-exponentiality of relaxation" is correlated with the fragility m (defined from the temperature dependence of the relaxation time at T_g as explained earlier). This is demonstrated in Fig. 17, in which complimentary data from studies on chain polymers [88] are included.

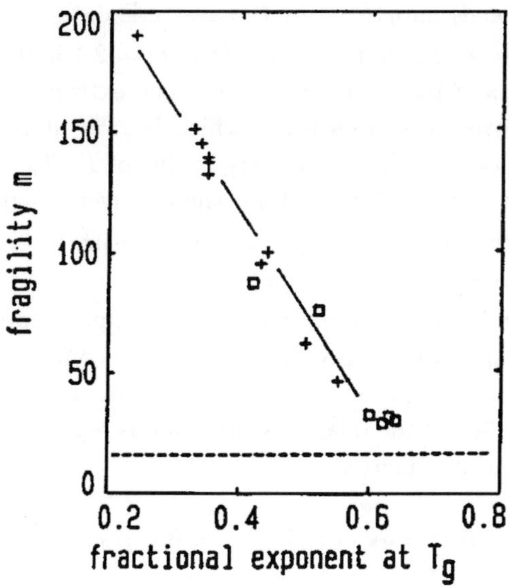

Fig. 17. Fragility m vs. fractional exponent b at the glass transition for polymers (circles, taken from Plazek and Ngai (Ref. [88]) and chalcogenides (squares, this work). The broken line indicates the minimum fragility.

Secondly, if the relaxation is monitored after dropping the temperature below T_g, so that the system is "out of equilibrium," i.e. non-ergodic, then the extent to which the relaxation time depends on the departure from equilibrium seems to depend on the fragility. This "non-linearity of relaxation" is shown in Fig. 18 for the extreme case of selenium in which we focus on the part of a complex relaxation function which is believed to be the visco-elastic part (there is a longer time part assigned to a quasi-chemical ring-chain motif equilibration, which is described elsewhere [65].

The diagram shows that the relaxation function depends on waiting time, and that it requires some 60 hrs of waiting at $T = 300$ K before the structure responsible for the viscoelastic relaxation, seen in this time window, fully

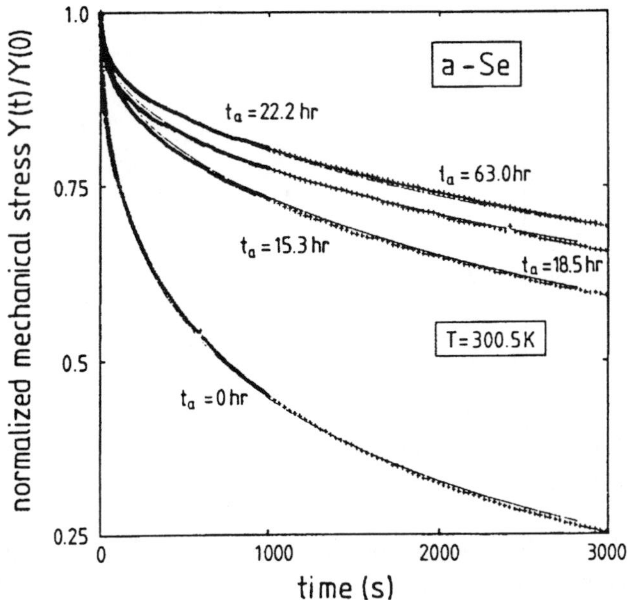

Fig. 18. Time dependence of normalized tensile stress $Y(t)/Y(t = 0)$ of amorphous selenium, which has previously been quenched from an equilibrium state obtained by annealing at 335 K, well above the glass transition temperature. The labels indicate the times at which the measurements have been started relative to the time of commencement of the first run. The solid lines are calculated using Eq. (11) with $\beta = 0.47$.

equilibrates. For compositions at $\langle r \rangle = 2.4$, no waiting time dependence could be observed [72]. This has the important consequence that glasses of this composition should be dimensionally stable over long periods of time near and below T_g.

In summary, a collection of measurements confirm that the rigidity percolation threshold of $\langle r \rangle \sim 2.4$ is of significance in covalently bonded glasses.

6. View from the Solid

While it is natural, in view of the way glasses are usually formed, to view the glass transition as an extension of the liquid state problem, it is also possible to treat it as a problem in solid state physics. The problem is to account for the transition from more or less harmonic behavior, well below T_g, to behavior so anharmonic at T_g that irreversible displacement of particles from their original sites occur and stresses can be relaxed by deformation on time scales of 100s

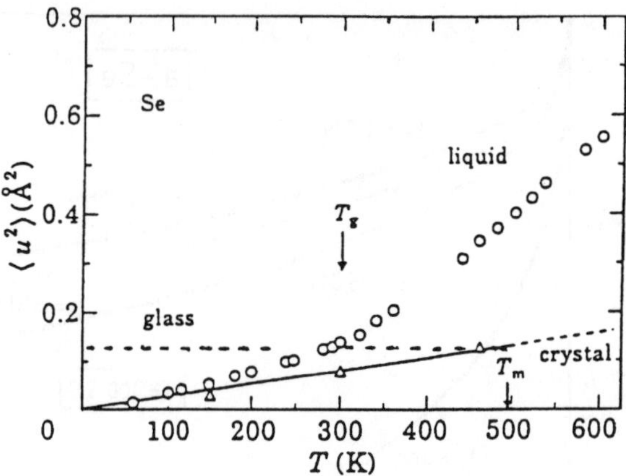

Fig. 19. Temperature effect on the mean square displacement of particles in selenium glass, crystal, and liquid (from Ref. [90]).

of seconds — something which does not happen in defect-containing crystals unless the crystals are very small.

Some sort of Lindemann criterion seems to apply because both neutron scattering [89, 90, 91] and computer simulation [2, 3, 92] studies show that the experimental glass transition tends to occur when the mean square vibrational amplitude reaches a value of about 1/3 of the atomic diameter. In the case of selenium where values at both glass transition and melting are known [89], they are coincident, as seen in Fig. 19. Clearly the rate of increase in $\langle r^2 \rangle$ per unit of thermal energy increase has to be much greater in the glass than in the crystal, which must be accountable in terms of the more random potentials in which the vibrations occur.

Whereas the limiting value of the mean square displacement in a crystal leads to the complete collapse of the structure in favor of the liquid, the consequence in the glass is only that very occasional local rearrangements of particles occur. Indeed the diffusivity of molecules in the crystal via defect diffusion near T_m is usually greater than that in the glass at the same $\langle r^2 \rangle$. However, the situation changes rapidly above T_g due largely to a change in structure which is usually (but not necessarily) accompanied by expansion, and is always accompanied by an increase in heat absorption per unit temperature increase. Evidently when the $\langle r^2 \rangle$ reaches $\langle 0.66 r_{\text{Se}} \rangle^2$, configurationally excited states of the system (the "configurons" of Sec. 4) can form at the expense of what would otherwise be excessive oscillatory excursions. The configuron states cannot be very extensive in space because micro-emulsion droplets of a glassformer (which exist at the bulk chemical potential) exhibit the same behavior (at least with respect to T_g) as the bulk [93].

The configurons must be coupled to the phonons by the anharmonic terms in the inter-particle potential. The more harmonic the effective potential, the weaker the coupling, hence the less probable the excitations, i.e. the greater the energy barrier to the phonon-configuron exchange [94], and the "stronger" the resulting liquid. The enhanced harmonicity in these cases is reflected in the overshoot in $\langle r^2 \rangle$ vs. t seen in computer simulations [92, 95, 96], which persists above T_g. This is the undamped remnant of harmonic "ringing" seen in simulations of crystals before dephasing destroys the correlations [97]. It is expected that, in network glasses, with longer intermediate range order, this damping/dephasing will be less marked so networks, intermediate range order, persistent boson peaks, and vibrational harmonicity, should all be linked together with "strong liquid" behavior [2].

Spin glasses, while showing a Vogel–Fulcher-like temperature dependence, do not depend on an anharmonicity coupling mechanism to establish equilibrium between phonon bath and spin system, so the basis for the spreading of systems in a strong vs. fragile pattern does not exist. This is presumably why a T_g/scaled presentation of spin glass behavior shows that most examples of the phenomenon are excessively fragile by comparison with structural glasses [98]. It is the high fragility which is the reason that most of the experimentation in spin glasses is carried out in the non-ergodic (non-linear) regime below T_g — the inverse of the situation with "structural glasses," which are mostly explored in their ergodic or viscous liquid states. Failure to recognize this essential difference is often a source of confusion when comparisons between the two phenomenologies are made.

7. Polyamorphism

The possibility that glassy systems in which the order is highly developed might be able to exist in more than one form, with fairly sharp structural barriers between forms, has been recognized for a long time [99], but clear evidence for the phenomenon has been obtained only recently [100, 101, 102, 103]. The best documented case is that of water which, interestingly enough, is very difficult to obtain in the vitreous state by normal liquid cooling methods. Mishima [101], following early work by Mishima *et al.* [100], has recently shown the quasi-reversible formation of alternately high density and low density forms of amorphous water in what amounts to close experimental reproduction of the slightly earlier computer simulations of Poole *et al.* [103]. The rather comparable behavior, in which a changeover from low density to high density forms occurs in a narrow range of pressures (although not in a strictly first order manner), has been seen also for vitreous silica and some other materials [102]. The phenomenon is seen most clearly in the simulation results of Poole and the experiments of Mishima, which are reproduced in Fig. 20. Poole's simulation study showed that for ST2 water and other simple pair potential cases, the transition could also, in principle, be seen in the supercooled *liquid* state at constant pressure but only if the pressure exceeded a critical value well above ambient. Equations of state for experimental water, on the other hand, all are consistent with the existence

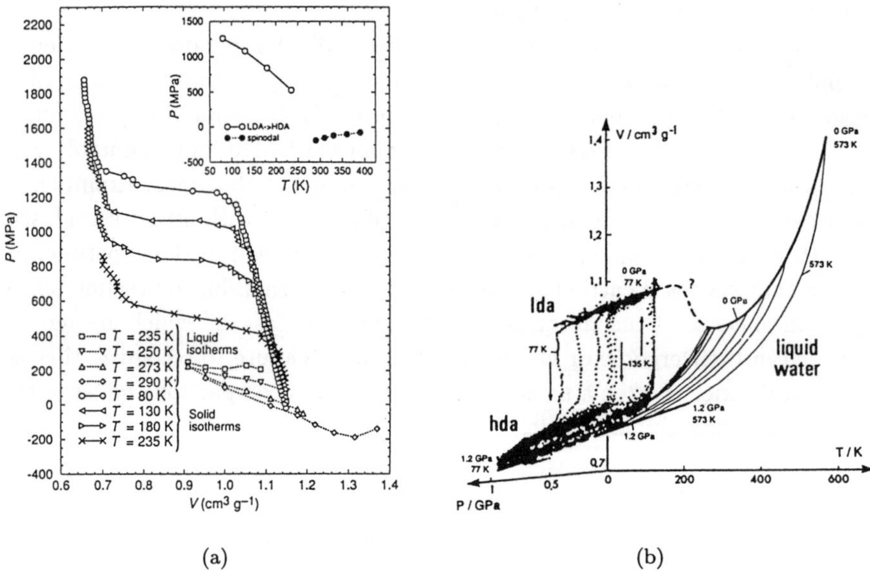

(a)   (b)

Fig. 20. (a) Variation of pressure with volume for simulated ST2 amorphous water. Note the sudden change in volume with minimum change of pressure which occurs at volumes below ~ 1.1 cm^3 s^{-1}. A hysteritic reversal of the density change (at negative pressure) during decompression is shown elsewhere in this reference (from Ref. [104], Nature, by permission). (b) Experimental equivalent of the results of part (a) from the work of Mishima [102] showing collapse of vitreous ice above 10 kb pressure to give high density form, and reversal of the density on decompression and warming. Behavior for various isotherms is demonstrated (from *J. Chem. Phys.* by permission).

of a re-entrant spinodal [104, 105] which implies [106] the existence of a first order liquid-liquid transition at ambient pressure (again under metastable conditions). The present author along with others [107] has claimed that the phenomenon is actually a transition from a fragile liquid to a strong liquid. We will produce some simulation evidence for this as a phenomenon common to a special class of liquid, later in this chapter.

In the simulation studies of water, direct observation of the first order liquid-liquid transition at higher pressures was frustrated by the slow dynamics in the region where the transition was expected [105]. The dynamics in the parallel case of liquid silicon seem to be rather faster, and the transition can be observed directly by simulation as well as in the laboratory. We will examine some relevant data in the next section.

7.1. *First order transitions in liquid silicon*

Although liquid silicon is a metal, it has proven possible to simulate many of its characteristics, including the generation of a tetrahedrally-coordinated phase at low temperatures, using the pairwise additive potential invented by Stillinger and Weber [108]. The potential leads, remarkably, to an accurate melting point, and although the coordination number in the liquid state is slightly smaller than that determined by experiment, it is large enough to be consistent with metallic behavior [109]. Extensive calculations of the supercooled state of silicon simulated with this potential have been carried out, and the occurrence of the transition from a relatively mobile liquid state to a rather immobile tetrahedrally-coordinated "amorphous" state has been reported [110]. The most extensive calculations, those by one of the authors, have remained largely unpublished until recently [83] and the important findings will be briefly summarized in this chapter.

The most significant features are summarized in Fig. 21 which shows the variation of the density during stepwise cooling and equilibration, using run lengths sufficient to ensure equilibrium at all temperatures down to the phase transition. The diagram shows the occurrence of a well-defined maximum at a temperature some 300 K below the melting point (cf. the case of SiO_2 [111, 112]), which is followed at a somewhat lower temperature (1060 K) by an abrupt change of density to a value 4 percent lower. As is seen in Fig. 22, this is associated with a relatively small decrease in coordination number. The change in potential energy accompanying this transition is found to be 20 percent of the heat of fusion.

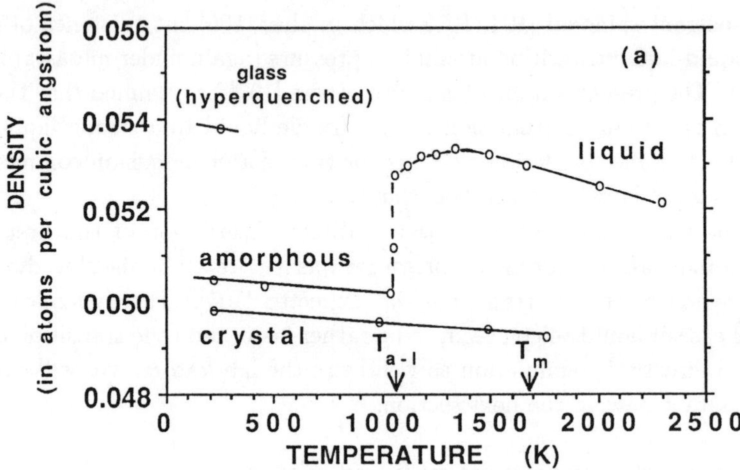

Fig. 21. Variation of the density of liquid and supercooled Stillinger-Weber liquid silicon during cooling at zero pressure, showing occurrence of density maximum at 1350 K followed by sudden change of density at 1060 K to an amorphous four-coordinated state of constant density. Densities of crystalline and hyper-quenched liquid states also shown.

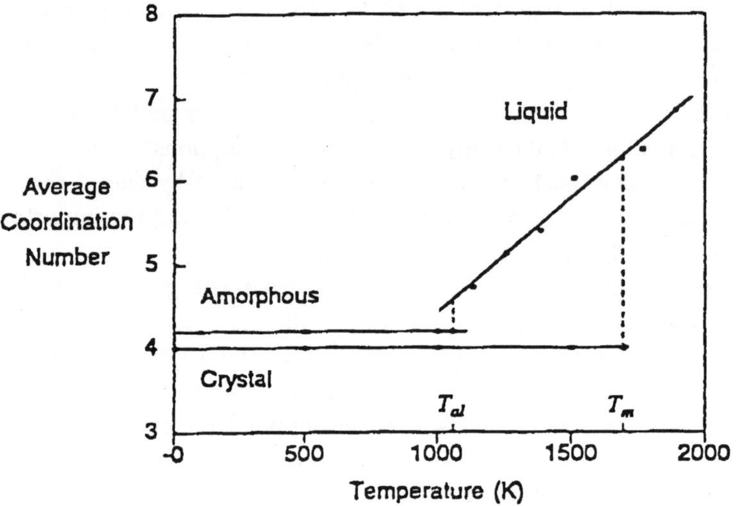

Fig. 22. Silicon atom coordination number in liquid S–W silicon during decrease of temperature, showing sudden small decrease at 1060 K where the first order phase transition in density and potential energy is observed.

The change of diffusion coefficient which accompanies this transition is too large to actually measure, but limits can be assigned. Certainly the diffusivity of silicon in the tetrahedral phase is at least two orders of magnitude less than that of the liquid before the transition occurs [83].

On the computer time scale, the tetrahedral phase is effectively an amorphous solid. This transition is the simulation analog of the event which has been reported on the basis of laser heating experiments by Thompson *et al.* [113]. The latter authors observed phenomena which they interpreted as due to a reversible first order liquid amorphous transition occurring at the temperature 1480 K, somewhat higher than that observed in the simulations.

Taking experiments and simulations together, there seems little question that a first order transition does occur in this system. The further observation of importance for the purposes of this paper is that the density maximum and the temperature of the transition both appear to decrease as the system is studied at smaller volumes, i.e. at higher pressures. In this respect, the system

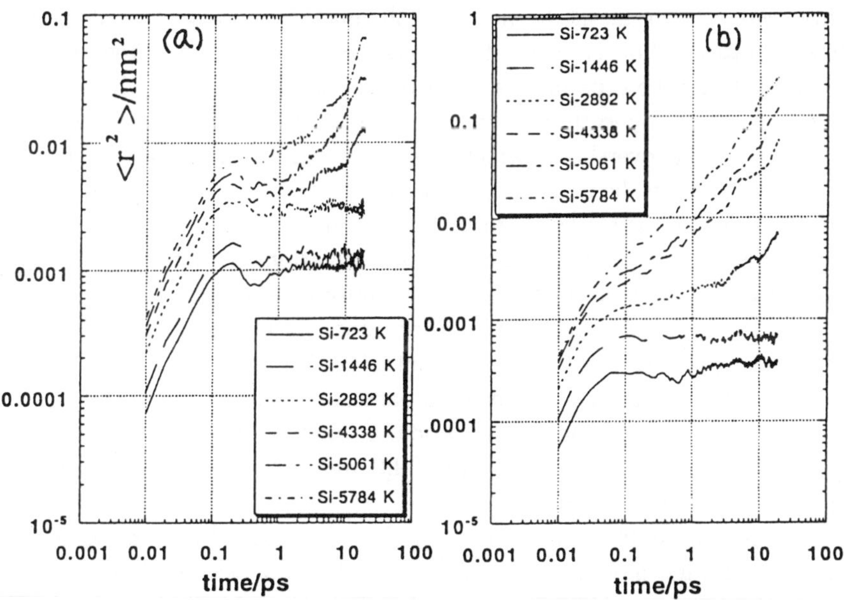

Fig. 23. Mean square displacement of Si^{4+} ions vs. time in (a) low density and (b) high density SiO_2 polyamorphs. Note (i) the pronounced overshoot in $\langle r^2 \rangle$ up to $\sim 3\,T_g$ (expt) in case (a) and its absence in case (b), and (ii) absence of diffusion up to 10 ps at 2892 K in case (a) compared with pronounced diffusion at 2892 K in case (b).

is very water-like. (Although it is controversial whether or not water passes through a first order transition at ambient pressure [2, 103, 105, 107], there is agreement that it should do so at higher pressures).

7.2. *Strong and fragile SiO_2*

We have seen in the foregoing the suggestion that liquids can transform isothermally from overcoordinated to lower coordinated states during decrease in temperature and the suggestion that the transition should be one from a liquid with fragile characteristics to a liquid with strong characteristics. So far we have not seen direct evidence of the difference in fragility between the open and the more close-packed amorphous structures. With silica polymorph simulations, these distinctions can be seen with some clarity [96].

We show in Fig. 23 the variations of the mean-square displacement (the Debye–Waller factor) for silicon in silica at two different densities. The densities correspond to the ambient pressure four-coordinated form and the high pressure six-coordinated form. We note firstly the strong overshoot in the Debye–Waller factor at the beginning of the plateau characteristic of the non-diffusive states and observe that this overshoot is present up to temperatures some three times the experimental glass transition temperature (which is much lower than the temperature where long-range diffusivity commences in the simulation). This overshoot (which is seen on careful examination of multiple averaged runs to be a "ringing" with period ~ 0.5 ps) is strikingly absent from the calculations for the high pressure form. Figure 24 shows the mean-squared displacement after a selection of simulation run times, for different temperatures. The figure includes data for both polymorphs.

From these "isochrones," a relaxation time can be calculated [96] on the basis of arguments [16c] and measurements [114] that show that the relaxation time is the time taken for the average particle to diffuse about one-half the molecular diameter. Using this criterion, relaxation times have been calculated for both polymorphs at a series of different temperatures, and these are displayed in Fig. 25.

It is clear from the data in Fig. 24 that the high density polymorph is gaining diffusional mobility much more rapidly with increasing temperature than is the low temperature polymorph. This is an indication of a higher fragility for the high-density polymorph which we now establish by comparing the behavior of the relaxation times. We obtain the relaxation times at different temperatures, as indicated above, by using different isochrones to establish the

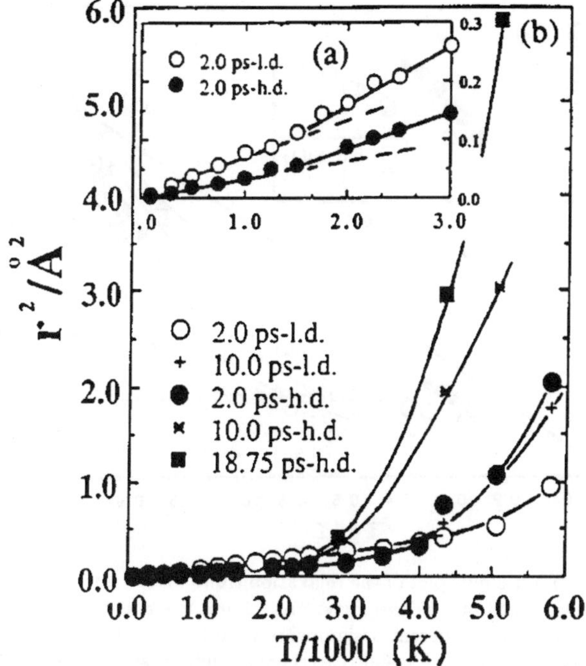

Fig. 24. (a) Mean square displacement of Si in low density (l.d.) and high density (h.d.) polyamorphs in pre-diffusion regime, showing breaks in slope near 1500 K (the laboratory T_g) more pronounced for the h.d. case. (b) $\langle r^2 \rangle$ for selected time windows vs. temperatures extending to values where diffusion commences on computational time scale. Note more rapid onset of diffusion in h.d. polyamorph notwithstanding higher packing density.

temperature at which the relaxation time has a certain value (the elapsed time of the isochrone). This is chosen as the temperature at which the mean-square displacement for the isochrone is $1 \sim$ Å in excess of the vibrational value. The results of this analysis based on Fig. 24 are summarized in Fig. 25 in which the relaxation times are plotted against the T_g-reduced reciprocal temperature T_g/T. Included in Fig. 25 are data for a variety of other substances.

Figure 25 makes clear the earlier assertion that the high density polymorph is much more fragile than the open network polymorph. While this does not prove that the earlier-observed first order transitions (observed during temperature change) are transitions from strong to fragile liquids, it certainly makes it much more plausible.

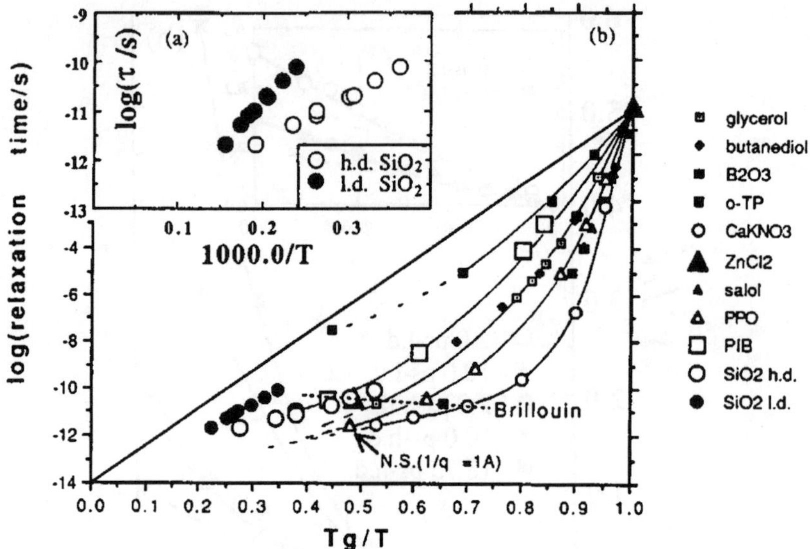

Fig. 25. (a) (insert) Arrhenius plot of the relaxation times for low and high density polyamorphic forms of SiO_2, based on the assignment of 0.1 nm average diffusional displacement of Si in one relaxation time. (Different assignments would change the vertical positions of the lines but not their slopes.) (b) T_g-scaled Arrhenius plots of structural relaxation times for a variety of liquids and two polymers, polypropylene oxide PPO and polyisobutylene PIB (segmental relaxation times), showing relation of relaxation times for the high and low density (six- and four-coordinated) forms of SiO_2 to the general pattern.

Acknowledgments

Support of this program by the National Science Foundation under grant number DMR9108028-002 is gratefully acknowledged. Much of this material was previously published as contributions to the proceedings of (1) the Enrico Fermi Summer School Varenna 1996 published as Proc. Int. School of Physics "Enrico Fermi" course CXXXIV ed. F. Mallamace and H.E. Stanley, 105 Press Amsterdam, 1997, (2) the NATO-ASI at Sozapol 1995 published as the volume "Amophorus Insulators and Semi Conductors" eds. M.F. Thorpe and M.I. Mitkova, NATO-ASI series, Plenum Press, 1997 and (3) the 1995 Pisa conference on complex systems published as "Non-Equilibrium Phenomena in Supercooled fluids, glasses, and amorphous materials," eds. M. Giordano, D. Leporini, and M.P. Tosi, by World Scientific, Singapore, 1996.

References

[1] D.R. Uhlmann, *J. Non-Cryst. Solids* **7**, 337 (1972).

[2] C.A. Angell, P.H. Poole and J. Shao, *Nuovo Cimento* **D16**, 993 (1994).

[3] C.A. Angell, *Science* **267**, 1924 (1995).

[4] R.J. Palmer and D.C. Stein, Relaxations in Complex Systems, eds. K. Ngai and G.B. Wright (National Technical Information Service, U.S. Department of Commerce, Springfield, VA 22161, 1985) p. 253.

[5] D.A. McQuarrie, *Statistical Mechanics* (Harper & Row, New York, 1973), p. 554. The *"ergodic hypothesis...*states that for a stationary random process, a large number of observations made on a single system at N arbitrary instants of time have the same statistical properties as observing N arbitrarily chosen systems at the same time from an ensemble of similar systems."

[6] A. Bondi, *Rheology*, Vol. 4, ed. F.R. Eirich (Academic Press, New York, 1958).

[7] C.A. Angell and E.J. Sare, *J. Chem. Phys.* **52**, 1058 (1970).

[8] C.T. Moynihan, A.J. Easteal, M.A. Debolt and J.C. Tucker, *J. Am. Ceram. Soc.* **59**, 16 (1976).

[9] K.J. Rao, C.A. Angell and D.B. Helphrey, *Phys. Chem. Glasses* **14**, 26 (1973).

[10] C.A. Angell, C. Alba, A. Arzimanoglou, R. Böhmer, J. Fan, Q. Lu, E. Sanchez, H. Senapati and M. Tatsumisago, *Am. Inst. Phys. Conf. Proc.* **256**, 3 (1992).

[11] C.A. Angell, *Current Opinion in Solid State and Materials Science* **1**, 578 (1996).

[12] B. Wünderlich, Assignment of the Glass Transition ASTM, STP 1249, ed. R.J. Syeler (ASTM Pub., Philadelphia 1994) p. 17.

[13] I.M. Hodge, *J. Non-Cryst Sol.* **131–133**, 435 (1991); and 169, 211 (1994).

[14] D.J. Plazek, C.A. Bero and I.C. Chay, *J. Non-Cryst. Solids* **172–174**, 181 (1994).

[15] A.Q. Tool, *J. Am. Ceram. Soc.* **29**, 240 (1946).

[16] C.A. Angell (a), Relaxations in Complex Systems, eds. K. Ngai and G.B. Wright (National Technical Information Service, U.S. Department of Commerce, Springfield, VA, 1985) p. 1.
(b) *J. Phys. Chem. Solids*, **49**, 863 (1988).
(c) *J. Non-Cryst. Sol.* **131–133**, 13 (1991).

[17] C.T. Moynihan, P.B. Macedo, C.J. Montrose, P.K. Gupta, M.A. DeBolt, J.F. Dill, B.E. Dom, P.W. Drake, A.J. Easteal, P.B. Elterman, R.P. Moeller, H.A. Sasabe and J.A. Wilder, *Ann. NY, Acad. Sci.* **279**, 15 (1976).

[18] M. Goldstein, *J. Chem. Phys.* **51**, 3728 (1969).

[19] H. Suga and S. Seki, *J. Non-Cryst. Solids* **16**, 171 (1974).

[20] H. Fujimori and M. Oguni, *J. Chem. Thermodynamics* **26**, 367 (1994).

[21] H. Fujimori and M. Oguni, *Solid State Commun.* **94**, 157 (1995).

[22] C. Alba, L.E. Busse and C.A. Angell, *J. Chem. Phys.* **92**, 617 (1990).

[23] J. Wong and C.A. Angell, *Glass: Structure by Spectroscopy* (Marcel Dekker, New York) (1976) Chap. 1.

[24] R.S. Pitzer and D.W. Scott, *J. Amer. Chem. Soc.* **65**, 803 (1943).

[25] H.S. Chen and D. Turnbull, *J. Chem. Phys.* **48**, 2560 (1968); *J. Appl. Phys.* **38**, 3646 (1967); *Acta. Metall.* **17**, 1021 (1969).

[26] L.E. Topol, S.W. Mayer and L.D. Ransom, *J. Phys. Chem.* **64**, 862 (1960).

[27] C.A. Angell and D.C. Ziegler, *Mat. Res. Bull.* **16**, 269 (1981).

[28] E. Thilo, C. Wieker and W. Wieker, *Silic. Tech.* **15**, 109 (1964).

[29] *Phase Diagrams for Ceramicists*, edited by E.M. Levin, C.R. Robbins and H.F. McMurdie, Diagram No 391 (American Ceramic Society, 1965).

[30] M. Poulain, M. Chantanasingh and J. Lucas, *Mater. Res. Bull.* **12**, 151 (1977).

[31] C.A. Angell, *J. Phys. Chem. Sol.* **49**, 863 (1988).

[32] D. Sidebottom and L. Torell, *Phys. Rev. Lett.* **71**, 2260 (1993).

[33] A.P. Sokolov, A. Kislink, M. Soltwisch and D. Quitmann, *Phys. Rev. Lett.* **69**, 1540 (1992).

[34] J. H. Gibbs, Modern Aspects of the Vitreous State, ed. J.D. McKenzie (Butterworths, London, 1960) Chap. 7, p. 230.

[35] J.H. Gibbs and E.A. Dimarzio, *J. Chem. Phys.* **28**, 373 (1958).

[36] G. Adam and J.H. Gibbs, *J. Chem. Phys.* **43**, 139 (1965).

[37] C.A. Angell, Nature (News & Views) **393**, 521 (1998).

[38] Myth of cathedral windows: This is the idea that glass flows slowly over hundreds of years because window panes from medieval churches tend to be thicker at the bottom. In view of measurements under stresses exceeding the gravitational stress by a factor of thousands which show no flow, the observations now seem to be better explained by supposing that the artisans of the time, when presented with hand-made panes of uneven thickness, preferred to set them with the thickest part at the bottom.

[39] F.H. Stillinger and T.A. Weber, *Science* **225**, 983 (1984); **267**, 1935 (1995).

[40] K.D. Ball, R.S. Berry, R.E. Kuntz, F.-Y. Li, A. Proykova and D.J. Wales, *Science* **271**, 963 (1996).

[41] R.J. Speedy and P.G. Debenedetti, *Mol. Phys.* **88**, 1293 (1996).

[42] (a) R.J. Speedy, *J. Phys. Chem.* **97**, 2723 (1993).
(b) R.J. Speedy and P.G. Debenedetti, *Mol. Phys.* **86**, 1375 (1995).

[43] (a) C.A. Angell and K.J. Rao, *J. Chem. Phys.* **57**, 470 (1972).
(b) C.A. Angell, B.E. Richards and V. Velikov, *J. Phys. Condensed Matter* **11**, 75–94 (1999).
(c) J.L. Green, K. Ito, K. Xu and C.A. Angell, *J. Phys. Chem.* **103**, 3991 (1999).
(d) R. Richert and C.A. Angell, *J. Chem. Phys.* **108**, 9016 (1998).

[44] J. Perez, *J. Physique* **46-C10**, 427 (1985).

[45] F.H. Stillinger, *J. Chem. Phys.* **38**, 7818 (1988).

[46] (a) C. Dasgupta, *Europhys. Lett.* **20**, 181 (1992).
(b) C. Dasgupta and O.T. Valls, *Phys. Rev.* **E50**, 3916 (1994).

[47] C.A. Angell, *J. Res. NIST*, **102**, 171 (1997); *Proc. 14th Citges Conf. on Theoretical Physics, 1966*, ed. M. Rubi (Springer), 1967, p. 1.

[48] A. Arzimanoglou and C.A. Angell, unpublished data, summarized in Ref. [16(b)].

[49] U. Mohanty, I. Oppenheim and C.H. Taubes, *Science* **266**, 425 (1994).

[50] C.A. Angell and C.T. Moynihan, Molten Salts; Characterization and Analysis, ed. G. Mamantov (Marcel Dekker, New York, 1969) p. 315.

[51] C.T. Moynihan, C.R. Smalley, C.A. Angell and E.J. Sare, *J. Phys. Chem.* **73**, 2287 (1969).

[52] F.S. Howells, R.A. Bose, P.B. Macedo and C.T. Moynihan, *J. Phys. Chem.* **78**, 639 (1974).

[53] J. Kawamura and M. Shimoji, *J. Non-Cryst. Sol.* **88**, 295 (1986).

[54] L. Boehm and C.A. Angell, *Proc. Fast Ion Transport in Solids Conf.*, Lake Geneva, Wisconsin, eds. P. Vashishta, J.N. Mundy and G.K. Shenoy (Elsevier, North Holland, 1979) p. 719.

[55] H. Fujimori and M. Oguni, *J. Chem. Therm.* **26**, 367 (1994).

[56] J.H. Magill, *J. Chem. Phys.* **47**, 2802 (1967). It is now known that the tri-napthyl benzene used in Macgill's work was not the tri-alpha isomer but the 1,3-bis(1-napthyl)-5 (2-napthyl) benzene (R.L. McMahon and C.M. Whitaker).

[57] Y. Privalko, *J. Phys. Chem.* **84**, 3307 (1980).

[58] C. Alba, L.E. Busse and C.A. Angell, *J. Chem. Phys.* **92**, 617 (1990).

[59] F. Stickel, E.W. Fischer and A. Schönhals, *Phys. Rev. Lett.* **73**, 2936 (1991).

[60] F. Stickel and E.W. Fischer, *Physica* **A201**, 263 (1993).

[61] F. Stickel, E.W. Fischer and R. Richert, *J. Chem. Phys.* **104**, 2043 (1996).

[62] C.A. Angell, C. Alba, A. Arzimanoglou, R. Böhmer, J. Fan, Q. Lu, E. Sanchez, H. Senapati and M. Tatsumisago, *Am. Inst. Phys. Conf. Proc.* **256**, 3 (1992).

[63] R. Böhmer and C.A. Angell, *Phys. Rev.* **B45**, 10091 (1992).

[64] R. Böhmer, K.L. Ngai, C.A. Angell and D.J. Plazek, *J. Chem. Phys.* **99**, 4201 (1993).

[65] R. Böhmer and C.A. Angell, *Phys. Rev.* **B48**, 5857 (1993).

[66] C.A. Angell, *Polymer* **38**, 6261 (1997).

[67] R.O. Davies and G.O. Jones, *J. Adv. Phys.* **2**, 370 (1953).

[68] R.O. Davies and G.O. Jones, *J. Proc. Roy. Soc.* (London) **A217**, 26 (1953).

[69] M. Goldstein, *J. Chem. Phys.* **39**, 3369 (1963); *J. Phys. Chem.* **77**, 667 (1973).

[70] C.T. Moynihan, A.V. Lesikar, *Ann. N.Y. Acad. Sci.* **371**, 151 (1981).

[71] Actually, the Adam–Gibbs equation predicts that the product TS_c, rather than Sc alone, should be constant at the transition. This leads to a different version of Eq. (6) which, however, does not change the prediction of Eq. (6) much beyond experimental error for fragile liquids. For $ZnCl_2$, a semi-strong liquid, a discrepancy in the correct direction was found (C.A. Angell, E. Williams, K.J. Rao and J.C. Tucker, *J. Phys. Chem.* **81**, 238 (1977); C.A. Angell and W. Sichina, *Ann. N.Y. Acad. Sci.* **279**, 53 (1976)). The constancy of TS_c at T_g has been directly demonstrated in the experiments of S. Takahara, O. Yamamuro and H. Suga, *J. Non-Cryst. Solids* **171**, 259 (1994).

[72] R. Böhmer and C.A. Angell, *Disorder Effects on Relaxational Processes*, eds. A. Blumen and R. Richert (Springer, Berlin, 1994) p. 11.

[73] I.M. Hodge, *J. Res. NIST* **102**, 195 (1997).

[74] H. Rawson, *Inorganic Glassforming Systems* (Academic Press, London) 1967, p. 216.

[75] R. Zallen, *The Physics of Amorphous Solids* (John Wiley & Sons, New York, 1983).

[76] S.R. Ovshinsky, *Phys. Rev. Lett.* **21**, 1450 (1968).

[77] J.C. Phillips, *J. Non-Cryst. Solids* **34**, 153 (1979).

[78] J.C. Phillips and M.F. Thorpe, *Solid State Commun.* **53**, 847 (1986).

[79] M.F. Thorpe, *J. Non-Cryst. Solids* **57**, 355 (1983).

[80] B.L. Halfpap and S.M. Lindsay, *Phys. Rev. Lett.* **57**, 847 (1986).

[81] B.L. Halfpap, PhD Thesis, Arizona State University (1990).

[82] M. Tatsumisago, B.L. Halfpap, J.L. Green, S.M. Lindsay and C.A. Angell, *Phys. Rev. Lett.* **64**, 1549 (1990).

[83] C.A. Angell, S. Borick and M. Grabow, *J. Non-Cryst. Solids* **205–207**, 463 (1996).

[84] (a) C.A. Angell and E.J. Sare, *J. Chem. Phys.* **49**, 4713 (1968); **52**, 1058 (1970).
(b) H. Kanno, *J. Phys. Chem.* **91**, 1967 (1987).
(c) D.R. MacFarlane, *Cryoletters* **6**, 33 (1985); **23**, 230 (1986).

[85] N.J. Kreidl, Glass: Science and Technology, Vol. 1 (Academic Press, New York, 1983) p. 105.

[86] U. Senapati and A.K. Varshneya, *J. Non-Cryst. Sol.* **185**, 289 (1995).

[87] J. Lucas, H.L. Ma, X.H. Zhang, H. Senapati, R. Böhmer and C.A. Angell, *J. Sol. State Chem.* **96**, 181 (1992).

[88] D.J. Plazek and K.-L. Ngai, *Macromolecules* **24**, 1222 (1991).

[89] U. Buchenau and R. Zorn, *Europhys. Lett.* **18**, 523 (1992).

[90] B. Frick and D. Richter, *Phys. Rev.* **B47**, 14795 (1993).

[91] B. Frick, D. Richter, W. Petry and U. Buchenau, *Z. Phys.* **B70**, 73 (1988).

[92] C.A. Angell, P.H. Poole and J. Shao, *Nuovo Cimento* **D16**, 993 (1994).

[93] J. Dubochet, M. Adrian, J. Teixeira, R.K. Kadiyala, C.M. Alba, D.R. McFarlane and C.A. Angell, *J. Phys. Chem.* **88**, 6727 (1984).

[94] C.A. Angell, *J. Am. Ceram. Soc.* **51**, 117 (1968).

[95] J. Shao and C.A. Angell, *Proc. XVIIth Int. Congress on Glass*, **ed. Not known**, **1**, (Beijing, 1995), p. 311.

[96] C.A. Angell, *Computational Mat. Sci.* **4**, 285 (1995).

[97] We are indebted to Sir Roger Elliott of Oxford University for pointing out the relation to crystalline state behavior. Simulations of crystals show that the corresponding ringing phenomenon persists for a very long time.

[98] J. Souletie, *J. Phys.* **51**, 883 (Paris, 1990).

[99] C.A. Angell and E.J. Sare, *J. Chem. Phys.* **52**, 1058 (1970).

[100] O. Mishima, L.D. Calvert and E. Whalley, *Nature* **310**, 393 (1984); **314**, 76 (1985).

[101] O. Mishima, *J. Chem. Phys.* **100**, 5910 (1994).

[102] G.H. Wolf *et al.*, *High-Pressure Research: Application to Earth and Planetary Sciences*, eds. Y.S. Manghani and M.H. Manghani (American Geophysical Union, Washington, DC, 1992), pp. 503–517.

[103] P.H. Poole, F. Sciortino, U. Essmann and H.E. Stanley, *Nature* **360**, 324 (1992); *Phys. Rev.* **E86**, 982 (1982).

[104] L. Haar, private communication, see plot of the Haar, Gallagher–Kell equation in Q. Zheng, D.J. Durben, G.H. Wolf and C.A. Angell, *Science* **254**, 829 (1991).

[105] S.B. Kiselev, J.M.H. Levelt Sengers and Q. Zheng, Physical Chemistry of Aqueous Solutions: Meeting the Needs of Industry, eds. H.J. White Jr., J.V. Sengers, D.B. Neumann and J.C. Bellows (Begell House: New York) 1995, p. 378.

[106] P.H. Poole, F. Sciortino, T. Grande, H.E. Stanley and C.A. Angell, *Phys. Rev. Lett.* **73**, 1632 (1994).

[107] C.A. Angell, *J. Phys. Chem.* **97**, 6339 (1993).

[108] F.H. Stillinger and T. Weber, *Phys. Rev.* **B31**, 5262 (1985).

[109] M. H. Grabow and G.H. Gilmer, *MRS Symposia Proc.* **141**, 341 (1989).

[110] W.D. Lüdke and U. Landman, *Phys. Rev.* **B37**, 4656 (1988); **40**, 1164 (1989).

[111] R.W. Douglas and J.O. Isard, *J. Soc. Glass Technol.* **35**, 206 (1951).

[112] R. Bruckner, *J. Non-Cryst. Solids* **5**, 281 (1971).

[113] M.O. Thompson, G.J. Galvin and J.W. Mayer, *Phys. Rev. Lett.* **52**, 2360 (1984).

[114] M. Lohfink and H. Sillescu, *AIP Conf. Proc.* **256**, 30 (1992).

2

Dual Nature of Molecular Glass Transitions

J. C. PHILLIPS

Bell Laboratories, Lucent Technologies,
Murrary Hill, N. J. 07974-0636
jcp@physics.bell-labs.com

Contents

1. Introduction

Most materials crystallize almost immediately when cooled slightly below their freezing points, so that the ability to form a glass is quite unusual. There are a number of mechanisms that can provide sufficiently strong barriers to crystallization to form glasses in a melt cooled at 10 K/s or less. We can identify three common kinds of glasses: two- and three-dimensional network formers (such as As_2S_3 and SiO_2); polymers (long chains like polybutadiene, PB), and large organic dipolar molecules (alcohols, such as glycerol). Among the polymers it is clear that the best glass formers (those with apparently negligible crystallization) are the ones with the longest chains and the smallest and least interactive side groups, which are most likely to solidify without microcrystalline ordering. For a long time the origin of glass formation among network formers was less obvious: ionic aluminosilicates (clays) have strong and very characteristic crystalline network structures, similar to ionic SiO_2, but they are poor glass formers, while the crystal structures of covalent As_2S_3 and ionic SiO_2 glasses seem to be quite different. The uniquely successful quantitative explanation for the glass-forming ability of network materials based on hierarchical valence force fields (principally bond-stretching and bond-bending) and constraint counting is discussed elsewhere in this volume by Thorpe. This theory is more accurate and detailed than the general theory of the visco-elastic properties of polymers developed by Flory, Doi and Edwards and De Gennes. At present there is no quantitative theory of alcoholic molecular glass formers, where electric dipolar disorder is the crystallization-inhibiting mechanism, but there is a large and rapidly growing body of data on molecular and ionic glass formers which is the focus of this paper.

2. Prototypical Molecular Glasses

Inert gases (such as argon) are stabilized in their condensed phases attractive Van der Waals and repulsive electron exclusion forces, but they crystallize easily and may actually be the worst glass-formers. In molecular dynamics simulations (MDS) glassy relaxation has been achieved [1] in soft sphere mixtures with a size ratio on the edge of phase separation on a time scale of 10^{-8} s, but even these systems would probably nucleate and begin to crystallize on a somewhat longer time scale (say μs or ms). There are only a few non-alcoholic molecular materials which do not crystallize on laboratory time scales: two of the most-studied are orthoterphenyl (OTP) and $(Ca,K_2)(NO_3)_2$ (CKN). The former is a planar hydrocarbon, while the latter is a fused salt. It seems likely that the reason OTP is a good glass former while *para* terphenyl crystallizes readily is that the former is strongly bent while the latter is linear. Similarly the glassforming ability of CKN can be traced to two combined effects, the obstacles to cation diffusion presented by planar NO_3 ions, and metastable three-center coordination configurations where a mixture of divalent Ca and monovalent K cations are shared between three (or more) non-coaxial nitrate anions. These two materials exhibit strongly curved (non-Arrhenius) viscosities $\log \eta(T)$ which are often fitted with the Vogel–Fulcher expression $-DT_0/(T-T_0)$, where T_0 is a kinetic glass transition temperature close to the thermal glass transition temperature T_g. Glasses with such large curvatures have been termed "fragile" to contrast them [2] with "strong" network-forming glasses with Arrhenius viscosities. Here we will see that a better description of fragile molecular glasses recognizes that there are two glass transition temperatures, T_g and T_k, and that these two temperatures are dual in several rigorous mathematical descriptions of these transitions in configuration space.

3. Kinetic Data

For many years Angell and coworkers have studied the glass transition in CKN in an effort distinguish the nature of its glass transition from those in network and polymeric glasses [2]. Their strong/fragile dichotomy is illustrated in Fig. 1. Network glasses, such as SiO_2, are mechanically strong and exhibit nearly Arrhenius viscosities when measured by a variety of methods. The two materials with the largest curvature of $\log \eta(T)$, on the lowest curve, are CKN and OTP. The methods traditionally used by chemists are mostly macroscopic, and if the samples are annealed long enough (typically 30 hrs for CKN with T near T_g) yield results for the relaxation time τ and the stretching fraction β which generally agree reasonably well with those measured by microscopic methods (better than 10% on β). Here β and τ refer to the parameters used to fit a measured quantity (such as stress I) which is relaxing according to the stretched exponential or Kohlrausch formula

$$I(t) = I(0) \exp[-(t/\tau)^{\beta}] \tag{1}$$

with $0 < \beta < 1$. The viscosity η is assumed to be proportional to the relaxation rate $1/\tau$ as determined indirectly from relations such as (1), and the values of $\tau(T)$ obtained in this way generally agree fairly well with those obtained more directly using, for example, tracer molecules with sizes close to that of the host supercooled liquid or glass.

Note that the range of $\eta(T)$ in plots such as shown in Fig. 1 is about 10^{15} which means that the data are usually obtained by several experimental methods. This naturally raises the question of whether the different methods yield the same or very similar values of τ and β. The answer is that there is usually enough overlap to obtain reasonably smooth curves for τ, but in

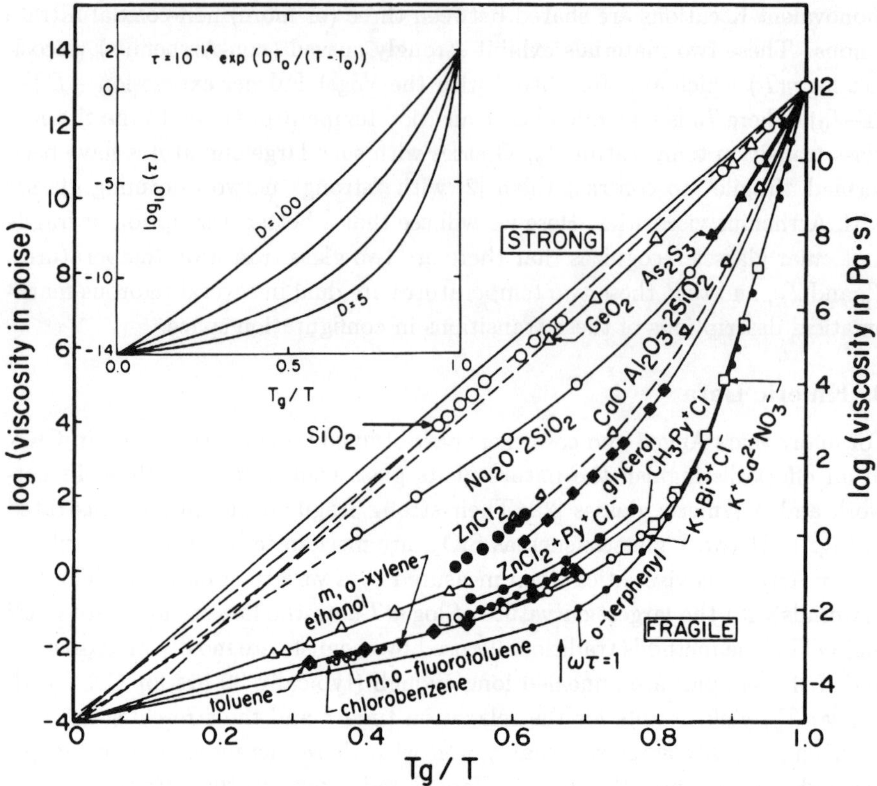

Fig. 1. Fragile glasses (molecular or ionic) have strongly curved $\log\eta(T)$ viscosities, in contrast to strong network glasses, where the same functional relation is essentially linear. The inset shows the Vogel–Fulcher fitting function $\log\eta(T) = \mathrm{const.} - DT_0/(T - T_0)$ for several values of D. The figure is from Ref. [2] and is reproduced here for the reader's convenience.

most cases the accuracy of macroscopic methods is insufficient to demonstrate convincingly, for example, that different relaxation mechanisms are involved, for instance, above and below the fragility knee near $T_k/T_g = 0.75$ in Fig. 1. The situation for β was formerly quite confused, but there has been great progress recently in understanding this parameter, as will appear in Sec. 7.3. In any case, a number of recent microscopic experiments have clarified the origin of the fragility knee, as we shall now see.

4. Diffraction Data: The Upper Glass Transition

The structure factors of good glass formers are never simple; this rule applies to both CKN and OTP. The first two peaks of $S(Q)$ in OTP, at $Q_1 = 1.4$ and $Q_2 = 1.9$ A^{-1}, confirm this rule [3]. At $T_g + 70$ K, the two peak heights are approximately equal, but near and below T_g, the height of the second peak has increased by about 10% and its position has shifted to 2.0 A^{-1}, while the height and position of the first peak are little changed. The origin of these peaks was a mystery [3] which we partially resolve here. OTP molecules are nearly planar, and so a layered local structure is an obvious possibility. Crystals of OTP have never been grown, but the crystal structure of biphenyl has been determined with high accuracy [4]. There are a large number of C–C interlayer spacings between 3.7 and 3.9 A so that an average of 3.8 A is reasonable. Similarly $(Q_1 + Q_2)/2 = 1.65$ A^{-1}, and the product of these two averages is essentially 2π. This means that the unsplit peak here has the same simple physical interpretation [5] as the first sharp diffraction peak in chalcogenide alloy glasses such as As_2S_3 and GeS_2. The actual peak splitting could be the consequence of subtle interference effects, and the magnitude of this splitting (FWHM = 0.25 A^{-1}) probably represents the breadth of the peak and cannot be used to obtain, for example, two different planar spacings. However, the magnitude of the splitting suggests that the correlation length for the structure must be larger than 20 A. Moreover, the fact that the splitting does not change much with temperature tells us that the short range order defined by this structure has not changed much between $T = T_g$ and their highest measuring temperature, $T_g/T = 0.77$. Referring to Fig. 1, we see that this temperature lies just above the fragility knee at $T_g/T_k = 0.75$. If neutron diffraction data could be obtained at higher temperatures below this knee (near $T_g/T = 0.62$, for example), they would be of great interest, for the slope of the splitting as a function of T might change just at $T = T_k$.

The dynamical intermediate scattering function $S(Q, t, T)$ of OTP has been studied extensively by spin-polarized neutron scattering [6] with $Q = Q_1$. The coherent Debye–Waller factor $f_Q(T)$ exhibits a singular term $(1 - T/T_c)^{1/2}$ with $T_g/T_c = 0.84$. We are tempted to call T_c the upper glass transition temperature, a temperature at which density fluctuations are arrested, while at lower temperatures rotational or angular degrees of freedom freeze out. However, because all the data were taken only for $Q = Q_1$, there are still many unanswered questions, such as what temperature dependence would be found if $f(T)$ were measured for a range of Q values between 1.3 and 2.2 A^{-1}: in other words, does this singularity describe the behavior of the unsplit peak (whose structural significance we understand in a simple way) or does it refer only to the lower peak? Answers to these questions might help us to understand why T_c falls halfway between the fragility knee at T_k and the thermal glass transition at T_g. The mean square displacement $\langle r^2(T)\rangle$ is linear (as in a crystal) for $T < T_g$ and increases rapidly between T_g and T_c with a value that extrapolates to $r_c = 0.8$ A. It is interesting to note that $r_c/3.8$ A $\sim 2(Q_2-Q_1)/(Q_1+Q_2)$, so that the displacements responsible for the peak splitting are essentially similar in magnitude to the thermal vibrational distortions at the upper transition temperature, that is, they may refer to angular or rotational degrees of freedom, rather than radial ones (density fluctuations), which in turn are related to the average value of Q.

The structural data for CKN obtained by neutron scattering [7] are quite different from those for OTP, as one would expect from the reduced planarity of its component ions. Instead of the density and rotational fluctuations of OTP we may expect to encounter density and charge fluctuations, and this indeed seems to be the case. While early experimental data were interpreted as containing evidence for $(1-T/T_c)^{1/2}$ singularity in static structure factor, on the grounds that this was a universal effect predicted by a heuristic theory, the more recent data have been interpreted more modestly, by studying the height of the largest structure factor peak (linear in T with a break in slope at T_g) and $\langle r^2(T)\rangle$ as determined from the same peak (linear in T, with a break in slope at "T_g", where $T_g/T_c = 0.91$). Again we notice that this procedure yields a value of T_g/T_c which is much too large to be associated with the fragility knee at $T_g/T_k = 0.75$. Apparently the phase sensitivity of diffraction data emphasizes medium-range order [5], while transport data are more sensitive to the onset of short-range order. According to the latter, CKN and OTP are quite similar as regards short-range order, but the former exhibits quite different medium-range order, reflecting large qualitative differences between the building blocks involved. At the same time it appears that the development of structural arrest really begins at T_k with the development of short-range order. This means that the non-universal nature of the structural data limits the insight into the nature of the glass transition that can be obtained in this way.

5. NMR Relaxation Data

Much more microscopic information on diffusion near the glass transition in OTP can be obtained from dynamical NMR measurements [8] in a homogeneous magnetic field (which gives a relaxation time τ_r associated with rotations of C–^{2}H bonds and a diffusion coefficient D_r) or in a magnetic field gradient (which gives a translational relaxation time τ_t and D_t). The length scales involved in these two measurements are unavoidably quite different, so the observed result that the two diffusion coefficients have different temperature dependencies is not surprising. The interesting point, however, is that for OTP the two appear to be the same above T_c, where they agree with extrapolated tracer measurements, and diverge only below it, which suggests an intrinsic effect related to the freezing of density fluctuations seen by dynamical spin neutron diffraction (see above). It was suggested [8] that this effect represents the formation around T_g of correlated "solid" (slowly relaxing) regions 50 A in diameter separated by "liquid" (fast relaxing) regions. We note that this solid domain length of 50 A near T_g is larger than the structural static correlation length of 20 A deduced above from the peak splitting for higher temperatures near the fragility knee, as it must be.

In CKN the ^{15}N static NMR spectrum is motionally narrowed at $x = T_g/T = 0.90$, but as T approaches T_g a spectrum characteristic of the uniaxial symmetry of the nitrate ion emerges [9]. The spectrum can be deconvoluted assuming that there are two relaxation modes between $x = 0.92$ and 0.95 with τ's that differ by about a factor 10. This region of x is very close to the glass transition and the presence of bimodal relaxation at such temperatures is strongly suggestive of the presence of at least two kinds of regions. Once again, however, we are far from the fragility knee, so that in CKN NMR may be most sensitive to solid-like regions only when these are rather large compared to the static structure which may already form, as in OTP, at temperatures close to the fragility knee ($T_g/T_k = 0.75$).

6. Dual Relaxation Modes

The efforts described above to identify the formation of "solid" clusters embedded in a liquid-like matrix involve direct methods whose details vary greatly from one material (OTP) to the next (CKN), even though the fragilities of these two materials, as described by the curvatures of $\eta(T_g/T)$ shown in Fig. 1, are virtually the same. In such circumstances one cannot hope to obtain much insight into the mechanism responsible for the crossover at T_k from a higher temperature region of modest $d\eta/dT$ with an activation energy of order a bond-breaking energy kT_b, where T_b is the boiling point, to a much larger activation energy of order $5kT_b$ near the glass transition. It is clear that what is needed to analyze fragility is a more general or universal method which can identify dual behavior universally at a microscopic level. At one time it was hoped that heuristic mode-coupling theory [10] might be such a method, but this has not proved to be the case even for only OTP (where freezing of density fluctuations was observed by diffraction [6], albeit not at T_k, but only at $T_c \sim (T_k + T_g)/2$) and CKN (where no freezing is observed [7] above T_g). Indeed, as pointed out above, the microscopic quantity most easily observed to have a break in slope at T_k, the splitting of the first peak in $S(Q)$, has yet to be even studied for $T > T_k$.

Some time ago Angell gave an excellent review [2] of the macroscopic aspects of fragility and their relation to various measures of dynamical relaxation, both microscopic and macroscopic. At that time it seemed as if a different relaxation time τ was associated with each probe, although the curved nature of $\log \eta(T)$ was very similar in all of them. It was also obvious a decade ago that by far the most satisfactory description of dynamical relaxation is provided by the stretched exponential (which is not contained in mode-coupling theory [10]), although no one understood the microscopic basis for this success in the context of molecular relaxation. It was also evident that the stretching fraction β correlated quite well with differences in relaxation times τ in certain experiments; for instance, shorter τ correlated with larger β. This is after all not so surprising, as for $T > T_m$ one has generally simple exponential (Debye) relaxation with $\beta = 1$. This idea has been extended [11] to define a fragility index $m = \lim(\mathrm{d}\log \tau/dx)_{x=1}$ and it is found that m correlates roughly linearly with β at T_g in two groups, (polymers and networks) vs. (simple and complex molecules and alcohols). Unfortunately these heuristic correlations are rather broad and do not provide microscopic insight into the nature of β itself, but the correlation is of considerable bibliographic value.

The most substantial progress has occurred because it has recently become possible to understand the origin of stretched exponential relaxation $I(t)$ at a microscopic level [12]. For a long time it was thought that $I(t)$ was only useful for fitting curves, and that β was no more than a felicitous curve-fitting parameter with no microscopic meaning. About twenty years ago, however, it was discovered (first in one dimension [13] and later more generally [14–16]) that the Kohlrausch function provided the asymptotic solution to a natural physical problem, that of dispersive current transport in a material with a high trap density as measured either with pulsed currents or through non-radiative exciton decay. In this model the origin of the stretched exponential is easy to understand. As the traps or sinks absorb excitations (which can be either real particles or entropy packets in the case of structural relaxation), the density of excitations near the traps is depleted, resulting in a decreasing rate of relaxation as particles must diffuse increasingly large distances before being absorbed at the traps or sinks. Once this had been demonstrated there was a tendency to declare the problem *solved* (at least among some theorists [17]), even while experimentalists still regarded it as one of the major *unsolved* problems of physics [18].

This divergence of viewpoints arose because the theory seemed to be incapable of predicting the values of β observed in experiments on different materials with different pumps and probes. When the relaxation takes place entirely through short-range interactions at randomly distributed static traps in a configuration space of restricted dimensionality d^*, the theory unambiguously predicts $\beta = d^*/(d^* + 2)$, and with the unrestricted value $d^* = 3$ this gives $\beta = \beta_{SR} = 0.6$. However, experimentally a wide range of values of β between 0.9 and 0.1 is observed, corresponding to apparently arbitrary fractal configuration space dimensionalities d^*, even at low temperatures where β has become temperature-independent. The natural explanation for this is that residual relaxation depends on the impurity structure in some way. However, if this is the case, then there is no reason for the observed invariance of the stretched exponential functional form either, and we are left no better off than Kohlrausch [19].

While the problem of connecting the theory of dispersive transport, with depletive effects near traps, to residual relaxation experiments quantitatively has long been deemed insoluble, a recent lengthy survey [12] of conceptual and mathematical models, computer simulations and a very wide range of experimental data has shown that the situation can be greatly clarified by combining fundamental understanding of the mathematical problems involved in defining d^* with a detailed analysis of material properties. One of the key steps was the realization that the randomly distributed sinks which are traps in the electronic case become relaxation centers in the molecular case. These relaxation centers act as "black holes" for the non-ergodic dissipation of structural excitations. The local nature of the relaxation centers is irrelevant to the value of β; all that is required is that the diffusion coefficient of these centers be small compared to that of medium excitations, and similarly for their decay rate. In this way it becomes possible to separate extrinsic and intrinsic effects, and to recognize that for dispersive transport there are predominantly two, and only two, possiblities for the dimensionality of configuration space d^*. These are either the physical dimensionality d when relaxation occurs through density fluctuations, or $d^* = d/2$, corresponding to fractal dispersive transport (FDT). When the relaxation occurs through charge fluctuations FDT was explained by an Ewald-like construction which assumed that collective plasma fluctuations are ineffective in linearly relaxing individual particles, much as only phonons, but not phasons, contribute to linear diffusion and relaxation in quasicrystals [20]. However, when the relaxation occurs in a hindered environment, such as near kinks in a polymer chain embedded in a bundle of such chains, a different explanation for FDT is possible in the context of dimensional regression [21].

The choices $d = 3$ or $d = 3/2$ correspond to $\beta = 3/5$ and $\beta = 3/7 = 0.43$, denoted [12] by β_{SR} and β_K, respectively. It is instructive to review [12] which values are obtained, within a few percent, with different experimental probes on OTP and CKN. For OTP β_{SR} is observed by both pulsed specific heat (0.60) and spin-polarized neutrons (0.62). For CKN β_K is observed with ultrasonic and Brillouin scattering, but β_{SR} is observed with spin-polarized neutrons. More recently with multidimensional NMR (fast and slow nested spin echoes) both values are found in OTP, β_K in the short time window, and β_{SR} in the long time window [22].

It is not surprising that even for a given material β varies with probe, but what is not only surprising but in fact unprecedented, is that only two values of β are observed, and these correspond quite simply to $d = 3$ or $3/2$. The microscopic significance of such an astonishingly simple result has already been discussed at length [12, 21], and it seems likely that the discussion has just begun and may not end for several decades. It may be, for instance, that the slow and fast relaxation modes observed [9] in CKN are associated with β_K and β_{SR}, respectively. One could also ask where these two modes are located spatially. Are the slow modes associated with solid-like regions [23] as suggested [9], or is another explanation possible? (For example, the liquid-like regions themselves could be of two types, interfaces between solid-like regions which are quasi-two-dimensional [21] and are associated with slow relaxation, and three-dimensional interstitial free volume [23] pockets associated with fast relaxation.) We will not pursue it further here, but will leave such analysis now to others who may well have new insights derived from further experiments.

References

[1] J.N. Roux, J.L. Barrat and J.-P. Hansen, *J. Phys.: Condens. Matter* **1**, 7171 (1989).

[2] C.A. Angell, *J. Non-Cryst. Sol.* **102**, 205 (1988); *ibid.*, **131 & 133**, 13 (1991).

[3]   E. Bartsch, H. Bertagnolli, P. Chieux, A. David and H. Sillescu, *Chem. Phys.* **169**, 373 (1993).

[4]   A. Hargreaves and S. Hasan Rizvi, *Acta Cryst.* **15**, 365 (1962).

[5]   J.C. Phillips, *J, Non-Cryst. Solids* **43**, 37 (1981).

[6]   W. Petry *et al.*, *Z. Phys.* **B83**, 175 (1991).

[7]   E. Kartini, M.F. Collins, B. Collier, F. Mezei and E.C. Svensson, *Phys. Rev.* **B54**, 6292 (1996).

[8]   I. Chang *et al.*, *J. Non-Cryst. Solids* **172–174**, 248 (1994).

[9]   S. Sen and J.F. Stebbins, *Phys. Rev. Lett.* **78**, 3495 (1997).

[10]   W. Gotze and L. Sjogren, *Rep. Prog. Phys.* **55**, 241 (1992).

[11]   R. Bohmer, K.L. Ngai, C.A. Angell and D.J. Plazek, *J. Chem, Phys.* **99**, 4201 (1993).

[12]   J.C. Phillips, *Rep. Prog. Phys.* **59**, 1133 (1996).

[13]   H. Scher and M. Lax, *Phys. Rev.* **B7**, 4491 (1973).

[14]   R. Friedberg and J.M. Luttinger, *Phys. Rev.* **B12**, 4460 (1975).

[15]   M.D. Donsker and S.R. Varadhan, *Commun. Pure and Appl. Math.* **28**, 525 (1975).

[16]   P. Grassberger and I. Procaccia *J. Chem. Phys.* **77**, 6281 (1982).

[17]   H. Scher, M.F. Shlesinger and J.T. Bendler, *Phys. Today* **44** (1), 26 (1991).

[18]   W.F. Brinkman *et al.*, *Physics Through the 1990's* (National Academic Science Press, Washington DC, 1986).

[19]   R. Kohlrausch, *Ann Phys. Lpz.* **12**, 393 (1847); *Poggendorf's Ann. Phys.* **91**, 56, 179 (1854); J. Jackle, *Phil. Mag.* **B56**, 113 (1987).

[20]   M. Dzugutov and J.C. Phillips, *J, Non-Cryst. Solids* **192 & 193**, 397 (1995).

[21]   J.C. Phillips and J.M. Vandenoorg, *J. Phys.: Cond. Matt.* **9**, L251 (1997).

[22]   R. Bohmer, G. Hinze, G. Diezemann, B. Geil and H. Sillescu, *Europhys. Lett.* **36**, 55 (1996).

[23]   G.S. Grest and M.H. Cohen, *Adv. Chem. Phys.* **48**, 454 (1981).

3

The Generic Phenomenology of Glass Formation

I. GUTZOW

Institute of Physical Chemistry, Bulgarian Academy of Sciences,
Sofia 1113, Bulgaria
gutzow@ipchp.ipc.bas.bg

Contents

1. Introduction and Historic Background

Most semiconducting and dielectric substances are typical glass formers and their technical applications depend greatly on the properties of matter in the vitreous state.

In the last years, the interest in the theoretical problems of glass science and of the process of glass formation (or vitrification, as it is better called), as well as in the very nature of glass as a physical state, and in the structure of diverse glasses has increased considerably. There are even statements in current literature (see e.g. [1, 2]) that vitrification is the greatest challenge in contemporary solid state physics. In an attempt to meet this challenge most of the efforts of present day investigators are focused on constructing more or less complicated atomistic models of the process of glass formation, e.g. in terms of mode coupling theory, bond fluctuation model (see [3, 4]) and related general approaches. An overview of these efforts and developments can be found in [5, 6] and in the already cited paper of Binder [2].

The atomistic approach has in every science the great advantage of the *ab initio* formulation of problems. However, in treating the kinetics of vitrification and glass relaxation the complicated atomistic models used to describe these processes gave but relatively limited results: as summarized in [2] and [6] in most cases only a qualitative coincidence with experiment and with already well known empirical dependences was obtained. Further progress in the framework of these and similar atomistic models of vitrification according to [2, 6] may be expected only by computer simulation, and this is a fairly tedious route.

As in other cases, however, in this field of science there exists still another possibility, another way alternative to atomistic and computer modeling: the phenomenological approach in describing and solving the problems of glass formation and of the kinetic stability of glasses. The present paper is devoted to this formal approach based on the thermodynamic analysis of both processes: glass formation and glass relaxation.

However, classical thermodynamics is only applicable to equilibrium states. It is valid only for different equilibrium forms of appearance of the same substance: with thermodynamic phases.

Following the traditions and the molecular and structural concepts of the 18th and the 19th centuries, present day molecular physics still divides matter into three states of aggregation: gaseous, liquid and solid (crystalline). Sometimes, in analogy to the beautiful schemes from Hellenistic times (see [7]), where four elemental forces constituting the Universe are combined (air, water, earth and fire), a fourth state of aggregation is introduced for the partly or totally ionized state of matter — the plasma state.

From the viewpoint of classical thermodynamics, the states of aggregation are thermodynamic phases and thus also equilibrium states of a system. Thermodynamic phases may coexist in equilibrium (as different states of aggregation or as various phases in the same state of aggregation), their coexistence being governed by Gibb's Phase Rule. By changing the external parameters phase transformations are initiated, their thermodynamic classification being first given by Ehrenfest [8].

The unusual properties of glasses, their twin position as amorphous solids and their liquid-like structures led in the early 1930's to the proposal (made by Berger [9] and then by Parker, for details see [10]) to characterize the vitreous state as the fourth state of aggregation of matter (beside gases, liquids and crystals). During the long years of debate on the nature of glass, described also in [10], several attempts were made to classify vitrification in line with Ehrenfest's thermodynamic scheme, i.e. as a first, second and even as a third order thermodynamic phase transition.

However, all these and similar proposals (some of them historically and emotionally interesting, others giving more a testimony of their author's limited thermodynamic knowledge) could not solve the problems related to the definition and classification of glasses and of the vitreous state as a particular physical state.

According to results accumulated at the beginning of the 1930's and mainly thanks to Simon [11], it was already known at those times that glasses are non-equilibrium, frozen-in systems in which the amorphous structure corresponding to the liquid state is arrested. This statement means, however, that glasses are non-thermodynamic systems and are thus excluded from any classical thermodynamic classification either as phases or as states of aggregation, and vitrification cannot be considered as a process of phase formation. In this sense, the historically introduced terms "glass formation" and "glass transformation temperature" T_g (and "glass transformation region") are misleading. As it will become evident from the discussion to follow, nothing new is formed upon vitrification and nothing is transformed from one structure to another. On the contrary, in the vicinity of temperatures close to T_g motion and structure and structural changes are arrested and the liquid melt stiffened into the rigid structure of the glass.

The thermodynamic treatment of frozen-in states is only possible in the framework of a phenomenological science that developed from classical thermodynamics also in the 1930's, mainly thanks to the efforts of de Donder and Prigogine (see [10, 12] and [13]): in the *Thermodynamics of Irreversible Processes*. It can even be stated that this further development and generalization of thermodynamics as a science was greatly called by the necessity to describe frozen-in states (with glasses as the most typical representatives) and steady-state-processes (e.g. life). Classical thermodynamics is capable of analyzing only equilibrium states (i.e. states which are single-valued functions of the external parameters of the system) and the possibility of changes in between them. Frozen-in states become thermodynamically defined according to de Donder's approach by the introduction of additional frozen-in internal parameters. This procedure, applied originally to the problems of glass science by Prigogine and Defay [12], Davies and Jones [14], and Kanai and Satoh [15] (for more details, see again [10]), turned out to be exceptionally efficient and examples illustrating its potential are:

- the explanation of one of the basic features in the kinetics of vitrification: the jump in the thermodynamic coefficients at T_g (see [10, 12]);
- the derivation of the remarkable expression (an analog to the Ehrenfest equations [8, 10]) now known as the Prigogine–Defay Ratio [10, 12], interconnecting the changes in the three thermodynamic coefficients at T_g;
- the definition of glass given by Cooper [16] and of its vapour pressure and solubility we proposed in [17];
- the analysis of the general kinetics of vitrification [18, 20] and the derivation of a kinetic criterion for vitrification [10];
- the dependence of vitrification temperature T_g established by Bartenev [22, 23] and of the thermodynamic properties of glass on the cooling rate q, we derived in [24];
- the geometry of the thermodynamics of vitrification as described in [25, 26].

No other theoretical approach has given such abundant results using so simple premises as the thermodynamics of irreversible processes and especially in its simplest, linear approach, mostly used here.

In the present article, an effort is made to give a generic description of theoretical Glass Science in terms of the phenomenology of the *Thermodynamics of Irreversible Processes*. Thus, we hope to produce not only a summary of results and of the history of a remarkable development, but also to give an outline of the *axiomatics* of glass science.

In order to make the necessary analysis and derivations, we need also a model to relate the kinetic properties of the vitrifying system (time of molecular relaxation, viscosity) to its thermodynamic functions. For equilibrium bodies, the ergodic hypothesis states that the frequency of fluctuations is determined by the statistical mean values characterizing the system (i.e. by the corresponding thermodynamic potential differences). This statement of statistical physics is used here in the derivations in the framework in the generalized form of the thermodynamic model of the Activated State Complex of the Absolute Rate Theory, well known from chemical reaction kinetics (see [27, 28]).

The essential features of this second phenomenological approach to be employed here, relating the kinetic properties of the system (e.g. of the undercooled melt) with its thermodynamic properties, may be traced in the subsequent literature (see [27–29]). Its employment is *a priori* guaranteed only for equilibrium states: it is assumed in the Activated State Model that both the initial system and the activated complex are in equilibrium.

Below T_g the frozen-in melt, the glass, is a non-equilibrium system. Thus it is non-ergodic and the quantitative treatment of kinetic processes possible in glasses requires, in principle, a more general theoretical approach (see [30–32]).

However, by a procedure we employed in [33, 34], the results obtained in the framework of the Activated State Model Approach at equilibrium (i.e. at temperatures T just above the corresponding temperature of vitrification T_g) can be extrapolated also for temperatures below vitrification (i.e. for $T < T_g$, see [35, 36]).

2. Basic Phenomenological Dependences

The foundations of the thermodynamics of irreversible processes were laid by de Donder as summarized in his remarkable little book [37]. Further developments may be followed in [10, 12, 13, 16] and, in an alternative way, also in [38], where earlier ideas of Mandelstam and Leontovich are employed (see [20]).

We exploit here, as we have done in [24], de Donder's basic idea in treating non-equilibrium states: the introduction of internal structural parameters, ξ_i (or reaction coordinates, see [10, 12, 13, 37]).

Let us assume that ξ is some generalized internal order parameter describing with sufficient accuracy the structure of a system and its changes when its external parameters are changed. Let ξ be defined within the limits $0 \le \xi \le 1$ (e.g. $\xi = 0$ denoting complete order, and $\xi = 1$ — complete structural disorder, see [24]). At equilibrium, ξ is a single valued function of the state variables, i.e.

$$\xi_e = \xi(P, T, Z_i) \tag{1}$$

where T and P stand for temperature and pressure, respectively, and Z_i denotes additional external parameters (e.g. electric or magnetic fields, E or M, tangential stress Π, etc., see [33, 34]). At equilibrium the relations

$$(\partial \Delta G/\partial \xi)_{T,P,Zi} = 0; \qquad (\partial^2 \Delta G/\partial \xi^2)_{T,P,Zi} > 0 \tag{2}$$

are fulfilled, ΔG being the configurational part of the Gibbs free energy of the system under consideration (in our case the undercooled melt).

However, when the state of the system is abruptly changed (e.g. at $P =$ const., and $Zi = 0$) by rapid quenching with a velocity

$$q = -(dT/dt) \tag{3}$$

its structure, described by the internal parameter ξ, cannot follow the alteration of temperature. In this case at lower temperatures $(T < \tilde{T})$, a frozen-in state of increased disorder remains in the melt which is characterized by the value ξ corresponding to the temperature T, i.e.

$$\tilde{\xi} = \xi(\tilde{T})_{P,Zi} \tag{4a}$$

Considering also abrupt changes (e.g. at $P, T =$ const.) of one of the additional state variables Z_j, e.g. of electric field strength E, we obtain

$$\tilde{\xi} = \xi(\tilde{Z}_i)_{P,T,Zj}, \qquad (j \neq i) \tag{4b}$$

when the corresponding velocity of change

$$w = -(dZ_i/dt) \tag{5}$$

is higher than the rate of structural alteration. In this way, with Eqs. (4), it becomes evident that the value $\xi(\tilde{Z})$ of the structural parameter ξ will be arrested in the frozen-in system.

At cooling run experiments we have to write for any temperature $T < \tilde{T}$, e.g. for $T = T_g$

$$d\xi/dt|_{T<T_g} = 0 \tag{6a}$$

and also

$$d\xi/dT|_{T<T_g} = 0 \tag{6b}$$

as far as below T_g we have $\xi = \tilde{\xi} = $ const.

An extension to vitrification processes at changing electric fields or pressure and stress may be found in [33, 34]. Here, however, we restrict our derivations to $P = \text{const.} = P_0$ and $Z_i = 0$ (absence of additional external fields and pressure).

The tilde sign ($\sim$) introduced above indicates the non-equilibrium value of the structural parameter ξ to which a corresponding frozen-in value of the state variable T (or Z, P) can be attributed via Eqs. (4). When the cooling rate q corresponds to some standard or conventional value the notation ξ_g is used. The corresponding state variable (e.g. temperature T) is denoted in such a case by the subscript g. Thus T_g denotes the conventional temperature of vitrification (the transformation temperature T_g) corresponding (at $P = \text{const.} = P_o$ and $Z_i = 0$) to cooling rates $q \cong 10^2$ K/sec (see [10, 25]).

For non-equilibrium states it is not Eq. (2) but the expression

$$(\partial \Delta G(P, T, \xi, X)/\partial \xi)_{P,T,Zi} > 0 \tag{7}$$

that holds with the aforementioned provision of ξ (for higher disorder, higher ξ values). The above procedure can also be applied, when the description of the state of the vitrifying system requires more (e.g. i) internal parameters ξ_i.

The thermodynamics of irreversible processes adds to the three basic principles of Classical Thermodynamics an additional Fourth Principle (see [10, 13]). This principle gives the two basic dependences which, as demonstrated here, can be used to derive a general condition for vitrification and to describe the kinetics of both glass formation and glass stabilization.

Usually this Fourth Principle is formulated in the form that thermodynamic fluxes, J (in our case the rates of structural change, $J = d\xi/dt$) are proportional to the respective thermodynamic driving forces, X.

Proceeding in the way as first done by de Donder [37] (see also [10, 13] and the literature given there), i.e. assuming that the force X is, in general, a (non-specified) function of the affinity A of the process of structural change, we can write this principle in the form

$$(d\xi/dt) = F(A) \tag{8}$$

The affinity of the reaction $\tilde{\xi} \to \xi_e$ defined as

$$A = -[\partial \Delta G(P, T, \xi, Zi)/\partial \xi]_{p,T,Zi} \tag{9}$$

plays a major role in the formalism of the thermodynamics of irreversible processes. In Eq. (8) we have to make only the provision that for equilibrium (i.e. for $A \equiv 0$, cf. Eqs. (2) and (8)), the unknown function in Eq. (8) has the value $F(A \equiv 0) = 0$.

Applying further a well known formalism (see [10] and especially [25]), i.e. using a truncated Taylor expansion of $F(A)$ for $A \to 0$ and of $[\partial \Delta G (T, P, \xi)/d\xi]_{P,T,E}$ (at $\xi \to \xi_e$), we arrive at de Donder's classical approximation

$$(d\tilde{\xi}/dt) \cong -[\tilde{\xi}(T) - \xi_e(T)]/\tau \tag{10}$$

Thus Eq. (8) is brought to a dependence stating that the rate of structural change is proportional to the deviation from equilibrium $(\tilde{\xi} - \xi_e)$ and reversely proportional to the characteristic time scale τ of the process under investigation. In the case of melt vitrification, τ in Eq. (9) corresponds to the time of molecular relaxation of the liquid under consideration (see [25]).

Introducing only the simplest linear approximations into the above briefly summarized derivation which is given in detail in [25], τ is defined as a function only of the equilibrium value ξ_e of the structural parameter (i.e. as a single valued function of the state variables) and has to be written as $\tau(T, P, Z_i)$ (or as $\tau(T)$ at $P = $ const., $Z_i = 0$). A more generalized approach, we have also given in [25], provides for relaxation and memory effects in the definition of τ. Thus, for non-equilibrium states (again for $P = $ const., $Z_i = 0$), this kinetic coefficient has to be written in the form $\tau(T, \tilde{\xi})$, where $\tilde{\xi}$ is defined via Eqs. (4). In this way, τ becomes dependent not only on T, but also on ξ and thus (via Eq. (10)) on the time t and on the previous history of the sample [25].

The solution of Eq. (10) reads

$$\tilde{\xi}(t) = \left\{ \tilde{\xi}(t=0) + \int_0^t (\xi_e, \tau) \left[\exp \int_0^t (dt/\tau)dt \right] \right\} \exp \left[- \int_0^t (dt/\tau) \right] \tag{11}$$

for τ independent of ξ, and (thus also of t) and thus, the well known result of a Maxwellian Relaxation Law follows immediately:

$$[\tilde{\xi}(t) - \xi_e(T)] = [\xi(t=0) - \xi_e(T)] \exp[-t/\tau(T)] \tag{12}$$

Different forms of Eq. (10) can be obtained by exploiting different $\tau(T)$ dependences to investigate the kinetics of relaxation at constant external parameters (e.g. at $P, T = $ const., see Fig. 1). However, and this is the essential point here, the same Eq. (12) also gives the possibility to analyse the kinetics of vitrification as τ-values change with state the variables such as temperature and structure, $\tilde{\xi}$.

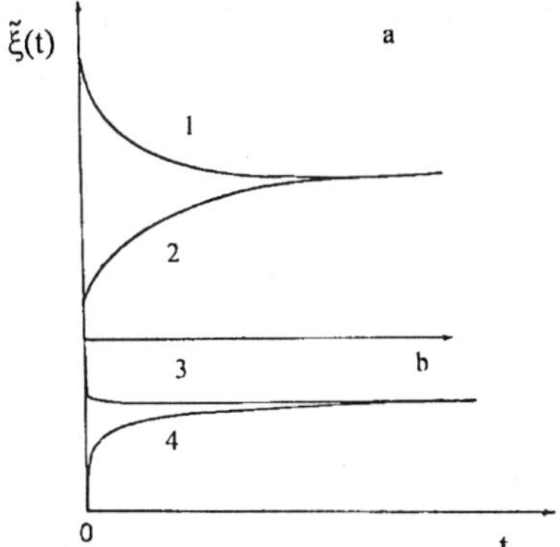

Fig. 1. Kinetics of glass stabilization according to Eqs. (10) and (12).
(a) Relaxation from "above" (curve 1) and from "below" (curve 2) is symmetrical in time,
when the time of molecular relaxation τ is a function only of temperature (Eq. (17)).
(b) Relaxation with τ dependent on ξ according to the non-linear solutions of Eq. (10): non-symmetrical course, when relaxation from "above" (i.e. from higher temperatures, curve 3)
and from "below" (i.e. from lower temperatures, curve 4) is compared.

Suppose that we change the temperature of a system in which a process
of structural relaxation from $\tilde{\xi}$ to ξ_e takes place according to Eq. (10). Considering only constant cooling and heating rates (i.e. in Eq. (3) we have to
write $q = \text{const.} = \pm q_o$) and following a procedure first applied by Bragg and
Williams [39]: to replace dt by (dT/q_o) (see also the book of Mott and Jones
[40] and [20, 41]), Eq. (10) can be written (at $p = \text{const}$, $Z_i = 0$) as

$$(d\tilde{\xi}/dT) = [\tilde{\xi}(T) - \xi_e(T)]/q_o\tau(T, \tilde{\xi}) \tag{13}$$

Another derivation of Eq. (13), adopted in [18], is to write the full differential of ξ in terms of both time t and temperature T (or any additional external
parameter: P or Z_i) in the form

$$d\xi = (\partial\xi/\partial T)_{t,P,Zi}\,dT + (\partial\xi/\partial t)_{(T,P,Zi)}dt \tag{13a}$$

or

$$d\xi/dT = (\partial\xi/\partial T)_{t,P,Zi} + (\partial\xi/\partial t)_{T,P,Zi}(dt/dT) \tag{13b}$$

The above expression refers to the case, when both T and t are independent variables. However, in any real or imaginary cooling (or heating) run
experiment, T and t are connected via Eq. (3). Thus in Eq. (13b) we have
to write $(dt/dT) = \pm 1/q$ and as far as for $T = f(t)$, we have to expect (for

$t = \text{const.}$) also $(\partial \xi / \partial T)_t = 0$, Eq. (13) is immediately obtained by a somewhat more general procedure which, however, involves an additional assumption.

The solutions of Eq. (13) can be written (again for τ independent of $\tilde{\xi}$) as

$$\xi(T) = \left\{ \xi(T = T_i) + \int_{T_i}^{Tt} (\xi_e / q\tau) \left[\exp \int_{T_i}^{T} (dT/q\tau) \right] dT \right\} \exp \left[- \int_{T_i}^{T} (dT/q\tau) \right]$$

(14)

where T_i is the initial temperature of the cooling (or heating) run (i.e. $T = Ti$ at $t = 0$).

The solutions of Eq. (13) offer for different $\xi_e(T)$ and $\tau(T)$ (or $\tau(T,\xi)$) functions a simple way of treating and illustrating the kinetics of vitrification processes (at $q_o < 0$) in vitrifying undercooled melts, as well as the changes occurring in the already frozen-in system (i.e. in the glass) at heating runs (at $q_o > 0$). The possibilities thus arising from Eq. (13) (mostly with (14)) have been exploited by several authors (see [18–20, 40–43] and the literature given in Mazurin's monographs [19, 43]); a somewhat modified version of Eq. (13) is also used in [44].

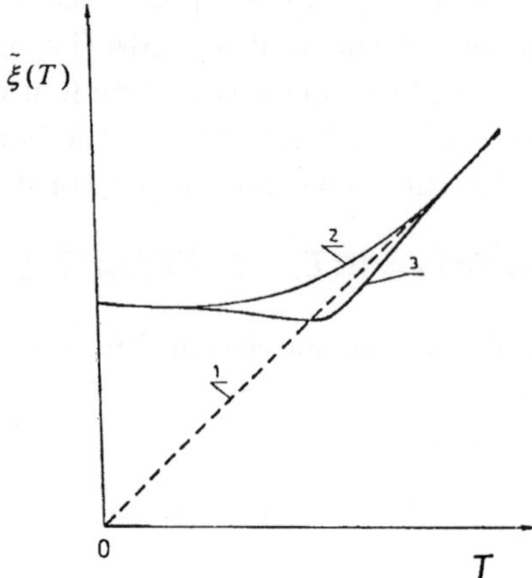

Fig. 2. The temperature course of the thermodynamic structural parameter ξ according to Bragg and Williams [39], and Mott and Jones [40] for heating and cooling run experiments according to Eqs. (13) and (14):
1 — temperature course of the equilibrium values $\xi_e(T)$ of the structural parameter;
2 — cooling run with quenching rate $-q_0$;
3 — heating experiment with heating rate $+q_0$. In both cases Eq. (23) with a constant value of the activation energy is anticipated in solving Eq. (13).

Moreover, as we have shown recently [21, 45], Eq. (13) also gives a direct possibility to formulate a very general criterion for vitrification. This criterion has been introduced up to now into the theory of glass formation either as an additional postulate (the Frenkel–Kobeko formula, see [10, 22, 46–48]) or via a very general kinetic argumentation (see the paper of Volkenstein and Ptizyn [20]) involving, however, again an additional assumption (see for more details [45]).

Here we have only to note that Eq. (13) gives as an immediate result the freezing-in of the structure (curve 2 in Fig. 2), when a cooling run is performed (see again Fig. 2, drawn according to Bragg and Williams' original figure in [39]), and a clear memory effect, when the system is reheated (curve 3 in Fig. 2).

In order to proceed further on we have to specify the $\tau(T, \xi)$ function in both Eqs. (10) and (13) determining the kinetics of isothermal and non-isothermal structural changes. Moreover, we have to introduce an appropriate temperature function for the equilibrium values of ξ, i.e. for $\xi_e(T)$.

3. Time of Molecular Relaxation in Glass Forming Liquids

The temperature dependence of $\tau(T)$ for undercooled glass forming liquids is usually described in the form of an exponential dependence of the type

$$\tau(T) = \tau_o \exp\left[U(T)/RT\right] \tag{15}$$

with the additional requirement for the activation energy $U(T)$ (see [10, 15])

$$d[U(T)]/dT < 0 \tag{16}$$

The Activated Complex Model approach specifies Eq. (15) in the framework of the Absolute Rate Theory of Reactions as (see [36])

$$\tau(T) = \tau_o \exp(U_o/RT) \exp[\beta^*/\xi(T)] \tag{17}$$

i.e. according to this phenomenological model

$$U(T) = [U_o + RT\beta^*/\xi(T)] \tag{18}$$

Here U_o is a nearly constant term of $U(T)$ (roughly proportional to the enthalpy of evaporation of the liquid, see [35]). For equilibrium systems (the undercooled melt), $\xi(T)$ in Eq. (18) is determined by $\xi_e(T)$. For non-equilibrium states, however, the $\xi(T)$ dependence following from Eq. (13) has to be inserted into

Eqs. (17) and (18). In both cases τ_o is the time of molecular oscillations which, according to the Absolute Rate Theory (see [27, 28]), can be estimated by

$$\tau_o \cong h/kT \tag{19}$$

Here, k is Boltzmann's constant, h is Planck's constant and the dimensionless numeric factor β^* has values $\beta^* \cong 1$ for simple (non-associating) liquids like metallic melts, while $\beta^* = 4$ to 8 is found for more complex structures, e.g. organic polymer glass forming melts (see [35, 49]).

The above dependences (Eqs. (17) and (18)), as discussed in more details in our recent publications [33, 34, 36], are obtained when it is assumed that the frequency ω of the flow-determining fluctuations in the undercooled melt is given by

$$\omega = (kT/h) \exp[-\Delta G^\# /kT) \tag{20}$$

where $\Delta G^\# = [G^* - G_f(T)]$ is the difference between the thermodynamic potentials of the activated complex state G^* and the thermodynamic potential of the ground state $G_f(T)$ (here: the undercooled liquid f). Following well known thermodynamic dependences, we can write (for $P = P_0 = $ const. and $Z_i = 0$)

$$\Delta G^\# = \Delta U^\# + P_0 \Delta V^\# - T\Delta S^\# \tag{21}$$

Here $\Delta U^\# = U^* - U_f(T)$, $\Delta V^\# = V^* - V_f(T)$ and $\Delta S^\# = S^* - S_f(T)$ are the corresponding changes in internal energy, free volume and entropy, when a building unit of the melt is brought from the initial state (f) into the activated state $(^*)$ of the flow determining fluctuation. It can be shown, in accordance with the basic assumptions of the Activated State Model (see [33]), that in Eq. (21) the values of $\Delta U^\#$ and $\Delta S^\#$ can be taken to be nearly constant. Expressing, however, P_0/RT via the caloric equation of state of the undercooled liquid, it can be shown, as we have done in [33], that Eqs. (20) and (21) lead directly to Eq. (17) when the meaning of relative free volume $\vartheta(T)$ is attributed to ξ

$$\xi(T) \cong [V_f(T) - V_c(T)]/V_f(T) \approx [V_f(T) - V_c(T)]/V_c(T) \tag{22}$$

where V_f and V_c are the molar volumes of the liquid and of the crystal, respectively.

Thus Eq. (17) can be in fact considered as representing the well known empirical Litowitz–Macedo equation [50] for the temperature dependence of the viscosity η of glass forming liquids.

We employ here Eq. (17) in two different approximations:

$$\tau \cong \tau_o \exp\{U_o/RT)$$ (23)

for substances for which in Eq. (18) $U_o \gg RT_g/\xi(T_g)$ (i.e. for liquids for which, according to Angell [51, 52], the term strong liquids has been proposed), and

$$\tau \approx \tau_o \exp[\beta^*/\xi(T)]$$ (24)

for liquids, which according to Angell's classification are termed fragile liquids (for them $\xi_e(T)$, is a steep temperature dependence and $U_o \ll RT_g/\xi(T)$, as we have established in [26].

In the above outlined formalism for equilibrium states, as already mentioned, the equilibrium values of ξ have to be introduced in Eqs. (17), (18) and (23). For non-equilibrium states the $\xi(T)$ values, as determined by Eq. (13), can be inserted into Eqs. (17) and (18) (see [36]).

4. Temperature Dependence of the Structural Parameter ξ

In order to avoid unnecessary mathematical difficulties in solving Eq. (13) we have to choose for $\xi_e(T)$ the simplest possible function — the linear dependence. This was done in all previous efforts to analyze the above mentioned equation (see [20, 39–43]). In order to account for the Kauzmann temperature T_o (where the free volume and the configurational entropy vanish, and thus also $\xi_e(T) \approx 0$ has to be expected, see [10, 53]) typical for glass forming substances, we have to specify $\xi_e(T)$ for undercooled melts (i.e. below the melting temperature T_m) as

$$
\begin{aligned}
\xi_e(T) &= g_0(T - T_0), &\quad \text{for } T_m \geq T \geq T_0 \\
\xi_e(T) &= 0 &\quad \text{for } T_0 > T \geq 0
\end{aligned}
$$ (25)

where $g_0 = (d\xi/dT)$ is taken to be a constant.

If $\xi(T)$ is considered as being a structural parameter related (e.g. via Eq. (22)) to the free volume of the liquids, g_0, has the physical meaning of a coefficient of thermal expansion (then according to Eq. (22), $g_0 \cong (dV_f(T)/dT)_p$) of the configurational volume of the melt.

More generally, it has to be recalled (see [10, 12, 13]) that according to the thermodynamics of irreversible processes the configurational specific heats $\Delta Cp(T) \cong Cp_f(T) - Cp_c(T)$ of a system are defined via any structural parameter ξ and the configurational part $\Delta H(T)$ of the enthalpy of the liquid as

$$\Delta C_p = [\partial \Delta H(\xi)/\partial \xi]_{P,T} \cdot (d\xi/dT) \tag{26}$$

(see [10, 12, 13]). In a first approximation (typical for the linear formalism of the thermodynamics of irreversible processes) it can be assumed that

$$\Delta H(T) \approx h_0.\xi(T) \tag{27}$$

where h_0 is a constant, and thus

$$\Delta C_p(T) = h_0.(d\xi/dT) \tag{28}$$

In defining the configurational part of the specific heats, the crystalline state (c) is used as a reference system.

The temperature dependence of the equilibrium value of the configurational specific heats of our simple model of an undercooled melt is thus defined with Eqs. (25) and (27) as

$$\begin{aligned}
\Delta C_p(T) &= g_0\, h_0 = \text{const.}, && \text{for } T_m \geq T > T_0 \\
\Delta C_p(T) &= 0, && \text{for } T_0 > T \geq 0
\end{aligned} \tag{29}$$

i.e. in a way that was used by us in [10, 55] to derive a number of useful dependences in the thermodynamics of glass forming melts.

For most fragile liquids, $T_0 \approx 1/2\,T_m$, as revealed from both caloric and volumetric measurements and from the temperature dependence of their viscosity (i.e. via Eq. (24), see [53, 54]).

It becomes evident from Eq. (28) that Eq. (13) defines in fact the temperature dependence of the configurational specific heats, $\widetilde{\Delta C_p}(T)$, of our model system in which a cooling or heating run is performed with a velocity q_0.

By employing well known thermodynamic dependences (see e.g. [10]) we can write

$$\Delta H(T) = \Delta H_0 + \int_0^T \Delta C_p dT \tag{30a}$$

$$\Delta S(T) = \int_0^T (\Delta C_p/T) dT \tag{30b}$$

$$\Delta G = \Delta g(0) - \int_0^T dT \int_0^T (\Delta C_p/T) dT \tag{30c}$$

The temperature dependence of the configurational part of both the thermodynamic functions $(\Delta H(T), \Delta S(T))$ and of the thermodynamic potential ΔG (the configurational part of Gibbs free energy) can be constructed for our model system in the whole temperature interval from T_m to the absolute zero of temperature $(T = 0)$. By using the equilibrium values of $\Delta C_p(T)$ obtained from experiment, by some atomistic theory or by an appropriate thermodynamic model (e.g. by the simple model indicated with Eq. (29)), we can construct via Eqs. (30) the thermodynamic picture of our imaginary model system, representing the undercooled glass forming melt, as this is done in Fig. 3.

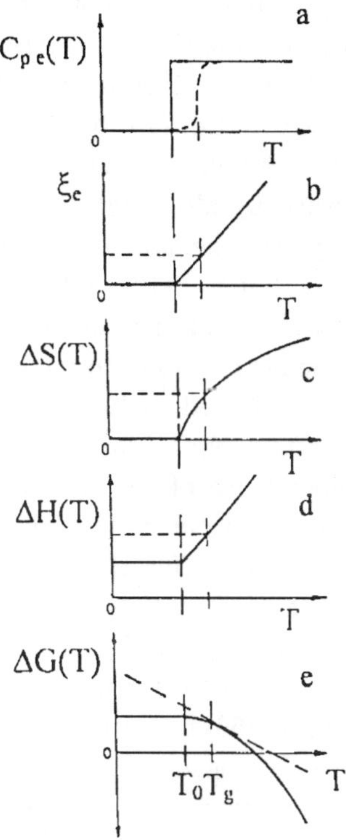

Fig. 3. The thermodynamic model of an undercooled melt employed in calculations:
(a) course of the configurational specific heats according to Eq. (29);
(b) $\xi_e(T)$ dependence following from Eq. (25);
(c) configurational entropy $\Delta S(T)$ determined by Eq. (30b) with ΔC_p values according to Eq. (29);
(d) and (e) configurational enthalpy ΔH and configurational Gibbs free energy ΔG, corresponding, according to Eqs. (30), to the ΔC_p dependence given with Eq. (29).

This picture is obtained in the framework of classical thermodynamics and here no vitrification is to be expected. The steep fall of ΔC_{pe} at temperatures $T = T_0$ is only a specific feature of the approximation employed (see [56, 57]). It is artificially introduced into the model (see Eq. (29)) in order to comply with the Third Principle of classical thermodynamics ($\Delta C_{pe}(T) = 0$ at $T \to 0$), see [24, 56, 57]) according to Nernst's Heat Theorem.

The steepness of the $\Delta C_p(T)$ course, as we established in [57], depends on the model employed. Especially for strong liquids a smooth, non-catastrophic $\Delta C_p(T)$ course is to be expected (and is, in fact, observed experimentally, see [56]), and even for fragile liquids a non-catastrophic $\Delta C_{pe}(T)$ course can be constructed for realistic thermodynamic models [56, 57].

When, however, the thermodynamics of irreversible processes is employed (i.e. by using its additional Fourth Principle given here with Eq. (8) and its approximative forms expressed by Eq. (10) for $T = $ const. and with Eq. (13) for cooling or heating runs), the picture schematically illustrated in Fig. 4 is observed. This was established by Moynihan [18] and Mazurin [19] from the solutions of Eq. (13), as they are schematically presented in this figure.

The essential point here is that Eq. (13), following from the basic principle of the thermodynamics of irreversible processes, leads to a $\Delta C_p(T)$ course corresponding, as given in the next section, to vitrification. The integration of this $\Delta C_p(T)$ dependence, as indicated by Eqs. (30), leads to $\Delta S(T)$, $\Delta H(T)$ and $V(T)$ functions of the system typical for glass formation (see [58]).

Such a temperature dependence of the thermodynamic functions of the configurational part of glass forming melts has been verified for the first time via caloric measurements by Simon (for the case of SiO_2 and glycerol, see [11, 59]) and it is discussed in more details in the subsequent literature and especially in [10, 24] and further on in Sec. 5. This course is quite different from the temperature dependences expected for thermodynamic phase transitions, as they have been first drawn in [60] by Justi and von Laue.

The differentiation of Eq. (13) leads via a thermodynamic argumentation to the general kinetic condition for vitrification (see [21, 45]).

5. The Kinetic Conditions for Vitrification and the Definition of the Vitreous State

At relaxation times for which $\tau \to \infty$, de Donder's Eq. (10) gives directly the condition (Eq. (6a)) which in combination with Eq. (7) has been used to give a definition of the vitreous state (see [10, 16, 25]).

In [10] we have formulated this definition in the following way: glasses are non-equilibrium systems (see Eq. 7) in which the state of disorder and the amorphous structure ($\tilde{\xi}$ corresponding to $\tilde{T}$) of a liquid are frozen-in.

This definition follows directly from Eq. (10), i.e. it is a consequence of Eq. (8) and of the aforementioned general principle of the thermodynamics of irreversible processes. Its generalization to include any state of frozen-in increased disorder (including spin glasses, etc.) is also given in [10] (see pp. 120/122 there).

However, Eq. (10) refers in fact to constant temperatures and can be used in defining the vitreous state only for temperatures $T < T_g$ below the temperature of vitrification T_g (at $\tau \to \infty$).

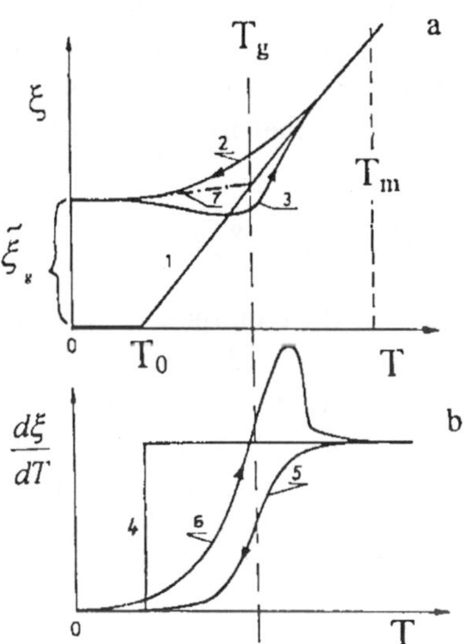

Fig. 4. Solutions of Eq. (13) and the course of the derivative $(d\xi/dT) \approx \Delta C_p(T)$ for a ξ-dependent course of the time of molecular relaxation τ.
(a) The temperature dependence of the structural parameter $\xi(T)$: curve 1 — for equilibrium according to Eq. (25); curves 2 and 3 — cooling and heating experiments with $-q_0$ and $+q_0$, respectively. Broken-dotted (dashed?) line 7 - extrapolation of frozen-in value of the structural parameter ξ_g from $T = 0$ to $T = T_g$, illustrating the commonly employed experimental definition of the temperature of vitrification $T = T_g$.
(b) The course of the derivation $d\xi/dT$ (i.e. of the specific heats $Cp(T)$) upon vitrification (curve 5) and in a reheating run (curve 6) performed with the same velocity as the quench. Line 4 indicates the temperature dependence of ΔC_{pe} at equilibrium. This figure is constructed as the schematic generalization of the results of several computer experiments with different, but realistic parameters of τ_0 and U_0.

The Bragg–Williams Eq. (13) on the other hand gives a possibility for a direct kinetic definition of glass and the vitreous state at changing temperatures, i.e. at the conditions of glass formation normally employed.

For $q = \infty$ it follows from Eq. (13) that $\Delta Cp = 0$, i.e. an abrupt change of the configurational specific heats at *any* temperature. For $q \to 0$ (i.e. practically at $T = $ const.) this equation gives, as expected, $\tilde{\xi}(T) \to \xi_e(T)$. Between these two extreme cases of temperature change, Eq. (13) leads, as we have shown in a recent investigation [21, 45], to the dependence

$$q_0.\tau(T, \xi)|_{T \to Tg} = [d(\ln \tau)/dT]^{-1}|_{T \to Tg} \approx \text{const.} = C_0 \tag{31a}$$

defining the temperature of vitrification T in dependence of the cooling rate q (Eq. (31) defines the case for standard cooling rates at $T = T_g$). The above result follows considering Fig. 4: for any temperature change performed at cooling rates ranging between $q = \infty$ and $q = 0$, the $\Delta C_p(T)$ curve has to pass in the vicinity of T_g (or at any other freezing-in temperature $\tilde{T}$) through an inflection point, where

$$[d^3\xi(T)/dT^3] = \partial^2 \Delta Cp(T)/dT^2 = 0 \tag{32}$$

This condition, combined with the requirement that at $T \to T_g$ the $\log(\tau) vs\ T$ curve has to change its course when $\xi(T)$ goes to $\xi_g = $ const. (cf. Eq. (17) and Figs. 5 and 6), leads (as shown in [21, 45]) directly to Eq. (31). A similar result is also obtained when $(U/RT)|_{T \to T_g} > 20$ to 30, i.e. for typical strong liquids. For the details of this derivation, we have to refer to the above mentioned study in [21, 45]. The essential point to be mentioned here is that Eq. (31), known as the Frenkel–Kobeko equation (see [22, 46, 47]), also written in the form given by Reiner [48, 61, 62],

$$q_0.\tau|_{Tg} = \Delta t \tag{31b}$$

where Δt is the observation time of the processes, is not an additional postulate deliberately introduced in glass science, but a direct consequence of Eq. (13) and thus of the principles of the thermodynamics of irreversible processes.

For the first time, a kinetic derivation of Eq. (31), based on an equation equivalent to Eq. (13), was given by Volkenstein and Ptizyn [20], although in a form not free of artificial assumptions (see [45]). The detailed analysis of the course of the $\xi(T)$ and $Cp(T)$ curves and of their derivatives, as is given in [21], makes the assumptions adopted in [20] superfluous.

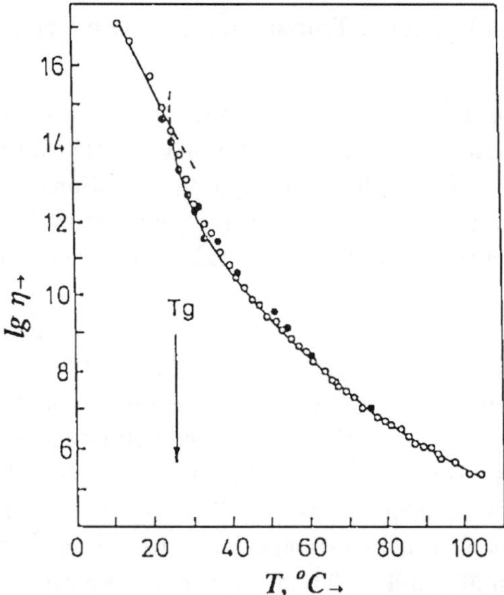

Fig. 5. Temperature course of the bulk viscosity η of undercooled selenium melts and of selenium glass (for $T < T_g$) according to Nemilov [66]. Note the peculiar change of slope of the $d\log\eta/dT$ curve at $T = T_g$.

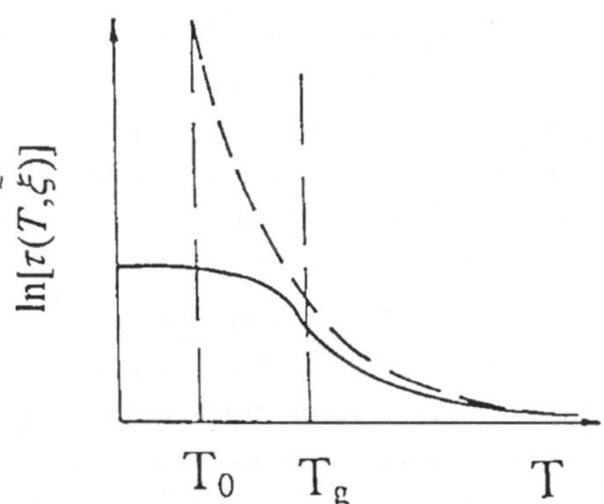

Fig. 6. Temperature dependence of the time of molecular relaxation $\tau(T,\xi)$ in the temperature interval from T_m through T_g to T_0 (Schematically).
Full line — the temperature dependence of $\log\tau$ according to Eq. (24) with $\xi(T)$ dependence given by the solution of Eq. (13).
Dashed line — the expected course of the $\log\tau(T,\xi)$ curve at equilibrium according to Eq. (24) for $\xi = \xi_e(T)$ given with Eq. (25).

6. Thermodynamic Phase Transitions and the Process of Vitrification

The analysis of the thermodynamics of phase transitions shows that the $G(T)$ curves of the metastable and of the stable phases either intercept at the equilibrium temperature T_e (for first order phase transitions, where $T_e = T_m$) or they coalesce at and above the equilibrium temperature T_e (for second order phase transitions, where $T_e = T_2$ corresponds to an order-disorder change) [10, 25, 60, 63].

However, at the glass transition temperature T_g, (or at any other vitrification temperature T), the Gibbs free energy curve of the frozen-in system (the glass) is a tangent to the $G(T)$ curve of the metastable phase (the undercooled liquid). This statement, essential for the understanding of the nature of the vitreous state, was given for the first time in our papers [17] (for the temperature course of the vapour pressure) as well as by Gupta and Moynihan [68] of glasses and undercooled melts (see also [10]). Its subsequent detailed analysis may be followed in [63] and in [25, 26] and more recently — in [45].

Following the theorems of differential geometry, an analysis of both thermodynamic phase transitions (first and second order) and of the process of vitrification has to be performed examining the temperature course of the first and second derivatives of the corresponding thermodynamic potentials and their differences, i.e. of the $\Delta S(T)$ and $\Delta Cp(T)$ curves. For the details of this analysis we refer to [25, 26, 58] and to [45], where the geometric approach in the thermodynamics of vitrification is developed in new lines.

Here, we would like only to point out that the above mentioned course of the thermodynamic potential also follows in a generic way from Eq. (13) — via an integration in terms of Eq. (30c). The result of such a procedure performed along lines 1 and 2 in Fig. 7 is given in Fig. 7a with lines 4 and 5.

For first order phase transitions below and above the temperature of phase equilibrium T_e, the difference $\Delta G(T)$ changes sign (the metastable undercooled liquid above T_m becomes a stable melt and the crystalline phase turns metastable). For second order phase transitions, one of the phases (the ordered one) is stable below T_2, the other one (the disordered one) is metastable. At the equilibrium point $(T_e = T_2)$, the potential difference $[G(T)_{\text{disordered}} - G_{\text{ordered}}]$ is nullified for $T > T_2$.

An analysis of the geometry of the thermodynamics of vitrification shows that the frozen-in system (the glass) is unstable both below and above T_g: for glasses $\Delta G = G(T)_{\text{glass}} - G_{\text{undercooled melt}}$ is always positive and this is the

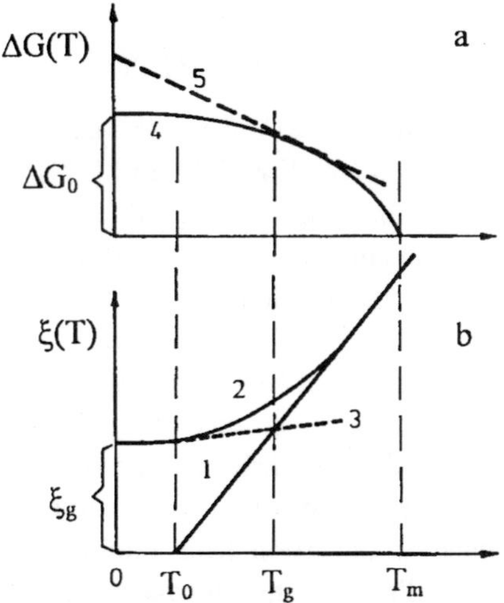

Fig. 7. The temperature dependence of the thermodynamic potential of a vitrifying system following from the integration of the solution of the Bragg–Williams Eq. (13) for $T < T_g$. (a) the $\Delta G(T)$ course upon vitrification according to [17, 18]; (b) the solution of Eq. (13) indicated with Eq. (14); 1 — equilibrium values of the structural parameter $\xi_e(T)$; 2 — cooling run according to Eqs. (13) and (14); 3 — thermodynamic potential difference of the metastable undercooled melt; 4 — temperature dependence of the thermodynamic potential of the glass; ξ_g^* — value of ξ frozen-in upon vitrification; ΔG_0 = thermodynamic potential difference corresponding to a fictive equilibrium melt at absolute zero temperature.

thermodynamic difference between vitrification and any form of thermodynamic phase transition. The consequences of this unusual behaviour are analyzed in detail in [25, 26] and in [58]. This is the thermodynamic difference between vitrification and thermodynamic phase transitions.

The *kinetic* difference between vitrification and any thermodynamic phase transition is that, according to Eq. (13) (and to its consequence, Eqs. (31)), the vitrification temperature becomes time dependent (more exactly: dependent on the cooling rate q_o). The corresponding dependence known as the Bartenev–Ritland formula (see [22], [47] and [62])

$$(1/T_g) = C_1 - C_2 \log q \tag{33}$$

with

$$C_1 = 2.3[k/U(T_g)].\log(C_0/\tau_0) \tag{33a}$$

and

$$C_2 = [2.3\,k/U(T_g)] \tag{33b}$$

can be derived directly from Eqs. (31), when the exponential character of the $\tau(T)$ dependence is considered (see Eqs. (15) and (17)). The simple derivation leading to Eq. (33) may be found in [10, 22, 24].

The essential point to be emphasized here is that both the thermodynamic features of vitrification and the kinetic nature of this process can be described in terms of the thermodynamics of irreversible processes; moreover, they follow directly from the basic principles of this thermodynamic phenomenology.

Classical thermodynamics is the science of equilibrium: it gives the classification of thermodynamic phase transitions and of changes possible between equilibrium systems. Thermodynamics of irreversible processes is the science of processes taking place at deviation from equilibrium, in non-equilibrium or in frozen-in systems. This is why the nature of the vitreous state can be understood only from the standpoint of this more general thermodynamic phenomenology.

The thermodynamics of irreversible processes gives both the kinetics of glass formation as well as of the stabilization of glass. This second process, the relaxation of the metastable state of the glass to that of the metastable liquid, is governed by the de Donder formula [Eq. (10)] and its solutions with a more or less complex $\tilde{\xi}$-dependent activation energy $\tau(T,\tilde{\xi})$. In this case, when $\tau(T,\tilde{\xi})$ is used, as we have shown in a recent publication [64], Eq. (12) (determining a Maxwellian type of relaxation at $\tau = \tau(T)$) turns into a more complicated dependence known as the Kohlrausch stretched exponent $\xi(t)$ formula (see also [19]).

The definition of glass can be attempted from different points of view: molecular, structural, kinetic, etc. However, thermodynamics is the science deciding upon the nature of states and transitions, and this is why Ehrenfest has chosen classical thermodynamics to classify phase transitions, and Prigogine indicated thermodynamics of irreversible processes as the basis for the definition of glass. His approach was used here and in [10, 25] to define glass as a particular physical state.

Other possibilities, following from the non-equilibrium nature of glass can also be used for this purpose, e.g. ergodicity and non-ergodicity of states. However, the non-ergodicity of frozen-in systems is only a consequence of their non-thermodynamic character, and ergodicity (or better quasiergodicity) is a term whose definition is still open to discussion in statistical physics.

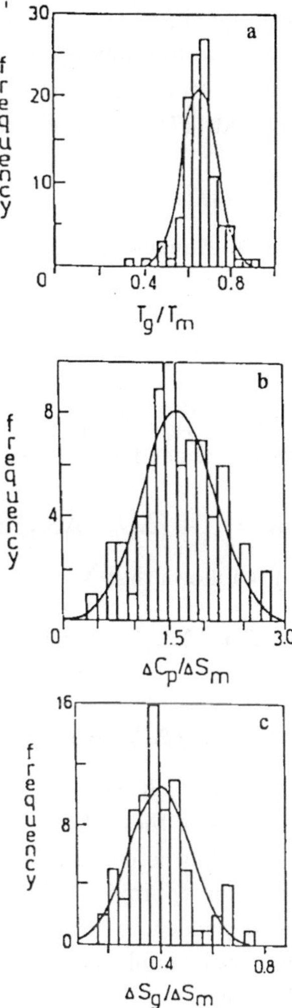

Fig. 8. The frozen-in thermodynamic properties of glasses, obtained at standard cooling rates. Experimental data from more than 80 single-component glass forming systems summarized in [24].

(a) the ratio T_g/T_m according to Eq. (36);

(b) configurational specific heats $\Delta C_p(T_g)/\Delta S_m$ ratio according to Eq. (36);

(c) frozen-in entropy values according to Eq. (36).

Note that the median value of the corresponding frequency curves lie at: $(T_g/T_m) \approx 0.65$, $(\Delta C_p/\Delta Sm) \cong 1.5$ and $(\Delta S_g/\Delta S_m) = 0.36$.

Therefore a purely thermodynamic approach in its more general formulation seems to be most appropriate to analyze, in a generic manner, both the nature of glass and its formation and relaxation.

It follows from Eqs. (31) and (17) that at equal, e.g. at standard cooling rates, we have to expect

$$(T_g/T_m) \approx \text{const.} \tag{34}$$

as also stated by the empirical Kauzmann–Beaman Rule. Introducing into Eq. (31) reasonable estimates for $U(T_g)/RT_g$, for the above indicated ratio values are obtained in the vicinity of 2/3 to 1/2 as predicted by this rule (for details see [10, 24] and especially [65], where for the first time a derivation of Eq. (34) was given).

With the constant value of ΔC_{pe}, anticipated according to Eqs. (29), for $T_m > T > T_0$ the temperature dependence of the confiurational entropy is defined as

$$\Delta S(T)/\Delta S_m = [1 + (\Delta C_{pe}/\Delta S_m)\ln(T/T_m)] \tag{35}$$

and with Eqs. (34) and (35) it follows that

$$[\Delta S_g/\Delta S_m] \approx \text{const.} \tag{36}$$

as far as

$$[\Delta C_p(T_g)/\Delta S_m] \approx \text{const.} \tag{37}$$

where ΔS_m denotes the entropy of melting.

The relatively good fulfillment of these three dependences [Eqs. (34), (36) and (37)] is illustrated in Fig. 8, where the results from an analysis of all existing thermodynamic data (for approximately 80 substances in 1990) are used (see [24]). These simple dependences, first established by us in [65], have a deeper molecular and structural significance. This, however, is beyond the scope of the present contribution.

7. Conclusions and Outlook

From the standpoint of thermodynamics and of statistical thermodynamics, undercooled melts and glasses belong to two different groups of objects: the former are equilibrium bodies, i.e. thermodynamic objects, ergodic systems, and they can be treated in the framework of classical thermodynamics and classical statistical thermodynamics. Glasses, however, are non-equilibrium, non-ergodic systems. Their phenomenological treatment is only possible in the framework of the thermodynamics of irreversible processes. It turns out that even the linear approximation of this science is sufficient, by introducing even one frozen-in structural parameter, to give at least a qualitative description of glass as a non-equilibrium system and of vitrification (or glass formation) as a process of structural arrest. Sometimes such a process is called a *kinetic* phase transition: thus its difference from *thermodynamic* phase transitions is underlined.

The quantitative description of vitrification in terms of the phenomenology of the thermodynamic of irreversible processes requires, however, not only new terms, but also a more general, non-linear extrapolation of this science itself, at least as far as the dependence of the time $\tau(T, \xi)$ of molecular relaxation is concerned. The introduction of more than one structural parameter improves further the quality of the description of vitrification (e.g. giving more realistic values of the Prigogine–Defay Ratio, see [10, 63]).

The two approaches: the classical thermodynamic description of the states of aggregation and of thermodynamic phases and the possible transitions in between them on the one hand, and of the vitreous state and of vitrification as a process in the framework of the thermodynamics of irreversible processes on the other hand, are the two possible ways, the two possible patterns of treatment: as changes between equilibrium states, and as transitions between equilibrium and non-equilibrium states. The difference between these two treatments is determined by the equilibrium character of coexistence between thermodynamic phases on the one hand and the irreversible nature of vitrification on the other hand. The notion kinetic phase transition, or diffuse phase transition, also proposed by some authors, requires the consideration of non-equilibrium phases and this is a difficult task. The definition of vitrification as a process of structural arrest, in which the state of disorder corresponding to undercooled liquid is frozen-in, seems to be sufficiently general and non-contradictory without introducing the term thermodynamic phases corresponding to another, equilibrium, type of bodies.

In one of the rooms of the Acropolis Museum in Athens, an archaic pedimental composition of porous limestone from the 6th Century B.C. is on exposition. It belonged to a large Hellenistic temple, probably Partenon II, which was the "archaios naos" of Athens (see its description in [67]).

On the right hand side of the pediment there is a picture of a three-bodied demon (it was originally four-bodied, one of the figures is lost), of three (originally four) male figures merging into a twisting snake like a tail of a dragon. Each of the four male figures holds in his hand one of the symbols of the four elements of Nature already mentioned in the Introduction: a bird (air), a wave (water), a crystal (solid) and a flame (fire).

These four symbols, according to the teaching of Hellenistic Philosophy (see [7]), are the four states, the elements of Matter and of the Universe, which have a common origin: equilibrium.

On the left hand side of the pediment, Heracles (the Hellenistic ordering force) is wrestling with the sea monster Triton — the Greek symbol of disorder. Both are petrified in their struggle as if symbols of frozen-in disorder. Is this the Hellenistic symbol of a frozen-in non-equilibrium state, of glass, different from any of the equilibrium states of aggregation, represented on the right hand side of the pediment? I do not know, but nevertheless I would recommend to every glass scientist to consider and reconsider those two beautiful symbols of frozen-in struggle of order and disorder on the one side and of the equilibrium states of the elements of Nature on the other side. It is really inspiring.

Further development in theoretical glass science should account for the differences and the generalizations, which this ancient picture seems to indicate. Simple phenomenological treatment has to precede complicated atomistic schemes and models in any quarter of theoretical thinking. Glass is a very ancient material, but theoretical glass science is a relatively young branch of solid state physics and its development in the nearest future should follow phenomenological lines and ways.

We have tried to show here, that the phenomenological treatment of vitrification, based on the thermodynamics of irreversible process has also the great advantage of treating and solving the problems of theoretical glass science in a simple and generic way without introduction of additional assumptions.

References

[1] R. Zallen, *The Physics of Amorphous Solids* (J. Wiley, New York, 1983).
[2] K. Binder, *Progr. Colloid. Polym. Sci.* **96**, 7 (1994).
[3] W. Goetze and L. Sjogren, *Rep. Progr. Phys.* **55**, 241 (1992).
[4] I. Carmesian and K. Kremer, *Macromolecules* **21**, 2819 (1988).
[5] R. Schilling, Mode Coupling Approach to the Glass Transition, *Disorder Effects on Relaxational Processes*, eds. R. Richert and A.B. Cumen (Springer Verlag Berlin, New York, 1990) p. 193.
[6] K. Binder, *Ber. Bunsenges. Phys. Chem.* **100**, 1381 (1996).
[7] J.D. Bernal, *Science in History* (A. Watts, London, 1957).
[8] P. Ehrenfest, *Proc. Amsterdam Acad.* **36**, 153 (1933).
[9] E. Berger, *Glast. Ber* **8**, 339 (1930).

[10] I. Gutzow and J. Schmelzer, *The Vitreous State: Thermodynamics, Structure, Rheology, and Crystallization* (Springer Verlag Berlin, New York, 1995).

[11] F. Simon, *Zs. anorg. allg. Chemie* **203**, 219 (1931).

[12] I. Prigogine and R. Defay, *Chemical Thermodynamics* (Longmans, London, 1954).

[13] R. Haase, *Thermodynamics of Irreversible Processes* (Steinkopf Verl., Darmstadt, 1962) (in German).

[14] R.O. Davies and G.O. Jones, *Proc. Roy. Soc.* (London) **A217**, 26 (1953).

[15] E. Kanai and T. Satoh, *J. Phys. Chem.* (Japan) **9**, 117 (1954); **10**, 1002 (1955).

[16] A.R. Cooper, Internal Parameters, Ordering Parameters, History and the Glass Transition, *Non-Crystalline Solids, Proc. Int. Conf.*, Clausthal 1976, ed. G.H. Frischat (Trans. Tech. Publishers, Acdermannsdorf, 1977) p. 384.

[17] E. Grantscharova and I. Gutzow, *J. Non-Cryst. Solids* **81**, 99 (1986). (see also I. Gutzow, Z. phys. Chemie (N.F.) **81**, 195 (1972) cited there).

[18] C.T. Moynihan, A.J. Eastel and J. Wilder, *J. Phys. Chem.* **78**, 2673 (1974).

[19] O.V. Mazurin, *Vitrification* (Nauka Publishers, Leningrad, 1986) (in Russian).

[20] M.V. Vol'kenstein and O.B. Ptizyn, *Proc. (Doklady) AN USSR* **103**, 795 (1955).

[21] I. Gutzow and F. Babalievski, A New Approach in the Thermodynamics and Kinetics of Vitrification, *Proc. 18th Int. Congress on Glass*, Paper No. 035, San Francisco, July 1998 (Americ. Ceram. Soc., Westerville, OH, 1998).

[22] G.M. Bartenev, *Proc. (Doklady) AN USSR* **76**, 227 (1951).

[23] A.R. Cooper and P.K. Gupta, *Phys. Chem. Glasses* **23**, 44 (1982).

[24] I. Gutzow and A. Dobreva, *J. Non-Cryst. Solids* **129**, 266 (1991).

[25] I. Gutzow, A. Dobreva and L.D. Pye, *J. Non-Cryst. Solids* **180**, 107, 117 (1995).

[26] I. Gutzow and A. Dobreva, Kinetics of Glass Formation, *in Amorphous Insulators and Semiconductors*, eds. M.F. Thorpe and M.I. Mitkova (Kluwer Academic Publishers, London, 1996) p. 21/43.

[27] S. Glasstone, H.J. Laidler and H. Eyring, *The Theory of Rate Processes* (Princeton University Press, New York, London, 1941).

[28] K.J. Laidler, *Reaction Kinetics, Homogeneous Gas Reactions — Vol. 1* (Pergamon Press, New York, London, 1963).

[29] J.O. Hirshfelder, Ch.F. Curtiss and R.B. Bird, Eyring's Theory of Reactions, Appendix A to Chapter 9, *Molecular Theory of Gases and Liquids* (Chapman & Hall, London, 1954).

[30] J. Bicerano, *J. Polymer Sci.* **B29**, 1329, 1345 (1991).

[31] I. Avramov and A. Milchev, *J. Non-Cryst. Solids* **104**, 253 (1988).

[32] I. Avramov, *J. Materials Sci.* **13**, 1367 (1994).

[33] I. Gutzow, A. Dobreva, C. Ruessel and B. Durschang, *J. Non-Cryst. Solids* **215**, 313 (1997).

[34] A. Dobreva and I. Gutzow, *J. Non-Cryst. Solids* **220**, 235 (1997).

[35] I. Gutzow, I. Avramov and K. Kaestner, *J. Non-Cryst. Solids* **129**, 266 (1991).

[36] I. Gutzow, V. Yamarov, D. Ilieva and L. D. Pye, Phys. and Chemistry of glasses, paper in print, 1999.

[37] Th. de Donder and P. van Rysselberghe, *Thermodynamic Theory of Affinity* (Stanford University Press, 1936).

[38] M.A. Leontowisch, *Introduction into Thermodynamics* (Deutsch Verl. Wiss., Berlin, 1953) (in German).

[39] W.L. Bragg and E.J. Williams, *Proc. Roy. Soc.* (London) **A145**, 699 (1934).

[40] N.F. Mott and H. Jones, *The Theory of Properties of Metals and Alloys* (Clarendon Press, Oxford, 1936) pp. 39–40.

[41] A.R. Cooper, *J. Non-Cryst. Solids* **95–96**, 1 (1987).

[42] C.T. Moynihan, A.J. Eastel, M.A. DeBolt and J. Tucker, *J. Americ. Ceram. Soc.* **59**, 12 (1976).

[43] O.V. Mazurin, *Vitrification and Stabilization of Inorganic Glasses* (Nauka Publishers, Leningrad, 1978) (in Russian).

[44] J.M. Hutchinson and A.J. Kovacs, The Relative Contributions of Temperature and Structural Parameters to Volume and Enthalpy Recovery of Glasses, *The Structure of Non-Crystalline Materials*, ed. P.H. Gaskell (Symp. Proc. Cambridge, 1977) p. 167.

[45] I. Gutzow, D. Ilieva, V. Yamakov, F. Babalievski and L.D. Pye, Glass Transition: An Analysis in a Differential Geometry Approach, in Nucleation: Theory and Applications, Eds. J. Schmelzer, G. Ropre and V. B. Priezzhev, JINR-Publishing House, Dubna (1999), pp. 368–409.

[46] P.P. Kobeko, *The Amorphous Substances: Physico-Chemical Properties of Simple and Polymer Amorphous Solids* (Academic Science USSR Publishers, Moscow, Leningrad, 1952) (in Russian).

[47] G.M. Bartenev, *Structure and Mechanical Properties of Inorganic Glasses*, (Building Materials Publishers, Moscow, 1966) p. 18–22 (in Russian).

[48] G.M. Stevels, *J. Non-Cryst. Solids* **6**, 307 (1971).

[49] I. Gutzow, D. Kashchiev and I. Avramov, *J. Non-Cryst. Solids* **73**, 477 (1985).

[50] B. Macedo and T.A. Litovitz, *J. Chem. Phys.* **42**, 245 (1965).

[51] C.A. Angell, *J. Non-Cryst. Solids* **102**, 205 (1988).

[52] C.A. Angell, Glass Formation and the Nature of the Glass Transition, Chapter 1, in: *Insulating and Semiconducting Glasses*, ed. P. Boolchand (World Scientific, 1998).

[53] C.A. Angell and K.J. Rao, *J. Chem. Phys.* **57**, 470 (1972).

[54] I. Gutzow, *Fizika i Khimiya Stekla* **1**, 431 (1975) (in Russian).

[55] I. Gutzow, *J. Non-Cryst. Solids* **45**, 301 (1981).

[56] B. Petroff, A. Milchev and I. Gutzow, *Macrom. Sci.* (*Phys.*) **B35**, 763 (1996).

[57] A. Milchev and I. Gutzow, *Macrom. Sci.* (*Phys.*) **B21**, 583 (1982).

[58] A. Milev and I. Gutzow, The Differential Geometry of the Thermodynamics of Phase Transition and of Vitrification, unpublished.

[59] F. Simon, 40th Gutrie Lecture, *Yearbook Phys. Soc.* (London) 1 (1956).

[60] E. Justi and M. von Laue, *Phys. Zs.* **35**, 945 (1934).

[61] M. Reiner, *Phys. Today* **17**, 62 (1964).

[62] H.N. Ritland, *J. Americ. Ceram. Soc.* **37**, 370 (1954).

[63] S.V. Nemilov, The Prigogine–Defay Ratio and the Structural Difference Glass/Liquid, *The Vitreous State, Proc. 8th All-Union Meeting*, ed. E.A. Porai-Koshitz (Nauka Publishers, Leningrad, 1988) p. 15.

[64] A. Dobreva, I. Gutzow and J. Schmelzer, *J. Non-Cryst. Solids* **209**, 257 (1997).

[65] I. Gutzow, The Thermodynamic Functions of Supercooled Glass Forming Liquids and the Temperature Dependence of their Viscosity, in *Amorphous Materials, Proc. 3rd Int. Conf.*, Sheffield, 1970, eds. R. W. Douglas and B. Ellis (Wiley, London, 1972) p. 159.

[66] S.V. Nemilov, *Zh. Priklad. Khimii* **37**, 1020 (1964).

[67] G. Papathanassopoulos, *The Acropolis: A New Guide of the Monuments and Museum* (Krane Editions, Athens, 1990), p. 46 and Fig. 35.

[68] P.K. Gupta and C.T. Moynihan, *J. Chem. Phys.* **65**, 4136 (1976).

4

The Structure and Rigidity of Network Glasses

M. F. THORPE, D. J. JACOBS and B. R. DJORDJEVIĆ

Department of Physics & Astronomy,
and Center for Fundamental Materials Research,
Michigan State University, East Lansing, MI 48824, USA
thorpe@pa.msu.edu

Contents

1. Introduction

In this chapter, we discuss continuous random networks and their mechanical properties. We have collected together most of the ideas that have been put forward over the past ten years on constraint counting and the resultant floppy modes in random networks. We show how model systems can help us better understand glasses via rigidity percolation. These ideas involving rigidity percolation have been tested experimentally in bulk glasses. We also examine marginal cases where the instability is caused by the surface, leading to the number of floppy modes scaling with the surface area. These effects may be important in porous silica, zeolites and possibly biological systems.

The study of network structures has fascinated scientists in many areas — ranging from engineering and mechanics to the material and biological sciences. Going back more than a century, Maxwell was intrigued with the conditions under which mechanical structures made out of struts, joined together at their ends, would be stable (or unstable) [1]. To determine the stability, without doing any detailed calculations (that would have been impossible then except for the simplest structures), Maxwell used the method of constraint counting. This counting is an approximate method that proves to be accurate for structures where the density (of struts or joints) is approximately uniform. Maxwell's constraint counting method is exact for some geometries that we will discuss in this paper. The idea of a constraint in a mechanical system goes back to Lagrange [2] who used the concept of holonomic constraints to reduce the effective dimensionality of the space. The difficult part is to determine which constraints are *linearly independent*. If the linearly independent constraints can be identified, then the problem is solved — however in most large systems this identification is not possible except using some numerical procedure on an actual realization.

The problem under consideration is a static one — given a mechanical system, how many independent deformations are possible without any cost in energy? These are the zero frequency modes, which we prefer to refer to as floppy modes because in any real system there will usually be some weak restoring force associated with the motion.

Often it is more convenient to look at the system as a dynamical one, and assign potentials or spring constants to deformations of the various struts (bonds) and angles. It does not matter whether these potentials are harmonic or not, as the displacements are virtual. However it is convenient to use harmonic potentials so that the system is linear. It is then possible to set up a Lagrangian for the system and hence define a dynamical matrix which is a real symmetric matrix and as a consequence has real eigenvalues. These eigenvalues are either positive or zero. The number of finite (non-zero) eigenvalues defines the rank of the matrix. Thus our counting problem is rigorously reduced to finding the rank of the dynamical matrix. The rank of a matrix is also the number of linearly independent rows or columns in the matrix. Neither of these definitions is of much practical help, and a numerical determination of the rank of a large matrix is difficult and of course requires a particular realization of the network to be constructed in the computer. Nevertheless the rank is a useful notion as it defines the mathematical framework for which the problem is well posed.

The genius of Maxwell [1] was to devise the simple constraint counting method that allows us to *estimate* the rank of the dynamical matrix and hence the number of floppy modes. In the next section, we discuss the application of these ideas to bulk covalent network glasses. In Sec. 5, we expand this work to look at the marginal case when surface floppy modes become important.

Until recently it has not been possible to improve on the approximate Maxwell constraint counting method, except on small systems using brute force numerical methods. Even these have had problems as will be described later. Recently a new powerful combinatorial algorithm, called the Pebble Game [3] has become available. This allows very large systems to be analyzed in two dimensional central force networks and three dimensional bond bending networks.

The layout of these notes is as follows. In the next section, we give a brief history of the construction of random network models for covalent glasses. In Sec. 3 we describe the powerful and simple ideas of constraint counting, and in Sec. 4 we discuss the exact algorithm, the pebble game and the insights that it has provided into rigidity percolation. In Sec. 5, we discuss *surface floppy modes*. In Sec. 6, we describe the current state of experiments on glasses that are relevant to this topic.

2. Continuous Random Networks

A perfect crystal is a structure in which the atoms, or groups of atoms, are arranged in a pattern that repeats itself periodically in three dimensions to form an infinite solid. For instance, crystalline diamond makes a lattice which is produced by the periodic repetition of the 8 atom diamond cubic cell (or the smaller 2 atom primitive unit cell) in all three spatial directions. A piece of the crystalline diamond structure is shown in Fig. 1.

Amorphous materials do not have the periodicity (long range order) characteristic of a crystal. Starting from a crystalline structure and slightly displacing every atom in a random manner, the periodicity is certainly destroyed, but the random structure obtained is not yet amorphous. An amorphous structure is

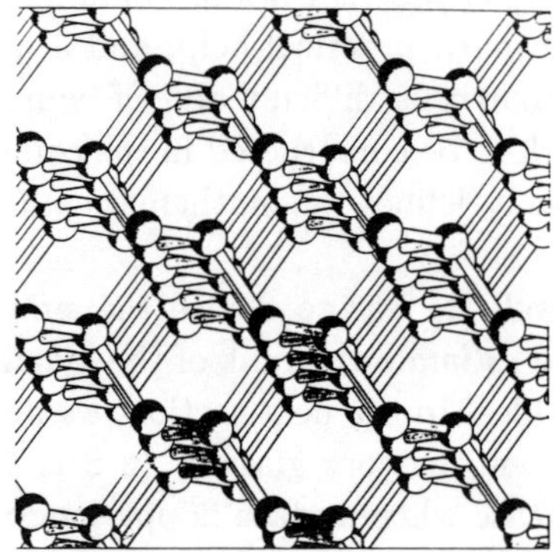

Fig. 1. A piece of a crystalline diamond structure. The characteristic six fold rings in a crystalline diamond structure can be seen.

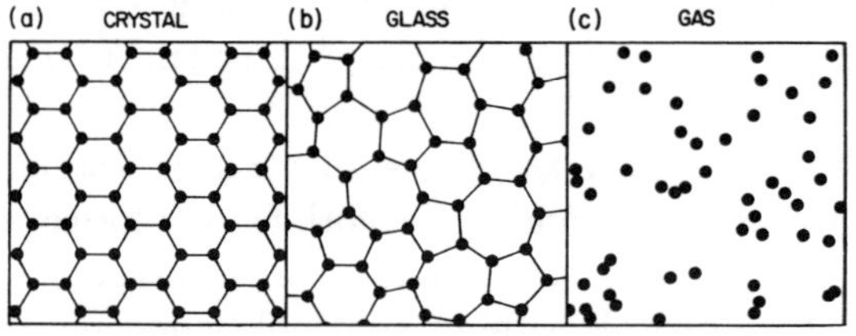

Fig. 2. Schematic sketches of the atomic arrangements in (a) a crystalline solid, (b) an amorphous solid, and (c) a gas. (Zallen [4])

topologically different from a crystalline one. Thus, to obtain an amorphous structure from a crystalline one, it is necessary not only to introduce randomness in the atomic positions, but also to change the *topology* of the original perfect lattice. We can schematically sketch atomic arrangements in three typical cases: a crystal, an amorphous solid, and a gas, as in Fig. 2.

In Fig. 3, we show a piece of the $3d$ *amorphous* diamond structure which was obtained in our computer simulations [5]. By amorphous diamond, we mean a structure that is 4-fold coordinated everywhere and therefore similar to a–Si and a–Ge, except for the relatively stronger bond bending forces.

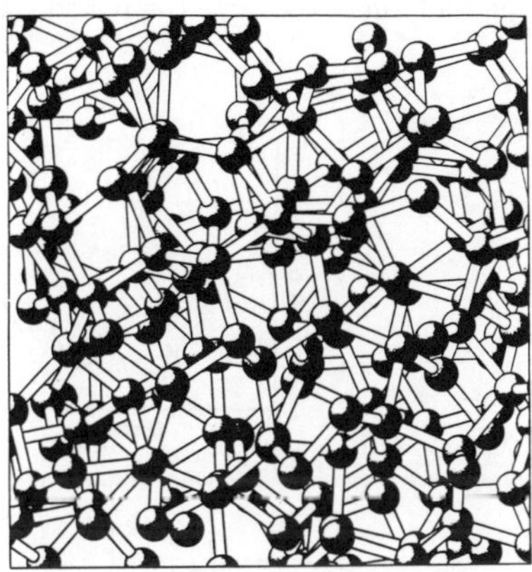

Fig. 3. A piece of an amorphous diamond structure.

Binary compounds like SiO_2, can be found both in the crystalline and in the amorphous state, the later being topologically different from crystalline SiO_2. Schematic two-dimensional representations of the crystalline and amorphous structures for a hypothetical A_2B_3 compound are shown in Fig. 4. Similarly, the crystalline diamond network possess only 6-fold rings of atoms, while the amorphous network also has 5-fold, 7-fold, and higher order rings, and consequently the number of 6-fold rings is reduced from its crystalline value.

The main experimental tool for verifying the structure of glasses is via the radial distribution function (RDF) that can be obtained by Fourier transforming X-ray or neutron diffraction data. This is illustrated in Fig. 5.

Modeling of amorphous structures can be divided in two major classes. The first one is the *Dense Random Packing* of hard spheres (DRP), in all its varieties, and the second is based on the *Continuous Random Network* (CRN),

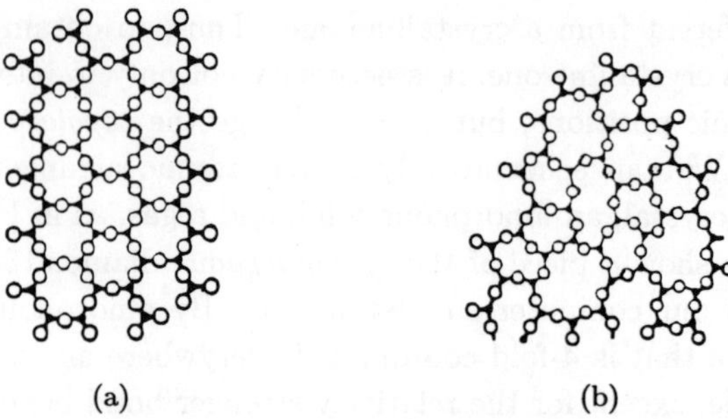

Fig. 4. Crystalline and amorphous form of a binary compound. (a) Hypothetical crystalline compound A_2B_3. (b) Amorphous form of the same compound. (Elliott 1984 [6])

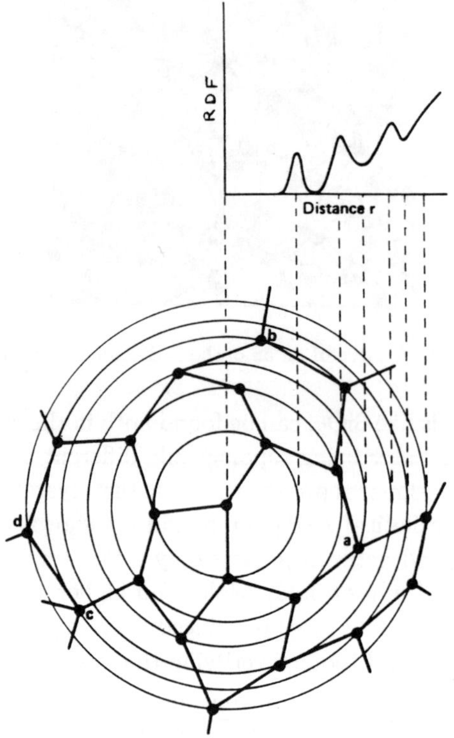

Fig. 5. Schematic illustration of the origin of the structural features in the RDF of an amorphous solid. Atoms are shown as lying on sharply defined rings for simplicity (Wooten and Weaire [7]).

concept. The DRP model was quite successful in explaining the amorphous structure of some liquids and also amorphous metals. Atoms in these systems are connected with each other predominantly by *non-directional* forces, which makes DRP a suitable model for the description of these materials. DRP modeling has been pioneered by Bernal [8, 9], and subsequently developed by Finney [10], Bennett [11] and others. For a review of modeling amorphous metals with DRP models see an excellent and comprehensive article of Cargill [12].

In this chapter, however, we shall focus our attention only on the *Continuous Random Network* models which are suitable for describing the structures of materials with predominantly covalent *directed* bonds, namely, covalent glasses. Covalent glasses have a much higher degree of short-range order than metallic glasses and are dominated by covalent bonding with a definite number of nearest neighbors; with distinct bond lengths and bond angles. The prototypical covalent amorphous solids are amorphous silicon (a–Si), and amorphous germanium (a–Ge) which belong to the semiconductors from Group IV. Every atom in these structures has four nearest neighbors, i.e. it is tetrahedrally coordinated.

We note here that the use of the term *continuous* in connection with random networks is somewhat unfortunate, because random networks are, in fact *discrete* in their structure. By *continuous* it is meant that one can build an amorphous network continuously, starting from the short-range order unit, up to an indefinite size, without including unsatisfied bonds or breaking the structure. In other words, *continuous* means that there are no identifiable boundaries in the structure, separating regions of distinctly different type of structure.

2.1. *Hand-built CRN models*

The first attempts to explain the structure of covalent glasses, around 1930, were based on the then very natural hypothesis, that amorphous materials consist of very large number of elemental microcrystals, randomly arranged into a very fine polycrystalline structure, which appears as amorphous. All attempts to fit the experimental data with such microcrystalline models failed, and the early idea of a *continuous random network* of Zachariasen [13] has become the only viable alternative. It was proposed by Zachariasen that *"... the atomic arrangement in glass is characterized by an extended three-dimensional network which lacks symmetry and periodicity."* In Fig. 6, we show

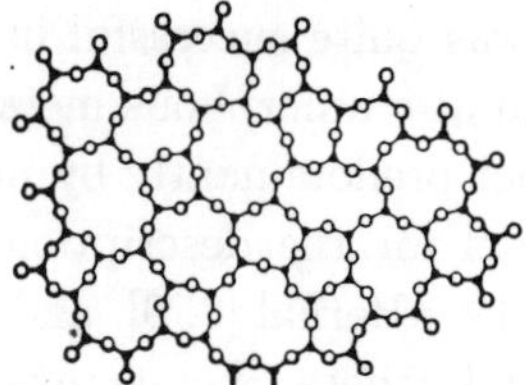

Fig. 6. Zachariasen's schematic 2d diagram represents his CRN model for an analogue of amorphous silica. Open circles are two-fold coordinated oxygen atoms, and black dots are Si atoms which are four-fold coordinated in 3d (Zachariasen [13]).

the historical schematic diagram of Zachariasen which represents his CRN model of an analogue of amorphous silica in 2d.

The first models based on the Zachariasen's ideas were built much later. One of those was the one for SiO_2, constructed by Bell and Dean [14, 15]. This was a hand-built model. This model was quite successful in explaining the diffraction data. In building the model, Bell and Dean used basic units consisting of a central Si atom and four rigid wires in a perfect tetrahedral arrangement. The O atoms were attached by inserting the wires into pre-drilled holes in the O atoms. The model was built to have a mean Si-O-Si bond angle close to 160°, but without any particular bond angle distribution. It maintained full coordination without dangling bonds in its interior. The model consisted of 614 atoms. Atomic coordinates were determined with the help of photographic techniques. From those coordinates the rms bond-length deviation was small in agreement with the experiment, while the bond-angle deviation was approximately ±30° around a mean value of 153°. The Radial Distribution Function was computed and was in fairly good agreement with the diffraction data of Mozzi and Warren [16].

Very soon, amorphous elemental semiconductors were intensively studied. The amorphous Group-IV semiconductors, a–Si and a–Ge, came to be regarded as prototypical covalent amorphous solids. It was not clear in the beginning whether tetrahedral bonds in such systems can be randomly connected *ad infinitum* in a disordered manner, or not. It was believed by Phillips [17, 18], that for instance, amorphous silicon, due to its *overconstrained* structure (a notion that will be explained in detail later), must necessarily have a discontinuous structure, in which regions of continuous random network are limited in size. But it is obvious that, in principle, there is a very close connection between a–Si and amorphous silica. A random network model for one can be easily converted into a corresponding model for the other, by adding or subtracting the oxygen atoms which link the silicon atoms and introducing a weak bond-bending force at oxygen sites.

The first CRN model for a tetrahedrally bonded amorphous solid was built by Polk [19]. There were no particular rules used in building Polk's model, except that there were no dangling bonds left within the interior of the structure. Whenever a choice about the bonding of a new atom should be decided, simply an arbitrary, random choice was made, just as in the work of Bell and Dean. The model contained a substantial number of five-, six-and seven-membered rings which ensured generation of a non-crystalline structure. There were no dangling bonds in the interior of the structure. Bond-length deviations were restricted within 1%, and bond-angle deviations were within $\pm 10°$ about the tetrahedral angle of 109°. The model consisted of 440 atoms. The original Polk model was physically extended to 519 atoms and refined by Polk and Boudreaux [20]. The refinement was in reducing the deviation of nearest-neighbor bond lengths from 1% to 0.2% using precise laser-beam measurements of the atomic coordinates and computer techniques. Steinhardt *et al* [21] and Duffy *et al* [22] used the 519 model of Polk and Boudreaux [20] and refined it further. This refinement consisted of introducing the Keating [23] potential (see later discussion) in the structure and relaxing the system toward the minimum of the elastic energy in terms of the bond lengths and bond angles.

The method of relaxation adopted by Steinhardt *et al* [21], is to move each atom separately and sequentially, while keeping the coordinates of the other atoms fixed, until there is no force on it. This process is repeated many times until convergence is achieved. In this way Steinhardt and coworkers found new, relaxed coordinates of the 519-atom model of Polk and Boudreaux. On the other hand, they also built a new 201-atom model starting from a 21-atom cluster with new atoms serially added. Whenever a small group of atoms was added to the growing cluster their positions were relaxed by computer to minimize the forces on them. Thus, the connectivity of this network is affected by the relaxation procedure. They did not find any significant structural difference between the two models. Duffy *et al* [22] used the same 519-atom model of Polk and Boudreaux but adopted a different relaxation procedure. Each atom was simultaneously moved in the direction of the force on it; a distance proportional to that force. After each move the forces are recalculated, the atoms moved again, and the process iterated until the forces and displacements become small. In this way, they obtained results which were pretty much the same as those of Steinhardt *et al* [21].

Provoked by the controversy involving the presence of five-membered rings in amorphous III-V compounds like GaAs, which would, of course imply the occurrence of energetically unfavorable, *wrong* bonds between like atoms, and to check if different ring statistics would produce different RDF's, Connell and Temkin [24] built a model of a–Ge with no odd-membered rings. The model consisted of 238 plastic and metal tetrahedra. This model did not produce any improvement of the agreement with experiment. Our own calculations of RDF for a–Si and a–C show that RDF is very similar for different ring statistics [5].

All these early hand-built models suffered from their finite sizes, or in other words, from the lack of periodic boundary conditions. In these relatively small models, a large fraction of the atoms were necessarily located close to the surface. This complicates the analysis of the bulk properties, and the calculation of the RDF. The first hand-built model which had periodic boundary conditions implemented was made by Henderson [25]. The model consisted of only 61 atoms.

2.2. *Computer-built CRN models*

Perhaps the very first computer modeling of this kind was done by Kaplow *et al* [26]. They tried to simulate the structure of amorphous selenium and to compare it to the experimental results. The method Kaplow *et al.* adopted in creating a computer model is a good example of a misconception that one can make an amorphous structure just by randomizing the atomic coordinates in a crystalline structure. This is exactly what the authors did in this early work. Their computer-generated coordinates of crystalline arrangement of about 100 atoms, and then used Monté Carlo procedure to randomize these coordinates. Only those which improved RDF were accepted. After about 10^5 moves a *reasonable* agreement with experiment was achieved. The problem with this procedure, as we have already mentioned before, is that obtained structure is topologically identical to the crystalline one, and thus cannot be expected to give good description of the amorphous selenium. This inverse Monté Carlo method has been considerably refined subsequently and is discussed at length elsewhere [27].

Henderson *et al* [25] experimented with computer algorithms for creating structures with periodic boundary conditions, but they did not succeed in getting good agreement with experiment. Nevertheless, the approach of

Henderson, in generating periodic structures and in using a simple computer algorithm, inspired later work on computer modeling. In contrast to this approach, Shevchik [28] and Polk and Boudreaux [20] tried to imitate in their algorithm the hand-building process but that proved to be a very cumbersome procedure.

As we have already stressed, a truly amorphous structure created by a successful randomization of the crystalline atomic positions, must be topologically different from the structure of its crystalline parent, and it must have perfect connectivity. One obvious technique to ensure these two requirements, for instance in creating amorphous silicon, is to start from the diamond cubic structure, remove one atom at random, and rejoin the four dangling bonds in pairs. In this way the number of atoms in the structure clearly decreases, but the topology is changed due to the fact that six-membered rings of the crystalline structure are destroyed, while five-, and seven-membered rings are introduced. Four-fold connectivity of the network is preserved. This method was used by Alben *et al* [29] who used 64-atoms crystalline supercell as a starting point, and ended up with a 58-atoms model of amorphous silicon. In an unpublished work of Wooten and Weaire (see Wooten and Weaire [7]) similar studies were done on models that contained 700–1000 atoms. Results were less then impressive. Calculated RDF's are in poor agreement with experiment. The structures obtained in this way have large bond-angle distortions, and it was concluded that they are not satisfactory as models of amorphous silicon.

The other technique which illustrates the topological connection between the structure of vitreous silica and amorphous silicon, was employed by Evans *et al* [30, 31]. They used a decoration transformation (see Wright *et al* [32]) of the vitreous silica random network model of Evans and King [33] which involved removing the oxygen atoms, rejoining the two dangling bonds and adjusting the new network to have the mean bond length appropriate for amorphous silicon. The model achieved relatively good agreement with experiment in the first two peaks, but the model pair correlation function exhibits too much structure at higher separations.

2.2.1. *Guttman model*

The first approach at computer generation of periodic random networks was developed by Guttman [34, 35]. His aim was to construct an ideal random network with four-fold coordination which would describe the structure of amorphous germanium. Guttman did not use a single computer algorithm, but otherwise his approach is quite similar to the Wooten and Weaire method [7]. Guttman starts from the crystalline super-cell which contains many unit cells, and imposes periodic boundary conditions on the supercell. There are no bonds in the model initially. Bonds are introduced in a stochastic way: four neighbors are selected at random from the given set of atoms which are assigned to each atom in the lattice. Thus four bonds are constructed at a time, and the process is repeated until, hopefully, the desired network is created. Guttman noticed that the probability of constructing a fully four-fold coordinated network in this way was *decreasing* with the size of the model. The size of the model varied from 64 atoms to 512 atoms. The structure was relaxed using a simple Hooke's-law bond-stretching, and bond-bending term, similar to that in Eq. (1). Guttman also varied the super-cell size to find the optimal density which minimizes the total strain energy. He found that the optimal model density was 15% greater than the observed crystalline density, while the density of a–Ge experimentally was found to be within the few percent of the crystalline density.

Furthermore, a large compressive stress was introduced by random bonding assignments, and the rms angular deviation from the perfect tetrahedral angle (109.47°) was found to be about 17°; much greater than the experimental value of abut 10° (see Etherington *et al* [36]). Guttman then used a specific topological rearrangement to generate the random network. In Fig. 7 we show two examples of Guttman's bond rearrangements or bond switches.

The switch is accepted if it lowers the energy of the network, and rejected if it does not. In this way the RDF was slightly improved, but the models

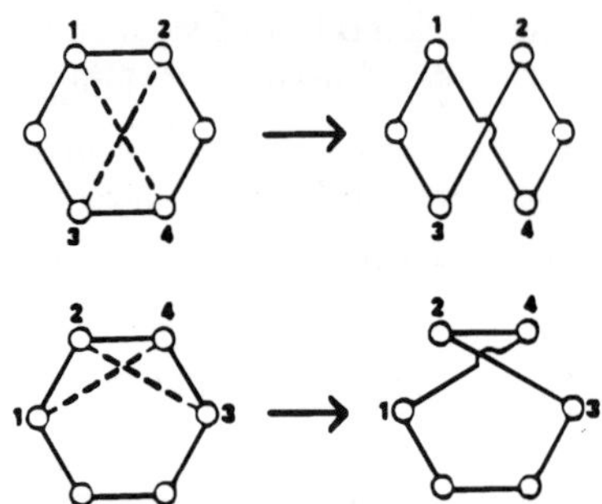

Fig. 7. Two examples of bond switches used by Guttman [35]. Six-membered rings of the crystalline lattice are shown, and four atoms are selected. A path 1-2-3-4 is defined and bonds 1–2 and 3–4 are exchanged with non-bonds 2–3 and 4–1 respectively.

still did not agree very well with experiments, because all Guttman's models contained many dangling bonds.

2.2.2. *Wooten–Weaire method*

In the Wooten and Weaire model of a–Si and a–Ge, the Keating potential is adopted to describe the interaction between the atoms. The Keating potential (1) was introduced earlier to fit the elastic and vibrational properties of Group-IV elements [23]. It represents a semiempirical description of bond-stretching and bond-bending forces and involves only few parameters. It consists of two terms, the first being the bond-stretching one, and the second the bond-bending.

$$V = \frac{3}{16}\frac{\alpha}{d^2} \sum_{l,i} (\vec{r}_{li} \cdot \vec{r}_{li} - d^2)^2 + \frac{3}{8}\frac{\beta}{d^2} \sum_{l\{i,i'\}} \left(\vec{r}_{li} \cdot \vec{r}_{li'} + \frac{1}{3}d^2 \right)^2 \qquad (1)$$

where α and β are bond-stretching and bond-bending force constants, respectively, and d is the strain-free equilibrium bond length in the diamond structure. For Si its value is $d = 2.35$ Å, and for Ge, we have $d = 2.46$ Å.

Central bond-stretching forces are not enough to stabilize the structure, because the number of constraints imposed by maintaining fixed bond lengths is smaller than the number of degrees of freedom. This will be explained in more detail later. At the moment it is enough to stress the importance of the bond-bending forces for creating a stable network. Wooten and Weaire followed Martin [37] by taking $\beta/\alpha = 0.285$ in building their model of a–Si and a–Ge. Due to the dominance of the α term, when the model is relaxed to minimize the energy, all bonds relax to within a few percent to their ideal unstretched length, but because this is not enough to stabilize the structure, the β term defines the stable structure which minimizes the bond-bending energy. Thus, the total energy is roughly proportional to β, and the relaxed structure is independent of β/α (Wooten, Winer and Weaire [38]; Wooten and Weaire [7]). The Keating potential is schematically show in Fig. 8.

Wooten, Winer and Weaire [38] created a new model of amorphous silicon and germanium which significantly improved the agreement with experiment. It consisted of 216 atoms. A new bond-switch was invented which does not produce large bond-length and bond-angle distortions, and is also simple. This new bond switch is shown in Fig. 9.

The bond switch shown in Fig. 9 is characterized by interchanging two second-neighbor bonds that are parallel to each other in the perfect diamond

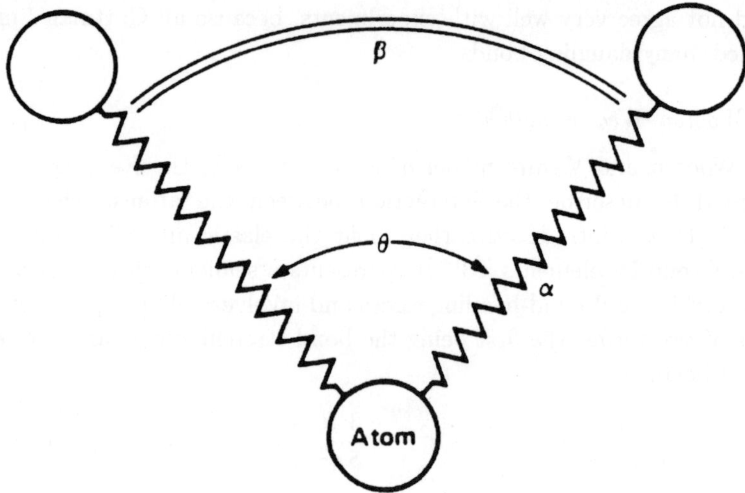

Fig. 8. The Keating potential bond-stretching α, and bond-bending β terms (Wooten and Weaire [7]).

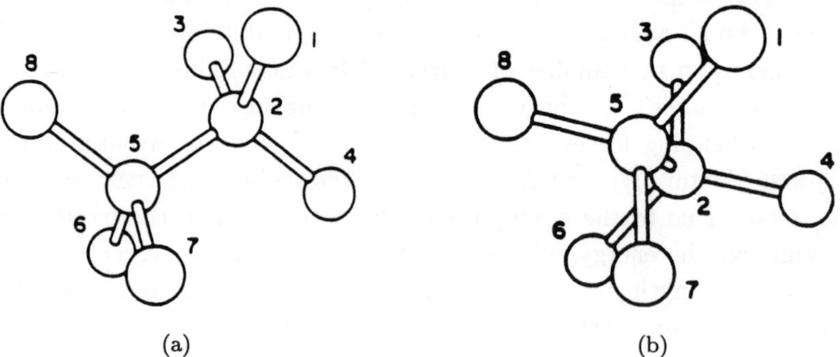

Fig. 9. Wooten–Weaire bond switch. (Left panel) Configuration of bonds in the diamond cubic structure; (Right panel) Relaxed configuration after switching two bonds.

structure. This is a local rearrangement of bonds that creates minimal strain into an originally strain-free perfect diamond structure, which was not the case for Guttman's switch (see Fig. 7). Atoms are numbered, their coordinates are listed, and a neighbor-table is created. The switch consists of breaking those two bonds [1–2 and 5–6] in Fig. 9, and creating two other bonds, [1–5 and 2–6]. The switch is not acceptable if new bonds are longer than 1.7 times the unstretched nearest-neighbor bond, which is a length slightly larger than the second-neighbor distance, as it should be if a successful bond switch is to be made in the original crystalline structure.

One can define a *ring* as any closed path of bonds. The number of n-membered rings defined in this way increases with n. An *irreducible ring* is defined as one that has no shortcuts across it. That is, given any two atoms (vertices) on the ring, there is no shorter path between any two atoms (as measured by the number of bonds along the path) than a path on the ring itself. One advantage of such rings for topological purposes is that the number of n-membered rings is zero for large n, but no topologically important rings are omitted, and the complete table of ring statistics remains finite (Wooten and Weaire [39]).

The perfect diamond structure has only 6-membered irreducible rings. The introduction of a bond switch changes the ring statistics by destroying even-membered rings and creating odd-membered rings, 5-, 7-, 9-, etc. membered rings. This process is schematically shown in the $2d$ case in Fig. 10. When the structure is already randomized, bonds that are switched are chosen to be approximately parallel by ensuring that they do not belong to the same 5-, 6-, or 7-membered ring.

One bond switch introduced into the crystalline diamond super-cell affects the topology of the structure by destroying 12 six-membered rings, and simultaneously creating 4 five-membered, and 16 seven-membered rings. During the process of initial randomization, the RDF becomes gradually less structured, reflecting the amorphous nature of the created structure.

After the initial randomization produces a disordered, but highly strained model, it is necessary to continue with creating new bond switches searching for those which will lower the strain energy. Of course, the lowest energy possible is zero, i.e. that of crystal. So one has to be careful to avoid returning to the crystalline state during the process of annealing. Wooten and Weaire [7] showed that the system indeed returns to the original crystalline diamond structure if in the initial process of randomization, the number of introduced bond switches was not large enough. If the initial temperature is too high the subsequent annealing is not able to bring the system down to an acceptably low energy, then the model obtained has too much strain in it. In other words, with too high an initial temperature, the system is trapped in one of the numerous metastable states of relatively high energy. From the other side, the number of bond switches has also to be chosen optimally; if there are too

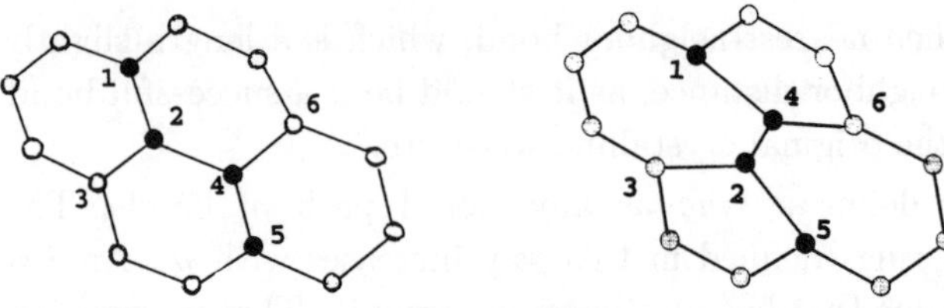

Fig. 10. Schematic representation of a Wooten–Weaire topological bond switch. There are four 6-membered rings initially (diamond structure), and after a bond switch is made, two 5- and two 7-membered rings are created.

many of them in the beginning, then again, the system will not be able to release a sufficient amount of energy during the annealing process, and the RDF of the final structure will show that the system has too large bond-length and bond-angle distortions. Wooten and Weaire found that for creating an a–Si model the optimal melting temperature for the initial process of randomization has to be $kT = 1$ eV.

The annealing process itself is based on Metropolis Monté Carlo algorithm [40] already applied in the literature to general problems of optimization by simulated annealing (Kirkpatrick *et al* [41]; Vanderbilt and Louie [42]). The Metropolis algorithm is generally used as an efficient simulation of a collection of atoms in equilibrium at a given temperature. In the process of creating a model of an amorphous structure, the introduction of topological rearrangements (bond switches) change the total strain energy of the lattice at a given temperature. If that change, ΔE, is negative, i.e. if the introduction of a bond switch lowers the total energy, this bond switch is accepted. On the other hand, if the ΔE is positive (increases the total energy), that bond switch may be accepted with probability $P(\Delta E) = \exp(-\Delta E/kT)$. In practice, a random number uniformly distributed in the interval (0,1) is chosen each time when a bond switch is introduced, and ΔE is calculated. If the picked random number is less than $P(\Delta E)$ then the new configuration of atoms is accepted. If the random number is greater than $P(\Delta E)$ the new configuration of atoms is rejected and the original one (before the bond switch was introduced) is used to start the next step. Of course, the next step always consists of introducing the new bond switch, i.e. in repeating the described basic step. When this is done many times, the procedure simulates the behavior of the system at given temperature T, as it reaches thermal equilibrium.

The introduction of the Boltzmann factor is essential in the process of annealing because it is able to lift the structure into a metastable state of higher energy, thus enabling it to find another path in the configurational space toward the lower-energy states. Without it the system would not be permitted to ever increase its energy, and thus it would be easily trapped in a metastable state with no way out. It has been noted (Wooten and Weaire [7]) that allowing the presence of the 4-membered rings in the annealing process facilitated the convergence toward the energy minimum, and helped to avoid trapping into high-energy metastable states.

Thus, after the structure has been initially randomized by the introduction of a large number of bond switches, the temperature is then reduced in small steps, and at each temperature the system is kept long enough by introducing new bond switches and using the Metropolis algorithm until the system is in equilibrium. As we already mentioned this is followed by the Keating relaxation of the local regions affected by a bond switch. A sequence of the temperatures is chosen, and at the end, when the temperature is zero, one hopes to obtain the structure with all characteristics of an amorphous material. Wooten, Winer and Weaire [38] obtained in this way a good model for a–Si and a–Ge. Their final 216-atom structure had an rms bond-angle deviation of about 12.6°. Consequently, the energy of their amorphous models was comparatively low. Our 4096-atom model of a–Si exhibits [5] even lower rms bond-angle deviation (10.5°–11°) which corresponds to a lower strain energy per atom than in the original Wooten and Weaire models. Nevertheless, some physicists object that there is still too much strain present in the amorphous structures obtained by using Wooten–Weaire method, and that a modification of it is needed if one wants to produce more realistic, less strained amorphous structures (Mousseau [43]).

The Wooten, Winer and Weaire model of a–Ge and a–Si [38] has a slight remnant of the original crystal visible in the structure factor, but nevertheless, there was no serious discrepancy with experiment (Wooten and Weaire [7]). The slight memory of the diamond cubic structure that remained in the amorphous structure is a consequence of the insufficient initial randomization and does not come from the annealing process. It should be noted here that the only essential difference between the a–Si and a–Ge is a simple scaling in proportion to the bond lengths ($d = 2.35$ Å for Si, and $d = 2.46$ Å for Ge). Otherwise, the experimental data are practically identical for a–Si and a–Ge. In Fig. 12, we show the pair correlation function obtained in Wooten and Weaire model for a–Ge compared with experiment.

The discrepancy in the RDF at high r are related to the presence of inhomogeneities in the real sample which are not taken into account in fully

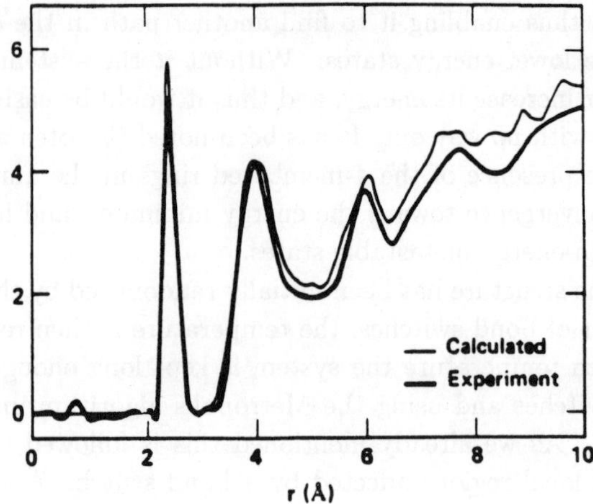

Fig. 11. Thin line is the RDF divided by r calculated from the Wooten and Weaire model (1985), and thick line is experimental result for a–Ge.

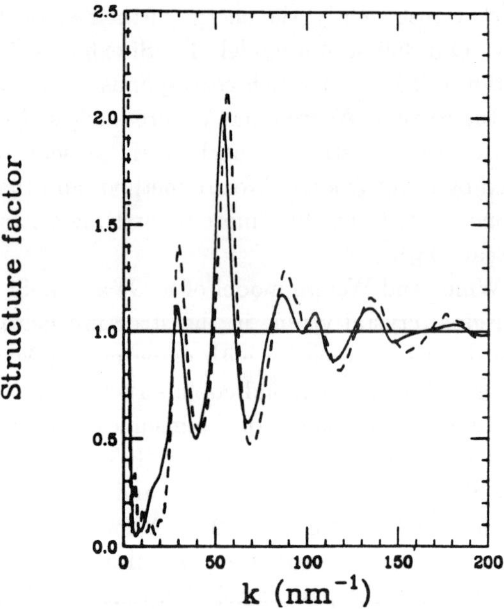

Fig. 12. A comparison of our calculated structure factor for a–C [5] (dashed line) with that experimentally observed [47]. Note that our model is 100% sp^3 whereas the experimental sample has from 10–15% sp^2 carbon in it.

tetrahedrally connected models, which always have slightly higher density than samples produced experimentally.

Since McKenzie and coworkers [44] have obtained 85–90% four-fold coordinated amorphous carbon by plasma-arc deposition, there has been an increasing interest in amorphous diamond-like carbon [45–48]. These diamond-like carbon films are found to be hard, optically transparent, and chemically inert, which makes them potentially important for applications in coating technology and also for use as wide band gap semiconductors. It is known that amorphous silicon and amorphous germanium exhibit a striking similarity in the shape of their radial distribution functions [7, 36]. In both of these materials the bond stretching forces are dominant compared to the weaker bond-bending forces. Amorphous carbon is different with respect to both a–Si and a–Ge because of very strong chemical bonding, and because the bond-bending forces are much larger relative to the bond-stretching forces. The ratio of bond-bending to bond-stretching forces β/α is 2.4 times larger for carbon than for silicon. One might have expected a slightly different amorphous structure for carbon (i.e. different ring statistics etc.) compared with silicon and germanium, because of the greater bond stiffness of carbon. Our results [5] show that the RDF of tetrahedral amorphous diamond is remarkably similar to the RDF's of amorphous silicon and amorphous germanium. The carbon first peak is broader than the silicon first peak, which is correlated with the larger rms bond-length deviation in carbon. This represents a different apportioning of the strain energy between bond length strain and bond angle strain in carbon and silicon. Nevertheless the overall features of RDF's are essentially unchanged between carbon and silicon. Using a Fourier transformation, the structure factor can be written as

$$I(k) = 1 + \frac{\rho_0}{k} \int_0^\infty \left[\frac{P(r)}{r} - 4\pi r \right] \sin(kr) dr \qquad (2)$$

where $P(r)$ is the RDF and ρ_0 is the average density. In Fig. 12, we show a comparison of our calculated structure factor for a–C with that observed in experiments of Gilkes *et al* [47].

Our large (4096 atom), and realistic structural model of amorphous diamond [5] is potentially very appealing for addressing some of the other important issues of the physics of amorphous solids. One of those is the nature of the band tails in the electronic density of states in amorphous materials (Mott and Davis [49]). Recently, Drabold [50] used our structural model of amorphous diamond [5] to study the nature of the electronic localization in three dimensions in the presence of topological disorder.

Another application of the Wooten–Weaire method was done by Mousseau and Lewis [51] in their computer simulations of amorphous silicon hydrides. The starting point is a model of amorphous silicon obtained by Wooten–Weaire method. Instead of the Keating potential the authors use the Stillinger–Weber potential (Stillinger and Weber [52]) to describe the interaction between atoms. They introduce pairs of H atoms in the a–Si structure in such a way that the fourfold coordination is preserved. The nearest-neighbor Si atoms were allowed to have a bond to each of the two H atoms. After each introduction of hydrogen pairs, the structure is Monté Carlo relaxed using a finite number of Wooten–Weaire bond switches, and relaxed using the Stillinger–Weber potential. The process is repeated until the desired concentration of H atoms is achieved. Mousseau and Lewis calculate Si–Si, Si–H, and H–H partial pair distribution functions, as well as corresponding static factors, and find that the agreement between the model and experiment improves with H concentration.

3. Constraint Counting

We start by examining a large covalent network that contains no dangling bonds. We can describe such a network by the chemical formula $Ge_xAs_ySe_{1-x-y}$ where the chemical element, Ge, stands for *any* fourfold bonded atom, As for *any* threefold bonded atom and Se for any twofold bonded atom. Every atom has its full complement of nearest neighbors and we consider the system in the thermodynamic limit where the number of atoms $N \to \infty$. There are no surfaces or voids and the chemical distribution of the elements is not relevant, except that we assume there are no isolated pieces, like a ring of Se atoms. The total number of atoms is N and there are n_r atoms with coordination $r(r = 2, 3$ or $4)$, then

$$N = \sum_{r=2}^{4} n_r \tag{3}$$

and we can define the mean coordination

$$\langle r \rangle = \frac{\sum_{r=2}^{4} r n_r}{\sum_{r=2}^{4} n_r} = 2 + 2x + y \tag{4}$$

We note that $\langle r \rangle$ (where $2 < \langle r \rangle < 4$) gives a partial but very important description of the network. Indeed when questions of connectivity are involved, it is the key quantity as we shall see.

In covalent networks like $Ge_x As_y Se_{1-x-y}$, the bond lengths and angles are well defined and small displacements from the equilibrium structure can be described by the Keating potential written in Eq. (1). The bond-bending force is essential to the constraint counting approach. The other terms in the potential are assumed to be much smaller and can be neglected at this stage. If floppy modes are present in the system, then these smaller terms in the potential will give the floppy modes a small finite frequency. For more details see Thorpe [53]. If the modes already have a finite frequency these extra small terms will produce a small uninteresting shift in the frequency. This division into large and small forces is absolutely essential if the constraint counting approach is to be of any use. It is for this reason that it is of little, if any, use in metals and ionic solids. It is fortunate that this approach provides a very reasonable starting point in many covalent glasses.

We will regard the solution of the eigenmodes of the potential of Eq. (1) as a problem in classical mechanics [1, 53, 54]. The dynamical matrix has a dimensionality $3N$ which corresponds to the $3N$ degrees of freedom. In a stable rigid network we would expect all the squared eigenfrequencies $\omega^2 > 0$ with six modes being at zero frequency. These six modes are just the three rigid translations and the three rigid rotations. We are assuming that our (large) network has free boundary conditions. Of course these 6 modes have no weight in the thermodynamic limit. The total number of zero frequency modes can be estimated by the Maxwell counting algorithm [1]. This approach was first done for glasses by J.C. Phillips [17, 18].

The constraint counting proceeds as follows. There is a single constraint associated with each bond which we can assign as $r/2$ constraints associated with each $r-$ coordinated atom. In addition there are constraints associated with the angular forces in Eq. (1). For a twofold coordinated atom there is a single angular constraint; for an $r-$fold coordinated atom there are a total of $2r - 3$ angular constraints. The total number of constraints is

$$\sum_{r=2}^{4} n_r [r/2 + (2r - 3)] \tag{5}$$

The fraction f of zero-frequency modes is given by

$$f = \left[3N - \sum_{r=2}^{4} n_r [r/2 + (2r - 3)] \right] / 3N \tag{6}$$

which can be conveniently rewritten in the compact form

$$f = 2 - \frac{5}{6}\langle r \rangle \tag{7}$$

where $\langle r \rangle$ is defined in Eq. (4). Note that this result only depends upon the combination $2x + y$ which is the relevant variable. When $\langle r \rangle = 2$ (e.g. Se chains), then $f = 1/3$; that is one third of all the modes are floppy. As atoms with higher coordination than two are added to the network as crosslinks, f drops and goes to zero at $\langle r \rangle = 2.4$. and the network becomes rigid, as it goes through the phase transition. This mean field approach has been quite successful in covalent glasses and helps explain a number of experiments as will be described later. Also in Sec. 5, we describe the results of computer experiments and show that they are rather well described by the results of this section.

4. Generic Rigidity Percolation

The elastic properties of random networks of Hooke springs has been studied over the past 12 years [53–58]. One of the most interesting findings has been that effective medium theory describes the behavior of the elastic constants and the number of floppy modes remarkably well [56, 58] except very close to the phase transition from a rigid to a floppy structure.

Unfortunately, attempts to study the critical behavior in central-force networks have not been very satisfactory and the question of the universality class of the rigidity transition [55, 57–61] has remained unresolved. This question is fundamental in understanding the nature of the rigidity transition, and may have important implications as to how the character of the glass transition is affected by the mean coordination, as has been discussed recently via fragile and strong glass formers [62]. We show here how substantial progress can be made in understanding the geometrical nature of generic rigidity percolation [3, 63].

There are two important differences between rigidity and connectivity percolation. The first difference is that rigidity percolation is a vector (not a scalar) problem, and secondly, there is an inherent long range aspect to rigidity percolation. For example, Fig. 13(a) shows four distinct rigid clusters consisting of two rigid bodies attached together by two rods connecting at pivot joints. Now the placement of one additional rod, as shown in Fig. 13(b), locks the previous four clusters into a single rigid cluster. This non-local character

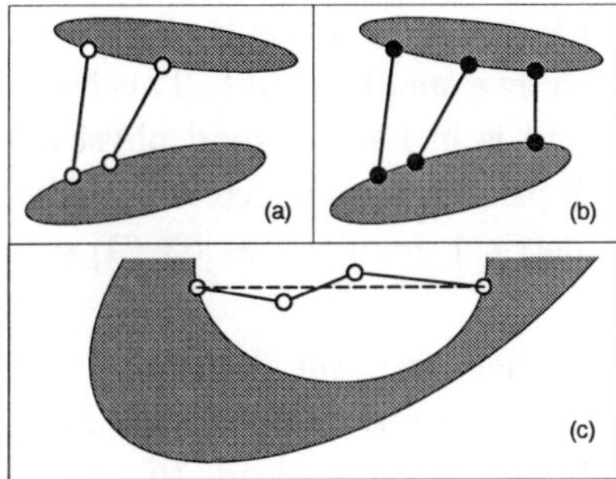

Fig. 13. The shaded regions represent 2d rigid bodies. The (closed, open) circles denote pivot-joints that are members of (one, more than one) rigid body. (a) A floppy piece of network with four distinct rigid clusters. (b) Three generic cross links between two rigid bodies make the whole structure rigid. If the bonds were parallel, the structure would not be rigid to shear [64]. (c) Three non-collinear connected rods connecting across a rigid body is generic and contains one internal floppy mode. If they were collinear (along the dotted line), then there would be two infinitesimal (not finite) floppy motions, and under a horizontal compression buckling would occur.

allows a single rod (or bond) on one end of the network to affect the rigidity all across the network from one side to the other.

Using concepts from graph theory, we set up generic networks where the connectivity or topology is uniquely defined but the bond lengths and bond angles are arbitrary. A generic network does not contain any geometric singularities [61] which occur when certain geometries lead to null projections of reaction forces. Null projections are caused by special symmetries, such as, the presence of parallel bonds or connected collinear bonds. Rather than these atypical cases their generic counterparts will be present as shown in Figs. 13(a) and (b). This ensures that all infinitesimal floppy motions carry over to finite motions [61, 65].

All previous studies on rigidity percolation have been on regular (non-generic) lattices which as we now know [3, 63] has inadvertently delayed a proper understanding of the rigidity transition. In non-generic (referred to as atypical) networks many geometrical singularities occur which lead to non-linear effects. For example, a diode-like problem frequently occurs in atypical networks where a string of collinear bonds can only be extended with a cost in energy but can be compressed with no cost in energy due to buckling [e.g. Fig. 13(c)]. The diode effect complicates studies because it leads to the breakdown of linear elasticity theory which must be reversible. A simple way to view a generic network is to take a regular lattice structure and randomly displace each site location by a small amount. This introduces local distortions throughout the lattice and is in itself a good physical model for amorphous and glassy materials. All prior studies which inevitably involve the non-linear effects arising from geometrical singularities [53–61] should be considered as a separate problem.

By considering generic networks, the diode effect and the problematic geometric singularities are completely eliminated. Therefore, the problem of rigidity percolation on generic networks leads to many conceptual advantages because all geometrical properties are robust, not depending on a multitude of special cases. Moreover, real glasses are modeled better by generic networks rather than regular atypical networks because of local distortions. In two dimensions, there exist efficient, exact, combinatorial algorithms allowing for the possibility of an in depth study of rigidity percolation.

The rigidity of a network glass is related to how amenable the glass is to continuous deformations that require very little cost in energy. A small energy cost will always arise from weak forces which are present in addition to the hard covalent forces that involve bond-lengths and bond-angles. These small energies can be ignored because the degree to which the network is deformable is well quantified by just the number of floppy modes [53] within the system. A mental picture of floppy and rigid regions within the network has led to the idea of rigidity percolation [53, 55]. An interesting model to consider is one with only central forces because its properties are most different from connectivity percolation which can be recovered if all pivot joints are welded fixed [66].

Much understanding of the general phenomena of central-force rigidity percolation can be obtained by studying a random network of Hooke springs. To be specific, we begin by considering a network of Hooke springs characterized by the potential

$$V = \frac{1}{2} \sum_{\langle ij \rangle} \alpha_{ij} \eta_{ij} (l_{ij} - l_{ij}^0)^2 \tag{8}$$

where the sum is over all bonds $\langle ij \rangle$ connecting sites i and j in the network. A bond connecting sites i and j is present if $\eta_{ij} = 1$ with probability p and absent if $\eta_{ij} = 0$ with probability $1 - p$. The spring constants, α_{ij}, and the equilibrium bond lengths, l_{ij}^0, are positive real numbers but are left arbitrary. In addition, the site locations are also left arbitrary as the network is generic.

Note that rigidity is a static concept, involving virtual displacements, so that while it is convenient to use harmonic potentials as done in Eq. (8), any set of pair potentials would give the same results for the geometric aspects of rigidity that is of interest here. A collection of sites form a rigid cluster when no relative motion within that cluster can be achieved without a cost in energy. Conversely, the floppy modes correspond to finite motions of the sample which do not cost energy. Therefore, the geometrical properties and the number of floppy modes can be determined by an equivalent bar and joint structure [61]. Note that a d-dimensional system always has at least $d(d+1)/2$ floppy modes due to d global translations and $(d-1)d/2$ global rotations.

The number of floppy modes in d dimensions is given by the total number of degrees of freedom for N sites minus the number of independent constraints. A redundant bond can only add additional reinforcement and/or cause internal stress in an existing rigid body. A key quantity is the number of floppy modes, F, in the network, or normalized per degree of freedom, $f = F/dN$. By defining the number of redundant bonds per degree of freedom as n_r, we can write quite generally,

$$f = \frac{dN - (\frac{1}{2}Nzp - dNn_r)}{dN} = 1 - \frac{p}{p^*} + n_r \qquad (9)$$

where $p^* = 2d/z$ and z is the lattice coordination. Neglecting the redundant bonds, as first done by Maxwell [1], we find that f is linear in the bond concentration, p, and goes to zero at the Maxwell approximation, p^*, for the threshold. The Maxwell approximation gives a very good account of the location of the phase transition and the number of floppy modes, but it ultimately fails since the number of independent constraints is not just the total number of bonds as some bonds are dependent. Redundant bonds occur in overconstrained regions of the networks. The labeling of such overconstrained regions is unique, as is the number of redundant bonds in a given overconstrained region. However which particular bonds are called redundant is arbitrary, although the number is fixed. If there is only a single redundant bond in an over constrained region, and it is cut, the region becomes isostatic and contains no redundant bonds. If a further bond is cut, this isostatic region will break up into two or more separate isostatic regions.

Short of the Maxwell constraint counting method, other ways that have been used for calculating the number of floppy modes exactly include rank determination of the dynamical matrix [53], relaxation methods [57, 59], transfer matrix methods [67], Gaussian elimination [60] and the equation of motion technique [58]. Hence, many numerical methods have been explored.

Here we focus on the geometrical aspects of rigidity percolation which have not been directly addressed before. Previous studies have used costly relaxation type methods [which use $O(N^3)$ floating point operations] so that networks containing $N \approx 10^4$ sites already present a difficult numerical challenge. Relaxation methods are well suited for calculating elastic constants, but not for characterizing the geometric structure. This is because numerically one cannot identify which bond has exactly zero stress or if a bond accidentally has zero stress. However, until now, this was the only approach available in determining the stress carrying backbone.

Many basic questions have remained open regarding the nature of the rigidity transition in spite of many years of research by many groups. We mention a few points regarding the triangular lattice. Hansen and Roux [59, 67] and later supported by Arbabi and Sahimi [60] have indicated that the fractal dimension of the stress carrying backbone is about $D_0 \approx 1.63$ which is very close to the current carrying backbone in connectivity percolation. Furthermore, the elastic moduli critical exponent (also denoted by f), or more precisely the ratio $f/\nu \approx 1.12$ was obtained for the random bond-dilution problem. Here the exponent ν controls how the correlation length diverges as the critical point is approached. Curiously, this value is the same as the full bond-bending model (all angular forces are present) which has an identical geometry to connectivity percolation. These two results gave strong evidence that the geometrical properties of central-force rigidity percolation is identical to connectivity percolation.

At first, the idea that the two types of percolation problems could share the same universality class is surprising since the vector and scalar character have different symmetry properties and because rigidity is a highly non-local phenomenon much different from connectivity. However, Hansen and Roux [59, 67] have argued that the reason why this is possible, is because at large length scales, the central forces are able to yield effective angular forces via lever arms. Thus, the central-force problem may renormalize into the full bond-bending model. Knackstedt and Sahimi Knackstedt [60] have suggested using a real space renormalization group calculation, the geometrical properties of site and bond rigidity percolation share the same universality class as far as geometry is concerned. However, the elastic moduli exponents were shown not to be the same.

4.1. *The pebble game*

We have been able to study networks containing more than 10^6 sites, using an integer algorithm which gives exact and unique answers to the geometric properties of generic rigidity percolation. Because of the non-local characteristic of rigidity percolation, [e.g. Figs. 13 (a) and 1(b)] burning-type algorithms [68] commonly used in connectivity percolation are useless.

A very efficient combinatorial algorithm, as suggested by Hendrickson [65], has been implemented to (i) calculate the number of floppy modes, (ii) locate over-constrained regions and (iii) identify all rigid clusters for $2d$ generic bar-joint networks. The crux of the algorithm is based on a theorem by Laman [69] from graph theory.

Theorem. A generic network in two dimensions with N sites and B bonds (defining a graph) does not have a redundant bond iff no subset of the network containing n sites and b bonds (defining a subgraph) violates $b \leq 2n - 3$.

By simple constraint counting it can be seen that there must be a redundant bonds when Laman's condition is violated. This necessary part generalizes to all dimensions such that if $b > dn - d(d+1)/2$ then there is a redundant bond for $n \geq d$. For $n < d$, it follows that if $b > n(n-1)/2$ then there is a redundant bond. Note that $n = 1$ is an excluded case in Laman's theorem since two sites are required for a bond to be present. The essence of Laman's theorem is that in two dimensions finding $b > 2n - 3$ is the only way redundant bonds can arise. This sufficient part does not generalize to higher dimensions [65].

The basic structure of the algorithm is to apply Laman's theorem recursively by building the network up one bond at a time. Only the topology of the network is specified, not the geometry. Because of the recursion, only the subgraphs which contain the newly added bond need to be checked. If each of these subgraphs satisfy the Laman condition, $b \leq 2n - 3$, then the last bond placed is independent, otherwise it is redundant. By counting the number of redundant bonds, the exact number of floppy modes is determined.

Searching over the subgraphs is accomplished by constructing a pebble game. Each site in the network has two pebbles tethered to it. A pebble is either free when it is on a site or *anchored* when it is covering a bond. A free pebble represents a single motion that a site can undertake. Consider a single site having two free pebbles, representing two translations. If two additional free pebbles can be found at a different site, then the distance between this pair of sites is not fixed. Placing a bond between this pair of sites will constrain their distance of separation. To record this constraint, one of the four free pebbles is anchored to the bond. Once the bond is covered, only three free pebbles can be shared between that pair of sites. After a bond is determined to be independent, it will always remain independent and covered.

We begin with a network of N isolated sites each having two free pebbles. The system will always have $2N$ pebbles; initially two free pebbles per site. We place one bond at a time in the network connecting pairs of sites. The topological placement of either the sites or bonds will depend on the model under study such as the site or bond diluted generic triangular lattice as done here. The independent bonds must be covered by a pebble, therefore, before a bond can be covered it must be tested for independence. For each bond placed in the network, four pebbles (two on each site at the ends of the bond) must be free for the bond to be independent. When a bond is determined to be independent, any one of the four pebbles can be anchored to that bond. In general, all four pebbles across an added bond will not be free because they are already anchored to other bonds. These anchored pebbles may possibly become free at the expense of anchoring a neighboring free pebble while keeping

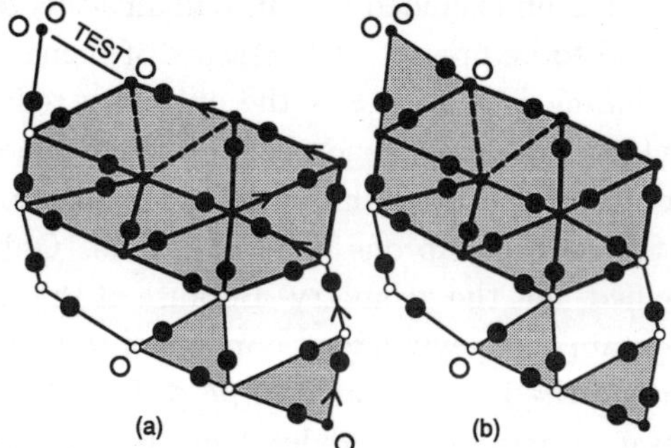

Fig. 14. A demonstration of the pebble game on a generic network. Independent (redundant) bonds are shown with solid (dashed) lines which are (are not) covered by a pebble. Large (filled, open) circles denote (anchored, free) pebbles on (bonds, sites). The two closest pebbles to a given site are tethered to that site. Small (filled, open) circles denote sites belonging to (one, more than one) rigid cluster. Over-constrained bonds are shown with heavy dark lines. Shaded regions denote $2d$ rigid bodies. (a) There are seven rigid clusters and the five free pebbles indicate 5 floppy modes until a new bond is added and tested for independence. A fourth free pebble is found via the path traced by arrows. (b) The added bond is independent and thus covered. There are now six rigid clusters and four floppy modes.

a particular independent bond covered. In other words, pebbles may be shuffled around the network provided all independent bonds remain covered.

It is always possible to free up three pebbles across a bond, since they correspond to its rigid body motion. When a fourth pebble across a bond cannot be found, then that bond is redundant and it is not covered. In Fig. 14 an example of how pebbles are shuffled is schematically shown on a small generic structure. Two distinct pebbles are associated with each site for which each pebble can either be used to cover a bond or is free to cover a bond. The two pebbles closest to a given site as schematically drawn in Fig. 14 are the pebbles that are tethered to that site. Thus a pebble may either be on a site (free pebble) or on a bond (anchored pebble) but it always remains tethered to a given site regardless of how the pebbles are shuffled. Note that a bond may be covered by a pebble from either of its end sites. Therefore, free pebbles can be moved across the network by exchanging the site from which a pebble is used to cover a bond.

Over-constrained regions are recorded each time a dependent bond is found. These regions correspond to the set of bonds that were searched in trying to free the fourth pebble but failed. These regions, called Laman subgraphs, violate the condition $b \leq 2n - 3$. An added bond onto a Laman subgraph will be redundant.

We identify all the rigid clusters after the network is completely built. First, we identify isolated sites. Then the rigidity of all other sites is tested with respect to a reference bond. If a test-bond between either one of the pair of sites forming the reference bond and the site in question is found to be (dependent, independent) then that site (is, is not) rigid with respect to the reference bond. The test-bond is actually never added to the network. Since a bond can only belong to one cluster (unlike sites), all the bonds within a rigid cluster are ascribed to a particular reference bond. A systematic search is made to map out all rigid clusters.

We show in Fig. 14(b) the end result of the pebble game applied to a simple structure. Many aspects of rigidity are displayed. It can be seen that:

(1) The exact number of floppy modes is determined by the number of free pebbles remaining. A depletion or excess of pebbles to cover a set of bonds distinguishes the over-constrained regions from the floppy regions; unlike the approximate global counting of Maxwell.

(2) This network is uniquely decomposed into a set of six distinct rigid clusters, although the clusters are not disconnected.

(3) The free pebble along the bottom edge cannot be shuffled over to the rigid body at the top which already has three free pebbles. This free pebble is shared among three bars and two triangles. Generally, free pebbles get trapped in floppy regions consisting of many rigid clusters giving arise to complex collective floppy motion.

(4) The number of redundant bonds is unique, whereas their locations are not unique since this depends on the order of placing the bonds. Nevertheless, each redundant bond belongs to a unique over-constrained region (Laman subgraph). For example, there are 19 over-constrained bonds in the rigid cluster at the top of the structure in Fig. 14(b), while having only two redundant bonds.

(5) A rigid cluster will generally have sub-regions that are over-constrained. If any bond that is over-constrained is removed, the rigidity of the network is unchanged.

A section of a large network on the bond-diluted generic triangular lattice is shown in Fig. 15 after the pebble game was applied. Here we see a typical topology and associated geometry of the set of rigid clusters in this section of network. Observe that nearly all individual rigid clusters form connected paths via pivots which are free joints shared by two or more rigid bodies. It can be seen that floppy regions form where the pivot sites cluster. Notice that within this section of the network, there is a spanning rigid cluster where rigidity forms a connected path from top to bottom and left to right. There are also clusters of over-constrained regions in the network which are not percolating in this section.

4.2. *Two dimensional central force networks*

In this section, we present some new results for central-force generic rigidity percolation on both the bond and site diluted **triangular** net. We begin by working out a better estimate for the bond and site rigidity threshold. The Maxwell counting prediction for the site rigidity threshold gives $q^* = 2/3$ which is the same as $p^* = 2/3$ for bond dilution. This comes about because the number of sites in a site diluted system is only qL^2 and the number of bonds is $q^2 z L^2/2$. When the number of floppy modes is normalized per degree of freedom, the final result for f can be expressed in the same way as the leftmost side of Eq. (9) with p replaced by q. To get a more accurate estimate for the rigidity threshold, the presence of redundant bonds and floppy inclusions must be accounted for.

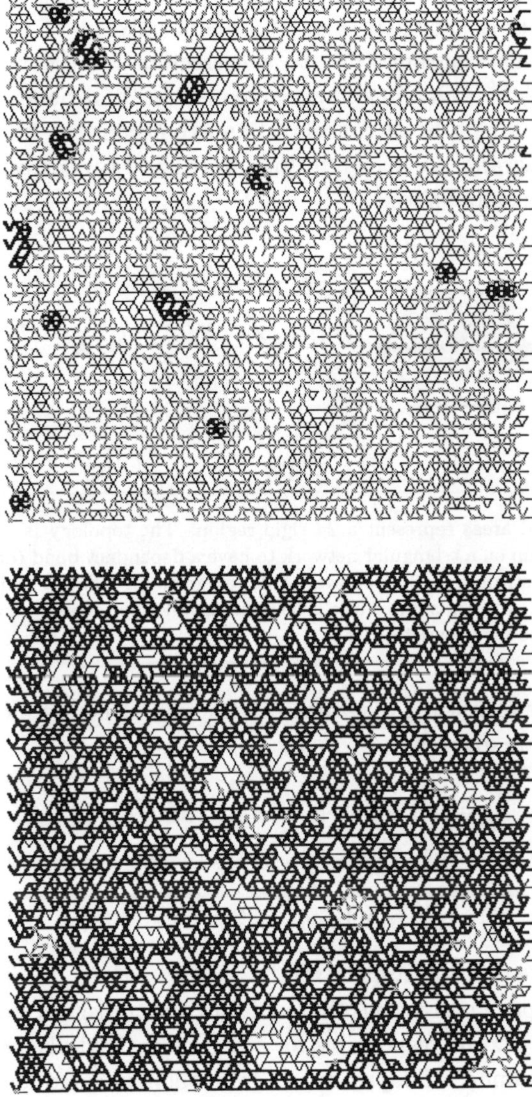

Fig. 15. The topology of a typical cut-out region from a bond-diluted generic network below percolation (top) and percolation (bottom). The heavy dark lines correspond to over-constrained regions. The open circles correspond to sites which are acting as pivots between two or more rigid bodies.

In a low bond concentration expansion the first diagram to contribute redundant bonds is shown in Fig. 16, where only 11 of the 12 bonds are independent. This diagram leads to a correction for the number of floppy modes as $n_r = \frac{1}{2}p^{12} + O(p^{18})$ and $n_r = \frac{1}{2}q^7 + O(q^{10})$ for bond and site dilution respectively. In a high bond concentration expansion, the first two dominant contributing diagrams are shown in Figs. 16(b) and (c) corresponding to dangling ends and isolated sites.

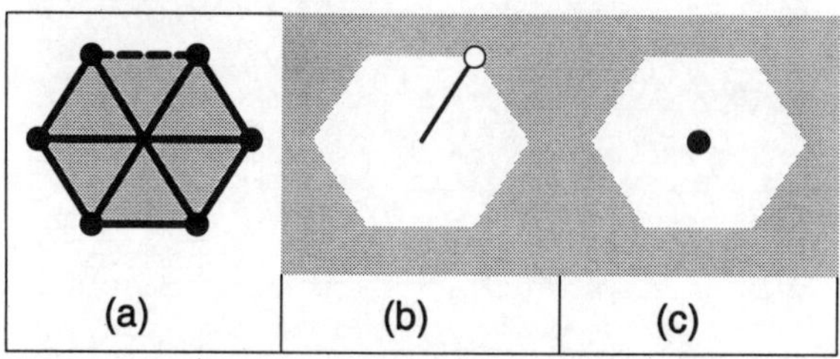

Fig. 16. All shaded areas represent a $2d$ rigid region. The topology is shown for: (a) The lowest order diagram on a triangular network to have a dependent bond (dashed line). All 12 bonds are over-constrained. (b) The lowest order diagram for a floppy inclusion corresponds to a dangling bond. (c) An isolated site is counted as a floppy inclusion and contributes two floppy modes at the next lowest order.

From these two leading diagrams the fraction of floppy modes, f, at high concentrations are given by

$$f = 3(1 - p)^5 - 2(1 - p)^6 + O((1 - p)^8) \tag{10a}$$

and

$$f = 3(1 - q)^5 - 5(1 - q)^6 + 2(1 - q)^7 + O((1 - q)^8) \tag{10b}$$

for bond and site dilution respectively.

It is these type of corrections that will shift the transition from the Maxwell threshold of 2/3 and be responsible for a non mean-field-like critical behavior. We equate the number of floppy modes from the truncated low and high concentration expansions. For bond and site dilution we estimate the thresholds to be 0.6622 and 0.6877 respectively. Thus we find a small downward shift for bond dilution and a somewhat larger upward shift for site dilution. The pebble game reveals that the rigidity thresholds shift about 50% more than the above estimates to $p_{\text{cen}} = 0.66020 \pm 0.0003$; and $q_{\text{cen}} = 0.69755 \pm 0.0003$;

for bond and site dilution respectively. The site diluted threshold is also in good agreement with that obtained by Duxbury and Moukarzel [70].

A sharp peak in the second derivative of the number of floppy modes shown in Fig. 17 appears without any signs of a discontinuity. The peak most resembles a simple cusp. As can be seen there is virtually no difference between the data for linear system sizes L= 680, 960 and 1150. Only very slight system size dependence has been observed. The trend is for smaller systems to show a cusp-like singularity in $f^{(2)}$ as well, but with the peak slightly shifted to the left with a smaller amplitude.

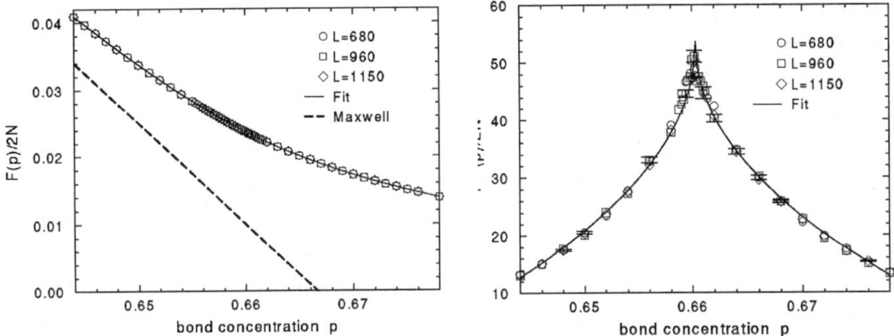

Fig. 17. (a) Simulation results for the fraction of floppy modes, $f = F/2N$, for a bond diluted generic network compared to the Maxwell prediction. All error bars are smaller than the symbols. Note that the non-generic threshold [60] is at 0.641. (b) The second derivative of the fraction of floppy modes for a bond diluted generic network as calculated from Monte Carlo sampling. The fitting results for the cusp yields $p_{\mathrm{cen}} = 0.6603 \pm 0.0003$ and an exponent of 0.48±0.05. Typical error bars are shown which reflect both the statistical errors in the Monté Carlo sampling and the ensemble averaging.

The behavior of the second derivative suggests that the number of floppy modes is analogous for rigidity and connectivity percolation. In the case of connectivity percolation, the number of floppy modes is simply equal to the total number of clusters, which corresponds to the free energy [71]. We find that the second derivative of the total number of clusters changes sign across the transition, thus violating convexity requirements. Noting that typically rigid clusters are not disconnected, we suggest that the number of floppy modes generalizes as an appropriate free energy. With this assumption, we have estimated the exponent α in the usual context of a *heat capacity* critical exponent. More work needs to be done to see how a free energy can be defined for rigidity percolation.

In addition to calculating the number of floppy modes, we analyzed the rigid cluster statistics [63] for bond dilution with free and periodic boundary conditions and site dilution with periodic boundary conditions. The free boundary condition data was generated mainly to check boundary effects. Finite size scaling techniques as used in percolation theory [68] are applied here assuming only a single relevant length scale exists.

We also look at the geometrical properties of the over-constrained regions within the network. These over-constrained regions are not necessary to sustain rigidity. An isostatic framework [61], for example, has just the right positioning of bonds to form a rigid cluster without a redundant constraint. However, when external forces are applied to a rigid isostatic framework, using rigid bus bars for example, over-constrained regions will be induced across it, which forms a percolating stressed region. Within a random environment there will be redundant bonds scattered throughout the network. Here the redundant bonds are essentially acting as external forces on an underlying isostatic framework. Physically, the resulting over-constrained regions characterize internal stress caused by bond mismatch. Thus, the over-constrained regions propagate stress (without externally applied forces), and are most closely analogous to the current carrying backbone in connectivity percolation. We find that monitoring the probability for a network of linear size L to contain either a spanning rigid cluster or spanning stressed backbone, leads to the same correlation length exponent, ν, and critical threshold.

From extrapolation using finite size scaling techniques [63] we find $p_{\text{cen}} = 0.66020 \pm 0.0003$ for the bond diluted triangular lattice which is in excellent agreement with that obtained from the cusp singularity in the second derivative of f as shown in Fig. 17. Similarly, for site dilution we find $q_{\text{cen}} = 0.69755 \pm 0.0003$. Moreover, the rigidity transition is second order, but in a different universality class than connectivity percolation, with the exponents; $\alpha = -0.48 \pm 0.05$, $\beta = 0.175 \pm 0.02$ and $\nu = 1.21 \pm 0.06$. The fractal dimension of the spanning clusters and the spanning stressed regions at the critical threshold are found to be $d_f = 1.86 \pm 0.02$ and $d_{BB} = 1.80 \pm 0.03$ respectively.

4.3. *Three dimensional bond bending networks*

Unfortunately it is not possible to extend the pebble game to $3d$ central-force networks, as this algorithm is based on Laman's theorem which does not generalize to $3d$. This is because of the existence of banana graphs as shown in Fig. 18.

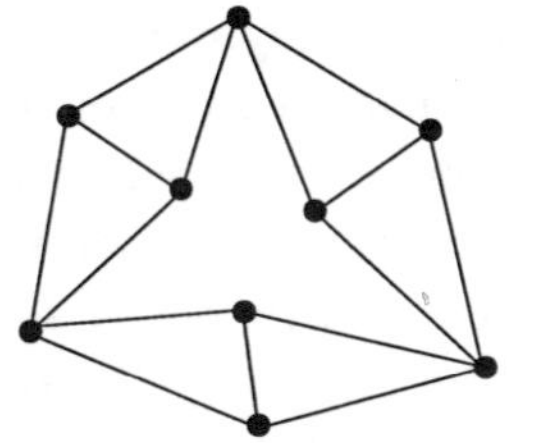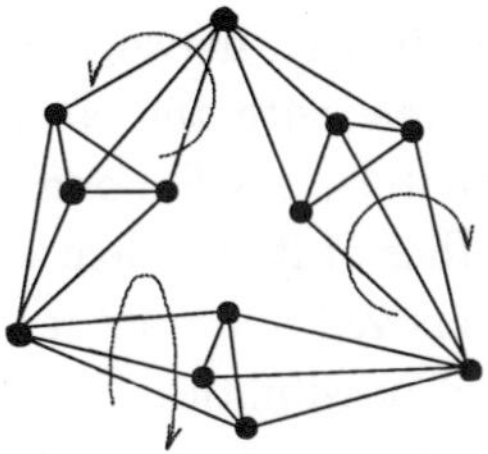

Fig. 18. (Left panel) A single rigid cluster in two dimensions. (Right panel): Showing 3 bananas in a $3d$ network. There are four rigid clusters here — three bananas plus an additional rigid cluster consisting of the three sites at the corners which are connected by implied bonds. Such non-contiguous rigid clusters cannot occur in $2d$ networks. Such clusters are responsible for the breakdown of Laman's theorem in $3d$.

While the pebble game described in this paper is only applicable in $2d$, we have generalized the rules for a certain class of $3d$ networks. Although the Laman condition is not generally sufficient [65, 69, 72] in three dimensions, we recently have shown [84] that it can be generalized within bond bending networks. The problem of the Laman bananas as shown in Fig. 18 is conveniently eliminated once angular forces are included as in the Keating potential Eq. (1). In particular, we have constructed a $3d$ pebble game for the bond-bending model, having nearest and next nearest neighbor forces. Fortunately, the bond-bending model is precisely the class of models that is applicable to the study of covalent glass-networks.

For the most part, the $3d$ pebble game rules come from naively generalizing the $2d$ rules (such as using three pebbles per site instead of two). However, there are some subtle differences caused by restricting the $3d$ networks to be in a certain class. The order of bond placement in building a network now becomes important, so that the bond-bending model at each step of the process is preserved and thereby insuring that at no time will banana structures form.

For purposes of testing rigidity in generic three dimensional bond-bending networks, it is only necessary to specify the network topology or connectivity of the central-force (CF) bonds, since the second nearest neighbors via CF bonds define the associated bond-bending constraints. Here, we have considered two test models. In the first model, a unit cell is defined from our realistic computer generated network of amorphous silicon [5] consisting of 4096 atoms having periodic boundary conditions. Larger completely four coordinated periodic networks containing 32,768, 262,144 and 884,736 atoms are then constructed from the basic amorphous 4096-atom unit cell.

The four coordinated network is then randomly diluted by removing CF bonds one at a time with the constraint that no site can be less than two coordinated. That is, a CF bond is randomly selected to be removed. If upon removal either of its incident atoms becomes less than two coordinated, then it is not removed and another CF bond is randomly selected from the remaining pool of possibilities. The order of removing CF bonds is recorded. This process is carried out until all remaining CF bonds cannot be removed, leading to as low an average coordination number as possible. All CF bonds that were successfully removed are marked. This method of bond dilution gives a simple prescription for generating a very large semi-realistic model of a continuous random $Ge_xAs_ySe_{1-x-y}$ type of network. For comparison sake, the second test model that was considered consisted of diluting in the same way a diamond lattice containing 32,768, 262,144 and 10^6 atoms.

In the application of the $3d$ pebble game, the network is built up one constraint at a time. The first step is to place all the unmarked CF bonds in the network. The order of placing these CF bonds in building the network is not important. However, once a CF bond is placed in the network, all its associated bond-bending constraints must also be placed before the next CF bond can be considered. In this way, the bond-bending form of the network is nearly conserved throughout the building process, which prevents the unwanted banana configurations, and thus enables us to test for generic rigidity using a $3d$ pebble game algorithm [84]. Then the marked CF bonds (including their associated bond-bending constraints) are placed in the network in the reverse order from that when they were removed as the network was randomly diluted. In this way, the rigidity properties of the network can be monitored as a function of the average coordination number which typically ranges from $(\langle r \rangle = 2.2$ to $4.0)$.

The ensemble average over many realizations of bond dilution for the fraction of floppy modes and its derivatives have been numerically obtained for different size networks. The fraction of floppy modes is calculated exactly for each realization, its first derivative is obtained by Monté Carlo sampling and the second derivative is determined by one numerical differentiation. The fraction of floppy modes and its second derivative for both the bond diluted amorphous silicon and diamond lattices are shown in Figs. 19 and 20 respectively. The bond-diluted generic diamond lattice behaves in nearly the same way as the bond-diluted amorphous silicon. They both have a rigidity transition slightly below the simple Maxwell constraint counting estimate of 2.4.

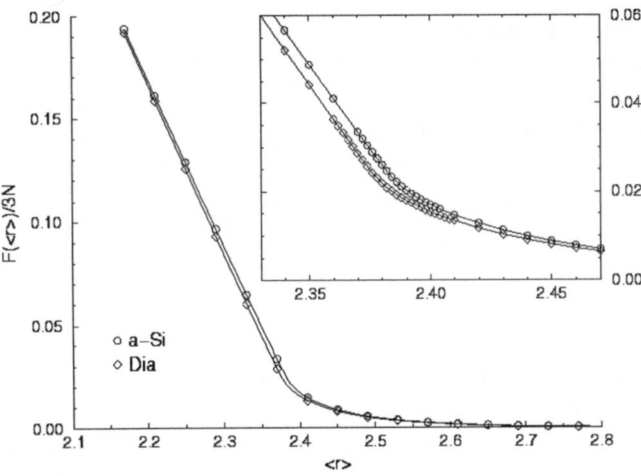

Fig. 19. Showing the number of floppy modes F as a function of the mean coordination $\langle r \rangle$. The insert shows a blow up of the region around $\langle r \rangle = 2.4$.

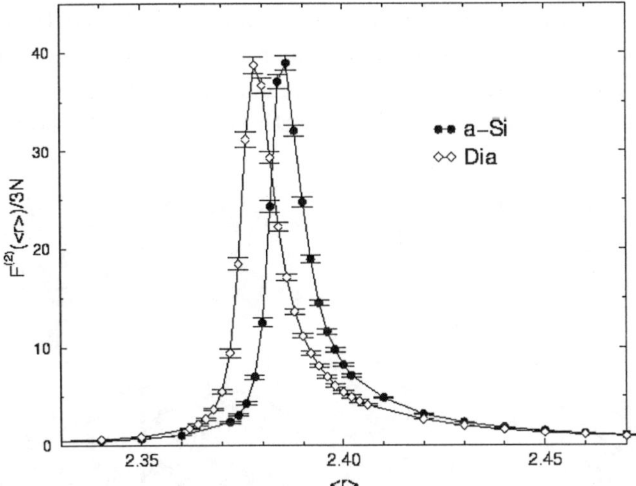

Fig. 20. Showing the second derivative of the number floppy modes $F^{(2)}$ as a function of the mean coordination $\langle r \rangle$.

The rigidity transition can be accurately found from the sharp peak in the second derivative in the fraction of floppy modes. In particular, for the diamond lattice $\langle r \rangle_c = 2.375 \pm 0.003$ and for a–Si $\langle r \rangle_c = 2.385 \pm 0.003$. Remarkably, the

Maxwell constraint counting estimate is accurate to about 1%. Furthermore, the small shift downward from the Maxwell estimate is expected since near the rigidity transition there are many floppy inclusions in the network which tends to lower the transition. On the other hand, the slightly higher transition coordination number for the a–Si compared to the generic diamond lattice can be qualitatively understood in that there exist five and four fold rings in a–Si, which tends to increase where the transition occurs.

In Fig. 21, we show the rigid and floppy regions of a typical section of the bond-diluted amorphous silicon model below and above the rigidity transition at an average coordination number of 2.37 and 2.40 respectively. It is useful to study these two figures, and compare them with Fig. 15 showing a similar map of rigid and floppy regions in a two dimensional central force network. It can be shown [84] that the only floppy element in a three dimensional bond-bending network is a hinge joint. Hinge joints can only occur through a CF bond and are always shared by two rigid clusters — allowing one degree of freedom of rotation through a dihedral angle.

In Fig. 21 it is seen that just below the rigidity transition, a–Si is still very floppy with small overconstrained regions sparsely scattered throughout the network – mostly due to the presence of four and five fold rings. Unlike in two dimensions, large isostatic rigid regions are not seen. Although most of the CF bonds are acting as hinge joints, it is worth noting that the average number

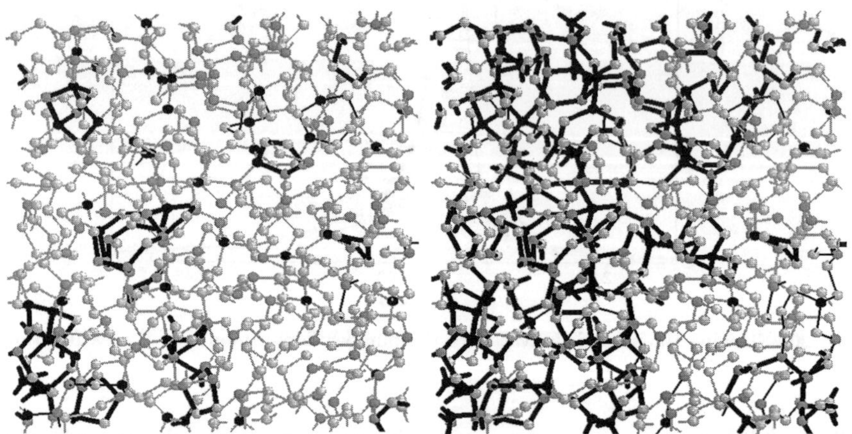

Fig. 21. Typical section of a random network with a low mean coordination, where rigidity has not percolated (left) and with a high mean coordination where rigidity has percolated (right). The dark lines represent overconstrained regions.

of hinge joints per floppy mode is about 15, indicating the existence of many collective motions, where one floppy mode is spread over many small rigid clusters. In Fig. 21, it is seen that with an increase in average coordination number of about 1.25%, a–Si becomes rigid with a spanning overconstrained region. There are still very few isostatic rigid regions found, although many floppy inclusions remain. The average number of hinge joints per floppy mode is again about 15 after passing through a maximum at the rigidity transition. Many of the floppy inclusions that remain are simple local motions involving one dihedral angle, but large floppy regions survive that involve many rigid clusters.

It has been suggested by Duxbury and co-workers [70] that the rigidity transition may contain some elements of a first order transition. In particular they claim that the infinite cluster percolation probability is discontinuous at the rigidity transition, whereas the backbone [overconstrained regions] probability changes continuously. This idea was partly generated through an elegant calculation of rigidity on a Bethe lattice, which they showed is a model that can be solved exactly. We have found an error in their solution — they identified the transition with the spinodal point rather than by computing the free energy equivalent [the number of floppy modes f], and making a Maxwell type equal area type construction to find the transition. Nevertheless the interesting result is that the rigidity transition on the Bethe lattice is of a conventional first order kind and not second order. To date there is no evidence that the rigidity transition on compact lattices is other than second order and continuous, although with larger sample sizes are studied, this possibility cannot be completely ruled out. These results are especially interesting as recently, Boolchand and co-workers [73] have found some dramatic evidence for small first order jumps in certain Raman frequencies in the vicinity of, or possibly at, the rigidity transition. This is a very exciting area of research that is not properly understood at the present time.

5. Surface Floppy Modes

The analysis in the previous section has all been in the thermodynamic limit ($N \to \infty$), and the number of floppy modes, F, was a thermodynamic quantity proportional to the number of atoms, N. Thus the floppy modes can be thought of as extended bulk modes. Because there are such modes, surface effects are normally irrelevant leading to totally negligible corrections. However, when the term proportional to N is zero, surface effects can become important and dominant. We will discuss this situation here.

5.1. *Basic counting techniques*

We use as an illustration the square net with $n_1 n_2$ sites forming a free standing rectangle. We can apply the Maxwell constraint counting method here also, taking careful account of the sides and corners. The basic algorithm is unchanged; the number of floppy modes equals the number of degrees of freedom (*twice the number of sites*) minus the constraints (*the number of bonds*).

$$F_2 = 2n_1 n_2 - \left[\frac{4}{2}(n_1 - 2)(n_2 - 2) + \frac{3}{2}2(n_1 - 2) + \frac{3}{2}2(n_2 - 2) + \frac{2}{2}4\right] \tag{11}$$
$$= n_1 + n_2$$

where the subscript 2 on F denotes two–dimensions. We have checked this computationally for a range of values of n_1 and n_2 and find that the formula (11) works *exactly*. The formula, Eq. (11) is based on topological considerations and has been found to work whether the network is generic or atypical. We have checked this computationally in both cases.

These planes can now be stacked up to form a simple cubic array and the result for F becomes

$$F_3 = n_3 F_2 + n_3(n_1 n_2) - (n_1 n_2)(n_3 - 1) \tag{12}$$

where the second term comes from the extra degrees of freedom and the last term is the additional constraints. Simplifying we find that

$$F_3 = n_1 n_2 + n_2 n_3 + n_3 n_1 \tag{13}$$

This is clearly a surface term, as it is proportional to the number of surface atoms in a large sample.

We can now generalize to hypercubical lattices d dimensions

$$F_d = n_d F_{d-1} + (n_d) \prod_{i=1}^{d} n_i - (n_d - 1) \prod_{i=1}^{d-1} n_i \tag{14}$$

which gives

$$F_d = n_d F_{d-1} + \prod_{i=1}^{d-1} n_i \tag{15}$$

We can solve this equation, using induction and the known answers for $d = 2$ and $d = 3$ to give

$$F_d = \left(\prod_{i=1}^{d} n_i\right)\left(\sum_{i=1}^{d} \frac{1}{n_i}\right) \tag{16}$$

This equation has been checked for various values of n_i for $d = 2$ and $d = 3$ and always works exactly for *both* generic and atypical networks. The reason why this counting is exact is because there are no redundant bonds in the network and thus Maxwell constraint counting is exact.

A nice way of obtaining the equation for F_2 is to count the number of *missing bonds* at the surface and divide by 2. An example is shown in Fig. 22. Here there are 29 sites and 24 surface bonds shown by dashed lines. A little thought will convince the reader that the number of floppy modes is $24/2 = 12$. This number clearly reduces to the previous answer $n_1 + n_2$ for a rectangular shape, but generalizes the answer to other less regular shapes. Note that the site in the indentation has all four bonds present, and so does not contribute, despite being a surface atom in some sense. This kind of construction also works if there are internal voids. Another way to look at this construction, is to dissect the network into distinct linear chains which decouple. Each linear chain has two (dashed) ends and a single zero frequency mode, leading immediately to the result above. This latter approach is equivalent to using Eq. (15) to solve for F_2 knowing that $F_1 = 1$.

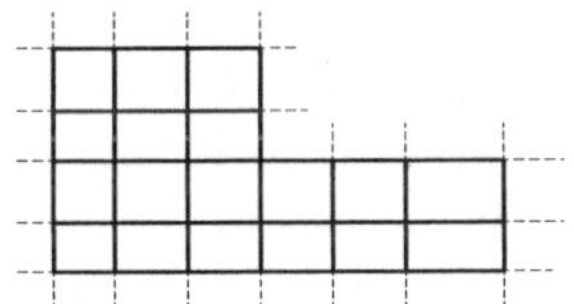

Fig. 22. A piece of a square net with 29 sites and 24 surface bonds shown by dashed lines. It is explained in the text that there are $24/2 = 12$ floppy modes of which 2 are the macroscopic rigid translations and 1 is the macroscopic rigid rotation.

5.2. *Problems with periodic boundary conditions*

The periodic case is more difficult to understand, as symmetry now seems to play a more important role, with clear differences appearing between distorted and undistorted networks [74]. The reader may complain that the periodic case is of no more than academic interest. This complaint may be justified, but we cannot claim to understand this whole area until we do understand the periodic case also. Constraint counting always give exactly zero for any d dimensional *hypercubic* lattice. This result is not exactly correct as there are always the d Goldstone modes, corresponding to the acoustic phonons in the long wavelength limit. For the *distorted or generic* periodic case, we find numerically that this is the total — there are indeed always exactly d modes. The *undistorted or atypical* periodic case can be solved using Bloch's theorem. For a periodic hypercubic lattice, the eigenfrequencies are given by

$$\omega_r^i = \sqrt{\frac{k}{M}} \left| \sin\left(\frac{\pi r_i}{n_r}\right) \right| \tag{17}$$

where $r_i = 0, 1, 2, \ldots n_r - 1$. Because of the decoupling, there are $\frac{1}{n_r}\left(\prod_{i=1}^{d} n_i\right)$ floppy modes associated with the direction r, so that the total number of floppy modes is again given by

$$F_d = \left(\prod_{i=1}^{d} n_i\right)\left(\sum_{i=1}^{d} \frac{1}{n_i}\right) \tag{18}$$

which curiously is the same answer we obtained for the square with free boundary conditions [74]. Again this is because the square can be dissected into periodic linear chains, each with a single floppy mode, as in the case with free boundary conditions.

The difference is that the distortions reduce the number of floppy modes from F_d to d in the periodic case, but the number of floppy modes for the non-distorted case is the same as F_d for free boundary conditions. Topologically, the periodic boundaries is equivalent to having long range bonds from end to end along the network. In the case of no distortions, the colinearity of these long range bonds with the $1d$ chains results in them not contributing as independent constraints.

6. Experiments

6.1. *Bulk materials*

The above findings have been confirmed using computations on model networks. Some of these results are shown in Fig. 23. The computed number of floppy modes (shown in the insert) closely follows Eq. (7) and the elastic constants approach zero from the high coordination side, also at $\langle r \rangle = 2.4$, which is referred to as the point at which *rigidity percolation* occurs. This is a phase transition from rigid to floppy, driven not by temperature, but rather by the mean coordination.

Measurements of the elastic constants [74] appear to be influenced considerably by the weak forces, and the phase transition is washed out as shown in Fig. 23. The best experimental confirmation of these ideas to date comes from

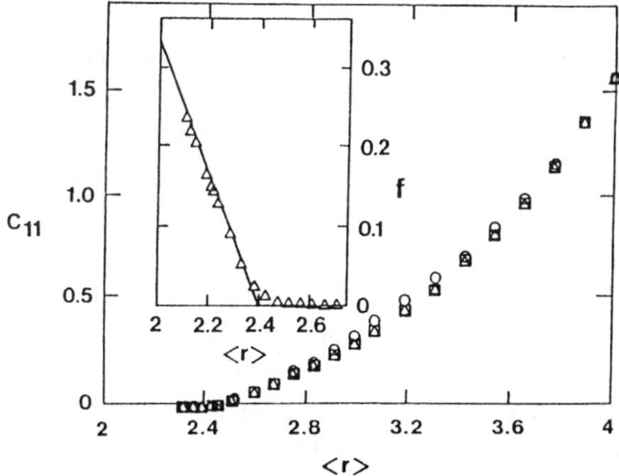

Fig. 23. Showing the elastic constant c_{11} for a model network as a function of the mean coordination $\langle r \rangle$ for three different series of random networks. In the insert the fraction of floppy modes, f, is shown. The points are from computer simulation [58], and the solid line in the insert is the straight line given by Eq. (7).

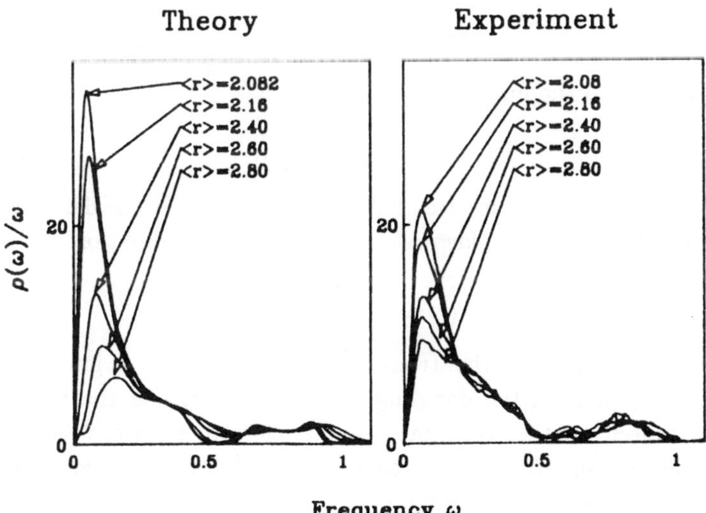

Fig. 24. The density of states divided by the frequency, $\rho(\omega)/\omega$, for various values of the mean coordination, $\langle r \rangle$. The left hand set of graphs are theoretical [54] and the right set of graphs are from inelastic neutron scattering experiments [74] on $Ge_x Se_{1-x}$. The frequency is in units where the maximum frequency is unity.

inelastic neutron scattering measurements of the density of states [74], shown in Fig. 24. The agreement between theory and experiment is excellent, even when the weak forces are included in a very simple way and adjusted to bring the zero-frequency modes to the correct (low) frequency. Note that the weight in the floppy modes, given by constraint counting, is unaffected by the weak forces, even though there is some background response to contend with.

6.2. *Correction for dangling bonds*

The previous section describes the situation when there are no dangling bonds present. Constraint counting fails when dangling bonds are present because the expression for the number of angular forces $2r - 3$ gives -1 when $r = 1$, instead of the correct answer of zero. Thus Eq. (7) needs correction. This has recently been done in a compact way by a number of authors [75–77] following earlier efforts [53, 78].

It is rather straightforward to extend Eq. (7) to include the summation over the dangling bonds ($r = 1$) and correct for the miscounting of the angular constraints to give

$$f = \left[3N - \sum_{r=1}^{4} n_r[r/2 + (2r - 3)] - n_1 \right] /3N \tag{19}$$

which now leads to the form

$$f = 2 - \frac{5}{6}\langle r \rangle - \frac{n_1}{3N} \tag{20}$$

where the definition of $\langle r \rangle$ is extended from (2) to include the dangling bonds. The transition now takes place at a lower mean coordination $\langle r \rangle$ which is given by

$$\langle r \rangle = 2.4 - 0.4\frac{n_1}{N} \tag{21}$$

It is not surprising that the transition takes place at a lower mean co-ordination $\langle r \rangle$, because the dangling ends play no role in the network connectivity. Indeed another conceptual approach is to strip the dangling ends away and define a *skeleton network* that has only 2, 3 and 4 coordinated atoms. The theory described in this section can then be applied to the skeleton network. Equation (21) has recently been applied to networks containing iodine, which forms a dangling end [76, 77].

These ideas have also been applied to amorphous carbon networks by Tamor and are described in [74]. Amorphous carbon networks can be thought of as consisting of three kinds of atoms: fourfold (diamond-like) carbon, threefold (graphitic) carbon, and often considerable amounts of atomic hydrogen that ties off dangling ends and so is singly coordinated. Suppose that there are N atoms in the network, with a fraction x_4 of fourfold coordinated carbon, a fraction x_3 of threefold coordinated carbon and x_1 of singly bonded atomic hydrogen. Then by definition, we have

$$x_4 + x_3 + x_1 = 1 \tag{22}$$

To illustrate the use of the skeleton network, we apply it to this case. The mean coordination $\langle r \rangle$ of the skeleton network, with the hydrogen removed is

$$\langle r \rangle = \frac{2(\text{Number of bonds})}{\text{Number of sites}} = \frac{(4Nx_4 + 3Nx_3 + Nx_1) - 2Nx_1}{Nx_4 + Nx_3} \tag{23}$$

which gives

$$\langle r \rangle = \frac{4x_4 + 3x_3}{1 - x_1} - \frac{x_1}{1 - x_1} \tag{24}$$

The important new term is the $2Nx_1$ in the numerator of Eq. (23), which removes the bonds between hydrogen and the rest of the network that are not present in the skeleton network. In deriving Eq. (24), we have assumed that there are no molecular fragments, like for example methane CH_4, that get detached from the network. If this were the case, then these fragments should also be eliminated for the purposes of counting. It appears that the elasticity and especially the hardness properties of carbon networks [77] follows Eq. (24) closely. Indeed the hardness goes all the way from close to that of crystalline diamond at the one extreme, to mush when too much hydrogen is present.

For the *skeleton network*, the number of floppy modes is still given by Eq. (7) as before. The total number of sites in the skeleton lattice is reduced to $N(1 - x_1)$, so that the total number of floppy modes F is given by

$$F = 3N(1 - x_1)\left(2 - \frac{5}{6}\langle r \rangle\right) \tag{25}$$

We now insert $\langle r \rangle$ from Eq. (24) into Eq. (25) and find that

$$\frac{F}{3N} = 2 - \frac{10}{3}x_4 - \frac{5}{2}x_3 - \frac{7}{6}x_1 \tag{26}$$

which of course reduces to the old result of Eq. (7) if there is no hydrogen present so that $x_1 = 0$. The total number of floppy modes is not changed when the hydrogen is recombined with the skeleton network to reconstruct the original network that existed before the hydrogen was stripped off. This occurs because adding back a single hydrogen atom, adds 3 extra degrees of freedom, but also adds 3 constraints (1 for the central force associated with the bond, and 2 for the angular forces associated with attaching this bond to an atom in the network). Thus there is no change and all the floppy modes are associated with the skeleton network and not specifically with the hydrogen. No direct measurements of these modes have yet been done in carbon networks. Inelastic neutron scattering experiments in the low frequency region, similar to those done for chalcogenide glasses and shown in Fig. 22, would be very interesting.

The transition from rigid to floppy occurs when the number of floppy modes, F, goes to zero. The most interesting experiment to date measures the hardness of networks containing carbon as a function of the mean coordination. The results seem to be in quite dramatic agreement with the ideas presented here.

It has recently been suggested that it is better to plot physical properties, like hardness, not against the mean coordination of the skeleton network, but rather to use the original network [77], which contains dangling ends. This may be connected with the fact that it takes energy to move the dangling

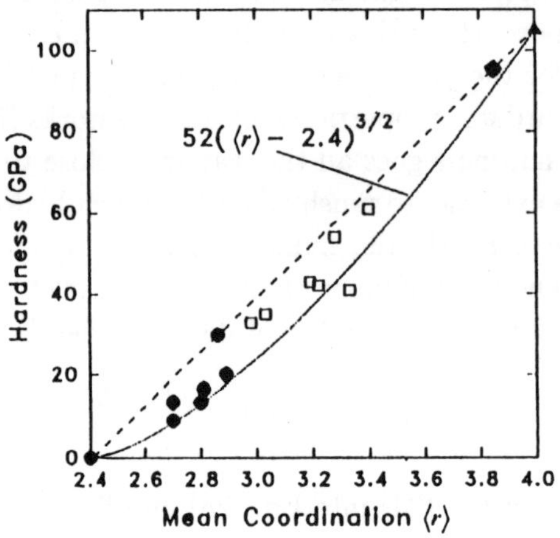

Fig. 25. The hardness (measured by nano-indentation) of various diamond films, some containing some hydrogen, as a function of the mean coordination as defined by Eq. (24). The data were complied and presented in this form by Tamor [79] based on previous work by Tamor and others [74], using various preparation techniques. The solid triangle is the result for crystalline diamond. The 3/2 power law is the computed behavior of the bulk modulus.

bonds apart in an experiment that measures hardness. It would be interesting to have measurements of other physical quantities to see if this observation still persists.

6.3. *Silicate networks*

Although the floppy modes that we have been discussing emanate from the surface, they are in no sense surface modes. The amplitude is not damped away from the surface. They are bulk in extent and involve the whole solid. It is only the total number that scales like the surface area. This concept is important in other marginal structures. Another $2d$ example is the kagome lattice which consists of triangles joined at the corners. The most important three dimensional example is provided by silicates $[SiO_2]$ where the SiO_4 tetrahedra are corner-sharing. This case has been examined extensively, both theoretically and experimentally recently [80]. These authors refer to these modes as rigid unit modes [RUM's] in which the SiO_4 tetrahedra are not distorted. These modes can be seen using inelastic neutron scattering. This experiment is not so straightforward as in the chalcogenide glasses, because the number of RUM's is proportional to the surface area $[N^{\frac{2}{3}}]$, rather than the bulk $[N]$, where N is the number of atoms.

It is not yet clear if the exact results given for the number of surface floppy modes in this section, are exactly or only approximately generalizable to the kagome and silicate structures.

7. Summary

Lord Kelvin has said *"I am never content until I have constructed a mechanical model of the subject I am studying. If I succeed in making one, I understand; otherwise I do not"* [81]. This has been the approach that we have found most useful to follow in trying to understand the mechanical aspects of the behavior of glasses.

In this review, we have tried to collect together some of the more important results regarding floppy modes and constraint counting in glasses that have emerged in the past decade. This work has provided a useful conceptual framework within which to discuss some of the physical properties of glasses. Some of the arguments in this area are subtle and still controversial. For example using a reduced dimensionality of 2 rather than 3 for layered materials like As_xSe_{1-x} does not seem to be correct, as these atoms are still embedded in a 3 dimensional space [82]. It is true that using this reduced dimensionality does seem to improve agreement with experiment where the discontinuity seems to be at $\langle r \rangle = 2.67$ rather than $\langle r \rangle = 2.4$, but perhaps there are other explanations for this.

Now that a $3d$ pebble game is at hand for bond-bending networks, we are in a good position to test the level of accuracy of Maxwell counting, and accurately pin point the location as well as study the nature of the rigidity transition. Our preliminary results suggest that the rigidity transition for generic $3d$ bond-bending networks is first order under some circumstances. It appears that Maxwell counting predicts remarkably well the location of this transition. Our preliminary results for a generic $3d$ bond-bending network with Poisson distributed bonds yields the actual location of the transition to be a bit down shifted to an average coordination of approximately 2.38, in comparison to 2.4. We are now just beginning to apply the $3d$ pebble game for a systematic study of rigidity on networks that more realistically model network glasses [83].

Recent work on surface floppy modes is beginning to look interesting as a way of understanding open structures. So far the only serious application has been to bulk silicate networks [80], but this approach looks promising for other more complex structures like porous silica, clays and zeolites, that contain internal surfaces or voids. Internal floppy modes may permit voids to alter their shape so as to facilitate chemical reactions and catalysis. This remains to be seen. An even more distant hope is that this work may find applications in biological systems, where the ability of molecules to move at little cost in energy probably has important implications for enzyme activity and other biologically important interactions.

Acknowledgments

This research was supported by the NSF under grant no. DMR–9632182 and CHE–9224102. A copy of the FORTRAN program *the pebble game* for analyzing two dimensional generic networks is available upon request from M.F. Thorpe.

References

[1] J.C. Maxwell, *Phil. Mag.* **27**, 294–299 (1864).
[2] J.L. Lagrange, *Mécanique Analytique*, Paris (1788).
[3] D.J. Jacobs and M.F. Thorpe, *Phys. Rev. Lett.* **75**, 4051 (1995).

[4] R. Zallen, *The Physics of Amorphous Solids* (John Wiley & Sons, New York, 1983).

[5] B.R. Djordjević, M.F. Thorpe and F. Wooten, *Phys. Rev.* **B52**, 5685 (1995); B.R. Djordjević, PhD Thesis, Michigan State University (1996).

[6] S.R. Elliott, *Physics of Amorphous Materials* (Longman, London and New York, 1984).

[7] F. Wooten and D. Weaire, H. Ehrenreich, F. Seitz and D. Turnbull (eds.), *Solid State Physics* **40**, 1 (Academic, New York, 1987).

[8] J.D. Bernal, *Nature* **183**, 141 (1959).

[9] J.D. Bernal, *Proc. Roy. Soc.* (London) **A280**, 299 (1964).

[10] J.L. Finney, *Proc. Roy. Soc.* **A319**, 479 (1970).

[11] C.H. Bennett, *J. Appl. Phys.* **43**, 2727 (1972).

[12] G.S. Cargill, H. Ehrenreich, F. Seitz and D. Turnbull (eds.), *Solid State Physics* **30**, (Academic, New York, 1975) p. 227.

[13] W.H. Zachariasen, *J. Am. Chem. Soc.* **54**, 3841 (1932).

[14] R.J. Bell and P. Dean, *Nature* (London) **212**, 1354 (1966).

[15] R.J. Bell and P. Dean, *Phil. Mag.* **25**, 1381 (1972).

[16] R.L. Mozzi and B.E. Warren, *J. Appl. Cryst.* **2**, 164 (1969).

[17] J.C. Phillips, *J. Non-Cryst. Solids* **34**, 153 (1979).

[18] J.C. Phillips, *J. Non-Cryst. Solids* **43**, 37(1981).

[19] D.E. Polk, *J. Non-Cryst. Solids* **5**, 365 (1971).

[20] D.E. Polk and D.S. Boudreaux, *Phys. Rev. Lett.* **31**, 92 (1973).

[21] P. Steinhardt, R. Alben and D. Weaire, *J. Non-Cryst. Solids* **15**, 199 (1974).

[22] M.G. Duffy, D.S. Boudreaux and D.E. Polk, *J. Non-Cryst. Solids* **15**, 435 (1974).

[23] P.N. Keating, *Phys. Rev.* **145**, 637 (1966).

[24] G.A.N. Connell and R.J. Temkin, *Phys. Rev.* **B9**, 5323 (1974).

[25] D. Henderson, *J. Non-Cryst. Solids* **16**, 317 (1974).

[26] R. Kaplow, T.A. Rowe and B.L. Averbach, *Phys. Rev.* **168**, 1068 (1968).

[27] L. Pusztai, *Amorphous Insulators and Semiconductors*, eds. M.F. Thorpe and M.I. Mitkova, [NATO ASI Series 3, High Technology — Vol. **23**, (Kluwer Academic Publishers], 1997) p. 215; N. Zotov, ibid, p. 225.

[28] N.J. Shevchik, *Phys. Status Solidi* **B58**, 111 (1973).

[29] R. Alben, D. Weaire, J.E. Smith, Jr. and M.H. Brodsky, *Phys. Rev.* **B11**, 2271 (1975).

[30] D.L. Evans, M.P. Teter and N.F. Borrelli, *A.I.P. Conf. Proc.* **20**, 218 (1974).

[31] D.L. Evans, M.P. Teter and N.F. Borrelli, *J. Non-Cryst Solids* **17**, 245 (1975).

[32] A.C. Wright, G.A.N. Connell and J.W. Allen, *J. Non-Cryst. Solids* **42**, 69 (1980).

[33] D.L. Evans and S.V. King, *Nature* **212**, 1353 (1966).

[34] L. Guttman, *A.I.P. Conf. Proc.* **20**, 224 (1974).

[35] L. Guttman, *A.I.P. Conf. Proc.* **31**, 268 (1975).

[36] G. Etherington, A.C. Wright, J.T. Wemzel, J.C. Dore, J.H. Clarke and R.N. Sinclair, *J. Non-Crystalline Solids* **48**, 265 (1982).

[37] R.M. Martin, *Phys. Rev.* **B1**, 4005 (1970).

[38] F. Wooten, K. Winer and D. Weaire, *Phys. Rev. Lett.* **54**, 1392 (1985).

[39] F. Wooten, D. Weaire and J. Kalivas (ed.), *Adaptation of Simulated Annealing to Chemical Problems* (Elsevier Science, 1996).

[40] N. Metropolis, A. Rosenbluth, M. Rosenbluth, A. Teller and E. Teller, *J. Chem. Phys.* **21**, 1087 (1953).

[41] S. Kirkpatrick, C.D. Gelatt, Jr. and M.P. Vecchi, *Science* **220**, 671 (1983).

[42] D. Vanderbilt and S. G. Louie, *J. Comp. Phys.* **56**, 259 (1984).

[43] N. Mousseau, private communication.

[44] D.R. McKenzie, D.A. Muller and B.A. Pailthorpe, *Phys. Rev. Lett.* **67**, 773 (1991).

[45] D.A. Drabold, P.A. Fedders and Petra, *Phys. Rev.* **B49**, 16415 (1994).

[46] H. Hirai, Y. Tabira, K. Kondo, T. Oikawa and N. Ishizawa, *Phys. Rev.* **B52**, 6162 (1995).

[47] K.W.R. Gilkes, P.H. Gaskell and J. Robertson, *Phys. Rev.* **B51**, 12303 (1995).

[48] N.A. Marks, D.R. McKenzie, B.A. Pailthorpe, M. Bernasconi and M. Parrinello, *Phys. Rev. Lett.* **76**, 768 (1996).

[49] N.F. Mott and E.A. Davis, *Electronic Processes in Non-Crystalline Materials* (Clarendon Press, Oxford, 1979).

[50] D.A. Drabold, *Amorphous Insulators and Semiconductors*, eds. M.F. Thorpe and M.I. Mitkova, NATO Series 3, High Technology — Vol. 23 (Kluwer Academic Publishers, 1997) p. 405.

[51] N. Mousseau and L.J. Lewis, *Phys. Rev.* **B41**, 3702 (1990).

[52] F.H. Stillinger and T.A. Weber, *Phys. Rev.* **B31**, 5262 (1985).

[53] M.F. Thorpe, *J. Non-Cryst. Solids* **57**, 355 (1983).

[54] Y. Cai and M.F. Thorpe, *Phys. Rev.* **B40**, 10535 (1989).

[55] S. Feng and P. Sen, *Phys. Rev. Letts.* **52**, 216 (1984).

[56] S. Feng, M.F. Thorpe and E.J. Garboczi, *Phys. Rev.* **B31**, 276 (1985).

[57] A.R. Day, R.R. Tremblay and A.-M. S. Tremblay, *Phys. Rev. Lett.* **56**, 2501 (1986).

[58] H. He and M.F. Thorpe, *Phys. Rev. Lett.* **54**, 2107(1985).

[59] A. Hansen and S. Roux, *Phys. Rev.* **B40**, 749 (1989), see especially Figs. 1 and 3.

[60] M.A. Knackstedt and M. Sahimi, *J. Stat. Phys.* **69**, 887 (1992); S. Arbabi and M. Sahimi, *Phys. Rev.* **B47**, 695 (1993).

[61] E. Guyon, S. Roux, A. Hansen, D. Bideau, J.-P. Trodec and H. Crapo, *Rep. Prog. Phys.* **53**, 373 (1990).

[62] M. Tatsumisago, B.L. Halfpap, J.L. Green, S.M. Lindsay and C.A. Angell, *Phys. Rev. Lett* **64**, 1549 (1990); R. Böhmer and C.A. Angell, *Phys. Rev.* **B45**, 10091 (1992).

[63] D. Jacobs and M.F. Thorpe, *Phys. Rev.* **E53**, 3682 (1996).

[64] V. Heine, private communication.

[65] B. Hendrickson, *SIAM J. Comput.* **21**, 65-84 (1992) and private communications.

[66] Y. Kantor and I. Webman, *Phys. Rev. Lett.* **52**, 1891 (1984); see also D. Bergman, *Phys. Rev.* **B31**, 1696 (1985).

[67] S. Roux and A. Hansen, *Europhys. Lett.* **6**, 301 (1988).

[68] D. Stauffer, *Indroduction to Percolation Theory* (Taylor and Francis, London, 1985).

[69] G. Laman, *J. Engrg. Math.* **4**, 331 (1970); see also L. Lovasz and Y. Yemini, On Generic Rigidity in the Plane, *SIAM J. Alg. Disc. Meth.* **3**, 91 (1982).

[70] C. Moukarzel and P.M. Duxbury, *Phys. Rev. Lett.* **75**, 4055 (1995); C. Moukarzel, P.M. Duxbury and P.L. Leath, *Phys. Rev. Lett.* **78**, 480 (1997); ibid. *Phys. Rev.* **E55**, 5800 (1997).

[71] C.M. Fortuin and P.W. Kasteleyn, *Physica* **57**, 536 (1972); P.W. Kasteleyn and C. M. Fortuin, *J. Phys. Soc. Japan* **26**, 11 (1969); see also J.W. Essam, *Rep. Prog. Phys.* **43**, 833 (1980).

[72] D.S. Franzblau, *Siam J. on Discrete Math.* **8**, 388 (1995); D.S. Franzblau and J. Tersoff, *Phys. Rev. Lett.* **68**, 2172 (1992); D.S. Franzblau, private communications.

[73] D.J. Jacobs. and M.F. Thorpe, unpublished.

[74] M.F. Thorpe, *J. Non-Cryst. Solids* **182**, 355 (1995). This mini-review contains references to many experimental results. Also see R. Cappelletti (Chapters 6A) and P. Boolchand (Chapter 6B) in the present volume.

[75] J.C. Angus and F. Jansen, *J. Vac. Sci. Technol.* **A6**, 1778 (1988).

[76] P. Boolchand and M.F. Thorpe, *Phys. Rev.* **B50**, 10366 (1994).

[77] P. Boolchand, M. Zhang and B. Goodman, *Phys. Rev.* **B53**, 11488 (1996).

[78] G.H. Döhler, R. Dandaloff and H. Bilz, *J. Non-Cryst. Solids.* **42**, 87 (1981).

[79] M. Tamor, private communication.

[80] M.T. Dove, A.P. Giddy and V. Heine, *Ameri. Crystal Assoc.* **27**, 65 (1993); A.P. Giddy, M.T. Dove, G S Pawley and V. Heine, *Acta Crystallographica* **Λ40**, 697 (1993).

[81] Quoted in *A Dictionary of Scientific Quotations* by A. Mckay (IOP Publishing, Bristol and Philadelphia, 1994).

[82] K. Tanaka, *Phys. Rev.* **B39**, 1270 (1988).

[83] D. Jacobs and M.F. Thorpe, to be published.

[84] D. Jacobs, *Generic rigidity in three dimensional bond-bending networks*, *J. Phys. A: Math. Gen.* **31**, 6653 (1998).

5

Glass Structure by Scattering Methods and Spectroscopy

A. X-RAY AND NEUTRON DIFFRACTION

ADRIAN C. WRIGHT

*J.J. Thomson Physical Laboratory, University of Reading,
Whiteknights, Reading, RG6 6AF, U.K.*
a.c.wright@reading.ac.uk

Contents

1. Introduction

Amorphous insulators and semiconductors[a] mainly exhibit predominantly directional bonding with significant covalent character and hence have network structures, which are traditionally described in terms of Zachariasen's random network theory [1]. This chapter will describe the general principles involved in studying the structure of amorphous network solids by X-ray and neutron diffraction and give examples of the use of structural modelling in the interpretation of diffraction data. A much more detailed account of the techniques involved and further examples, together with a comprehensive list of references, can be found in Ref. [2] and two earlier reviews by the present author [3, 4].

---

[a]In this chapter the term "glass" will be reserved for materials covered by the ASTM definition of a glass as "an inorganic product of fusion which has cooled to a rigid condition without crystallizing."

An amorphous solid is one with a structure which lacks periodicity, extended symmetry and long range order and it is the absence of the last of these that leads to a diffraction pattern which is a continuous and relatively slowly varying function of the magnitude of the scattering vector, Q, in contrast to the sharp Bragg peaks of a polycrystalline material. As with the latter, X-ray and neutron diffraction are the major direct structural probes but, due to the inherent differences between the two classes of materials, the problems encountered in diffraction studies of amorphous solids are somewhat different from those experienced in crystallography. The existence of a periodic structure means that, given good diffraction data over a reasonable region of reciprocal space, it is in practice possible to determine the structure of simple crystalline solids absolutely. The same is not true for amorphous solids. Because they are normally isotropic on a macroscopic scale, the maximum that can be obtained from a diffraction experiment on an amorphous solid is a one-dimensional correlation function, from which the regeneration of the underlying three-dimensional structure can never be unique. It is for this reason that modelling plays such an important role in structural studies of amorphous solids and why the choice between possible models involves a wide range of experimental techniques and not just X-ray and/or neutron diffraction.

1.1. *The random network theory*

The random network theory was first introduced by Zachariasen [1] to describe the structure of oxide glasses and is illustrated schematically in two dimensions, for a pure glass forming oxide A_2O_3, in Fig. 1. As pointed out by Zachariasen, the atoms in a conventional melt-quenched glass are linked together by forces essentially the same as those in the corresponding crystalline materials and, if the internal energy of the glass is to be comparable to that of the crystal, then it is necessary that the oxygen polyhedra in the vitreous and crystalline states are similar. As in the latter, extended three dimensional networks are formed and the atoms vibrate about definite equilibrium positions, but in the glass the structural units (AO_3 triangles in Fig. 1) are linked together randomly to form a non-periodic structure which lacks long range order. It is clear from Zachariasen's original diagram that he expected the structural units (s.u.) in a network glass to be as regular as those in the crystal and that, for a three dimensional structure, disorder would be introduced via a distribution of the torsion angles α_1 and α_2 and the oxygen bond angle β_O, as defined for vitreous silica in Fig. 2, a view supported by modern diffraction data.

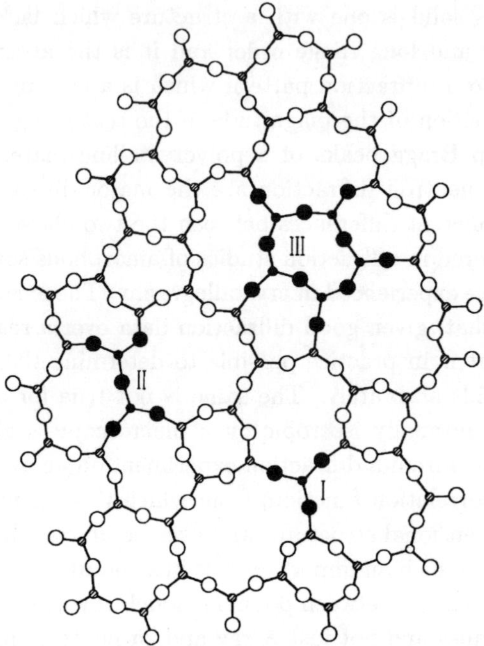

Fig. 1. Zachariasen's original random network diagram (retraced) for a glass of composition A_2O_3 [1], with structural units shaded to indicate range I, II and III order. Small atoms, A and large atoms, O.

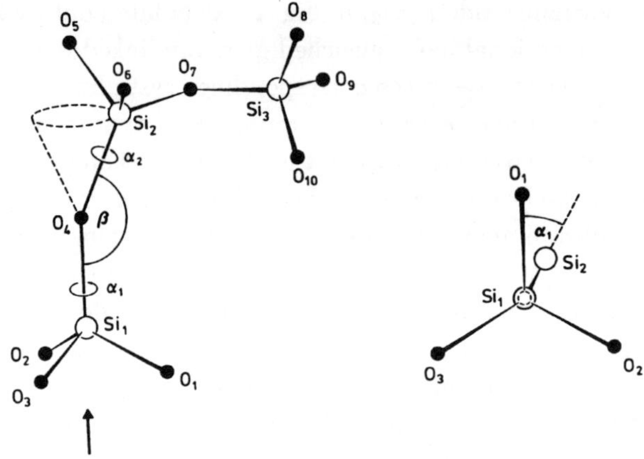

Fig. 2. Definition of the bond angle β_O and the torsion angles a_1 and a_2 for vitreous silica.

Two important points should be noted about the random network theory. First, the perfect (continuous) random network plays the same role in amorphography as does the idealized crystal structure, represented by the unit cell, in crystallography. In real materials there will be "defects," but these will, in general, be undetectable in an X-ray or neutron diffraction study. Second, the random network theory was formulated to describe conventional oxide glasses quenched from the supercooled melt at relatively slow cooling rates. Thus, although the random network theory has been extended to cover other amorphous network solids, such as amorphous semiconductors (e.g. amorphous germanium and hydrogenated amorphous silicon), Zachariasen's arguments, based on internal energy considerations, concerning the relationship between the structural units found in the crystalline and vitreous states are not necessarily applicable to materials which have been prepared by techniques involving highly non-equilibrium processes such as vapour deposition or glow discharge. For example, the tetrahedral bonding in evaporated amorphous Si and Ge (cf. Sec. 5.1) is very much more distorted than that in the ambient crystalline modification and both materials contain a much greater fraction of broken bonds than do the corresponding oxide glasses SiO_2 and GeO_2.

1.2. *Chalcogenide systems*

Chalcogenide materials are, in general, structurally much more complicated than their oxide counterparts, in that even two element systems, such as As–S and Ge–Se, exist as amorphous solids over a wide range of composition and contain homopolar bonds. Models for the numbers of each type of bond present range from the random covalent model to complete chemical ordering. The former places no restriction on the bonds which may be formed, the number of each type of bond being determined solely from the $8 - N$ rule and the concentration of each atomic species. A chemically ordered system, on the other hand, contains the maximum number of bonds between unlike atoms. In the vitreous state, the relevant bond strengths are a useful guide to the relative number of each type of bond likely to be formed but, for example, in vapour deposited materials the balance may be significantly affected by the molecules present in the vapour phase (cf. Sec. 7.4).

1.3. *Multi-component glasses*

In multi-component oxide glasses, the individual constituents are usually classified as network formers or network modifiers, although some intermediate materials can act in either capacity and this distinction is only applicable for glasses in which the network modifying cation bonding is significantly more ionic than that involving the network forming cations. The addition of a further network former to a single component glass will result in a mixed network of two different structural units, whereas the conventional view of a network modifier is that it leads to the formation of negatively-charged non-bridging oxygen atoms. However, the addition of a network modifier may also result in an increase in the network-forming cation coordination number, $n_{A(O)}$, such as occurs preferentially in alkali borate glasses at low alkali contents, where $n_{B(O)}$ increases from 3 to 4.

An extremely controversial aspect of the stuctures of glasses containing network modifiers concerns the spatial distribution of the network modifying cations. Traditionally, the network modifying cations have been envisaged as occupying holes in the random network close to the non-bridging oxygen atoms, but modern diffraction and EXAFS data strongly suggest that the network-modifying cations adapt their local environment to obtain a more desirable co-ordination polyhedron, similar to that found in related crystalline materials. In addition, the negatively-charged non-bridging oxygen atoms tend to cluster, since they are attracted to the positively-charged network modifying cations and vice versa. This has led to a modified random network theory for silicate glasses [5] in which the network-modifying cations are concentrated in channels between regions of silica-rich network.

1.4. *Superstructural units*

Crystalline borate systems contain a rich variety of so-called *superstructural units*, which comprise well-defined arrangements of the basic BO_3 and BO_4 structural units (cf. Fig. 1 of Ref. [6]) and there is increasing evidence that the same units are also present in the corresponding glasses, as revealed by sharp lines in their Raman spectra. This has led Bray [7] to suggest a structural model in which it is the superstructural units that are linked together randomly to form the vitreous network, in addition to the basic BO_3 and/or BO_4 units.

2. Quantification of Amorphous Solid Structures

Although the concept of a random network is easy to describe qualitatively, there is no way to completely define the resulting structures quantitatively, except by specifying the coordinates of every atom present, which is clearly impossible for any real material. Traditionally the structural order within an amorphous solid has been divided into three ranges: short, intermediate (medium) and long. However, for systems with well-defined directional bonding, it is much more convenient to discuss the analysis of diffraction data using four ranges [2], the first three of which are indicated schematically in Fig. 1.

2.1. *Range I: The structural unit*

The most important fact to establish for an amorphous network solid is the identity of the structural unit(s) present. The structural parameters in range I specify the detailed geometry of the structural unit(s), including the distribution of bond lengths and angles. Superstructural units lead to well-defined interatomic distances which are larger than those found in the basic structural unit(s).

2.2. *Range II: The interconnection of adjacent structural units*

The range II order involves the relative orientation and, where appropriate, the interconnection of adjacent structural units. The number of parameters required depends on the number (if any) of shared atoms. For example the interconnection of two corner-sharing SiO_4 tetrahedra requires one ($Si–\hat{O}–Si$) bond angle and two torsion angles as defined in Fig. 2.

2.3. *Range III: The network topology/order beyond the adjacent structural unit*

A useful concept in discussing the so-called intermediate range order for an amorphous network solid is that of an underlying topological network, which can be decorated using various atomic motifs to represent different amorphous solids [2]. As discussed in Ref. [2], the topology of the network can only be fully specified by a connectivity matrix or a near neighbour table, which again is impossible for a real material, and hence it is usual to characterize a network in terms of the ring statistics by shortest path analysis.

2.4. *Range IV: Long range density fluctuations*

Although amorphous solids lack long range order, there may nevertheless be longer range fluctuations in density and/or composition arising from phase separation, etc. The characteristic distances involved in such fluctuations are such that they must be studied using small angle scattering techniques, which are beyond the scope of the present chapter.

2.5. *Experimental parameters*

In general the parameters just outlined are not those which are obtained directly from experiment, diffraction or otherwise. The isotropic nature of most amorphous solids means that in real space the result of a single diffraction experiment is a weighted sum of component correlation functions,

$$t_{ij}(r) = 4\pi r \rho_{ij}(r) = d_{ij}(r) + t_j{}^\circ(r) \tag{1}$$

$$d_{ij}(r) = 4\pi r[\rho_{ij}(r) - \rho_j{}^\circ] \tag{2}$$

$$t_j{}^\circ(r) = 4\pi r \rho_j{}^\circ \tag{3}$$

in which $\rho_{ij}(r)$ is on average the number density of atoms of type j (usually elements, but it may be important to distinguish between chemically distinct atoms of the same element) a distance r from the ith atom in the composition unit (c.u.) and $\rho_j{}^\circ$ is the average number density of j atoms. For a sample containing n elements/atom types, there are $n(n+1)/2$ independent component correlation functions.

3. Theoretical Outline

In this section, only a very brief outline of the background theory of X-ray and neutron diffraction by amorphous solids will be given. A more detailed account, using the same formalism, can be found in Refs. [2–4] and in a series of papers discussing neutron diffraction studies of vitreous silica [8–11].

For an amorphous solid, the scattered intensity, $I(Q)$, is a function only of the magnitude of the scattering vector,

$$Q = \frac{4\pi}{\lambda} \sin\theta \tag{4}$$

λ being the incident wavelength and 2θ the scattering angle. This arises because any interatomic vector, r_{pq}, has equivalent vectors elsewhere in the sample in all possible orientations with respect to Q. Hence for such materials it is possible to perform an isotropic average to yield the Debye equation

$$I_N(Q) = \sum_p \overline{a_p^2(Q)} + \sum_p \sum_q \overline{a_p(Q)}\ \overline{a_q(Q)} \frac{\sin r_{pq}Q}{r_{pq}Q} \tag{5}$$

in which $\overline{a(Q)}$ represents either the X-ray form factor, $f(Q)$, or the neutron scattering length, $\bar{b}$, and the p and q summations are taken over all the atoms in the sample with $p \neq q$. The Debye equation may be used to calculate the scattered intensity if all the interatomic vectors are known or can be calculated, as in the case of a structural model (cf. Sec. 6.1).

For a real sample, on the other hand, the interatomic vectors are unknown and hence it is necessary to work in terms of the component correlation functions, $t_{ij}(r)$. The double summation over all atoms in Eq. (5) may thus be replaced by an integration over the component correlation functions to give

$$I(Q) = \sum_i \overline{a_i(Q)} + \sum_i \sum_j \overline{a_i(Q)}\, \overline{a_j(Q)} \int_0^\infty t_{ij}(r)\frac{\sin Qr}{Q}\, dr \qquad (6)$$

the intensity, $I(Q)$, now being normalized to one composition unit. The i summation is taken over the atoms in one composition unit, while that for j is over atom types/elements. The final corrected X-ray or neutron diffraction pattern takes the general form

$$I(Q) = I^\circ(Q) + I^S(Q) + i(Q) \qquad (7)$$

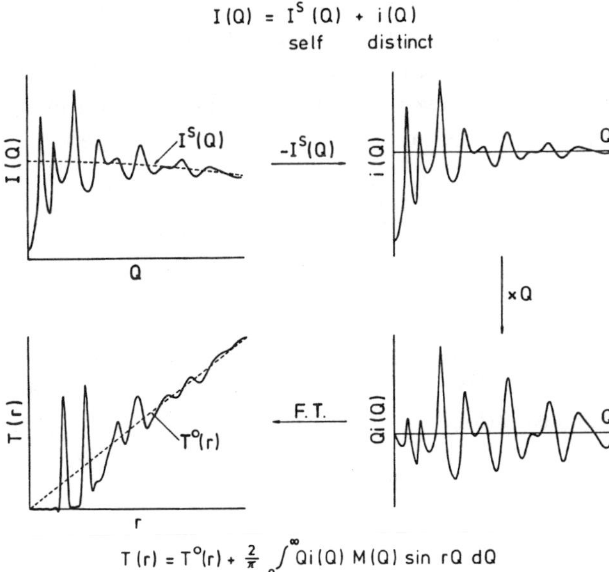

Fig. 3. The relationship between the corrected, normalized diffraction pattern $I(Q)$ and the real space correlation function $T(r)$. The data are for vitreous silica.

$I^\circ(Q)$ is the experimentally inaccessible scattering at $Q \sim 0$, due to the average sample density, $I^S(Q)$ is known as the self (independent) scattering and $i(Q)$ as the distinct (interference) scattering. The required structural information contained within $i(Q)$ may be extracted by means of a Fourier sine transformation of the interference function, $Qi(Q)$, as outlined for neutrons in Fig. 3. The experimental total neutron correlation function is given by

$$T^N(r) = T^\circ(r) + D^N(r) \tag{8}$$

in which $D^N(r)$ is the corresponding differential correlation function

$$D^N(r) = (2/\pi) \int_0^\infty Qi^N(Q)M(Q)\sin(rQ)\,dQ \tag{9}$$

and

$$T^\circ(r) = 4\pi r\rho^\circ \left(\sum_i \bar{b}_i\right)^2 \tag{10}$$

ρ° is the average composition unit number density and $M(Q)$ is a modification function to allow for the fact that $Qi^N(Q)$ can only be determined for Q less than or equal to some maximum value $Q_{\max}$ and is zero for $Q > Q_{\max}$.

In the case of X-rays, it is conventional to divide $Qi^X(Q)$ by the *sharpening function $f_e^2(Q)$*, before Fourier transformation, to yield an "atomic" correlation function, $f_e(Q)$ being the average form factor per electron for the sample in question. This yields the X-ray differential correlation function

$$D^X(r) = (2/\pi) \int_0^\infty \frac{Qi^X(Q)}{f_e^2(Q)} M(Q)\sin(rQ)\,dQ \tag{11}$$

The expression for $T^X(r)$ is analogous to Eq. (8) for $T^N(r)$ and, in $T^\circ(r)$ {Eq. (10)}, the atomic number Z_i replaces $\bar{b}_i$.

The fact that data can only be obtained for $Q \leq Q_{\max}$ means that the relationship between $T^N(r)$ or $T^X(r)$ and the component correlation functions $t_{ij}(r)$ is one of convolution,

$$T(r) = \sum_i \sum_j t'_{ij}(r) \tag{12}$$

where

$$t'_{ij}(r) = \int_0^\infty t_{ij}(r')[P'_{ij}(r - r') - P'_{ij}(r + r')]\,dr' \tag{13}$$

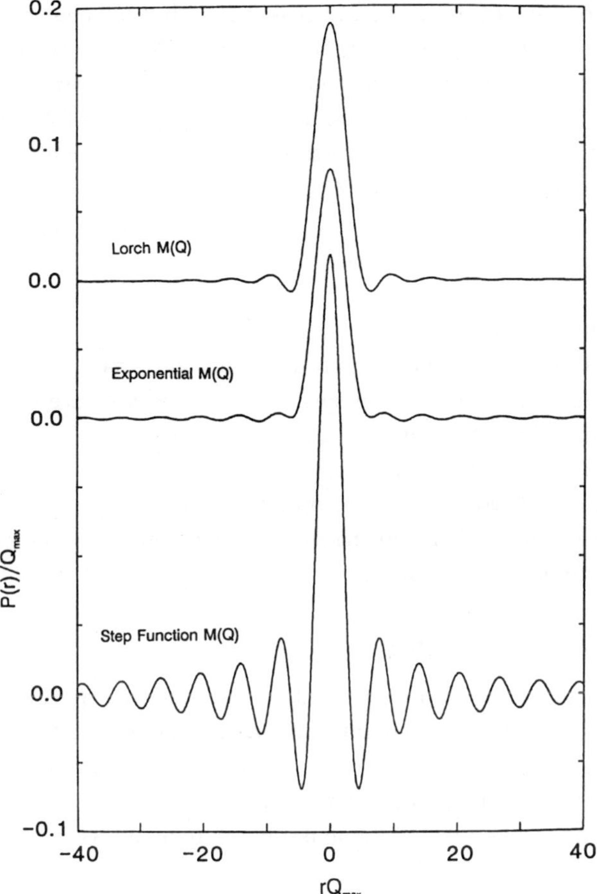

Fig. 4. Reduced neutron peak functions: (a) Lorch [12] function, $M(Q) = \sin(\Delta rQ)/(\Delta rQ)$, with $\Delta r = \pi/Q_{\max}$; (b) artificial temperature factor, $M(Q) = \exp(-BQ^2)$, with $B = \ln 10/Q_{\max}$, and (c) Step function, $M(Q) = 1$.

r' is a dummy convolution variable and the prime indicates N or X, for neutrons or X-rays respectively. The corresponding peak functions,

$$P_{ij}^N(r) = \frac{\bar{b}_i \bar{b}_j}{\pi} \int_0^\infty M(Q) \cos(rQ) \, dQ \tag{14}$$

and

$$P_{ij}^X(r) = (1/\pi) \int_0^\infty \frac{f_i(Q)f_j(Q)}{f_e^2(Q)} M(Q) \cos(rQ) \, dQ \tag{15}$$

define the experimental resolution in real space, that for neutrons being given for the common modification functions in Fig. 4. Note that the experimental real space resolution (width of the central maximum) is inversely proportional to $Q_{\max}$ and that, for X-rays, the different Q dependence of $f(Q)$ for each element means that the shape of $P_{ij}^X(r)$ changes for each independent component. (See, for example, Fig. 14 of Ref. [2].)

In principle, $Qi(Q)$ and $T(r)$ contain precisely the same information, albeit expressed in a different form, although in practice the information content of $T(r)$ is slightly reduced by the use of a modification function other than a step function. The advantage of $T(r)$ is that the information for a particular interatomic distance is concentrated around the appropriate value of r, whereas it is spread throughout reciprocal space. Both functions are one-dimensional representations of a three-dimensional structure and are a gross average over the whole irradiated volume. In addition, as shown by Eqs. (12) and (13), the experimental correlation function is broadened by $P'_{ij}(r)$ and a single diffraction experiment on a multi-element sample yields only a weighted sum of the individual components, $t'_{ij}(r)$.

4. Experimental Techniques

As indicated above, an X-ray or neutron diffraction experiment involves a determination of the scattered intensity as a function of the magnitude of the scattering vector Q. In order to achieve the required variation in Q, it is possible to scan the scattering angle, 2θ, at fixed incident wavelength, λ, (conventional technique) or to make measurements as a function of λ at fixed 2θ.

4.1. *Neutrons*

The two techniques are compared for neutrons in Fig. 5 and examples of the conventional and time-of-flight diffractometers used for amorphous solids are given in Ref. [2]. In the conventional steady state reactor twin-axis experiment (e.g. the D4b diffractometer at the Institut Laue-Langevin), the flux, $\Phi(\lambda)$, of thermal neutrons extracted from the moderator/reflector of a steady-state reactor has a Maxwellian distribution of velocities and is time independent. A beam of wavelength λ, selected by the monochromator crystal, is incident on the sample and scattered into the detector through a variable angle 2θ. The variable λ time-of-flight technique is usually employed with a pulsed

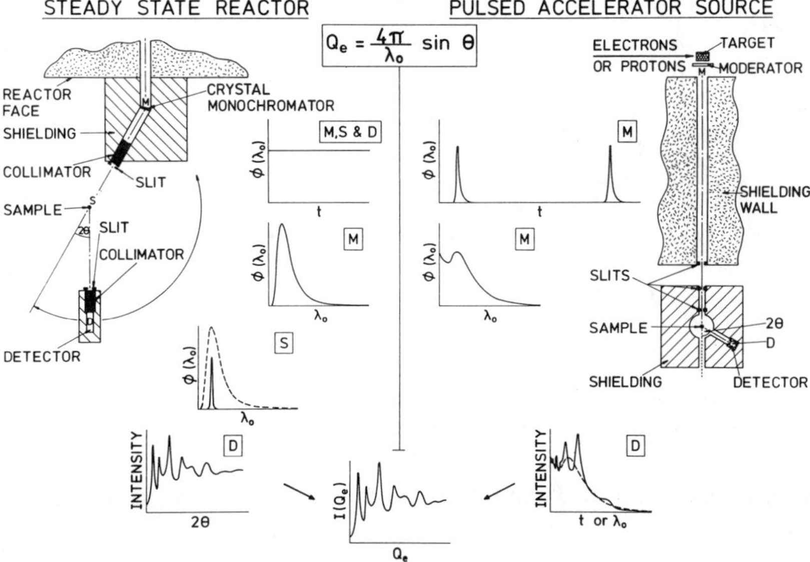

Fig. 5. A schematic comparison of reactor and pulsed source techniques for measuring the neutron diffraction pattern from an amorphous sample.

accelerator source. In a spallation neutron source, such as ISIS at the Rutherford Appleton Laboratory, protons strike a heavy metal target producing pulses of fast neutrons which are partially moderated before being incident on the sample and scattered into a detector situated at a fixed angle 2θ. The detector records the scattered intensity, as a function of the time-of-flight for the distance from the moderator via the sample to the detector. For any neutron, the time-of-flight is simply related to λ through the velocity and De Broglie's relationship and the experimental diffraction pattern, $I^E(Q)$, may be extracted by dividing the measured intensity by the incident neutron spectrum shape. The main difference between the real instruments and the schematic diagrams of Fig. 5 is that the former have multiple counters to increase the counting efficiency and, for the time-of-flight spectrometer, to provide coverage of the full range of Q. The great advantage of a pulsed accelerator source over a steady-state reactor is that the former is undermoderated, which gives rise to a strong epithermal component in the incident spectrum, $\Phi(\lambda)$ (cf. Fig. 5). These short wavelength neutrons allow data to be obtained to much higher values of Q, resulting in a corresponding increase in real space resolution, and,

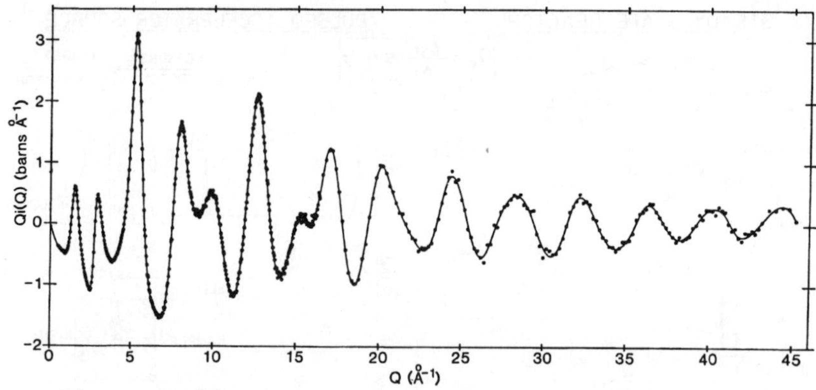

Fig. 6. The interference function for vitreous silica obtained from time-of-flight data ($\bullet$, experimental points and ———, cubic spline fit).

as an example, a time-of-flight interference function for vitreous silica is shown in Fig. 6.

4.2. *X-rays*

In the case of X-rays, the situation for amorphous solids is complicated by the presence of incoherent Compton scattering, which at high values of Q can be of much higher intensity than the required coherent contribution. Hence, for accurate quantitative work on amorphous solids, a very much more sophisticated instrument is required, which employs both an incident beam monochromator and an analyser in the diffracted beam. The former is required to ensure monochromatic radiation since, unlike a polycrystalline powder, an amorphous solid does not separate the elastic scattering into clearly defined Bragg peaks. The analyzer is required because neither the wavelength distribution nor the integrated intensity of the Compton scattering from a given sample can be satisfactorily calculated and therefore it is necessary either to remove this contribution or to measure the energy distribution of the scattered radiation at each angle and then extract the desired coherent elastic scattering using peak fitting techniques, e.g. by the use of a solid state detector with pulse height (X-ray energy) analysis or an analyzing crystal in conjunction with a position sensitive detector.

For laboratory systems, many authors have used only a fixed analyzer in the diffracted beam, tuned to the incident K_α wavelength, but the problem with this arrangement is that "white" background radiation from the

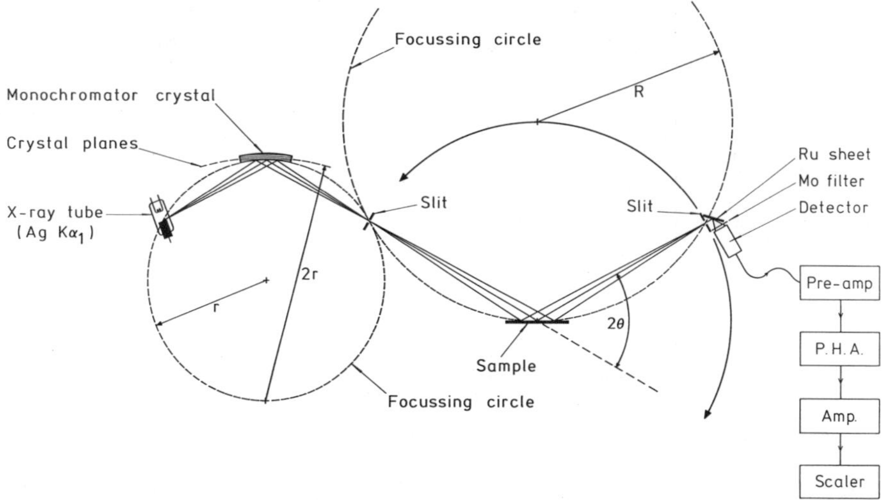

Fig. 7. The fluorescent foil X-ray technique [13].

X-ray tube can be Compton scattered into the monochromator envelope. A much better arrangement is that due to Warren and Mavel [13], shown in Fig. 7, which employs a conventional curved crystal monochromator in the incident beam (Ag $K_{\alpha 1}$: $\lambda = 0.5594$ Å), but the analyzer in the diffracted beam comprises a foil with an absorption edge at a wavelength slightly longer than the characteristic line of the X-ray tube (Ru K edge: $\lambda = 0.5605$ Å) such that the coherent intensity will excite fluorescence whereas the Compton scattering will not. The fluorescent radiation is then recorded by the detector. Despite the obvious superiority of this technique, however, it is very rarely used with the result that X-ray data quality is frequently not up to that of its neutron counterpart. Mozzi and Warren's X-ray data for vitreous silica [14] are shown in Fig. 8.

The increasing use of synchrotron radiation sources in X-ray diffraction studies of amorphous solids is one of the most important recent developments. Provided X-rays of sufficiently short wavelength (high $Q_{\max}$) can be generated using a wiggler, this type of source has several advantages over a standard laboratory X-ray generator and when used with an appropriate diffractometer, optimized for amorphous materials, is capable of yielding high quality data. The continuous spectrum of a synchrotron radiation source allows the selection of the optimum wavelength for anomalous dispersion experiments (cf. Sec. 8.2)

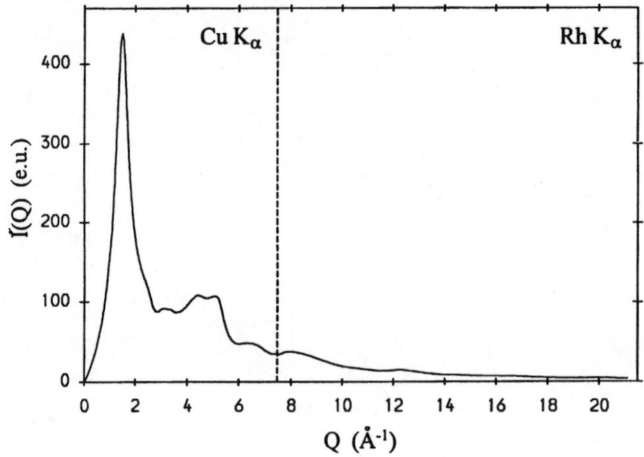

Fig. 8. Mozzi and Warren's X-ray data for vitreous silica [14], which was obtained with Cu K_α and Rh K_α radiation. The vertical dashed line indicates the change from Cu K_α to Rh K_α radiation.

and the highly collimated beams produced make it possible to perform grazing incidence experiments to investigate very thin films or surface layers. As with laboratory systems, a major factor limiting accuracy is the elimination of the Compton scattering and a relatively recent development has been the adaption of the Warren and Mavel technique for use on amorphous materials diffractometers at synchrotron sources [15].

4.3. *Data reduction*

Any measurement involves a certain number of corrections to the raw data, but the good experiment minimizes these corrections or puts them in a form in which they are easily handled. A detailed account of the corrections applied to both X-ray and neutron data from amorphous solids can be found in Refs. [2, 3, 9] and [11]. Normally corrections are included for:

  (i) counter paralysis time,
 (ii) instrumental background and sample container scattering,
(iii) absorption,
 (iv) multiple scattering and self-shielding,
  (v) polarization (X-rays),
 (vi) residual Compton scattering (X-rays) and
(vii) Static Approximation Distortions (Placzek Corrections, neutrons).

Following correction, it is necessary to normalize the data to absolute units. This may be achieved either by the use of self consistent (Krogh–Moe and Norman) integration techniques [3] or, for neutrons, by measuring the (incoherent) scattering from a standard vanadium sample.

4.4. *Fourier transformation*

One of the most controversial aspects of the analysis of X-ray and neutron diffraction data for amorphous solids is the Fourier transformation of the interference function to give the real space correlation function, mainly because the Fourier transform is a clear indication of data quality. The Fourier transformation may be performed using either a quadrature or a fast Fourier transformation algorithm, the latter being more economical on computer time, but much less flexible. The data may also be smoothed before Fourier transformation but this has very little effect on the resulting transform since the frequency range of the noise removed mainly corresponds to distances in excess of those of interest.

4.5. *Assessment of accuracy*

The statistical accuracy of diffraction data in reciprocal space can be simply calculated from the total number of counts at each intensity point, but a much better indication of the overall accuracy, including systematic errors, is given by the structure in the correlation function at low r, below the first true peak. If the transform is well behaved in this region, then the data are of reasonable quality. Great care is needed in making this assessment, however, as some authors either plot the radial distribution function, $rT'(r)$, which has the effect of reducing the relative amplitude of the error ripples at low r, or use these false oscillations as a criterion for "massaging" their data before publication and do not include the original unadulterated transform. In the absence of other information, such massaged data must be treated with the utmost suspicion. Similar techniques are sometimes used to "remove" the effects of terminating the data at finite $Q_{\max}$. Information theory, however, indicates that it is impossible to replace the unmeasured data without making some assumption about the material under investigation so that the resulting correlation function merely becomes one possible model which "fits" (or not, as the case may be) the results, rather than an unbiased Fourier transform. Similar objections can be raised concerning the maximum entropy technique, which also has the additional problem that it is impossible to know what to do to the correlation function from a structural model in order to be able to directly compare it with the experimental maximum entropy version. An alternative method of assessing data quality is to perform two independent sets of measurements on the same material, as indicated in Sec. 5.

5. Methods of Interpretation

There is an unfortunate tendency in the literature to regard the correlation function, or its individual components, as the final result of a structural study and to merely describe its general form and list a few peak positions. However, the correlation function is closely related to the crystallographer's Patterson function and no crystallographer would ever end a paper with just a Patterson function. The experimental interference and correlation functions should thus be considered the starting point for the real science, once the data reduction is complete, the object being to extract the maximum information concerning the structure of the material in question. The methods for doing this vary for the different ranges of order. In range I, the parameters for the structural unit are best extracted using peak fitting techniques whereas, in ranges II and III, the non-unique, one-dimensional nature of diffraction data leads to an extensive use of modelling techniques. Before discussing each range in detail, however, it is first appropriate to make a few general comments concerning the form of the interference and correlation functions for an amorphous solid and the comparison of peak fits and structural models with experimental data.

Diffraction data are frequently held to be insensitive to the exact nature of the structure of an amorphous solid but, if performed correctly, modern diffraction experiments are able to yield accurate data with high real space resolution, which in practice provide a stringent test for the various structural models proposed in the literature. The main problem with the analysis of diffraction data from amorphous solids is one of uniqueness in that, even if a perfect fit is obtained with experiment, this is no guarantee that there are not other models which would fit equally well. Agreement with diffraction data is thus **a necessary but not sufficient criterion** for any valid structural model. What can be stated with absolute certainty, however, both for diffraction data and for the results from any other experimental technique, is that **any model which is not consistent with the data is wrong and must be rejected**. Indeed many models may be rejected on the basis of quite simple measurements, the average number density, ρ°, being an obvious and most important example. Too often diffraction data are published, along with the author's favoured model, totally ignoring other published models and/or the results of complementary techniques which would invalidate the model in question.

The neutron correlation function for vitreous silica at ambient temperature is shown in Fig. 9, marked with the approximate extent of ranges I, II and III. The high real space resolution ($Q_{max} = 45.2$ Å^{-1}) means that the characteristics of the three ranges of order can be clearly seen. The two peaks from the intrastructural unit order (range I) are extremely sharp and, even at this value of Q_{max}, there is still a significant contribution to their width from the peak function $P_{ij}^{N}(r)$ (cf. Sec. 3, Fig. 4). Ranges II and III overlap but the main features between 3.0 and 5.5 Å are due to distances between atoms in adjacent tetrahedra (range II). Their increased breadth compared to the peaks in range I arises from the wide distribution of intertetrahedral Si–$\hat{\text{O}}$–Si and bond torsion angles. Finally range III is characterized by the broadest features in the correlation function which extend out to $\sim 10 - 20$ Å, the limit of the intermediate range order.

The general form of the diffraction pattern, or interference function, for an amorphous solid is also worthy of note since many structural models, particularly those derived from crystalline structures, fail to get this correct. As may be seen from Fig. 9, the absence of long range order leads to a

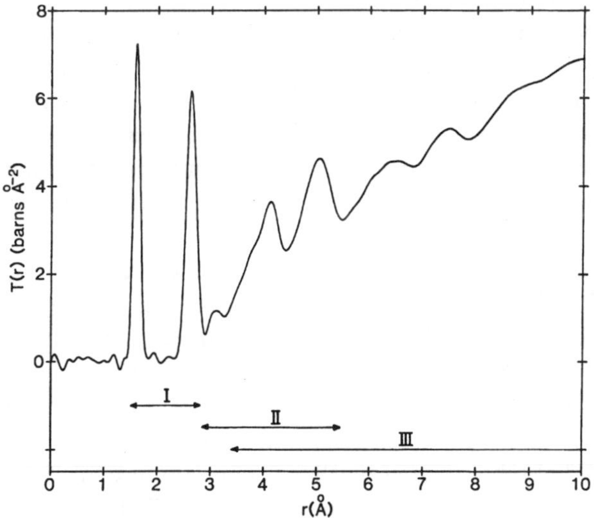

Fig. 9. The neutron correlation function for vitreous silica. The roman numerals indicate the extent of the ranges of order discussed in Secs. 2.1 to 2.3.

correlation function in which the peaks tend to increase in width and decrease in amplitude with increasing r. This means that, in reciprocal space, the highest frequency Fourier components decay the most rapidly and hence the interference function for an amorphous solid is in general characterized by a succession of peaks which get broader with increasing Q until only the Fourier component arising from the sharpest real space peak, usually due to the first interatomic distance, remains (cf. Fig. 6). Thus it is the first few peaks in reciprocal space which are particularly sensitive to the range III order and hence a disagreement between a structural model and experiment in this region ($Q \lesssim 5 \, \text{Å}^{-1}$) is predominantly indicative of incorrect intermediate range order. Similarly, the rapid decay of the highest frequency Fourier components means that the first peak in the diffraction pattern from an amorphous material is almost invariably the sharpest (cf. Sec. 9).

There is considerable misunderstanding in the literature as to what constitutes (good) agreement of a structural model with experiment and the comparison of models with experiment is frequently grossly inadequate. Indeed, for most of the models so far reported, a rigorous comparison reveals serious discrepancies between the model and diffraction data. Many authors term good agreement with experiment as getting the peaks in either $Qi'(Q)$ or $T'(r)$ in the right place. Only very poor models fail to achieve this. What is necessary is also to get peak shapes and areas correct, which involves including the effects of thermal vibration for static models (e.g. computer relaxed random networks) and in real space folding the model component correlation functions with the correct component peak functions $P'_{ij}(r)$ for the experiment in question. Otherwise, the comparison between model and experiment is meaningless. Note that $P'_{ij}(r)$ has satellite features on either side of the central maximum which depend on the exact form of $M(Q)$ (cf. Fig. 4) and themselves lead to features in $T'(r)$. Thus a Gaussian approximation to $P'_{ij}(r)$ is simply not adequate, especially when the width is arbitrarily chosen to give the best agreement with experiment. Resolution effects may also be important in reciprocal space and, if so, the model interference function must be broadened by the experimental reciprocal space resolution function. The functions $Qi'(Q)$ and $T'(r)$ emphasize different aspects of a given structure and hence it is vital, when comparing models with experiment, to make such comparisons in both real and reciprocal space. It should also be stressed that any valid structural model must not only be consistent with diffraction data but also with all the other available data for the material in question. For example, a reverse Monte Carlo simulation (cf. Sec. 7.2) may yield a structure which is consistent with diffraction data but has to be rejected due to the presence of too many over or under coordinated atoms as determined by magnetic resonance measurements.

One reason why so many inaccurate and irrelevant structural models and computer simulations are accepted for publication in the literature is that authors are not required to include any quantitative measure of the (dis) agreement between their model and experiment. This would never be allowed in a crystallograpic communication, where authors are expected to quote the reliability (R) factor for their structure and compare it with the theoretical minimum R factor for the experimental data. Since the functions used in amorphography have both positive and negative regions, it is convenient to use a slightly different reliability factor [16], which in real space takes the form

$$R_\chi = \left(\frac{\sum_i [T_{\mathrm{exp}}(r_i) - T_{\mathrm{fit}}(r_i)]^2}{\sum_i T_{\mathrm{exp}}^2(r_i)} \right)^{1/2} \tag{16}$$

As an indication of the accuracy of modern neutron diffraction data, the discrepancy between two completely independent data sets for vitreous silica, over the range $1 \leq r \leq 8$ Å, is characterized by an R_χ factor of 1.5% [17], this being the value which should be approached by any structural model which claims to be in agreement with these data.

5.1. *Range I*

The modern method of extracting range I parameters from diffraction data is via peak fitting techniques. Such fits may be performed either in reciprocal or real space, but the latter has the advantage that the Fourier transformation has the effect of at least partially decoupling the parameters for different peaks. Assuming a Gaussian peak in $t_{ij}(r)$ about a mean interatomic distance r_{ij}, with an rms deviation of $(\overline{u_{ij}^2})^{1/2}$, the contribution to the interference function is given by

$$Qi'_{ij}(Q) = n_{i(j)} \overline{a_i(Q)}\, \overline{a_j(Q)} \frac{\sin r_{ij}Q}{r_{ij}} \exp(-\overline{u_{ij}^2}Q^2/2) \tag{17}$$

in which $n_{i(j)}$ is the coordination number of j type atoms about the ith atom in the composition unit.

Peak fits to the X-ray and neutron correlation functions for vitreous SiO_2, with the Lorch [12] modification function, are illustrated in Fig. 10, together

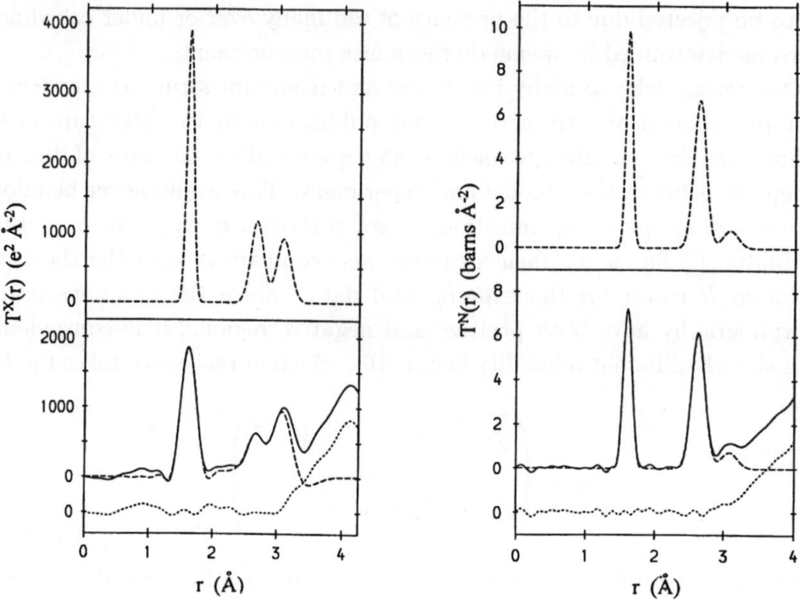

Fig. 10. Peak fits (lower curves) to the X-ray (left) and neutron (right) correlation functions for vitreous silica, obtained using the Lorch [12] modification function, together with the corresponding unbroadened peaks (upper curve). The peak parameters are given in Table 1 (————, experiment; - - - -, fit and . . . ., residual).

Table 1. X-ray and neutron peak parameters for vitreous silica.

Radiation	R_χ	$i-j$	$n_{i(j)}$	r_{ij} (Å)	$(\overline{u_{ij}^2})^{1/2}$ (Å)
X-rays	7.3	Si–O	3.77 ± 0.12	1.626 ± 0.004	0.053 ± 0.009
		O–O	6.15 ± 0.38	2.657 ± 0.012	0.102 ± 0.016
		Si–Si*	4.0	3.077	0.111
Neutrons	3.8	Si–O	3.85 ± 0.16	1.608 ± 0.004	0.047 ± 0.004
		O–O	5.94 ± 0.23	2.626 ± 0.006	0.091 ± 0.005
		Si–Si*	4.0	3.077	0.111

*From Ref. [18].

with the appropriately weighted unbroadened distributions, and the peak parameters are summarized in Table 1. In both fits, parameters for the first Si–Si distance are taken from the work of Konnert and Karle [18] and this distribution is included since its low r tail extends under the O–O peak. For X-rays, the unbroadened peaks are weighted according to the atomic numbers of Si and O and it is interesting to note that the relative heights of the O–O and Si–Si peaks in the experimental correlation function are not what would be expected from the atomic numbers of Si and O, due to the satellite features on either side of the central maximum of the component peak functions (Fig. 14 of Ref. [2]). The areas under the O–O and Si–Si peaks are $768e^2$ and $784e^2$, respectively.

The increase in the width of the second peak in the correlation function over that for the Si–O bond length reflects the variation in the O–$\widehat{\text{Si}}$–O intratetrahedral bond angle and hence is a measure of the tetrahedral distortion. Assuming that the bond length and bond angle variations are uncorrelated, it is possible to use a Monte Carlo technique to extract the mean bond angle and the rms bond angle variation from the widths of the first two peaks [19]. Using this procedure with the neutron peak parameters yields values of $109.7 \pm 0.4°$ and $4.5 \pm 0.4°$, respectively, the latter including both thermal and static disorder. The agreement between the former and the expected tetrahedral angle of $109.5°$ is well within the experimental uncertainty. A similar fit to the first two peaks for amorphous Ge indicates that the distribution of intra-tetrahedral angles (rms variation $10.5 \pm 0.7°$) is much broader than that for vitreous silica, reflecting the very much higher level of network strain associated with its more extreme method of preparation.

5.2. *Range II*

In addition to interatomic distances and bond angles, the quantification of the order for an amorphous network solid in range II involves the distribution of

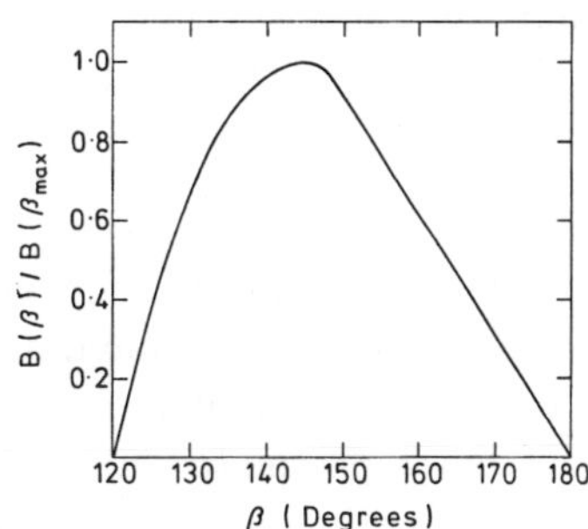

Fig. 11. The Si–$\widehat{\text{O}}$–Si bond angle distribution for vitreous silica as determined by Mozzi and Warren [14].

bond torsion angles, $A(\alpha)$, which cannot be obtained directly from diffraction data without assuming some model. For materials such as SiO_2, the difference between the structures of the crystal and glass are that, in the latter, there is a broad distribution of Si–$\widehat{O}$–Si and bond torsion angles whereas in general the former is characterized by one, or at most a few, discrete values. One simple model for the interconnection of structural units in a random network assumes a uniform distribution of torsion angles, except where excluded by steric effects, and has been used by Mozzi and Warren [14] to obtain the Si–$\widehat{O}$–Si bond angle distribution in Fig. 11. The structural models discussed in the next section on range III order must also, by definition, contain some inherent bond and torsion angle distribution and indeed often fail in this respect, making a comparison of the range III order with experiment pointless.

5.3. *Range III*

The usual method of investigating the order in range III is via modelling studies, which fall into two groups, those of an analytical nature and those leading to actual atomic coordinates, where an attempt is normally made to generate a "typical region" of the structure that can be used both to compare to diffraction data and in the calculation of other properties. The problem with the first group is that there is no guarantee that it is possible to build a real structure corresponding to the analytical formalism and hence the second group is of very much greater use in understanding the structure of amorphous solids. A wide range of structural models of amorphous insulators and semiconductors has been presented in the literature and the most important types are summarized in Table 2. Within the space available, it is only possible to give a few examples of the use of structural models in interpreting diffraction data for amorphous network solids and these are chosen to illustrate specific points. Further examples can be found in Refs. [2] and [17].

6. Structural Modelling

The importance of modelling techniques in the study of amorphous solid structures cannot be overstated. The simple act of building a model by hand gives a "feel" for the structure which it is difficult to obtain in any other way and may lead to the conclusion that a postulated structure is sterically impossible, even if no quantitative information is extracted from the finished model. The simplest models are of the hand-built variety, which are constructed using suitable units to reflect the experimental parameters such as the structural unit

Table 2. Structural models of amorphous network solids.

Model	Example(s)
Random Network (a) Hand built (b) Computer generated (c) Geometric transformation	As, As_2O_3, B_2O_3, Ge, SiO_2 Ge, SiO_2 B_2O_3, H_2O, P, P_2O_5
Polymer Models (a) Random coil (b) Bundled coil	S, Se
Crystal Based Models (a) Limited range of order (b) Strained crystal	$GeSe_2$, SiO_2 SiO_2
Molecular Model	CCl_4, P_4Se_3, $As_{1-x}S_x$
Layer Model	$As_{1-x}S_x$, $Ge_{1-x}Se_x$
Amorphous Cluster	Ge, SiO_2
Monte Carlo (a) Energy minimization (b) χ^2 minimization	BeF_2, Ge $As_{1-x}Se_x$, SiO_2, $ZnCl_2$
Molecular Dynamics	BeF_2, SiO_2, Na_2O–SiO_2

regularity and the bond angle distribution. The disadvantage of hand building is that the construction and quantitative measurement of large models is extremely tedious and it is difficult to vary the model parameters in a systematic way. The use of computer relaxation techniques (molecular mechanics), however, reduces the accuracy with which it is necessary to measure the atomic coordinates and also removes any anisotropy which might result from the effects of gravity on a large model. Alternatively, models may be completely computer generated which removes both the tedium and any unintentional bias associated with hand-built models.

An important aspect of coordinate models, as opposed to the analytical variety, is the problem of finite model size, which limits the correlation function to interatomic distances less than or equal to some maximum value, r_{max}. This problem may be approached in two different ways, cluster models and those which employ periodic boundary conditions, and it is vital to understand the restrictions which these place on the model in both real and reciprocal space. It is important to note that most structural modelling is performed in real space, whereas diffraction data are collected in reciprocal space, and hence, to compare model and experiment, it is necessary to perform a Fourier transformation, either explicitly or implicitly. However, while the effect of the finite range of Q on diffraction data is reasonably well understood (cf. Sec. 3), it is not generally appreciated, particularly by those who build structural models but do not themselves perform diffraction experiments, that exactly the same considerations apply when transforming a structural model to reciprocal space.

6.1. *Cluster models*

The simplest method of calculating the component correlation functions, $t_{ij}(r)$, for a cluster model is only to use those atoms as centre which are at distances in excess of $r_{\max}$ from the model surface. In this case, there is no effect of the finite model size on $t_{ij}(r)$ for $r \leq r_{\max}$ and, on transformation to reciprocal space, the corresponding interference functions, $Qi_{ij}(Q)$, are simply broadened by the reciprocal space peak function, $P(Q)$, appropriate to the modification function used in the Fourier transformation; e.g. for a step function cut-off, $P(Q)$ is of the same form as Fig. 4(C). The problem with this approach is that only a small fraction of the total atoms are used as centre and hence the statistical accuracy of $t_{ij}(r)$ is very much poorer than if all the whole model had been used. Consequently, it is normal to use all of the atoms in a spherical cluster, of radius $R\,(r_{\max} = 2R)$, taken from the centre of the model, for which the component correlation functions are reduced relative to those for an infinite model by the factor [20]

$$F(r) = \left(\frac{r}{R-2}\right)^2 \frac{\frac{r}{R+4}}{16} \qquad r \leq 2R$$
$$F(r) = 0 \qquad\qquad\qquad r \geq 2R \tag{18}$$

The corresponding intensity and interference functions in reciprocal space are calculated using the Debye equation {Eq. (5)}. For a spherical model, the reciprocal space peak function is the Fourier cosine transform of $F(r)$ and takes the form [2]

$$P(Q) = \frac{3}{4\pi RQ^2}\left[1 - 2\frac{\sin 2RQ}{2RQ} + \frac{\sin^2 RQ}{(RQ)^2}\right] \tag{19}$$

which is identical to the particle size broadening function for a polycrystalline material with spherical crystallites. The finite model size also leads to small

Q scattering appropriate to a uniform sphere of the same size and average scattering density as the original model,

$$I^\circ(Q) = \frac{12\pi\rho^\circ}{R^3 Q^6} \left[\sum_i \overline{a_i(Q)}\right]^2 [\sin(RQ) - RQ\cos(RQ)]^2 \qquad (20)$$

This is equivalent to the average density term, $I^\circ(Q)$, for a real sample (cf. Sec. 3). Thus, for a cluster model, the Fourier transformation to reciprocal space leads to a broadening of the interference function and the introduction of satellite features, in exactly the same way as for an experimental correlation function in real space. Hence, in order to compare model and experiment in reciprocal space, it is necessary to fold the *experimental* interference function with the reciprocal space peak function, $P(Q)$, appropriate to the model calculation. Similarly the model interference function must be broadened by the reciprocal space resolution function for the instrument used in the experiment, if this is significant.

6.2. *Periodic boundary conditions*

The use of periodic boundary conditions in the modelling of amorphous materials is extremely controversial, in that a periodic boundary model is in reality a crystal. The lack of internal symmetry within the unit cell means that the structure belongs to the triclinic $P1$ space group, although the unit cell itself is frequently pseudo-cubic. Any calculation which uses periodic boundary conditions, therefore, is strictly appropriate to a $P1$ crystal and not an amorphous solid. For a periodic boundary model, $r_{\max}$ is usually taken as half the smallest unit cell dimension and the periodic images are used in calculating $t_{ij}(r)$ to avoid any effects of the finite cell for $r \leq r_{\max}$.

In reciprocal space, the application of periodic boundary conditions means that, with the exception of (inelastic) thermal diffuse scattering, the diffraction pattern for such a structure, after averaging over all orientations of the unit cell with respect to the scattering vector Q, is only sampled at discrete values of Q (Q_{hkl}, the δ-function Bragg peaks) and comprises a Debye–Scherrer polycrystalline powder diffraction pattern, not the continuous distribution typical of an amorphous solid. Assuming a pseudo-cubic unit cell, with lattice parameter a_0, the positions of the Bragg peaks are given by

$$Q_{hkl} = \frac{2\pi}{a_0}(h^2 + k^2 + l^2)^{1/2} = \frac{2\pi}{a_0}m^{1/2} \qquad (21)$$

where m is an integer, and there is no elastic scattering below the Bragg cut-off at $Q = 2\pi/a_0$. The correct method of calculating the interference function for a periodic boundary model is thus via the square of the isotropically averaged crystalline structure factor, $|F_{hkl}|^2$.

In Fig. 12, the spacing of the Bragg peaks at low Q for a model of amorphous Ge by Wooten and Weaire [21] (cf. Sec. 7.2) is shown, relative to the experimental neutron diffraction pattern for amorphous Ge [22]. Each Bragg peak is represented by a vertical line in the appropriate position, the height of which is proportional to the peak intensity for the static model. The prominent Bragg peak at 1.87 Å^{-1} is close to the 111 Bragg reflection for crystalline Ge, suggesting that the model may contain a small amount of residual crystallinity. Conventionally, such a diffraction pattern is artificially broadened, e.g. by summing the δ-function Bragg peaks into relatively broad histogram columns, in which case it is imperative that, before intercomparison, the experimental data are treated identically.

Many workers, using periodic boundary models, incorrectly calculate reciprocal space functions for their model by Fourier transforming the component real space differential correlation functions, $d_{ij}(r)$, over the range $0 \le r \le r_{\mathrm{rmax}}$ (cf. Sec. 6.1), but this results in a modified model, for which the range of

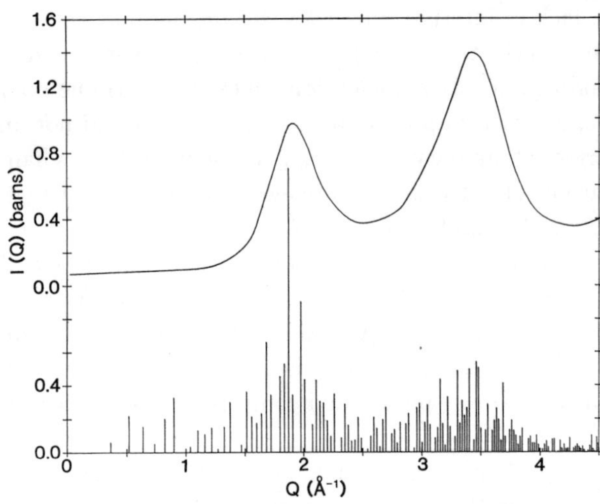

Fig. 12. The positions and intensities of the Bragg peaks (vertical bars) for a model of amorphous Ge, generated by Wooten and Weaire [21] (Model C in Fig. 15), relative to the experimental neutron diffraction pattern for amorphous Ge [22] (solid line).

order is limited by an *extra* multiplicative factor, $F(r)$, and does *NOT* yield the true interference function appropriate to their model (i.e. the factor of $F(r)$ introduces extra disorder, which is not present in the original model). Similarly, any calculation which truncates the structure (correlation function) at some distance less than a_0 is simply incapable of distinguishing between the crystalline and amorphous states and the limitation of the range of order is a property of the calculation, not the underlying structural model.

7. Examples of Structural Models for Amorphous Network Solids

7.1. *Random network models*

Two important hand-built structural models, which have been much used in the literature, are the Bell and Dean model for vitreous silica [23] and the Polk and Boudreaux model for amorphous Si and Ge [24]. The correlation function for the Bell and Dean model, after energy minimization [25] and thermal broadening, is compared to neutron and X-ray data in Fig. 13. The peaks at

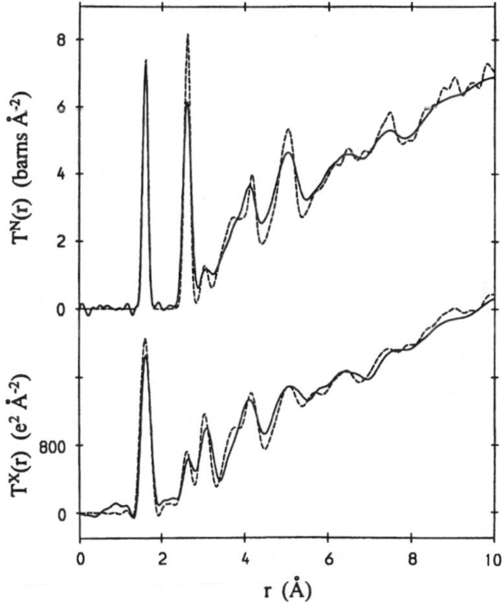

Fig. 13. Neutron (upper curves) and X-ray (lower curves) correlation functions for the Bell and Dean model of vitreous silica [23, 25] (dashed lines) compared to experiment (solid lines).

4.1 Å (mainly Si–O interactions) and 5.0 Å (O–O) arise mainly from the range II order and it can be seen that the shoulder on the low r side of the 4.1 Å peak is much too pronounced compared to experiment, indicating that the range II order in the model is incorrect. This is a common failing of models of vitreous silica and a similar discrepancy between model and experiment occurs for almost all of the other random network models of this material discussed in the literature [17]. The Polk and Boudreaux model [24] is compared to neutron data for amorphous Ge [22] in Fig. 14 and is interesting because it was constructed from rigid plastic tetrahedra joined together with rigid aluminium sleeves (cf. Fig. 39 of Ref. [2]), which had the effect of constraining the Ge–$\widehat{\text{Ge}}$–Ge bond angle distribution to be much too narrow, as may be seen from the second peak in the correlation function. Hence it is a classic example of using inappropriate units in a modelling study, which lead to an incorrect structure.

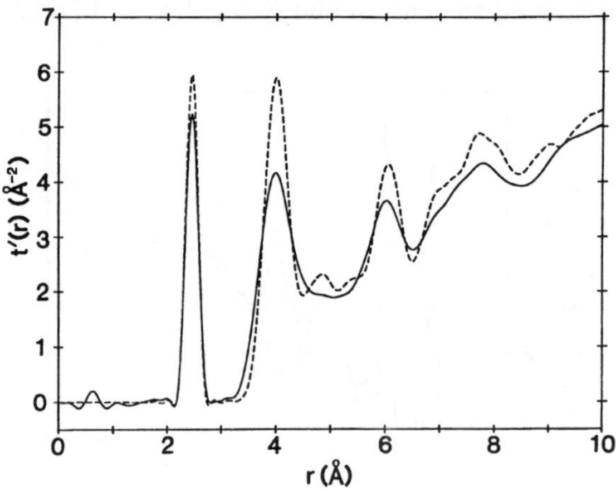

Fig. 14. The correlation function for the Polk and Boudreaux [24] model of amorphous Ge compared neutron data [22] (———, experiment and - - - -, model).

7.2. *Monte Carlo techniques*

The time-consuming nature of hand building and the difficulty of controlling the distribution of the various structural parameters during construction means that the future must reside with computer generated models, using either an algorithm which mimics the hand-building process or some form of Monte Carlo technique. The latter are of two main types, according to the quantity which is minimized. In conventional Monte Carlo simulations, the energy is minimized while, in the so-called reverse Monte Carlo technique, the algorithm optimizes the agreement with the experimental data. Although usually termed reverse Monte Carlo simulation, this technique does not in practice involve a simulation, since it does not try to generate the structure and properties of the material in question starting from some suitable interatomic potential. Instead, it moves the atoms in a periodic unit cell or cluster to yield the best agreement with experiment, as indicated by a reliability factor such as that given in Eq. (16), and is thus the amorphous solid equivalent of the Rietveld technique used in powder crystallography.

A novel and highly successful Monte Carlo technique, for generating random network models of amorphous Si and Ge ("Sillium" models), has been evolved by Wooten and Weaire [21]. This starts with the ambient crystalline network and then introduces disorder via a series of topological transformations which are tested against a Maxwell–Boltzmann factor as in the case of conventional Monte Carlo simulations. Correlation functions for three models due to Wooten and Weaire are compared to neutron diffraction data in Fig. 15. Models A and B are topologically equivalent, but have been relaxed with different interatomic potentials, while model C contains 4-membered rings which were specifically excluded from A and B. It can be seen that all three models have a bond angle distribution very much closer to that of the real material than the Polk and Boudreaux model [24] (Fig. 14) and in general are in very good agreement with experiment, except that the average model number densities are too high. The additional structure in $t'(r)$, at higher r, is almost certainly statistical in nature, due to the small model size (216 atoms).

The reverse Monte Carlo technique was originally devised by Rechtin *et al* [26] but the most recent advances have been made by McGreevy and co-workers [27], who have introduced an exponential probability for accepting unfavourable moves and used very much larger systems. Clearly, given adequate computer time, reverse Monte Carlo modelling is capable of yielding excellent agreement with experiment, but this is to be expected since a model with N atoms is equivalent to fitting the experimental data with $3N$ adjustable parameters (typically, $1000 \leq N \leq 10000$). In comparison, a cubic spline fit to the same data requires only a few hundred parameters and hence the number of excess parameters in the reverse Monte Carlo case is extremely large. For this reason it is imperative, when using reverse Monte Carlo techniques, to examine the resulting configuration to ensure that it is chemically

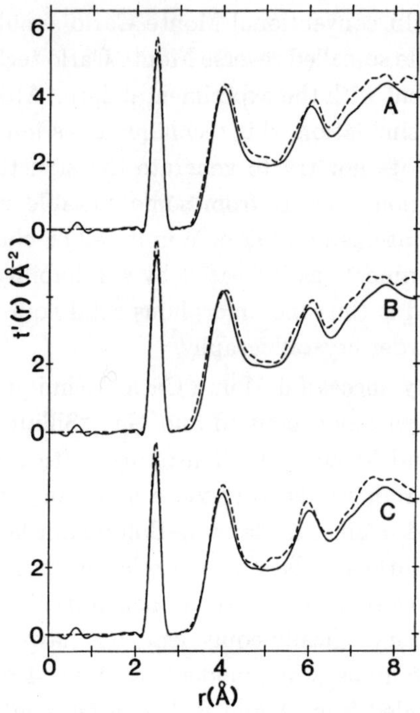

Fig. 15. A series of periodic boundary models of amorphous Ge generated by Wooten and Weaire [21] (dashed lines) compared to experiment (solid lines). (A), model relaxed with Keating potential; (B), as model A but relaxed with Weber potential and (C), model containing 4-membered rings.

reasonable and that it is in accord with all other available data for the material in question. Thus, for example, the vitreous silica model of Keen *et al* [28] must be rejected in that it contains far too many silicon atoms with coordination numbers other than 4. Further constraints, such as the maximum percentage of incorrectly coordinated atoms, may of course be included in the modelling procedure, but at the expense of considerable extra computational time. In addition to checking that the model is chemically reasonable and does not contain too many defects, it is also very important that it be stable, when relaxed with a reasonable interatomic potential, since an unstable structure is similarly unacceptable as a model on which to base further calculations. The structure must be stable to small atomic (e.g. thermal) displacements (i.e. not critically metastable) and kinetically accessible.

Instead of starting from random coordinates, the reverse Monte Carlo technique may also be employed to refine models generated in other ways. For example, an extremely promising algorithm has been developed by Gladden which mimics the hand building process and, in a recent study of vitreous silica, Gladden [29] has further refined such a model using reverse Monte Carlo techniques. The final model is compared to neutron data in Fig. 16. The agreement between the model and experiment is excellent, yielding an R_χ factor of 2.0%, and the number of "defects" in the model is very much less than that for the model of Keen *et al* [28] which started from random atomic coordinates.

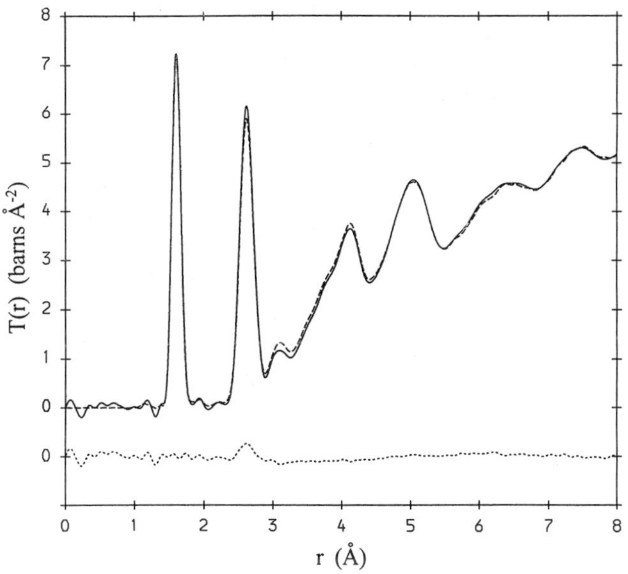

Fig. 16. The neutron correlation function for a vitreous silica model of Gladden [29], refined using the reverse Monte Carlo technique (——, experiment; - - - -, model and ...., residual).

7.3. *Molecular dynamics simulations*

An alternative approach is to use molecular dynamics simulations, which are discussed in very much greater detail elsewhere in this volume [30]. However, it is worthwhile including a few comments here on the comparison of such simulations with diffraction data. Early molecular dynamics simulations of amorphous network solids employed 2-body (ionic) potentials, with no bond angle restoring forces, but led to structural units that are far too distorted and, as a result, more recent simulations have used 3-body forces. Such a simulation of vitreous silica by Vessal *et al* [31] is shown in Fig. 17, together with the difference curve (dotted line). The R_χ factor of 9.1% is somewhat lower than that for the relaxed Bell and Dean model ($R_\chi = 11.1\%$; cf. Sec. 7.1).

Even though the simulation in Fig. 17 employs 3-body potentials, it is obvious from the difference curves that the SiO_4 tetrahedra are still too distorted. The intratetrahedral O–O peak is narrower than for 2-body potentials, although still too broad, but the reduction in the spread of O–$\widehat{Si}$–O bond

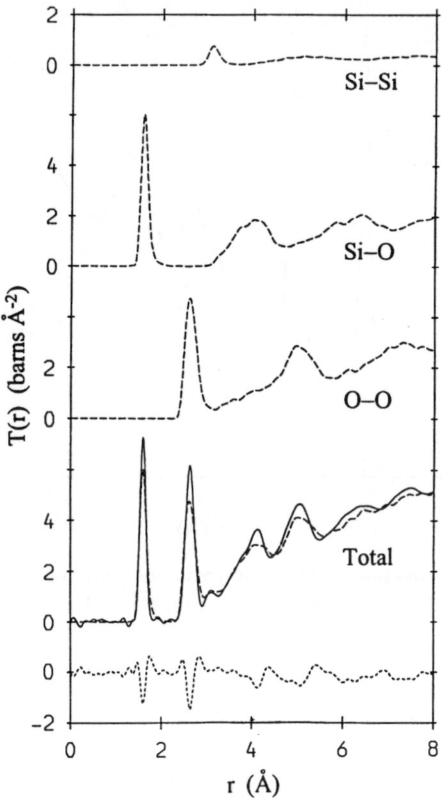

Fig. 17. A comparison of the molecular dynamics simulation of vitreous silica by Vessal *et al* [31] with the neutron data from Fig. 9 (———, experimental data; - - - -, simulation and ...., difference curve).

angles has been obtained at the expense of an increased spread of bond lengths, such that the Si–O peak is also too broad. The structure in the correlation function at higher r is similarly reduced, relative to that for the real glass. A major problem with molecular dynamics simulations of glasses is that the quench rates from the liquid state are many orders of magnitude faster than that for the real glass and hence the simulated structure is appropriate to a much higher fictive temperature. Thus it is to be expected that the latter will be more disordered, as indeed is the case. The way in which the thermal energy is removed during quenching is also different in that, for the simulation, this is usually done homogeneously by reducing the energy of each atom by the same fraction, whereas the quenching of a laboratory glass relies on conduction to the outer surfaces. An interesting possibility would be to try to refine the structure from a molecular dynamics simulation using reverse Monte Carlo techniques, as Gladden [29] has done for her computer generated model.

One important application of molecular dynamics simulations in the analysis of diffraction experiments, even those using 2-body potentials, is in the interpretation of data for complex multi-component systems, particularly when coupled with a versatile molecular graphics package. For example, a number of simulations have been performed of alkali silicate systems and a frame from a simulation of a Na_2O–SiO_2 glass [32] is illustrated in Fig. 18. In the case of vitreous Na_2O–SiO_2, the Na–O distance distribution is asymmetric and this is predicted by the Na–O contribution from the molecular dynamics simulation, shown as the dashed upper curve in Fig. 19, together with the separate Na–O distributions for non-bridging and bridging oxygen atoms. The disorder

Fig. 18. A molecular graphics representation of the structure of vitreous Na_2O–SiO_2, generated by molecular dynamics simulation [32].

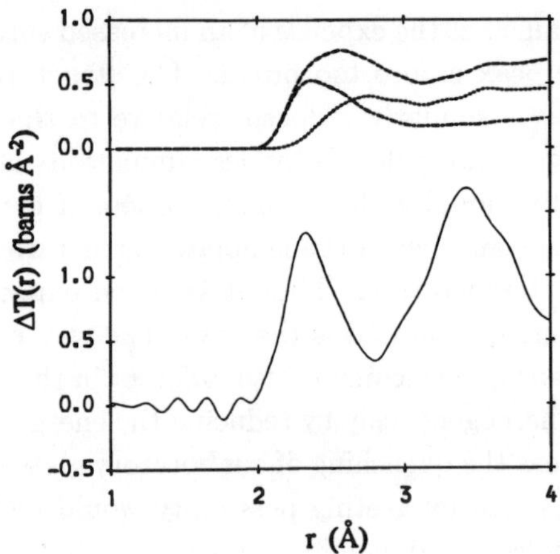

Fig. 19. An analysis of the Na–O distribution for vitreous Na_2O–SiO_2, obtained using the difference technique [33] (c.f. Sec. 8). Bottom, difference correlation function (________); top, molecular dynamics simulation showing the total Na–O distribution (- - - -) and the separate contributions from non-bridging (- · - · -) and bridging (. . . . .) oxygen atoms.

inherent to the vitreous state means that not all of the oxygen atoms in the first neighbour shell around the sodium ions will be non-bridging and, similarly, that the cavities in which some of the Na^+ ions reside will not be sufficiently regular for them to be at the ideal distance from every oxygen atom. In this situation, the positively-charged Na^+ ions will tend to adopt the optimum configuration with respect to the negatively-charged non-bridging oxygen atoms, rather than the bridging oxygen atoms which only carry the much smaller effective charge associated with heteropolar covalent bonding. Thus, it might be expected that the Na–O distribution will comprise an approximately symmetric peak at the non-bridging oxygen atom distance plus a higher r contribution arising mainly from the bridging oxygen atoms, which is consistent both with the molecular dynamics simulation and with the Na(O) coordination number of 3.0 ± 0.5 extracted from a symmetric peak fit [32], since in an oversized cavity the Na^+ ion will normally only be "in contact" with three oxygen neighbours.

7.4. *Analytical models*

Instead of generating atomic coordinates, some modelling techniques generate real or reciprocal space functions analytically, examples being the molecular model and many crystal based models. As indicated in Sec. 5.3, the problem with this approach is that there is no guarantee that it is possible to generate a real structure (i.e. atomic coordinates) consistent with the analytical formalism. For example, many crystal based models calculate $Qi'(Q)$ and/or $D'(r)$ for crystal-like regions, but completely ignore the necessary interconnecting material which in any real structure is likely to comprise the major fraction of the total volume.

Examples of vapour deposited amorphous solids include arsenic sulphide, where the molecules have undergone at least partial polymerisation. Fits to the first peaks in $T'(r)$ for vitreous ($As_{0.38}S_{0.62}$) and vapour-deposited ($As_{0.43}S_{0.57}$) arsenic sulphide indicate that, while the bulk glass is chemically ordered, the vapour deposited film contains far more As–As bonds than the minimum required by stoichiometry as revealed by a high r shoulder [34]. Moreover, the mean As–As distance (2.57 Å) exceeds the normal covalent bond length (2.49 Å) and is much nearer that (2.59 Å) found in the As_4S_4 realgar molecule, which is known from mass spectrometric studies to be present in the vapour phase. Hence it can be concluded that vapour deposited arsenic sulphide contains a significant fraction of As_4S_4 molecules and/or molecular fragments. A random orientation molecular model [34, 35], based on the realgar molecule, is compared to X-ray data in Fig. 20. At high Q, $Qi^X(Q)$ is dominated by intra-molecular contributions and in this region the model is in best agreement with experiment. The differences between model and experiment at low Q are the result of polymerization and orientational effects arising from the fact that the As_4S_4 molecule is in reality not spherical.

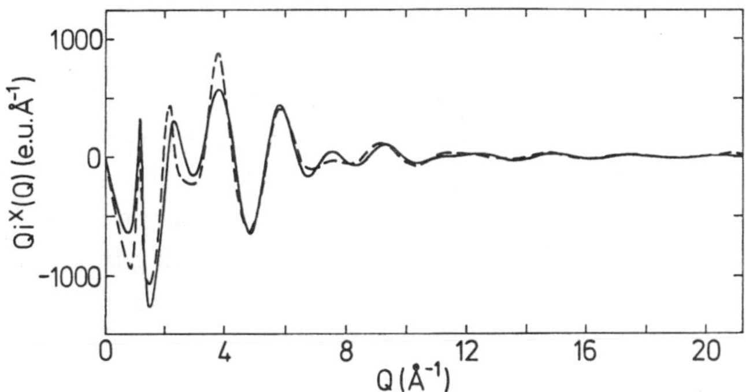

Fig. 20. The X-ray interference function for an As_4S_4 molecular model (solid line) compared with experimental data for thin film $AsS_{1.38}$ (dashed line) [35].

8. Separation of Individual Component Correlation Functions

Various methods exist for the separation of individual component correlation functions, or linear combinations of subsets of these components, which involve the variation of the scattering length of one or more elements. The most important of these are isotopic substitution (neutron diffraction) and the anomalous dispersion technique (X-ray and neutron diffraction), but it is also possible to utilize neutron magnetic diffraction and approximate methods such as isomorphous substitution. A first order difference correlation function, involving a variation in the scattering length of element A via either isotopic substitution or anomalous dispersion, is the equivalent of an EXAFS experiment at the absorption edge of A.

Unfortunately, for many amorphous insulators and semiconductors, none of these techniques is feasible, but even in such cases valuable extra information can be obtained from a combination of neutron and X-ray diffraction, for which the component weighting factors are different ($\bar{b}_i \bar{b}_j$ and $\sim Z_i Z_j$, respectively). In addition, for multi-component systems, the information obtainable from diffraction experiments is greatly increased if a systematic study is carried out as a function of composition. A possible approach, in such cases, is to use the difference technique in which the correlation function for the base amorphous solid is subtracted from that of a more complex material [33].

8.1. *Isotopic substitution*

The neutron scattering length of an element with more than one isotope may be varied by altering the relative isotopic abundances. Samples may thus be prepared with different values of $\bar{b}_A$. For a few elements (H, Li, Ti, Cr, Ni, Sm, Dy and W), isotopes exist with both positive and negative scattering lengths, the latter corresponding to scattering without the normal phase change of π. This leads to a special form of isotopic substitution known as the null technique in which a combination of isotopes is used such that element A has a zero scattering length and hence does not contribute to the measured interference or correlation functions. A measurement of the scattering from such a sample thus yields the X–X component directly, where X indicates any element present other than A.

Figure 21 shows the Dy–Dy + Dy–X (X = Na, Be or F) difference correlation function for vitreous NaF–DyF$_3$–BeF$_2$, together with the results of a molecular dynamics simulation using 2-body potentials [36]. Due to slightly

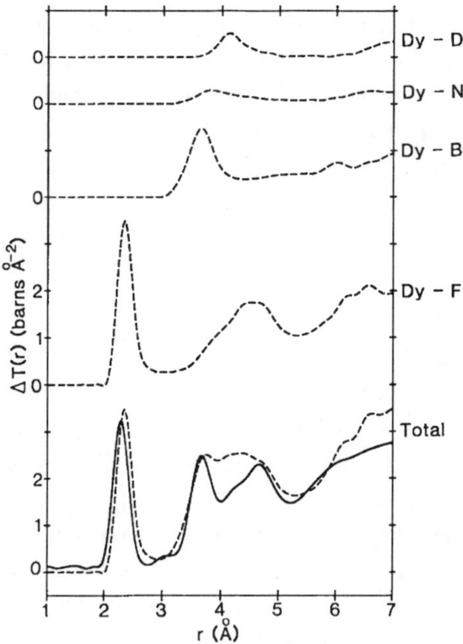

Fig. 21. The Dy–F + Dy–X (X = Na, Be, or F) isotopic difference correlation function for vitreoous NaF–DyF$_3$–BeF$_2$ (solid line) compared to the predictions of a molecular dynamics simulation (dashed lines) [36].

too large an ionic radius for Dy^{3+}, the first neighbour Dy–F peak for the simulation is displaced to higher r but its width is in agreement with experiment. The structure for the simulation between 3.3 and 5.0 Å is in qualitative agreement with experiment but is more smeared out. The simulation does, however, identify the experimental peak at 3.7 Å as being due to the first Dy–Be coordination shell, while the feature between 4 and 5 Å arises from higher order Dy–F interactions with a small contribution from Dy–Dy correlations.

8.2. *Anomalous dispersion*

The X-ray anomalous dispersion technique [37] is based on the fact that, near an absorption edge, the X-ray form factor has both real and imaginary parts which are wavelength dependent

$$f(Q) = f^\circ(Q) + \Delta f'(\lambda) + i\Delta f''(\lambda) \qquad (22)$$

The utilization of anomalous dispersion to separate component correlation functions has, however, only really become feasible with the advent of suitable diffractometers on synchrotron X-ray sources, which allow the wavelength to be continuously varied to obtain the optimum values of $\Delta f'(\lambda)$ and $\Delta f''(\lambda)$. Anomalous dispersion experiments are also possible with neutrons, for which a variation in scattering length occurs close to an absorption resonance and is much larger for the isotope in question than is the case for X-ray scattering.

9. First Diffraction Peak

For the reasons outlined in Sec. 5, the rapid decay of the highest frequency Fourier components means that the first peak in the diffraction pattern of an amorphous solid is almost invariably the sharpest and as such it has attracted a great deal of attention in the literature. Before discussing the structural origins of this often named "first sharp diffraction peak" (FSDP), however, it is perhaps worth making a few general comments to dispel some of the mysticism and misunderstanding with which this feature has been associated. First, and this seems to be completely lost on many authors who claim special importance for the first diffraction peak, any diffraction pattern with peaks must by definition have a first peak and the fact that this is usually the sharpest also means that it is the most complicated since it has contributions from the reciprocal space Fourier components arising from all of the peaks in $T(r)$. Second, the first diffraction peak does not arise from a *distance* in real space but from the longest period, most slowly decaying (assuming that the first peak is indeed the sharpest) real space *Fourier component*, the period being given by $2\pi/Q_1$ where Q_1 is the peak position in reciprocal space. As for crystalline Bragg peaks, therefore, the first diffraction peak for a glass is evidence of a *periodicity* in real space, the much greater width relative to crystalline materials indicating that the associated Fourier component decays after relatively few periods. There is of course absolutely no reason why the origin of the longest period Fourier component should be the same for all amorphous solids, which indicates the futility of searching for some universal theory for the first diffraction peak, with which many workers seem so obsessed. A review of the theories of the first diffraction peak for amorphous materials is given by Moss and Price [38].

Many workers, in interpreting diffraction data from amorphous chalcogenides, have placed great emphasis on the first diffraction peak, which it is claimed is anomalous in that it occurs at very low Q and is very sharp.

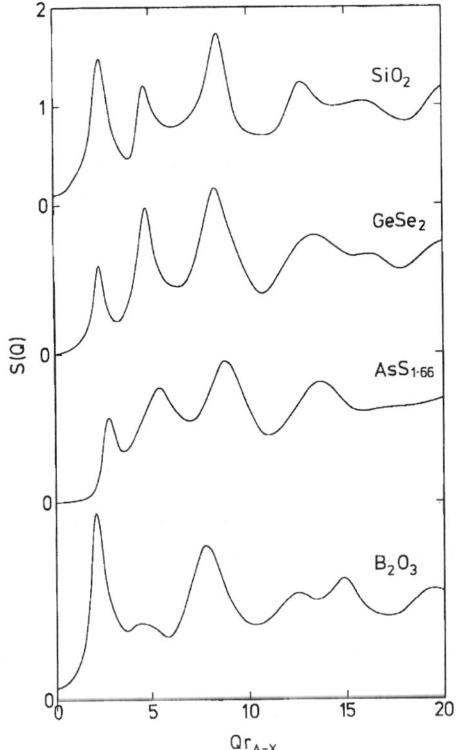

Fig. 22. Reduced diffraction patterns for vitreous SiO$_2$, GeS$_2$, AsS$_{1.66}$ and B$_2$O$_3$ [35].

In the case of vitreous arsenic sulphide (Fig. 22), this peak is close to the 020 reflection for the mineral orpiment, which has a layer structure, the 020 reflection corresponding to the interlayer spacing. A similar situation exists for other chalcogenides (e.g. As$_2$Se$_3$, GeS$_2$ and GeSe$_2$) and has consequently led to the suggestion that the glasses also have layer structures. The first neutron diffraction peak for vitreous GeSe$_2$ is indeed at a lower value of Q than that for vitreous SiO$_2$, but this would be expected, since the Ge–Se bond is much longer than the Si–O bond (2.370 Å as opposed to 1.608 Å). If the vitreous SiO$_2$ structure were expanded to give the Ge–Se bond length, the Q scale would contract by the same factor. Thus, in order to compare the neutron diffraction patterns for the two materials, it is necessary to plot the structure factor $S(Q)$ against $r_{A-X}Q$ where r_{A-X} is the A–X bond length (A = Si or Ge; X = O or Se). This is done in Fig. 22 and it can be seen that the first three neutron diffraction peaks for GeSe$_2$ are almost exactly coincident with those of vitreous SiO$_2$, which certainly does not have a layer structure. The situation for vitreous AsS$_{1.66}$ is also ambiguous in that its first diffraction peak can equally well be predicted by a molecular model as demonstrated in Sec. 7.4. For amorphous network solids, the first diffraction peak has been associated with the periodicity arising from the boundaries between a succession of the cages which comprise the structure of a 3-dimensional covalent network (cf. Fig. 1). In the case of 3-fold coordinated structural units, the cages tend to become flattened and a few such adjacent cages, flattened in the same direction, will approximate a set of layers and give rise to the necessary Fourier component to generate the first diffraction peak, without the need to resort to what is traditionally understood by a layer structure.

10. Conclusions

The examples discussed in this chapter, and in Refs. [2–4] and [17], demonstrate the role of X-ray and neutron diffraction techniques in investigations of the structure of amorphous insulators and semiconductors. The best modern diffraction experiments are capable of providing accurate data with high real-space resolution, which if used correctly are an extremely fine filter for the various models proposed in the literature. The greatest barrier to progress in understanding the structure of these materials lies not so much with the diffraction technique itself but in the development of modelling procedures, in which model parameters can be varied in a systematic way until agreement with experiment is obtained, in both real and reciprocal space, within the known experimental uncertainties. This situation is unique to amorphous materials and has no parallel in corresponding studies of the crystalline state.

References

[1] W.H. Zachariasen, *J. Amer. Chem. Soc.* **54**, 3841 (1932).
[2] A.C. Wright, Neutron and X-Ray Amorphography, *Experimental Techniques of Glass Science*, eds. C.J. Simmons and O.H. El-Bayoumi (Amer. Ceram. Soc., Westerville, 1993) Chap. 8.
[3] A. C. Wright, *Adv. Struct. Res. Diffr. Meth.* **5**, 1 (1974).
[4] A.C. Wright and A.J. Leadbetter, *Phys. Chem. Glasses* **17**, 122 (1976).
[5] G.N. Greaves, A. Fontaine, P. Lagarde, D. Raoux and S.J. Gurman, *Nature* **293**, 611 (1981).
[6] A.C. Wright, N.M. Vedishcheva and B.A. Shakhmatkin, *J. Non-Cryst. Solids* **192 & 193**, 92 (1995).

[7] P.J. Bray, *J. Non-Cryst. Solids* **75**, 29 (1985).

[8] R.N. Sinclair and A.C. Wright, *J. Non-Cryst. Solids* **57**, 447 (1983).

[9] P.A.V. Johnson, A.C. Wright and R.N. Sinclair, *J. Non-Cryst. Solids* **58**, 109 (1983).

[10] A.C. Wright and R.N. Sinclair, *J. Non-Cryst. Solids* **76**, 351 (1985).

[11] D.I. Grimley, A.C. Wright and R.N. Sinclair, *J. Non-Cryst. Solids* **119**, 49 (1990).

[12] E.A. Lorch, *J. Phys.* **c2**, 229 (1969).

[13] B.E. Warren and G. Mavel, *Rev. Sci. Instrum.* **36**, 196 (1965).

[14] R.L. Mozzi and B.E. Warren, *J. Appl. Crystallogr.* **2**, 164 (1969).

[15] G. Bushnell-Wye, J.L. Finney, J. Turner, D.W. Huxley and J.C. Dore, *Rev. Sci. Instrum.* **63**, 1153 (1992).

[16] A.C. Wright, *J. Non-Cryst. Solids* **159**, 264 (1993).

[17] A.C. Wright, *J. Non-Cryst. Solids* **179**, 84 (1994).

[18] J.H. Konnert and J. Karle, *Acta Crystallogr.* **A29**, 702 (1973).

[19] A.C. Wright, R.A. Hulme, D.I. Grimley, R.N. Sinclair, S.W. Martin, D.L. Price and F.L. Galeener, *J. Non-Cryst. Solids* **129**, 213 (1991).

[20] G. Mason, *Nature* **217**, 733 (1968).

[21] F. Wooten and D. Weaire, *Solid State Phys.* **40**, 1 (1987).

[22] G. Etherington, A.C. Wright, J.T. Wenzel, J.C. Dore, J.H. Clarke and R.N. Sinclair, *J. Non-Cryst. Solids* **48**, 265 (1982).

[23] R.J. Bell and P. Dean, *Phil. Mag.* **25**, 1381 (1972).

[24] D.E. Polk and D.S. Boudreaux, *Phys. Rev. Lett.* **31**, 92 (1973).

[25] P.H. Gaskell and I.D. Tarrant, *Phil. Mag.* **B42**, 265 (1980)

[26] M.D. Rechtin, A.L. Renninger and B.L. Averbach, *J. Non-Cryst. Solids* **15**, 74 (1974).

[27] R.L. McGreevy and L. Pusztai, *Mol. Simulation* **1**, 359 (1988).

[28] D.A. Keen and R.L. McGreevy, *Nature* **344**, 423 (1990).

[29] L.F. Gladden, Structure and Dynamics of 4–2 Coordinated Glasses, *The Physics of Non-Crystalline Solids*, eds. L.D. Pye, W.C. LaCourse and H.J. Stevens (Taylor and Francis, London, 1992) p. 91.

[30] P. Boolchand (ed.), Insulating and Semiconducting Glasses, *Molecular Dynamics Simulations of Network Glasses*, ed. D.A. Drabold, Chapter 9 (World Scientific, Singapore, 2000) pp. 607–651.

[31] B. Vessal, M. Amini and C.R.A. Catlow, *J. Non-Cryst. Solids* **159**, 184 (1993).

[32] N.M. Vedishcheva, B.A. Shakhmatkin, M.M. Shultz, B. Vessal, A.C. Wright, B. Bachra, A.G. Clare, A.C. Hannon and R.N. Sinclair, *J. Non-Cryst. Solids* **192 & 193**, 292 (1995).

[33] A.C. Wright, A.G. Clare, B. Bachra, R.N. Sinclair, A.C. Hannon and B. Vessal, *Trans. ACA* **27**, 239 (1991).

[34] M.F. Daniel, A.J. Leadbetter, A.C. Wright and R.N. Sinclair, *J. Non-Cryst. Solids* **32**, 271 (1979).

[35] A.C. Wright, R.N. Sinclair and A.J. Leadbetter, *J. Non-Cryst. Solids* **71**, 295 (1985).

[36] A.G. Clare, G. Etherington, A.C. Wright, M.J. Weber, S.A. Brawer, D.D. Kingman and R.N. Sinclair, *J. Chem. Phys.* **91**, 6380 (1989).

[37] J. Krogh-Moe, *Acta. Chem. Scand.* **20**, 2890 (1966).

[38] S.C. Moss and D.L. Price, Random Packing of Structural Units and the First Sharp Diffraction Peak in Glasses, *Physics of Disordered Materials*, eds. D. Adler, H. Fritzsche and S.R. Ovshinsky (Plenum, New York, 1985) p. 77.

B. MÖSSBAUER SPECTROSCOPY

P. BOOLCHAND

*Department of Electrical & Computer Engineering and Computer Science,
University of Cincinnati, Cincinnati, OH 45221-0030, USA*
punit.boolchand@uc.edu

Contents

1. Introduction

In 1932, W.H. Zachariasen [1] proposed the structure of SiO_2 glass as a "*continuous random network*" (CRN). With the profound developments in both theoretical and experimental methods to probe glass structure over the past 65 years, such as neutron scattering [2], ^{17}O NQR^3, ^{29}Si solid state NMR [4], molecular dynamic simulations [5], and extended constraint theory [6], one has begun to see a convergence of ideas on the rather special role of a CRN as a structural description of SiO_2 glass. In SiO_2 glass, the weakness of bond-bending forces about bridging oxygen atoms allows angular disorder to remain when melts are cooled past the glass transition temperature (T_g). The underlying O-bond-bending *constraint* is thus *broken* [6]. A network of corner-sharing $Si(O_{1/2})_4$ tetrahedra with wide variation in bridging oxygen bond angles $(110°-180°)$ can then fill space with the accumulation of minimal internal strain. The number of mean-field constraints per atom in such a CRN structure of SiO_2 glass then approaches 3, and equals the degrees of freedom per atom in a $3d$ network. This rather unique set of conditions [6] accounts for the propensity of glass formation in this prototypical oxide.

In analogy to the case of silica, it has been fashionable to describe the molecular structure of stoichiometric chalcogenide glasses (As_2Se_3, $GeSe_2$, $SiSe_2$) also in terms of a CRN. Such a structural description of these stoichiometric glasses appears to us to be simplistic today. Diffraction methods [7] provide an essential but limited description of the non-crystalline state of matter because of the spatial averaging intrinsic to such measurements. Local probes such as Mössbauer spectroscopy [8], Raman scattering [9], solid state NMR [10] and EXAFS [11] have provided crucial insights in refining structural models of insulating and semiconducting glasses. In sharp contrast to diffraction methods, local probes can, in favorable cases, provide a multimodal distribution of sites and bonds, thus imposing limits on acceptable structural models. In the chalcogenides, a perusal of the available structure results reveals a pattern; glasses that possess a low mean-coordination number ($\bar{r} < 2.4$) are, by and large, floppy homogeneous [12] random networks (although some exceptions occur), but at medium- ($r = 2.4$) and high mean-coordination number ($r > 2.4$), glasses become generally rigid and heterogeneous [12] as structural correlations on a scale of 5 Å and larger (molecular nanophases) appear. In many cases, these molecular phases bear a connection to closely lying crystalline structures while in others, the molecular phases populated in glasses are *low-pressure conformations.* Our use of the term molecular phases in this context is to denote distinct nanophase morphologies that may either coexist or exist separately to constitute the glass network backbone. In this chapter, we shall provide several examples to illustrate the underlying ideas focusing on applications of Mössbauer nuclear hyperfine structure as a probe of short and medium range structure of glasses.

The power of Mössbauer spectroscopy as a probe of glass structure derives in large part from its potential to quantitatively characterize local environments of cations or anions in a network through *local fields* and the *local vibrational density of states.* The local fields are directly accessible from the Mössbauer hyperfine structure. The quadrupolar fields are usually well defined because the nearest-neighbor structure in glasses is rather well developed, as shown by diffraction methods [7]. In covalent materials, the local fields can be correlated in many cases to those in corresponding crystalline phases. The power and elegance of the spectroscopy also derives from its potential to characterize the vibrational behavior of these local environments in a network, through a measurement of the Lamb–Mössbauer factor [13]. One is thus able to probe details of both the static local structure, and some aspects of the dynamical behavior of a glass network. The local- and medium-range structure of a glass network influences both the static hyperfine structure and the vibrational density of states. Taken together, these results in a glass serve to provide a consistency check on interpretation of each other. In Chapter 6(B), we describe [13] Mössbauer spectroscopy as a local probe of vibrational density of states in glasses, and also provide details of a typical spectrometer. Next, we introduce the elements of hyperfine structure and the extraction of local fields from the observed lineshapes as a means to characterize glass structure. In the sections following, we provide results of investigations on selected oxide, chalcogenide and chalcohalide glasses. We conclude the chapter with a summary of the principal findings on network glasses and some remarks on possible future directions of the method within the context of glass science.

2. Hyperfine Interactions and Local Structure

The discovery of nuclear resonant emission and absorption of gamma rays in solids by R.L. Mössbauer [14] made possible the study of electron-nucleus interaction (hyperfine interaction) in an excited nuclear state (in the energy domain) for the first time. Prior to the discovery of the Mössbauer effect, hyperfine interactions in a nuclear ground state were accessible by NMR and

NQR in the energy domain. The recoil-free, i.e. zero phonon nature of the nuclear emission and absorption process, makes it possible to obtain the natural linewidth of a nuclear transition determined solely by the lifetime ($\bar{\tau}$) of the excited state $\Delta E = \hbar/\bar{\tau}$ (Heisenberg's uncertainty principle). For a nuclear excited state, such as the 23.8 keV state in ^{119}Sn, possessing a lifetime (τ) of 17.75 nsec, the natural width of the 23.8 keV γ-ray is 2.57×10^{-8} eV, yielding an energy resolution of one part in 10^{12}. At this level of precision, it becomes possible to probe the interaction of a nucleus with its overlapping and non-overlapping electron charge cloud in a solid, since the magnitude of the hyperfine interactions are typically in the $10^{-8} \sim 10^{-12}$ eV range. There are three principal interactions observed in Mössbauer spectroscopy [15], isomer shift, electric quadrupole interaction, and magnetic dipole interaction.

2.1. *Isomer shift*

The Coulomb interaction between the nuclear charge and its overlapping electron cloud directly (and the non-overlapping electron cloud indirectly) contribute to the isomer shift. In Mössbauer spectroscopy, one detects the nuclear resonance by comparing the nuclear transition energy in an emitter matrix to that in an absorber matrix. By imparting a precise motion to the emitter matrix relative to a stationary absorber matrix (with a linear velocity transducer), one scans the nuclear resonance profile using the Doppler effect according to: $\delta = E_a - E_e = \hbar\nu\, v/c$ where $E_{e,a}$ are the γ-ray transition energies in the emitter and absorber, respectively. The isomer (or chemical) shift is given by

$$\delta = E_a - E_e = \frac{2\pi}{5} Ze^2 \Delta|\psi(0)|^2 \Delta\langle r^2 \rangle \tag{1}$$

where $\Delta\langle r^2 \rangle$ represents the change in the nuclear charge radius between the ground state and excited state, while $\Delta|\psi_0|^2$ represents the change in the contact electron density at the nuclear-site between the two matrices in question. Thus, the nuclear charge radius in the excited state must differ from that in the ground state ($\Delta\langle r^2 \rangle \neq 0$) to observe an isomer shift. It is customary to measure the isomer shift relative to a standard matrix which, for the case of ^{119}Sn, is taken to be Sn^{4+} as in $CaSnO_3$. Figure 1 displays spectra of metallic Sn (β-Sn), ^{119}Sn as a substitutional dopant in crystalline-Si, and $BaSnO_3$ each taken with an emitter of ^{119m}Sn in Vanadium metal. In practice, the isomer shift provides the formal charge state of the resonant atom, which in insulators and semiconductors also yields the local coordination, as we shall illustrate next for the case of the ^{119}Sn nuclear resonance.

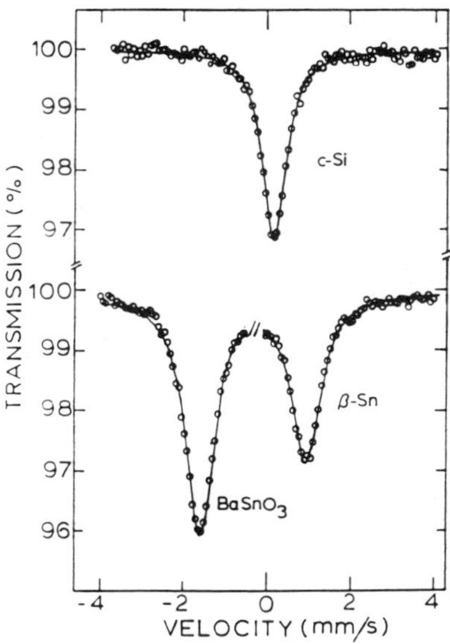

Fig. 1. ^{119}Sn spectra of indicated absorbers taken with a source of ^{119m}Sn in Vanadium revealing a positive isomer shift for β-Sn and Sn in c–Si relative to BaSnO$_3$. Figure is taken from Ref. [8].

Figure 2 gives a ^{119}Sn isomer shift scale for selected Sn bearing crystals as well as glasses. We find that shifts characteristic of tetrahedral Sn reside in the region of +1.3 mm/s to +2.0 mm/s, which is between the shifts of Sn^{4+} and Sn^{2+}. The large positive shifts of Sn^{2+} are understood in terms of the two 5s-like valence electrons on Sn. These contribute overwhelmingly to the contact charge density $|\psi(0)|^2$. Just the reverse is true for Sn^{4+} which has the lowest $|\psi(0)|^2$ because of the absence of 5s electrons. Sn present in a local tetrahedral symmetry, as in c–Si, is described in terms of sp^3–like covalent bonds. This configuration has a shift that resides in-between Sn^{4+} and Sn^{2+}, primarily because only one 5s-like electron contributes to $|\psi(0)|^2$.

α–Sn and Sn as an impurity in the group IV element semiconductors represent some of the few examples of tetrahedrally coordinated species found in crystalline hosts. Because of its more metallic character, Sn tends to choose octahedral over tetrahedral coordination in most Sn-bearing crystals. In covalent glasses, which are less dense than their crystalline analogues, just the reverse is the case, and one finds, for example, that Sn is predominantly tetrahedral in GeX$_2$ glasses (X = S, Se, and Te). As discussed later, spectra of

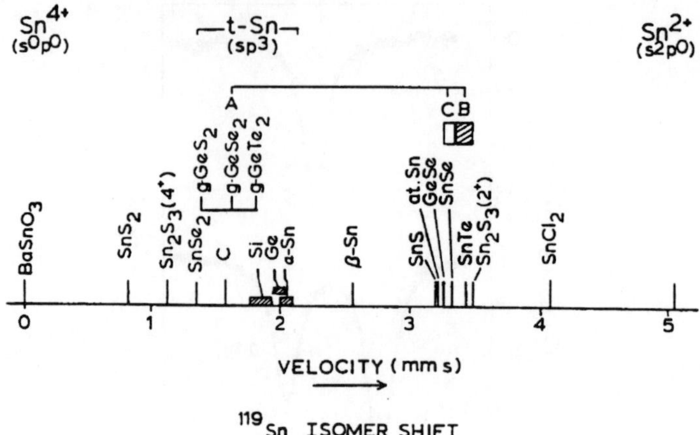

Fig. 2. ^{119}Sn isomer shifts of selected crystals and glasses plotted with respect to BaSnO$_3$. Figure is taken from Ref. [8].

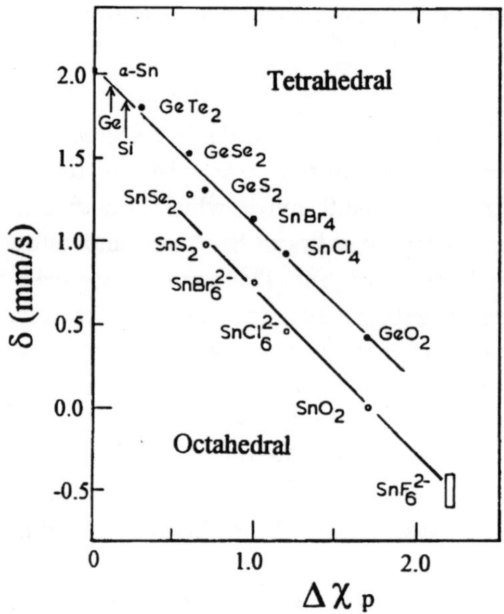

Fig. 3. ^{119}Sn isomer shifts of selected tetrahedral (SnX$_4$) and octahedral (SnX$_6$) species found in inorganic glasses and crystals plotted as a function of the Pauling electronegativity difference $\Delta\chi_p = \chi_x - \chi_{Sn}$. Figure is taken from Ref. [16].

these glasses display a prominent Sn resonance which is characterized by a narrow width (*A*-site). Plotting the lineshifts of these (*A*-sites) resonances along with the shifts of several Sn tetrahalides (where the cation is also fourfold coordinated in a tetrahedral symmetry), against the Pauling electronegativity difference $\Delta\chi_p = \chi_x - \chi_{Sn}$, gives the universal curve in Fig. 3 [16]. This correlation can be quantitatively understood in terms of covalently bonded interactions which are modified by charge transfer effects [16]. This correlation serves as conclusive evidence that the *A*-site resonances seen in GeX_2 glasses represent geometrically and chemically tetrahedrally coordinated Sn species. (This is a point that was not recognized by early workers [17] in the field, who instead ascribed this resonance to Sn^{4+}, an octahedrally coordinated species, as found in corresponding crystals.) A parallel correlation between δ and $\Delta\chi_p$ occurs for octahedrally (sp^3d^2) coordinated species. The latter correlation is displaced to lower δ values in relation to the former (sp^3) largely because of the shielding effect of two additional d electrons.

The isomer shift of Sn^{2+} species in covalent networks usually tends to be in the vicinity of 2.9 mm/s to 3.4 mm/s and, furthermore is correlated with the quadrupole splitting that invariably accompanies such species. This correlation has been discussed by a number of previous workers in the field and the reader is referred to Ref. [19] for a more complete discussion.

2.2. *Electric quadrupole interaction*

Nuclear states of spin $I \geq 1$, in general can possess a finite quadrupole moment eQ, associated with the asphericity of nuclear charge distribution. Interaction with the electric field gradient tensor (EFG) leads to nuclear-level splittings, described by the following Hamiltonian [19],

$$H_Q = \frac{e^2 Q V_{zz}}{4I(2I-1)} \left\{ 3I_z^2 - I(I+1) + \frac{\eta}{2}(I_+^2 + I_-^2) \right\} \qquad (2)$$

A prolate nuclear charge distribution implies a positive quadrupole moment eQ. The asymmetry parameter, η, of the tensor provides a measure of the departure of the electronic charge distribution from cylindrical symmetry, i.e. $\eta = |(V_{xx} - V_{yy})/V_{zz}|$. For a cylindrically symmetric local environment, $\eta = 0$ or $V_{xx} = V_{yy}$. I_+ and I_- are the step-up and step-down nuclear spin operators. The local site symmetry of a probe atom in a solid determines in a direct manner the sign and magnitude of the EFG tensor. The EFG tensor vanishes if the probe atom occupies a site of cubic or tetrahedral symmetry.

The EFG tensor at a probe atom in a solid generally derives from two contributions. Thus

$$V_{zz} = e(V_{zz})_{\text{ion}}(1 - R) + e(V_{zz})_{\text{host}}(1 - \Upsilon_\infty) \tag{3}$$

where the first-term represents the contribution from electrons of the probe-atom (ion), while the second-term represents a contribution from more distant atoms comprising the host-solid. In (3) R and Υ_∞ are the shielding and anti-shielding coefficients, respectively, of the two gradients by the core electrons on the probe atom. For probe atoms with open p-valence shells, the contribution of $(V_{zz})_{\text{ion}}$ in general dominates and, for this reason, the nature of probe chemical bonding with its first- and second-nearest neighbors largely determine the quadrupole coupling (e^2QV_{zz}) in glasses and crystals.

Nuclear quadrupole resonance (NQR) spectroscopy [21] deals with measurements of EFGs in nuclear ground states [see Chapter 5(C)]. Mössbauer spectroscopy, on the other hand, opens the possibility to measure EFGs using excited states of nuclei as well. In working with chalcogen based materials, this is particularly attractive because there are no suitable Te nuclei available as NQR probes. On the other hand, Te-based compounds [22] can be probed using Mössbauer resonances in ^{125}Te and ^{129}I, as we shall discuss later.

The 23.8 keV resonance in ^{119}Sn involves a gamma transition from an excited state of spin 3/2 (possessing a finite quadrupole moment $eQ = -0.08$ barns) to a ground state spin of 1/2 ($eQ = 0$). In SnTe which possesses a rocksalt structure, the quadrupole interaction vanishes, and one observes a narrow absorption line because of the cubic symmetry. On the other hand, in SnS which possesses a distorted rocksalt environment, one observes a doublet. The doublet results from the presence of a finite quadrupole interaction in the $3/2^+$ state. The quadrupole splitting in $I = 3/2$ state is given by solving the quadrupole Hamiltonian given in Eq. (2):

$$\Delta = \frac{e^2 V_{zz} Q}{2}(1 + \eta^2/3)^{1/2} \tag{4}$$

Since the splitting Δ depends on the product of the quadrupole coupling $(e^2 V_{zz} Q)$ and the factor $(1 + \eta^2/3)^{1/2}$, it is not possible to determine η and the magnitude and sign of the coupling constant from a measurement of Δ alone. The uncertainty due to the lack of an *a priori* knowledge of η, introduces a maximum error of 17% in the coupling since the range spanned by η is $0 < \eta < 1$.

An example of quadrupole interaction in *crystalline* samples is illustrated in Fig. 4(a) for the case of $Ge_{2-2x}Sn_{2x}S_3$ compounds [23–25] at several values of x. The stoichiometric crystalline compound at $x = 1/2$, i.e. $GeSnS_3$ consists of globally parallel chains of corner-sharing $Ge(S_{1/2})_4$ tetrahedra, which are laterally coupled by Sn^{2+} sites. The doublet structure seen in Fig. 4(a) results from the non-cubic coordination of Sn^{2+} local environments in this structure. At $x = 1$, i.e. Sn_2S_3, the structure consists of infinitely long, thin, and narrow molecular ribbons that produce two types of cations sites [23, 24], which occur with equal frequency [see Fig. 4(c)]: an octahedrally coordinated Sn^{4+} site in the ribbon interior giving a single line and an asymmetrically coordinated 3-fold coordinated Sn^{2+} site at ribbon edges giving rise to a doublet in Fig. 4(a). Spectra at the intermediate compositions unequivocally demonstrate that the alloy at $x = 0.60$ takes on the zigzag chain structure [Fig. 4(b)] while the one at $x = 0.75$ takes on the molecular ribbon structure [Fig. 4(c)] as discussed elsewhere [23]. In these spectra the power of Mössbauer spectroscopy to elucidate the structure of the crystals by merely probing the local order has been confirmed by X-ray crystallography. However, in glasses one does not have the luxury to check the interpretation of Mössbauer spectroscopy with X-ray crystallography because of the absence of long range order. However, it is possible to confirm or complement results of Mössbauer spectroscopy on glasses with Raman scattering measurements as we shall illustrate later. Returning to our discussion of the quadrupole interaction, in principle in dealing with crystalline materials, η and the sign of the quadrupole coupling can be established by applying an external magnetic field [15] or the use of textured samples, such as single crystals [26].

With gamma-transitions involving nuclear states of spin $I \geq 3/2$, such as for ^{129}I $(5/2 \rightarrow 7/2)$, quadrupole interaction leads to a 12-line multiplet structure as shown in Fig. 5. The resulting lineshapes for the known quadrupole moment ratio $R = Q_e/Q_g$ (where Q_e and Q_g designate quadrupole moments of the excited and ground state), and different η values are sketched in Fig. 6. Note that when $\eta = 0$, the spectrum displays a rather asymmetric lineshape revealing a triplet lineshape structure at negative velocities. On the other hand, when $\eta = 1$, spectrum becomes symmetric about the center of gravity. It is possible to deduce the sign and magnitude of the quadrupole coupling, and η from a measurement of the Mössbauer spectrum on a polycrystalline or non-textured material like a glass. ^{129}I lineshapes observed in elemental chalcogens and binary glasses will be discussed in Sec. 4.1.2.

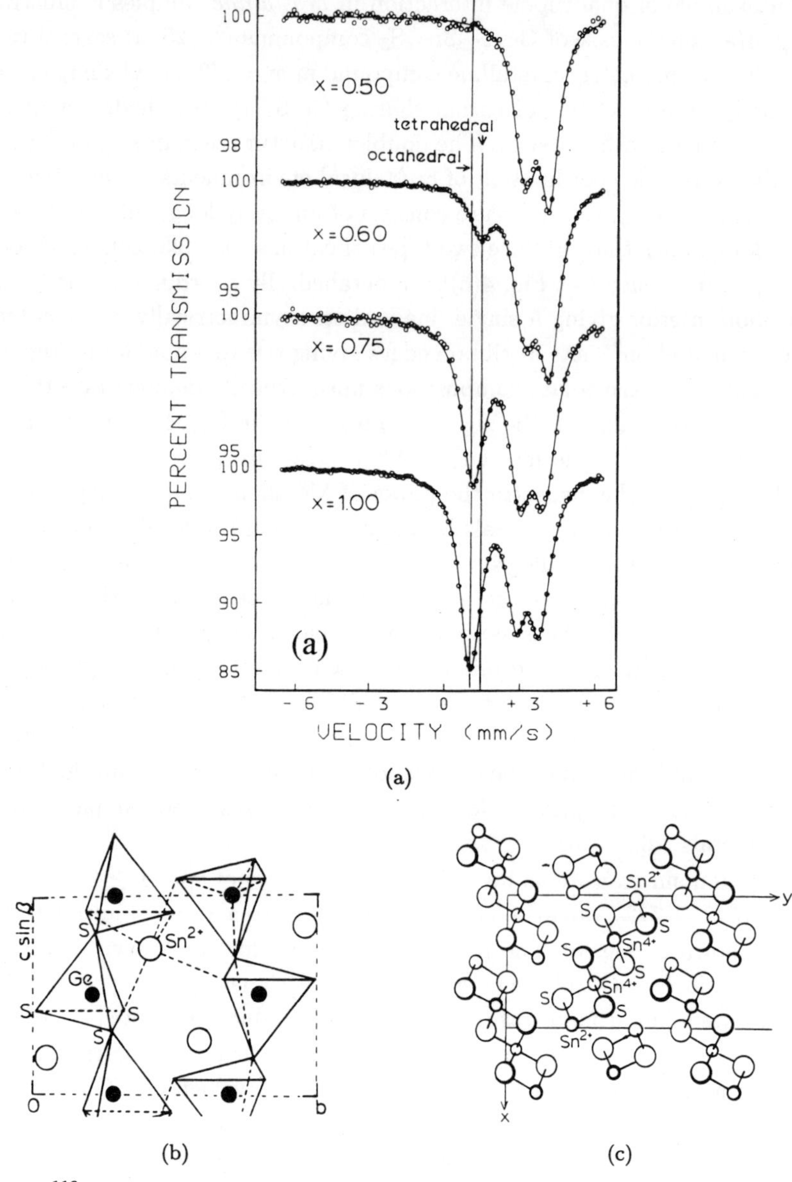

Fig. 4(a) ^{119}Sn spectra of $Ge_{2-2x}Sn_{2x}S_3$ crystalline compounds at indicated compositions x. (b) Zigzag chain molecular structure of c–SnGeS$_3$ consisting of corner-sharing tetrahedra of $Ge(S_{1/2})_4$. (c) Structure of c–Sn$_2$S$_3$ consisting of infinitely long, thin molecular ribbons running along the z-axis.

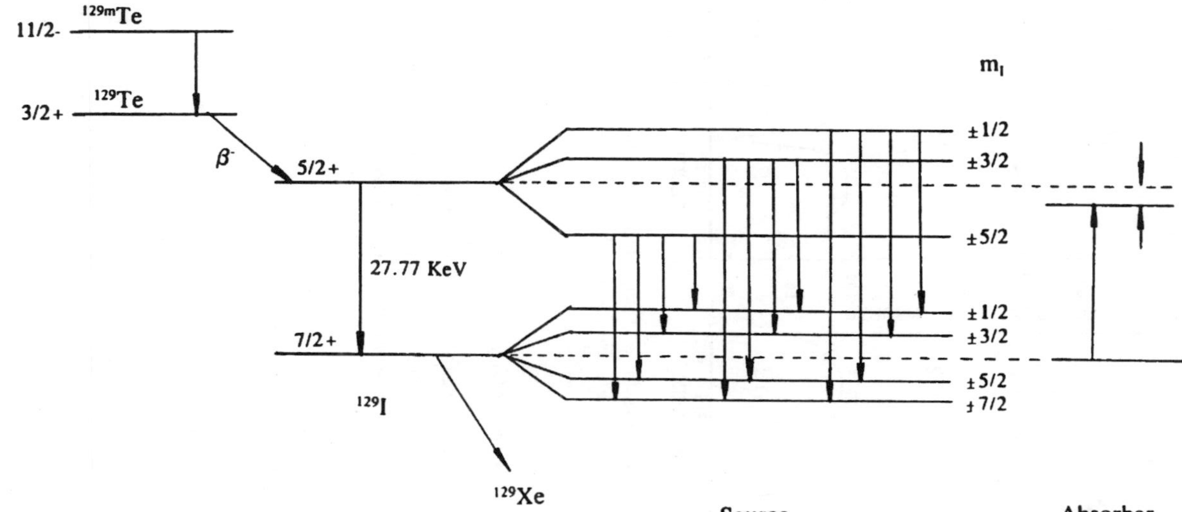

Fig. 5. Nuclear scheme of decay of ^{129m}Te. The figure shows nuclear quadrupole splitting of the $5/2^+$ and $7/2^+$ levels in ^{129}I revealing a 12-line multiplet structure in emission.

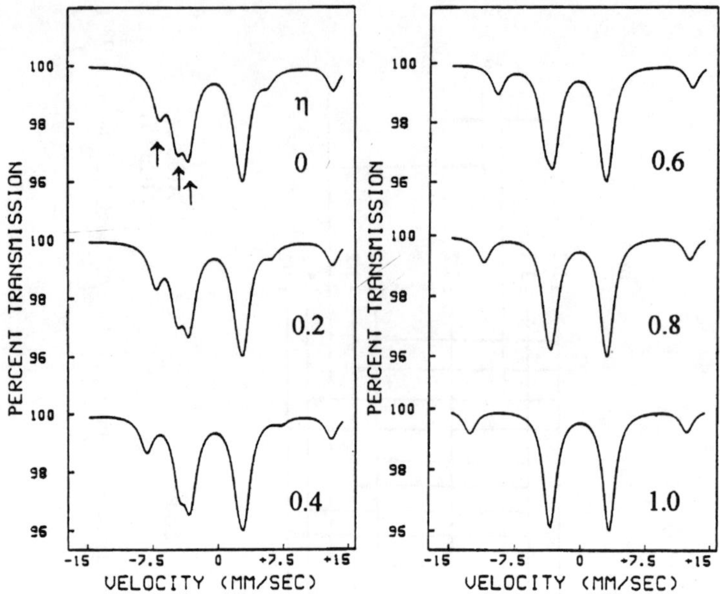

Fig. 6. Simulations of ^{129}I lineshapes for a static quadrupole interaction. Changes in line-shape due to a finite η value of the EFG are displayed.

2.3. *Magnetic dipole interaction*

The third interaction often accessible in Mössbauer spectroscopy is the magnetic dipole interaction. This interaction has served as a powerful probe of magnetism [27] in bulk crystalline solids, thin-films, mesoscopic systems like nanograined systems, and glassy systems. Among the latter, metallic glasses based on metal-metalloid such as on Fe–B and Ni–P have been examined extensively and the interested reader is steered to several reviews of the subject [28]. The Zeeman interaction

$$H = -\boldsymbol{\mu} \bullet \mathbf{H} = -g\mu_n \mathbf{I} \bullet \mathbf{H} \tag{5}$$

between a nuclear magnetic moment ($\boldsymbol{\mu}$) and a magnetic field ($\mathbf{H}$) splits the $\pm m$ degenerate magnetic levels and results in complete removal of the $2I + 1$ fold degeneracy of a nuclear level. In Eq. (5), g and μ_n represent the nuclear g-factor and the nuclear magneton ($= 5.049 \times 10^{-27}$ Joules/Tesla). To illustrate the principle, we show in Fig. 7 the magnetic hyperfine splitting of the

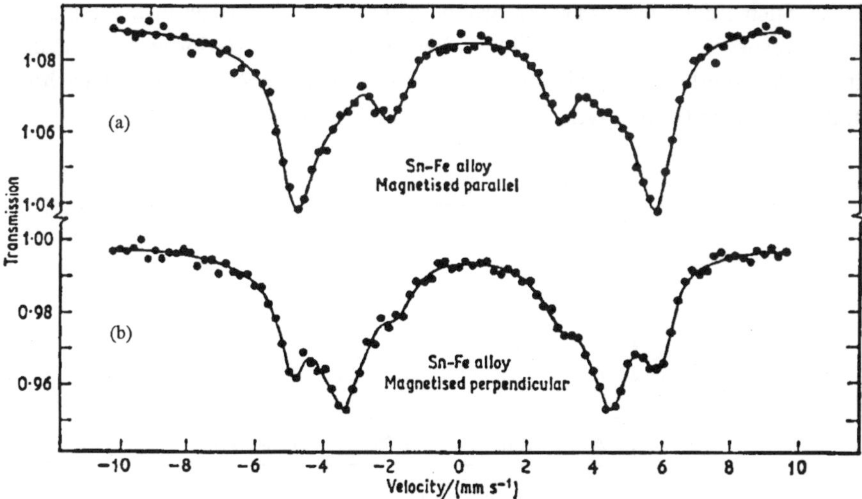

Fig. 7. ^{119}Sn magnetic hyperfine structure in α-Fe revealing the 6-line pattern changes in line intensities upon magnetizing the sample parallel and perpendicular to γ-ray wave vector direction can be seen. Figure is taken from Ref. [29].

23.8 keV gamma resonance into a 6-line multiplet structure [29] for the case of Sn as a dilute impurity in metallic Fe. The local magnetic field experienced by the dopant at room temperature is found [29] to be 78.5 KOe. The Fe-foil magnetization direction in relation to the γ-ray wave vector leads to changes in intensity ratios of the magnetic components. Specifically, one expects the intensity ratio of the 6-line multiplet to change from 3:0:1:1:0:3 to 3:4:1:1:4:3 as the foil magnetization is changed from a direction parallel to perpendicular to the foil plane. Magnetic fields of much larger value (> 200 kOe) at Sn are documented [30] in the Heusler alloy Pd$_2$MnSb, leading to a completely resolved 6-line magnetic hyperfine structure. In our discussion so far, the hyperfine interactions were assumed to be static on the scale of nuclear lifetimes. But this is not always the case as in magnetic oxides where spin relaxation effects can manifest and lead to unusually rich lineshapes from which relaxation times can be deduced as discussed in a review [28] by Wickman and Wertheim.

2.4. *Nuclear probes*

There are about 30 nuclei that possess a low energy (< 100 keV) nuclear transition to the ground state, that have been used to observe the Mössbauer effect. Such nuclei can serve as probes of condensed matter in a nuclear resonant absorption experiment. The relevant part of the periodic table dealing with chalcogenide (and chalcohalide) glasses is reproduced in Table 1 with the number in parenthesis giving the covalent radius of the elements.

Table 1.

IV	V	VI	VII
Si	P	S	Cl
(1.11 Å)	(1.06 Å)	(1.02 Å)	(0.99 Å)
Ge	As	Se	Br
(1.22 Å)	(1.20 Å)	(1.16 Å)	(1.14 Å)
Sn	Sb	Te	I
(1.41 Å)	(1.40 Å)	(1.36 Å)	(1.33 Å)

Binary alloys of group IV (Ge) and group V (As) elements with the chalcogens (group VI) over wide concentrations form the prototypical chalcogenide glasses. Such binary glasses when alloyed with a halogen (Group VII) form the prototypical chalcohalide glasses. Although neither the $3sp$, nor the $4sp$ elements offer the prospect of a suitable Mössbauer probe, the fact is that each of the $5sp$ elements in the 3rd row have suitable Mössbauer probe nuclei. These include ^{119}Sn, ^{121}Sb, ^{125}Te, ^{129}I in *absorption spectroscopy*. In *absorption spectroscopy* the glass of interest is doped with traces of the resonant isotope and used as an absorber in conjunction with a single line emitter. On the other hand, in *emission spectroscopy*, the Mössbauer parents ^{119m}Sn, ^{121m}Sn, ^{125m}Te, ^{129m}Te are alloyed in the glass of interest and a spectrum of the emitter is recorded using a single line absorber. In many instances emission spectroscopy offers unique opportunities to examine solids by the appropriate choice of Mössbauer parent nucleide incorporated in minute concentrations either by melt quenching or by ion-implantation using an isotope separator.

In principle, one can probe the cation (anion) chemistry of Si- or Ge-based chalcogenide glasses or crystals by alloying traces of the isovalent ^{119}Sn (^{125}Te) dopant. Since the covalent radii of the $5sp$ elements are in general larger than those of their $4sp$ and $3sp$ counterparts, the solubility and dopant site occupancies will in general be restricted. Our experience with the Ge-based chalcogenides has revealed that the solubility of Sn in the glasses is in general higher than in corresponding crystals. This is the expected result given that molar volumes of glasses in general exceed those of corresponding crystals by up to 10%. In practice, whether Sn dopant substitutes for a Ge cation in a chalcogenide glass or a system phase separates into a Sn-rich phase and a Ge-rich phase depends to a large extent on the melting point difference between the Ge- and Sn-chalcogenide compounds existing in the respective phase diagrams.

Isovalent atom substitution utilizing ^{129m}Te Mössbauer parent has proved to be a particularly powerful probe of chalcogen-site chemistry in glasses as we shall illustrate in the next section. Because of the 15 percent larger covalent radius of Te in relation to that of covalent Se (see Table 1), the oversized dopant does not *randomly* substitute for Se or S local environments, but instead selects those chalcogen environments associated with more free volume so as to permit the oversized probe atom to relax locally.

3. Oxide Glasses

Mössbauer spectroscopy offers the prospect only *indirectly* to examine network structures of prototypical binary glasses, such as B_2O_3, SiO_2, GeO_2 and P_2O_5 since none of the network forming cations in these glasses are suitable nuclear probes. Fortunately, other local structural methods such as NMR and NQR, have been particularly useful in this regard as discussed by P. Bray [in Chapter 5(C)] and H. Eckert [in Chapter 5(D)] of this volume. An exception to the above is the case of TeO_2 glass, where the network forming cation, Te, can be used to probe the glass molecular structure both in emission and absorption spectroscopy. Pure TeO_2 is a marginal bulk glass former. Addition of the usual network modifiers such as alkali oxide (Ak_2O), alkaline-earth oxide (AeO), or even the transition metal oxides (vanadates, tungstate, molybdates) readily yield bulk glasses. Glasses produced by alloying traces of Fe_2O_3 and SnO_2 in prototypical binary oxide glasses (B_2O_3, SiO_2, GeO_2, P_2O_5) with the usual modifiers have been examined extensively using the ^{57}Fe or ^{119}Sn resonance. References [31] and [32] review these studies and elucidate the role of the chemical state of the dopant (Fe^{2+} or Fe^{3+}, Sn^{2+} or Sn^{4+}) in glass processing. Such studies may be of relevance to controlling the physical behavior of glasses by alloying dopants. However, such measurements have provided little insight into the molecular structure of the network backbone, or even how the backbone is modified by the presence of the usual modifiers since these probe atoms do not form part of the backbone or the principal modifier regions. There is a general consensus that alkali-oxide or alkaline-earth oxide alloying in prototypical binary oxide glasses leads in most cases but not all, to depolymerization of the backbone as bridging oxygen atoms are replaced by non-bridging ones. The non-bridging oxygen sites serve to charge compensate for the mono- or divalent cation (Na^+, Ca^{2+}) additives in the network. Two of the glass systems, Fe-phosphate and alkali tellurate, have been the subject of more recent studies. In both these cases, the network forming cations (Fe in Fe-phosphate and Te in alkali-tellurates) are also suitable nuclear probes. It is therefore not surprising that new insights on glass backbone structure have evolved using Mössbauer spectroscopy.

3.1. *Alkali-tellurate glasses* $(Ak_2O)_x$ $(TeO_2)_{1-x}$; $Ak = Li$, Na, K

Although pristine TeO_2 melts crystallize upon a water quench, alloying 5 molar percent of Ak_2O in such melts promotes glass formation. Glass transition temperatures $T_g(x)$ in the titled glasses reveal a systematic reduction [33] with alkali oxide (Ak_2O) alloying (Fig. 8) which suggests a decrease in global connectivity of the backbone. The glass forming tendency of $(Na_2O)_x(TeO_2)_{1-x}$ melts displays pronounced maxima near $x = 0.15$ and $x = 0.27$ as revealed by the

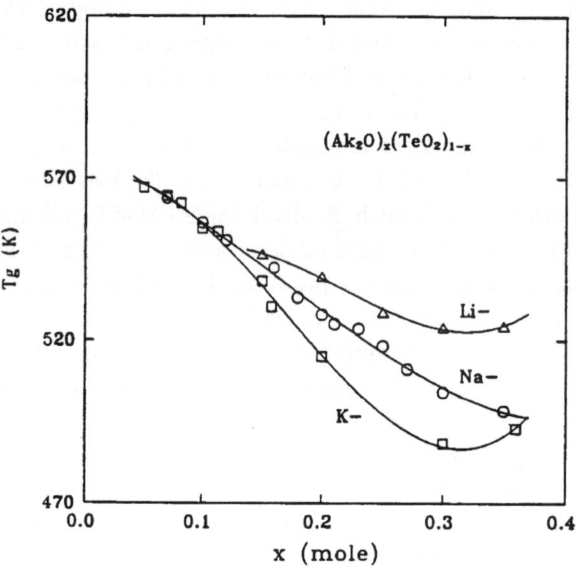

Fig. 8. Glass transition temperature $T_g(x)$ variation in alkali-tellurate glasses taken as a function of alkali-oxide content for indicated alkali. Note that in each case $T_g(x)$ decreases with x. Figure is taken from Ref. [34].

XRD traces of alloyed samples [33] with different Na_2O content that have received the same undercooling. The enhancement of glass forming tendency near $x = 0.18$ is an effect traced to the rigid to floppy transition and discussed [13] in Chapter 6(B). The enhancement of glass formation near $x = 0.27$ on the other hand is related to the presence of a eutectic.

^{125}Te absorption spectra of these glasses display a quadrupole doublet, with the magnitude of the average splitting $\langle \Delta \rangle$ increasing nearly linearly [34] with x as shown in Fig. 9. These results are understood in terms of growth of Te $(O_{1/2})_3 O^-$ local units at the expense of the more symmetric Te $(O_{1/2})_4$ units emerging in the network as non-bridging O^- sites are formed upon Na_2O alloying. The increased quadrupole splitting at Te $(O_{1/2})_3 O_{ax}^-$ and Te $(O_{1/2})_3$ O_{eq}^- units in relation to the trigonal bipyramidal Te $(O_{1/2})_4$ units is ascribed to the axial or equatorial non-bridging oxygen (O^-) back donating some of valence charge to the central Te cation. The compositional trends of $T_g(x)$ and those in ^{125}Te quadrupole splittings $\Delta(x)$ with increasing x are opposite, i.e. a T_g decrease is associated with a $\langle \Delta \rangle$ increase. We believe such a negative correlation should be observed generally for the network forming cation (Te), assuming that the role of the network modifier (Ak_2O) is to

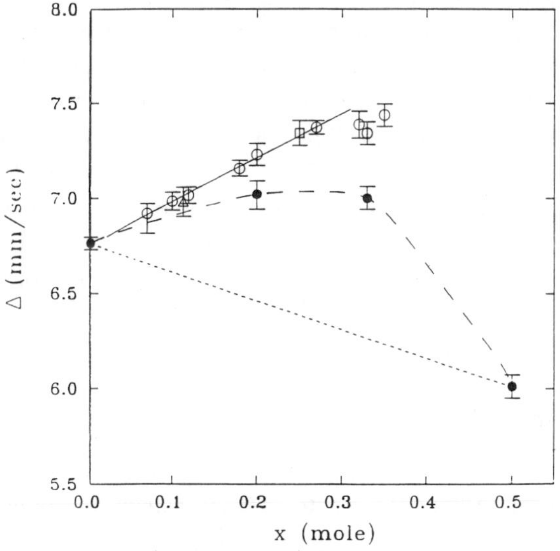

Fig. 9. ^{125}Te quadrupole splitting $\Delta(x)$ variation in $(Na_2O)_x(TeO_2)_{1-x}$ glasses revealing an increase with x. The filled circles represent results for crystalline compositions.

depolymerize the network backbone by forming non-bridging oxygen sites. The $T_g(x)$ reduction in the alloys is a direct manifestation of the reduced global connectivity [35] or mean-coordination number of the network backbone. The underlying physical idea has been made quantitative by development of a stochastic matrix model of agglomeration [35], a subject we return to later.

A T_g vs. ^{57}Fe quadrupole splitting $|\Delta|$ correlation has also been observed [36] in sodium tellurate, barium gallate, calcium aluminate, potassium silicate glasses as a function of alkaline-earth or alkali oxide content. For the Mössbauer measurements typically 5 to 10 molar percent of Fe_2O_3 was alloyed in each of the samples. Surprisingly, a positive correlation of T_g with $\langle\Delta\rangle$ (^{57}Fe) was noted, in other words, the magnitude of the Fe^{3+} quadrupole splitting decreases as the network is modified by alkali-oxide alloying. The linear correlation of T_g with $\Delta(^{57}$Fe) has been ascribed [36] to Fe dopant forming part of the network backbone to acquire a local configuration characteristic of the network forming cations like Te (tbp), Ga (tetra), Al (octa) and Si (tetra) in the indicated oxides. In our view, perhaps a more likely interpretation of these results on the alkali tellurate glasses [36] is that Fe_2O_3, is in a modified region of the network, but not in the backbone. As the mean coordination of the network decreases upon Ak_2O addition, the modified regions of the network become less distorted. The reduced distortion is detected by the tracer atoms and leads to a reduced mean ^{57}Fe quadrupole splitting, which thus correlates with T_g reduction (due to Ak_2O alloying). On chemical and thermal considerations, it is unlikely that Fe will substitute for Te in alkali-tellurate glasses to acquire a trigonal bipyramidal (tbp) coordination and serve as a network former. In sodium tellurate glasses, the T_g vs. $\Delta(^{125}$Te) negative correlation on the one hand, and the T_g vs. $\Delta(^{57}$Fe) positive correlation on the other reflects the fact that ^{125}Te nuclei probe the chemical behavior of the network backbone, while ^{57}Fe nuclei probe specific modified regions of the glass network. It would be desirable to measure EFGs in the gallates (using Ga NQR) as a function of alkaline-earth alloying to establish the behavior characteristic of the backbone.

3.2. *Fe phosphate glasses*

Pseudobinary $(Fe_2O_3)_x(P_2O_5)_{1-x}$ glasses have attracted interest as possible hosts for storage of radioactive waste [37] as well as for biological applications. Most phosphate glasses are hygroscopic, however pseudobinary glasses of Fe_2O_3 with P_2O_5 are known to be remarkably resilient, a behavior that is traced to the presence of the chemically more stable Fe–O–P signatures

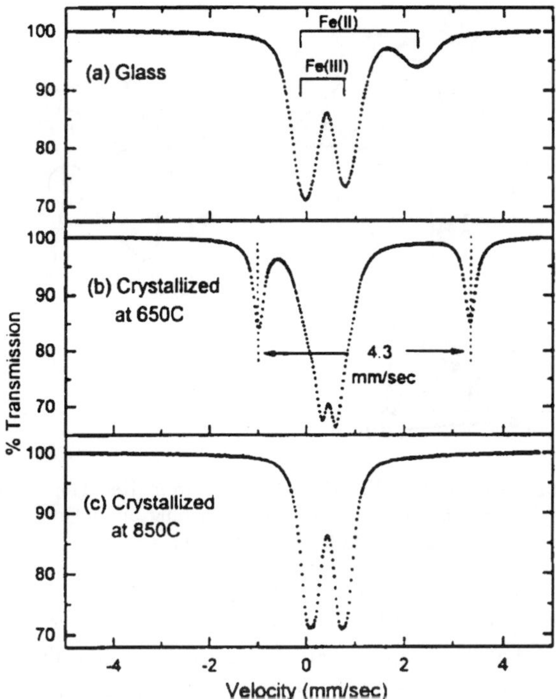

Fig. 10. ^{57}Fe spectra of a $(Fe_2O_3)_{0.4}(P_2O_5)_{0.6}$ bulk glass in (a) its pristine state, (b) upon thermal cycling through 650°C and (c) upon a second thermal cycling to 850°C. Figure is taken from Ref. [38].

replacing the P–O–P signatures in the pristine glasses. Binary glasses of the composition $x \sim 0.40$ can be synthesized by reacting respective oxides in air at 1200°C, and are found [38] to display a glass transition of about 500°C. Figure 10 displays room temperature spectra [38] of a glass in (a) its pristine state, (b) after heat treatment of the glass at 650°C (when $Fe_3(P_2O_7)_2$ is known to crystallize as revealed from XRD measurements) and (c) after heat treatment to 850°C (when the crystalline product is transformed to $Fe_4(P_2O_7)_3$). The observed lineshape in the pristine glass reveals the presence of both Fe^{3+} and Fe^{2+} cations with typically a quarter of the Fe present as Fe^{2+}. The crystalline structure of $Fe_3(P_2O_7)_2$ consists of Fe^{3+} cations present in an octahedral coordination and Fe^{2+} cations present in the body of trigonal prisms, with the latter prisms sandwiched between a pair of octahedral units as shown in Fig. 11.

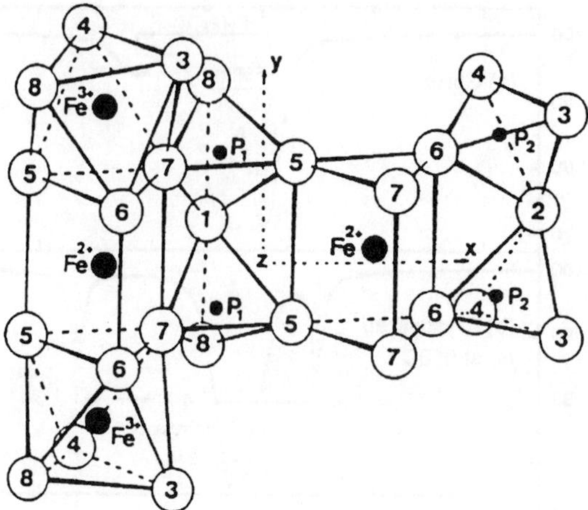

Fig. 11.   Crystalline structure of $Fe_3(P_2O_7)_2$ taken from Ref. [38] showing octahedrally coordinated Fe^{3+} cations and trigonal prismatic coordination of Fe^{2+} cations. The latter cation experience a large quadrupole splitting of 4.3 mm/s.

The strikingly large quadrupole splitting of 4.3 mm/s in c–$Fe_3(P_2O_7)_2$ (Fig. 10) is identified with Fe^{2+} cations present in trigonal prisms. The Mössbauer lineshapes clearly show that Fe^{2+} in trigonal prisms do not constitute a building block of the glass. The backbone structure in these Fe-phosphate glasses largely consists of Fe^{3+} and Fe^{2+} cations in octahedra and tetrahedra with diphosphate tetrahedra units intermixed. Recently, Raman scattering results on these glasses have been reported [39] and the richness of the observed lineshapes which change with glass composition and upon crystallization, has provided useful information on the structure of the network backbone. The presence of medium range structure describing connectivity of the building blocks remains a subject of continued discussion, however. Nevertheless, important clues in understanding the local structure of Fe-phosphate glasses have resulted from these Mössbauer and Raman scattering results, which suggest that c–$Fe_3(P_2O_7)_2$ forms a good starting point to model the glass. It is possible that progress in decoding the medium-range structure of these phosphate glasses can emerge from compositional studies using these local probes. These spectroscopic results underscore the fact that the *structure of Fe-phosphate glasses* is not really *random* and derives from or bears a connection to that of a *nearby crystalline compound*. This theme will recur when we examine the spectroscopic landscape on the chalcogenide glasses.

4. Chalcogenide Glasses

The structures of chalcogenide glasses have been examined by diffraction methods, resonance methods (NQR, NMR) and vibrational spectroscopy as discussed in Chapters 5 and 6 of this volume. In this section, we will peruse the Mössbauer spectroscopy results globally (over wide compositions) and correlate these results with Raman scattering and Differential Scanning Calorimetry. When the mean coordination number, $\bar{r}$, of these glasses exceeds about 2.4, chalcogenide glasses appear generally to separate into distinct nanophases, and the observed multi-modal Mössbauer site distribution a quantitative measure of the different nanophases populated, as we shall illustrate in the next few sections.

4.1. *Binary glasses* ($Ge_x S_{1-x}$; $Ge_x Se_{1-x}$)

Stoichiometric compositions ($x = 0, 1/3$) in the titled binaries have attracted special attention in the past because crystalline compounds are known to occur in the equilibrium phase diagram at these compositions. For this reason, one can directly compare the molecular structure of the bulk glasses with their crystalline counterparts. However, an interpretation of the spectroscopic differences between a stoichiometric glass and its crystalline counterpart experience has shown not to be all that straightforward. Eventually, glass structural models at stoichiometric compositions must smoothly connect with models of these glasses at non-stoichiometric compositions. In this respect, Mössbauer spectroscopy has been remarkably quantitative in decoding the molecular structure of these chalcogenide glasses. Figure 12 provides a summary of the Ge cation site-intensity distribution inferred [40] using [119]Sn absorption spectroscopy. Specifically, in these compositional studies, a minuscule amount of [119]Sn dopant is introduced in $(Ge_{0.99}Sn_{0.01})_x Se_{1-x}$ and $(Ge_{0.99}Sn_{0.01})_x S_{1-x}$ melts, which are quenched in water to yield bulk glasses and used as absorbers in Mössbauer absorption experiments.

Before addressing glass structure issues, it may be pertinent to pose the following question. Does a Sn guest atom alloyed in a Ge–Se melt or a Ge–S melt substitute for Ge local environments in these glasses in a random fashion, or does the guest create its own Sn-like local environment in these glasses? Figure 13 illustrates an example of a study [43] that addresses some of the issues. The observed [119]Sn Mössbauer lineshapes remain unchanged upon cycling a bulk GeS_2 glass doped with traces of Sn through T_g, but the lineshape changes qualitatively upon cycling through T_x (crystallization temperature) as

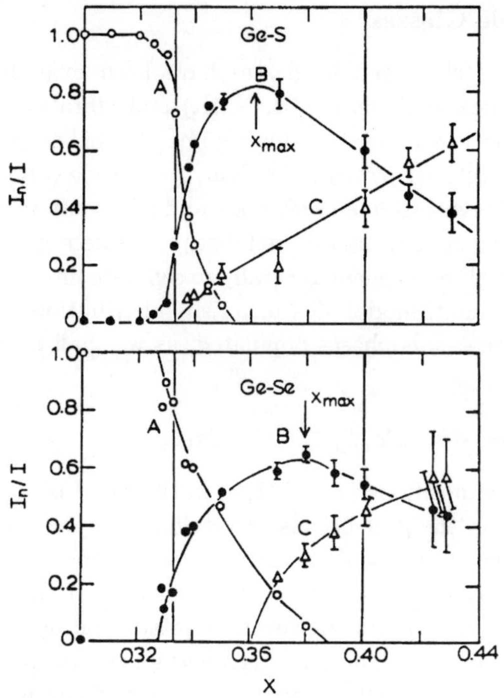

Fig. 12.   Molecular nanophases A, B and C observed in Ge_xS_{1-x} (top panel) and in Ge_xSe_{1-x} (bottom panel) bulk glasses. Note that the intruding B-phase consisting of ethane-like Ge_2S_3 or Ge_2Se_3 is largely responsible for glass formation at $x > 1/3$ in both glass systems. See text for details.

one would expect. In the latter instance, half of the dopant continues to be substitutionally incorporated in the layer form $(\alpha-)$ of GeS_2 while the rest of it precipitates as a Sn-rich phase of c–SnS, suggesting that the solubility of Sn in α–GeS_2 is about 1/2 atomic percent. One would expect the solubility of Sn dopant in the glasses to exceed that in the crystalline phase, largely because of the more open and flexible structure of the glass. Parallel results are documented [16] for the Ge_xSe_{1-x} binary.

There are plausible reasons to suggest that the trimodal Sn(Ge) cation site distribution (A, B and C) represents Sn randomly replacing Ge in *three distinct nanophases* [40, 44] present in the glasses. The case for random substitution of Sn dopant for Ge cation environments in the nanophases derives from quantitative correlation of Mössbauer spectroscopy with Raman

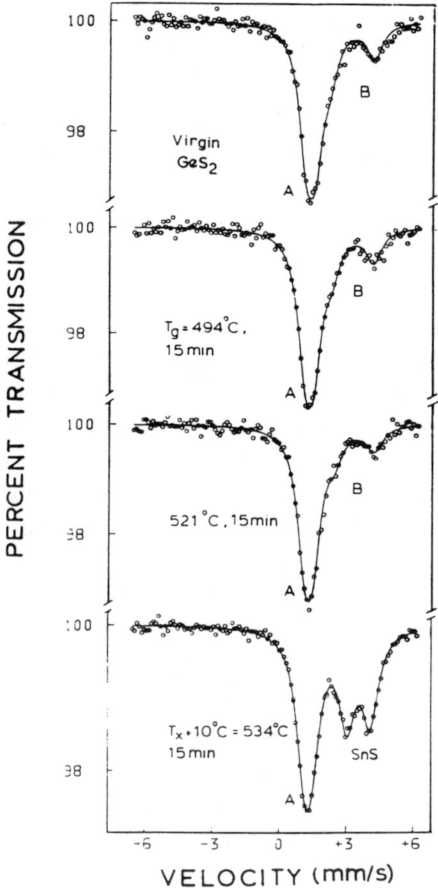

Fig. 13. ^{119}Sn spectra of $Ge_{0.99}Sn_{0.01}S_2$ bulk glass in its virgin state obtained by a water quench of a melt (topmost), after thermal cycling through T_g for 15 min., after thermal cycling just below T_x, and after thermal cycling above T_x, the crystallization temperature respectively. Note that thermal cycling through T_g leaves the lineshape unchanged while a thermal cycle through T_x precipitates c–SnS and leaves some of the Sn substitutional in α-GeS$_2$. Figure is taken from Ref. [43].

scattering results where possible. At low x, in analogy to a Ge atom, a Sn atom also cross-links chains of S or Se to acquire a tetrahedral (A) local configuration. Such a local environment is consistent with the observed value of the isomer shift, and the absence of a quadrupole interaction for site A. With increasing x and in the $0.20 < x < 0.30$ composition range, as suggested by Raman scattering results [43], site A represents the signature of a characteristic cluster consisting of edge-sharing fragments cross-linked by corner-sharing $Ge(Se_{1/2})_4$ units, in which Sn replaces Ge at the *more flexible corner-sharing units*. As x increases to $1/3$, the appearance of a quadrupole doublet (site B) in the line-shape represents the signature of a non-tetrahedral species that is thought to emerge from a second phase. In Ge_xSe_{1-x} glasses, the *concommitant* growth of site B integrated intensity with the 180 cm^{-1} Raman mode scattering strength (Fig. 14) at $x \geq 0.31$ strongly suggests that the underlying molecular phase consists of face-sharing quasi one-dimensional ethane-like $Ge_2(Se_{1/2})_6$ chain-fragments. The concentration of such fragments increases monotonically in the composition range $0.31 \leq x \leq 0.37$ for the case of the Ge_xSe_{1-x} binary [Fig. 14(b)]. At higher Ge concentrations, a third molecular phase (site C) appears in the network in which Sn is present as Sn^{2+} experiencing a sizeable EFG. The C-phase is based on the distorted rocksalt structure of c–GeSe, in which Ge cation acquires three short and three more distant Se near-neighbors [43]. Such a local environment is characteristic of an ionically stabilized structure.

These Mössbauer spectroscopy results are more quantitative than Raman scattering ones because the nuclear resonant cross-sections are independent of chemical coordination while the Raman scattering cross-sections can vary by factors of two or more from mode to mode. Furthermore, lineshape deconvolution and phase identification in Mössbauer spectroscopy is relatively straightforward since it involves deconvoluting the observed lineshape in terms of either a doublet or a singlet with a Lorentzian profile having a FWHM that is close to the natural linewidth. On the other hand, in Raman scattering [45] a specific molecular phase may yield *several vibrational modes* of varying scattering strengths with linewidths that vary substantially from mode to mode. In some cases mode identification becomes rather challenging when vibrational mode frequencies of two molecular species overlap, as in the case of the A- and B-molecular phases in the Ge_xS_{1-x} binary. Specifically, the 340 cm^{-1} mode in these glasses is generally identified [45] with the symmetric stretch of a $Ge(S_{1/2})_4$ unit in the A-phase, an identification that certainly appears plausible at $x \leq 0.35$. However, this particular mode not only persists in Ge_xS_{1-x} glasses at $x \geq 0.38$, but actually becomes the most important feature in the Raman spectra of Ge-rich glasses, while in corresponding Mössbauer spectroscopy results, there is no evidence of the A-phase (in which Sn is tetrahedral). Indeed, the inability to distinguish between the A- and B-molecular phases in Ge-S glasses by Raman scattering has lead early workers

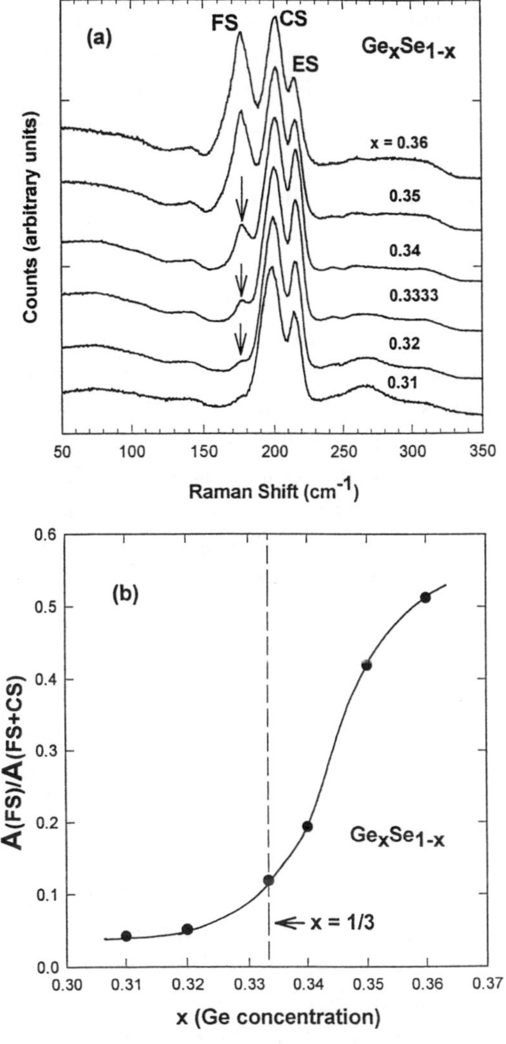

Fig. 14. Raman scattering in bulk Ge_xSe_{1-x} glasses at indicated Ge concentrations x, showing evolution of the ethane-like B-molecular phase (top panel). The bottom panel shows a plot of the scattering strength ratio A_{fc}/A_{cs} of the 180 cm^{-1} ethane-like mode to the 202 cm^{-1} corner-sharing mode with x, revealing a value of 0.12 at $x = 1/3$ ($GeSe_2$). The scattering was excited by 647 nm line using a Kr-ion laser. Perusal of the observed lineshapes also reveals that as $x \rightarrow 0.37$, the scattering strength ratio of the ES/CS modes decreases from its value at $x = 1/3$. This suggest that the ethane-like units must also contribute a mode at the CS mode of about 200 cm^{-1}.

[45] to suggest that GeS_2 glass is largely chemically ordered in contradiction to the Mössbauer results [43]. In spite of these shortcomings there is little doubt that Raman scattering in glasses has continued to provide valuable insights on subtle aspects of glass structure through the vibrational density of states [45].

The molecular phase diagram of Fig. 12 suggests presence of only a *partial chemical order* in the glasses at the stoichiometric compositions $x = 1/3$ and 2/5. In the equilibrium phase diagram [46] although there exist two polymorphs [47] (α- and β-) of $GeSe_2$ ($x = 1/3$) consisting of tetrahedral building blocks, a crystalline compound of Ge_2Se_3 ($x = 2/5$) stoichiometry consisting of face-sharing tetrahedra is not known to exist. The B-nanophase is thus *peculiar* to the glasses in both these binaries and its presence was first identified by chemical dissolution and thermal studies [48]. A central theme to emerge from the results of Fig. 12 is that for both the Ge_xSe_{1-x} and Ge_xS_{1-x} binaries, although two (A- and C-) of the nanophases present have crystalline counterparts, the B-nanophase is unique to the glasses and melts and is not stable as a crystalline compound. Apparently, the B-phase is populated in such melts and glasses because it represents a low-pressure molecular conformation. Indeed, upon crystallization, the B-phase disproportionates into the more compact A- and C-phases. Thus, the extensive bulk glass formation witnessed in these binaries at $\bar{r} > 2.67$ is, in fact, largely due to the presence of the marginally rigid B-molecular phase. The bond-bending constraints about the Se atoms would necessarily be broken if the face-sharing chains were not to be linear, as one would expect it to be the case in glasses. Constraint counting procedures under these circumstances show [49] that the mechanically effective coordination of the B-nanophase equals 2.56, only marginally larger than the optimum value of 2.40. The loss of network connectivity as the B-nanophase emerges at $x > 1/3$ is primarily responsible for the reduction of T_g in the Ge-rich glasses.

Support for these molecular phase diagrams of Ge_xS_{1-x} and Ge_xSe_{1-x} glasses has recently emerged from unusual quarters. There are two sets of unrelated doping experiments that have been reported recently which provide quantitative support for the proposed phase separation model of the Ge-rich glasses. Rare-earth dopants in chalcogenides have attracted interest for laser applications [50]. In particular, the solubility of Pr as a dopant in Ge_xS_{1-x} glasses has been established by a rather precise optical method [51]. The optical method consists of measuring the integrated absorption coefficient from the ground state to the $^3F_4$, $^3F_3$, $^3F_2$ and $^3H_6$ levels of Pr^{3+} and studying its variation with Pr doping concentration. Departure of the absorption coefficient from linearity with doping concentration sets in due to Stark splitting of the levels when dopant clustering occurs, and thus provides a precise means to optically establish solubility of the rare-earth guest atom in the chalcogenide glass host. These experiments reveal [51] that the Pr^{3+} solubility in $Ge_xS_{1-x-y}I_y$ ($y \simeq 0.10$) glasses increases almost linearly with x at $x > 1/3$ as shown in Fig. 15. This observation has a fairly straightforward interpretation. PrS crystallizes [52] in a rocksalt structure, and for that reason, Pr can be expected to replace Ge in the C-molecular phase and to acquire a distorted rocksalt configuration compatible with its local bonding requirement. It is for this reason that Pr^{3+} solubility in Ge_xS_{1-x} glasses *quantitatively* tracks (Fig. 15) the concentration of the distorted rocksalt C-molecular phase established earlier by Mössbauer spectroscopy results (Fig. 12). Perhaps what is most remarkable in this correlation is the fact that two seemingly independent and rather precisely measured observables on the same glass system can be traced to a common molecular origin.

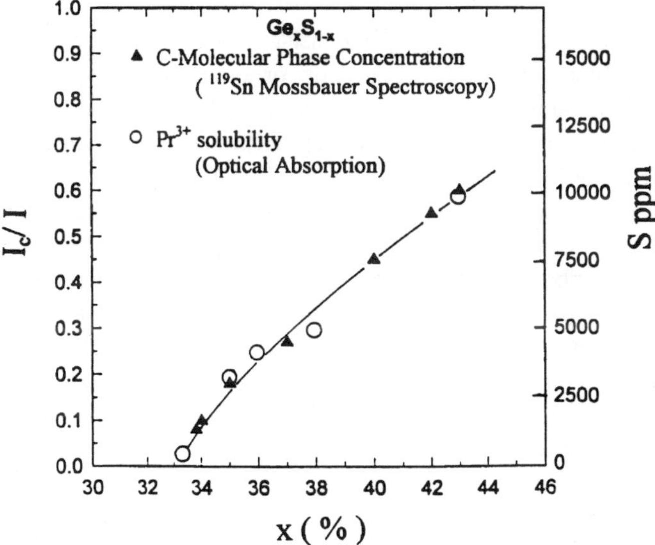

Fig. 15. $I_C/I(x)$, concentration variation of C-molecular phase (▲) in Ge_xS_{1-x} glasses taken from Fig. 12, and solubility of Pr^{3+} in $Ge_xS_{1-x-y}I_y$ ($y = 0.08$) glasses (O) reported in Ref. [51]. The continuous line through the data is a guide for the eye. The scaling underscores that the rare-earth dopant acquires a distorted rocksalt configuration by substituting in the C-molecular phase.

The second example comes from glass formation in the Ag–Ge–(S or Se) ternaries that reveal [53] two distinct compositional regions (region I and II). Region I comprises chalcogen-rich $[Ge_y (S\ or\ Se)_{1-y}]_{1-x} Ag_x$ glasses corresponding to $y < 1/3$, in which Ag bonds with the excess chalcogen to phase separate into a glassy $Ag_{2+\delta}Se$ phase and a chalcogen deficient $Ge_t (S\ or\ Se)_{1-t}$ backbone with $t > y$. Bimodal $T_g s$ are observed in region I corresponding to these two phases. Region II comprises Ge-rich glasses, i.e. $y > 1/3$ in which unimodal $T_g s$ are observed and Ag is known to acquire a 3-fold coordination. These two regions are separated by a narrow corridor along the $[Ge_{0.33} (S\ or\ Se)_{0.66}]_{1-x} Ag_x$ tie-line, with no bulk glass formation taking place at low x ($x < 0.15$). Propensity for glass formation in region II, at least at low x, can be traced to Ag replacing Ge in the C-molecular phase and to acquire a distorted rocksalt configuration which satisfies the 3-fold bonding requirement of the Ag dopant. Thus the glass forming tendency in Ge–(S or Se)–Ag ternaries in region II can be microscopically traced to the existence of the C-molecular phase at Ge concentration greater than 1/3.

4.1.1. *Stoichiometric GeS$_2$ and GeSe$_2$ glasses*

As noted before, these glasses largely consist of tetrahedral $Ge(X_{1/2})_4$, $X = S$ or Se building blocks in which the ratio of corner-sharing (CS) to edge-sharing (ES) tetrahedral units is close to two, as found in 2d form (high-T form) of corresponding crystalline phases [47]. Diffraction experiments [41] and Raman scattering [54–57] experiments broadly support this conclusion. [119]Sn Mössbauer spectroscopy results on these glasses reveal that there exists a small but reproducible concentration of Ge(Sn) non-tetrahedral sites. In addition, [119]Sn Lamb–Mössbauer factors reveal that the non-tetrahedral sites (B) and tetrahedral sites (A) possess significantly different *vibrational temperatures*, suggesting that B-sites are *not part of the tetrahedral network*, but are phase separated from it on a molecular scale [see Chapter 6(B)]. Phase separation in $Ge_x Se_{1-x}$ glassy melts apparently initiates at $x \geq 0.30$ and leads to a degree of broken chemical order (I_B/I) of 14(2)% in GeSe$_2$ and of 29(2)% in GeS$_2$ bulk glasses. Raman scattering in the $Ge_x Se_{1-x}$ binary glass system near the concentration $x = 1/3$ are displayed in Fig. 14(a). Noteworthy in the spectra is the growth in scattering strength of the face-sharing (FS) ethane-like $Ge_2(Se_{1/2})_6$ mode at 180 cm^{-1}. The Raman scattering strength for FS mode at 180 cm^{-1} normalized to the scattering strength of the CS A_1 mode at 202 cm^{-1} is 0.12(1) in GeSe$_2$ glass [Fig. 14(b)]. Since there are two face-sharing tetrahedra in each ethane-like unit, one must double the observed ratio of 0.12(1) to obtain the ratio of FS to CS tetrahedra in the network. Further, if allowance is made for the Raman scattering cross-section [58] of the FS mode to be 35% larger than the CS mode, one obtains a value of 0.15(1) for the concentration ratio of FS/(FS + CS). This should be compared to the Mössbauer site intensity ratio $I_B/I = 0.14(2)$. Thus there is *remarkable agreement* about the broken chemical order in $GeSe_2$ glass from Raman and Mössbauer spectroscopy. (In these calculations we have assumed that the oversized Sn replaces Ge in FS and CS units but not in ES tetrahedra.) From the above phase separation model we can estimate the concentration of homopolar bonds in $GeSe_2$ to be [59] about 1.78%. This number is just at the limit of detection for diffraction measurements [41]. These numbers unequivocally demonstrate the sensitivity of Raman and Mössbauer spectroscopy as a probe of network chemical and structural order in that these methods probe respectively modes and local fields of network building blocks, i.e. a group of correlated atoms rather than counting network bonds at random.

On general grounds of stoichiometry, the presence of a finite concentration of Ge–Ge bonds in a GeS_2 (or $GeSe_2$) glass [40, 41, 44, 60, 61] implies a finite concentration of chalcogen-chalcogen (S–S or Se–Se) bonds. The chemistry of chalcogens in these glasses has been examined in detail [8, 60, 61] by doping ^{129m}Te tracer in the melts and obtaining ^{129}I Mössbauer emission spectra of the bulk glasses. These Mössbauer experiments on stoichiometric glasses are *unusually sensitive* to the small fraction of chalcogen-chalcogen bonds as is discussed next.

The principal results are displayed in Fig. 16, which compares ^{129}I spectra of the elemental chalcogens with those of the stoichiometric glasses, $GeSe_2$ and GeS_2. As noted earlier in Sec. 2.2, the $5/2 \to 7/2$ nuclear spin sequence leads to 12 quadrupolar components, and the observed lineshapes can be least squares fit to deduce the quadrupole coupling (e^2QV_{zz}), η and δ and thus to characterize the local environment of the ^{129}I daughter atom formed in these glasses and crystalline hosts when the ^{129m}Te parent β-decays. Since the nuclear lifetime (20 nsec) of the 27.8 keV Mössbauer level in ^{129}I is orders of magnitude larger than the typical chemical bond rearrangement time (a few psec), the ^{129}I daughter experiences a well defined static EFG. In general, when the parent Te atom is 2-fold coordinated, the I-daughter is found to be σ-bonded to a nearest neighbor atom, thus possessing a 1-fold coordination as, for example, in solid or molecular I_2. In the lighter elemental chalcogens (S, Se), the ^{129}I spectra reveal formation of a relaxed I–S or I–Se σ bond with a nearly cylindrically symmetric ($\eta = 0$) positive electric field

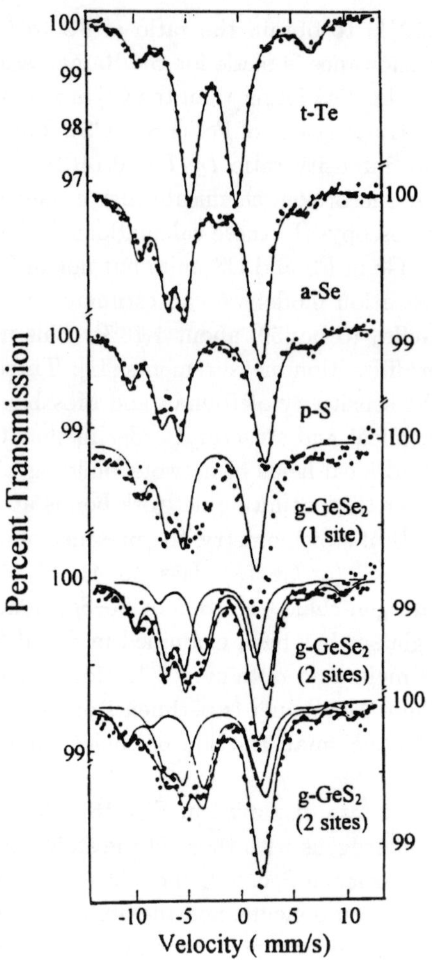

Fig. 16. ^{129}I emission spectra of indicated elemental chalcogens and binary glasses. Note that a 2-site deconvolution of the lineshape in the binary glasses reveals one site (A) to be common to both GeSe₂ and GeS₂ glass and the second site (B) in the glass to be common to respective elemental chalcogen. Figure is taken from Ref. [60].

gradient (electronic charge distribution). And since the nuclear quadrupole moment is negative, the signature of 1-fold coordination leads to a negative quadrupole coupling $(-e^2qQ)$. In c–Te, the results reveal that although I continues to be σ-bonded to a Te near-neighbor in the trigonal phase, interactions of the I-probe also occurs with a second Te near-neighbor in the Te helical chain. The value of $\eta = 0.76$ constitutes direct evidence for the presence of these π interactions.

In the stoichiometric glasses (GeS_2 and $GeSe_2$), the central result to emerge from the observed lineshapes [60, 61] is that two chemically inequivalent ^{129}I sites are populated. One of the sites, labelled A, is common to both these glasses in that the quadrupole coupling parameters (η, δ, e^2qQ) are strikingly similar to each other. This particular ^{129}I local environment (site A) is thus identified with an I–Ge σ-bond, involving the cation common to both glasses. The second site, labelled B, is found to have quadrupole coupling parameters that closely match those of respective elemental chalcogens. In other words, the B-site parameters in $GeSe_2$ (GeS_2) glass are similar to the ^{129}I site found in elemental Se(S). The interpretation of these results is that the A-site in the glasses results from a Te parent replacing Se or S at the bridging location between two $Ge(S_{1/2}$ or $Se_{1/2})_4$ tetrahedra, i.e. at the *chemically ordered site* expected in a network composed of tetrahedral building blocks. The B-site, on the other hand, is the signature of $\equiv$ Ge–Se–Se– or $\equiv$ Ge–S–S–linkages present in these glasses. The I-site in question (B) is formed when the Te parent replaces Se or S in the homopolar bonds (Se–Se, S–S) to yield a one-fold coordinated ^{129}I daughter bonded as in the elemental chalcogens. In these experiments it would be well to recall that the Te $\rightarrow$ I transformation occurs in the solid glass cooled to 4.2 K. The repulsive lone pair interactions between I and Se (or S) completely overwhelm the charge transfer interactions between I and Ge, and result in I bonding with the Se (or S) rather than the Ge [8, 60] nearest-neighbor. But perhaps the most remarkable discovery of these experiments is that the relative intensities $I_B \gg I_A$, showing that the Te dopant selects [8] the homopolar bonds over the heteropolar ones by a factor of 20 or more. The most natural interpretation of these results is that the Se–Se bonds *dress surfaces or edges* of characteristic molecular clusters and that the over-sized dopant (Te) *segregates to these cluster edges*. Such substitution minimizes strain by permitting Te to bond with its two near-neighbors and relax in the intercluster region where van der Waals interaction rather than the stronger covalent interactions occur. Indeed, as shown elsewhere [8], when allowance is made for the high selectivity of Te to replace Se or S, the degree of broken chemical order deduced from the ^{129}I measurement is in *quantitative* accord with the concentration of Ge–Ge bonds deduced from the ^{119}Sn Mössbauer experiments.

Given the wealth of Raman and Mössbauer spectroscopic results on the compositional trends of Ge–Ge signatures in binary Ge_xSe_{1-x} glasses, it would be inappropriate to view in isolation the results at $x = 1/3$ and to suggest that these represent defects in an otherwise ordered bond network [42]. B-sites which appear abruptly at $x > 0.30$, and grow in concentration (Figs. 12 and 14) with x to peak near $x = 0.37$, are manifestation of a metastable nanophase formed in melts and glasses.

An objection to a cluster model of these stoichiometric glasses ($GeSe_2$, GeS_2) is that simulations of the structure factor for such a model yield sharp features in the interference function which are not to be found in the measured diffraction results of the glasses. This is valuable information, and suggests that $GeSe_2$ glass cannot be viewed as consisting of van der Waals coupled characteristic Se-rich clusters as suspected, for example, in P_2Se glass [62] composed of P_4Se_3 monomers. On the other hand, diffraction measurements cannot exclude *spatially inhomogeneous* structures consisting of Se-rich and Ge-rich regions where the concentration of homopolar bonds is only about 2%. The presence of internal surfaces leading to a partial instead of complete polymerization of the network appears to be an essential ingredient of $GeSe_2$ and GeS_2 glass structure. In fact, the observed reversible photo-microcrystallization effects [63] on bulk $GeSe_2$ glass (when irradiated by a photon pump in the Tauc edge are persuasive in demonstrating that the pristine bulk glass network reconstructs when light absorption occurs in those parts of the network where homopolar bonds are localized and these are converted to heteropolar ones. The underlying reconstruction is an athermal process akin to photoinduced fluidity demonstrated in recent years for As_2S_3 fibers [64] where viscous flow of the material occurs well below the actual glass transition temperatures.

As_xS_{1-x} and As_xSe_{1-x} binary glasses have been examined by [129]I emission spectroscopy [22, 65] and the results also reveal a bimodal distribution of chalcogen sites even at the stoichiometric composition ($x = 2/5$) suggesting that some degree of phase separation is also present in these binary glasses.

A final comment on the molecular structure of Si-chalcogenides glasses. Si_xTe_{1-x} glasses have been examined in [125]Te Mössbauer spectroscopy and show evidence of a change in glass structure [66] with x near $x = 0.20$. The molecular structure of Si_xSe_{1-x} glasses has been extensively examined by Raman scattering [56, 67, 68] and also produce unambiguous evidence of a morphological change in structure from a floppy random structure at $x \leq 0.20$ to a rigid heterogeneous structure consisting of several molecular phases at $x \geq 0.27$. In sharp contrast to $GeSe_2$, $SiSe_2$ glass appears to be chemically ordered; and in the Si-rich phase (Si_xSe_{1-x}, $x > 1/3$) signature of face-sharing tetrahedra appear [69]. In $Ge_x Se_{1-x}$ glasses $T_g(x)$ shows a maximum near $x = 0.34$. On the other hand, in $Si_x Se_{1-x}$ glasses no such maximum is observed [68], and $T_g(x)$ continue to show an increase at $x > 1/3$, as do melt viscosities. Such a result constitutes evidence for the face-sharing tetrahedra (B-nanophase) forming part of the $Si_x Se_{1-x}$ network backbone (A-nanophase) in contrast to the situation in $Ge_x Se_{1-x}$ glasses where the face-sharing tetrahedra form a separate nanophase from the backbone (A-nanophase).

4.1.2. $Ge_x Te_{1-x}$ glasses

Interest in these glasses developed in the mid 1960's when Stan Ovshinsky and collaborators at ECD used these materials to demonstrate threshold and memory switching [70]. Such glasses with As and Si additives have become the phase change materials in rewritable compact disk. The subject is reviewed in Chapter 12(A) of this volume. In the past five years several new chalcogenide glasses have been discovered [70] that display threshold and memory switching, transitions that appears to be correlated with network connectivity. This is an area that will undoubtedly receive more attention in the future. Bulk glass formation by water quenching melts occurs over a rather narrow compositional window $0.14 < x < 0.25$ in the $Ge_x Te_{1-x}$ binary. However, J.P. deNeufville and collaborators [71] were able to prepare amorphous thin-films of $Ge_x Te_{1-x}$ over wide compositions $0.07 < x < 1.0$ by vapor deposition onto a variety of substrates such as glass slides and Al-foils using RF diode sputtering. The molecular structure of such amorphous thin-films and bulk glasses has now been examined in Raman scattering and in ^{125}Te and ^{119}Sn Mössbauer spectroscopy measurements [71].

In contrast with the rich molecular structure details on $Ge_x Se_{1-x}$ and $Ge_x S_{1-x}$ bulk glasses shown in the phase diagram of Fig. 12, one may ask why do bulk $Ge_x Te_{1-x}$ glasses only form over a narrow compositional window? There are several factors that apparently enhance glass formation in this binary while several others suppress it. An examination of the equilibrium phase diagram of the $Ge_x Te_{1-x}$ binary reveals that stoichiometric crystalline compounds occur at $x = 0$ (c–Te) and at $x = 1/2$ (c–GeTe). Thus, neither the A-molecular phase corresponding to $x = 1/3$, GeTe$_2$, nor the B-molecular phase corresponding to $x = 2/5$, Ge$_2$Te$_3$ are known to form stable crystalline phases. The C-molecular phase corresponding to $x = 1/2$ GeTe does form, but it is rocksalt-like, with a cubic structure in which the mean coordination is 6. Thus the network is highly overconstrained. The possibility of distorting the rocksalt phase and lowering the mean coordination to 3, as is apparently the case for C-phase in Ge_xSe_{1-x} and Ge_xS_{1-x} glasses, is therefore not feasible in the case of Ge_xTe_{1-x} glasses. Finally, liquid Te is metallic and 3-fold coordinated in contrast to liquid S or Se that are semiconducting and 2-fold coordinated. Bulk glassy Te therefore cannot be synthesized by supercooling melts. A striking parallel exists between elemental Si and elemental Te in this regard in that neither element forms bulk glasses upon melt-quenching, although amorphous thin-films can be readily prepared by vapor deposition. Given these factors which suppress glass formation, it is remarkable indeed that bulk glasses in the Ge_xTe_{1-x} binary can be formed at all in the narrow compositional window around $x = 0.20$. There are two factors which actually promote bulk glass formation in this binary. First, the existence of a eutectic at $x = 0.16$ leads to a lowering of melt temperatures and an underlying incipient molecular phase separation. But the second factor, and perhaps more significant one contributing to glass formation, is the *rigidity percolation threshold* at $x = 0.20$ for which direct evidence [72] has now been obtained from ^{125}Te Lamb–Mössbauer factors, and is discussed in Chapter 6(B). Raman and ^{119}Sn Mössbauer spectroscopy on bulk $Ge_{20}Te_{80}$ glasses provide evidence for both face-sharing $Ge_2(Te_{1/2})_6$ and ES tetrahedra which serve to cross-link Te_n-chain segments in the bulk glasses. ^{119}Sn Mössbauer spectra reveal not only a narrow single line from tetrahedrally coordinated Sn (in ES units), but also a quadrupole doublet feature, presumably from Sn present in FS units with a non-tetrahedral coordination. The mean-field count of valence force field atomic constraints for such a network, in which Ge is 4-fold and Te is 2-fold coordinated, yields a value of 3 at $x = 0.20$. The *onset of rigidity* near $x = 0.20$ is perhaps the most important factor promoting *bulk glass formation* in this poorly ordered binary glass system. Ideas on onset of rigidity in random networks are developed in Chapter 6(B).

4.2. *Ternary glasses*

The popularity of structure studies on binary glasses, such as AB_2 or A_2B_3 glasses which consist of tetrahedral $A(B_{1/2})_4$ or pyramidal $A(B_{1/2})_3$ building blocks, derives in part from the commercial interest in these materials (SiO_2, B_2O_3, As_2S_3, $GeSe_2$), and also in part from the advent of computer simulations of random networks of these materials. Numerical simulations [73–75] has gained wide acceptance as a means to gain insight into the physical behavior of complex systems, even though such an approach currently has its own limitations [75].

In ternary glass systems, random network simulations are less straightforward and for that reason such systems are less studied. This is a rather unfortunate circumstance because, as we shall illustrate next, the most direct demonstration of the molecular origin of glass forming tendency in the chalcogenides emerges from thermal analysis and Mössbauer spectroscopic results on ternary glass systems.

4.2.1. $(Ge_2Z_3)_{1-x}(Sn_2Z_3)_x$; $Z = S$ or Se

We noted in Sec. 2.2 that Sn_2S_3 crystallizes into long, thin and narrow molecular ribbons [Fig. 4(c)]. The ribbons provide for two types of cations, Sn^{4+}:Sn^{2+} that occur with equal frequency 1:1. The molecular ribbons have two interior Sn^{4+} sites that are octahedrally coordinated, and two edge Sn^{2+} sites that are 3-fold coordinated to S near-neighbors in a ribbon. Although binary Sn_2S_3 melts crystallize upon a water quench, remarkably quenching ternary $(Ge_2S_3)_{1-x}(Sn_2S_3)_x$ alloy melts readily yield bulk glasses over wide molar concentrations, $0 < x < 0.62$ of Sn_2S_3. In the equilibrium phase diagram, there also exists a molecular conformation of zig-zag chain structure at $x = 1/2$ or

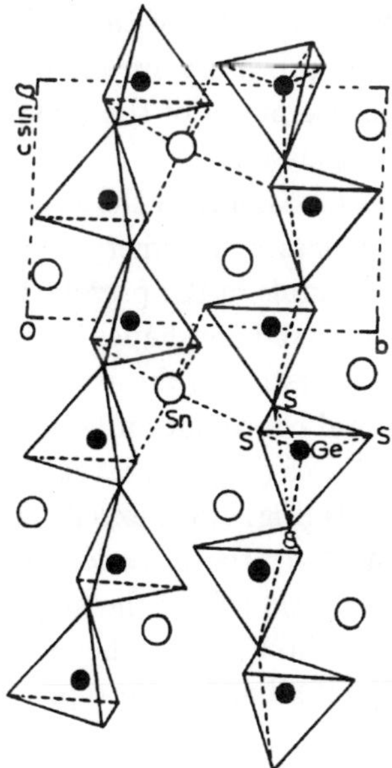

Fig. 17. Zigzag polymeric chain structure of c–GeSnS₃.

GeSnS$_3$. This phase consists of corner-sharing Ge(S$_{1/2}$)$_4$ tetrahedral units that are laterally coupled by Sn^{2+} cations (Fig. 17). The latter cation forms two short intrachain and three longer interchain Sn–S bonds, with the former bonds providing for the zig-zag nature locally and the latter bonds for the parallelism of the chains globally as shown more clearly in Fig. 17. The floppy zig-zag molecular chain conformation, as we shall see next, is central to understanding the glass forming tendency in these ternaries.

The glass transition temperatures decrease with increasing x in the $0 < x < 0.20$ composition range, and thereafter display a local maximum that is centered at $x = 1/2$. The latter behavior is akin to a T$_\ell$ (liquidus temperature) maximum when compound formation occurs in the equilibrium phase diagram and suggests that the local maximum in $T_g(x)$ at $x = 1/2$ could represent formation of the zigzag molecular chain-structure in both melts and bulk glasses. While such chains are parallel in the crystalline phase, these can be expected to bend and twist in glasses and glassy melts. The crystallization behavior of such glasses has been examined [75] by Time-Temperature-Transformation curves over a wide composition range and reinforce the molecular structure results outlined here. Specifically, these measurements reveal that both nucleation and growth processes occur upon crystallization of glasses at low x ($x < 0.2$). At higher x, the nose of the TTT curve becomes conspicuously absent and only growth processes are involved in crystallization of the glasses, understandably because nuclei of the crystalline phase (zigzag chains of SnGeS$_3$) are native to the glasses.

Direct evidence [23] for the existence of the $x = 1/2$ zig-zag chain molecular conformation in the ternary glasses has emerged from ^{119}Sn Mössbauer spectroscopy of the glasses and crystals in the present ternary. In the composition range $0.50 < x < 0.62$, the observed lineshapes in the crystals [Fig. 4(a)] and in the glasses (Fig. 18) can be understood if the excess Sn replaces Ge in the zigzag chains to acquire a tetrahedral coordination (single line in the spectra). The observed lineshapes in the glasses (Fig. 18) reveal changes with x that parallel those in corresponding crystals [Fig. 4(a)], except that the observed linewidths are larger in the glasses. We infer the presence of the zigzag chain structure in the glasses by plotting the ratio [76] of the singlet to doublet site-intensity ratio $I_s/I_d(x)$ as a function of x in Fig. 19. The plot reveals an average slope of 2 which is the expected slope if all excess Sn replaces available Ge-sites in the corner-sharing chains. At $x > 0.62$, the Sn for Ge substitution builds enough strain (since tetrahedral covalent radius of Sn (1.41 Å) exceeds the tetrahedral covalent radius (1.22 Å) of Ge) that it is energetically favorable to nucleate

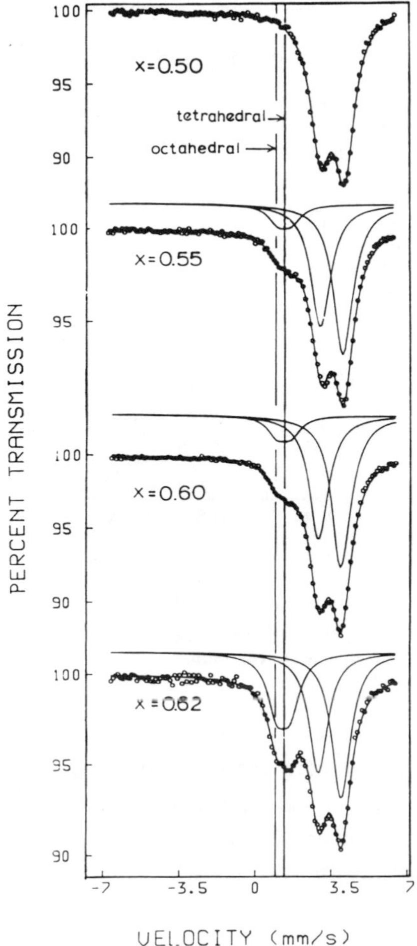

Fig. 18. ^{119}Sn spectra of $Ge_{2-2x}Sn_{2x}S_3$ glasses at indicated values of x. Note presence of a doublet at $x = 1/2$. Note evolution of a singlet (satellite peak) at $x > 1/2$, whose integrated intensity ratio, I_s/I_d, increases with x. I_s-integrated intensity of singlet, I_d-integrated intensity of doublet. Figure is taken from Ref. [23].

the ionically bonded and rigid Sn-rich molecular ribbons of Sn_2S_3 in the melts with consequent loss of bulk glass formation. The sharp threshold in the glass forming tendency of the present ternary can thus be traced to nucleation of a specific floppy molecular fragment in the melts which promotes, and rigid (Sn_2S_3) molecular ribbons in the melts which suppress bulk glass formation.

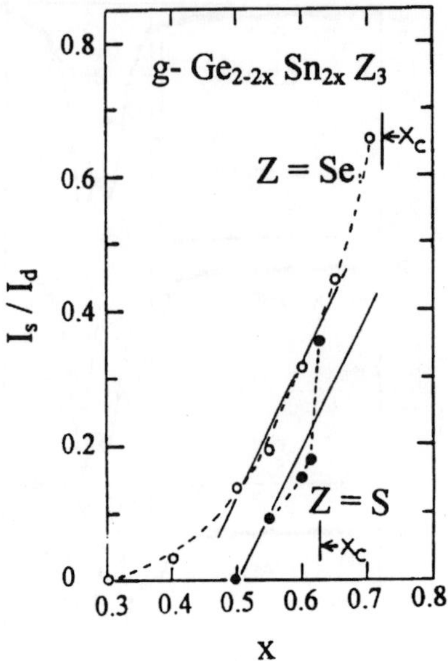

Fig. 19. ^{119}Sn Mössbauer site intensity ratio variation, $I_s/I_d(x)$, in indicated glasses reveals a slope of 2 (continuous line). This variation suggests that the medium-range structure in the glasses consist of the zigzag chain fragment shown in Fig. 17. Figure is taken from Ref. [76].

The floppiness of the zig-zag chains deserves a comment. In a ternary SnGeS$_3$ glass or melt, in contrast to the stoichiometric crystal, it is reasonable to expect the three longer and weaker Sn^{2+}–S bonds to display a wider distribution of lengths than the two intrachain Sn^{2+}–S bonds as chains cease to remain globally linear. If this is the case then the *mechanically effective* coordination of Sn is 2, and it is this coordination that essentially controls the thermal behavior of melts and glasses. This is not to say that the 3 interchain Sn–S bonds do not exist, but we are suggesting that the constraints underlying these weaker interactions are intrinsically *broken* in the melts and glasses. Specifically, bond-lengths and bond-angles associated with these weaker interactions must display wide variations in the glasses. Remarkably, if one calculates the valence force field constraints/atom (n_c) for the zigzag molecular chain structure of SnGeS$_3$, assuming Ge to be 4-fold and Sn and S to be 2-fold, one obtains a mean coordination of

$$\bar{r} = (4 + 2 + 2 \times 3)/5 = 2.4 \tag{6}$$

This is precisely the condition for n_c to equal 3 and for a network of such chains to promote glass formation. We are thus lead to the idea that the isolated distorted molecular chains are optimally constrained in the constraint counting sense, and that the microscopic origin of glass forming tendency reflects this optimal constraint. Constraint counting algorithms have proved to be unusually insightful in understanding the molecular origin of glass formation and are discussed in Chapter 6(B).

The glass forming tendency in corresponding selenides, i.e., in the $(Ge_2Se_3)_{1-x}(Sn_2Se_3)_x$ ternary displays [76] close parallels to those of its S-counterpart. Specifically, the Se-bearing glasses can be easily processed by melt-quenching in water in the concentration range $0 < x < 0.70$. $T_g(x)$ are found to decrease monotonically in the $0 < x < 0.2$ concentration range, and to display a broad local maximum centered [23, 76] near $x = 0.50$ as shown in Fig. 20.

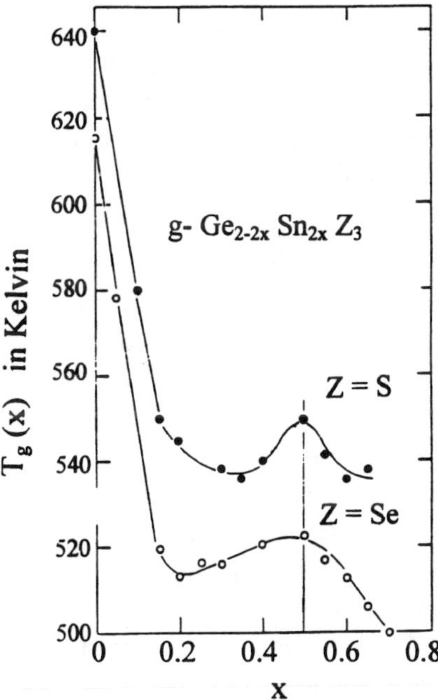

Fig. 20. Glass transition temperature $T_g(x)$ variation in indicated glasses revealing a local maximum near $x = 1/2$, suggestive of the formation of the Fig. 17 fragment in melts and glasses. Figure is taken from Ref. [76].

The zigzag chain molecular conformation of Fig. 17 is not stable when S is fully replaced by Se. Samples of c–GeSnSe$_3$ consist of the mono- and diselenides of Sn and Ge in which the cations are octahedrally coordinated. One is thus led to conclude that while the open zig-zag chain structure with Se anion is unstable in the crystalline state, that molecular conformation appears to readily form in melts and glasses because of their lower density in relation to the crystalline state. Thus, the molecular structure of GeSnSe$_3$ *glass* bears no relation to that of *crystalline* GeSnSe$_3$. In melts and glasses, the low pressure molecular conformation is populated, while in corresponding crystals that conformation is unstable against compaction, and disproportionates into binary selenides.

The $T_g(x)$ results and Mössbauer site intensity ratio $I_s/I_d(x)$ variation with x in $(\mathrm{Ge_2Se_3})_x(\mathrm{Sn_2Se_3})_{1-x}$ glasses are compared to those of their S-counterpart in Figs. 19 and 20. The close parallels in the thermal and spectroscopic results unequivocally demonstrate that the floppy zigzag chain structure form the elements of medium-range order in both S- and Se-bearing glasses near $x = 1/2$, and that this optimally coordinated fragment is primarily responsible for the extensive glass forming tendency in the present ternaries. At $x > 0.70$, for both $\mathrm{Z} = \mathrm{S}$ and Se more compact and overconstrained molecular structures apparently emerge in melts in which cations are octahedrally coordinated and suppress bulk glass formation.

4.2.2. $(GeZ_2)_{1-x}(SnZ_2)_x$; $Z = S$ or Se

We noted in Sec. 4.1.1 the evidence for broken chemical order in a GeSe$_2$ glass provided by [119]Sn Mössbauer spectroscopy. The structural evolution of ternary $\mathrm{Ge_{1-x}Sn_xSe_2}$ alloy glasses as a function of Sn concentration x has also been probed [77, 78] by Mössbauer spectroscopy. At low x (< 0.10), these measurements show how Sn replaces Ge in the host network and at higher x (> 0.10), elucidate how the host network is altered and re-formed as more Sn replaces Ge. Such ternary melts are at the limit of bulk glass formation and can be synthesized in small sizes (1/4 gm). The glass transitions $T_g(x)$ are found to decrease systematically with x.

The fraction $T(x)$ of tetrahedrally coordinated $\mathrm{Sn(Se_{1/2})_4}$ units obtained by [119]Sn spectroscopy over the glass forming range ($0 < x < 0.6$) is displayed in Fig. 21. One sees that $T(x)$ displays a rather curious two peak structure, with a maximum at $x_{\mathrm{max}} = 0.07$ and a more spectacular one at $X_{\mathrm{max}} = 0.35$ in the selenide glasses.

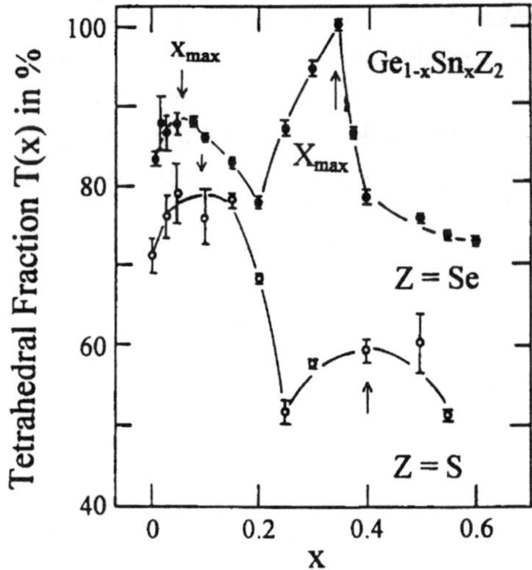

Fig. 21. ^{119}Sn tetrahedral fraction $T(x)$ in Ge$_{1-x}$Sn$_x$S$_2$($\bullet$) and Ge$_{1-x}$Sn$_x$S$_2$(o) bulk glasses revealing a 2-peak structure at $x = x_{\max}$ and $x = X_{\max}$. The second peak at $X_{\max}$ in the selenide glasses represent the stiffness transition, and leads to the formation of a CRN consisting of Sn(Se$_{1/2}$)$_4$ and Ge(Se$_{1/2}$)$_4$ tetrahedra in the approximate ratio of 1:2.

The monotonic increase in $T(x)$ in the narrow $0 < x < 0.07$ composition range, starting from GeSe$_2$ glass ($x = 0$), implies a decreasing concentration of ethane-like units. This trend is *qualitatively incompatible* with the predicted growth of homopolar (Ge–Ge) bonds from the Nerst relation $e^{-\Delta H_f/kT_g}$ as T_g decreases, if indeed such bonds are responsible for the non-tetrahedral defects in an ordered bond network [42] of GeSe$_2$. Here ΔH_f is the bond enthalpy difference $H(\text{Ge–Ge}) + H(\text{Se–Se}) - 2H(\text{Ge–Se})$. On the other hand, this $T(x)$ trend would be expected in the phase separated molecular cluster model if the oversized Sn substituent replaces Ge atoms in CS units at cluster edges [78] in the Se-rich regions. Such substitution may relieve some of the surface stress and promote network polymerization by a pair of homopolar bonds (Ge–Ge and Se–Se) recombining to form heteropolar ones. At higher x and in the alloying regime, $0.07 < x < 0.20$, large scale rearrangement of the network backbone must occur. Currently discussions of such rearrangements must remain largely speculative in the absence of detailed Raman results. However, the spectacular second peak in $T(x)$ at $X_{\max} = 0.35$, invites a special comment. There are reasons to think that it represents a *structural transition of a morphological nature* as we discuss below.

Unlike in the case of Ge, the bond-bending forces for the more metallic Sn are weak [79] and could be intrinsically broken in these ternary alloys. In the present ternary, constraint counting algorithms reveal that if the β-constraints of Sn cation are broken, then the number of mean-field Lagrangian constraints/atom, will equal 3 when the mean coordination number of the network is given by

$$\bar{r} = 2.4 + 2(m_4/N) \tag{7}$$

In Eq. (7), (m_4/N) represents the fraction of 4-fold coordinated atoms of the network that have their β-constraint broken. In the present $Ge_{1-x}Sn_xSe_2$ ternary, Eq. (7) is fulfilled when

$$[4(1-x) + 4x + 4]/3 = 2.4 + 2(x/3)$$

$$\text{or } 8/3 = 2.4 + 2(x/3) \tag{8}$$

$$\text{or } x_c = 0.4$$

The proximity of the predicted composition $x_c = 0.4$ to the observed transition at $x = 0.35$ is *remarkable*. It reflects the power and elegance of constraint counting algorithms, pioneered by J.C. Phillips [79], to understand glass forming systems. It strongly suggests that the second peak at $X_{\max} = 0.35$ represents the *stiffness transition* [77–79] in this case realized by cation tunning of valence force fields on the Sn dopant. Two additional reports on the $Ge_{1-x}Sn_xSe_2$ ternary have been the subject of more recent publications, one on bulk glasses and a second on thin films [81] wherein the molecular structure and optical properties of thin-films have been studied [80, 81].

Parallel results [82] have been obtained on corresponding S-bearing glasses, as illustrated in Fig. 21. $T(x)$ in this ternary also displays a two peak structure, although only partial chemical order is restored at $X_{\max} = 0.40$, probably because in these S-bearing glasses the β-constraint about Sn cations may only be partially broken at T_g.

5. Chalcohalide Glasses

Alloys of the group IV and group V elements with the chalcogens (group VI) and the halogens (group VII) represent chalcohalides. The glass forming tendency in this class of materials was recently reviewed [83]. There is a remarkable correlation between the predictions of constraint counting algorithms and the observed glass forming regions. Te halides occur in a variety of oxidation states; ($TeCl_4$, $TeCl_2$, TeF_6), but to our knowledge none of these halides form bulk glasses. On the other hand, the subhalide Te_3Cl_2 does display an unusual glass forming tendency [84] and also an unusual chain molecular structure. Chalcohalides are thus interesting systems for probing the molecular origin

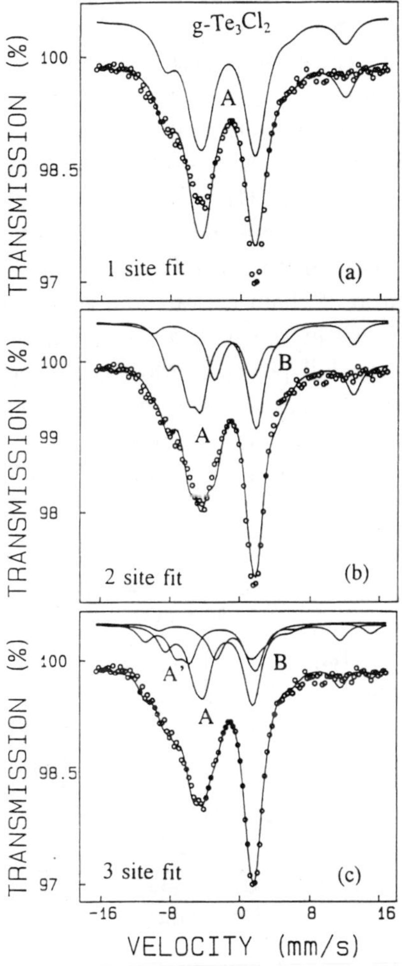

Fig. 22. ^{129}I emission spectra of Te_3Cl_2 glass. The observed lineshape is deconvoluted in terms of one site (a), two sites (b) and three sites (c). The microscopic origin of the three sites (A, A' and B) is traced from the measured quadrupole coupling parameters in Fig. 23. Figure is taken from Ref. [86].

of glass formation. Furthermore, the high infrared transmission [85] of these glasses has led to important applications in thermal imaging and optical communications, a subject that is discussed in Chapter 11 of this book.

5.1. *Binary Te_3X_2; $X = Cl$ or Br glasses*

Rich details on the local structure of g–Te_3Cl_2 (and g–Te_3Br_2) have emerged [86, 87] from Mössbauer experiments with ^{129m}Te tracer in the bulk glasses. ^{129}I emission spectra of these glasses display (Fig. 22) a trimodal distribution of local environments (A, A', B), whose nature can be established from isomer-shift and nuclear quadrupole coupling parameters (Fig. 23). Two of the sites (A, B) represent intra-chain sites characteristic of the chain structure of crystalline Te_3Cl_2 (Fig. 24). The third site (A') is unique to the glass and is attributed to a chain-end Te cation terminated by a halogen dangling end. The site-intensity ratio I'_A/I_A or I'_A/I_B, provide a measure of the edge/interior ratio of chain cations and serve to characterize the average length of the chains in the bulk glass. These results are striking in glass science because the full power of Mössbauer spectroscopy as a probe of both the *local or short-range structure* from the hyperfine structure, and the *medium range structure* from the variation of site-intensity ratio of a chain-glass has been realized for the first time. The Mössbauer spectroscopy

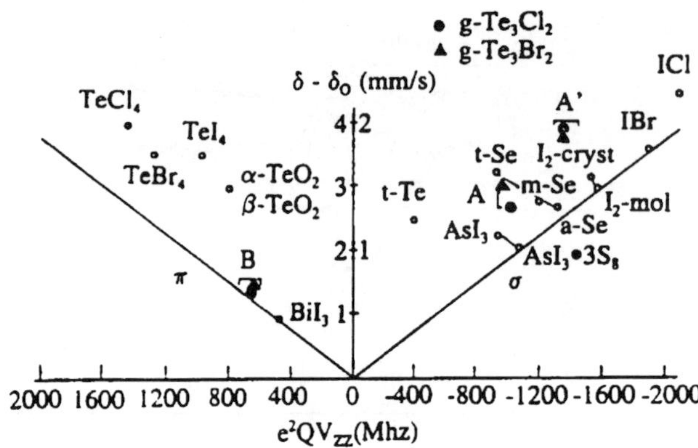

Fig. 23. ^{129}I isomer shift ($\delta - \delta_0$) vs. quadrupole coupling (e^2QV_{zz}) plot revealing domains of π and σ bonding. Note that site A and A' parameters in Te_3Cl_2 glass (•) and Te_3Br_2 glass (▲) are suggestive of I species σ bonded to Te and a halogen (Cl, Br), respectively. Site B parameters, on the other hand, reveals an I site which is π-bonded, and resulting from a parent Te coordination greater than two.

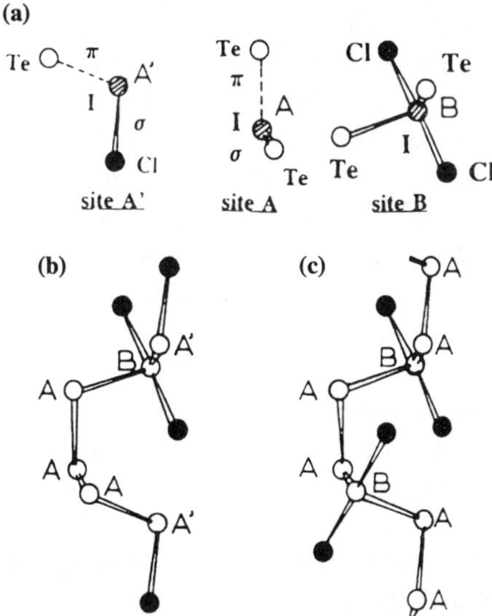

Fig. 24. Local structure of various I sites (A, A', B) shown in (a), as these relate to (b) chain fragments composing the structure of the glass, which is based on infinitely long chains of c–$\mathrm{Te_3Cl_2}$ shown in (c). Figure is taken from Ref. [86].

findings reveal that both $\mathrm{Te_3Cl_2}$ and $\mathrm{Te_3Br_2}$ glasses are characterized by short chain segments about 1.5 nm in length. Thermal measurements show a glass transition temperature of 357 K for $\mathrm{Te_3Cl_2}$ and of 348 K for $\mathrm{Te_3Br_2}$, consistent with a mean coordination of the glassy networks of $\bar{r} = 2$.

There are several implications of the above molecular structure findings on these stoichiometric glasses. Previously, the role of halogens as a chain terminator in a Se glass was inferred from viscosity measurements. However, the model calculations [88] of the observed T-dependence of viscosity give a wide variation (4 order of magnitudes) in Se chain-length assignment. The mean chain-length estimate for $\mathrm{Te_3Cl_2}$ glass from the Mössbauer intensity ratios represents an important quantitative step and, in conjunction with viscosity measurements of $\mathrm{Te_3Cl_2}$ melts, may assist in refining model parameters to reliably ascertain details of chain lengths in other inorganic polymer glasses.

The finding of *short-chain segments* as the structural elements of $\mathrm{Te_3Cl_2}$ glass is not that surprising in itself, given that c–$\mathrm{Te_3Cl_2}$ consists of infinitely long polymeric chains. On the other hand, the present finding of c–$Te_3\,Cl_2$ *like* chain-segments for $\mathrm{Te_3Br_2}$ glass is unusual because c–$\mathrm{Te_3Br_2}$ does not have the chain structure of c–Te_3Cl_2. Here again we have an example of an exception to the Ioffe-Regel rule, in that the molecular structure of the glass and its crystalline counterpart, bear no direct correlation to each other. In Te_3Br_2 glasses a *low-pressure molecular conformation* is apparently frozen in from the melt. In c–Te_3Br_2, a more compact molecular structure [84] emerges in which Te cations are alternately quasi octahedral and quasi-tetrahedral, characteristic of a *high-pressure molecular conformation*. Support for this interpretation comes from the crystallization exotherm for a Te_3Br_2 glass obtained by differential scanning calorimetry which display a much larger heat of transformation than found in Te_3Cl_2 glass.

Thus, in unravelling the structure of glasses, one encounters not too infrequently a circumstance when the molecular structure of a glass and its corresponding crystal have little or no bearing to each other. In such instances the molecular structures prevailing in the melt and glass usually represent a *low-pressure molecular conformation* while, in the crystalline phase, a *high-pressure molecular conformation* is stablized.

5.2. *Ternary* $Te_{3-z}Se_zX_2$; $X = Cl$ *or* Br *glasses*

The glass forming regions in the Te–Se–Cl and Te–Se–Br ternaries span a wide range of compositions [85] as illustrated in Fig. 25. The line from Te_3Cl_2 to

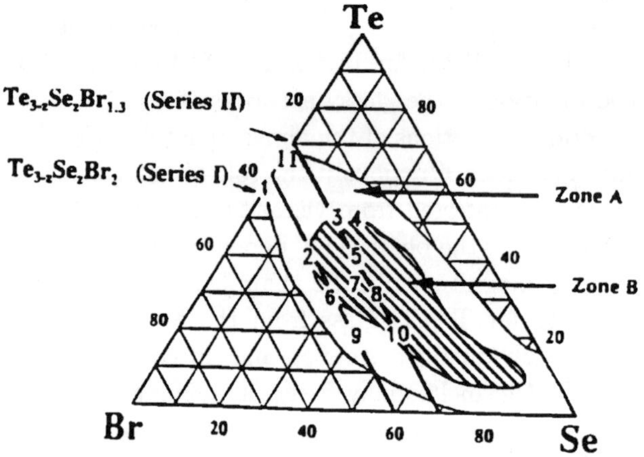

Fig. 25. Glass forming region in the Te–Se–Br ternary taken from Ref. [85]. Glasses in two series (I and II) indicated by the two dark lines taking cuts of the glass forming region were studied. Glasses in the shaded region reveal no T_x while those in the unshaded region do reveal a T_x.

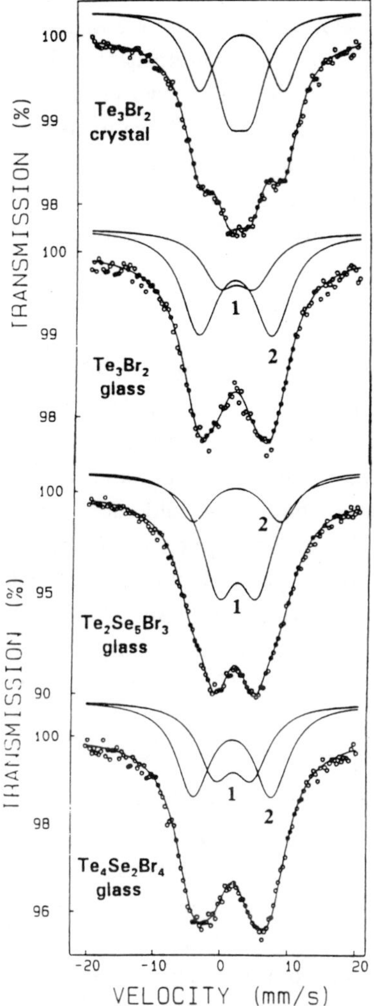

Fig. 26. ^{125}Te spectra of indicated glasses deconvoluted into two doublets, one of the doublets is labelled "1" while the other labelled "2". Figure is taken from Ref. [87].

Se passing through the center composition ($Te_{1/3}Cl_{1/3}Se_{1/3}$) cuts through the heart of the glass forming region. To gain insight into the possible role of molecular structure on the glass forming tendency, we have used ^{125}Te Mössbauer spectroscopy to probe [87] the titled ternary glasses. In these experiments, spectra of the glass samples were recorded using a monoenergetic source of

60 days ^{125}Te^m in Mg_3TeO_6, with the source at 300 K and the glass samples cooled to 12 K. Two sets of glass compositions, $Te_{3-z}Se_zBr_2$ (Series I) and $Te_{3-z}Se_zBr_{1.3}$ (Series II), representing two cuts of the glass forming region (indicated in Fig. 25) were examined. These measurements elucidate the role of replacing Te by Se and, independently, the role of changing the Br content of the starting Te_3Br_2 glass composition.

In spite of the intrinsically broad ^{125}Te natural linewidth ($\Gamma = 5.2$ mm/s), from the observed lineshapes in the glasses we could uniquely deconvolute [87] two distinct types (1,2) of Te local quadrupole field environments; a type (2) environment with a quadrupole splitting of 5.5 mm/s attributed to a Te atom that is 4-fold coordinated (2 to Te and 2 to Br as in Te_3Br_2), and a type (1) Te environment characterized by a quadrupole splitting of 12 mm/s (see Fig. 26). The latter quadrupole field increases systematically with the Se/Te ratio $z/(3-z)$ of the glasses and thus suggests that the type (1) local environment represents a Te atom that is two-fold coordinated. In the stoichiometric glass Te_3Br_2, type (1) local environments provide an integrated average of contributions from A and A' type of local environments. As discussed earlier, signals from these local environments could be resolved from each other in ^{129}I emission spectroscopy measurements largely because of the intrinsically narrower natural linewidth of the 27.8 keV nuclear resonance.

5.2.1. *Role of Se- for Te-replacement in Te_3Cl_2-like chains*

One of the central results to emerge from the present glass compositional studies [87] is that Se readily replaces Te in the Te_3Br_2 chain fragments that are coordinated to two chalcogen near-neighbors. One can thus also understand the systematic increase of $\Delta_1(z)$, as 2-fold coordinated Te environments with 2 Se nearest-neighbors (nn) or one Te and one Se nn emerge at the expense of Te environments with 2 Te nn (characterized by a splitting of 9.6 mm/s). Parallel results were seen in c–Se_xTe_{1-x} alloys where similar local units occur in copolymeric $Te_{1-x}Se_x$ chains). The Se for Te replacement in type (1) local environments eventually saturates when the intensity ratio I_1/I_2 deduced from the Mössbauer lineshapes approaches zero; i.e. when $z \to 2.5$ (Fig. 27). Actually, our Mössbauer results along with the known glass forming compositions also reveal that, once z increases to $z_c = 2.5$, and $I_1/I_2 \to 0$, the glass forming tendency also rapidly disappears. This correlation *underscores* the fact that once Se is forced to substitute for the 4-fold coordinated Te local environments (i.e. B-sites) in the Te_3Br_2-like chain fragments, the chains become chemically

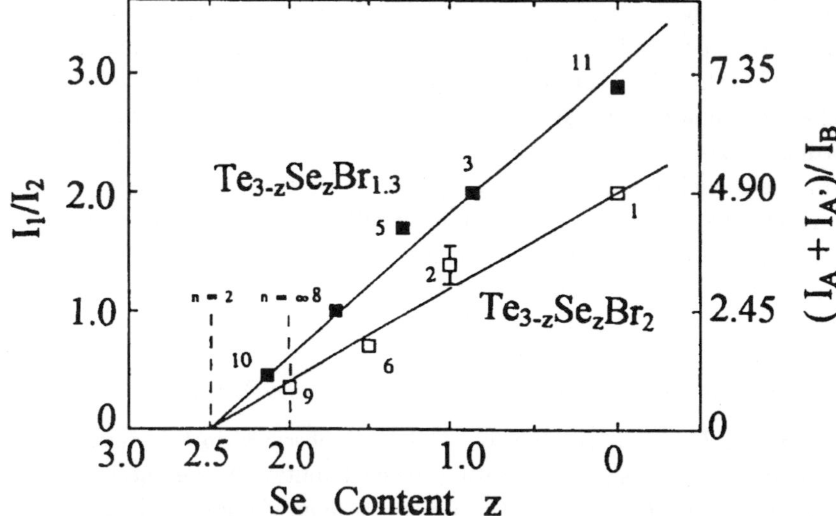

Fig. 27. Mössbauer site-intensity ratio $I_1/I_2(z)$ variation as a function of Se content z in indicated glasses. Figure is taken from Ref. [87].

unstable. Se is a smaller atom than Te and it is also less polarizable than Te. Te replacement by Se at the 4-fold coordinated sites requires a pair of Se-halogen dangling bonds to form in addition to the pair of intra-chain bonds with Se near-neighbors along a chain. Such a local bonding configuration with the less polarizable Se may not be energetically cost effective. The loss of glass formation, once $z > z_c$, can thus be traced to the chemical instability of the $Se_3 Cl_2$-*like* chain-molecular conformation.

5.2.2. *Role of Br content*

The variation of I_1/I_2 as a function of Se content shown in Fig. 27 for glass samples in Series II is quite similar to the one in Series I. For Series II, the Br-content of the samples is smaller than for samples in Series I and, for that reason, comparatively more A- than B-sites are populated in the chains since there are not as many halogen atoms available to form B-sites. It is for this reason the starting value of I_1/I_2 in sample 11 in Fig. 27 (corresponding to Te_2Br composition) is larger than in sample 1 (Te_3Br_2). The ^{125}Te Mössbauer results for sample 11 are taken from the work of M. Takeda and N.N. Greenwood [89]. Remarkably, we see that I_1/I_2 extrapolates to 0 at $z = 5/2 = z_c$

for samples in both series. The essential difference between the two series of samples is that the initial populations of A- and B-sites in the chains are different. In both series I and II, the halogen atoms serve a dual role, these act as chain terminators and also bond to intra-chain Te-sites to form the 4-fold coordinated B-sites.

In these chalcohalide glasses the presence of 1-fold coordinated halogen atoms serve to decouple the Te_n chains globally and to mechanically constrain them locally. The latter behavior follows directly from constraint counting algorithms [90]. These dual functions explain why alloying a halogen in elemental Te promotes glass formation.

5.2.3. *Average length of chain fragments in $Te_{3-z}Se_zBr$ glasses*

The bimodal I_1, I_2 distribution of ^{125}Te sites observed in this ternary glass system fortunately permits separating the contribution in the lineshape of the 2-fold coordinated Te-sites from the 4-fold coordinated ones. In the Se-alloying experiments (Fig. 27) we were thus able to establish the critical Se concentration z_c at which all 2-fold coordinated Te-sites in a chain are fully occupied by Se. Experimentally this is realized when $I_1/I_2 \to 0$, or $I_1 \to 0$. Consequently, the ratio $z_c/(3-z_c)$ has another physical meaning, it provides a direct measure of the concentration of 2-fold to 4-fold coordinated chalcogen sites in a chain fragment and this information permits estimating the length of the chains as we show next.

If the chains constituting the glasses were infinite in extent, i.e. $n = \infty$, (here n is the number of Te_3Cl_2 formula units as in crystalline Te_3Cl_2) then one expects $z_c/(3-z_c) = 2$ since there are twice as many 2-fold coordinated Te-sites as 4-fold coordinated ones. The previous relation yields $z_c = 2.0$. On the other hand, if the chains were much smaller in length, say $n = 4$ formula units long, then simple counting of sites reveals $z_c = 2.25$. Finally, for a chain fragment that is only $n = 2$ formula units long, a count of the sites populated reveals $z_c/(3-z_c) = 5$ or $z_c = 2.5$. Thus z_c spans a range of $2.0 < z_c < 2.5$ as the chain length varies in the range as $\infty > n > 2$.

The value of $z_c = 2.5$ deduced from the Mössbauer intensity ratio I_1/I_2 measurements (Fig. 27) in the present ternary glass system, thus implies that the average length of the chains is given by $n = 2$. Exactly the same average length of the chain fragments was inferred quite *independently* from earlier ^{129}I Mössbauer emission spectroscopy measurements on the stoichiometric glasses $Te_3(Br$ or $Cl)_2$. It appears that the average length of $n = 2$ of the chain fragments is independent of Se and Br contents of the glasses for the region of the phase diagram presently studied. While a chain-length of $n = 2$ can be easily rationalized for Series I samples, it is less obvious to understand such a result for Series II samples. A possible interpretation is that a glass of Te_7Br_3 composition (Series II) is composed of two types of $n = 2$ chain fragments. One of these fragments contains only halogens as terminations (Te_6Br_2) while the other fragment of Te_6Br_4 stoichiometry contains a pair of intra-chain halogens in addition to a pair of halogens acting as terminations as shown in Fig. 24. A glass stochiometry close to Te_7Br_3 would then result if on an average, for every Te_6Br_4 chain-fragment one had a pair of Te_6Br_2 fragments in the network.

6. Conclusions and Future Work

The judicious choice of Mössbauer probe atoms has led to the quantitative aspects of glass structure in several chalcogenides, chalcohalides and selected oxides. Experimental results on several families of IV–VI glasses reveal that at a low mean coordination number $\bar{r} \leq 2.4$, glasses usually consist of floppy random networks of Se or S chains that are marginally crosslinked. As the coordination number of glasses increases to $\bar{r} \geq 2.40$, the morphological structure undergoes a change with correlations on a scale of 5 Å and larger appearing as the backbone becomes percolatively rigid. Stoichiometric $GeSe_2$ and GeS_2 glass are intrinsically phase separated into Ge-rich and S- or Se-rich nanophases for which unambiguous spectroscopic evidence is documented from Mössbauer and Raman scattering measurements. Networks of Ge_2Se_3 or Ge_2S_3 stoichiometry are also not chemically ordered and owe their glass forming tendency to the presence of two Ge-rich-nanophases of $Ge_2(S \text{ or } Se)_3$ and $Ge(S \text{ or } Se)$ stoichiometry, respectively. Alloying Sn_2Se_3 in Ge_2Se_3 extends bulk glass formation in the $(Ge_2Se_3)_{1-x}(Sn_2Se_3)_x$ pseudobinary up to $x = 70$ molar percent of Sn_2Se_3. Glass formation in this pseudobinary is traced to the appearance of a zigzag molecular chain conformation of $GeSnSe_3$ stoichiometry for which rather direct and striking Mössbauer spectroscopic evidence is documented. In all instances, the experiments reveal that bulk glass formation of melts ceases when the underlying molecular phases are mechanically overcoordinated ($\bar{r} = 3$) such as the case for the Ge-monochalcogenides (GeS, GeSe and GeTe) and $Sn_2(S \text{ or } Se)_3$.

The present molecular structure studies on bulk Ge–Sn–S or Se glasses can be extended to amorphous thin-films prepared by vapor deposition, using either evaporation or laser-ablation or RF sputtering. The quantitative aspects of

Mössbauer and Raman scattering have yet to be fully exploited in decoding the structure of chalcogenide thin-film. Thin-film growth involves a multitude of complex factors such as formation of metastable local configurations controlled by the mobility of deposited species on the substrate and by the effluent species emitted by the source. Also coming into play are the formation kinetics of the network backbone and its subsequent relaxation and polymerization upon thermal annealing after deposition [91]. To date, studies on chalcogenide thin-films have largely used optical absorption edge shifts, small angle neutron scattering, and refractive index as probes, all providing macroscopic average properties. In contrast local spectroscopies like Raman and Mössbauer can provide the details of structural inhomogeneities as recently demonstrated [92] for obliquely deposited $GeSe_2$ films displaying a columnar structure.

The existing Mössbauer spectroscopy data base on Ge–Sn–chalcogenides can be extended to include other families of binary (Si–Se, As–Se, Sb–Se) and ternary (Ge–As–Se, Ge–P–Se) chalcogenides including pnictides and the lighter tathogen. There are new and interesting possibilities to probe pnictide local environments in chalcogenides with ^{119}Sn emission (using ^{119}Sb radioactive parent), and ^{121}Sb absorption spectroscopy. Intrinsic differences of chemical bonding between As- and Sb-cations in chalcogenide and chalcohalide glasses can strongly influence glass forming tendencies in binary and ternary systems.

Rare-earths dopants in IR transmitting chalcogenides have attracted interest as active emitters for lasers and optical amplifiers. There are new and interesting possibilities of using Rare-earth Mössbauer probes such as ^{151}Eu, ^{149}Sm and ^{169}Tm to examine the chemical bonding configurations chosen by rare-earths in chalcogenides.

Lamb–Mössbauer factors in glasses are sensitive to low-frequency and particularly floppy modes in glasses [see Chapter 6(B)]. The unique potential of Mössbauer spectroscopy as a site-specific vibrational probe of network glasses within the context of connectivity induced rigidity transition promises to complement inelastic neutron scattering measurements.

Acknowledgments

Many students have participated in the glass science projects at University of Cincinnati and some of the recent ones include Min Zhang, Xingwei Feng and Deepak Selvanathan. It is also a pleasure to acknowledge the participation of several colleagues including W.J. Bresser, R.N. Enzweiler and J. Wells in the experiments. I have benefited from numerous discussions with B. Goodman,

D.H. McDaniel and J.C. Phillips. The present work is supported by the National Science Foundation grant DMR-97-01289.

References

[1] W.H. Zachariasen, *J. Am. Chem. Soc.* **54**, 3841 (1932).

[2] A.C. Wright and A.J. Leadbetter, *Phys. Chem. Glasses* **17**, 122 (1976); also see E.A. Porai–Koshits, *J. Non-Cryst. Solids* **25**, 87 (1977); Ke Tanaka, *Japan J. Appl. Phys.* **37**, 1747 (1998).

[3] A.E. Geissberger and P.J. Bray, *J. Non-Cryst. Solids* **54**, 121 (1983).

[4] R. Dupree and R.F. Pettifer, *Nature* **308**, 523 (1984).

[5] S.K. Mitra, M.N. Amini, D. Fincham and R.W. Hockney, *Phil. Mag.* **B43**, 365 (1981).

[6] M. Zhang and P. Boolchand, *Science* **266**, 1355 (1994). Also see J.C. Phillips, *Solid State Comm.* **47**, 203 (1983). Ideas on constraint counting were developed by J.C. Phillips, *J. Non-Cryst. Solids* **34**, 153 (1979).

[7] A.C. Wright, R.N. Sinclair and A.J. Leadbetter, *J. Non-Cryst. Solids* **71**, 295 (1985). Also see A.C. Wright, Chapter 5(A) of this volume. S.C. Moss and D.L. Price, *Physics of Disordered Materials*, ed. D. Adler, H. Fritzsche, S.R. Ovshinsky (Plenum Press, New York) p. 77. The microscopic origin of the first sharp diffraction peak in glasses continues to be an open question. Also see S. Elliot, Ref. [42].

[8] P. Boolchand, *Physical Properties of Amorphous Materials*, eds. D. Adler, B.B. Schwartz and M.C. Steele (Plenum Press, New York, 1985) p. 221.

[9] M. Murase, Chapter 6(C), this volume.

[10] H. Eckert, Chapter 5(D), this volume.

[11] G.N. Greaves, *Glass Science and Technology*, eds. D.R. Uhlmann and N.J. Kreidl (Academic Press Inc., New York, 1990) p. 1.

[12] P. Boolchand, X. Feng, D. Selvanathan and W.J. Bresser, *Rigidity Theory and Applications*, eds. M.F. Thorpe and P.M. Duxbury (Plenum Press, New York, 1999) (in press).

[13] P. Boolchand, Chapter 6(B), this volume.

[14] R.L. Mössbauer, *Z. Naturforsch* **14**, 211 (1959).

[15] N.N. Greenwood and T.C. Gibb, *Mössbauer Spectroscopy* (Chapman and Hall Ltd., London, 1971).

[16] P. Boolchand and M. Stevens, *Phys. Rev.* **B29**, 1 (1984).

[17] P.P. Seregin, A.R. Regel, A.A. Andreev and F.S. Nasredinov, *Phys. Stat. Sol.* (a) **74**, 373 (1982).

[18] P.A. Flinn, *Mössbauer Isomer Shifts*, eds. G.K. Shenoy and F. Wagner (North Holland, Amsterdam, 1978) p. 595. Also see J.D. Donaldson and B.J. Senior, *J. Inorg. Nucl. Chem.* **31**, 881 (1969).

[19] T.P. Das and E.L. Hahn, *Nuclear Quadrupole Resonance Spectroscopy, Supplement I of Solid State Physics*, eds. W. Seitz and D. Turnbull (Academic Press Inc., New York, 1958).

[20] R.M. Sternheimer, *Phys. Rev.* **84**, 244 (1951), ibid **95**, 736 (1954).

[21] P. Bray, Chapter 5(C), this volume.

[22] P. Boolchand, *Z. Naturforsch* **51a**, 572 (1996). Also see H. deWaard, *Mössbauer Effect Data Index*, eds. J.G. Stevens and V.E. Stevens (Plenum Press, New York, 1975) p. 447.

[23] D. Ruffolo and P. Boolchand, *Phys. Rev. Lett.* **55**, 242 (1985).

[24] G. Amthauer, J. Fenner, S. Hafner, W.B. Holzapfel and R. Keller, *J. Chem. Phys.* **70**, 4837 (1979).

[25] A. Feltz, B. Voight and E. Schlenzig, *Amorphous and Liquid Semiconductors*, eds. J. Stuke and W. Brenig (Taylor and Francis, London, 1974) p. 261.

[26] P. Boolchand, B.L. Robinson and S. Jha, *Phys. Rev.* **B2**, 3463 (1970).

[27] S.S. Hanna, J. Heberle, C. Littlejohn, G. Perlow, R.S. Preston and D.H. Vincent, *Phys. Rev. Lett.* **4**, 177 (1960). Also see *Nuclear Zeeman Effect and Recent Advances in Mössbauer Spectroscopy*, ed. P. Boolchand, *Hyperfine Interactions* **72**, 1–3 (1992).

[28] An early review of work on metallic glasses appeared in *J. Nucl. Instrum. Methods* **199** (1982). Also see M.M. Abd-Almeguid, H. Micklitz and I. Vincze, *Phys. Rev.* **25**, 1 (1982). For spin relaxation effects in solids, see H.H. Wickman and G.K. Wertheim, *Chemical Applications of Mössbauer Spectroscopy*, eds. V.I. Goldanskii and R.H. Herber (Academic Press, New York, 1968) p. 548.

[29] O.C. Kistner, A.W. Sunyar and J.B. Swan, *Phys. Rev.* **123**, 179 (1961).

[30] M. Tenhover and P. Boolchand, *Phys. Rev.* **B18**, 6292 (1978).

[31] C.R. Kurkjian, *J. Non-Cryst. Solids* **3**, 157 (1970). Also see W. Müller-Warmuth and H. Eckert, *Phys. Reports* **88**, 93 (1982).

[32] G. Tomandl, *Glass Science and Technology* **B4**, 273 (1990), eds. D.R. Uhlmann and N.J. Kreidl (Academic Press, Boston, 1990) p. 273.

[33] M. Zhang, S. Mancini, W.J. Bresser and P. Boolchand, *J. Non-Cryst. Solids* **151**, 149 (1992). Also see P. Boolchand and M. Zhang, *Science* **268**, 1510 (1995).

[34] M. Zhang, PhD Thesis, University of Cincinnati, 1994 (unpublished).

[35] G.G. Naumis and R. Kerner, *J. Non-Cryst. Solids* **231**, 111 (1998).

[36] T. Nishida, M. Yamada, H. Ide and Y. Takashima, *J. Mater. Sci.* **25**, 3546 (1990). Also see T. Nishida, *Z. Naturforsch* **51a**, 620 (1995).

[37] B.C. Sales and L.A. Boatner, *Science* **226**, 45 (1984). Also see *Radioactive Waste Forms for the Future*, eds. W. Lutze and R.C. Ewing (North Holland, Amsterdam, 1988) p. 193.

[38] G.K. Marasinghe, M. Karabulut, C.S. Ray, D.E. Day, M.G. Shumsky, W.B. Yelon, C.H. Booth, P.G. Allen and D.K. Shuh, *J. Non-Cryst. Solids* **222**, 144 (1997).

[39] A. Mogus-Milankovic, B. Pivac, K. Furie and D.E. Day, *Phys. Chem. Glasses* **38**, 74 (1997).

[40] P. Boolchand, J. Grothaus and J.C. Phillips, *Solid State Commun.* **45**, 183 (1983).

[41] I.T. Penfold and P.S. Salmon, *Phys. Rev. Lett.* **67**, 97 (1991). Also see P. Boolchand, J.C. Phillips, *Phys. Rev. Lett.* **68**, 252 (1992) and response thereafter.

[42] S.R. Elliott, *Glasses and Amorphous Materials*, Vol. 9, ed. J. Zarzycki (VCH, New York, 1991) p. 377.

[43] P. Boolchand, J. Grothaus, M. Tenhover, M.A. Hazle and R.K. Grasselli, *Phys. Rev.* **B33**, 5421 (1986).

[44] P. Boolchand, *Comments in Cond. Mat. Phys.* **12**, 163 (1986).

[45] G. Lucovsky, J.P. deNeufville and F.L. Galeener, *Phys. Rev.* **B9**, 1591 (1974); G. Lucovsky, F.L. Galeener, R.C. Keezer, R.H. Geils and H.A. Six, *Phys. Rev.* **B10**, 5134 (1974).

[46] H. Ipser, M. Gambino and W. Schuster, *Monatsh. Chem.* **113**, 389 (1982).

[47] V.G. Dittmar and H. Schäfer, *Acta Cryst.* **B31**, 2060 (1975); V.G. Dittmar and Schäfer, *Acta Cryst.* **B32**, 1188 (1976).

[48] A. Feltz, K. Zickmuller and G. Pfaff, *Proc. 7th Int. Conf. on Amorphous and Liquid Semiconductors*, eds. W.E. Spear and G.C. Stevenson (Institute of Physics, Bristol, 1978) p. 125.

[49] For a Ge_2Se_3 unit, taking Ge to be 4-fold ($n_c = 7$) and Se to be 2-fold coordinated ($n_c = 1$), one obtains $\bar{n}_c = 3.40$ if the bond-bending constraint of Se is broken. The mechanically effective coordination number, $\bar{r}$, is obtained by requiring $3.40 = (5/2)\bar{r} - 3$, giving $\bar{r} = 2.56$.

[50] D.P. Machewirth, K. Wei, V. Krasteva, R. Dutta, E. Snitzer and G.H. Sigel Jr., *J. Non-Cryst. Solids* **213 & 214**, 295 (1997).

[51] V. Krasteva, D. Hensley and G. Sigel Jr., *J. Non-Cryst Solids* **222**, 235 (1997).

[52] M. Picon and M. Patrie, *Compt. Rend.* **242**, 1321 (1956).

[53] M. Mitkova, Y. Wang and P. Boolchand, *Phys. Rev. Lett.* **83** 3848 (1999). Also see Z.U. Borisova, *Glassy Semiconductors* (Plenum, New York, 1981) p. 227.

[54] P. Tronc, M. Bensoussan, A. Brenac and C. Sebbene, *Phys. Rev.* **B8**, 5947 (1973).

[55] P.M. Bridenbaugh, G.P. Espinosa, J.C. Phillips and J.P. Remeika, *Phys. Rev.* **B20**, 4140 (1979).

[56] S. Sugai, *Phys. Rev.* **B35**, 1345 (1987).

[57] K. Murase and T. Fukunaga, Defects in Glasses, *Mat. Res. Soc. Symp. Proc.* **61**, 101 (1986).

[58] Koblar Jackson, private communication. The Raman cross-sections using density-functional cluster predictions for modes of fs:cs:es units = 54:39:32. Thus the observed scattering strength ratios of A_{fs}/A_{cs} leads to a concentration ratio, $C_{fs}/C_{cs} = 0.17$, or $C_{fs}/(C_{cs}+C_{fs}) = 0.13$. Also see K. Jackson, M.R. Pederson, D. Porezag, Z. Hajnal and T. Frauenheim, *Phys. Rev.* **B55**, 2549 (1997).

[59] In $GeSe_2$, there are 12 ethane-like units for 100 CS ones and 50 ES ones. Since there is one homopolar (Ge–Ge) bond for 6 heteropolar ones in an ethane-line unit, the concentration of heteropolar/homopolar bonds in the network is given by $12/[(100 + 50)4 + 12]$ or $12/[672]$ or 1.78%.

[60] W.J. Bresser, P. Boolchand, P. Suranyi and J.P. deNeufville, *Phys. Rev. Lett.* **46**, 1689 (1981). Also see C. Kim and P. Boolchand, *Phys. Rev.* **B19**, 3187 (1979).

[61] W.J. Bresser, P. Boolchand and P. Suranyi, *Phys. Rev. Lett.* **56**, 2493 (1986).

[62] D.L. Price, M. Misawa, S. Susman, T.I. Morrison, G.K. Shenoy and M. Grimsdith, *J. Non-Cryst. Solids* **66**, 443 (1984). D.J. Verrall, S.R. Elliott, *Phys. Rev. Lett.* **61**, 974 (1988).

[63] J. Griffiths, G.P. Espinosa, J.P. Remeika and J.C. Phillips, *Phys. Rev.* **B25**, 1272 (1982).

[64] H. Hisakuni and Ke Tanaka, *Science* **270**, 975 (1995).

[65] For probe of chalcogen local environments, see P. Boolchand, W.J. Bresser and P. Suranyi, *Hyp. Int.* **27**, 385 (1986). For probe of As local environments, see P. Craig Taylor, *Z. Naturforsch* **51a**, 603 (1996).

[66] B. Norban, D. Pershing, R.N. Enzweiler, P. Boolchand, J.E. Griffiths and J.C. Phillips, *Phys. Rev.* **B36**, 8109 (1987).

[67] M. Tenhover, M.A. Hazle and R.K. Grasselli, *Phys. Rev. Lett.* **51**, 404 (1983).

[68] D. Selvanathan, W.J. Bresser and P. Boolchand, *Solid State Commun.* **111**, 619 (1999).

[69] J.E. Griffiths, M. Malyj, G.P. Espinosa and J.P. Remeika, *Phys. Rev.* **B30**, 6978 (1984).

[70] S.R. Ovshinsky, *Phys. Rev. Lett.* **21**, 1450 (1968). S.R. Ovshinsky and H. Fritzsche, *IEEE Trans. on Electron Devices* **ED20**, 91 (1973). More recent work on threshold and memory switching has appeared from the Institute of Science in Bangalore. See for example in S. Murugavel and S. Asokan, *Phys. Rev.* **B58**, 3022 (1998).

[71] J.P. deNeufville, *J. Non-Cryst. Solids* **8–10**, 85 (1972). Also see P. Boolchand, B.B. Triplett, S.S. Hanna and J.P deNeufville in *Mössbauer Effect Methodology* **9**, 53 (1974) eds. J. Gruverman, C.W. Seidel and D.K. Dieterly (Plenum, New York).

[72] R.N. Enzweiler, W.J. Bresser, J. Selvanathan and P. Boolchand (unpublished).

[73] D.S. Franzblau and J. Tersoff, *Phys. Rev. Lett.* **68**, 2172 (1992). Also see D. Drabold, Chapter 9 of this volume.

[74] Carlo Massobrio, A. Pasquarello and Robert Car, *Phys. Rev. Lett.* **80**, 2342 (1998).

[75] G. Lemon and P. Boolchand, *J. Non-Cryst. Solids* **91**, 1 (1987).

[76] R.N. Enzweiler and P. Boolchand, *Solid State Commun.* **62**, 197 (1987).

[77] Mark Stevens, J. Grothaus, P. Boolchand and J. G. Hernandez, *Solid State Commun.* **47**, 199 (1983).

[78] M. Stevens, P. Boolchand and J.G. Hernandez, *Phys. Rev.* **B31**, 981 (1985).

[79] J.C. Phillips, *Solid State Commun.* **47**, 203 (1983).

[80] M.J. Peters and L.E. McNeil, *J. Non-Cryst. Solids* **139**, 231 (1992).

[81] D. Islam and R.L. Cappelletti, *Phys. Rev.* **B44**, 2516 (1991).

[82] J. Grothaus and P. Boolchand, *J. Non-Cryst. Solids* **72**, 1 (1985).

[83] M. Mitkova and P. Boolchand, *J. Non-Cryst. Solids* **240**, 1 (1998).

[84] J. Lucas, Chapter 11 of this volume.

[85] J. Lucas and X.H. Zhang, *J. Non-Cryst. Solids* **125**, 1 (1990).

[86] J. Wells, W.J. Bresser, P. Boolchand and J. Lucas, *J. Non-Cryst. Solids* **195**, 170 (1996).

[87] W.J. Bresser, J. Wells, M. Zhang, and P. Boolchand, *Z. Naturforsch* **51a**, 378 (1996).

[88] W.C. Cooper and R.A. Westburg in *Selenium*, eds. R.A. Zingaro and W.C. Cooper (Van Nostrand Reinhold, New York, 1974) p. 108.

[89] M. Takeda and N.N. Greenwood, *JCS Dalton* **631** (1976).

[90] P. Boolchand, M. Zhang and B. Goodman, *Phys. Rev.* **B53**, 11488 (1996).

[91] For some early work on thin-films, see P. Boolchand, R.N. Enzweiler and M. Tenhover, *Diffusion and Defect Data* **53–54**, 415 (1987), Trans. Tech. Publications Ltd., Switzerland.

[92] P. Boolchand, K.L. Chopra and W.J. Bresser, *Bull. Am. Phys. Soc.* **44**, 1434 (1999).

C. NUCLEAR QUADRUPOLE RESONANCE (NQR) STUDIES OF GLASS STRUCTURE

PHILIP J. BRAY

*Department of Physics, Box 1843, Brown University Providence,
Rhode Island, USA 02912*
bray@physics.brown.edu

Contents

1. Introduction

NMR has been used for some 40 years [1] to study structure, chemical bonding and other phenomena in glasses. (See the accompanying Chapter by Hellmut Eckert entitled "Solid State NMR as a structural tool in Glass Science"; other reviews of this work are also available [2–6].) A significant portion of that work has involved the use of quadrupolar effects [7] in the NMR spectra of various nuclei (e.g. ^{7}Li, ^{9}Be, ^{11}B and ^{10}B, ^{17}O, ^{27}Al, etc). NQR (nuclear quadrupole resonance) studies of glasses are more recent [8, 9], but provide substantial increases in the resolution of the spectra and major gains in the precision of the data. Since the less accurate NMR results are regularly used to determine the areas of frequency that should be swept for the NQR responses, a brief synopsis of NMR studies involving quadrupolar effects is appropriate and will introduce the quadrupolar parameters that are measured in both experiments (NMR and NQR). The NMR spectra and data will also be referenced repeatedly in discussing the results of the NQR investigations.

2. NMR Spectroscopy with Quadrupolar Effects

2.1. *Zeeman and quadrupole interactions*

Figure 1 displays the energy levels for the case of the pure Zeeman interaction (i.e. no quadrupole interaction present) for which the single transition frequency (the Larmor frequency) is ν_0, and the shifting of those levels when

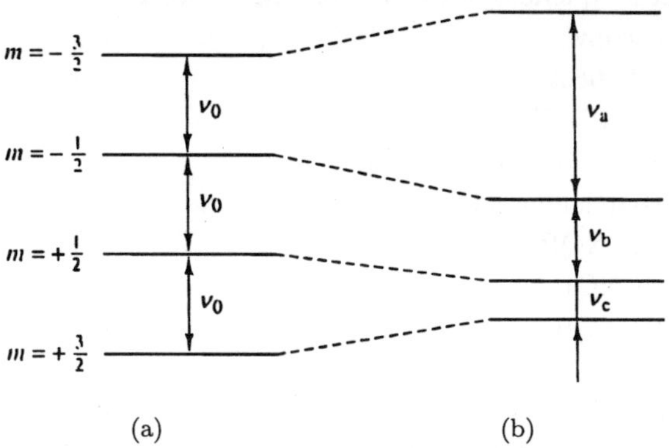

Fig. 1. Energy levels arising from the interaction of the nuclear magnetic dipole moment with a magnetic field; (a) no electrical quadrupole interaction; (b) small quadrupole interaction present (the levels shown are appropriate for the ^{11}B nucleus).

a relatively small quadrupole interaction is present. The shifts depend on the coupling of the nuclear electrical quadrupole moment (Q) with the components of the electric field gradient (EFG) tensor at the nuclear site. These are V_{xx}, V_{yy}, and V_{zz} in the system of principal axes for the EFG. Here V is the electrical potential at the nuclear site arising from charges outside the nucleus, and $V_{xx} = \partial^2 V/\partial x^2$, etc. The shifts also depend on the angles θ and ϕ that specify the orientation of the magnetic field $\mathbf{H_0}$ with the principal axes of the EFG. Since all angles θ and ϕ are equally probable in a finely divided polycrystalline powder or a glass, the NMR spectra for those cases will be the envelope of responses observed for a single crystal as it is placed in all possible orientations in the magnetic field. The envelope is labeled a "powder pattern".

2.2. *First-order quadrupolar effects*

The theoretical "powder pattern" for a quadrupole interaction that is sufficiently small to be treated with first-order perturbation theory is displayed in Fig. 2 for the case of a nuclear spin $I = 3/2$. (No dipolar or other broadening mechanisms have been taken into account.) Here $Q_{cc} = e^2 V_{zz} Q/h = e^2 q Q/h$ is the quadrupole coupling constant that measures the strength of the quadrupole interaction, and $\eta = (V_{xx} - V_{yy})/V_{zz}$ is the asymmetry parameter that measures the departure of the EFG from axial symmetry. (With $|V_{zz}| \geq |V_{yy}| \geq |V_{xx}|$, one finds that $0 \leq \eta \leq 1$.) Note that the "central" transition $(m = +1/2 \leftrightarrow -1/2)$ is unaffected in first order and remains at the Larmor frequency ν_0, while the "satellite" transitions $(m = +3/2 \leftrightarrow +1/2$ and $m = -3/2 \leftrightarrow -1/2)$

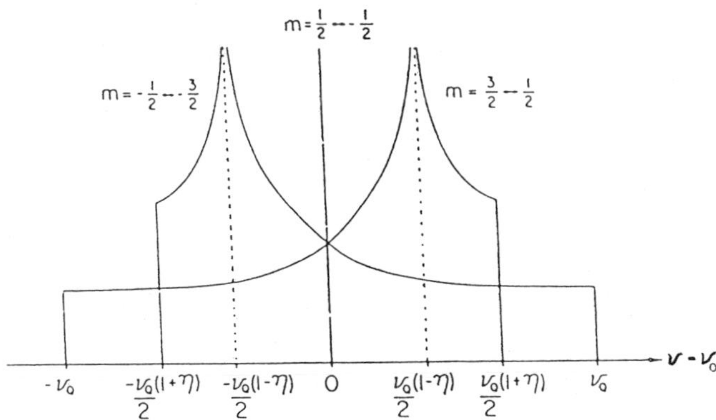

Fig. 2.   First-order quadrupolar spectrum for $I = \frac{3}{2}$ nuclei in a powder or glass. Here $\nu_Q = 3Q_{cc}/2I(2I-1)$ and 0 on the horizontal (frequency) axis marks the location of ν_0.

are spread out into the symmetric "powder pattern". (In Fig. 2 and subsequent figures, 0 marks the location of the Larmor frequency ν_0.)

Broadening mechanisms (e.g. dipolar interactions, distributions in the values of the EFG components) smooth the divergences in the theoretical powder pattern and give breadth to the central transition peak. This is exemplified in Fig. 3 for the particular case of $\eta = 0$. When use is made of older CW (continuous wave) NMR spectrometers, the recorded resonance is usually the first derivative of the absorption curve. This is displayed in Fig. 4 for the ^{11}B ($I = 3/2$) response in polycrystalline boron phosphate (BPO$_4$) for which the value of $Q_{cc} = 50.4$ kHz if η is assumed to be zero [10]. (The boron atom is at the approximate center of a B0$_4$ tetrahedron that is only slightly distorted so the components of the EFG are quite small. The EFG vanishes for perfect tetrahedral and octahedral symmetry.)

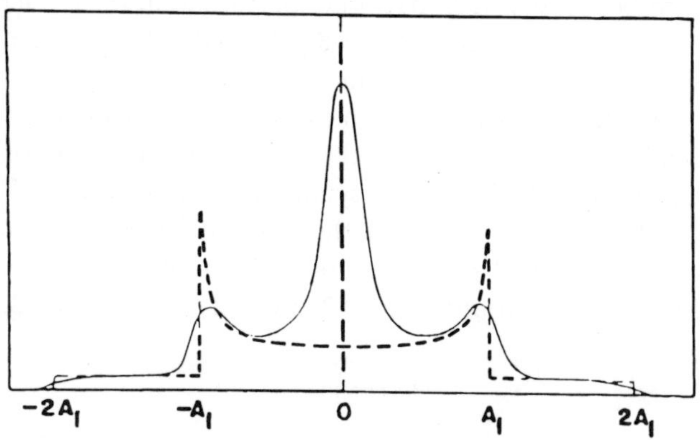

Fig. 3. Resonance lineshape (solid curve) predicted for a nucleus such as ^{11}B with a small quadrupole interaction in a glass or polycrystalline powder. This is the $\eta = 0$ case. Here $A_1 = \frac{\nu_Q}{2}(1-\eta)$. The dashed curve is the lineshape in the absence of broadening mechanisms.

Figure 5 displays the derivative response [11] for ^{7}Li ($I = 3/2$) in a glass containing 25 mol% Li$_2$O and 75 mol% B$_2$O$_3$. If $\eta = 0$, then $Q_{cc} \approx 185$ kHz. (The electrical quadrupole moment Q for ^{7}Li is relatively small.) The substantial "smearing" of the divergences in the powder pattern arises from the distribution in the EFG components caused by the disorder of the vitreous state. See Krämer *et al* [12] for other examples of ^{7}Li NMR spectra for glasses.

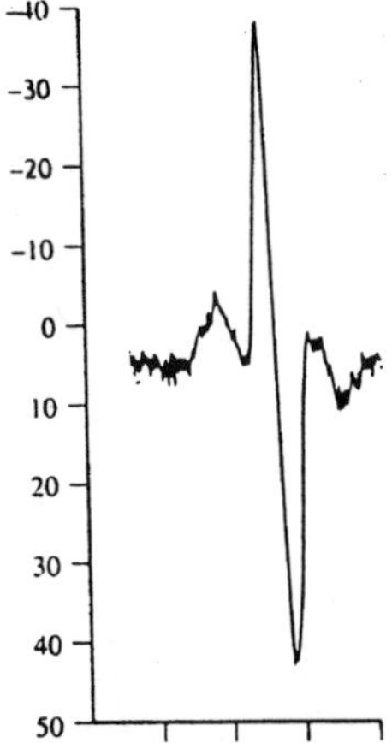

Fig. 4. The NMR of ^{11}B in boron phosphate (BPO$_4$); $\nu_0 = 7.177$ MHz.

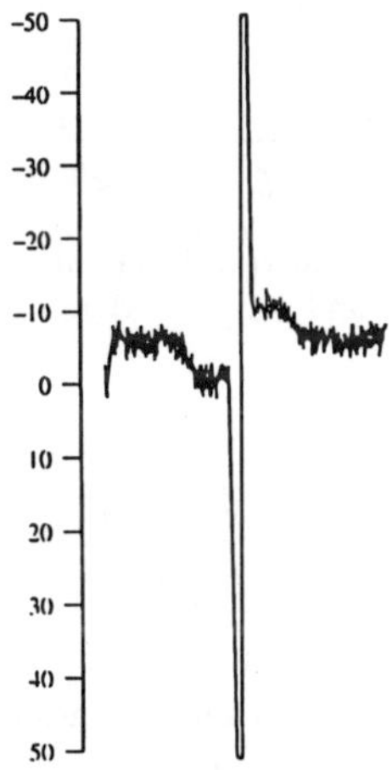

Fig. 5. ^{7}Li NMR spectrum for a glass of composition Li$_2$O $\cdot$ 3B$_2$O$_3$.

2.3. *Second-order quadrupolar effects*

When the quadrupole interaction is larger (but still treatable as a perturbation of the Zeeman interaction), the central ($m = +1/2 \leftrightarrow -1/2$) transition is affected in second order. Figure 6 displays the theoretical pattern for this case, both for $\eta < 1/3$ and $\eta > 1/3$.

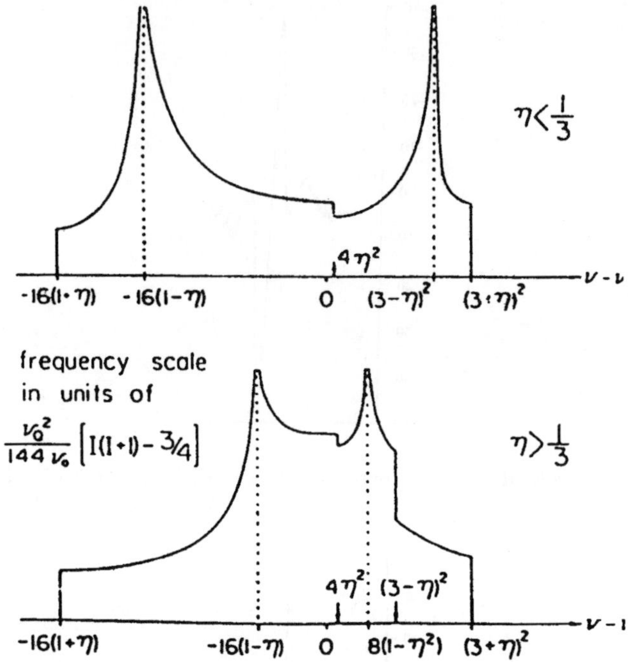

Fig. 6. Powder patterns for the central transition ($m = -\frac{1}{2} \leftrightarrow m = +\frac{1}{2}$) of half-integer spin nuclei in the regimes: $\eta < 1/3$ and $\eta > 1/3$. Here $\nu_Q = 3Q_{cc}/2I(2I-1)$ and ν_0 is the Larmor frequency.

2.3.1. *Case of a small asymmetry parameter*

The particular case of small or zero η is illustrated in Fig. 7 where the dashed lines indicate the theoretical pattern without broadening and the solid line indicates the expected lineshape when broadening is included. The NMR spectrum (absorption curve itself, not the derivative) for ^{11}B in vitreous boron oxide (B_2O_3) is displayed [13] in Fig. 8. In this case, in which all of the boron atoms are in the (approximate) center of planar triangles of oxygens (i.e. BO_3 units) and all of the oxygens are bridging (i.e. bonded to two borons), one finds $Q_{cc} = 2.76$ MHz and $\eta = 0.10$. The departure from perfect trigonal symmetry about the axis perpendicular to the BO_3 units and passing through the boron is small or moderate. (η would be zero if the trigonal symmetry were perfect.) One finds for such BO_3 units that 2.4 Mhz $\leq Q_{cc} \leq 2.9$ MHz and $0 \leq \eta \leq 0.2$. (In the case of vitreous B_2O_3, most of the BO_3 units are

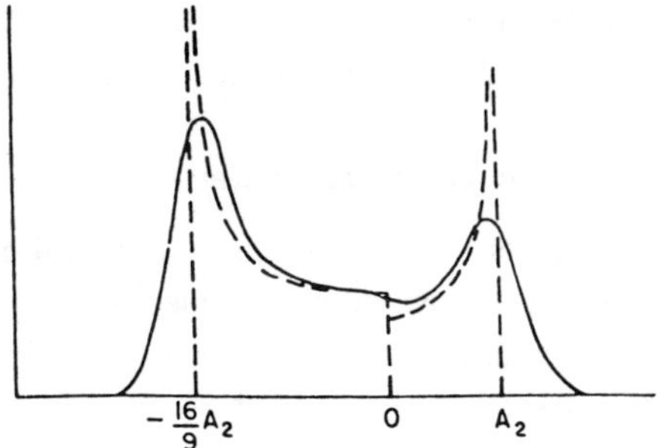

Fig. 7. Resonance lineshape (solid curve) for a nucleus with $I = \frac{3}{2}$ and a moderately large quadrupole interaction in a glass or polycrystalline powder. This is the $\eta = 0$ case. Here $A_2 = \frac{9}{64} \frac{2I+3}{4I^2(2I-1)} \frac{Q_{cc}^2}{\nu_0}$.

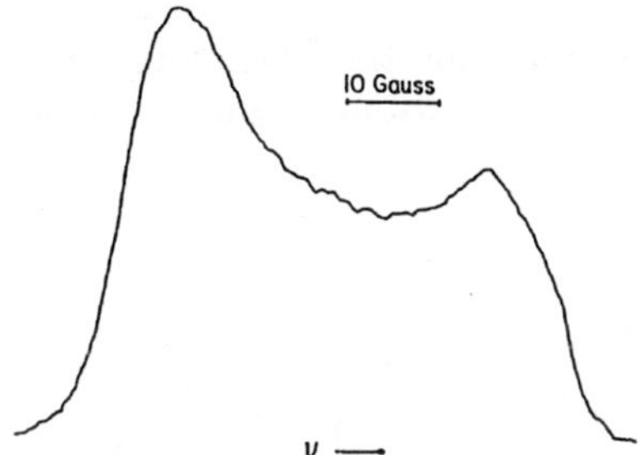

Fig. 8. The ^{11}B NMR spectrum for vitreous B_2O_3 at $\nu_0 = 16$ MHz.

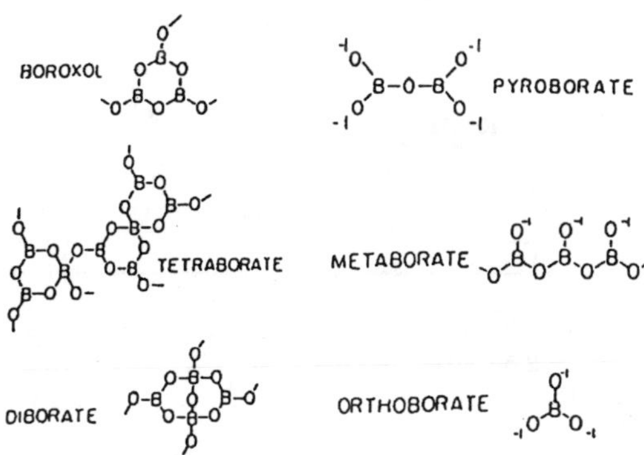

Fig. 9. Some of the structural groupings found in alkali borate compounds and glasses.

portions of the boroxol structural groupings depicted in Fig. 9, but they occur also in other borate structural groupings shown there.)

2.3.2. *Case of a large(r) asymmetry parameter*

In alkali borate compounds of 50 mol% or more of alkali oxide, one finds BO_3 units that have one or more nonbridging oxygens (NBO). (That is, the oxygen is bonded to only one boron and is negatively charged.) If the BO_3 unit has one or two of these, the trigonal symmetry noted above is lost and the asymmetry parameter has been found from NMR spectra to rise into the range $0.4 \leq \eta \leq 0.9$. (However, it is still the case that $2.4\text{ MHz} \leq Q_{cc} \leq 2.9\text{ MHz}$.) An example is calcium metaborate ($CaO \cdot B_2O_3$) whose chain structure is depicted in Fig. 9. The NMR spectrum [14] for this material is displayed in Fig. 10.

2.3.3. *Use of computer simulation to extract values of Q_{cc} and η*

These asymmetric BO_3 units, and the BO_4 and symmetric BO_3 units, are all present in many borate glasses. Responses from all three units are present in the ^{11}B NMR spectrum displayed in Fig. 11, where the narrow structureless

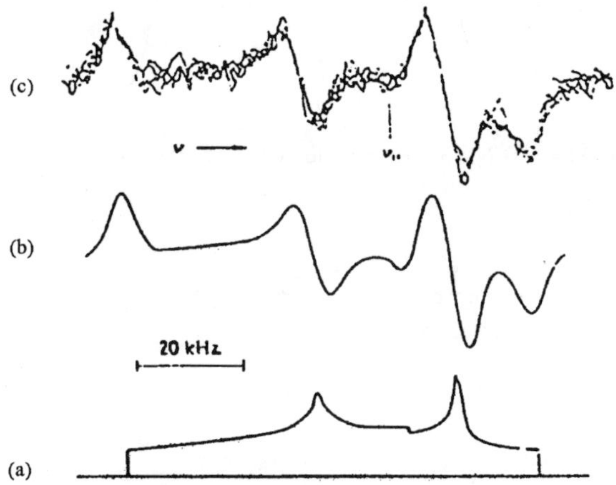

Fig. 10. (a) Theoretical powder pattern for the transition ($m = +\frac{1}{2} \leftrightarrow m = -\frac{1}{2}$), with $I = \frac{3}{2}$, $\nu_0 = 16$ MHz, $Q_{cc} = 2.56$ MHz, and $\eta = 0.54$.
(b) First derivative of the theoretical powder pattern after convolution with a Gaussian curve of linewidth $2\sigma = 5$ kHz.
(c) Superposition of four experimental traces for ^{11}B in polycrystalline calcium metaborate, at a resonance frequency of $\nu_0 = 16$ MHz.

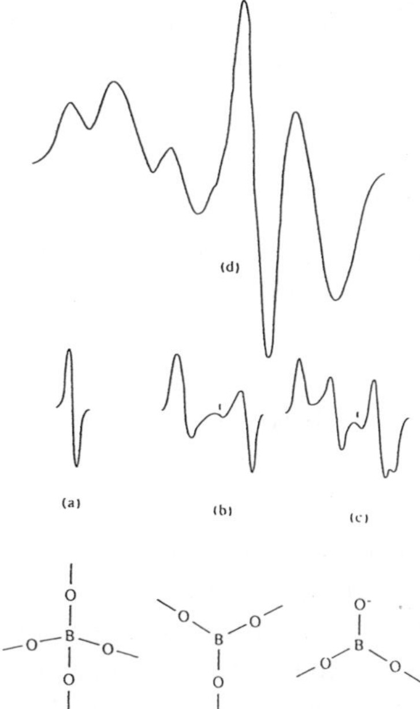

Fig. 11. The ^{11}B NMR spectrum of (a) a BO_4 unit, (b) a BO_3 unit, (c) a BO_3^- unit (with one NBO) and (d) a sample containing all three units.

line from BO_4 units is near the center of the pattern and the symmetric BO_3 give rise to the middle positive peak on the left and part of the negative peak on the right. Computer simulation (as displayed in Fig. 10(b) for $CaO \cdot B_2O_3$) permits choice of the values of Q_{cc} and η that yield the observed quadrupolar perturbation in the NMR spectrum. When the spectrum exhibits more than one response, as in Fig. 11, the computer simulation must be generated for each response. Then the weightings of those computed spectra required to reproduce the observed spectrum give the amounts of each nuclear site in the material under study.

2.3.4. *Consequences of overlap of NMR spectra*

As seen in Fig. 11, the responses from all nuclear sites are — of course — centered around ν_0 (where the center of the strong narrow response crosses the baseline), and they overlap. Smaller responses may, in fact, be lost under the stronger resonances, and the precision of the Q_{cc} and η values is generally no better than 3 and 2 significant figures, respectively. As will be been in the following sections, NQR can yield much higher resolution and substantial improvement in precision.

3. NQR Spectroscopy

3.1. *Genesis of spectra*

The quadrupolar interaction by itself generates a set of energy levels in the absence of a magnetic field. Transitions between these levels will yield resonance spectra if the spectrometers employed are sufficiently sensitive in the relevant frequency regions. These zero-field NMR studies are labeled pure nuclear quadrupole resonance (NQR) spectroscopy.

3.2. *Discovery of NQR*

Pure NQR spectra in solids were first detected by Dehmelt and Kruger in 1950 [15]. A large body of NQR studies of chemical bonding and atomic arrangements in crystalline solids, using quadrupolar nuclei, has followed from that discovery and continues to expand. (See, for example, the summaries by Lucken [16], Schempp and Bray [17], Semin *et al* [18], and Rubinstein and Taylor [19]).

3.3. *Instrumentation and procedures*

The apparatus used for NQR studies is generally the same pulsed or continuous wave equipment used for NMR. (But the frequency must be swept since there is no applied magnetic field to sweep through the resonances). Discussions of standard instruments and procedures can be found in various books [20–23] and journal articles [24–25]. Instruments also used, but less frequently, are of the superregenerative [15, 26–28] and SQUID-based type [29–31].

3.4. *NQR of particular nuclei in inorganic solids*

The following sections summarize the application of NQR to glass studies employing particular nuclei.

3.4.1. ^{75}As NQR

The first detection of NQR in a glass was reported by Rubinstein and Taylor[RT] in 1972 [8] and extended by Taylor and colleagues [19, 32]. The RT work employed the spin $I = 3/2$ nucleus ^{75}As in vitreous arsenic trisulfide

(As$_2$S$_3$). The extended studies [19, 32] also encompassed vitreous arsenic telluride and elemental arsenic.

The Hamiltonian for the pure quadrupole interaction is [33]

$$H = \frac{Q_{cc}}{4I(2I-1)}\left[3I_Z^2 - I(I+1) + \frac{1}{2}\eta(I_+^2 + I_-^2)\right] \tag{1}$$

where $I_\pm = I_x \pm I_y$. When $I = 3/2$, the secular equation (with E in units of $Q_{cc}/4$) is

$$E^2 - \left(1 + \frac{1}{3}\eta^2\right) = 0 \tag{2}$$

and the frequency ν_o of the single allowed transition between the $m = \pm 3/2$ and $m = \pm 1/2$ states, both of which are degenerate in the absence of a magnetic field, is

$$\nu_o = \frac{1}{2}Q_{cc}\sqrt{1 + \frac{1}{3}\eta^2}\,. \tag{3}$$

(Clearly, the values of Q_{cc} and η cannot be independently extracted from this one transition. Methods to determine η — and thus Q_{cc} — independently are discussed later in this article.)

The transition at frequency ν_o given in Eq. 3 was observed by Rubinstein and Taylor [8] for crystalline (orpiment) and vitreous As$_2$S$_3$ at 4.2 K using a variable frequency pulsed NQR/NMR spectrometer and observing the frequency dependence of the spin-echo amplitude. Figure 12 displays the data points and fitted curves for the glass; the two narrower lines in the 70–73 MHz region are data for crystalline As$_2$S$_3$. The latter resonances occur at 70.38 and 72.86 MHz with respective half-widths at half-maximum amplitude (HWHM) of 87 and 44 kHz. (There are two inequivalent As sites in the known structure of orpiment.) But the glass yields only one very broad response centered at 71.6 MHz and having a HWHM of 3.5 MHz.

Orpiment has a layer structure in which each As atom is bonded to three S atoms in a triangular pyramidal arrangement with an As atom at the apex and three S atoms at the base. Rubinstein and Taylor [8] assumed that the large breadth of the lineshape in the As$_2$S$_3$ glasses arises from a distribution of distortions in the apex angles α. Since a difference of 1.3° between the α values for the two sites in orpiment produces a difference of 2.48 MHz between the NQR responses, a distribution of apex bonding angles α with a half-width of 2° will produce a distribution of frequencies with the observed half-width of

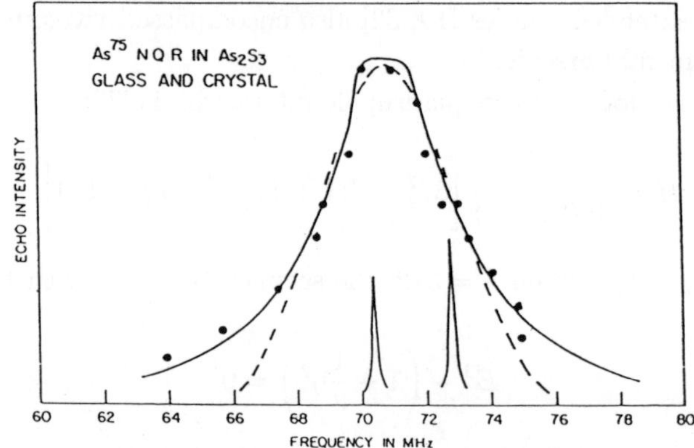

Fig. 12. ^{75}As NQR spectrum of vitreous As_2S_3 at 4.2 K. The experimental data are shown by the points. The two narrow lines are data for crystalline As_2S_3 (see Rubinstein and Taylor [19]). Attempted fitting of the data for the glass involves superposition of two separated Gaussian (dashed line) or Lorentzian (solid line) responses [19]. (This figure is taken from Ref. [19] which contains later and more accurate data than Ref. [8]).

3.5 MHz for the glass. (The analysis assumes that the sites present in orpiment, and yielding the observed NQR frequencies, are present in the As_2S_3 glass. This has been placed in question by the work of Szeftel and Alloul that is discussed later in this article.)

In a later study Rubinstein and Taylor [19] detected and analyzed the pure NQR spectra for crystalline and vitreous As_2Se_3 and obtained a better resolved spectrum for vitreous As_2S_3 than that obtained in the original study [8]. They also measured the spin-lattice relaxation time [21] T_1 at various temperatures. From the near equality of the As_2S_3 crystalline and amorphous initial relaxation behavior, Rubinstein and Taylor [19] conclude that the near-neighbor environments of As atoms in the crystal and the glass are quite similar (i.e. the two sites in orpiment, and two-dimensional correlations between As_2S_3 pyramidal units in the layer-structured crystalline phase, are retained in the glass.)

Application of a small magnetic field at a series of orientations of an orpiment single crystal yielded data that, based on calculations by Bersohn [33] and Dean [34], yielded values of the asymmetry parameter for the two sites in the crystal that Rubinstein and Taylor conclude are also present in the glass. This case of a Zeeman perturbation of an NQR spectrum is discussed in a later section of this chapter. The values of η for the two sites are 0.343 and 0.374.

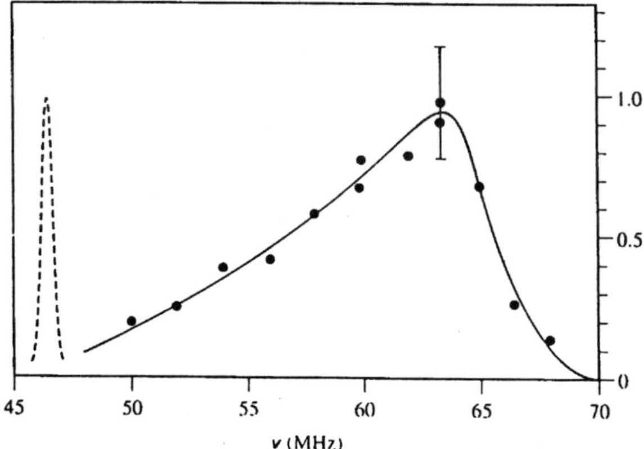

Fig. 13. [75]As NQR absorption in amorphous As (circles and solid line) and orthorhombic As (dashed line) at 4.2 K. (See Jellison *et al* [34]).

Jellison, Petersen, and Taylor [32] reported in 1980 an NQR study of amorphous, orthorhombic, and rhombohedral arsenic. Figure 13 displays the highly asymmetric response for the arsenic glass, and the much more narrow line for orthorhombic arsenic. Jellison *et al.* found that the asymmetric pattern found for the glass could be reproduced in a calculation by varying the dihedral angle using distributions in the angle suggested in the literature. (The dihedral angle is the angle of rotation along the As–As bond required to bring the pair into an eclipsed position.) The temperature dependence of the NQR frequencies for the crystalline forms of arsenic were obtained and could be well fit by the customary Bayer model [35], but the NQR lineshape for the arsenic glass is too broad for observation of shifts in the peak frequency with temperature.

In more recent studies [36], Taylor and colleagues have used [75]As NQR and [63]Cu NMR spectroscopy to study chalcogenide glasses in the systems Cu–As–S and Cu–As–Se.

3.4.2. [11]B NQR

The NQR resonance frequencies, predicted from the quadrupolar effects observed in NMR studies, for many other quadrupolar nuclei in glasses (e.g. [7]Li, [10]B, [11]B, ^{14}N, [17]O, [23]Na, [27]Al, [39]K, [51]V) lie below 10 MHz and even below 1 MHz in some cases. Sufficient signal-to-noise (S/N) values for detection of those responses can be very difficult to achieve. However, a modified Robinson circuit [24], placing the entire LC tank circuit at 77 K and utilizing

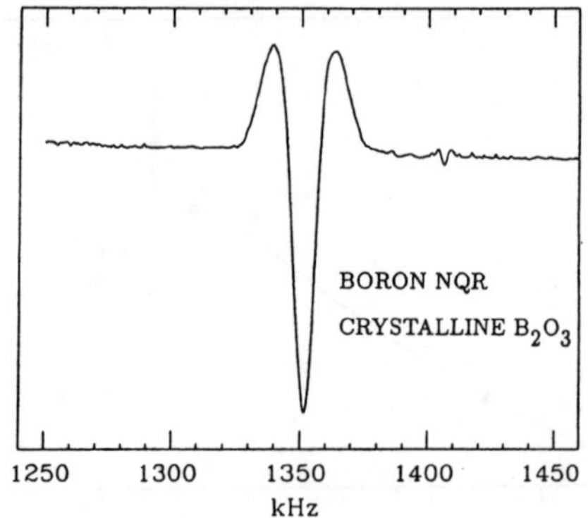

Fig. 14. Boron NQR spectrum for crystalline B_2O_3.

state-of-the-art solid state components and Zeeman modulation (in this case, an antisymmetric square wave of magnetic field), has achieved high S/N at low frequencies. Figure 14 from a 1990 Gravina and Bray paper [9] displays the ^{11}B NQR spectrum for crystalline boron oxide (B_2O_3). ^{11}B has a spin $I = 3/2$ and Eq. 3 relates the NQR frequency to Q_{cc} and η. A S/N value between 100 and 200 for the ^{11}B response is easily obtained at this frequency of 1.358 MHz. The small response at a higher frequency than that for the large ^{11}B response is discussed later in this chapter.

In light of the extreme width of the ^{75}As NQR spectra in amorphous As and vitreous As_2S_3 and As_2Se_3, one might anticipate failure in seeking the ^{11}B NQR spectrum in borate glasses near 1.36 MHz. But, if those widths of some MHz are caused by distributions in the EFG components, they will scale with the resonance frequency. Figure 15 displays the ^{11}B NQR spectrum for vitreous B_2O_3. The full width at half maximum (FWHM) of the larger response is 21 kHz. This is only a moderate increase in width from the FWHM of 15 kHz, arising from dipolar broadening, in crystalline B_2O_3.

There are clearly two ^{11}B responses in the spectrum displayed in Fig. 15, one having approximately four times the intensity of the other. The NMR spectrum for the glass [1, 11, 13, 37–39] displays only one response when conventional low-field spectrometers are employed. (See Fig. 8.) Clearly, the weaker NQR response seen in Fig. 15 is obscured under the NMR signal (Fig. 8) arising from the boron nuclei that produce the stronger NQR resonance in Fig. 15. (This possibility was discussed earlier in relation to Fig. 11.) The strong response arises [38] from borons in boroxol structural groupings (see Fig. 9);

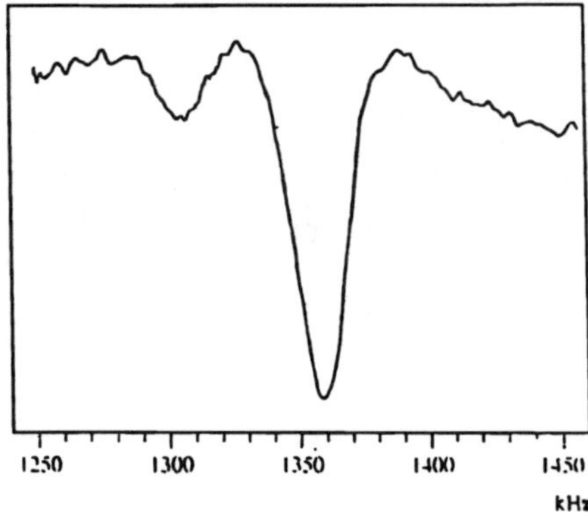

Fig. 15. ^{11}B responses from vitreous B_2O_3.

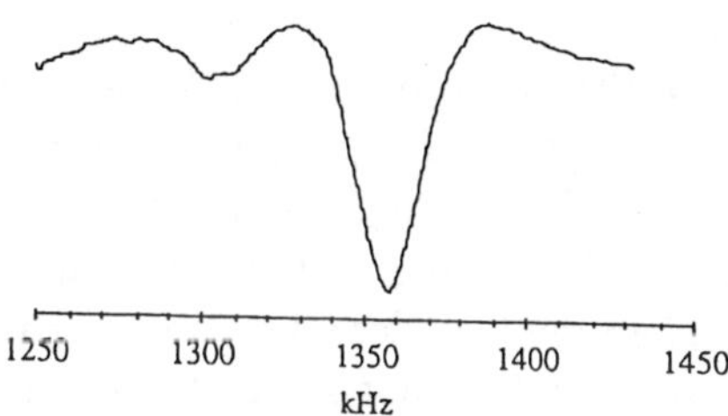

Fig. 16. ^{11}B NQR spectrum of quenched (i.e. rapidly cooled) B_2O_3 glass.

identification of the origin of the weaker signal has not yet been accomplished, but it might arise from BO_3 units that connect the boroxol range [38].

The spectrum in Fig. 15 for vitreous boron oxide was obtained from a sample that had been allowed to cool without quenching. Figure 16 displays the NQR spectrum for a sample quenched between brass blocks. Note the structure that is present in the weaker response. Clearly, from comparison of Figs. 15 and 16, there are thermal history effects on the structure of the glass, and the NQR spectrum from a glass fiber of B_2O_3 (which involves extremely rapid quenching) will be of interest. Substantial thermal history effects have already been detected [40] by ^{11}B NMR in a glass of the system $CaO \cdot SiO_2 \cdot B_2O_3 \cdot Al_2O_3$.

Some work has been completed in ^{11}B NQR studies of structural arrangements and chemical bondings in alkali borate glasses. According to the model proposed by Jan Krogh-Moe [41] and supported (but not proved) by the ^{11}B

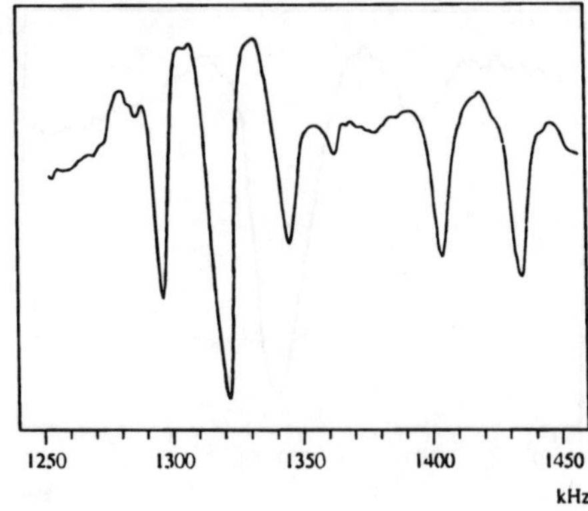

Fig. 17.   [11]B responses from a polycrystalline sample having the sodium tetraborate composition ($Na_2O \cdot 4\,B_2O_3$).

and [10]B NMR studies of Bray and co-workers [42–44], those glasses should be composed of mixtures of the structural groupings presented in Fig. 9. As an indication of the sensitivity of the [11]B NQR resonance frequency to the particular structural grouping in which the BO_3 or BO_4 units are located, it is instructive to study Fig. 17. The sample under study was prepared as polycrystalline sodium tetraborate ($Na_2O\cdot4B_2O_3$), but X-ray diffraction showed the material to be a mixture of sodium borate compounds. [11]B NQR responses from at least six sites are resolved in the spectrum displayed in Fig. 17, and they span a range of some 150 kHz. There is hope, then, of identifying the particular structural groupings present in any borate glass, and even gaining a measure of the amount of each grouping present.

Figure 18 displays the [11]B NQR spectrum for a glass of molar composition $0.15Na_2O$–$0.85\,B_2O_3$ with the spectrum for vitreous B_2O_3 superposed. It is clear that there are at least three different structural groupings present in the sodium borate glass, and that one of them is probably the boroxol unit found in vitreous B_2O_3. On the basis that the structural groupings from the crystalline compounds nearest in composition to the glass should predominate, the other responses may arise from the tetraborate group or the pentaborate and triborate groups that can join to form the tetraborate structural grouping (see Fig. 9).

A problem with [11]B NQR is that the single resonance frequency (see Eq. 3) does not permit extraction of the values of Q_{cc} and η: both parameters occur

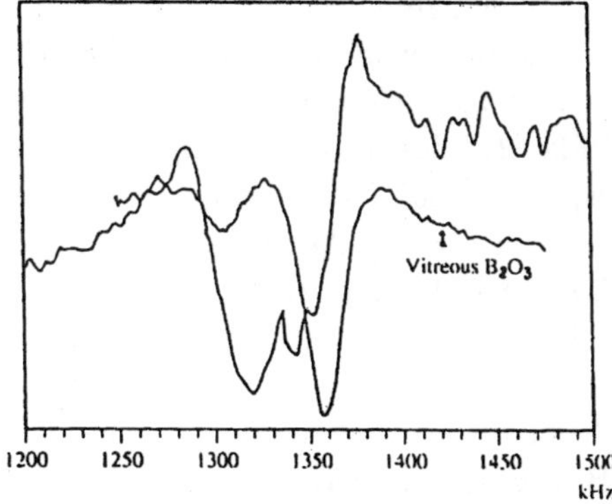

Fig. 18. ^{11}B NQR for a glass of composition 0.15 Na$_2$O–0.85 B$_2$O$_3$ referenced to vitreous B$_2$O$_3$.

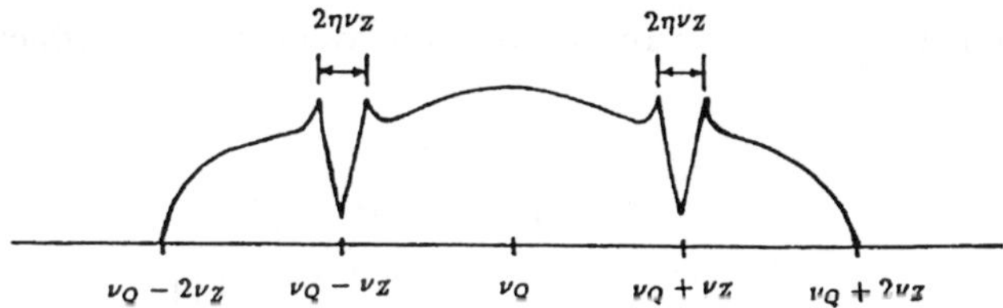

Fig. 19. NQR powder pattern for $I = \frac{3}{2}$ nuclei in the presence of a small external magnetic field ($\mathbf{H_0}$) parallel to the axis of the coil (inductance L) containing the sample; $\nu_z = \gamma H_0$ (see text).

in the single equation. But fortunately, there is a simple modification of the NQR experiment that does allow one to obtain Q_{cc} and η from the single ^{11}B response. This involves applying a small (e.g. 30 gauss) magnetic field along the axis of the coil containing the sample. (The coil is also the inductance in the LC tank circuit of the spectrometer.) When the field is present, the NQR response changes from the simple lineshape of Fig. 14 to the structured pattern displayed in Fig. 19 where ν_Z is just the product of the known gyromagnetic ratio (γ) of the nucleus and the strength H$_0$ of the applied magnetic field. (The drawing in Fig. 19 is inverted from the actual spectrum in Fig. 14.) Clearly, one can easily first obtain ν_Z and then η from this pattern. Then Eq. 3 yields Q_{cc}. Values of Q_{cc} and η accurate to 3 and 2 significant figures, respectively, can be obtained by this method. These are the limits of accuracy previously noted when Q_{cc} and η are obtained from NMR spectra.

In actual practice, an antisymmetric square-wave of magnetic field (Fig. 20) is employed rather than a static field. While the field is off, the response pattern is a simple bell-shaped curve (broken line in Fig. 20); when the field is on, the response is that of Fig. 21 (dotted line in Fig. 20). The electronics of the system place the responses 180° out of phase so that the recorded signal is given by the solid curve in Fig. 20. An actual response [45, 46] for crystalline $CaO \cdot B_2O_3$ at 77 K is displayed in Fig. 21. The values of $Q_{cc} = 2.59 \pm 0.01$ MHz and $\eta = 0.52 \pm 0.02$ obtained are in agreement with the results of earlier NMR studies [47].

This discussion has so far considered only boron in BO_3 units. Since the values of Q_{cc} for ^{11}B in BO_4 units have been found from NMR measurements to be less than 900 kHz, regions of frequency below about 500 kHz would have to be searched for the boron NQR responses for these 4-coordinated borons. Success [48] has been achieved (Fig. 22) for the single ^{11}B response at about 275 kHz for 4-coordinated borons in polycrystalline lithium diborate ($Li_2O \cdot 2B_2O_3$) at 77 K. This realtively large value of Q_{cc} (which would be

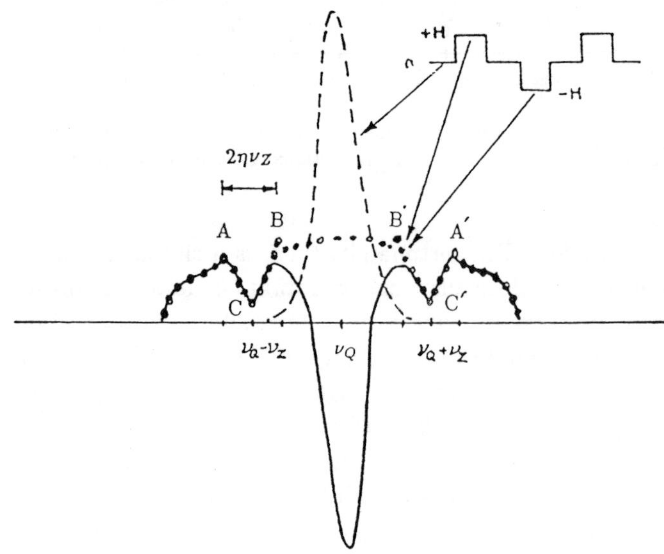

Fig. 20. NQR lineshape obtained using a square wave Zeeman modulation. The resonance obtained when the modulation field is off is indicated by the broken line; the powder pattern appears (circles) when the field is on; the detected response (solid line) is the difference between the broken line and the line of circles.

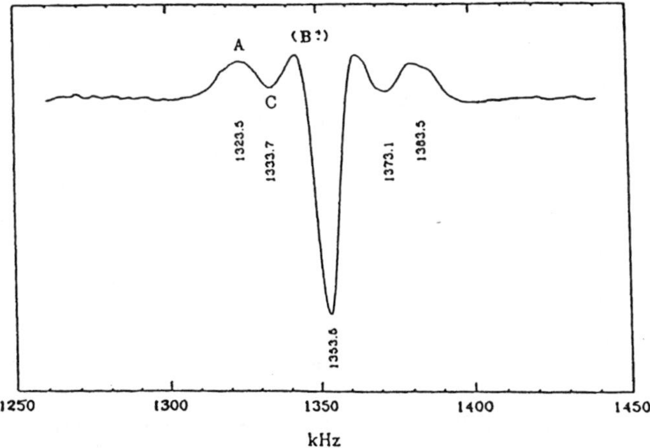

Fig. 21. ^{11}B NQR pattern for polycrystalline CaO·B$_2$O$_3$ obtained using the procedure diagramed in Fig. 20; the sharp features A, B, and C of Fig. 20 are smoothed by line broadening mechanisms (e.g. dipolar interactions).

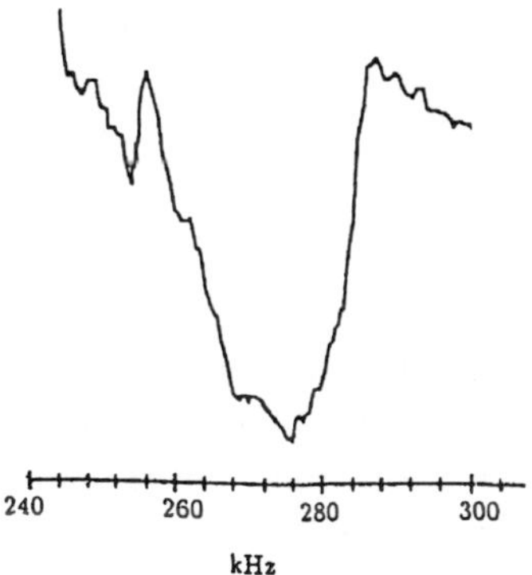

Fig. 22. ^{11}B NQR spectrum for four-coordinated borons in crystalline Li$_2$O·2B$_2$O$_3$.

550 kHz if $\eta = 0$) probably arises from distortion of the BO$_4$ units in the diborate grouping (Fig. 9). (The value of Q_{cc} would be zero for perfect tetrahedral symmetry.) The concept of "bent bonds" may be applicable in achieving the geometry of this grouping.

The ^{11}B NQR spectrum [49] for crystalline lead diborate (PbO·2B$_2$O$_3$) is shown in Fig. 23. Two responses were detected at 406.7 ± 0.5 and 482.5 ± 0.5 kHz, indicating the existence of two distinct types of BO$_4$ units in

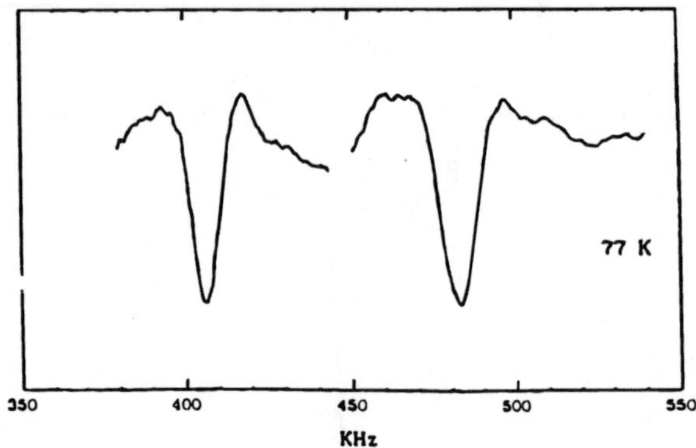

Fig. 23. ^{11}B NQR responses for crystalline PbO·2B$_2$O$_3$.

PbO·2B$_2$O$_3$. This result is in agreement with an X-ray study of PbO·2B$_2$O$_3$ by Perloff and Block [50], who reported the existence of two types of BO$_4$ tetrahedra with different B-O bond lengths and O-B-O angles. The existence of two different tetrahedra may be a consequence of the fact that, in order for all of the borons to be four-coordinated in this compound, some of the oxygens are three-coordinated while the rest are in the usual two-coordinated state.

3.4.3. ^{10}B NQR

Fortunately, there is a way to carry out NQR of boron that greatly increases the accuracy of the data obtained; that is to use the NQR responses of the ^{10}B isotope. The small response in Fig. 14 at a frequency a little higher than that of the single ^{11}B resonance arises from ^{10}B. (Figure 24 displays the responses in Fig. 14 but with a greatly expanded vertical scale.) The ^{10}B isotope has a spin $I = 3$ and its quadrupolar interaction gives rise to as many as thirteen NQR transitions [51]. Figure 25 displays a plot of the values of ν/Q_{cc} vs. η where ν is the measured frequency for a ^{10}B NQR response. In principle, it is only necessary to measure the frequencies of any two ^{10}B responses, form their ratio, and determine at what value of η in Fig. 27 that ratio of two values of ν/Q_{cc} is found. Then Q_{cc} is obtained from the measured value of ν and the value of ν/Q_{cc} in Fig. 27 at the applicable value of η. In practice, many responses are identified and their frequencies are fed into a computer program that has been designed [45, 52] to yield from the entered data the values of Q_{cc} and η. Figure 26 displays another ^{10}B response, near 883 kHz, for crystalline

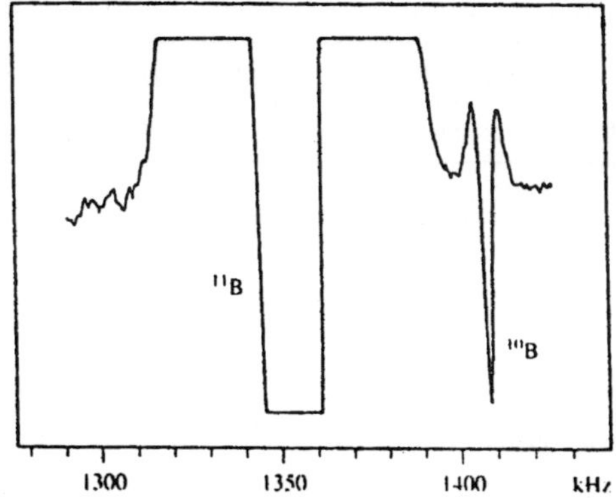

Fig. 24. The features of Fig. 14 displayed with a greatly expanded vertical scale.

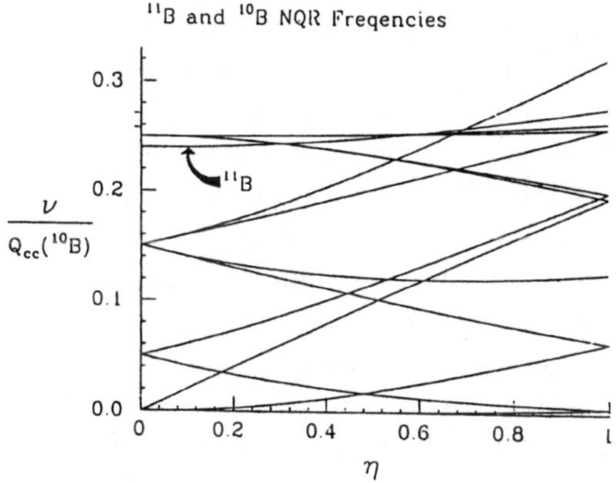

Fig. 25. NQR frequencies (divided by pertinent value of Q_{cc}) for the ^{10}B nucleus ($I = 3$) plotted as a function of the asymmetry parameter η. For ^{11}B nucleus ($I = 3/2$) there is only a single transition, the frequency variation is indicated by an arrow.

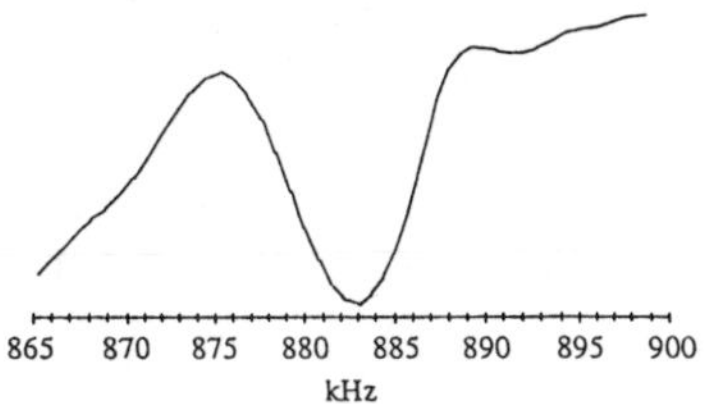

Fig. 26. A ^{10}B resonance from the spectrum of crystalline B_2O_3 at 77 K.

Table 1. ^{11}B quadrupole parameters in crystalline B_2O_3.

^{11}B	NMR 300 K	NQR 300 K	NQR 77 K
Q_{cc} (kHz)	2690 ± 30	2682.6 ± 0.5	2701.1 ± 0.2
η	0.06 ± 0.02	0.071 ± 0.001	0.067 ± 0.001

B_2O_3. Use of this and other responses at 77 K, including the one in Figs. 14 and 24, yields the values [45] for the ^{11}B quadrupolar parameters displayed in Table 1. Characteristically, NQR using the ^{10}B isotope yields values of Q_{cc} and η with 5 and 3 significant figures, respectively, as compared to the 3 and 2 figures obtained in NMR studies [53, 54].

Another example of the use of ^{10}B NQR is given by Fig. 27 where the ^{10}B responses [45, 46] for polycrystalline calcium metaborate at 77 K are displayed. (The weaker responses were confirmed by many repetitions of the scan of the relevant region of frequencies.) Analysis of the data yields the following values of the quadrupolar parameters for ^{11}B: $Q_{cc} = 2594.30 \pm 0.05$ kHz and $\eta = 0.515 \pm 0.001$. These values should be compared with those given above that were obtained using the ^{11}B NQR response and a small magnetic field.

A problem arises, however, in attempting to secure the ^{10}B NQR responses for glasses. The ^{10}B lines in crystalline compounds are quite narrow (see Fig. 24) due to two circumstances [21]: (1) dipolar broadening depends on the product of the gyromagnetic ratios of the interacting nuclei, and that of ^{10}B is much smaller than that of ^{11}B; (2) the broadening interaction between unlike nuclei such as (^{10}B, ^{11}B) is less than that between like nuclei such as (^{10}B, ^{11}B). (Note that the natural abundances of ^{10}B and ^{11}B are 19.6 and 80.4% respectively so that ^{10}B and ^{11}B nuclei both interact predominantly with ^{11}B nuclei.) But the nuclear electrical quadrupole moment Q for ^{10}B is about twice as large as that of ^{11}B. Since the linewidth of the NQR spectra for glasses is determined by the spread in the value of the EFG components, not by dipolar broadening, the width of ^{10}B responses in glasses will be about twice that of the ^{11}B lines. The 21 kHz FWHM for ^{11}B in vitreous B_2O_3 (Fig. 15), as compared to the 15 kHz FWHM for crystalline B_2O_3, indicates that one can expect a FWHM value of some 42 kHz for the ^{10}B lines in the glasses. In the crystal those lines are less than 3 kHz broad. The increase by a factor of perhaps 15 for ^{10}B linewidths in going from crystal to glass will make detection of the ^{10}B responses in the latter very difficult. (Note in Fig. 14

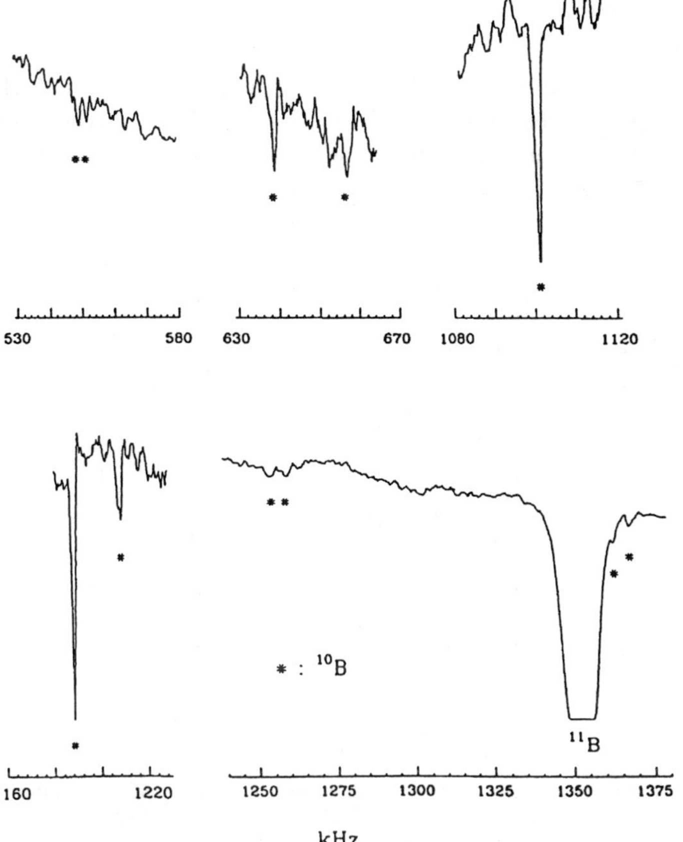

kHz

Fig. 27. ^{11}B and ^{10}B NQR responses for $CaO \cdot B_2O_3$ at 77 K; the ^{11}B line is driven off scale in order to show the weaker ^{10}B responses.

the relative sizes of the ^{10}B and ^{11}B responses in the crystalline material.) No ^{10}B responses in glasses have been detected so far. But improvements in spectrometer sensitivity, use of much longer periods of signal averaging, enhancement of the ^{10}B fraction of boron from the naturally occurring 19.8% to 90–95%, and/or work at temperatures well below that of liquid nitrogen (77 K) should permit detection of the resonances. Fortunately, the less accurate quadrupolar parameters obtained from application of a small magnetic field to the very strong ^{11}B NQR response will permit identification of structural groupings in the glasses in many cases.

3.4.4. $^{27}Al\ NQR$

The Robinson-type CW spectrometer has been used successfully [55] to detect ^{27}Al NQR spectra in polycrystalline powders and glasses at relatively low frequencies. Figure 28 displays the NQR responses at 77 K from the 5-coordinated and 6-coordinated aluminum sites in polycrystalline andalusite [56]. ^{27}Al has a spin $I = 5/2$ and the secular equation for the energy values E (with E in units of $Q_{cc}/20$) is

$$E^3 - 7(3 + \eta^2)E - 20(1 - \eta^2) = 0 \,. \tag{4}$$

Numerical solution via a computer program yields the plot of the resonance frequencies divided by Q_{cc} vs. η shown in Fig. 29. The values of Q_{cc} and

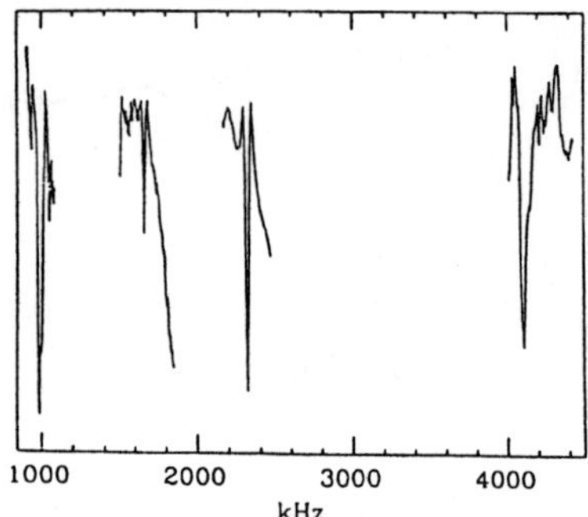

Fig. 28. ^{27}Al NQR responses from aluminum sites in polycrystalline andalusite at 77 K: below 2 MHz, five-coordinated site; above 2 MHz, six-coordinated site.

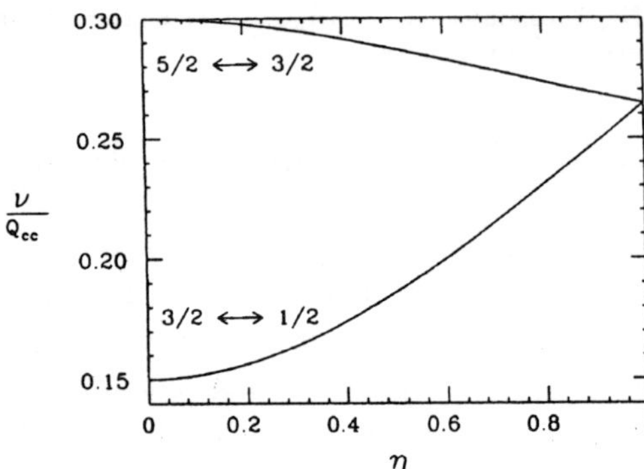

Fig. 29. Plot of ν/Q_{cc} vs. η for an $I = \frac{5}{2}$ nucleus. Here ν is the transition frequency.

η extracted from the responses in Fig. 28 are displayed in Table 2 along with values determined from two NMR rotational studies of single crystals [57, 58], and an earlier NQR study [59] of a sample that was not of good quality, containing inclusions of other material. The results from NQR study 2 for the polycrystalline andalusite agree well with the (much less accurate) single-crystal study by Hafner *et al* [57]. The major gain in accuracy obtained with the much simpler NQR detection using a polycrystalline powder is apparent.

Table 2. ^{27}Al quadrupole parameters for the mineral andalusite.

Studies	AlO$_5$		AlO$_6$	
	Q_{cc}(MHz)	η	Q_{cc}(MHz)	η
NQR study 1 (59)	5.696 ± 0.018	0.398 ± 0.008	14.112 ± 0.095	0.300 ± 0.019
NQR study 2 (56)	5.960 ± 0.007	0.700 ± 0.002	15.261 ± 0.003	0.106 ± 0.001
NMR study 1 (57)	5.9	0.7	15.6	0.08
NMR study 2 (58)	5.7	0.7	10.5	1.0

Note. The values of Q_{cc} and η for two different aluminum-oxygen coordinations (i.e. five and six coordinations) obtained from this NQR work are compared with those obtained from NMR studies (57, 58) of single crystals. The second and third columns are the Q_{cc} and η values obtained from the five-coordinated aluminum site, and the fourth and fifth columns are those obtained from the six-coordinated aluminum site.

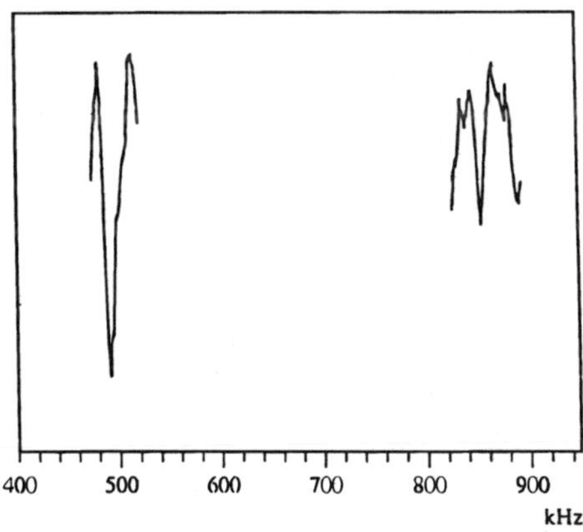

Fig. 30. The two ^{27}Al NQR responses for vitreous 16.8 Al$_2$O$_3$–83.2 TeO$_2$ in molar %.

Figure 30 displays the two NQR responses [60] for ^{27}Al in a glass of composition (mol %) of 16.8 Al $_2$O$_3$–83.2TeO$_2$ at 77 K. The frequencies are 490.1 and 853.3 kHz and the widths (FWHM) are approximately 15 kHz. Values of $Q_{cc} = 2.909 \pm 0.014$ MHz and $\eta = 0.346 \pm 0.014$ were obtained from the data. An NMR measurement yielded an isotropic chemical shift of 36.7 ppm which identifies a pentahedrally coordinated aluminum site.

It is clear that NQR studies of aluminum in 4, 5, and 6-coordination in glasses are feasible and very accurate in measurements of the quadrupolar parameters.

3.4.5.　*NQR of polymers*

Although organic polymers are often not included in discussions of inorganic glasses, they do share many characteristics with the latter. In fact, the more disordered regions of polymers are labeled "glassy" while the more ordered regions are called "crystalline". An example is one of the nylons, nylon-6, in which the "crystalline" content can approach some 70% depending on the thermal history [61].

Semin and co-workers [18, 62–64] have published ^{35}Cl NQR spectra for a number of polymers, including poly (vinylidene chloride) and its copolymers with polystyrene and the polyester of hydroxyenanthic acid. Whereas there are four responses in the ^{35}Cl NQR spectrum for the monomeric vinylidine chloride (average frequency = 36.544 MHz), there are 8 broader lines for the

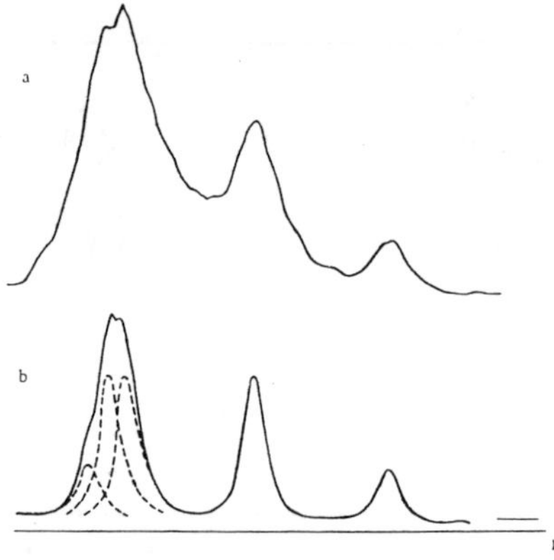

Fig. 31. NQR spectra of ^{35}Cl in (a) pure poly (vinylidene chloride); (b) in the copolymer of poly (vinylidene chloride) with the polyester of hydroxyenanthic acid.

polymer, with an average frequency of about 37.65 MHz). Those lines clump into three broad, but distinct, responses having values of the full-width at half-maximum (FWHM) of 160, 200, and 400 kHz. (See Fig. 31.) Values of the FWHM up to 800 kHz, with widths of about 2 MHz at the baseline, have been observed in other chlorine-containing polymers. These large widths arise, of course, from the distribution in the values of the EFG components caused by the disorder in the polymer. Such broad responses complicate and reduce the ability to gain structural and bonding information from the NQR spectra.

Since the width should scale with the frequency (if the degree of disorder is the same), detection of ^{14}N NQR spectra at frequencies in the 2 to 4 MHz region (with anticipated widths of perhaps only a few to 20 or even 80 kHz) should prove useful in structural and dynamic studies of nitrogen-containing polymers complementing and extending the insights gained from extensive NMR studies [65–67] of ^{15}N chemical shifts. Other nuclei (e.g. ^{17}O) yielding relatively low-frequency NQR spectra (and consequently relatively small linewidths) may also prove useful in studying a variety of polymeric materials.

3.5. *NQR with a Zeeman perturbation*

We survey now the case in which a small magnetic field is present during the NQR experiment. This Zeeman splitting of a pure NQR spectrum was first treated by Bersohn [32] using a perturbation procedure. Dean [33] subsequently treated the theory of Zeeman splitting rigorously and discussed the case of nuclear spin $I = 3/2$ in detail. The applied field lifts the degeneracy of the $m = \pm 3/2$ and the $m = \pm 1/2$ states, yielding the energies [68] and transition frequencies displayed in Fig. 32.

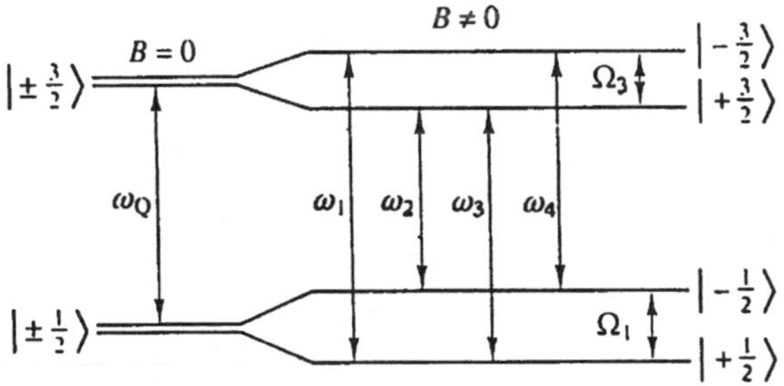

Fig. 32.   Energy-level diagram for a nuclear spin $I = 3/2$ in the presence of a strong quadrupole coupling and, respectively, zero and nonzero Zeeman perturbation (see Szeftel and Alloul [69]).

Rubinstein and Taylor [19] (RT) applied the expressions for the four transition frequencies $\omega_i = \nu_I$, $i = 1, 2, 3, 4$ to the three cases of the field $\mathbf{H}_0$ applied along the mutually orthogonal $\mathbf{a}$, $\mathbf{b}$, and $\mathbf{c}$ axes of the single crystal (orpiment) of As_2S_3, using a magnetic field of about 875 gauss. (The Zeeman interaction in this field would be 638 kHz whereas the NQR frequencies are in the 70–72 MHz range. A first-order perturbation treatment of the Zeeman interaction is adequate.) The values of η extracted for the two inequivalent As sites in orpiment are 0.343 and 0.374.

The RT procedure involved sweeping the frequency of the NQR spectrometer while maintaining the constant value of the magnetic field. This is also the procedure relevant to the earlier discussion in this chapter of obtaining both Q_{cc} and η from the single ^{11}B NQR response. (See Figs. 19, 20 and 21 and the text surrounding those figures.)

Szeftel and Alloul [69] employed the frequency Ω_1 in Fig. 32 to determine the value of the asymmetry parameter for ^{75}As in several vitreous chalcogenides. (In theory, Ω_3 could also be used, but the transition probability is lower than for Ω_1. No response was detected [69] at Ω_3.) The frequency Ω_1 is given by

$$\Omega_1 = \Omega \left[\left(1 - \tfrac{2}{\rho}\right)^2 \cos^2\theta + \left[\left(1 + \tfrac{1}{\rho}\right)^2 + \tfrac{\eta^2}{\rho^2} - 2\tfrac{\eta}{\rho}\left(1 + \tfrac{1}{\rho}\right)\cos 2\phi \right] \sin^2\theta \right]^{1/2}$$

$$(5)$$

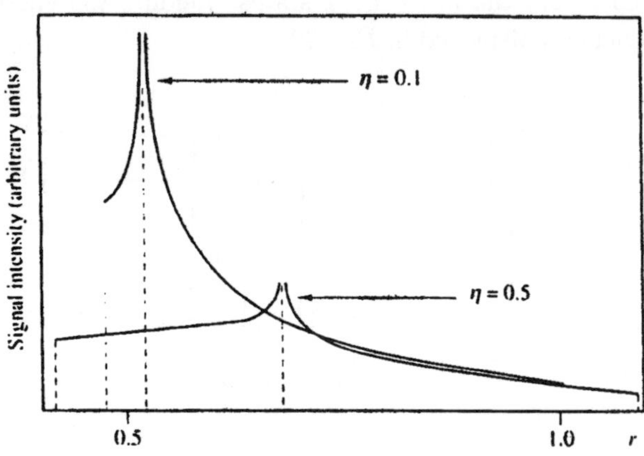

Fig. 33. Spectra (solid lines) calculated for two values of η (see Szeftel and Alloul [69]).

where $\Omega = \gamma H_0$, γ is the gyromagnetic ratio, H_0 is the strength of the applied magnetic field, and $\rho = 1 + \frac{\eta^2}{3}$. Since Eq. (5) does not include the quadrupole coupling constant Q_{cc}, this transition can be used to obtain the value of η quite decoupled from that of Q_{cc}.

Prediction of the observable lineshape for a polycrystalline powder or glass requires averaging over all θ and ϕ including the dependence of the state functions and transition probabilities on those angles. Figure 33 displays the

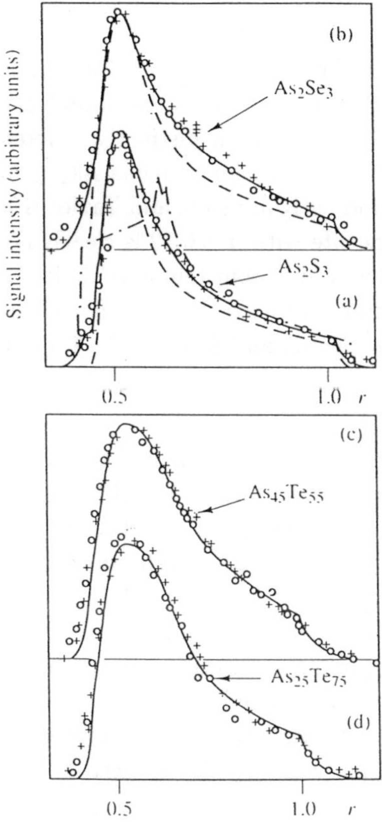

Fig. 34. ^{75}As spectra recorded for four vitreous chalcogenides at 5.9 MHz (circles) and 10.3 MHz (crosses) as a function of the parameter r, as defined in text. The powder spectrum calculated with η values of orpiment is plotted as dashed-dotted lines. Solid lines indicate the best fit. When two Gaussian distributions are needed to obtain the best fit, the contribution of the distribution of higher weight is plotted as dashed lines (see Szeftel and Alloul [69]).

calculated lineshapes for two values of η. The parameter $r = \Omega/\nu = \gamma H_0/\nu$, where ν is the frequency of the spectrometer, has the limits $r_m \leq r \leq r_M$ and a logarithmic divergence occurs at r_p, where

$$
\left.
\begin{aligned}
r_m &= \left[1 + \frac{1+\eta}{\rho}\right]^{-1}, \quad r_M = \left(\frac{2}{\rho} - 1\right)^{-1} \\
r_p &= \left[1 + \frac{1-\eta}{\rho}\right]^{-1}
\end{aligned}
\right\}.
\tag{6}
$$

The ^{75}As spectra obtained at 4.2 K for four vitreous chalcogenides are displayed in Fig. 34. Analysis of the spectra was carried out using Gaussian distributions in η about the value determined at the logarithmic divergence (r_p). For vitreous As_2S_3 and As_2Se_3, it was found necessary to include the contributions from two As sites with distinctly different values of η. These were determined as 0.12 and 0.36 for vitreous As_2S_3, whereas the values for crystalline orpiment found by Rubinstein and Taylor [19] using a single crystal are 0.343 and 0.374. The site with $\eta = 0.12$ is clearly not present in orpiment, and is a site not considered by Rubinstein and Taylor [19] in their discussions of the ^{75}As NQR spectrum for vitreous As_2S_3. For vitreous As_2Se_3, the values of η were determined by Szeftel and Alloul to be 0.14 and 0.45.

Acknowledgments

The boron NQR studies reported here have been carried out at Brown University by Samuel J. Gravina, Degen Mao, and Donghoon Lee while they were graduate students in my research group, and I am most grateful for their work as my colleagues. I am indebted to literally dozens of former graduate students who were involved with the earlier NMR studies of borate materials. Dr. Gary Petersen [70] has been a major source of help and information on electronic circuits and equipment, encouragement, and useful discussions. Both the National Science Foundation and U.S. Borax, Inc. [71] have supported the research, and that financial support is deeply appreciated, as is the extensive advice and provision of materials by Dr. Robert A. Smith and others at U.S. Borax.

References

[1] A.H. Silver and P.J. Bray, *J. Chem. Phys.* **29**, 984 (1958).

[2] H. Eckert, Progress in Nuclear Magnetic Resonance Spectroscopy, eds. Emsley, Feeney and Sutcliffe, Pergamon Press Ltd. England **24**, 3 (1992) 159–293.

Page 163 of that work contains a table of 39 earlier reviews, listing their main topics.

[3] P.E. Stallworth and P.J. Bray, *Glass Sci. Technol.* **B4**, 77 (1990).

[4] A.R. Grimmer, *Exp. Tech. Physik* **36**, 81 (1988).

[5] P.J. Bray, S.J. Gravina, D.H. Hintenlang and R.V. Mulkern, *Magn. Reson. Rev.* **13**, 263 (1988).

[6] P.J. Bray, *J. Non-Cryst. Solids* **95 & 96**, 45 (1987).

[7] M.H. Cohen and F. Reif, Solid State Physics, eds. Seitz and Turnbull (Academic Press, New York, 1958) **5**, p. 321.

[8] M. Rubinstein and P.C. Taylor, *Phys. Rev. Lett.* **29**, 119 (1972).

[9] S.J. Gravina and P.J. Bray, *J. Magn. Reson.* **89**, 515 (1990).

[10] P.J. Bray, J.O. Edwards, J.G. O'Keefe, V.F. Ross and Tatsuzaki, *J. Chem. Phys.* **35**, 435 (1961).

[11] P.J. Bray, Classe des Memoirs, de l'Academie Royale de Belgique, Sciences XXXIII **3**, 37 (1961).

[12] F. Krämer, W. Müller-Warmuth, J. Scheerer and H. Dutz, *Z. Naturforsch.* **82a**, 1338 (1973).

[13] P.J. Bray, Interaction of Radiation with Solids, ed. A. Bishay (Plenum Press, 1970) pp. 11–40. The spectrum was obtained by S.G. Bishop, PhD Thesis, Brown University, 1969.

[14] H.M. Kriz, S.G. Bishop, and P.J. Bray, *J. Chem. Phys.* **49**, 557 (1968).

[15] H.G. Dehmelt and H. Kruger, *Naturwiss* **37**, 111 (1950); **38**, 921 (1951).

[16] E.A.C. Lucken, Nuclear Quadrupole Coupling Constants (Acadmic Press, New York, 1969).

[17] E. Schempp and P.J. Bray, Physical Chemistry (Academic Press, New York, 1970) **4**, p. 521.

[18] G.K. Semin, T.A. Babushkina and G.G. Yakobson, Nuclear Quadrupole Resonance in Chemistry (John Wiley and Sons, New York, 1975).

[19] M. Rubinstein and P.C. Taylor, *Phys. Rev* **B9**, 4258 (1974).

[20] E. Fukushima and S.B.W. Roeder, Experimental Pulse NMR, A Nuts and Bolts Approach (Addison-Wesley, New York, 1981).

[21] A. Abragam, The Principles of Nuclear Magnetism (Oxford University Press, 1961).

[22] P.E. Stallworth and P.J. Bray, Advances in Structural Analyses, eds. by D.R. Uhlmann and N.J. Kreidl (Academic Press, Inc., 1990) **B4**, Chap. 2, entitled "Nuclear Magnetic Resonance".

[23] J.F. Emerson and P.J. Bray, Experimental Techniques of Glass Science, eds. K.J. Simmons and O.H. el-Bayoumi (American Ceramic Society, Westerville, OH, 1993), Chap. 3.

[24] F.N.H. Robinson, *J. Phys. E: Sci. Instrum.* **13**, 861 (1980) and **15**, 814 (1982).

[25] G.L. Petersen, P.J. Bray and R.A. Marino, *Naturforsch* **A49**, 65 (1994).

[26] J.R. Whitehead, Superregenerative Receivers (Cambridge University Press, New York, 1950).

[27] C. Dean, PhD Thesis, Harvard University (1952).

[28] H.G. Dehmelt and H. Krüger, *Z. Physik* **129**, 401 (1951).

[29] C. Connor, J. Chang and A. Pines, *Rev. Sci. Instrum.* **61**, 1059 (1990).

[30] N.Q. Fan and J. Clarke, *Rev. Sci. Instrum.* **62**, 1453 (1991).

[31] C. Connor, Advances in Magnetic and Optical Resonance (Academic Press, Inc., 1990) **15**, p. 201.

[32] G.E. Jellison, Jr., G.L. Petersen and P.C. Taylor, *Phys. Rev.* **B22**, 3903 (1980).

[33] R. Bersohn, *J. Chem. Phys.* **20**, 1505 (1952).

[34] C. Dean, *Phys. Rev.* **96**, 1053 (1952).

[35] H. Bayer, *Z. Phys.* **130**, 227 (1951).

[36] Z.M. Saleh, G.A. Williams and P.C. Taylor, *Phys. Rev.* **B46**, 10557 (1989).

[37] P.J. Bray and J.G. O'Keefe, *Phys. Chem. Glasses* **4**, 37 (1963). J. Zhong and P.J. Bray, *J. Non-Cryst. Solids* **111**, 67 (1989).

[38] G.E. Jellison, Jr. and P.J. Bray, *Solid State Commun.* **29**, 517 (1976) and *J. Non-Cryst. Solids* **29**, 187 (1978).

[39] G.E. Jellison, Jr., L.W. Panek, P.J. Bray and G.B. Rouse, Jr., *J. Chem. Phys.* **66**, 802 (1977).

[40] P.K. Gupta, M.L. Lui and P.J. Bray, *J. Am Ceram. Soc.* **68**, C-82 (1985).

[41] J. Krogh-Moe, *Phys. Chem. Glasses* **3**, 101 (1962).

[42] P.J. Bray, *J. Non-Cryst. Solids* **75**, **29** (1985) and references therein.

[43] G.E. Jellison, Jr. and P.J. Bray, *J. Non-Cryst. Solids* **29**, 187 (1978).

[44] P.J. Bray, S.A. Feller, G.E. Jellison, Jr. and Y.H. Yun, *J. Non-Cryst. Solids* **38–39**, 93 (1980).

[45] Degen Mao, PhD Thesis, Brown University (1991).

[46] Degen Mao and Philip J. Bray, *Solid State Nuc. Magn. Res.* **1**, 255 (1992).

[47] H.M. Kriz, S.G. Bishop and P.J. Bray, *J. Chem. Phys.* **49**, 557 (1968).

[48] S.J. Gravina, P.J. Bray, and G.L. Petersen, *J. Non-Cryst. Solids* **123**, 165 (1990).

[49] D. Mao and P.J. Bray, *J. Non-Cryst. Solids* **144**, 217 (1992).

[50] A. Perloff and S. Block, Acta Crystallogr. **20**, 274 (1966).

[51] R.B. Creel, *J. Magn. Reson.* **50**, 82 (1982).

[52] S.J. Gravina, PhD Thesis, Brown University (1989).

[53] D. Kline, P.J. Bray and H.M. Kriz, *J. Chem. Phys.* **48**, 5277 (1968).

[54] C. Rhee and P.J. Bray, *J. Chem. Phys.* **56**, 2476 (1972).

[55] D. Lee, PhD Thesis, Brown University (1991).

[56] D. Lee and P.J. Bray, *J. Magn. Reson* **94**, 51 (1991).

[57] S.S. Hafner, M. Raymond and S. Ghose, *J. Chem. Phys.* **52**, 6037 (1970).

[58] G.W. Schultz, W. Mueller-Warmuth, W. Poch and J. Scheerer, *Glasstech Ber.* **41**, 435 (1968).

[59] R.V. Pound, *Phys. Rev.* **79**, 685 (1950).

[60] D. Lee and P.J. Bray, *The NQI Newsletter* **1**, 22 (1994).

[61] V.R. Gowariker, N.V. Viswanathan and Jayadev Sreedhar, *Polymer Science*, Halsted Press (a division of John Wiley and Sons, Inc. New York) (1986).

[62] T.A. Babushkina, V.I. Robas and G.K. Semin, Doklady AN SSSR **159**, 164 (1964).

[63] G.K. Semin, V.I. Robas and T.A. Babushkina, Readiospektroskopiya Tverdogo Tela Atomizdat (1967) p. 218.

[64] A.P. Suprun, A.S. Shashkov, T.A. Soboileva, G.K. Semin, T.T. Vasileva, G.P. Lopatina, T.A. Babushkina and R. Kh.Freidlina, Doklady AN SSSR **173**, 1356 (1967).

[65] L.J. Mathias, Solid State NMR of Polymers (Plenum Press, New York, 1991).

[66] L.J. Mathias, D.G. Powell and A.M. Sikes, *Polym. Commun.* **29**, 192 (1988).

[67] D.G. Powell and L.J. Mathias, *J. Am. Chem. Soc.*, **112**, 669 (1990) and references therein.

[68] T.P. Das and E.L. Hahn, Solid State Physics, eds. F. Seitz and D. Turnbull (Academic Press, New York, 1958). Supplement 1.

[69] J. Szeftel and H. Alloul, *Phys. Rev. Lett.* **42**, 1691 (1979).

[70] G.L. Petersen, President, RITEC, Inc., 60 Alhambra Road, Suite 5, Warwick, Rhode Island, USA 02886.

[71] U.S. Borax, Inc., P.O. Box 926, Valencia, California 91380-9026.

D.   SOLID STATE NMR AS A STRUCTURAL TOOL IN GLASS SCIENCE

HELLMUT ECKERT

*Institut für Physikalische Chemie, Westfälische Wilhelms-Universität Münster,
Schloßplatz 7, D48149 Münster, Germany
eckerth@mail.uni-muenster.de*

Contents

1. Introduction

Fifty years after its invention, nuclear magnetic resonance (NMR) spectroscopy has evolved tremendously as a structural and analytical tool in solid state chemistry and materials science. Today, the richness and detail available on structure and dynamics is unparalleled by any other spectroscopic technique carried out on condensed matter. For the investigation of the solid state, NMR often complements diffraction techniques (X-ray and neutron) nicely, by virtue of being element-selective and locally sensitive. These features become particular important in the study of highly disordered, amorphous, and compositionally complex systems, which are generally less accessible to diffraction methods. Here, solid state NMR is able to provide information about coordination numbers, bonding partners, local symmetry, and internuclear distance distributions. In addition to the structural information provided by NMR methods, the direct proportionality of the signal intensity to the number of nuclei giving rise to this signal makes NMR potentially superior to many other commonly used spectroscopic methods. Furthermore, the techniques used are non-destructive and require little sample preparation. Although the low intrinsic detection sensitivity and hence the need for comparatively large sample sizes is a possible disadvantage, for compositionally complex and heterogeneous systems and materials the study of large samples is often necessary to ensure results that are representative. Finally, unlike with any other technique, the inherent time window of NMR can be altered by many orders of magnitude. This makes it possible to study molecular and chemical dynamics over many decades of timescales.

Due to these opportunities, solid state NMR has become one of the most important structural tools in glass science [1–4]. The complexity of the glassy state demands that the information obtained by NMR be combined with evidence from complementary structural techniques. At the same time, it is equally important to exploit each technique to its fullest potential. This is particularly fruitful in the field of solid state NMR, which has remained a vital and attractive research area in its own right. Close attention to technique-oriented research is therefore particularly important and helps to anticipate future applications. Therefore, this review has a three-fold responsibility: (1) To summarize the fundamentals of the technique, (2) to review current high-impact applications of established NMR methodology to important glass issues, and (3) to discuss new developments and ideas likely to influence future directions.

2. Fundamentals of Solid State NMR

2.1. *Nuclear magnetism and resonance*

Naturally, within the confines of this chapter, it is not possible to review the entire theory of solid state NMR spectroscopy. However, the main facts relevant to the applications to glasses must be re-stated. For more information, a number of excellent monographs are available [5–8]. In brief, NMR spectroscopy exploits the fact that certain nuclei possess angular momentum, *spin*. Subject to the laws of quantum mechanics, magnitude and orientation of the nuclear spin are quantized according to:

$$|\hat{\mathbf{I}}| = \sqrt{I(I+1)} \cdot \hbar \tag{1a}$$

$$|\hat{\mathbf{I}}_z| = m\hbar \tag{1b}$$

where $\hat{\mathbf{I}}$ and $\hat{\mathbf{I}}_z$ are the vector operators of the nuclear spin and its z component, I is the spin quantum number and $m(I \geq m \geq -I)$ is the orientational quantum number.

Furthermore, the relationship

$$\mu_z = \gamma m\hbar \tag{2}$$

relates each spin state to a magnetic moment μ_z. Here, γ, the *gyromagnetic ratio*, is a constant specific to the nucleus under consideration. To detect nuclear magnetic moments, an external magnetic field B_o is applied. The interaction is represented by the Zeeman Hamiltonian

$$\mathcal{H}_Z = -\mu_z B_o \tag{3a}$$

with eigenvalues

$$E_m = -m\gamma\hbar B_o \tag{3b}$$

Thus, the Zeeman interaction causes energy splittings between the various nuclear spin orientational states. In the simplest case ($I = 1/2$) two distinct eigenstates exist. By application of electromagnetic waves fulfilling the Bohr condition, $\Delta E = \hbar\omega$, transitions between adjacent quantized spin states can be effected and observed spectroscopically. To a first approximation, the angular resonance frequency ω is given by:

$$\omega = \gamma B_o \tag{4}$$

Since the values of γ differ greatly for different types of nuclei, NMR spectroscopy is intrinsically element-selective. Table 1 summarizes important characteristics of the nuclei frequently employed for NMR measurements on solids and glasses in particular.

2.2. *Spectroscopic technique*

To explain how the NMR signal is detected, it is useful to appeal to the classical prediction that the interaction of the nuclear magnetic moments with the magnetic field forces the nuclei to precess around the field direction with the angular Larmor frequency

$$\omega_p = \gamma B_{\text{loc}} \tag{5}$$

Modern NMR methods measure this precession frequency in the time domain, following pulsed irradiation of the spin system on resonance. Since, due to their differences in energy, the spin eigenstates are not equally populated, the vector sum of the associated magnetic moments produces a resultant magnetization in a macroscopic sample, according to the Boltzmann distribution. This magnetization is the target of the pulsed NMR experiment. The principle is illustrated with vector diagrams in the rotating frame, i.e. in a co-ordinate system that rotates with the carrier frequency ω_o about the magnetic field axis (see Fig. 1). A short (1–10 μs length), intense (100–1000 W power) radiofrequency pulse is applied that tips the magnetization into the plane perpendicular to the magnetic field direction (90° pulse). Following this flip, the magnetization precesses with the ω_p components of the nuclei present and induces an a.c. voltage in the detector coil of the NMR probe. This voltage signal decays in time because the magnetization vectors lose their phase-coherence due to their differences in ω_p as well as due to spin-spin interactions. Following further signal processing and storage of this time-domain signal, the frequency domain spectrum is obtained by Fourier-transformation.

2.3. *Internal interactions*

While the measured precession frequencies are dominated by the Zeeman interaction, they are additionally influenced by a variety of internal interactions,

Table 1. Nuclear magnetic properties, applications and problems encountered in NMR studies of certain isotopes.

Nucleus	Spin	Natural abundance (%)	Resonance frequency (MHz)[*]	Most commonly measured parameter (and experiment)	Typical problems
^{1}H	1/2	100	300.1	δ_{iso} (CRAMPS)[a]	strong dipole-coupling limits resolution
^{2}H	1	0.016	46.07	C_Q, η (Quad. Echo)	isotopic labelling required, 90° pulse lengths $< 2\mu s$ needed
^{6}Li	1	7.42	44.15	δ_{iso} (MAS)	long T_1^{e}, small range of δ_{iso}
^{7}Li	3/2	92.58	116.59	δ_{iso} (MAS), C_Q, T_1	small range of δ_{iso}, e^2qQ/h
^{9}Be	3/2	100	42.18	δ_{iso} (MAS)	small range of δ_{iso}, e^2qQ/h
^{10}B	3	19.85	32.25	C_Q, η (cw)[b]	low sensitivity, broad lines
^{11}B	3/2	80.42	96.25	$C_Q, \eta, \delta_{\text{iso}}$ (MAS)	small range of δ_{iso}
^{13}C	1/2	1.11	75.47	δ_{iso} (CPMAS or MAS)	long T_1 in H-free samples, quantitation difficult
^{14}N	1	99.63	(21.68)	C_Q, η (NQR)[d]	low frequency, broad lines
^{15}N	1/2	0.37	30.41	δ_{iso} (CPMAS or MAS)	isotopic labelling required
^{17}O	5/2	0.037	40.68	$C_Q, \eta, \delta_{\text{iso}}$ (MAS, cw)	isotopic labelling required
^{19}F	1/2	100	282.32	δ_{iso} (CRAMPS)	strong dipole coupling limits resolution
^{23}Na	3/2	100	79.38	δ_{iso} (MAS, MQMAS[g] nutation)	lineshape interpretation; difficult quantitation
^{27}Al	5/2	100	78.20	δ_{iso} (MAS, MQMAS nutation)	lineshape interpretation; difficult quantitation
^{29}Si	1/2	4.70	59.70	δ_{iso} (MAS)	long T_1, labelling often required
^{31}P	1/2	100	121.65	$\delta_{\text{iso}}, M_{2d}$ (MAS, echo)	interpretation of δ_{iso} unclear
^{45}Sc	7/2	100	72.90	C_Q, η (NQR)	strong quadrupole coupling

Table 1. (*Continued*)

Nucleus	Spin	Natural abundance (%)	Resonance frequency (MHz)[*]	Most commonly measured parameter (and experiment)	Typical problems
[51]V	7/2	99.76	78.87	$\delta_{iso}, \delta_{ii}, C_Q, \eta$, (wideline, MAS, nutation)	lineshape deconvolution
[63]Cu	3/2	69.09	(79.54)	C_Q, η (NQR)	strong quadrupole coupling
[71]Ga	3/2	39.60	91.60	C_Q, η, δ_{iso} (MAS)	strong quadrupole coupling
[75]As	3/2	100	(51.38)	C_Q, η (NQR)	strong quadrupole coupling
[77]Se	1/2	7.58	57.20	δ_{iso} (MAS)	long T_1, large csa[f]
[113]Cd	1/2	12.20	66.71	δ_{iso} (MAS)	long T_1, large csa
[119]Sn	1/2	8.58	111.85	δ_{iso} (MAS)	long T_1, large csa, wide δ-range
[125]Te	1/2	7.00	94.79	δ_{iso} (MAS)	long T_1, large csa, wide δ-range
[133]Cs	7/2	100	39.36	δ_{iso}, C_Q, η (MAS)	lineshape deconvolution
[199]Hg	1/2	16.84	53.50	$\delta_{iso}, \delta_{ii}$ (MAS)	long T_1, large csa, wide δ-range
[205]Tl	1/2	70.50	173.12	$\delta_{iso}, \delta_{ii}$ (cw, spin echo)	strong exchange couplings, wide δ-range
[207]Pb	1/2	22.60	62.79	δ_{iso} (cw, MAS)	long T_1, wide δ-range

[*]At a field strength of 7.05 T

[a]CRAMPS = combined rotation and multiple pulse spectroscopy

[b]cw = continuous wave spectroscopy

[c]CPMAS = cross polarization with MAS

[d]NQR = nuclear quadrupole resonance

[e]T_1 = spin-lattice relaxation time

[f]csa = chemical shift anisotropy

[g]MQMAS = multiple quantum excitation with MAS

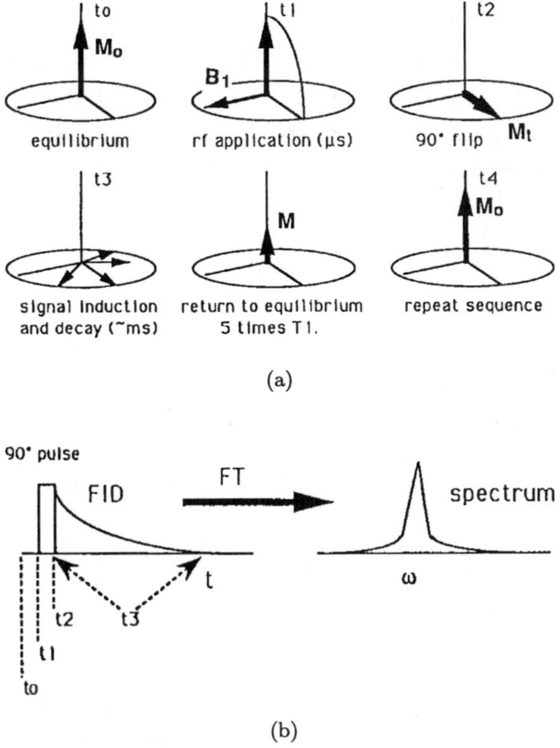

Fig. 1. (a) Detection of NMR signals by pulsed spectroscopy, shown in the rotating coordinate system associated with the oscillating magnetic field component at the applied radio frequency ω_o at various stages $(t_0 - t_4)$ of the experiment: t_0-spin system with magnetization (fat arrow) at equilibrium, t_1-irradiation of the B_1 field orthogonal to the magnetization direction tips the magnetization, t_2-the system after a 90° pulse resulting in transverse magnetization M_t, t_3-off-resonance precession and free induction decay in the signal acquisition period following the pulse, t_4-return to spin equilibrium due to spin-lattice relaxation. (b) Timing diagram of the experiment, followed by Fourier transformation.

whose parameters reflect the details of the local structural environment. In the solid state, three principal internal interaction mechanisms must be considered: (1) magnetic interactions of the nuclei with the surrounding electrons (*chemical shielding interaction*), (2) magnetic dipole-dipole interactions among nuclei, and (3) interactions between the electric quadrupole moment of spin $> 1/2$ — nuclei with electrostatic field gradients surrounding these nuclei (*nuclear electric quadrupolar coupling*). The effects of these interactions on the energy levels can be calculated using standard perturbation theory.

2.3.1. *Chemical shielding interaction*

Magnetic polarization effects induced in the electronic environment of the nuclei under the influence of B_o affect the local field at the site of the nuclei. Since this effect is particularly sensitive to the valence electron configuration, it has been labelled the *chemical shielding interaction*. In solids the secular chemical shielding Hamiltonian (corresponding to the application of first order perturbation theory)

$$\mathcal{H}_{\mathrm{CS}} = \gamma \hat{\mathbf{I}}_z \overset{\leftrightarrow}{\sigma} \hat{\mathbf{B}}_o \tag{6}$$

is parametrized by a symmetric second rank (3×3) tensor $\overset{\leftrightarrow}{\sigma}$. This tensor is diagonal in a specific principal axis system, which can be related to local symmetry. The diagonal components σ_{11}, σ_{22}, and σ_{33} can also be cast in terms of the parameters:

$$\sigma_{\mathrm{iso}} = \frac{1}{3}(\sigma_{11} + \sigma_{22} + \sigma_{33}) \quad \text{(isotropic component)} \tag{7a}$$

$$\sigma_{\mathrm{ax}} = \sigma_{33} - \frac{1}{2}(\sigma_{22} + \sigma_{11}) \quad \text{(axial component)} \tag{7b}$$

$$\eta = \frac{\sigma_{22} - \sigma_{11}}{\sigma_{33} - \sigma_{\mathrm{iso}}} \quad \text{(asymmetry parameter)} \tag{7c}$$

As a convention $|\sigma_{33} - \sigma_{\mathrm{iso}}| > |\sigma_{22} - \sigma_{\mathrm{iso}}| > |\sigma_{11} - \sigma_{\mathrm{iso}}|$. Due to the chemical shielding anisotropy, the nuclear precession frequency in the solid state depends on orientation:

$$\omega_p(\theta, \phi) = \gamma B_o \{ (1 - \sigma_{11})^2 \cos^2 \phi \sin^2 \theta + (1 - \sigma_{22})^2 \sin^2 \phi \sin^2 \theta$$

$$+ (1 - \sigma_{33})^2 \cos^2 \theta \}^{1/2} \tag{8a}$$

For the case of axial symmetry ($\eta = 0$) this relation reduces to:

$$\omega_p(\theta) = \gamma B_o \left(1 - \sigma_{\mathrm{iso}} - \frac{1}{3}\Delta\sigma(3\cos^2\theta - 1) \right) \tag{8b}$$

Here the angles θ and ϕ specify the orientation of the principal axis system with respect to the magnetic field direction.

Experimentally accessible are not absolute shielding values (which would require comparisons with bare nuclei) but rather *chemical shifts* measured in relationship to the resonance frequency of a reference compound:

$$\delta = \frac{\omega_p - \omega_{\mathrm{ref}}}{\omega_{\mathrm{ref}}} \tag{9}$$

Both isotropic and anisotropic components of the chemical shift are sensitive to the nature of chemical bonding and to the coordination numbers of the atom investigated. In addition, the chemical shift anisotropy can be related to the symmetry of the local environment.

2.3.2. *Direct magnetic dipole-dipole coupling*

The effective magnetic fields, which dominate the resonance frequencies of the nuclei under study, are also influenced by the magnetic moments of nuclei in the neighborhood. The secular dipolar Hamiltonian $\mathcal{H}_D$ describing this interaction can be separated into a homo- and a heteronuclear term, associated with the contributions from like and unlike nuclei, respectively.

$$\mathcal{H}_D + \mathcal{H}_{Dhomo} + \mathcal{H}_{Dhetero} \tag{10a}$$

$$\mathcal{H}_{Dhomo} = -\left(\frac{\mu_o}{4\pi}\right)\frac{\gamma^2\hbar^2}{r_{ij}^3}\left(\frac{3\cos^2\theta - 1}{2}\right)(3\hat{I}_z^2 - \hat{I}^2) \tag{10b}$$

$$\mathcal{H}_{Dhetero} = -\left(\frac{\mu_o}{4\pi}\right)\frac{\gamma_I\gamma_S\hbar^2}{r_{IS}^3}\left(\frac{3\cos^2\theta - 1}{2}\right)(\hat{I}_z\hat{S}_z) \tag{10c}$$

Note in particular the direct scaling with the inverse cube of the internuclear distance. For this reason, evaluation of the dipolar interaction has important structural significance. The orientational dependence is governed by the term $(3\cos^2\theta - 1)$, where θ is the angle between the internuclear distance vector and the magnetic field direction.

In systems characterized by multiple spin-spin interactions, the dipolar couplings are characterized by the average mean square of the local field (the *second moment*), which can be related to internuclear distances by the van Vleck formulae:

$$M_2 = M_{2homo} + M_{2hetero}$$

$$M_{2homo} = \frac{3}{5}\left(\frac{\mu_o}{4\pi}\right)^2 I(I+1)\gamma^4\hbar^2\sum_{i\neq j}r_{ij}^{-6} \tag{11a}$$

$$M_{2hetero} = \frac{4}{15}\left(\frac{\mu_o}{4\pi}\right)^2 S(S+1)\gamma_I^2\gamma_S^2\hbar^2\sum_{S}r_{IS}^{-6} \tag{11b}$$

2.3.3. *Nuclear electric quadrupolar interaction*

For nuclei with $I > 1/2$, the charge distribution is non-spherically symmetric. This asymmetry can be described by an electrical quadrupole moment superimposed upon a charged sphere. Classical physics predict that electrical quadrupole moments interact with spatially inhomogeneous electric fields, i.e. an *electric field gradient*, EFG, generated by the local bonding environment. The EFG is a symmetric second rank tensor, which can be diagonalized in a molecular axis system. The diagonal sum of these EFG components $q_{zz} + q_{yy} + q_{xx}$ vanishes (Laplace equation), so that the quadrupolar interaction is completely defined by two Hamiltonian parameters: the nuclear electric quadrupolar coupling constant C_Q specifying the maximum cartesian component q_{zz} and the asymmetry parameter η_Q, representing its deviation from cylindrical symmetry. The quadrupolar interaction competes with the Zeeman interaction for the spin alignment, and easy solutions to the eigenvalue problem are only obtainable if one interaction dominates. The case $\mathcal{H}_Q \ll \mathcal{H}_z$ is particularly simple, and first-order perturbation theory suffices to compute the energy corrections to the Zeeman levels. In this case the first-order approximation for the quadrupolar Hamiltonian in the presence of a large Zeeman interaction is given by:

$$\mathcal{H}_Q^{(1)} = \frac{C_Q}{2I(2I-1)} \cdot \frac{1}{2}(3\cos^2\theta - 1 - \eta_Q \sin^2\theta \cos 2\phi) \cdot \frac{1}{2}[3\hat{I}_z^2 - I(I+1)]$$

$$(12)$$

where $\eta_Q = \frac{q_{yy}-q_{xx}}{q_{zz}}$ and θ and ϕ specify the orientation of the magnetic field direction in the molecular axis system.

If the condition $\mathcal{H}_Q \ll \mathcal{H}_Z$ is not fulfilled, the perturbation calculation has to be extended to second order. In this case the Hamiltonian can be separated into two contributions $\mathcal{H}_Q^{(1)}$ and $\mathcal{H}_Q^{(2)}$.

Similarly to the chemical shift anisotropy, the nuclear electric quadrupolar coupling parameters sensitively reflect the local coordination geometry and symmetry.

2.4. *Experimental separation strategies*

Considering all of the above interactions, the total Hamiltonian in the solid state is given by:

$$\mathcal{H}_{(\text{total})} = \mathcal{H}_Z + \mathcal{H}_{\text{CS}}(\text{iso}) + \mathcal{H}_{\text{CS}}(\text{aniso}) + \mathcal{H}_{\text{Dhomo}} + \mathcal{H}_{\text{Dhetero}} + \mathcal{H}_Q^{(1)} + \mathcal{H}_Q^{(2)}$$

$$(13)$$

In the most general case, the NMR spectrum of a solid is influenced by all of these parameters simultaneously, making the analysis a completeley hopeless task. The special value of solid state NMR for the structural analysis of materials arises from the fact, however, that the individual contributions to the Hamiltonian can be separated from each other, using suitable experimental strategies. There exist a plethora of such *selective averaging experiments*, which eliminate the effect of certain interactions while preserving others. These techniques generally utilize special geometrical sample manipulations or application of radiofrequency pulse trains in many cases combined with two-dimensional data processing.

2.4.1. *Magic-angle spinning*

The simplest and most popular selective averaging experiment involves recording the spectrum while rotating the sample at an angle α of 54.7° (*the magic angle*) relative to the magnetic field direction. This operation is designed to eliminate all of the anisotropic contributions to $\mathcal{H}$(total). This is possible because under fast sample rotation about an angle α from the magnetic field direction the anisotropic energy level shifts caused by the dipole-dipole coupling, the chemical shift anisotropy, and (to first order) by the quadrupolar

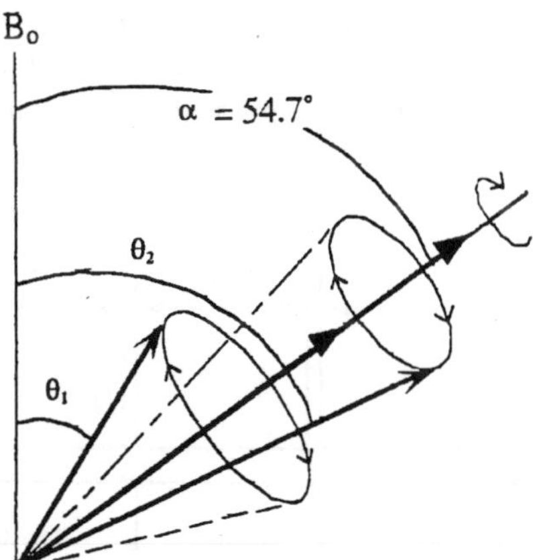

Fig. 2. Principle of the MAS experiment: at sufficiently fast spinning rates, the NMR interaction tensor orientations with initial angles of θ_1 or θ_2 relative to B_o, have rotational averages of 54.7°, resulting in an average term $(1/2)\overline{(3\cos^2\theta - 1)} = 0$.

interaction, are scaled with the factor $3\cos^2\alpha - 1$. As illustrated in Fig. 2, this time average is zero for $\alpha = 54.74°$, the *magic angle*. Thus, for spin-1/2 nuclei, the Hamiltonian under fast MAS is given by:

$$\mathcal{H}(\text{MAS}) = \mathcal{H}_Z + \mathcal{H}_{\text{CS}}(\text{iso}) \tag{14}$$

resulting in high-resolution spectra, in which individual chemical environments can be distinguished on the basis of their corresponding isotropic chemical shifts (and scalar spin-spin couplings). For quadrupolar nuclei, the averaging by MAS is incomplete, and highly characteristic lineshapes can be obtained, which allow the simultaneous determination of chemical shift and quadrupolar coupling parameters.

2.4.2. *Multi-dimensional NMR*

An alternative approach towards enhancing the informational content of solid state NMR is to separate different interactions into different time-domains. Figure 3 illustrates the basic idea of a two-dimensional experiment: during the preparation period transverse magnetization is created. The experiment is then designed such that the internal Hamiltonian affecting the spin system during the evolution period t_1 (which is systematically incremented), is different from that affecting the spin system during the detection period t_2. Fourier-Transformation with respect to both time domains yields a plot in which the effects of different internal interactions are displayed separately onto two distinct frequency axes and correlated with each other in a two-dimensional representation. Various applications of this concept will be given in this article.

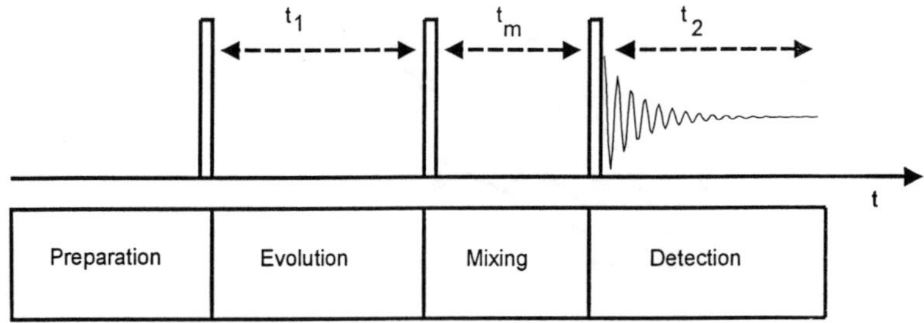

Fig. 3. Basic idea of a two-dimensional NMR experiment, consisting of a preparation period, an evolution period t_1, a mixing time t_m, and the detection period t_2.

3. Structural Issues in Non-crystalline Solids and Glasses

Materials that lack long-range structural order and possess a viscosity exceeding 10^{13} Poise are considered non-crystalline (amorphous) solids [10]. The absence of long-range structural order is usually operationally defined by the technique of X-ray diffraction. The term "glass" is reserved for non-crystalline solids that exhibit structural relaxation. For a glass there exists a well-defined temperature interval, the *glass transition region*, where its thermodynamic properties (enthalpy, entropy, molar volume etc.) show abrupt changes in temperature coefficient. Microscopically this effect is explained by the onset of long-range molecular motion with correlation times comparable to the inherent time scales of physical measurements. This implies a dramatic increase of the heat capacity, which is easily detected by differential scanning calorimetry (DSC). Due to the opportunities of fine-tuning the physical-chemical properties by composition and processing, glasses have attained great importance in materials science and technology. To shed light on the relationship between composition and macroscopic properties, we seek knowledge about the microscopic structural organization of glasses. Since glasses lack translational symmetry, diffraction experiments, which ordinarily yield the complete distance geometry on crystalline materials, are only of limited value for the glassy state. To describe the structure of a glass, we need to elucidate different aspects of ordering on various length and time scales. Specifically the following questions are of interest:

(1) On the atomic scale, one is interested in two-atom correlations, concerning the presence of *short-range and chemical order*. Are there reproducible nearest-neighbor atomic environments? What can be said about the atomic coordination numbers and -symmetries, bond distances and bond angles? In general these parameters are not single-valued, but rather defined by distribution functions. From this (incomplete) point of view, the glass is an assembly of coordination polyhedra, to be identified and quantified.

(2) At the next stage of sophistication, one is interested in spatial correlations involving atoms in higher coordination spheres. This quest for *intermediate range order* concerns the connectivity of individual types of atomic polyhedra and the presence of larger units, such as clusters, chains or rings.

(3) At the 10–100 nm scale, the question of chemical homogeneity arises. Many glasses cannot be considered homogeneous continua, but are segregated

into chemically distinct micro-phases. The compositions and domain sizes of these micro-phases are an important part of the structural description.

(4) Finally, at the most general level the question arises *why* the glass structure, however incompletely we succeed in describing it, exists the way it does. In this context one needs to characterize chemical equilibria present in the liquid precursor phase and the dynamic processes at and below the glass transition region. Such studies might also provide the connection between microscopic structure and macroscopic properties.

4. Short-Range Order in Oxide Glasses

4.1. *Local coordination number and symmetry*

X-ray and neutron diffraction work have shown that the framework atoms of oxide glasses (Si, B, P, Al) possess well-defined nearest-neighbor environments that constitute short-range order. The identification and quantification of such framework (*network former*) sites is the most basic step in the structural analysis of a glass. Towards this goal, magic angle spinning NMR has proven very instrumental. The dominant chemical shift effect arises from the coordination number: the higher the coordination number, the lower the isotropic chemical shift value. In addition, the local symmetry possesses a considerable influence on the chemical shift. Figure 4 shows a glass prepared under high-pressure conditions, containing four-, five-, and six-coordinated silicon, which are separated by approximately 40–50 ppm each in the spectrum [11]. Glasses prepared at ambient pressures usually contain only four-coordinated silicon. Depending on cation content, five different $Q^{(n)}$ sites, differing in the number n of bridging oxygen neighbors, can be formed. As shown in Fig. 5, these five types of sites are resolvable in ^{29}Si MAS–NMR spectra [12–17]. Using spectral deconvolution techniques, it is possible to estimate their quantitative contributions to the glass structure. These results show that alkali or alkaline earth oxides (*network modifiers*) depolymerize the silicate network, successively creating non-bridging oxygen species. Analogous analyses and conclusions are possible from ^{31}P MAS–NMR studies of phosphate glasses [18–23]. In borate glasses, the dominant sites are the symmetric trigonal BO_3 units present in glassy B_2O_3. With successive addition of network modifier, however, negatively charged tetrahedral BO_4^-, and distorted trigonal BO_3^- units are formed [24]. These sites can be well-discerned in high-field ^{11}B MAS–NMR spectra on the basis of their different chemical shift and quadrupolar coupling parameters

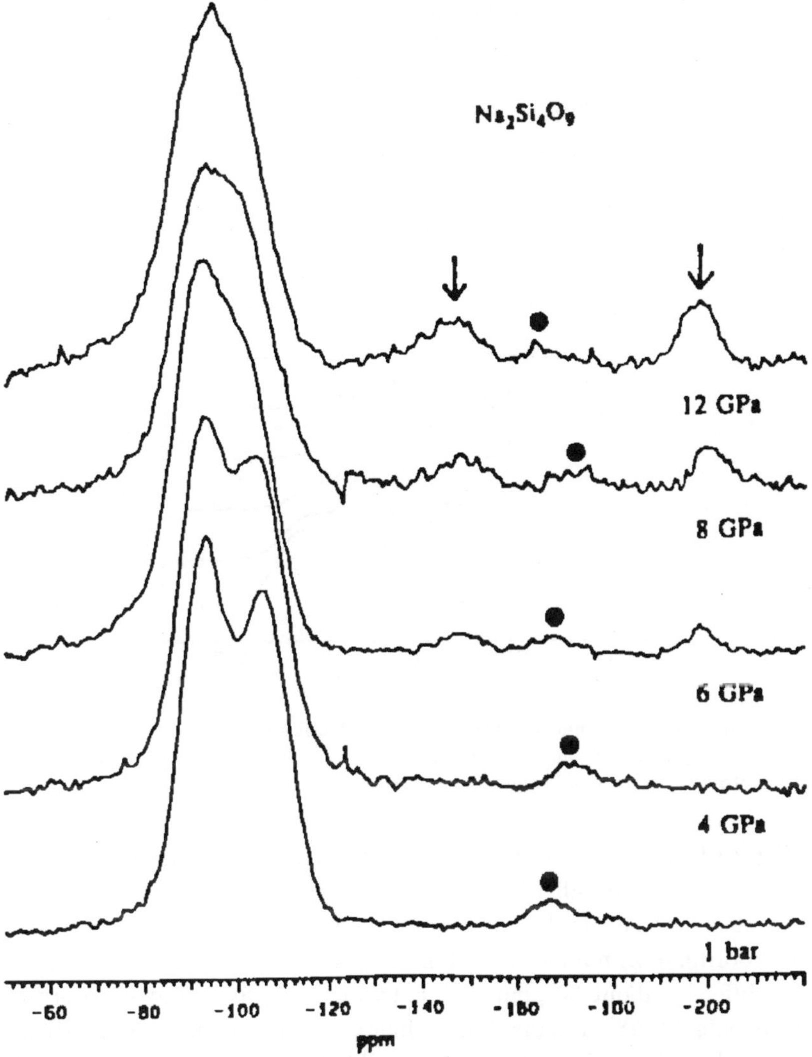

Fig. 4. ^{29}Si MAS–NMR spectra of glassy Na$_2$Si$_4$O$_9$ samples quenched from high temperatures and different pressures (listed in the figure) and decompressed at room temperature. The formation of five and six-coordinated Si is indicated by the gradual emergence of peaks at −150 and −210 ppm (shown by arrows) as the pressure is increased. The dot marks a spinning sideband. From Ref. [11].

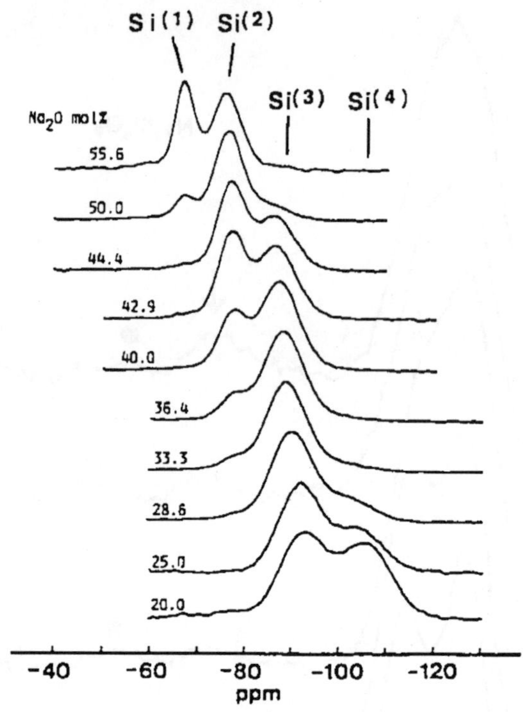

Fig. 5. Compositional dependence of the ^{29}Si MAS–NMR spectra in the glass system Na_2O–SiO_2. Sample compositions and peak assignments are indicated. The number in the superscript denotes the number of bridging oxygen species. From Ref. [12].

(see Fig. 6) [25]. Note in particular, that the axially symmetric BO_3 and the asymmetric BO_3^- sites differ in the asymmetry parameter-η_Q characterizing the nuclear electric quadrupolar interaction. MAS–NMR also yields information on the coordination number of aluminum. The ^{27}Al chemical shifts in oxide glasses differ by about 30 ppm each between four-, five-, and six-coordinated environments [26–28]. Finally, Fig. 7 illustrates further that ^{17}O NMR in isotopically labelled samples differentiates well between bridging and non-bridging oxygen sites in silicate glasses [29]. Since ^{17}O nuclear electric quadrupolar coupling constants are moderately large (up to 5 MHz), advanced sample reorientation techniques are required to eliminate the effect of second-order quadrupolar interactions on the lineshape. Thus, the spectrum shown in Fig. 7 was obtained by dynamic angle spinning (DAS), a two-dimensional experiment that involves switching of the spinner axis in the middle of the evolution period

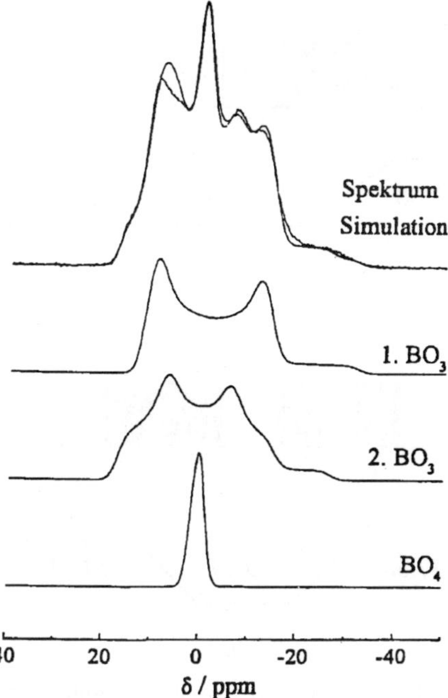

Fig. 6. Typical 7.0 T ^{11}B MAS–NMR spectra of a typical sodium aluminoborate glass and its simulation. Resonance assignments to individual spectral components are indicated.

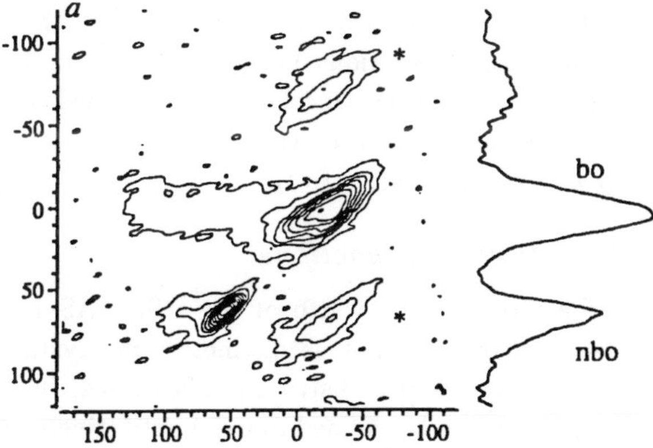

Fig. 7. ^{17}O DAS–NMR spectrum of glassy K$_2$Si$_4$O$_9$ enriched with ^{17}O isotope. The one-dimensional projection shows clear differentiation between bridging oxygen (bo) and non-bridging oxygen (nbo) species. From Ref. [29].

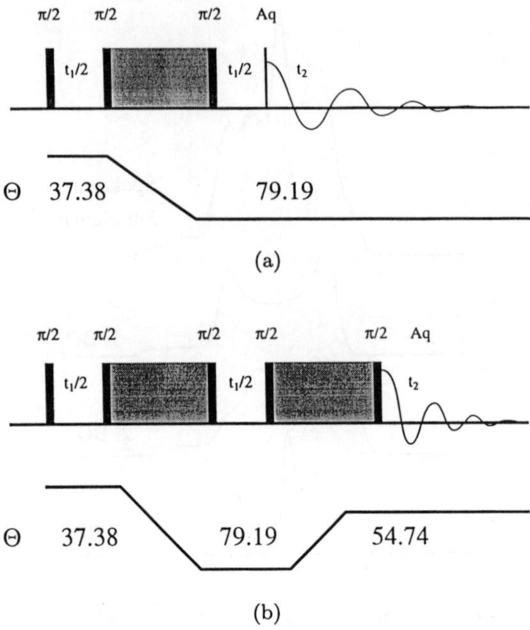

Fig. 8. Pulse sequence used in the DAS experiment. (a) Simplest two-angle version, (b) with detection at the magic angle. During the period characterized by the shaded areas the rotor angle is flipped. From Ref. [32].

(see Fig. 8) [30–32]. Judicious choice of the two rotation angles (typically 37.38° and 79.19° relative to the direction of B_o) ensures that the effects of the anisotropic quadrupolar interactions on the spin evolution before and after flipping have opposite signs and thus cancel each other. The isotropic spectrum is available by a two-dimensional Fourier Transform and subsequent projection onto the frequency axis corresponding to t_1.

4.2. *Bond angle distribution functions*

As illustrated by the intrinsic line width of the ^{29}Si MAS–NMR spectra of silicate glasses, the local order is not as distinct as in crystalline materials, indicating the presence of topological distribution functions. In favorable cases, such distribution functions can be defined more quantitatively using advanced solid state NMR experiments. For example, *ab initio* calculations [33] and experiments on crystalline model compounds [34] have shown that the nuclear electric quadrupolar coupling constant and the asymmetry parameter of ^{17}O nuclei in silicates depend sensitively on the value of the Si–O–Si bond angle ξ. To a first approximation the relationships are:

$$C_Q \sim \frac{\cos\xi}{1-\cos\xi} \tag{15a}$$

$$\eta = 1 + \cos\xi \tag{15b}$$

more detailed discussion is given in Ref. [34].

Thus, with increasing ξ value ($> 90°$) C_Q increases, whereas η decreases, the latter approaching the value of zero for the cylindrically symmetric environment where $\xi = 180°$. Obtaining a distribution of Si–O–Si angles for a silicate glass requires that the distribution functions of C_Q and/or η be determined from the ^{17}O NMR lineshape parameters associated with the bridging oxygen atoms. This task is complicated by a fundamental dilemma: while the MAS–NMR lineshape can be calculated rigorously from distribution functions of C_Q and η, the reverse process, namely extracting the distribution function of these

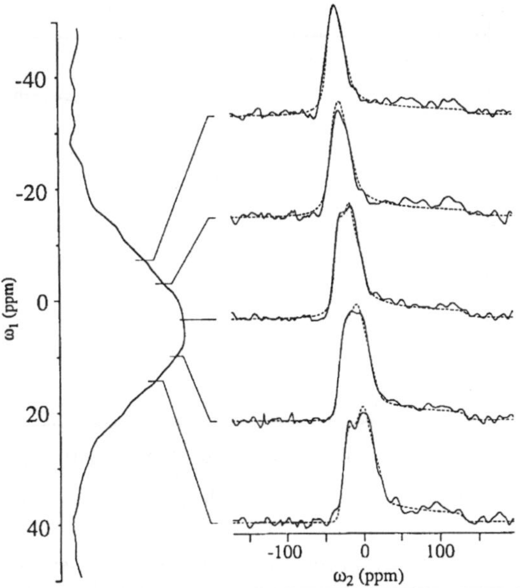

Fig. 9(a). ^{17}O DAS–NMR results on glassy $K_2Si_4O_9$. Indicated are the individual anisotropic slices for various points across the broadened DAS spectrum in the isotropic dimension. Note the systematic change in the asymmetry parameter. Dashed curves are simulated spectra. From Ref. [29].

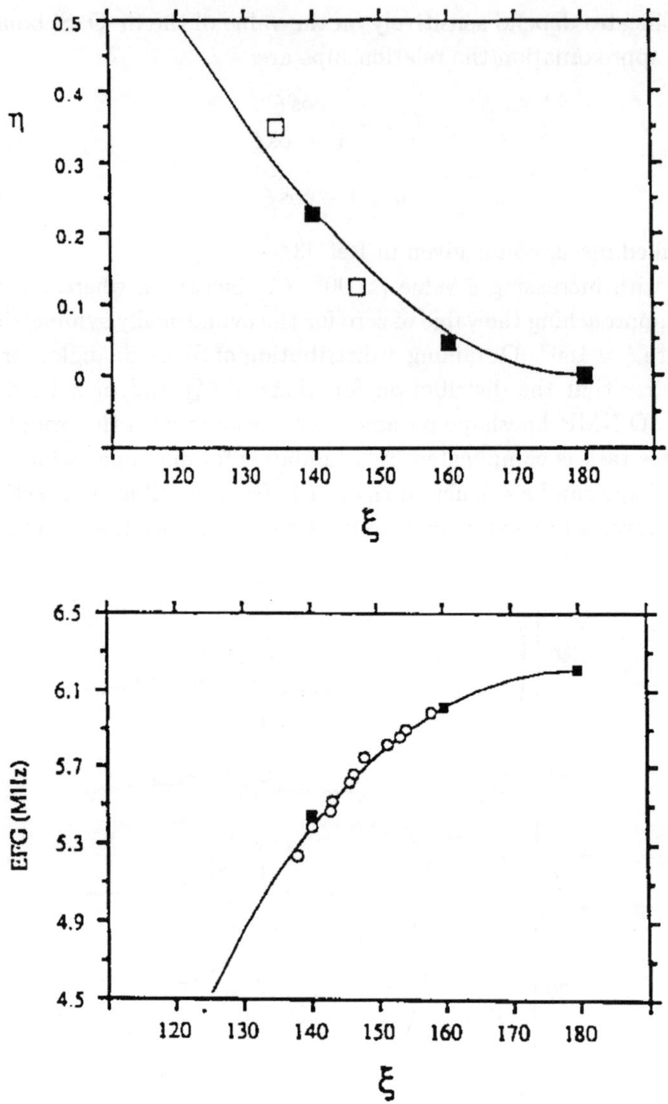

Fig. 9(b). Variation of (1) ^{17}O asymmetry parameter and (2) ^{17}O electric field gradient with various functions relating to the Si–O–Si bond angle ζ. Filled squares are results from *ab initio* calculations. From Ref. [29].

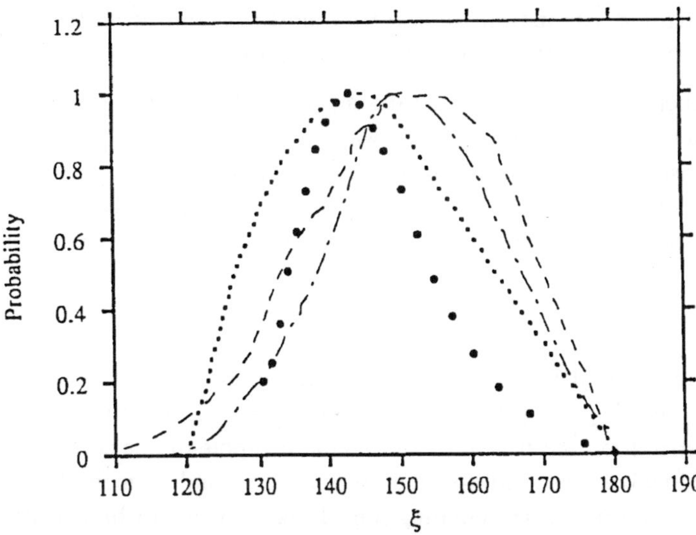

Fig. 9(c). Comparison of the bond angle distribution function derived from ^{17}O DAS–NMR for glassy $K_2Si_4O_9$ (circles) with the respective function derived from X-ray data on glassy SiO_2 (dots) and various molecular dynamics simulations (dashed curves). From Ref. [29].

parameters from the experimental ^{17}O NMR lineshape in an unambiguos manner is impossible by standard mathematical methods [20]. The situation is complicated further by the fact that the ^{17}O MAS–NMR spectrum is affected by distributions of both quadrupole coupling parameters and isotropic chemical shifts. A rather elegant solution to those problems emerges from the two-dimensional aspects of dynamic angle spinning NMR. Figure 9 shows an application to $K_2Si_4O_9$ glass [29]. Each point on the isotropic ^{17}O chemical shift distribution curve measured for the bridging oxygen site by DAS can be correlated with the asymmetry parameter of the ^{17}O nuclear electric quadrupolar interaction. Progressing from low to high values of isotropic shift C_Q decreases while η increases, in good agreement with the predictions made by Eqs. (15a) and (15b). The distribution function of ξ derived on the basis of this 2–D correlation plot is shown in Fig. 9(c). This landmark study has shown that NMR can be used to great advantage for elucidating topological distribution functions in glasses. Aside from the need for isotopic labelling, the methodology portrayed here is general and can be extended to many other glass compositions.

4.3. *Spatial distribution of modifier cations and structural implications of the mixed-alkali effect*

Besides their effect on the framework structure, the network modifier ions in glasses are of interest, because knowledge of their spatial distribution may provide insight into the mechanism of ionic conduction [35]. One of the most mysterious phenomena in this regard is the *mixed alkali effect* [36, 37]. Mixed-alkali glasses, i.e. systems containing two different types of modifier cations show appreciably reduced ionic mobility compared to single-alkali glasses with the same overall cation contents. In the past, numerous proposals have been made, all of which explain the effect based upon the interactions and spatial distributions of the two types of cations present [38–41]. On this topic, NMR spectroscopy can provide vital quantitative information, since the spatial cation distribution is reflected by the magnetic dipole-dipole couplings between the spins involved. For multi-spin systems, the second moment characterizing this interaction in a corresponding mixed alkali glass can be calculated from the internuclear distance distributions r_{ij} according to Eq. (11b). Experimentally, these second moments can be measured selectively by the spin echo double resonance (SEDOR) NMR technique [42–44] illustrated in Fig. 10 and applied to glasses for the case ^{23}Na/^{7}Li. The experiment consists of two parts: in the first part the intensity of a regular ^{23}Na spin echo is measured as a function of evolution time t_1. The normalized intensity of this echo $F(2t_1)/F_o$ decays only as a result of homonuclear dipole-dipole couplings, whereas the magnetization decay attributable to the heterodipolar ^{7}Li–^{23}Na interaction is refocused during the t_1 period following the 180° pulse. In the second part

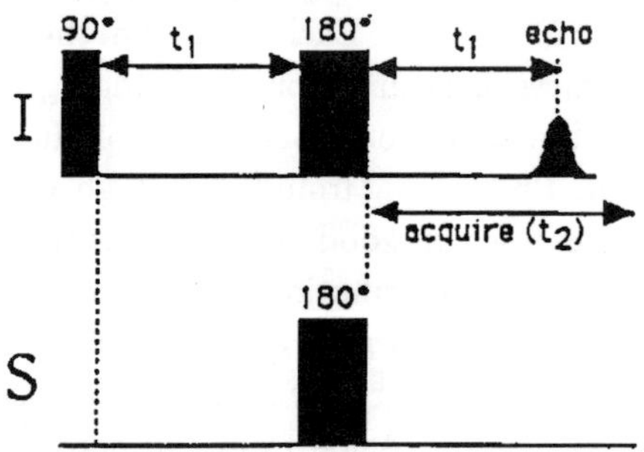

Fig. 10.   Pulse sequence used for ^{23}Na{^{7}Li} spin echo double resonance (SEDOR) experiments.

of the experiment, however, the additional 180° pulse applied to the ^{7}Li spin species inverts the sign of the ^{7}Li–^{23}Na dipole-dipole coupling constant after t_1, and thereby reintroduces this coupling into the spin evolution. As a result, the spin echo decay is accelerated if a significant heterodipolar coupling is present. A typical example is shown in Fig. 11 for a mixed-alkali silicate glass. Qualitatively, the strong effect of the ^{7}Li irradiation on the ^{23}Na spin echo signifies strong dipole-dipole interactions between both spin species. Clearly this result rules out cation segregation models previously invoked in the explanation of mixed alkali effects. In more quantitative terms the ^{23}Na{^{7}Li} SEDOR decay can be analyzed according to:

$$\frac{I(2t_1)}{I_0} = \frac{F(2t_1)}{F_0} \cdot \exp - \left\{ (2t_1)^2 \cdot \frac{M_{2\text{hetero}}}{2} \right\} \tag{16}$$

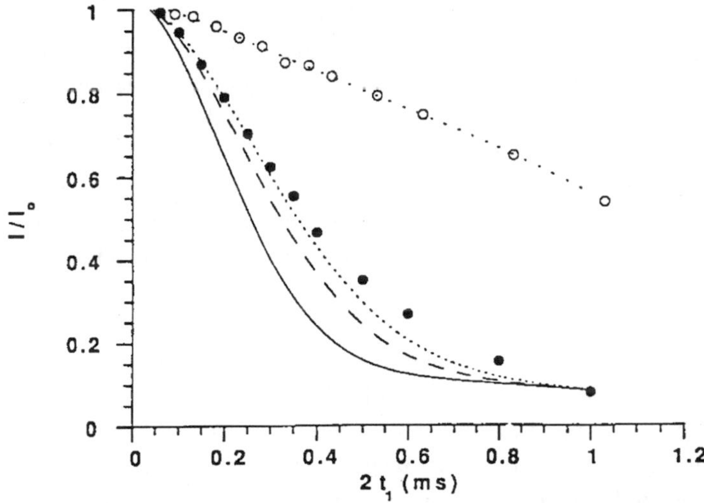

Fig. 11. ^{23}Na{^{7}Li} spin echo double resonance experiment on a mixed alkali glass with composition indicated. The normalized ^{23}Na spin echo intensity is plotted vs. dipolar evolution time. Open circles: no ^{7}Li 180° pulse; closed circles: with ^{7}Li 180° pulse. The various curves shown are calculated for various scenarios for the Na vs Li spatial distribution. The best agreement with experimental data is found for a scenario assuming statistical Li–Na mixing within an overall homogeneous cation distribution (dotted curve). From Ref. [46].

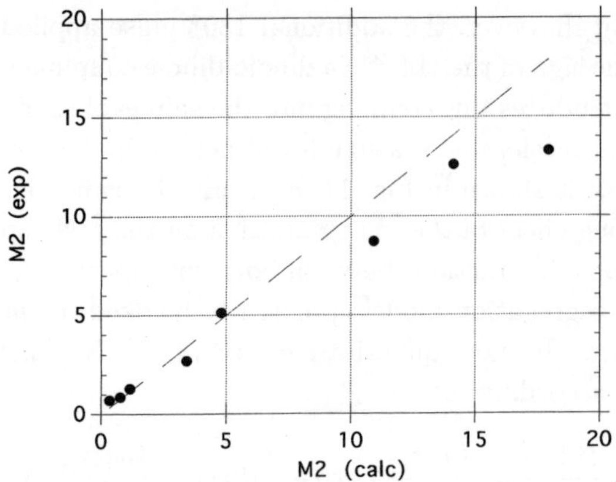

Fig. 12. Heterodipolar second moments from ^{23}Na$\{^{7}$Li$\}$ and ^{23}Na$\{^{6}$Li$\}$ spin echo double resonance experiments on various mixed alkali silicate glasses vs. calculated values for the scenario assuming statistical Li–Na mixing within an overall homogeneous cation distribution. The solid curve represents the identity for the simulations.

yielding second moment values characterizing the average strength of heterodipolar ^{7}Li–^{23}Na interactions. These second moments can serve as quantitative criteria against which hypothetical atomic distribution models can be tested. During the past few years the mutual lithium-sodium distribution in mixed alkali silicate glasses has been studied by SEDOR spectroscopies [45–48]. Figure 12 summarizes such results from both ^{23}Na$\{^{6}$Li$\}$ and ^{23}Na$\{^{7}$Li$\}$ SEDOR studies carried out in our laboratory. The data are in good agreement with a model in which the overall cation distribution is close to homogeneous, and where the two types of cations are statistically mixed. This conclusion is also consistent with recent ^{17}O DAS–NMR spectra of Na, K silicate glasses [49]. As Fig. 13 demonstrates, the ^{17}O isotropic chemical shifts of the non-bridging oxygen species are greatly influenced by the type of cation coordinated to it. Although the resonances assigned to bridging and non-bridging oxygen sites are partially overlapped at higher Na contents, a deconvolution is still possible. Assuming each non-bridging oxygen species to be surrounded by 4 alkali cations, five configurations are possible: Si–O–Na$_4$, Si–O–Na$_3$K, Si–O–Na$_2$K$_2$, Si–O–NaK$_3$ and Si–O–K$_4$. The ^{17}O DAS–NMR spectra are well-reproduced by assuming binomial population statistics. The results preclude significant like-cation segregation and are generally supportive of models of the mixed alkali effect that involve blocking of cation migration pathways by foreign cations.

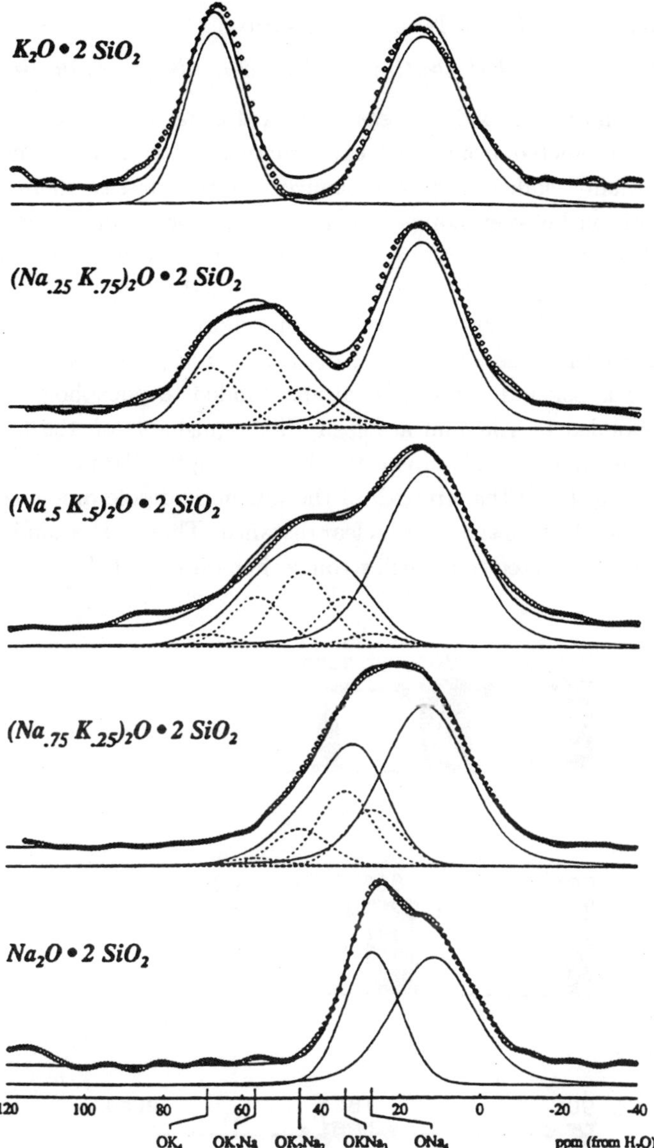

Fig. 13. One-dimensional isotropic 9.4 T ^{17}O DAS spectra of mixed-alkali disilicate glasses with the compositions indicated. The experimental data are shown as diamonds, the simulation as straight line, and each component as dashed lines, assuming a binomial distribution. From Ref. [49].

5. Short-Range Order in Non-Oxide Glasses

5.1. *Chemical bond distribution and intermediate range order*

Unlike the situation in oxide glasses, the local order in non-oxide glasses is not as easily predicted from the local structures of crystalline model compounds. In many chalcogenide and semiconducting glasses, there is an efficient competition between homo- and heteropolar bond formation, resulting in a great deal of *chemical disorder*. This effect results in an extremely wide distribution of chemical shifts, producing very broad MAS–NMR lines [50]. A case in point is the binary glass system phosphorus–selenium, where chemical disorder concerns the quantitative distribution of phosphorus–phosphorus and phosphorus–selenium bonds. *P* atoms bonded to phosphorus and those exclusively bonded to selenium are easily distinguishable on the basis of the homonuclear magnetic dipole-dipole coupling among the ^{31}P nuclei [51, 52]. As expressed by Eq. (11a) the strength of the magnetic dipole-dipole coupling is sensitively dependent on the internuclear distance. Thus, a *P*-atom involved in P–P bonding is subjected to a much stronger homodipolar field (corresponding

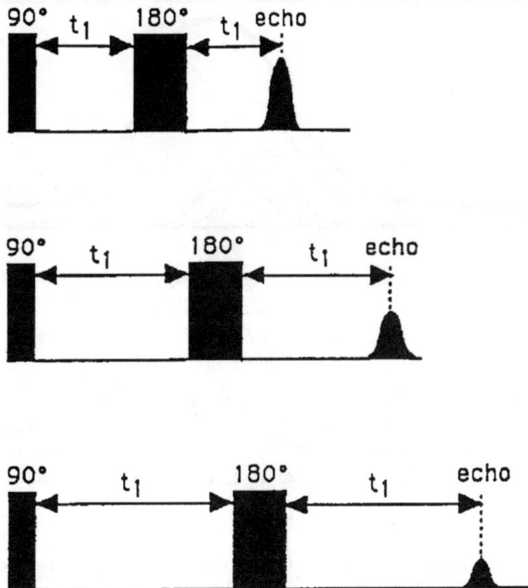

Fig. 14.   Pulse sequence used in spin echo decay spectroscopy for the measurement of homodipolar interactions.

to a much larger homodipolar second moment) than phosphorus atoms bonded exclusively to selenium. Experimentally, M_{2homo} characterizing the homonuclear dipole-dipole coupling among the ^{31}p spins can be measured by a spin echo pulse sequence, as illustrated in Fig. 14. Following an initial 90° preparation pulse, the transverse magnetization decays during the evolution period t_1

^{31}P Spin Echo NMR: P-bonded P ^{31}P Spin Echo NMR: non-P-bonded P

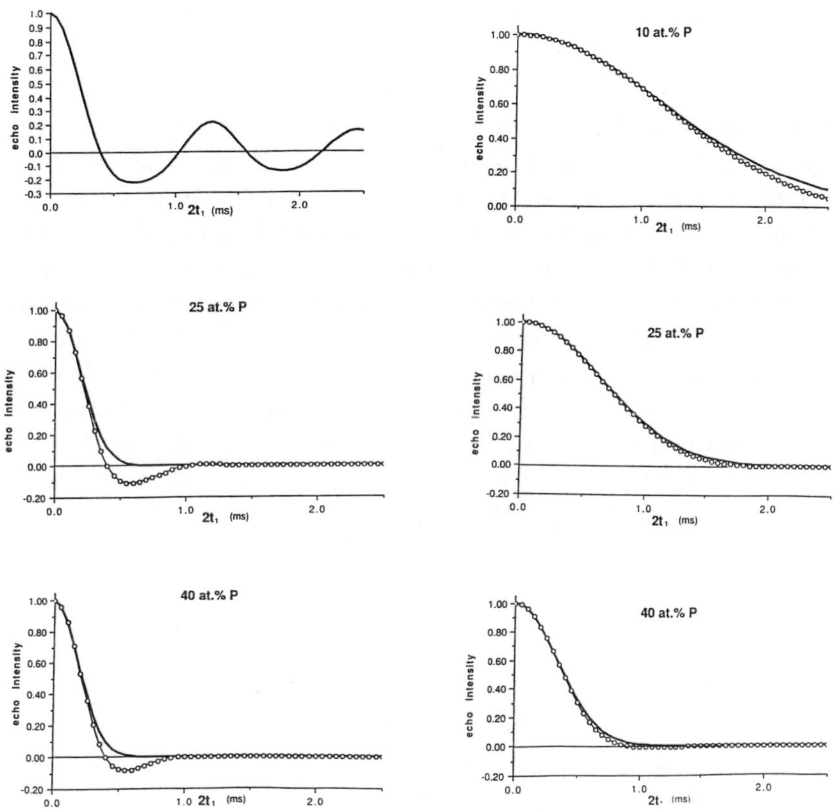

Fig. 15. Simulated ^{31}P spin echo decays in P–Se glasses with various P contents. Shown are the decays expected for P atoms involved in a single P–P bond, as well as for P-atoms only coordinated to Se and not involved in P–P bonds. Results from rigorous calculations and Gaussian approximations (solid curves) are shown. The effect of secondary interactions to ^{31}P nuclei at larger distances is taken into account by Gaussian broadening, assuming a random P distribution in space.

due to the combined effects of dipolar interactions and chemical shift dispersion. Application of the 180° pulse after t_1, however, reverses the part of the decay due to the chemical shift contributions and heteronuclear dipole-dipole couplings, resulting in a spin echo at the time $2t_1$. Thus, the intensity of the spin echo is attenuated only by homonuclear dipolar couplings during the dipolar evolution time $2t_1$. The evolution time is systematically incremented, giving rise to an experimental spin echo decay curve $I(2t_1)/I_o$. For a system with multiple interactions, a Gaussian decay is expected:

$$\frac{I(2t_1)}{I_o} = \exp\left\{-(2t_1)^2 \frac{M_{2\text{homo}}}{2}\right\} \tag{17}$$

As discussed elsewhere, a number of caveats need to be borne in mind when applying this technique to systems with complex spin dynamics. Using a range of suitable crystalline model compounds, it was demonstrated, however, that correct homodipolar second moments for ^{31}P can be obtained with $\pm10\%$ accuracy [53]. As expected, the experimental spin echo decay curves observed in the phosphorus–selenium system reveal approximately bimodal behavior: the rapidly decaying initial part (large M_2 value) of the curve is due to P-bonded phosphorus, while the slowly decaying part (small M_2 value) belongs to P atoms exclusively bonded to selenium. Typical simulations, modelling the contributions for both types of species in the glasses are shown in Fig. 15. A typical ^{31}P spin echo decay and its bimodal fit are shown in Fig. 16. Analysis of the experimental data in terms of this bimodal distribution thus

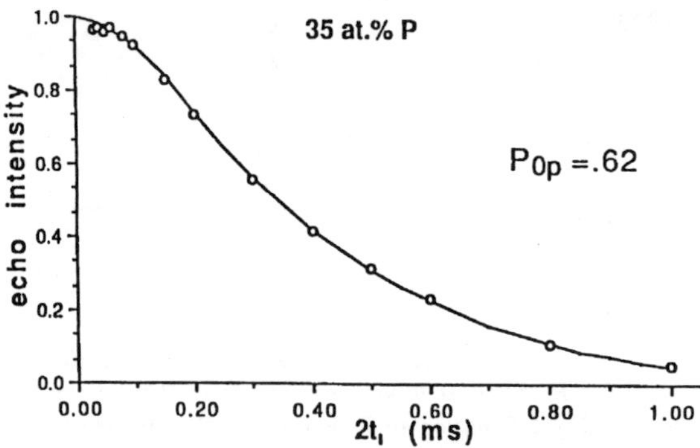

Fig. 16.   Bimodal analysis of the spin echo decay curve for a P–Se glass containing 35 at. % P. The fit implies 62% of the P atoms not to be P-bonded.

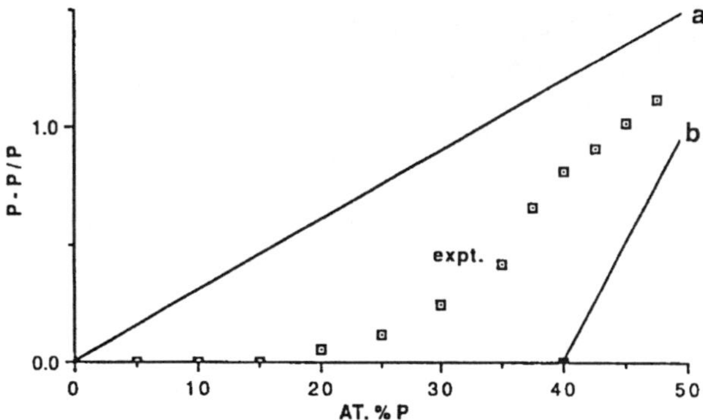

Fig. 17. Average number of P–P bonds per *P* atom in Phosphorus–Selenium glasses as a function of composition. Solid lines indicate the expected values (a) based on a statistical distribution of P–P and P–Se bonds, and (b) based on a chemically ordered glass model excluding P–P bonds below 40 at. % P and minimizing the number of P–P bonds at higher phosphorus contents.

allows one to quantify the number of P-bonded *P* atoms for each glass composition. The results are summarized in Fig. 17. P–P bonds appear in the structure of these glasses at P contents above 25 at. % P, however, their contribution to the glass structure remains sub-statistical over the entire range of compositions. Thus, there is a definite, quantifiable preference for the formation of P–Se over the formation of P–P bonds in these glasses.

In a similar manner, the ^{77}Se$\{^{31}$P$\}$ SEDOR decay curves of these glasses are found to be bimodal, consisting of a rapid initial decay due to P-bonded selenium atoms and a much slower decaying component arising from selenium atoms not directly bonded to phosphorus (see Fig. 18) [54]. By extrapolating the slowly decaying part of these curves to zero evolution time, it is possible to estimate the fraction of Se atoms that are exclusively bonded to selenium. Again, these numbers turn out to be sub-statistical. Most recently, ^{31}P spin echo decay NMR studies have shown that the tendency towards P–P bond formation is greatly enhanced in phosphorus containing telluride glasses [55]. In the system Si–P–Te, P–P bonds occur with statistical probability, whereas in Ge–P–Te glasses, the local bonding of phosphorus is dominated by other *P* atoms altogether.

A second example concerns the atomic distribution in glassy CdGeP$_2$ [56]. The crystalline counterparts of these materials are strictly chemically ordered with P^{3-} ions occupying exclusively the anionic sites of the tetragonal

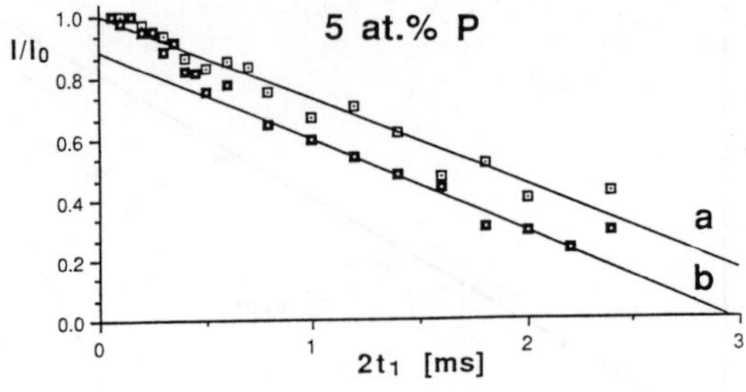

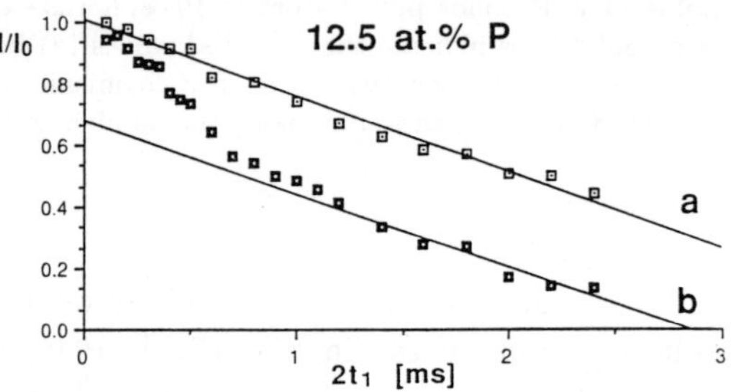

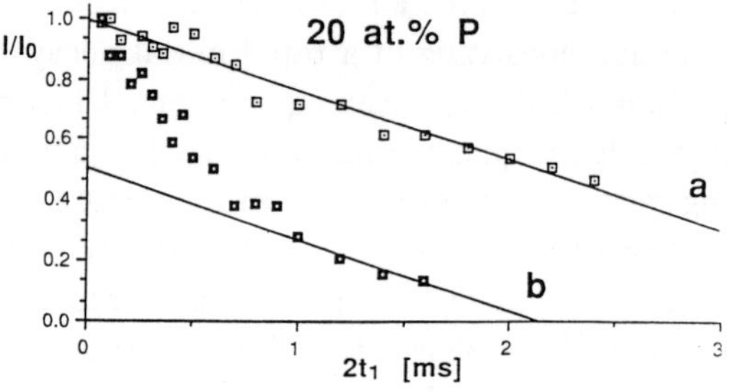

Fig. 18. ^{77}Se$\{^{31}$P$\}$ spin echo double resonance (SEDOR) NMR results for P–Se glasses with low phosphorus contents. Shown are normalized ^{77}Se spin echo intensities as a function of evolution time $2t_1$ (a) without and (b) with application of the ^{31}P pulse. The straight lines are linear least squares fits to the whole data of curve (a) and to the long-time behaviour of curve (b), respectively.

chalcopyrite structure, while the Cd^{2+} and Ge^{4+} ions occupy the cation sites. This arrangement minimizes the strength of ^{31}P–^{31}P homodipolar couplings, because it maximizes the phosphorus–phosphorus distances and excludes direct phosphorus–phosphorus bonds. An alternative hypothetical model for the glass is a tetrahedral lattice where all of the sites are randomly occupied (complete chemical disorder). Since this model produces a statistical number of direct P–P bonds the average dipole-dipole coupling is greatly increased. Figure 19 compares the experimental ^{31}P spin echo decay for the glassy material with corresponding decays calculated for both scenarios; the data are well-simulated with an approximately equal-par superposition of both.

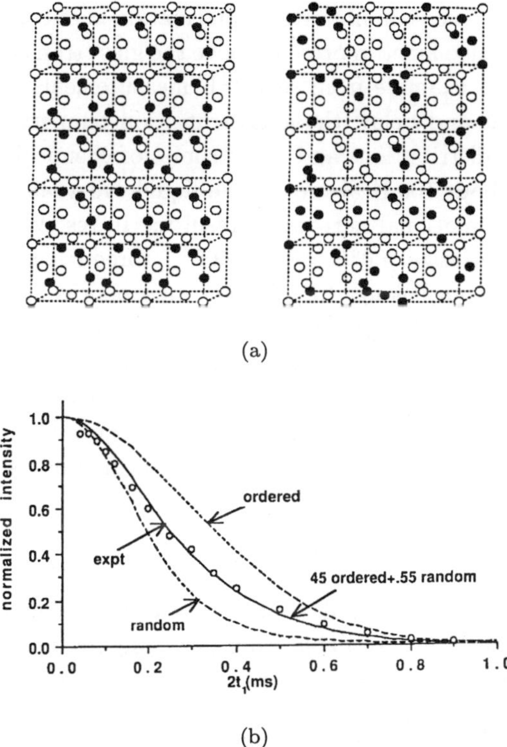

(b)

Fig. 19. (a) Possible distributions of P in glassy $CdGeP_2$: left: chemically ordered distribution, right: random distribution. (b) ^{31}P spin echo decay NMR spectroscopy of glassy $CdGeP_2$. Included is a comparison of the experimental data with data calculated based on a chemically ordered and a random distribution of P atoms as shown in (a). The best fit to the data turns out to be a superposition of both models.

These results illustrate the presence of homopolar P–P bonds at a substantial level, although heteropolar bonds still appear to be favored. Analogous conclusions are valid for glasses within the entire compositional region $CdGeP_2$–$CdGeAs_2$ [57].

5.2. *Chemical equilibria and kinetics in glassforming liquids*

As the facts and details on glass structures are being gathered, questions concerning the *how* and *why* are emerging. The structural speciations in the glassy state reflect liquid-state equilibria that are frozen in at the glass transition temperature. *In-situ* high-temperature NMR spectra can serve to identify these equilibria and (in suitable cases) to measure the corresponding equilibrium constants. They also contain important information about motional dynamics and chemical exchange kinetics present above the glass transition temperature. Figure 20 shows an example from such ^{31}P measurements in the P–Se glass system [58, 59]: at and below the glass transition temperatures the spectra are dominated by distributions and anisotropies of chemical shifts. However, at temperatures about 100 K above T_g, the liquid eventually attains a viscosity consistent with motional narrowing, resulting in resolved ^{31}P resonances. In the temperature range $300°C < T < 400°C$ three resonances are evident: the peak near 140 ppm belongs to polymeric phosphorus units, while the resonances near 70 and -70 ppm belong to the chemically distinct P sites of molecular P_4Se_3. The spectral contribution of P_4Se_3 grows with increasing temperature, signifying a successive depolymerization of the molten network. As discussed in detail elsewhere [59], this process is modelled by an equilibrium in which these molecules are formed from precursor fragments with P–P bonds according to the scheme

$$3\,Se_{2/2}\,P\text{–}P\,Se_{2/2} \Leftrightarrow P_4\,Se_3 + 2\,P\,Se_{3/2}$$

leaving $PSe_{3/2}$ fragments behind. Temperature dependent equilibrium constants for this process are available from peak integration of the resonances at 70 and -70 ppm, and a plot of ln K vs. T^{-1} yields a reaction enthalpy of 50 kJ/mol.

The spectra of Fig. 20 not only serve to characterize molten-state thermodynamics, but also contain interesting kinetic information: while up to a temperature of $375°C$ the spectra reveal three clearly resolved resonances, coalescence phenomena are evident at higher temperatures, indicating that there is chemical exchange on the NMR timescale. With the help of detailed computer

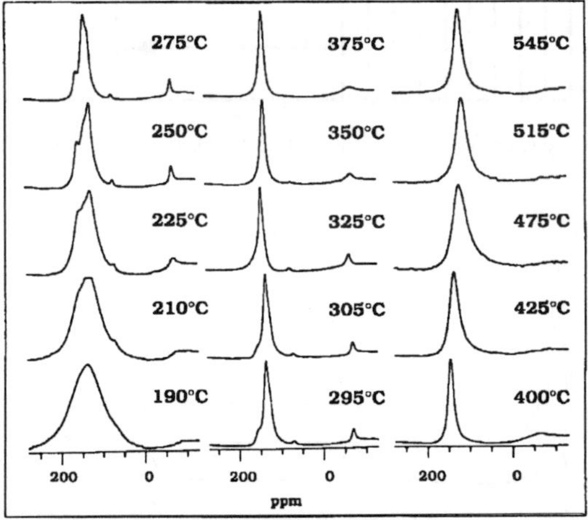

Fig. 20(a).   Variable temperature ^{31}P NMR spectra of a P–Se glass melt containing 45 at. % P.

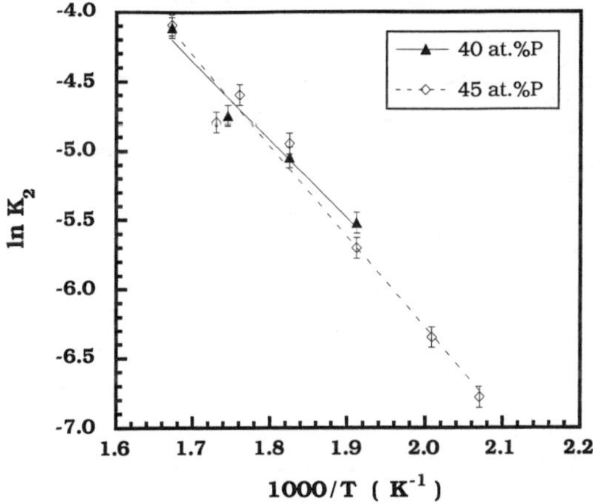

Fig. 20(b).   Temperature dependence of the equilibrium constant for the reaction 3PSe $\Leftrightarrow$ 2PSe$_{0.75}$ + PSe$_{3/2}$ in P–Se glasses.

simulations, it is possible to extract temperature-dependent rate constants for this process [58]. The NMR timescale can be extended even further with the help of magnetization transfer experiments. Figure 21 shows the NMR experiment used for such studies: using a frequency selective DANTE pulse train

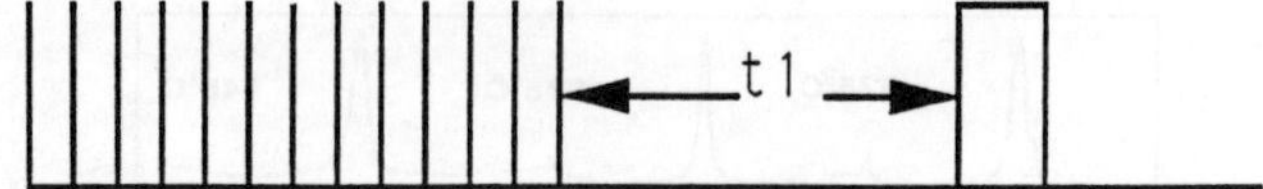

Fig. 21. Pulse sequence used for measuring exchange rates via the one-dimensional saturation transfer method.

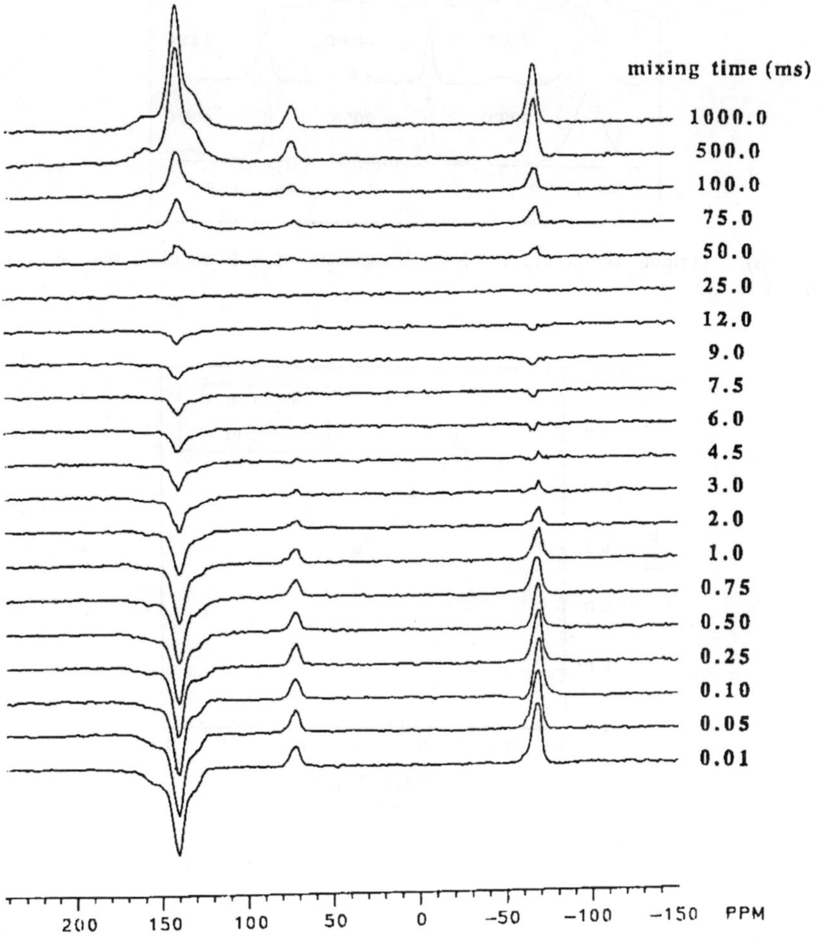

Fig. 22.   Representative data set of a saturation transfer experiment carried out on a phosphorus–selenium glass containing 48 at. % phosphorus, at 300°C. Selective inversion of the resonance at 140 ppm, assigned to P atoms in the polymeric melt.

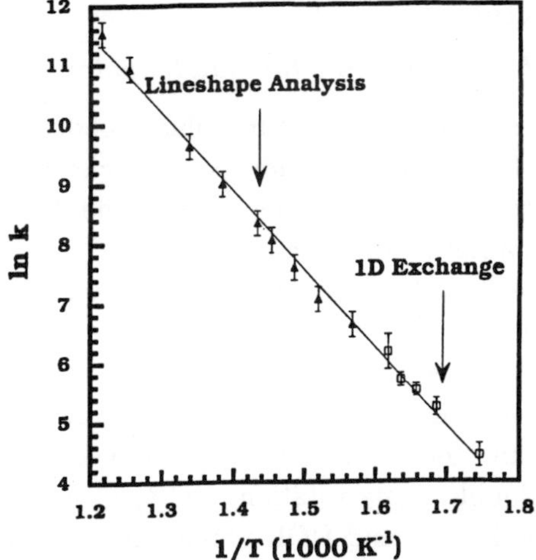

Fig. 23. Arrhenius plot of the chemical exchange rate constants extracted from the saturation transfer experiments (open squares) and comparison with the rate constants extracted from simulations of the variable temperature NMR lineshapes at higher temperatures (triangles) for the same sample.

[60], the spin populations of one resonance are inverted selectively. During the *mixing time*, t_1, relaxation of this perturbed system back to equilibrium will affect all of the site populations if chemical exchange is present. These changes are detected by the final non-selective 90° read pulse [61]. Corresponding rate constants can then be extracted by experiments under systematic variation of the mixing time. Figure 22 shows a typical set of experimental data [62]. An Arrhenius plot of rate constants extracted from both lineshape analysis and saturation transfer experiments is displayed in Fig. 23, yielding an activation energy of 109 kJ/mol. Other types of chemical exchange processes, occurring at lower temperatures, have been identified in the P–Se glass system [63]. In addition, high-temperature ^{77}Se NMR studies have shown that in chalcogenide glasses, unlike to the situation in silicate glasses, the timescales of segmental mobility and chemical exchange are distinct from each other. Thus within a distinct temperature range it is possible to observe resolved, motionally narrowed, spectra while chemical exchange is still slow on the NMR timescale [63, 64]. Overall, *in-situ* liquid state NMR studies of these systems have revealed their potential for characterizing motional and chemical exchange dynamics in glassforming liquids.

6. Future Perspectives

6.1. *Towards higher resolution for quadrupolar nuclei*

The elimination of second order quadrupolar broadening from the central transition spectra of half-integer quadrupolar nuclei remains an important frontier in solid state NMR methodology. While the dynamic angle spinning (DAS) technique represents an elegant approach, intrinsic limitations arise from the time requirements for flipping the spinner axis and from the fact that during t_1 the spin system evolves at spinning angles different from the magic angle. Thus, DAS works best with samples having moderate-to-weak dipolar interactions and fairly long spin-lattice relaxation times. If these conditions are not met, a second method, involving simultaneous sample reorientation about two angles (double angle rotation, DOR) gives superior performance [65]. Nevertheless both DAS and DOR remain non-standard techniques that require specialized equipment. Recently, an elegant multiple-quantum (MQ) excitation method has been developed that allows high-resolution central transition NMR studies with standard MAS–NMR equipment [66–69]. Figure 24

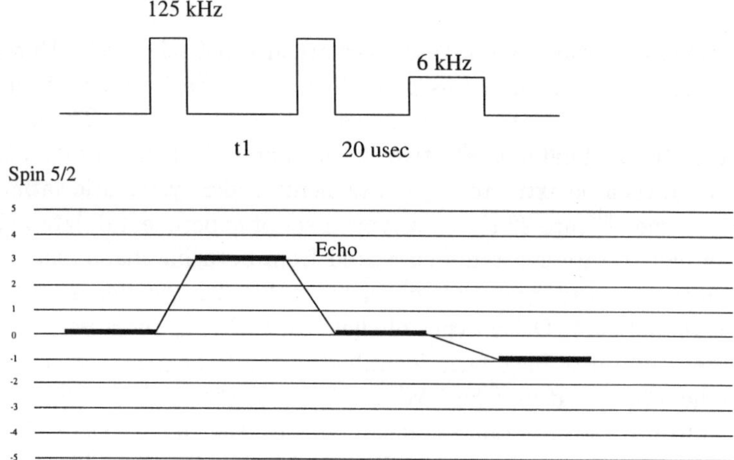

Fig. 24. Pulse sequence and corresponding coherence level diagram for triple-quantum (TQ) MAS–NMR spectra of quadrupolar nuclei. During the period t_1, evolution of a triple quantum coherence takes place. Evolution is stopped by the second pulse. The third pulse creates observable magnetization. The specific coherence path shown is selected via appropriate phase cycling. Typical delay times and nutational frequencies used for the various pulses are indicated in the figure.

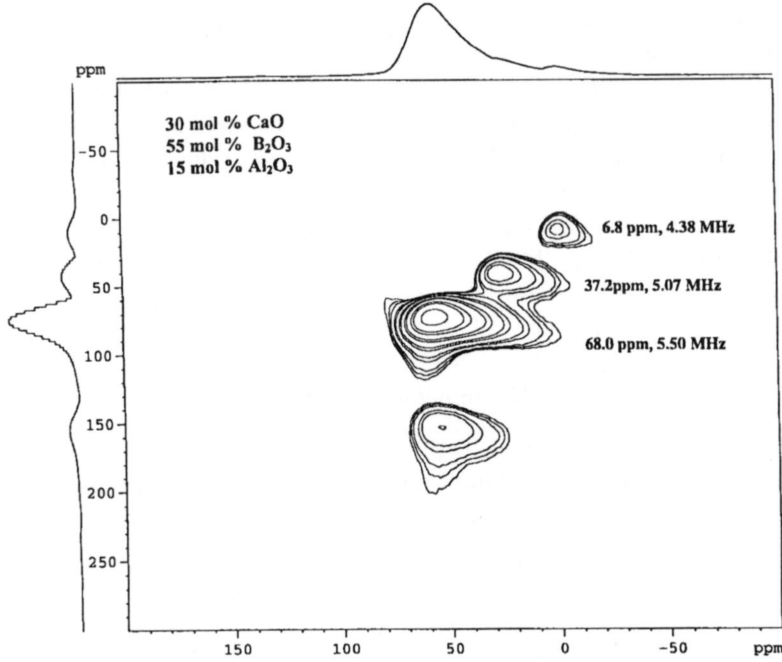

Fig. 25. ^{27}Al TQ–MAS–NMR spectrum of a calcium aluminoborate glass. Note the enhanced resolution present in the projection onto the frequency axis corresponding to the evolution time t_1. The intensity shown at (150, 50 ppm) arises from a spinning sideband.

shows the pulse sequence and a corresponding coherence level diagram. In the meantime, the utility of this method has been demonstrated for a variety of crystalline model compounds while applications to glasses are still at an early stage. A representative ^{27}Al MQ–MAS–NMR result from our laboratory is shown in Fig. 25 for a calcium aluminoborate glass. The spectrum illustrates the facile distinction between four-, five-, and six-coordinated aluminum sites [70]. Since the efficiency of triple-quantum excitation depends on the magnitude of C_Q, no direct site quantification is possible from these spectra.

6.2. *Recovery of dipolar interactions in MAS–NMR: site connectivities*

As discussed above, MAS–NMR frequently affords a facile quantitative distinction between local structural units in glasses. Towards an improved understanding of the glassy state, we need to know how these local structural units are connected to each other to form the overall glassy network. In principle, information about these *site connectivities* is contained in internuclear dipole-dipole couplings, which are sensitively distance dependent. On the other hand, to obtain the high resolution necessary for site distinction, we require magic-angle spinning, which by design eliminates the very dipole-dipole couplings of interest. Research progress made during the past ten years has demonstrated that this dilemma can be overcome with the help of advanced pulsed techniques applied under MAS conditions. The design of these techniques depends on whether the dipole-dipole interactions to be recovered are homo- or heteronuclear in origin.

6.2.1. *Double resonance NMR in heteronuclear systems*

For heteronuclear spin systems, it is possible to re-introduce the dipolar interaction by the rotational echo double resonance ("REDOR") experiment [71, 72]

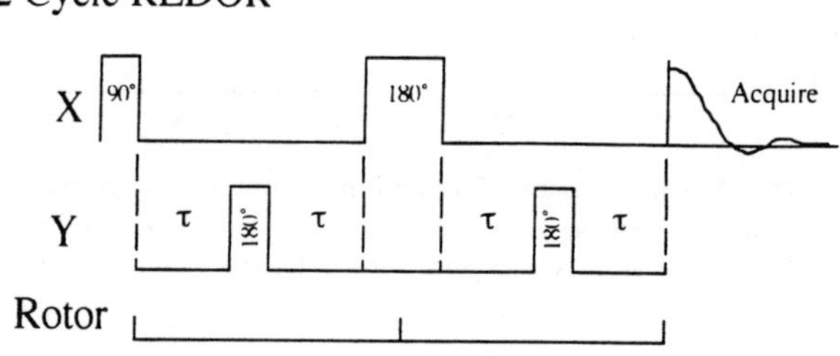

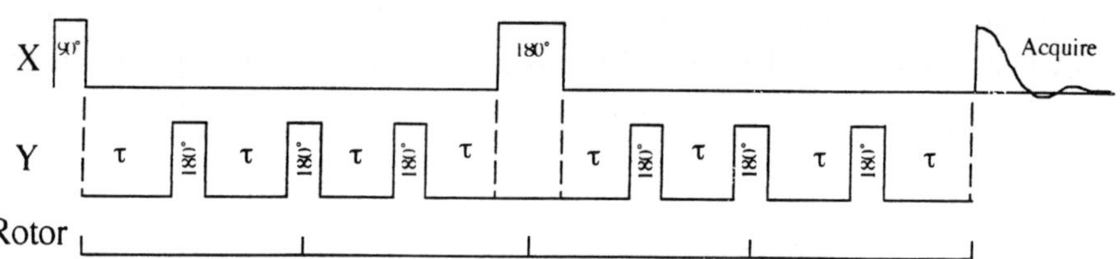

Fig. 26. Pulse sequence used for REDOR experiments on inorganic materials. 2-Cycle REDOR emphasizes the strongest dipolar couplings. For measurements of D, the number of rotor cycles is systematically incremented.

shown in Fig. 26. The experiment is conducted in two parts. In the first part, a MAS spin echo with intensity S_o is produced. In the second part, the 180° pulses applied to the non-resonant S-spins in the middle of the rotor period reverse the averaging of the heteronuclear I–S interaction by MAS, resulting in a reduction of the signal intensity of the observed I-spins. The magnitude of the REDOR difference signal depends (1) on the I–S dipole-dipole coupling constant and (2) on the overall evolution dime during which the interaction is active. This timescale is under experimental control and given by $N \cdot T_r$, the total number of rotor cycles multiplied by the period of one rotor revolution. Thus, a plot of the normalized REDOR difference signal $\Delta S/S_o$ yields the dipolar coupling constant D, from which the internuclear distance can be directly calculated:

$$D = \frac{\gamma_I \gamma_S \hbar}{2\pi r^3} \left(\frac{\mu_o}{4\pi} \right) \tag{18}$$

In multi-spin systems, the specific shape of the $\Delta S/S_o$ vs. $N \cdot T_r$ curve depends on the number of coupled nuclei, the distances involved, and on the detailed geometry with which the distance vectors are oriented relative to each

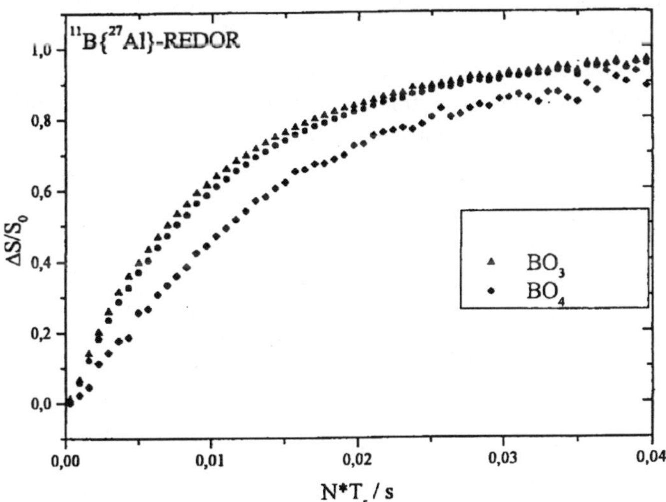

Fig. 27. ^{11}B$\{^{27}$Al$\}$ REDOR curve of a sodium aluminoborate glass of composition: 30 mole % Na_2O–50 mole % B_2O_3–20 mole % Al_2O_3. Note that the effect of the heterodipolar coupling to ^{27}Al is significantly stronger for the trigonal boron site (triangles), compared to the tetrahedral boron site (diamonds). Circles denote the effect on the total boron resonance.

other [73, 74]. Detailed simulations show that the slope of this curve at short evolution times is linearly correlated with the second moment of the I–S interactions [75]. Figure 23 shows recent ^{11}B–^{27}Al REDOR studies of sodium aluminoborate glasses obtained in our laboratory [76, 77]. From MAS–NMR it is known that the various types of framework sites present in these glasses include trigonal $BO_{3/2}$, tetrahedral $BO_{4/2}$ and tetrahedral $AlO_{4/2}$ units, however, the information about connectivities and spatial correlations between these sites is not available. ^{11}B$\{^{27}$Al$\}$ REDOR curves of the kind as those shown in Fig. 27 reveal that the trigonal boron sites interact much more strongly with aluminum than the tetrahedral boron sites. These results are quantitatively consistent with a model where $AlO_{4/2}$–$BO_{4/2}$ connectivities are essentially excluded from the glass structure. This appears to be the case for glasses in the entire composition range.

6.2.2. *Zero- and double quantum NMR in homonuclear systems*

The existence of dipole-dipole couplings in homonuclear spin systems is conveniently established via the process of magnetization transfer (*zero-quantum NMR, spin diffusion*). To detect and quantify spin diffusion, standard two-dimensional NMR experiments can be utilized. Figure 28 shows a typical pulse sequence employed successfully to detect spin diffusion among phosphorus sites in phosphate glasses [78]. Following the initial 90° preparation pulse, the magnetization evolves under the influence of the regular MAS–Hamiltonian (the isotropic chemical shift) during the evolution time t_1. The second 90° pulse then stores the magnetization along the field direction and spin diffusion is

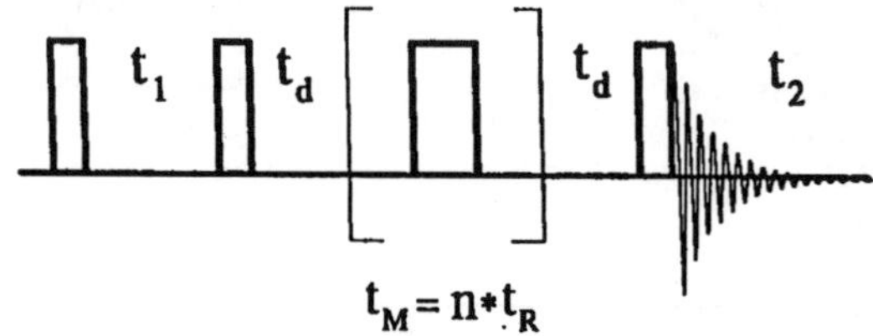

Fig. 28. 2D exchange NMR pulse sequence used for probing ^{31}P–^{31}P dipole-dipole interactions under MAS conditions. The narrow pulses are 90° pulses whereas in the mixing time rotor-synchronized 180° pulses are applied for recoupling of the dipole interaction. High speed MAS is applied throughout the entire experiment. The delays, t_d, before and after the mixing time are inserted to destroy transverse magnetization and are in the vicinity of 50 μs. From Ref. [78].

allowed to occur during the mixing time t_m. During this mixing time, the dipole-dipole interaction is reintroduced into the experiment by a train of 180° pulses applied in synchrony with the rotor period. The third pulse produces observable magnetization, which then evolves again under the regular MAS–Hamiltonian during the detection period t_2. As a result of spin diffusion during t_m, part of this magnetization now precesses at a frequency different from that present during t_1. This part gives rise to cross-peaks in the two-dimensional correlation plot that is obtained after Fourier Transformation with respect to both time axes.

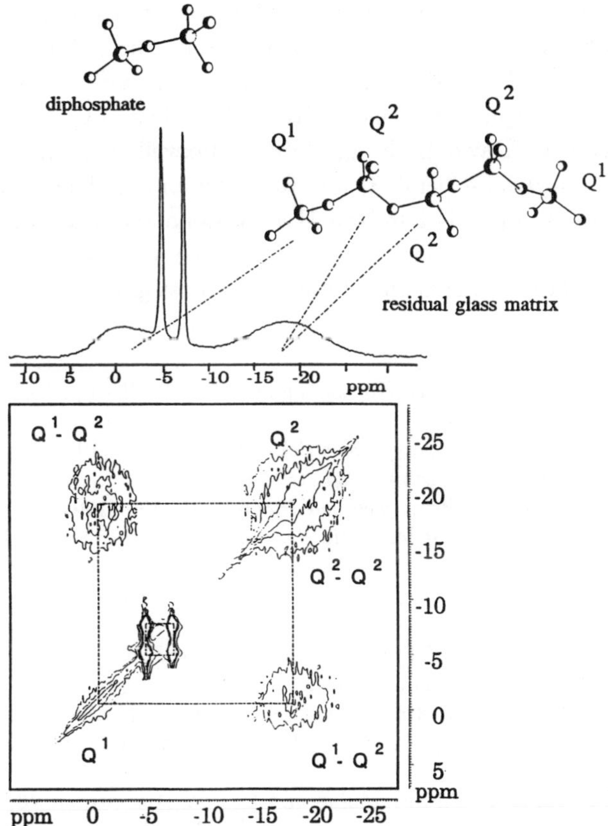

Fig. 29. ^{31}P 2D NMR spectrum of a CaO–Na$_2$O–P$_2$O$_5$ ceramic. The narrow resonances and cross peaks of the crystalline phosphate phase are due to a diphosphate anion. Connectivities inferred are shown in the figure. Reprinted from Ref. [79].

Figure 29 illustrates a typical example for a $CaO-Na_2O-P_2O_5$ ceramic [79]. The crosspeaks observable clearly establish site connectivity among $Q^{(1)}$ and $Q^{(2)}$ sites of the glass component of this ceramic. In addition, the substantial broadening of the diagonal peaks is indicative of spin diffusion among $Q^{(2)}$ sites with slightly different environments. Most recent developments in this research field suggest that excitation of double quantum coherences is a promising alternative to magnetization transfer experiments for probing homonuclear connectivities [80].

Acknowledgments

I wish to thank the past and present members of my research groups at the WWU Münster and the University of California, Santa Barbara, for many ideas and fruitful collaborations during the past ten years. In particular this review contains excerpts from the creative works of Drs. J.C.C. Chan, Deanna Franke, Becky Gee, David Lathrop, Robert Maxwell, Leo van Wüllen, and Lars Züchner, and Dipl.-Chem's Guido Regelsky and Michael Witschas. Our research has been supported by the U.S. National Science Foundation, by the Deutsche Forschungsge meinschaft and by the Wissenschaftsministerium Nordrhein–Westfalen. This financial support is most gratefully acknowledged.

References

[1] P.J. Bray, S.J. Gravina, D.H. Hintenlang and R.V. Mulkern, *Magn. Reson. Rev.* **13**, 263 (1988).

[2] P.E. Stallworth and P.J. Bray, *Glass Sci. Technol.* **4**, 77 (1990).

[3] H. Eckert, *Prog. NMR Spectrosc.* **24**, 159 (1992).

[4] H. Eckert, *NMR Basic Principles and Progress* **33**, 125 (1994).

[5] C.P. Slichter, *Principles of Magnetic Resonance* (Springer Verlag, Berlin 1989).

[6] C.A. Fyfe, *Solid State NMR for Chemists* (CRC Press, Guelph, Ontario 1984).

[7] M. Mehring, *Principles of High Resolution NMR in Solids* (Springer Verlag, Berlin 1983).

[8] K. Schmidt–Rohr and H.W. Spiess, *Multidimensional Solid State NMR and Polymers* (Academic Press, London 1994).

[9] R. Zallen, *The Physics of Amorphous Solids* (John Wiley, New York 1983).

[10] A. Feltz, *Amorphous Inorganic Materials and Glasses* (Verlag Chemie, Weinheim 1993).

[11] X. Xue, J.F. Stebbins, P.F. McMillan and B. Poe, *Am. Miner.* **76**, 8 (1991).

[12] H. Maekawa, T. Maekawa, K. Kawamura and T. Yokokawa, *J. Non-Cryst. Solids* **127**, 53 (1991).

[13] J. F. Stebbins, *Nature* **330**, 465 (1987).

[14] R. Dupree, D. Holland and D.S. Williams, *J. Non-Cryst. Solids* **68**, 399 (1984).

[15] A.R. Grimmer, M. Mägi, M. Hähnert, H. Stade, A. Samoson, W. Wieker and E. Lippmaa, *Phys. Chem. Glasses* **25**, 105 (1984).

[16] C. M. Schramm, B.H.W.S. DeJong and V.E. Parziale, *J. Am. Chem. Soc.* **106**, 4396 (1984).

[17] E. Schneider, J.F. Stebbins and A. Pines, *J. Non-Cryst. Solids* **89**, 371 (1987).

[18] R.K. Brow, R.J. Kirkpatrick and G.L. Turner, *J. Non-Cryst. Solids* **116**, 39 (1990).

[19] P. Losso, B. Schnabel, C. Jäger, U. Sternberg, D. Stachel and D.O. Smith, *J. Non-Cryst. Solids* **143**, 265 (1992).

[20] M. Villa, M. Scagliotti and G. Chiodelli, *J. Non-Cryst. Solids* **94**, 101 (1987).

[21] R.K. Sato, R.J. Kirkpatrick and R.K. Brow, *J. Non-Cryst. Solids* **143**, 257 (1992).

[22] R.K. Brow, R.J. Kirkpatrick and G.L. Turner, *J. Am. Ceram. Soc.* **76**, 919 (1993).

[23] S. Hayashi and K. Hayamizu, *J. Solid State Chem.* **80**, 195 (1989).

[24] P.J. Bray, *J. Non-Cryst. Solids* **95 & 96**, 45 (1987).

[25] L. van Wüllen and W. Müller–Warmuth, *Solid State Nucl. Magn. Reson.* **2**, 279 (1993).

[26] B.C. Bunker, R.J. Kirkpatrick, R.K. Brow and G.L. Turner, *J. Am. Ceram. Soc.* **74**, 1430 (1991).

[27] S.C. Kohn, R. Dupree, M.G. Mortuza and C.M.B. Henderson, *Am. Mineral* **76**, 309 (1991).

[28] C. Jäger, W. Müller–Warmuth, C. Mundus and L. van Wüllen, *J. Non-Cryst. Solids* **149**, 209 (1992).

[29] I. Farnan, P.J. Grandinetti, J.H. Baltisberger, J.F. Stebbins, U. Werner, M.A. Eastman and A. Pines, *Nature* **358**, 31 (1992).

[30] K.T. Mueller, B.Q. Sun, C.G. Chingas, J.W. Zwanziger, T. Terao and A. Pines, *J. Magn. Reson.* **86**, 470 (1990).

[31] B.F. Chmelka and J.W. Zwanziger, *NMR Basic Principles and Progress* **33**, 1 (1994).

[32] J.H. Baltisberger, S.L. Gann, E.W. Wooten, E.H. Chang, K.T. Mueller and A. Pines, *J. Am. Chem. Soc.* **114**, 7489 (1992).

[33] J.A. Tossell and P. Lazzeretti, *Phys. Chem. Miner.* **15**, 564 (1988).

[34] P.J. Grandinetti, J.H. Baltisberger, I. Farnan, J.F. Stebbins, U. Werner and A. Pines, *J. Phys. Chem.* **99**, 12341 (1995).

[35] J.W. Zwanziger, *Solid State Nucl. Magn. Reson.* **3**, 219 (1994).

[36] M.H. Ingram, *Phys. Chem. Glasses* **28**, 215 (1987).

[37] (a) J.O. Isard, *J. Non-Cryst. Solids* **1**, 235 (1969).
     (b) D.E. Day, *J. Non-Cryst. Solids* **21**, 343 (1976).

[38] J.R. Hendrickson and P.J. Bray, *Phys. Chem. Glasses* **13**, 43, 107 (1972).

[39] W.C. LaCourse, *J. Non-Cryst. Solids* **95 & 96**, 905 (1987).

[40] A. Bunde, M.D. Ingram and P. Maass, *J. Non-Cryst. Solids* **172–174**, 1222 (1994).

[41] S. Balasubramanian and K.J. Rao, *J. Non-Cryst. Solids* **181**, 157 (1995).

[42] C.D. Makowka, C.P. Slichter and J.H. Sinfelt, *Phys. Rev. Lett.* **49**, 379 (1982).

[43] B. Boyce and S.E. Ready, *Phys. Rev.* **B38**, 11008 (1988).

[44] S.E. Shore, J.P. Ansermet, C.P. Slichter and J.H. Sinfelt, *Phys. Rev. Lett.* **58**, 953 (1987).

[45] A.T.W. Yap, H. Förster and S.R. Elliott, *Phys. Rev. Lett.* **75**, 3946 (1995).

[46] B. Gee and H. Eckert, *J. Phys. Chem.* **100**, 3705 (1996).

[47] B. Gee and H. Eckert, *Ber. Bunsenges. Phys. Chem.* **100**, 1610 (1996).

[48] L. van Wüllen, B. Gee, L. Züchner, M. Bertmer and H. Eckert, *Ber. Bunsenges. Phys. Chem.* **100**, 1539 (1996).

[49] P. Florian, K.E. Vermillion, P.J. Grandinetti, I. Farnan and J.F. Stebbins, *J. Am. Chem. Soc.* **118**, 3493 (1996).

[50] D. Lathrop and H. Eckert, *J. Phys. Chem.* **93**, 7895 (1989).

[51] D. Lathrop and H. Eckert, *J. Am. Chem. Soc.* **111**, 3536 (1989).

[52] D. Lathrop and H. Eckert, *Phys. Rev.* **B43**, 7279 (1991).

[53] D. Lathrop, D. Franke, R. Maxwell, T. Tepe, R. Flesher, Z. Zhang and H. Eckert, *Solid State Nucl. Magn. Reson.* **1**, 73 (1992).

[54] D. Lathrop and H. Eckert, *J. Am. Chem. Soc.* **112**, 9017 (1990).

[55] M. Witschas, G. Regelsky and H. Eckert, *J. Non-Cryst. Solids* **215**, 226–235 (1997).

[56] D. Franke, R. Maxwell, D. Lathrop and H. Eckert, *J. Am. Chem. Soc.* **113**, 4822 (1991).

[57] D. Franke, R. Maxwell, D. Lathrop, K. Banks and H. Eckert, *Phys. Rev.* **B46**, 8109 (1992).

[58] R. Maxwell and H. Eckert, *J. Am. Chem. Soc.* **115**, 4747 (1993).

[59] R. Maxwell and H. Eckert, *J. Am. Chem. Soc.* **116**, 682 (1994).

[60] G. Bodenhausen, R. Freeman and G. A. Morris, *J. Magn. Reson.* **23**, 171 (1976).

[61] H. Rumpel and H. H. Limbach, *J. Am. Chem. Soc.* **111**, 5429 (1989).

[62] R. Maxwell, H. Erickson and H. Eckert, *Z. Naturforsch* **50a**, 395 (1995).

[63] R. Maxwell and H. Eckert, *J. Phys. Chem.* **99**, 4768 (1995).

[64] R. Maxwell, D. Lathrop and H. Eckert, *J. Non-Cryst. Solids* **180**, 244 (1995).

[65] A. Samoson, E. Lippmaa and A. Pines, *Mol. Phys.* **65**, 1013 (1988).

[66] A. Medek, J.S. Harwood and L. Frydman, *J. Am. Chem. Soc.* **117**, 12779 (1995).

[67] C. Fernandez and J.P. Amoureux, *Solid State Nucl. Magn. Reson.* **5**, 315 (1996).

[68] D. Massiot, B. Touzo, D. Trumeau, J.P. Coutures, J. Virlet, P. Florian and P.J. Grandinetti, *Solid State Nucl. Magn. Reson.* **6**, 73 (1996).

[69] J.H. Baltisberger, Z. Xu, J.F. Stebbins, S.H. Wang and A. Pines, *J. Am. Chem. Soc.* **118**, 7209 (1996).

[70] J.C.C. Chan, M. Bertmer and H. Eckert, *J. Am. Chem. Soc.* **121** 5238 (1998).

[71] T. Gullion and J.S. Schaefer, *J. Magn. Reson.* **81**, 196 (1989).

[72] T. Gullion and J.S. Schaefer, *Chem. Phys. Lett.* **194**, 423 (1992).

[73] A. Naito, K. Nishimura, S. Tuzi and H. Saito, *Chem. Phys. Lett.* **229**, 506 (1994).

[74] K. Herzog, B. Thomas, D. Sprenger and C. Jäger, *J. Non-Cryst. Solids* **190**, 296 (1995).

[75] M. Bertmer and H. Eckert, *Solid State Nucl. Magn. Reson.* **15**, 139 (1999).

[76] M. Bertmer, L. Züchner, J.C.C. Chan and H. Eckert, *J. Phys. Chem.*, submitted.

[77] L. Van Wüllen, L. Züchner, W. Müller–Warmuth and H. Eckert, *Solid State Nucl. Magn. Reson.* **6**, 203 (1996).

[78] C. Jäger, M. Feike, R. Born and H.W. Spiess, *J. Non-Cryst. Solids* **180**, 91 (1994).

[79] C. Jäger, P. Hartmann, G. Kunath–Fandrei, O. Hirsch, P. Rehak, J. Vogel, M. Feike, H.W. Spiess, K. Herzog and B. Thomas, *Ber. Bunsenges. Phys. Chem.* **100**, 1560 (1996).

[80] M. Feike, R. Graf, I. Schnell, C. Jäger and H.W. Spiess, *J. Am. Chem. Soc.* **118**, 9631 (1996).

6

Vibrational Excitations in Glasses

A. INELASTIC NEUTRON SCATTERING

RONALD L. CAPPELLETTI

*Department of Physics and Astronomy and
Condensed Matter and Surface Sciences Program,
Ohio University, Athens, Ohio 45701-2979
cappltti@helios.phy.ohiou.edu*

Contents

1. Introduction

Inelastic scattering of neutrons has played an essential role in revealing atomic dynamics in glasses and their connection to structure and to the glass-transition itself. Over the past dozen or so years, there have been improvements in instrumental resolution and range and a development of the use of the dynamical structure factor to study dynamics in detail. There have also been important conceptual developments such as the role of bond constraints in network vibrational dynamics and computational developments such as *ab initio* molecular dynamics. This chapter gives a brief introduction to the use of inelastic neutron scattering in the study of glasses and offers several selected illustrations of what we have learned from it.

2. Inelastic Neutron Scattering (INS)

The uncharged, slowly-moving, 1.0087 amu thermal neutron is a very useful probe of condensed matter. A key reason for this is that in many vibrational spectroscopies of atoms in glasses the intensities are weighted by excitation-energy-dependent "matrix elements" which, because of atomic disorder, are very difficult to calculate or account for in some other way. The nuclear scattering cross-section for thermal neutrons is relatively weak and essentially constant for each isotope, independent of the relative energy of the neutron and the scattering nucleus. It is weak enough for the neutrons to penetrate deeply into the system under study with the scattered intensity dominated by single scattering events. Neutrons are produced by some nuclear reaction (in a reactor or at a spallation source) and are slowed down by inelastic collisions in a light-atom moderating material to kinetic energies characteristic of the moderator temperature. The wavelengths of $\sim$ 300 K neutrons are $\approx$ 2 Å, i.e. of the order of interatomic spacing, which thus makes them very suitable for diffraction studies. For neutrons having a room-temperature spectrum the energies and momenta of the neutrons are *both* reasonably well-matched with those of characteristic atomic motions in solids. This allows the investigation of the spatial character of excitations in a manner which is unparalleled by other spectroscopies. It is clear that thermal neutrons have much to recommend them as probes of the structure and atomic dynamics of glasses.

Important techniques in measuring the vibrational density of states by inelastic neutron scattering are time-of-flight (TOF) methods and low-pass (e.g. beryllium) filter methods. Neither of these will be described in much detail here since there are extensive descriptions elsewhere [1, 2]. Basically, in a TOF measurement a pulse of monochromatized neutrons impinges on a sample and is scattered to a bank of detectors arrayed at various angles at a fixed distance from the sample. Neutrons losing (gaining) energy in the collision travel slower (faster) than those which are scattered elastically. By measuring the time-of-flight over the scattering path at a given angle one can measure the energy and momentum transferred to the sample. There is a description of a TOF instrument given in Sec. 2.4.

2.1. *Formalism*

Here, for want of space, we present a very brief account of the formalism required to understand what is measured in inelastic neutron scattering from glasses. The reader is urged to consult very thorough developments of the subject available in several excellent textbooks [3, 4, 5, 6] and articles [7]. Below we follow, with some variations, the notation of Ref. [7].

An incoming neutron of selected momentum $\hbar k_0$ scatters from a sample. It leaves with momentum $\hbar k_1$. Energy $E = \hbar^2 (k_0^2 - k_1^2) / 2\,m$ and momentum $\hbar Q = \hbar(k_0 - k_1)$ are transferred to the sample. The double differential scattering cross-section is:

$$\frac{d^2\sigma}{d\Omega dE} = \frac{k_1}{Nk_0} \sum_{d,d'} \sum_{\substack{i \in d \\ i' \in d'}} \overline{b_i^* b_{i'}} S_{ii'} \tag{1}$$

where

$$S_{ii'} = \frac{1}{2\pi\hbar} \int_{-\infty}^{\infty} dt e^{\frac{-iEt}{\hbar}} \langle e^{-iQ\cdot\hat{R}_i(0)} e^{iQ\cdot\hat{R}_{i'}(t)} \rangle \tag{2}$$

The brackets indicate an ensemble average and the bar indicates an average of the products of scattering lengths $b_i^* b_{i'}$ over nuclear spin and isotope distributions. $\hat{R}_i(0)$ is the quantum mechanical position operator of nucleus i at time 0 and $\hat{R}_{i'}(t)$ is the position operator of nucleus i' at time t. d and d' refer to the different elements in the sample. N is the total number of scattering centers (nuclei) exposed to the incoming beam. For the purposes of our discussion we may ignore incoherent scattering which arises from the rms deviation of the scattering lengths of a given element from the average for that element. Incoherent scattering is negligible for SiO_2 glass and small for the chalcogenide glasses we discuss below. In that case, we are left with coherent scattering which has a double differential cross-section given by

$$\frac{d^2\sigma}{d\Omega dE} = \frac{1}{N}\frac{k_1}{k_0} \sum_{d,d'} \bar{b}_d \bar{b}_{d'} \sum_{\substack{i \epsilon d \\ i' \epsilon d'}} S_{ii'}(\mathbf{Q}, E) \tag{3}$$

Note that now only the average element neutron scattering lengths appear in the formula. The coherent partial scattering function is then introduced:

$$S_c^{d,d'}(\mathbf{Q}, E) = \frac{1}{\sqrt{N_d N_{d'}}} \sum_{\substack{i \epsilon d \\ i' \epsilon d'}} S_{ii'} \tag{4}$$

where N_d is the number of atoms of element d in the beam. The coherent double differential scattering cross-section is given by

$$\frac{d^2\sigma}{d\Omega dE} = \frac{k_1}{k_0} \sum_{d,d'} \sqrt{c_d c_{d'}}\, \bar{b}_d \bar{b}_{d'} S_c^{d,d'}(\mathbf{Q}, E) \tag{5}$$

For elemental glasses or for the fortunate "pseudo-elemental" case where we can ignore differences among the average scattering lengths and masses of the elements in the glass (which is approximately the case in the Se–As–Ge glasses, for example) this formula reduces to

$$\frac{d^2\sigma}{d\Omega dE} = \frac{k_1}{k_0}\bar{b}^2 S_c(\mathbf{Q}, E) = \frac{k_1}{k_0}\bar{b}^2 \frac{1}{2\pi\hbar N} \sum_{ii'} \int_{-\infty}^{\infty} dt\, e^{\frac{-iEt}{\hbar}}$$

$$\langle e^{-i\mathbf{Q}\cdot\hat{\mathbf{R}}_i(0)} e^{i\mathbf{Q}\cdot\hat{\mathbf{R}}_{i'}(t)} \rangle \tag{6}$$

The reader will notice that the case of glassy alloys which are not pseudo-elemental will generate formulas complicated by the presence of partial structure factors. This case can provide very useful structural and dynamic information provided one can separately measure the partial structure factors. This requires the ability to vary the average elemental scattering lengths by, for example, isotopic substitution. The subject is beyond the scope of this article and treatments can be found elsewhere [8].

Let us bypass the use of the formulas introduced so far in the study of diffraction measurements which are considered in the chapter by A.C. Wright in this volume. Here we concern ourselves only with atomic dynamics of a pseudo-elemental glass in which each atom vibrates around an equilibrium position **i**. Then $\hat{\mathbf{R}}_i(t) = \mathbf{i} + \hat{\mathbf{u}}_i(t)$ with the displacement $\hat{\mathbf{u}}$ now being the dynamical variable operator. We may expand the scattering function in terms of orders of phonon scattering: $S_c(\mathbf{Q}, E) = S_{c,el}(\mathbf{Q}, E) + S_{c,1}(\mathbf{Q}, E) + S_{c,2}(\mathbf{Q}, E) +$ .... The first term represents elastic scattering:

$$S_{c,el}(\mathbf{Q}, E) = \delta(E) \sum_{i,i'} e^{i\mathbf{Q}\cdot(\mathbf{i}'-\mathbf{i})} e^{-(W_{i'}+W_i)} \tag{7}$$

W_i is a Debye–Waller (DW) factor which, for an isotropic system, is

$$W_i = Q^2 \langle u_i^2 \rangle / 6 \tag{8}$$

where $\langle u_i^2 \rangle$ is the mean square time averaged displacement of atom i.

The next term in the expansion (which is the important one in this chapter section) is, for the neutron-energy-gain case of interest here,

$$S_{c,-1}(\mathbf{Q}, E) = \frac{1}{N} \sum_j \frac{\hbar Q^2}{2\omega_j} \langle n_j \rangle \delta(E + \hbar\omega_j) \left| \sum_i^N e^{-W_i} (\hat{Q} \cdot \mathbf{e}_i^j) \frac{1}{\sqrt{M_i}} e^{i\mathbf{Q}\cdot\mathbf{i}} \right|^2 \tag{9}$$

where ω_j is the frequency of the jth normal mode, $\mathbf{e}_i^j$ is its polarization eigenvector at atom i, and $\langle n_j \rangle = [\exp(+\hbar\omega_j/k_B T) - 1]^{-1}$ is its population factor. The remaining terms in the expansion arise from scattering in which several vibrational quanta are excited in each event. Generally these multiphonon terms contribute a relatively featureless background to the inelastic scattering, and corrections need to be made for them in extracting the one-quantum process of Eq. (9).

2.2. *Dynamical structure factor, GVDOS*

Apart from the scattering length and DW factor, the structure factor term (in vertical brackets) in Eq. (9) is essentially a Fourier transform of the spatial vibration pattern of the jth mode. If we separately make appropriate averages of the DW and M_i factors and integrate over a small energy window ΔE about E, we get

$$\Delta E S_{c,+1}(\mathbf{Q}, E) \equiv \langle n_E \rangle e^{-2\bar{W}} \frac{\hbar^2 Q^2}{2\bar{M}} F(Q, E) \tag{10}$$

with

$$F(Q,E) = \left\{ \frac{\Delta E Z(E)}{NE} \left| \sum_i^N \frac{e^{-W_i}}{e^{-\bar{W}}} (\hat{Q} \cdot \mathbf{e}_i^j) \left(\frac{\bar{M}}{M_i} \right)^{1/2} e^{i\mathbf{Q}\cdot\mathbf{i}} \right|^2 \right\} \qquad (11)$$

Equation (10) defines the dynamical structure factor accessible from experiment, $F(Q,E)$ described by Eq. (11), which represents an average over all modes j having energies in the integration window. $\Delta E Z(E)$ is the number of such modes in that window. $F(Q,E)$ allows us to make contact with computations based on a model. Typically such a model involves a small number of atoms so that the computed dynamical structure factor must be orientationally averaged over $\hat{Q}$ to make a comparison.

A dynamic correlation function, basically a Fourier transform on the Q variable of the dynamical structure factor, has been introduced which provides a spatial representation of vibrating modes in the energy window [9]. This can give important insight into the nature of the modes, although, as is true of the Fourier transform of the static structure factor, the information is projected onto one dimension, namely a radial variable, and so cannot render a conclusive description.

To proceed further we need another approximation. In particular, returning to Eq. (9), if we imagine that in averaging over a sufficiently wide range of Q the interference effects arising from $e^{i\mathbf{Q}\cdot(\mathbf{i}'-\mathbf{i})}$ cancel to zero excepting when $\mathbf{i} = \mathbf{i}'$ (the "self" terms) then we arrive at the "incoherent approximation." Again, with suitable averaging of several factors we may define an effective generalized vibrational density of states (GVDOS) $G(E)$ accessible from experiment [10], even for the case of a (non-pseudo-elemental) alloy:

$$\frac{d\sigma}{dE}\Big|_{c,-1} \equiv e^{-2W_{av}} \langle \bar{b}^2 \rangle \frac{\hbar^2(Q_{\max}^4 - Q_{\min}^4)}{2M_{av}k_0^2} \langle n_E \rangle \frac{G(E)}{E} \qquad (12)$$

where

$$\frac{G(E)}{E} = \sum_j \left(\sum_i \frac{\bar{b}_i^2}{\langle \bar{b}^2 \rangle} \frac{M_{av}}{M_i} (\hat{Q} \cdot \mathbf{e}_i^j)_{or}^2 \frac{\delta(E - \hbar\omega_j)}{N\hbar\omega_j} \right) \qquad (13)$$

Here "or" stands for orientational averaging over $\mathbf{Q}$ occurring in the experiment. For a monatomic glass, assuming complete orientational averaging, this reduces to the usual vibrational density of states (VDOS) $Z(E)/E = \sum_j \delta(E - \hbar\omega_j)/3N\hbar\omega_j$. For a compound glass the GVDOS has weighting by vibrational mode square atomic amplitudes, scattering cross sections and inverse masses and so can differ substantially from $Z(E)$. In Eq. (10) $M_{av} = \sum_d (\alpha_d M_d)$ where $\alpha_d = (c_d \bar{b}_d^2 / M_d)/(\sum_l (c_l \bar{b}_l^2 / M_l))$, and $\langle \bar{b}^2 \rangle = \sum_l (c_l \bar{b}_l^2)$. $Q_{\min}$ and $Q_{\max}$ are the experimentally accessible extrema. The DW factor in Eq. (9) can be calculated self-consistently from Eq. (6) using $G(E)$ instead of $Z(E)$ in:

$$\overline{\langle u^2 \rangle} = \frac{3\hbar^2}{2M_{av}} \int \frac{\langle 2n_E + 1 \rangle}{E} G(E) dE \qquad (14)$$

Multiphonon corrections to $G(E)$ appear as a smooth slowly rising contribution as a function of E. $G(E)$ and multiphonon corrections can be calculated self-consistently. A code to do this, MUPHOCOR, has been developed at Karlsruhe [11] based on an analysis by Sjölander [12]. Typically the incident flux of the neutrons is too low at an incoming energy required to cover the sample vibrational spectrum in energy loss, so the measurement must be made on the neutron-energy-gain side of the elastic peak. If it can be arranged, however, there is an advantage of a neutron-energy-loss experiment, namely that the energy resolution is better on the loss side of the elastic scattering peak. In addition, the sample can be held at low temperatures (since thermal excitation of the quanta is unnecessary to the measurement) which reduces multiphonon contributions.

Corrections for multiple scattering of the incoming neutrons may be important and can be estimated by a Monte Carlo calculation as discussed by Copley [13]. To reduce these corrections, a vertical cylindrical thin annular shell extending along the beam aperture is a useful sample configuration.

2.3. *Facilities and instruments*

There is a great deal of information available from the Internet with regards to neutron scattering facilities, instruments, and methods, including pictures, schematic drawings, detailed specifications about instrumental resolution, Q range and E range, for example, and instructions and forms for proposals. The URL's for some of the major ones are listed here:

http://ncnr.nist.gov/
http://pnsjph.pns.anl.gov/ipns.html
http://neutron.chm.bnl.gov/HFBR/
http://www.lansce.lanl.gov/
http://www.missouri.edu/
http://neutrons.ornl.gov/
http://cu17.accl.ca

http://www.risoe.dk/
http://ndafleetwood.nd.rl.ac.uk/ISISpublic/
http://www.kfa-juelich.de/iff/iff_ns1.e.html
http://www.llb.cea.fr/index_e.html
http://www.ill.fr
http://nfdfn.jinr.dubna.su

Reference [1] gives an expert account of instrumental configurations and important parameters. As an example of a TOF instrument we describe the Fermi-chopper TOF spectrometer at the NIST Center for Neutron Research [14]. Figure 2.1.1. taken from Ref. [14] is a schematic of the instrument. Neutrons incident from the guide tube G on the double Bragg monochromator M_1/M_2 are low-pass filtered to remove higher order Bragg reflections by F to produce a beam of wavelength λ_0 incident on the chopper C. The chopper cuts the monochromatic beam into pulses which fly towards the sample S where they are scattered along an evacuated flight path towards ^{3}He-filled detectors D in a bank distributed over angles from 5° to 140° at a distance $\sim$ 2.3 m from the sample. The oscillating radial collimator (RC) reduces the flux of neutrons scattered towards the detectors from vacuum, furnace

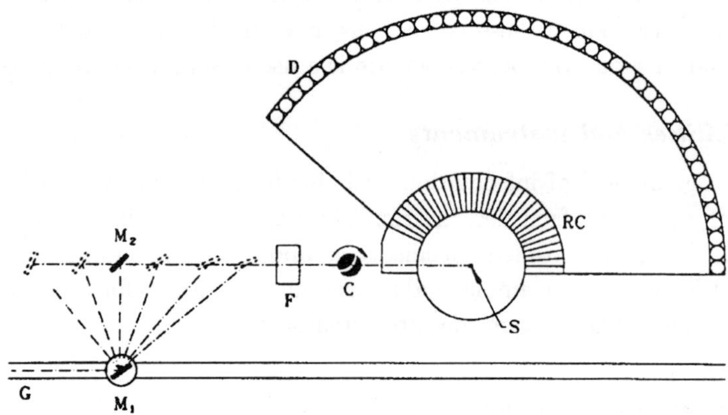

Fig. 2.1.1. Simplified plan view of the Fermi chopper spectrometer at the NIST Center for Neutron Research. The figure is reproduced from Fig. 14 of Ref. [14]. The letters G, F, C, S, and D, denote the guide, filter, chopper, sample and detectors respectively; M_1 and M_2 are the monochromator crystals and RC is the (oscillating) radial collimator. Incident wavelengths between 0.23 and 0.61 nm are obtained by modifying the Bragg diffraction angle of the double monochromator, as suggested by the dashed lines in the figure.

or cryostat wall which may be surrounding the sample. After appropriate electronic processing and storage, the signals from each detector may be displayed as counts vs. time channel. From this information one may obtain a histogram representing $S(Q, E)$ over the energy and momentum transfer regions allowed by the parameters of the experiment. This information is then used to derive the GVDOS as described in Sec. 2.2.

3. Selected Results

3.1. *General features of the vibrational spectrum of glasses*

3.1.1. *The boson peak*

In addition to acoustic-like modes having a Debye $\sim E^2$ spectrum, a feature seen in many glasses is a bump in $G(E)$ in the range of a few meV, see Fig. 3.1.1 [15]. These "extra" states, the low energy density of states (LEDOS), are apparently the same as those giving rise to a feature observed in Raman scattering called the "boson peak". They have been shown to be related to

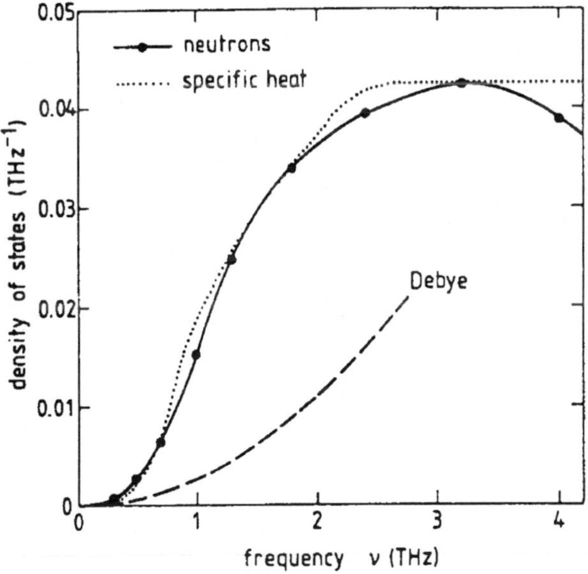

Fig. 3.1.1. Density of states $G(\nu)$ in **vitreous silica** reproduced from Fig. 1 of Ref. [15] which gives details of how it was derived. The dashed curve is the Debye contribution calculated from the measured speeds of sound. The dotted curve is calculated from the heat capacity.

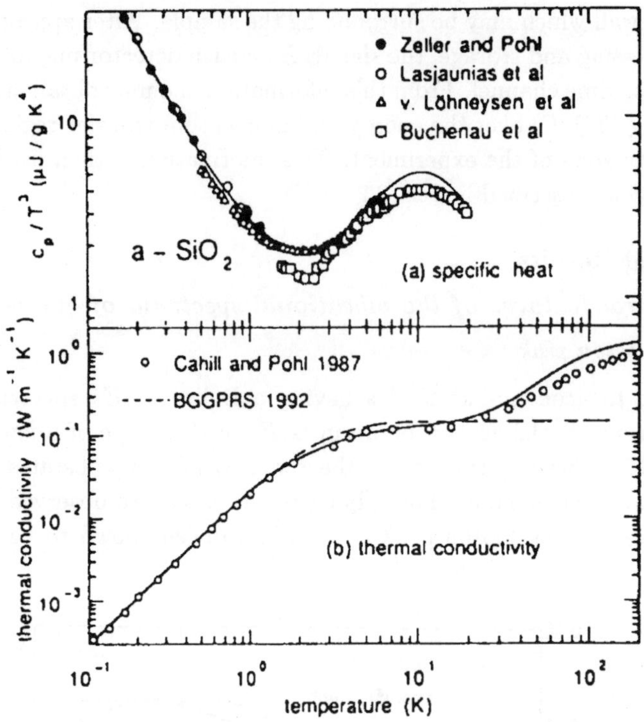

Fig. 3.1.2. Top: specific heat [75, 76, 77, 15] plotted as C_p/T^3 vs. T. Bottom: thermal conductivity [78] vs. T. The dashed and solid lines are model calculations as described in Ref. [79] from which this figure was reproduced.

a peak (Fig. 3.1.2) in $C(T)/T^3$, where C is the heat capacity, and a plateau (Fig. 3.1.2) and subsequent rise in the thermal conductivity in the temperature range between 1 and $\sim$ 10 K. The plateau has been interpreted in terms of an interaction between the acoustic modes and the extra modes [16].

Moreover, in several cases (though not all [17]) there appears to be a relation between these vibrational modes and a feature observed in the static structure factor of neutron and X-ray diffraction, the anomalously-behaved "first sharp diffraction peak" (FSDP), which is the signature of intermediate range order on the length scale of a few nm [18]. An interesting semi-quantative equality has been suggested between a correlation length ξ which can be defined from the width of the FSDP, ΔQ, $\xi = 2\pi/\Delta Q$, and a length d related to the wave-number of the boson peak (Fig. 3.1.3) [19]. It is argued that the elasto/optical

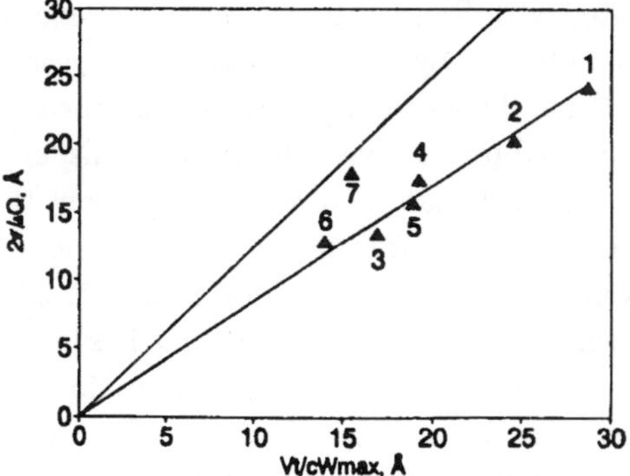

Fig. 3.1.3. Correlation length as obtained from FSDP vs. that from ω_{max} for several glasses (1–7). The figure was reproduced from Ref. [79] which gives the details of the samples and the two lines which are model curves.

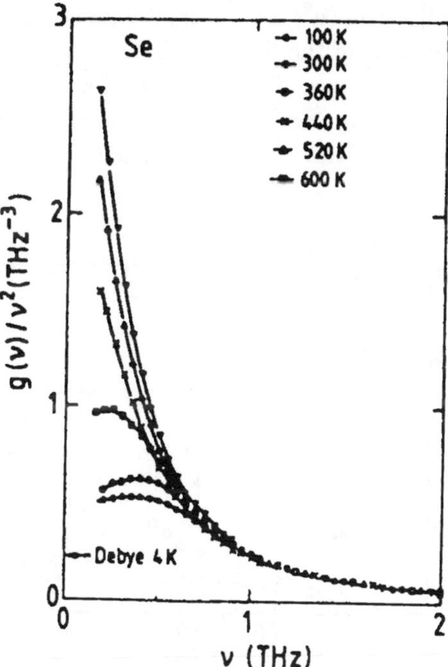

Fig. 3.1.4. Quasi-harmonic vibrational density of states fitted to neutron data [80] in glassy and liquid **selenium** below and above the glass temperature $T_g = 305$ K. The figure is reproduced from Ref. [16].

fluctuations associated with Raman scattering over a length-scale d are related to the peak wave-number ν_p by $d \simeq v/(c\nu_p)$, where v is the speed of sound and c the speed of light. This equality is then construed as evidence that the boson modes are localized (or at least coherent) on this length scale. However, it should also be mentioned that there is evidence from the Q-dependence of the dynamical structure factor for extended acoustic mode behavior of a least some of the modes in the region of the boson peak [15, 20].

An interesting series of experiments [16, 21] has demonstrated that the boson peak persists *above* T_g into the undercooled liquid. Moreover, it does not immediately give way to diffusional motion (Fig. 3.1.4). As T increases the peak red-shifts (softens) and eventually goes over to a quasi-elastic line associated with relaxational motion, but not until well above T_g.

Since the boson peak represents excess LEDOS it contributes (see Eq. 14) to an increase in the average atomic displacement $\langle u^2 \rangle$ which can be determined from neutron scattering. An example for Se is shown in Fig. 3.1.5 from the work of Buchenau [22] where $\langle u^2 \rangle$ for the crystal is shown for comparison. One remarkable feature evident in the figure is that $\langle u^2 \rangle_{T_g} \simeq \langle u^2 \rangle_{T_m}$, where T_m is the crystal melting temperature. This suggests that the Lindemann criterion [23], which says that melting occurs when atomic oscillations reach a certain (universal) small fraction of the interatomic spacing, is operable.

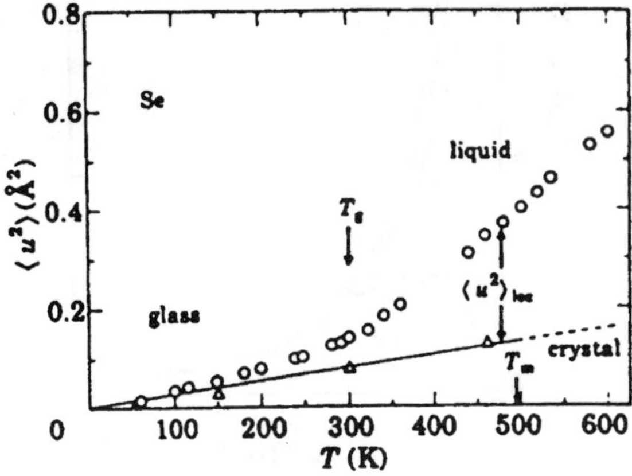

Fig. 3.1.5. Mean-square displacements in glassy, liquid and crystalline selenium determined from neutron data for motions with frequencies above 10^{11} Hz. The figure is reproduced from Ref. [22].

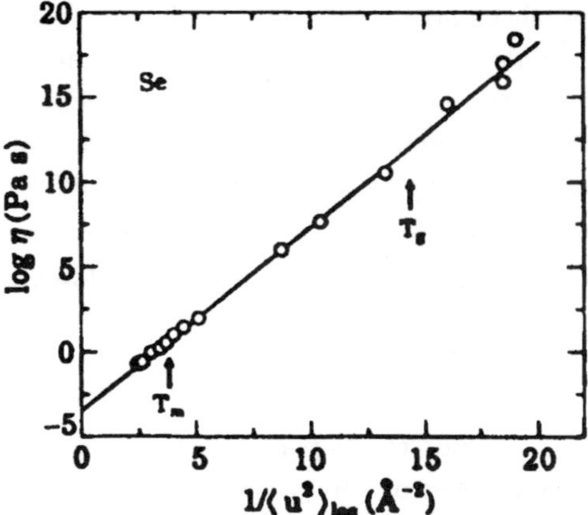

Fig. 3.1.6. Linear relation between the logarithm of the viscosity [81] and the inverse of the soft mode or β-relaxation mean-square displacement in selenium. The figure is reproduced from Ref. [22].

On this basis, Malinovski and Novikov [24] have described the glass transition as follows: beginning at high temperatures, the presence of LEDOS in the liquid phase implies a $\langle u^2 \rangle$ which is too large to satisfy the Lindemann criterion, and in this sense prevents crystallization unless the cooling rate is sufficiently slow. Finally, at T_g, $\langle u^2 \rangle_{T_g} \simeq \langle u^2 \rangle_{T_m}$ and solidification occurs. These authors use a phenomenological model to show how parameters describing the LEDOS can predict that $T_m/T_g \approx 1.5$, which is empirically true in many glasses. They further remark that the presence of LEDOS in the liquid is thus a predictor of glassforming ability.

A related connection has been made by Buchenau who showed that in **Se** one can account for the behavior of the viscosity through the glass transition over the entire range of measurement in terms of $\langle u^2 \rangle_{\text{local}}$, the part associated with the LEDOS, as shown in Fig. 3.1.6, thus providing a link between "fast" and "slow" motions characterizing the transition.

3.1.2. *The "floppy" modes*

The concept of "floppy" modes in network glasses, idealized as zero-frequency modes, arises in connection with constraint counting arguments introduced by Phillips [25], developed by Thorpe and co-workers [26], and recently extended by Boolchand and co-workers [27]. Phillips conceived of the bonds in network glasses as acting as mechanical constraints (which he then used as a basis for a theory of glassforming ability). It was proposed that when the number of constraints in the system is less than the number of mechanical degrees of freedom there could be atomic displacements which do not provoke restoring forces — hence, zero-frequency or mechanically "floppy" modes.

Since in network glasses the relative strengths of the forces go like: bond-stretching > bond-bending $\gg$ dihedral angle > "interchain", there is some ambiguity in deciding which forces to count as constraints. For example, if one can ignore the last two forces and one treats all atoms as having the average coordination number $\langle r \rangle$ (a "mean field"-like approximation), then each bond stretch contributes $\langle r \rangle/2$ constraints per atom (as each bond is shared by two atoms). Similarly, the number of bond angles per atom, each of which imposes a constraint, goes as $2\langle r \rangle - 3$, i.e. 1 for $\langle r \rangle = 2$, 3 for $\langle r \rangle = 3$, etc. (For a glass alloy $\langle r \rangle = \sum_d c_d r_d$, where c_d and r_d are, respectively the concentration and coordination number of element d.) Since in three dimensions the number of degrees of freedom per atom is three there is a "rigidity threshold" when the number of constraints per atom, N_c, is also three. This occurs for a value of $\langle r \rangle$ given by $5\langle r \rangle/2 - 3 = 3 \Rightarrow \langle r_c \rangle = 2.4$. For smaller $\langle r \rangle$ there are floppy modes. The fraction of the spectrum expected to be floppy is thus given by $f = 1 - N_c/3 = 1 - 5\langle r \rangle/6$. For example, for Se, $\langle r \rangle = 2$ and $f = 1/3$.

Thorpe showed by a computer model using a potential embodying only stretching and bending forces that the mean field results above works quite well with fluctuations being important only very near $\langle r_c \rangle$ [28]. In real network glasses the dihedral angle and interchain forces are small but not negligible. In effect, these forces act as weak constraints and so the floppy modes are shifted up to finite frequencies. A later computer calculation containing terms to model the latter forces showed that the rigidity threshold arising from bond stretching and bending forces is smeared out but is still apparent as a gentle break in slope at $\langle r \rangle = 2.4$ in the first inverse moment of the vibrational density of states [29].

The **Se–As–Ge** system forms network glasses over a broad range of compositions from Se to well beyond the rigidity threshold. One can make glasses of different compositions sharing the same $\langle r \rangle$ value. Moreover there is only a modest variation in masses and coherent scattering lengths among the elements so that $G(E)$ is expected to be a good representation for $Z(E)$. Kamitakahara *et al* [20] determined $G(E)$ at room temperature from neutron TOF measurements made at IN4 of the Institute Laue–Langevin (ILL). The

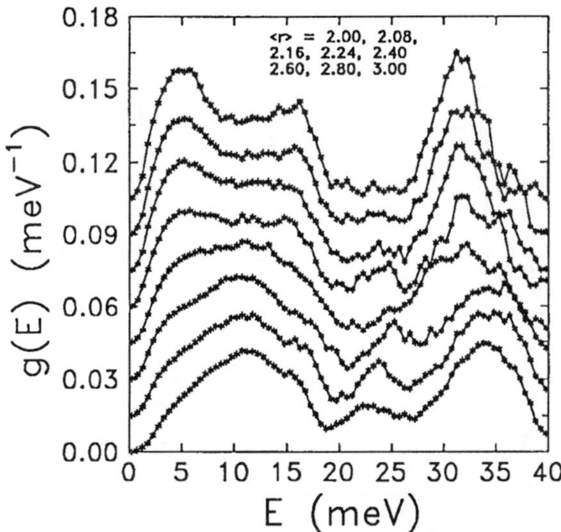

Fig. 3.1.7. VDOS for Se–As–Ge alloy glasses. Starting from the bottom ($\langle r \rangle = 3.00$), successive curves have been shifted up by 0.015 units on the vertical scale. The top curve ($\langle r \rangle = 2.00$) is for pure Se glass. The other alloy glasses in order are: $\langle r \rangle = 2.08$: $Se_{96}Ge_{04}$; $\langle r \rangle = 2.16$: $Se_{89}As_{06}Ge_{05}$; $\langle r \rangle = 2.24$: $Se_{88}Ge_{12}$; $\langle r \rangle = 2.40$: $Se_{675}As_{250}Ge_{075}$; $\langle r \rangle = 2.60$: $Se_{63}As_{14}Ge_{23}$; $\langle r \rangle = 2.80$: $Se_{51}As_{18}Ge_{31}$; $\langle r \rangle = 3.00$: $Se_{395}As_{210}Ge_{395}$. The peak at ≈ 5 meV is identified as the "floppy" mode peak. The figure was reproduced from Ref. [20].

results corrected for multiphonon scattering are shown in Fig. 3.1.7 for $\langle r \rangle$ values running from 2.0 to 3.0. The peak at 5.0 meV in the top spectrum (Se) has an integrated intensity of approximately 1/3 and is identified as the "floppy" mode peak — shifted up by dihedral angle forces (which give rise to torsional modes). As the Se chains (and rings) are cross-linked by As or Ge atoms, spectral weight is shifted to other regions of the spectrum and the fraction of floppy modes decreases. A plot of the integrated first inverse moment of the spectra (Fig. 3.1.8) is compared to theory [29] and also to low-temperature Mössbauer measurements [30] using as a probe a small amount of ^{119}Sn which replaces Ge. The data show a gradual change in slope in the vicinity of the rigidity threshold region in agreement with theory.

Figure 3.1.9 shows that below ~ 20 meV pairs of alloys of different composition having the same $\langle r \rangle$ value have essentially identical spectra. This is especially dramatic for $\langle r \rangle = 2.4$ where the middle and upper regions of

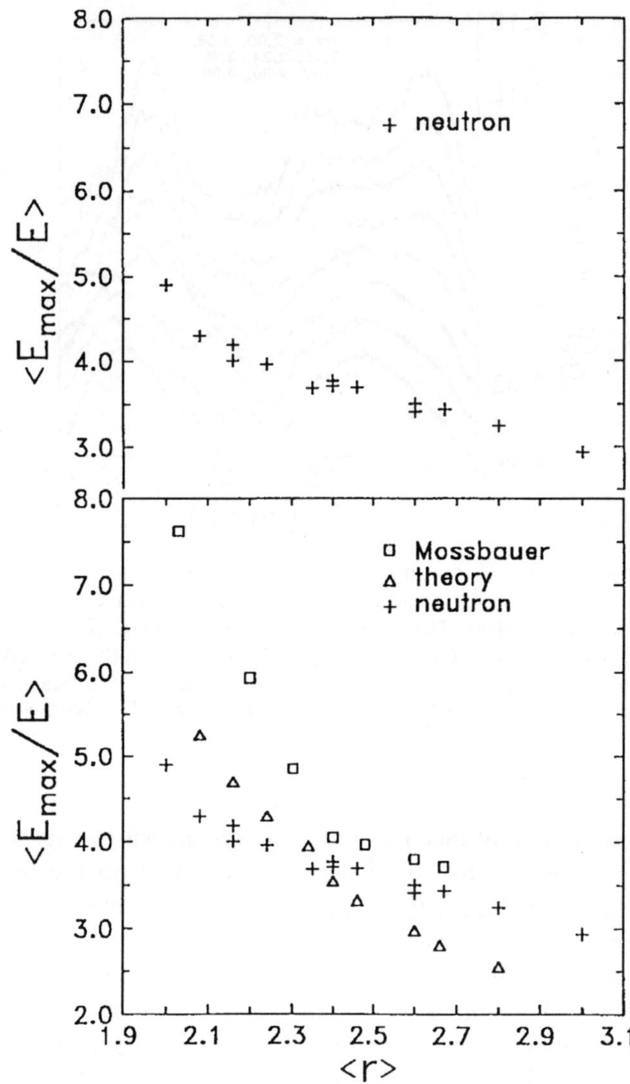

Fig. 3.1.8. The first inverse moment of the VDOS for Se–As–Ge glasses normalized to $E_{\mathrm{max}} = 40$ meV. Theory if from Ref. [29], Mössbauer results are from Ref. [30]. According to simple constraint-counting arguments the rigidity threshold occurs at $\langle r \rangle = 2.40$. The figure was reproduced from Ref. [20].

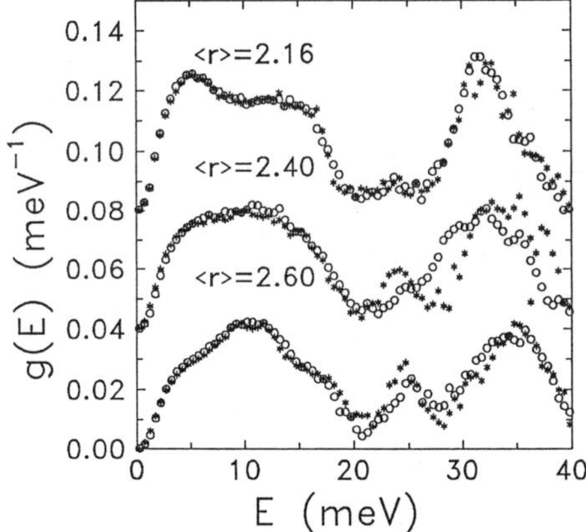

Fig. 3.1.9. VDOS comparing pairs of alloy glasses having the same $\langle r \rangle$ value. $\langle r \rangle = 2.16$: * $Se_{92}Ge_{08}$, ∘ $Se_{89}As_{06}Ge_{05}$, $\langle r \rangle = 2.40$: * $Se_{80}Ge_{20}$, ∘ $Se_{675}As_{250}Ge_{075}$, $\langle r \rangle = 2.60$: * $Se_{70}Ge_{30}$, ∘ $Se_{63}As_{14}Ge_{23}$. Starting from the bottom, successive data set pairs are shifted up by 0.04 units. The figure was reproduced from Ref. [20].

the spectrum are quite distinct. The remarkable insensitivity to composition has been seen in other properties in these glasses including the glass transition temperature, T_g and the viscosity of the liquid near T_g [31]. It has also been seen in spectral hole burning measurements [32]. While not completely understood, it appears to be an indication that where bond strengths and masses are roughly the same in these compound glasses, connectivity emerges as the dominant factor. The boson peak regime in Se–Ge glasses appears to be near 2 meV [33] as shown in Fig. 3.1.10.

The constraint counting argument presented above would indicate that SiO_2 is an "overconstrained" glass with no floppy modes. However, very recently there have been developments indicating that a special class of floppy modes, "rigid unit modes" (RUMs) should exist in *crystalline* SiO_2 because symmetry renders some constraints degenerate. It is suggested that some of these RUMs should survive as floppy modes in the glass [34]. Neutron scattering measurements of $G(E)/E^2$ in α–cristobalite, g–SiO_2 and g–$K_2O.2SiO_2$ indeed show the presence of peaks at 5 meV which the authors identify as a RUM in the crystal and a floppy mode in the glasses.

This raises the question of the appearance of various kinds of modes in the same frequency regime. The original Thorpe–Phillips floppy mode concept offers little detailed information about the nature of the modes (whether they extended or localized, for example) and so does not conflict with the usual

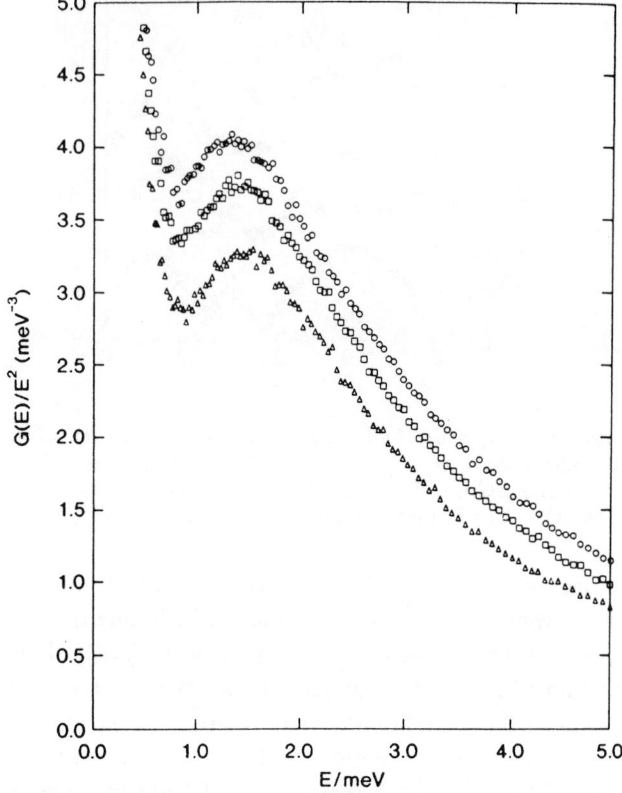

Fig. 3.1.10. $G(E)/E^2$ for a–$\mathbf{Ge_x Se_{1-x}}$ alloys with $x = 0.14(\circ)$, $x = 0.22(\square)$ and $x = 0.30(\triangle)$. The figure was reproduced from Ref. [33].

interpretation of the boson peak as modes which are localized on the nm length scale. However RUMs are quite particular kinds of modes which should present a signature in the Q-dependence of the dynamical structure factor.

3.2. *Modes of special structural clusters in glasses*

This section addresses special modes throughout the vibrational spectrum of particular glasses. Attention is drawn to specific examples where neutron scattering (sometimes combined with modelling) gives reasonably strong evidence for modes associated with specific molecular units or clusters of them within the glass.

3.2.1. g–SiO_2

The first example comes from the work of Buchenau [15] in the boson peak region of the spectrum. For ordinary sound waves the dynamical structure factor of Eq. (10) is expected to be a featureless function of Q. Features arise from vibrational modes in which the $\mathbf{e}_i^j$ vary significantly in direction on a length scale $\sim 2\pi/Q$. A plot of (essentially) $F(Q,E)$ in the 1 to 7 meV region of the spectrum of g–SiO_2 is shown in Fig. 3.2.1. Panels i and ii show what is to be expected for one and two tetrahedra librations. The error bars in panel iii represent the data, and curve represents what is to be expected for the (isolated) five-tetrahedron cluster librating as shown in Fig. 3.2.2. The

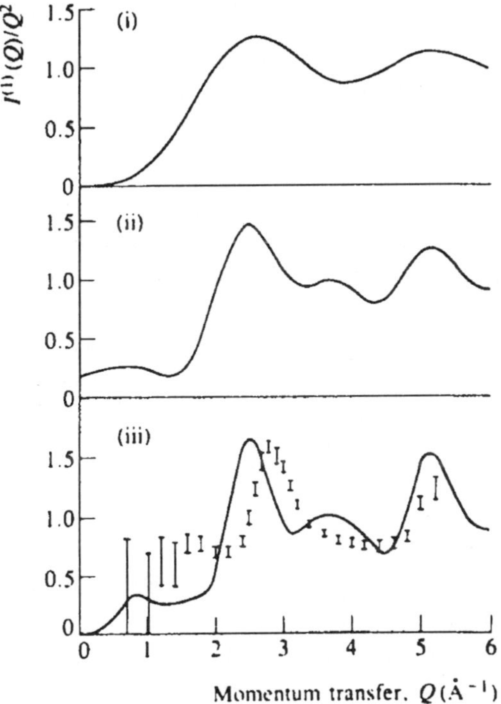

Fig. 3.2.1. Inelastic factor $I^{(1)}(Q)/Q^2$ [essentially the quantity of Eq. (10)] normalized to 1 for $Q \to \infty$ for (a) rotation of a single tetrahedron, (b) selective rotation of two tetrahedra, and (c) the relative rotation of five. Also shown are experimental point with error bars. The figure was reproduced from Ref. [15].

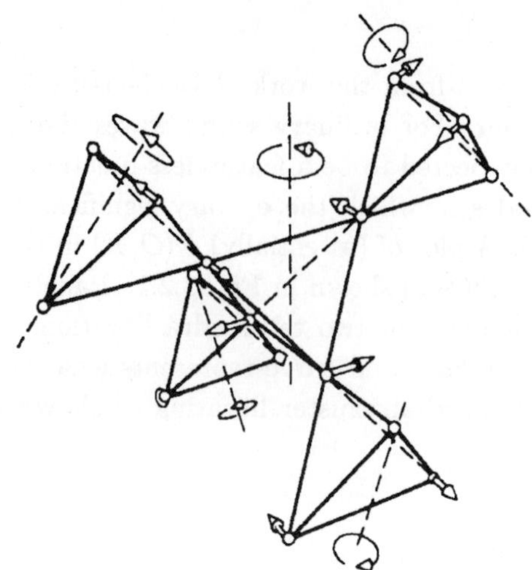

Fig. 3.2.2. Five tetrahedra structural model used in the calculation of the line in panel (c) of Fig. 3.2.1. The figure was reproduced from Ref. [15].

qualitative resemblance to the data is suggestive. However detailed calculations on such a cluster embedded in the glass remain to be done. As Buchenau *et al.* suggested, the proposed librations should be related to motions in crystalline silica. In particular, it would be interesting to know whether an α-cristobalite-like RUM (mentioned in Sec. 3.1.2) can account for the observed Q-dependence of the dynamical structure factor observed by Buchenau *et al.*

The second example is from the work of Arai *et al* [35] and it illustrates the level of analysis currently available for g–SiO_2. In Fig. 3.2.3 is shown the multiphonon-corrected GVDOS measured in neutron-energy-loss at the MARI TOF spectrometer at ISIS of the Rutherford–Appleton Laboratory. The results are in substantial agreement with earlier measurements by Price and Carpenter [7] made at the LRMECS TOF spectrometer at IPNS of Argonne National Laboratory. The frequencies marked by vertical lines are the results of calculations using the central-force Sen–Thorpe [36] model based on corner-sharing tetrahedra with force constant 562 N/m and $\langle \theta \rangle = 140°$. Figure 3.2.4 shows two corner-sharing tetrahedra and defines the angles. Figure 3.2.5 shows $S(Q, E)\Delta E$ [the quantity of Eq. (9)] as a function of Q for energy cuts around the major features near ω_1, ω_3, and ω_4 in Fig. 3.2.3. The lines through the data are the results of fits to a phenomenological model for motion of the atoms in a SiO_4 tetrahedron based on the following simplifying assumptions.

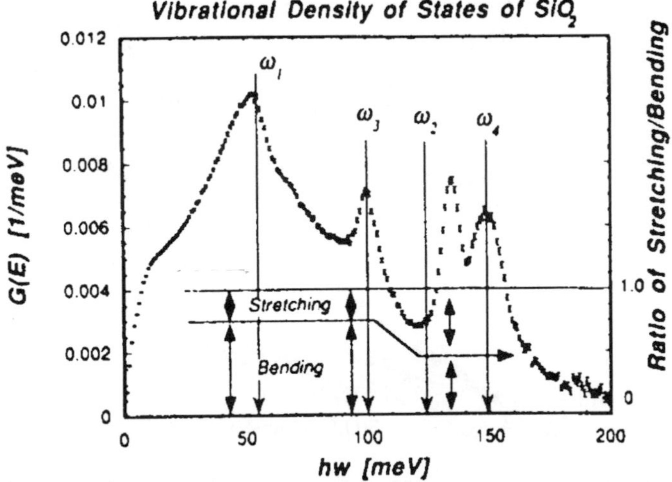

Fig. 3.2.3. The vibrational density of states $G(E)$ for g–SiO_2 (measured with $E_i = 220$ meV), indicating the normal modes ω_1, ω_2, ω_3 and ω_4 for the central-force-only calculation [36]. Also shown is the relative importance of bending and stretching in different energy regions, estimated by the model calculation discussed in the text. The figure was reproduced from Ref. [35].

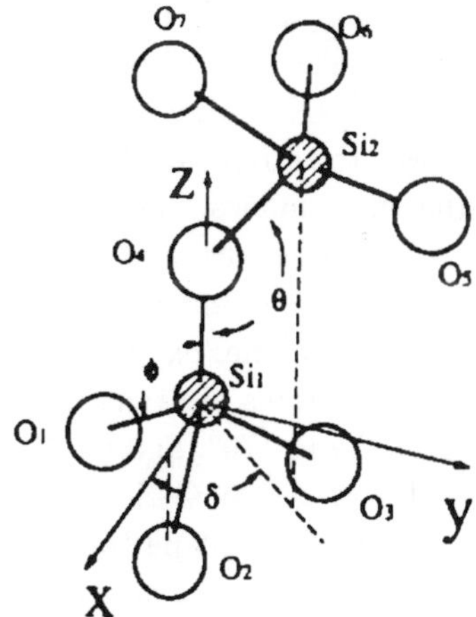

Fig. 3.2.4. Corner-sharing tetrahedron model for g–SiO_2. Each Si atom is bonded to four oxygen atoms which are shared with surrounding tetrahedron units. The Si–O–Si bond angle is θ and the dihedral angle is δ and together these give the angular orientation of the tetrahedron relative to the neighboring tetrahedra. The figure was reproduced from Ref. [35].

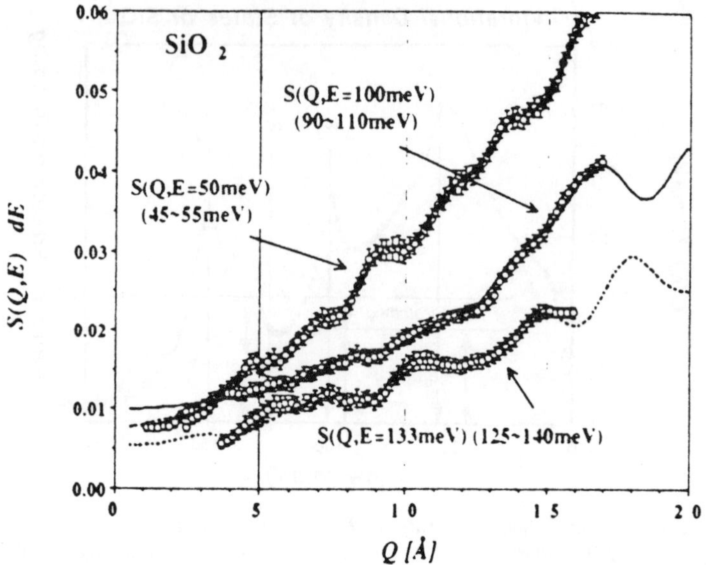

Fig. 3.2.5. The Q-dependence of $S(Q,E)$ for g–SiO_2 at specific energies of 50, 100, and 133 meV. The integrated ranges are specified in the parentheses. The lines are fit using the SiO_4 tetrahedron model. The figure was reproduced from Ref. [35].

(1) The tetrahedra in the network (apart from vibrations) are undistorted with Si–O bondlength of 1.62 Å. (2) The displacements of the atoms of the tetrahedron are free parameters. (3) Each individual atomic DW factor is replaced by the average one. In this way the atomic displacement vectors are calculated from the fit for each of the three modes. The resulting atomic motions are illustrated in Fig. 3.2.6. With some further assumptions the model is then used to estimate the Si–O–Si angles θ and dihedral angles δ associated with neighboring tetrahedra sharing the corner oxygens (only the neighboring Si are shown in the figure). The assumptions are that (a) the Si–O–Si bond angles are assumed to lie between 110° and 170°, and (b) the eigenvector is assumed to lie either along the bisector of the Si–O–Si angle (pure "bending", B) or perpendicular to it and in the Si–O–Si plane (pure "stretching", S). Having thus determined the possible Si–O–Si planes for each corner oxygen of the central tetrahedron, one can now estimate the angles θ and δ presumably in a way consistent with the tetrahedra linked to each corner not hindering each other. This method gives a broad distribution of angles $\langle\theta\rangle = 143°$ with FWHM $\sim 33°$ in reasonable agreement with early work by Mozzi and Warren [37],

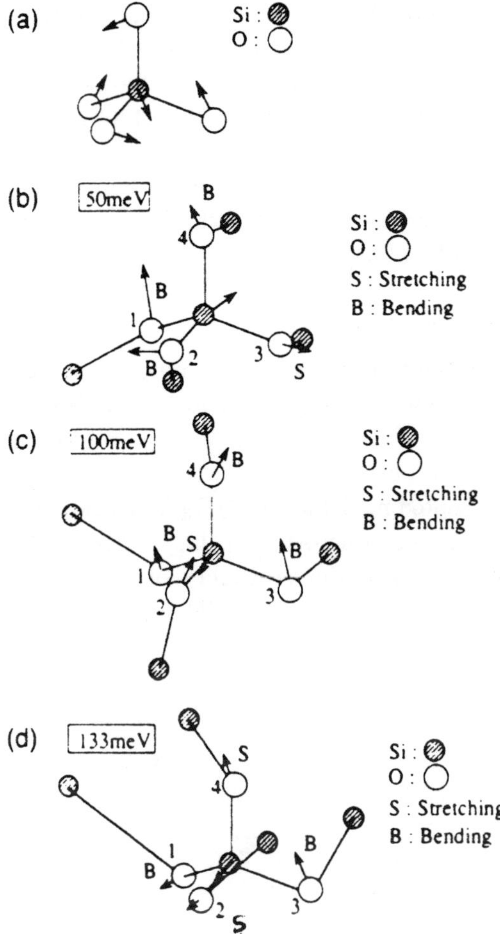

Fig. 3.2.6. (a) The SiO_4 tetrahedral unit with the free displacement vectors used for model fitting, together with the displacement vectors and position of the Si atoms in neighboring tetrahedra estimated by fitting the modes at 50 (b), 100 (c), and 133 meV (d). The figure was reproduced from Ref. [35].

phenomenological molecular dynamics (MD) models [38], and fitting by Reverse Monte-Carlo methods [39]. The resulting dihedral angle distribution seems to weakly prefer 0° and 120° which places neighboring Si atoms over the oxygens of the central tetrahedron rather than between them. Finally, one obtains the fraction of the mode which has B (bending) or S (stretching) character as shown in Fig. 3.2.3. For comparison, analysis of NMR measure-ments gives $\langle\theta\rangle$ in the 140° to 150° range [40]. Also, the result of *ab initio* molecular dynamics (aiMD) calculations on a 72 atom cell by Sarnthein *et al* [41] gives θ 's over the same range but with $\langle\theta\rangle = 136°$ and a somewhat sharper distribution.

In summary, the method described by Arai, though simplified, appears to hold some promise in getting at the structural parameters $\langle\theta\rangle$ and $\langle\delta\rangle$ which are difficult to obtain by other methods.

3.2.2. *Breathing boroxol ring mode in* g–B_2O_3

This next example taken from the work of Hannon *et al* [42] demonstrates the power of high-resolution neutron spectroscopy to address long-standing issues in glasses. Hannon gives a thorough review of the subject, so only the salient points will be touched on here.

The generally accepted model for g–$\mathbf{B_2O_3}$ is Zachariasen's celebrated con-tinuous random network of corner-sharing planar trigonal BO_3 structural units [43]. This structure permits the possibility of a boroxol group, namely a B_3O_6 planar ring of three corner-sharing BO_3 units shown in Fig. 3.2.7. One

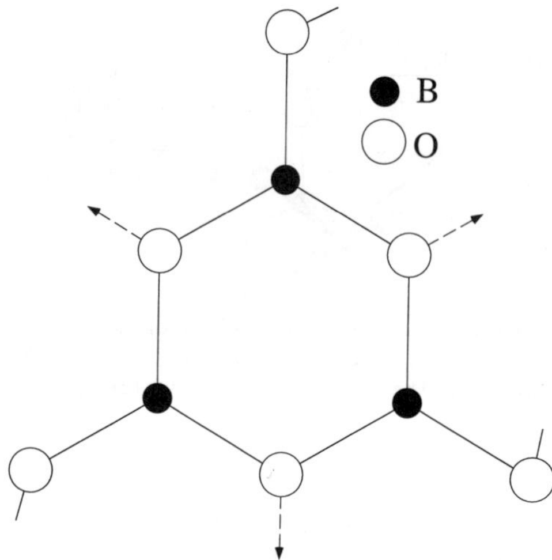

Fig. 3.2.7. A B_3O_6 boroxol group showing the motion (dashed arrows) involved in the breathing mode. The outer oxygens are linked to other BO_3 units in the glassy B_2O_3 network.

vibrational mode of such a ring occurs with the ring oxygens moving radially (the "breathing" mode) and, as the figure suggests, the B atoms and O atoms linking the ring to the remaining structure hardly move. For this reason the ring breathing mode is sharply localized in the ring, i.e. is a quasi-molecular mode, and thus occurs as a sharp feature in Raman spectroscopy at 808 cm^{-1} with a FWHM of ~ 15 cm^{-1}. At issue is the fraction f of B atoms in boroxol groups in the glass structure.

NMR experiments yield a value of $f = 0.82 \pm 0.08$ [44], earlier neutron diffraction experiments gave $f = 0.6 \pm 0.2$ [45], NQR results gave $f = 0.85 \pm 0.02$ [46]. An estimate from Raman scattering gives $f \simeq 0.75$ [47]. On the other hand, several MD computations based on various phenomenological potentials show no evidence of boroxol groups [48]. A MD calculation including a three-body force, i.e. allowing for bond-bending forces, gave $f = 0.23$ of B in puckered (not planar) rings [49]. The low fraction may be an artifact of the unphysical cooling rates in the MD simulations. Previous INS measurements of lower resolution near the boroxol breathing mode had also not provided convincing evidence for the existence of a peak [50].

The Hannon experiment consisted of high accuracy diffraction measurements using the D4 diffractometer at the ILL and the LAD instrument at ISIS, and of INS measurements using the HET spectrometer at ISIS. The samples were enriched to 99.57% ^{11}B to reduce absorption of neutrons by ^{10}B. The INS measurements reduced to a GVDOS are shown in Fig. 3.2.8 where the boroxol ring contribution is clearly visible at 100.2 meV (808 cm^{-1}). The calculated instrumental resolution at this peak was 1.63 meV FWHM, rather close to the observed width of 1.74 meV of the fitting Gaussian indicating a natural linewidth of the boroxol ring mode in g–B_2O_3 of ~ 0.6 meV (~ 5 cm^{-1}) assuming the widths add in quadrature. Such a narrow linewidth is consistent with a long mode lifetime and strong localization on single boroxol rings.

The diffraction results will not be discussed here in any detail. They were of sufficient accuracy to be able to compare them to a model which minimized R_χ at $f = 0.80 \pm 0.05$, in excellent agreement with the aforementioned nuclear resonance experiments. Thus several quite different experiments converge to challenge MD calculations on the structure of g–B_2O_3.

3.2.3. *Molecular modes in g–GeSe₂*

g–**GeSe$_2$** differs from g–SiO$_2$ in several respects. The zero-order structural model is one of linked tetrahedra of GeSe$_4$ but there is neutron diffraction

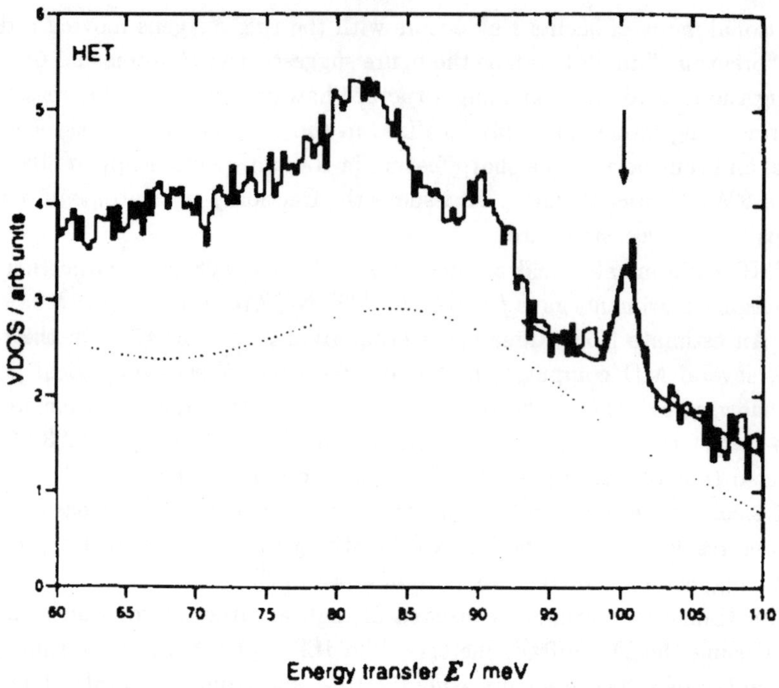

Fig. 3.2.8. The VDOS of vitreous v–B_2O_3 in the region of the boroxol ring breathing mode as measured on HET. The arrow indicates the energy at which the boroxol ring breathing mode is observed in Raman spectra, while the dotted line indicates the estimate of the multiple scattering contribution described in Ref. [42] from which the figure was reproduced. The continuous line indicates a Gaussian fit.

evidence for almost half of these tetrahedra being linked at two corners, i.e. they share an edge, in both the glass [51] and the liquid [51, 52]. This might be expected on the basis of the high-temperature form of the crystalline phase (β–$GeSe_2$) in which such edge-sharing tetrahedra link chains of corner-sharing tetrahedra to form layers [53]. There is also some evidence from Mössbauer studies of broken chemical order, i.e. homopolar bonds [54]. MD studies based on model atomic potentials which are in quite good agreement with diffraction data give a bimodal distribution for the Ge–Se–Ge angle with peaks at 111° and 88°, the latter corresponding to the edge-sharing tetrahedra. In the Sen–Thorpe [36] thoery 111° is less than the critical angle 136° in this glass at which molecular modes are merged into the bands [55], so one expects some molecular-like features in the Raman spectra. Indeed, a fairly sharp feature at

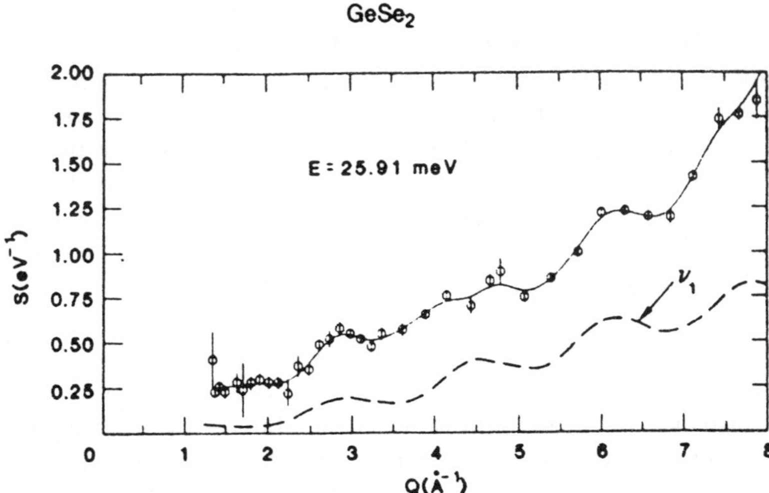

Fig. 3.2.9. $S(Q, E)$ measured around $E = 25.9$ meV (209 cm^{-1}). The dashed curve marked ν_1 represents a calculation based on Ge(Se$_{1/2}$) tetrahedra. The figure was reproduced from Ref. [57].

~ 201 (FWHM $= 17$) cm^{-1} has been identified as a symmetrical breathing tetrahedral mode (A$_1$) [56]. This identification (as well as that of two of other three tetrahedral modes) has been confirmed in a neutron scattering study at LRMECS at the IPNS of $S(Q, E)\Delta E$ by Walter *et al* [57] as shown in Fig. 3.2.9. The dotted lines are calculated using a valence force model on an isolated tetrahedron. "ν_1" is the breathing mode.

A second feature seen in Raman scattering, the "companion" mode at ~ 218 (FWHM $= 10$) cm^{-1} (A_{1c}) with about one half the intensity of the A_1 mode has been harder to interpret [56]. A detailed study and analysis of g–Se$_{1-x}$Ge$_x$ by Sugai interpreted the A_{1c} as arising from a breathing motion of edge-sharing tetrahedra (see Fig. 3.2.12 for an example) [56]. An interpretation requiring the presence of broken chemical order in the form of β–GeSe$_2$ crystalline fragments terminated along their chain edges by Se–Se (wrong bond) dimers has also been made, the companion being identified with a dimer mode [58]. This interpretation has been controversial because convincing structural evidence in support of the existence of the fragments has been hard to come by.

While many INS experiments have seen a broad peak in the A_1 region [20, 57, 59], some with a shoulder, the companion mode was not resolved. A recent energy-gain measurement at the Fermi Chopper TOF spectrometer in the Cold Neutron Research Facility at NIST with an energy resolution of 1.6 meV in the region of this peak on a portion of the same sample studied by Walter *et al* [57] has been analyzed for a multiphonon-corrected GVDOS which is shown in Fig. 3.2.10 [60]. Although they overlap heavily, the splitting of the A_1 and A_{1c} peaks is evident. Taking energy cuts around these peaks, $S(Q, E)\Delta E$ is displayed for both of them in Fig. 3.2.11. The statistics are adequate to make the point that the dynamic structure factors in these cuts do not differ much. This is borne out by aiMD calculations on a 216 atom cell the results of which are shown as solid lines for comparison [60].

One method for studying the vibrational power spectrum from MD simulations is the "frozen phonon" method. After following algorithms for making a glass model and relaxing it, the system is reduced to zero temperature. At this point, one of the advantages of the aiMD method becomes useful — it does not

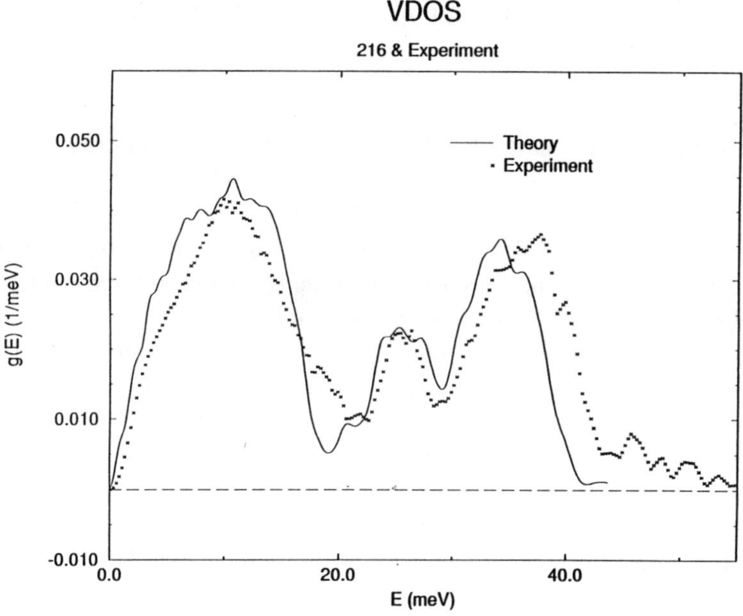

Fig. 3.2.10. VDOS of g–GeSe$_2$ measured at the Fermi Chopper TOF Spectrometer at NIST. The A_1 and A_{1c} modes are visible near 25 meV where the instrumental resolution is 1.6 meV. The solid line represents the results of an aiMD calculation on a 216 atom model. The figure was reproduced from Ref. [60].

rely on empirical interatomic potentials. One finds the force components acting on each atom by displacing it a small amount along each Cartesian coordinate and solving the quantum mechanical problem from which comes the change in total system energy. Thus the quantum mechanical origin of the interatomic forces is accounted for. In this way the elements of the dynamical matrix are determined and one proceeds to solve for the eigenvalues and eigenvectors in the harmonic approximation. The VDOS can then be calculated and is shown as the line in Fig. 3.2.10. In addition one has the atomic mode eigenvectors e_i^j so that the dynamical structure factor can also be calculated for each mode in an energy cut, as shown in Fig. 3.2.11. Even if one gets excellent agreement with data, it may be difficult to characterize the atomic motions, so the further step described below has been used.

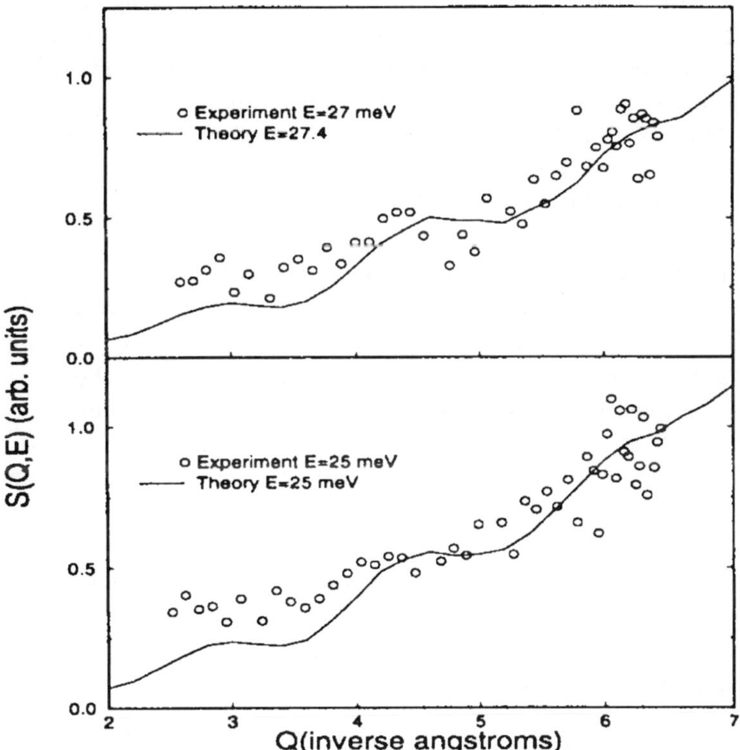

Fig. 3.2.11. $S(Q, E)$ for A_1 and A_{1c} regions. The lines are the results of an aiMD calculation on a 216 atom model. The figure was reproduced from Ref. [60].

One can animate each normal mode by oscillating the atoms in phase along their eigenvector directions and simply visualize the system on a computer screen, rotating the cell about to watch what is going on. In the case of g–GeSe$_2$ this has been helpful in gaining a qualitative understanding, especially in a region like the A_1–A_{1c} peaks where the dynamical structure factors do not differ very much. As expected, pure molecular modes are not really seen, but something approximating them is evident. For example, in the A_1 region a given mode may have eigenvectors quite well localized on a few separated tetrahedra in the cell which are exhibiting more-or-less radial Se motions. A neighboring mode separated by ~ 0.1 meV will look similar except that the eigenvectors will now be large on other tetrahedra and small on the previous ones. If one imagines the isolated molecule symmetric breathing as a "basis function" then the nearby states are reminiscent of splitting the degeneracy of this isolated molecular mode through the molecules being coupled in the network. In this same spectral region, edge sharing tetrahedra typically do not have large eigenvectors, but do participate in the motion. In addition, one finds localized modes on defects (e.g. 3-fold coordinated Se) which occur with greater frequency in the model than in the real glass. No doubt, one could develop a parameter to measure the "A_1" character of a given mode, but that might oversimplify the situation. It seems reasonable to simply compute the dynamical structure factor for comparison with the data.

A study of the modes in the A_{1c} region of the spectrum does indeed show more localization on edge-sharing tetrahedra including a breathing motion. Again, there is the resonance-like phenomenon of different edge-sharing tetrahedra "lighting up" in nearby modes. An example from a preliminary study of a 63 atom cell (in which it is easier to visualize the motions than in the 216 atom cell) is shown in Fig. 3.2.12 [61]. Modes in the lower band (below 20 meV) are typically extended throughout the cell and exhibit stronger Se motion than Ge motion. The opposite is true for the band above 29 meV: Ge motion dominates there.

Localization can be studied by computing the inverse participation ratio for each mode. A 216 atom cell probably contains too few modes in the boson region to study it properly in terms of localization. In general one finds more localized modes near the band edges, as expected.

4. Comparison to Molecular Dynamics

Increasingly in the literature one finds that neutron scattering data are compared to the results of MD computations. The approach based on the use of

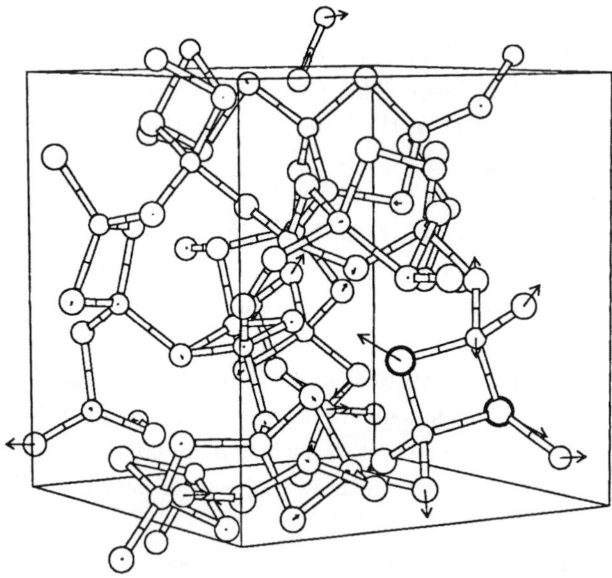

Fig. 3.2.12. A vibrational mode in the A_{1c} region of the spectrum of g–GeSe$_2$. The figure is the result of a frozen phonon calculation on a 63 atom model produced by aiMD. The arrows are the calculated eigenvectors. The figure shows a quasi-breathing mode substantially localized on an edge-sharing tetrahedron in the network. The figure was reproduced from Ref. [61].

empirical interatomic potentials has been available for more than twenty-five years. Modern large-scale computations of this form can involve up to $\sim 10^7$ atoms. Chapter 9 describes this approach and its successes in glasses in detail.

Since that chapter deals with classical MD, here we discuss the application of aiMD to glasses, a subject relatively in its infancy. This method has been described recently by Drabold [62]. Briefly, in 1985 R. Car and M. Parrinello (CP) combined density functional theory for calculating the electronic energy for a given configuration of ions with the simulated annealing technique thereby creating aiMD [63]. It uses one-electron density functional theory employing the local density approximation for the exchange-correlation term (LDA) in a clever scheme to compute total energies and forces directly from the variational principle of quantum mechanics. The ionic coordinates of the system evolve under the influence of these forces and ionic trajectories are thereby generated. (A review of the CP method which addresses important technical problems has been given recently by Payne *et al* [64].) As noted earlier, this scheme makes no assumptions about the form of the interatomic potential. There are no adjustable parameters. The forces arise from the solution of the underlying quantum mechanical problem. This gives the method a predictive power not available to empirical MD methods in studying the behavior of novel compounds. In principle the many-body forces acting on each atom are included. Moreover, the chemistry involved in bonding is not imposed but rather arises from quantum mechanics which can be an important advantage particularly in investigating multi-component systems which can admit homopolar bonds either as defects or as intrinsic to the system.

The original CP method involves the use of plane waves for basis functions and to date is limited in practical applications to systems of less than two hundred atoms solved on a supercomputer. In addition to studies of important amorphous elemental semiconductors such as a–Si [65] and molecular clusters [66], it has been been applied to the network glass-formers g–Se [67] and g–SiO$_2$ [41].

The study of l–Se and g–Se by Hohl and Jones [67] generates structural, vibrational and electronic information all of which is compared to experiment. For example, the VDOS computed from the Fourier transform of the velocity autocorrelation function has peaks at 52, 114, and 269 cm^{-1} compared to experiment: 44, 129, and 266 cm^{-1} respectively. The model is then used to provide structural information not yet available to experiment such as correlations between twofold and threefold coordinated Se atoms, bond-angle distributions for twofold and threefold coordinated atoms, and dihedral angle distributions and that of their phases. Finally, an extensive study is made of the electronic/structural coordination defects in the model, finding that the leading candidates are single threefold coordinated atoms.

The study of liquid and g–SiO$_2$ and α-quartz by Sarnthein *et al* [41], apart from providing the usual structural information such as partial pair correlation functions and bond-angle distributions for the liquid and glass, finds that their 72 atom cell is a perfectly chemically ordered network. Again, a key point about aiMD illustrated by this work is that the electronic properties can be studied. The electronic density of states compares well with experiment in structure, but the energy gap for the glass and crystal is too low (a well-understood characteristic discrepancy for the LDA). The significance of studying the electronic states is that they may be correlated (as mentioned for Se above) with structure (e.g. bonding) and with defects in the structure.

Sankey *et al* [68] developed an alternative approach to the plane wave method of CP which employs local atomic basis functions and a non-self-consistent version of density functional theory (the "Harris functional" [69)]. This method allows for larger systems ($\approx$ 500 atoms) to be computed using modern workstations instead of supercomputers. It has been applied to amorphous elemental semiconductors [70, 71] and recently to amorphous compounds (e.g. a–GaN [72]) and the network glass of interest here, $GeSe_2$ [60, 61]. The limitation of non-self-consistency has been lifted recently by Demkov *et al* [73]. Comparison of these aiMD results with experiment for vibrations has already been made for g–$GeSe_2$ in a previous section. Here we comment on the behavior of the electronic states in the liquid which show promise in connection with understanding photostructural phenomena characteristic of chalcogenide glasses (see Chapter 12).

Let us begin by focusing on a key finding of the computational liquid which is very likely relevant in the glass as well, namely a clear relationship between a static property, the localization of the LDA electronic eigenstates,

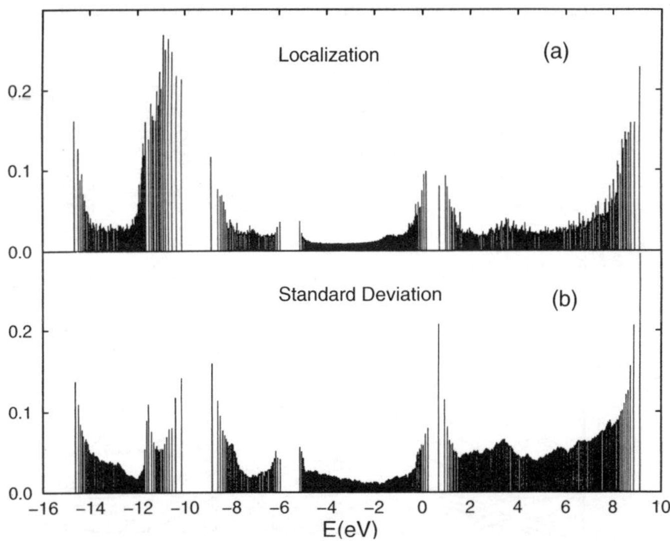

Fig. 4.1. (a) The average localization of the LDA electronic eigenstates as measured by the inverse participation ratio and (b) the root mean square deviation of the LDA electronic eigenvalues. The plot is a result of an aiMD calculation on a 216 atom model of liquid $GeSe_2$ over a 3.5 ps trajectory at $\sim$ 1144 K. The states whose eigenvalues are most sensitive to thermal fluctuations are those which are localized. The figure was reproduced from Ref. [74].

with a dynamic property, the RMS time fluctuation of the LDA eigenvalues. This trend is illustrated in Fig. 4.1 which is a plot of the inverse participation ratio, a measure of localization [74], in the top panel vs. energy. The bottom panel is a plot of the RMS deviation of the LDA energy eigenvalues vs. energy. The remarkable similarity of these plots establishes the above relationship. As the localization of an eigenstate increases, changes in the local network topology affect it strongly, and conversely as an eigenstates becomes more delocalized, thermal effects are averaged out, decreasing fluctuations in the eigenenergy. This is clearly an example of the interaction between the electronic and ionic dynamics, in other words, the electron-phonon interaction, but it emphasizes those unstable electronic states related to local structural rearrangements which are most likely to be the ones involved in photo-phenomena.

It is of particular interest that the liquid has *transient electronic states in the middle of the optical gap*. This is illustrated in Fig. 4.2 which is a plot of LDA eigenvalues in the gap region as a function of computational time step. Four gap-crossing events are observed. In each event either the valence band ascended or the conduction band descended into the optical gap creating a transient midgap state. It is possible to track the LDA energy eigenstates through

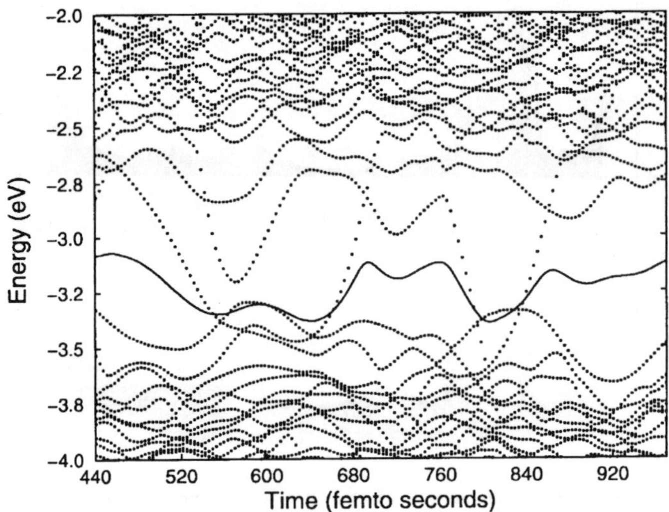

Fig. 4.2. LDA electronic eigenstates as a function of time displaying band-gap crossing events (dotted lines). The Fermi level is the solid line. The curves are the result of an aiMD calculation on a 216 atom model of liquid $GeSe_2$. The figure was reproduced from Ref. [74].

the atoms on which the states localize. In each event between two and five atoms had significant localization and all of these atoms were nearest or next nearest neighbors.

From a detailed visualization of these events and their analysis, as described in [74], the authors were able to observe that the transient midgap states arise from Se atoms which form weak bonds with their Se and Ge neighbors. A simple interpretation of these events is that lone pair states of Se atoms are being unoccupied as the energy level crosses from the valence band to the conduction band and vice-versa. When a lone pair state becomes unoccupied an antibonding state is created and a bonding state is created as well. This interpretation is consistent with their observation that the broadening of the conduction band edge (in the liquid compared to the glass) is associated with a high percentage of transient gap states which involve threefold coordinated Se atoms. Moreover, they speculate that these midgap transient states in the liquid phase, highly localized on structural defects and clearly sensitive to temperature changes, are related to corresponding states in the glass phase induced by light irradiation which are possibly responsible for light-induced electron spin resonance, photoluminescence, and photo-darkening.

To summarize, aiMD, although dealing with relatively small numbers of atoms at this stage, has several important advantages in simulating glasses. First, no assumptions are made about the interatomic potentials or bonding. Hence the method can be applied to make predictions about novel materials. Secondly, electronic information, possibly relevant to structural and atomic dynamical issues, is simultaneously available.

5. Conclusions and Future Work

In conclusion, several examples have been given of the important role played by INS in advancing our understanding of glasses. INS has been used to study the boson peak and to reveal a relation between the fast and slow dynamics of the glass transition. The "floppy" modes in chalcogenide glasses have been observed in approximate accordance with constraint-counting ideas. Special modes associated with molecular-like clusters have been observed and identified by high resolution INS measurements.

One can expect that the extension of the momentum and energy ranges and resolutions of neutron scattering instruments at major facilities will continue to have a strong impact on glass studies. The new field of aiMD is proving to be a powerful and discerning tool in the study of glasses, especially when used in conjunction with neutron scattering results, and will increase in significance with the growth of computational power.

Acknowledgments

I would like to express my appreciation to my colleagues W.A. Kamitakahara, P. Boolchand, D.A. Neumann, J.J. Rush, J.S. Lannin, and especially D.A. Drabold for many fruitful discussions and productive collaborations. I would also like to thank the National Science Foundation for support under Grant DMR-96-04921.

References

[1] C.G. Windsor, *Methods of Experimental Physics*, eds. by R. Celotta and J. Levine, Neutron Scattering **23** (Academic, Orlando, 1986) Chap. 3.

[2] C.G. Windsor, *Pulsed Neutron Scattering* (Taylor and Francis, London, 1981).

[3] L. Dobrzynski and K. Blinowski, *Neutrons in Solid State Physics*, eds. Malcom Cooper, (Ellis Horwood Limited, London, 1994).

[4] S.W. Lovesey, *Theory of Neutron Scattering from Condensed Matter* **1** (Clarendon, Oxford, 1984).

[5] G.E. Bacon, *Neutron Diffraction* (Clarendon, Oxford, 1975).

[6] P.A. Egelstaff, *Thermal Neutron Scattering* (Academic Press, New York, 1971).

[7] D.L. Price and J.M. Carpenter, *J. Non-Cryst. Solids* **92**, 153 (1987); D.L. Price and K. Sköld, *Methods of Experimental Physics*, eds. R. Celotta and J. Levine, *Neutron Scattering* **23-A** (Academic, Orlando, 1986) Chap. 1.

[8] K. Suzuki, *Methods of Experimental Physics*, eds. by R. Celotta and J. Levine, Neutron Scattering **23-A** (Academic, Orlando, 1986).

[9] A.C. Hannon, M. Arai, R.N. Sinclair and A.C. Wright, *J. Non-Cryst. Solids* **150**, 239 (1992).

[10] M.M. Bredov, B.A. Kotov, N.M. Okuneva, V.S. Oskotskii and A.L. Shakh-Budagov, *Sov. Phys. Solid State* **9**, 214 (1967); V.S. Oskotskii, *Sov. Phys. Solid State* **9**, 420 (1967).

[11] W. Reichardt, private communication.

[12] A. Sjölander, Arkiv för Fysik **14**, 315 (1958).

[13] J.R.D. Copley, *Comp. Phys. Commun.* **7**, 289 (1974); ibid. **9**, 59 (1975); ibid. **9**, 64 (1975); ibid. **21**, 431 (1981).

[14] J.R.D. Copley and T.J. Udovic, *J. Res. Nat. Inst. of Standards and Technology* **98**, 71 (1993).

[15] U. Buchenau *et al.*, *Phys. Rev.* **B34**, 5665 (1986).

[16] U. Buchenau, *J. Non-Cryst. Solids* **172–174**, 391 (1994).

[17] C. Levelut *et al.*, *Phys. Rev.* **B51**, 8606 (1995).

[18] D.L. Price, *Current Opinion in Solid State and Materials Science* **1**, 572 (1996).

[19] H.R. Schober, *Physica* **A201**, 14 (1993).

[20] W.A. Kamitakahara, R.L. Cappelletti, P. Boolchand, B. Halfpap, F. Gompf, D.A. Neumann and H. Mutka, *Phys. Rev.* **B44**, 94 (1991).

[21] U. Buchenau, *Phil. Mag.* **B71**, 793 (1995).

[22] U. Buchenau and R. Zorn, *Europhys. Lett.* **18**, 523 (1992).

[23] F.A. Lindemann, *Phys. Z.* **11**, 609 (1911).

[24] V.K. Malinovsky and V.N. Novikov, *J. Phys.: Condens. Matter* **4**, L139 (1992).

[25] J.C. Phillips, *J. Non-Cryst. Solids* **34**, 153 (1979); ibid. **43**, 37 (1981).

[26] M.F. Thorpe, *J. Non-Cryst. Solids* **57**, 355 (1983).

[27] P. Boolchand and M.F. Thorpe, *Phys. Rev.* **B50**, 10366 (1994); M. Zhang and P. Boolchand, *Science* **266**, 1355 (1994).

[28] H. He and M.F. Thorpe, *Phys. Rev. Lett.* **54**, 2107 (1985).

[29] Y. Cai and M.F. Thorpe, *Phys. Rev.* **B40**, 10535 (1989).

[30] P. Boolchand, R.N. Enzweiler, R.L. Cappelletti, W.A. Kamitakahara, Y. Cai and M.F. Thorpe, Solid State Ionics **39**, 81 (1990); W. Bresser, P. Boolchand and P. Suranyi, *Phys. Rev. Lett.* **56**, 2493 (1986).

[31] M. Tatsumisago, B.L. Halfpap, J.L. Green, S.M. Lindsay and C.A. Angell, *Phys. Rev. Lett.* **64**, 1549 (1990).

[32] S.P. Love, A.J. Sievers, B.L. Halfpap and S.M. Lindsay, *Phys. Rev. Lett.* **65**, 1792 (1990).

[33] R.N. Sinclair et al., *J. Non-Cryst. Solids* **150**, 219 (1992).

[34] M.T. Dove et al., *Phys. Rev. Lett.* **78**, 1070 (1997).

[35] M. Arai et al., *J. Non-Cryst. Solids* **192 & 193**, 230 (1995).

[36] P.N. Sen and M.F. Thorpe, *Phys. Rev.* **B15**, 4030 (1977).

[37] R.L. Mozzi and B.E. Warren, *J. Appl. Crystallogr.* **2**, 164 (1969).

[38] P. Vashishta, R.K. Kalia, R.P. Rino and I. Ebbsjö, *Phys. Rev.* **B41**, 12197 (1990).

[39] R.L. McGreevy, M.A. Howe, D.A. Keen and K.N. Clausen, *Inst. Phys. Conf. Ser.* **107**, 165 (1990).

[40] R. Dupree and R.F. Pettifer, *Nature* **308**, 523 (1984).

[41] J. Sarnthein, A. Pasquarello and R. Car, *Phys. Rev.* **B52**, 12690 (1995).

[42] A.C. Hannon et al., *J. Non-Cryst. Solids* **177**, 299 (1994).

[43] W.H. Zachariasen, *J. Am. Chem. Soc.* **54**, 3841 (1932).

[44] G.E. Jellison Jr., L.W. Panek, P.J. Bray and G.B. Rouse Jr., *J. Chem. Phys.* **66**, 802 (1977).

[45] P.A.V. Johnson, A.C. Wright and R.N. Sinclair, *J. Non-Cryst. Solids* **50**, 281 (1982).

[46] S.J. Gravina, P.J. Bray and G.L. Petersen, *J. Non-Cryst. Solids* **123**, 165 (1990); P.J. Bray, S. Gravina and D. Lee, *Boron-rich Solids*, eds. D. Emin, T.L. Aselage and B. Morosin, *AIP Conf. Proc.* **231**, 271 (1991); P.J. Bray et al., *J. Non-Cryst. Solids* **129**, 240 (1991); P.J. Bray et al., *The Physics of Non-Crystalline Solids*, eds. L.D. Pye, W.C. LaCourse and H.J. Stevens (Taylor and Francis, London, 1992) 713.

[47] A.C. Hannon et al., *J. Non-Cryst. Solids* **177**, 299 (1994) Sec. 2.3.

[48] A.C. Hannon et al., *J. Non-Cryst. Solids* **177**, 299 (1994) Sec. 2.8.

[49] H. Inoue, N. Aoki and I. Yasui, *J. Am. Ceram. Soc.* **70**, 622 (1987).

[50] A.C. Hannon *et al.*, *J. Non-Cryst. Solids* **106**, 116 (1988).

[51] S. Susman *et al.*, *J. Non-Cryst. Solids* **125**, 168 (1990).

[52] I.T. Penfold and P.S. Salmon, *Phys. Rev. Lett.* **67**, 97 (1991).

[53] Von G. Dittmar and H. Schäfer, *Acta Crystallogr. Sect.* **B32**, 2726 (1976).

[54] M. Stevens, P. Boolchand and J.G. Hernandez, *Phys. Rev.* **B31**, 981 (1985).

[55] See discussion of S.R. Elliott, *Physics of Amorphous Materials* (John Wiley & Sons, Inc., New York, 1990) 194.

[56] S. Sugai, *Phys. Rev.* **B35**, 1345 (1987).

[57] U. Walter, D.L. Price, S. Susman and K.J. Volin, *Phys. Rev.* **B37**, 4232 (1988).

[58] P.M. Bridenbaugh, G.P. Espinosa, J.E. Griffiths, J.C. Phillips and J.P. Remeika, *Phys. Rev.* **B20**, 4140 (1979); J.C. Phillips, *J. Non-Cryst. Solids* **43**, 37 (1981); J.E. Griffiths *et al.*, *Phys. Status Solidi* **B122**, K11 (1984).

[59] L.F. Gladden, S.R. Elliott, R.N. Sinclair and A.C. Wright, *J. Non-Cryst. Solids* **106**, 120 (1988).

[60] M. Cobb, D.A. Drabold and R.L. Cappelletti, *Phys. Rev.* **B54**, 12162 (1996).

[61] R.L. Cappelletti, M. Cobb and D.A. Drabold, *Phys. Rev.* **B52**, 9133 (1995).

[62] D.A. Drabold, *Amorphous Insulators and Semiconductors*, 3. eds. M.F. Thorpe and M.I. Mitkova (Kluwer Academic, Boston, 1996) 405.

[63] R. Car and M. Parrinello, *Phys. Rev. Lett.* **55**, 2471 (1985).

[64] M. Payne *et al.*, *Rev. Mod. Phys.* **64**, 1045 (1992).

[65] F. Buda *et al.*, *Phys. Rev* **B44**, 5908 (1991); D.A. Drabold, P.A. Fedders and M. Grunbach, *Phys. Rev.* **B54**, 5480 (1996).

[66] D. Hohl, R.O. Jones, R. Car and M. Parrinello, *Chem. Phys. Lett.* **139**, 540 (1987); D. Hohl, R.O. Jones, R. Car and M. Parrinello, *J. Chem. Phys.* **89**, 6823 (1988); R.O. Jones and D. Hohl, *J. Chem. Phys.* **92**, 6710 (1990); R.O. Jones and D. Hohl, *J. Am. Chem. Soc.* **112**, 2590 (1990); D. Hohl, R.O. Jones, R. Car and M. Parrinello, *J. Am. Chem. Soc.* **111**, 825 (1989); R.O. Jones and D. Hohl, *Proc. 10th General Meeting of the Condensed Matter Division* (European Physical Society, Lisbon, Portugal, 1990); R.O. Jones and D. Hohl, *Int. J. Quantum Chem. Symp.* **24** (1990).

[67] D. Hohl and R.O. Jones, *Phys. Rev.* **B43**, 3856 (1991).

[68] O.F. Sankey and D.J. Niklewski, *Phys. Rev.* **B40**, 3979 (1989); O.F. Sankey, D.A. Drabold and G.B. Adams, Bull. *Am. Phys. Soc.* **36**, 924 (1991).

[69] A. Nakano *et al.*, *Phys. Rev.* **B49**, 9441 (1994).

[70] D.A. Drabold, P.A. Fedders, S. Klemm and O. Sankey, *Phys. Rev. Lett.* **67**, 2179 (1991); P.A. Fedders, D.A. Drabold and S. Klemm, *Phys. Rev.* **B45**, 4048 (1992).

[71] D.A. Drabold, P.A. Fedders and P. Stumm, *Phys. Rev.* **B49**, 16415 (1994).

[72] P. Stumm and D.A. Drabold, *Phys. Rev. Lett.* **79**, 677 (1997).

[73] A.A. Demkov *et al.*, *Phys. Rev.* **B52**, 1618 (1995).

[74] M. Cobb and D.A. Drabold, *Phys. Rev.* **B56**, 3054 (1997).

[75] R.C. Zeller and R.O. Pohl, *Phys. Rev.* **B4**, 2029 (1971).

[76] J.C. Lasjaunias, A. Ravex, M. Vandorpe and S. Hunklinger, *Solid State Commun.* **17**, 1045 (1975).

[77] H.V. Löhneisen, H. Rüsing and W. Sander, *Z. Phys.* **B60**, 323 (1977).

[78] D.G. Cahill and R.O. Pohl, *Phys. Rev.* **B35**, 4067 (1987); D.G. Cahill, PhD Thesis (Cornell University, Cornell, 1989).

[79] A.P. Sokolov, A. Kisliuk, M. Soltwisch and D. Quitmann, *Physica* **A201**, 295 (1993).

[80] W.A. Phillips, U. Buchenau, N. Nücker, A.J. Dianoux and W. Petry, *Phys. Rev. Lett.* **63**, 2381 (1989).

[81] K. Ueberreiter and H.J. Orthmann, *Kolloid Z.* **123**, 84 (1951).

B. VIBRATIONAL EXCITATIONS IN GLASSES: RIGIDITY TRANSITION AND LAMB-MÖSSBAUER FACTORS

P. BOOLCHAND

*Department of Electrical, Computer Engineering and Computer Sciences,
University of Cincinnati Cincinnati, OH 45221–0030 USA
punit.boolchand@uc.edu*

Contents

1. Glass Forming Tendency and Rigidity Transition

The normal course of an inorganic melt upon cooling is a first order crystallization into a solid at the liquidus (T_l) temperature. Crystallization involves a latent heat and a first order change in specific volume with a discontinuity. *Select* melts can be supercooled past T_l into a glassy solid at T_g, the glass transition temperature. Glass formation also involves a latent heat but a second order change in specific volume near T_g as a solid with well defined local structural correlations emerges at $T < T_g$. Historically, discussions of the glass forming tendency (GFT) of melts have focused on avoidance of crystallization [1]. Threshold quenching rates were identified that would preclude large ($\sim 1\mu$m) grain growths in different types of melts (such as metallic, semiconducting and insulating). The focus of these discussions understandably, therefore, has dealt with the kinetic aspects of the glass transition, as also reviewed in Chapter 3 of this volume by I. Gutzow. More recently, attention has focused on the formation of distinct glassy states produced by different quench rates of melts or even by different processing methods. The trapping of a complex system like a glass in different local minima of an energy hypersurface is now popularly addressed by building landscape paradigms [2]. Furthermore, the temporal relaxation of a glass system between a pair of quasi-equilibrium states has been addressed by a class of experiments in which, surprisingly, observables display a common functional dependence on time, a stretched exponential $(e^{-t/\tau})^\beta$ with $0 < \beta < 1$, for which a common physical basis appears to be emerging [3, 4]. In 1847, Kohlrausch first noted [5] the decay of charge in a Leyden jar to follow such a function.

A novel and microscopic approach to glass formation emerged in the early 1980's when the notion of inter-atomic valence forces in a covalent solid serving as Lagrangian constraints was introduced. The idea pioneered by J.C. Phillips led to the suggestion [6] that the glass forming tendency of a melt is optimized when the mean number of Lagrangian constraints per atom, $\bar{n}_c$, equals the number of degrees of freedom per atom. For point-like atoms in a $3d$ network

$$\bar{n}_c = 3 \tag{1}$$

Overconstrained ($\bar{n}_c > 3$) and underconstrained ($\bar{n}_c < 3$) networks, he reasoned, would spontaneously crystallize, while optimally constrained ($\bar{n}_c \sim 3$) ones would polymerize with accumulation of minimal strain and thus form continuous networks. There is now substantial experimental evidence to suggest that the mean-field glass condition (1) does indeed describe optimization of the GFT in a variety of systems, as we shall illustrate in this chapter.

The algorithm for counting Langrangian constraints is *"r/2" for bond-stretching (α) forces and "2r − 3" for bond-bending (β) forces in 3d* networks provided atoms bond with a coordination number $r \geq 2$. For such continuous random networks with no dangling ends (singly bonded atoms), Eq. (1) takes on the form,

$$\bar{n}_c = 5\bar{r}/2 - 3 = 3 \qquad (2)$$

and leads to a critical mean coordination $\bar{r}_c = 2.40$. Upon quenching a melt of this critical composition or connectivity, glass formation instead of crystallization is predicted to be the normal course. This is observed in a variety of chalcogenides such as $As_2(S \text{ or } Se)_3$ and $Ge(S \text{ or } Se)_4$ where the mean coordination number $\bar{r}$ acquires the critical value of 2.4 if the assumed covalences are 2, 3 and 4 for S or Se, As and Ge, respectively.

Ternary $Ge_x As_y Se_{1-x-y}$ glasses provide ideal examples of covalently bonded networks in which the connectivity can be changed continuously by compositional tunning. Since over the glass forming range, as noted earlier, Ge, As and Se atoms largely possess coordination numbers of 4, 3 and 2, respectively, the mean coordination number $\bar{r} = 2 + 2x + y$. Physical properties such as the low-frequency ($0 < \omega < 10 \text{ cm}^{-1}$) vibrational density of states (VDOS) deduced from inelastic neutron scattering, activation energy for viscosity variation with temperature [8], glass transition temperatures ($T_g s$) [8], vibrational lifetime of a H_2O guest molecule [9], relaxation of persistent infrared spectral holes burned in the Se-H vibrational absorption band [10], are each found to correlate with the mean coordination $\bar{r}$ of a glass sample. These results may be related to the low-frequency vibrational excitations which are controlled strictly by global connectivity of the backbone i.e., $\bar{r}$. These observations make plausible that $\bar{n}_c$ in practice would depend linearly on $\bar{r}$ [Eq. (2)] and not on some higher power of $\bar{r}$.

Whether constraints are intact or broken in a specific system depends on how underlying bond strengths compare to thermal energies at the glass transition temperature T_g. Since β-forces are weaker than α-forces (by a factor of 3 or more), constraints associated with β-forces are more likely to be broken. Constraint counting algorithms have also been extended to networks that possess a finite fraction (m_2/N) of 2-fold coordinated atoms that have their β-constraint *broken*. Zhang and Boolchand showed [11] that for such networks Eq. (2) takes on a correction term, i.e.

$$\bar{r}_c = 2.4 + 0.4(m_2/N) \qquad (3)$$

with the correction term explicitly dependent on the m_2/N atom fraction.

The celebrated example of a broken bond-bending constraint in a glass network is the case of SiO_2 glass. For a binary $Si_x O_{1-x}$ network in which Si and O atoms possess a coordination number (CN) of 4 and 2, respectively, and in which the β-constraint about each O atom is intrinsically broken, Eq. (3) requires

$$4x + 2(1 - x) = 2.4 + 0.4(1 - x)$$

$$\text{or } x = 1/3 \tag{4}$$

leading to SiO_2 stoichiometry. One can thus understand the pronounced glass forming tendency [11, 12] of SiO_2, since the condition $x = 1/3$ satisfies the glass condition (1) exactly in a mean-field sense.

In chalcogenide glasses, Eq. (3) has also served as means to extend constraint counting to glasses that are *inhomogeneous* in character. In the $Ge_x Se_{1-x}$ binary glass system, for example in the Se-rich region ($x < 1/3$), an intrinsic phase separation occurs between Se-rich and Ge-rich regions. At finite $x(> 0.10)$ as $T_g s$ increase above $200°C$, it appears that the bond-angle constraint associated with Se_n-chain fragments joining Ge-rich regions is broken, causing the glass condition (1) to be realized at a higher Ge concentration than $x = 0.20$, as is actually observed in Raman scattering measurements [13]. The underlying upshift of the glass forming tendency to a higher mean coordination number $\bar{r}_c$, then also appears to be quantitatively described by Eq. (3).

Constraint-counting algorithms have also been extended to networks possessing dangling ends or 1-fold coordinated atoms, and it has been shown by Boolchand and Thorpe [14] that the glass condition (1) for such networks is satisfied when

$$\bar{r}_c = 2.40 - 0.4(n_1/N) \tag{5}$$

The GFT in several families of chalcohalide glasses that contain halogens and chalcogens σ bonded to group IV and separately to group V elements conforms to the prediction of Eq. (5) as shown in a recent review [15]. These correlations lend further credence to the notion that mechanical (Lagrangian) constraints in network glasses indeed play a *pivotal* role in determining the glass forming tendency.

From a normal mode analysis of covalently bonded $3d$ networks M.F. Thorpe recognized [16] that the number f of cyclical (floppy) or zero frequency solutions/atom of the dynamical matrix, is

$$f = 3 - \bar{n}_c \tag{6}$$

The glass condition (1) thus also becomes a condition for cyclical modes to vanish, $f \to 0$, i.e. for the network to first become rigid. Underconstrained networks ($\bar{n}_c < 3$) are easily deformable, while overcoordinated ones ($\bar{n}_c > 3$) are intrinsically rigid. On the other hand, optimally coordinated networks ($\bar{n}_c = 3$), are the ones that most readily form glasses. Thus, the condition for the *GFT to be optimized coincides with the onset of rigidity*. For this reason, the nature of the *rigidity transition* in network glasses has evoked particular attention in glass science. The manner in which a glass network becomes rigid upon crosslinking apparently has a direct bearing on the nature of molecular rearrangements and relaxation processes that ensue upon cooling a melt past T_g to form a glass. Glass transitions become thermally reversing near the rigidity transition as revealed by temperature modulated differential scanning calorimetry (MDSC). The results described by Eqs. (1)–(6) are mean-field results and represent averages over statistical variations in structural and dynamical properties. The mean-field result of a solitary transition describing rigidity onset in a 3*d network* at $\bar{r}_c = 2.40$ is remarkably well reproduced by numerical simulations [17] on *random networks* that identify a transition at $\bar{r}_c = 2.385$. It thus appears that the mean-field constraint-counting algorithms are a *good approximation* for describing the mechanical or elastic response of *continuous random networks*. Their great value is that they permit *specific predictions* for actual glasses that can be *tested in the laboratory*, thus providing an important guide to thinking about these systems. To the extent that glasses synthesized in the laboratory can be described as random networks, the predictions of mean-field constraint counting or numerical simulations on random networks have worked remarkably well. Notable exceptions have now appeared in binary chalcogenide glasses [18, 19]. The rigidity transition in these binary glasses occurs in two steps, with an onset point and a completion point defining a *transition region* rather than a solitary step. It appears that one of the challenges to understand the nature of the *transition region* in glasses is the need to identify aspects of medium range structure that apparently influence the nature of the rigidity transition.

Experiments have shown that the *rigidity transition* manifests directly in the *vibrational* and *thermal behavior* of network glasses. Inelastic neutron scattering measurements provide a means to probe the global vibrational density of states (VDOS) including the cyclical modes [20]. In the next section, we discuss the connection between *Lamb-Mössbauer factors* (f) and the *VDOS*, and their relation to the *rigidity transition* in glasses. Thermal reversibility of the glass transition probed in MDSC experiments has provided a new dimension as a probe of the transition region, as discussed in the last section of this chapter.

2. Vibrational Density of States and Lamb-Mössbauer Factors

Debye first showed [21] that the strength of recoil-free (Bragg) scattering of X-rays depends on the Debye-Waller factor. Recoil-free emission and absorption of γ-rays is governed by the closely related Lamb-Mössbauer [22] factor f. This factor represents the probability for a recoil-free event to occur and is given by

$$f = e^{-\overline{(\vec{k}\bullet\vec{u})^2}} \tag{7}$$

where $\vec{k}$ represents the γ-ray wave-vector, $\vec{u}$ the displacement vector of the Mössbauer active nucleus, and the bar over the dot product squared represents a motional average. For an isotropic solid (7) can be rewritten as

$$f = e^{-\frac{1}{3}k^2\overline{u^2}} \tag{8}$$

where $\overline{u^2}$ is the mean-square displacement. For example, in the Einstein model where all atoms vibrate with the same frequency ω_E

$$\overline{u^2} = \frac{3\hbar}{2M\omega_E}\coth\frac{\hbar\omega_E}{2k_BT} \tag{9a}$$

Then

$$\overline{u^2}_{T\to0} = \overline{u_o^2} = \frac{3\hbar}{2M\omega_E} \tag{9b}$$

and at high-T, $(T > \theta_E = \hbar\omega_E/k_B)$, one reaches the classical limit,

$$u^2_{T>\theta_E} = 3k_BT/M\omega_E^2 \tag{9c}$$

For a harmonic solid with a normalized VDOS $g(\omega)$, Eq. (9a) is replaced by the ω-integral:

$$\overline{u^2} = \frac{3}{2}\frac{\hbar}{M}\int_o^{\omega_{\max}} d\omega\,\frac{g(\omega)}{\omega}\coth\left(\frac{\hbar\omega}{2k_BT}\right) \tag{10a}$$

The $T \to 0$ value is

$$\overline{u_o^2} = \frac{3}{2}\frac{\hbar}{M}\overline{\omega^{-1}} \tag{10b}$$

where

$$\overline{\omega^{-1}} = \int_o^{\omega_{\max}} d\omega g(\omega)\omega^{-1} \tag{10c}$$

is the first inverse moment of the VDOS; and the high-T replacement of (9c) is

$$\overline{u^2}_{T\geq\theta_E} = \frac{3k_BT}{M}\overline{\omega^{-2}} \tag{11a}$$

in terms of the second inverse moment

$$\overline{\omega^{-2}} = \int_{o}^{\omega_{\max}} d(\omega)g(\omega)\omega^{-2} \tag{11b}$$

It is sometimes helpful to relate via Eq. (8) the inverse frequency moments directly with the temperature variation of the Lamb-Mössbauer factor $f(T)$

$$-\ln f(T) = \frac{1}{3}k^2\overline{u_T^2} = \frac{1}{3}\frac{E_{\gamma^2}}{\hbar^2 c^2}\overline{u_T^2} \tag{12a}$$

Then

$$\overline{\omega^{-1}} = \frac{2\hbar M c^2}{E_\gamma^2}(-\ln f_o) \tag{12b}$$

and, since $-\ln f_{T>\theta_E}$ is proportional to T,

$$\overline{\omega^{-2}} = \frac{Mc^2}{E_\gamma^2 k_B}\frac{d}{dT}(-\ln f) \tag{12c}$$

Equation (10a) and its limiting forms (10b, 11a) permit us, in principle, to correlate inelastic neutron scattering results on $g(\omega)$ with measurements of Lamb-Mössbauer factors. A useful reference for the temperature dependence of $f(T)$ is a Debye solid ($g(\omega) \sim \omega^2, \omega < \omega_D - k_B\theta_D/\hbar$), namely

$$f(T) = \exp\left[\frac{-6E_R}{k_B\theta_D}\left\{\frac{1}{4} + (T/\theta_D)^2\int_0^{\theta_D/T}\frac{xdx}{e^x-1}\right\}\right] \tag{13}$$

where $x = \hbar\omega/k_B T$, $E_R = $ recoil energy $= \hbar^2 k^2/2M = E_\gamma^2/2Mc^2$, and $\theta_D = $ Debye temperature. As $T \to 0$, f saturates to a value somewhat less than 1 on account of the finite zero point motion (zpm) given by the first term in Eq. (13). At high-T ($T > \theta_D$), $\ln f$ decreases linearly with T, and the slope $d(\ln f)/dT$ is found to be inversely proportional to θ_D. The latter feature is often used to characterize solids by a Debye temperature (θ_D) from $f(T)$ results. Furthermore, if a solid has an excess of VDOS at low-frequencies ($\omega_f \leq 5$ meV) such as due to cyclical modes in a network glass, then one expects the moments $\overline{\omega^{-1}}$ and $\overline{\omega^{-2}}$ (which carry increasingly larger weights as $\omega \to 0$) to acquire additional strength and result in a softening of the Lamb-Mössbauer factor $f(T)$ as T is lowered below a characteristic softening temperature $T_S \sim \hbar\omega_f/k_B$, where ω_f is the frequency of cyclical modes. The non-zero value of ω_f is presumably due to the longer range forces not considered in the nearest neighbor bond constraint-counting. For these reasons, Lamb-Mössbauer factors provide an elegant probe of low-frequency vibrational excitations in solids in general, and through cyclical modes in network glasses, a probe of the rigidity transition in particular.

Evidence for cyclical modes in ternary $Ge_x As_y Se_{1-x-y}$ chalcogenide glasses and particularly g-Se has emerged from inelastic neutron scattering measurements [20]. Lamb-Mössbauer factor measurements on such glasses are not directly feasible, since neither of the three elements offers the prospect of a suitable Mössbauer probe. However, it is possible to dope traces of isovalent Sn atoms for Ge in binary $Ge_x Se_{1-x}$ glasses, and deduce from [119]Sn Lamb-Mössbauer measurements, the first-and second-inverse moment of the *local* VDOS. Since the covalent radius of Sn is larger than Ge, it appears that the dopant replaces Ge at the more flexible corner-sharing tetrahedra, but not at the more constrained edge-sharing tetrahedra. Furthermore, since the mass of Sn (119 amu) is greater than the masses of Ge (72.59 amu) and Se (78.96 amu) atoms that comprise the network backbone, the *local* VDOS at [119]Sn displays subtle differences from the neutron VDOS in the glasses, a point we shall return to discuss in Sec. 5.

Anharmonicity of vibrations largely sets in at $T \geq T_g$ when a glassy solid transforms to a glassy liquid. The slope of mean-square displacement of atoms with temperature increases qualitatively [23, 24] near T_g and likewise viscosities and underlying relaxation times change by orders of magnitude [25]. The Lamb-Mössbauer factor measurements discussed here are performed at $T < T_g$ wherein the harmonic approximation under which Eqs. (9)–(13) are written to analyze the results is still a good one. Anharmonicity of lattice vibrations in crystalline solids has been addressed by Mössbauer spectroscopy elsewhere [26].

A general feature of glasses is the appearance of an excess density of vibrational states over and above Debye-like excitations in the low-frequency region ($0 < \omega < 40 \, cm^{-1}$). Low-$T$ anomalies in thermal conductivity and specific heat of glasses are identified [27] with the presence of these excess VDOS. Raman and inelastic neutron scattering measurements on glasses reveal a broad peak in the indicated frequency region which has been broadly termed the Boson peak [28]. Some have identified [29] the peak with the high frequency limit of acoustic modes that can propagate in a glass, a limit apparently set by structural inhomogeneities (molecular clusters) intrinsic to a glass. Furthermore, not all excess excitations in glasses observed in inelastic neutron scattering are found to be Raman active, as is the case for example with cyclical modes in an undercoordinated glass. In this context the correlation between low-T softening of Lamb-Mössbauer factors and cyclical modes in chalcogenide glasses suggests that the Mössbauer effect experiments clearly do probe these low-frequency excitations. In this sense, Lamb-Mössbauer factors complement Raman scattering as probes of low frequency vibrational excitations in glasses.

3. Experimental Considerations

3.1. *Mössbauer spectrometer*

A Mössbauer spectrometer can be assembled using (a) a PCAII card from Oxford Instruments in a 486 based PC, (b) an electromechanical transducer (K4 motor) and drive electronics (such as the Austin Science Associates model S-700), (c) a vibration-free model DE202 He cryocooler with a DMX-20 interface by APD Cryogenics, Inc., with suitable T-sensors and T-controller and (d) a vacuum pumping system to evacuate the vacuum shroud of the He cryocooler. Such a system can be assembled for about \$50,000 and used for low-T measurements from 8.5 K $< T <$ 320 K without the need of either liquid helium or liquid nitrogen. Such a spectrometer can be used in the standard constant acceleration mode or the constant velocity mode. The configuration of the system is schematically illustrated in Fig. 1.

The vibrations intrinsic to the expander in a He cryo cooler system have been addressed by a suitable mount of the cold head as discussed elsewhere

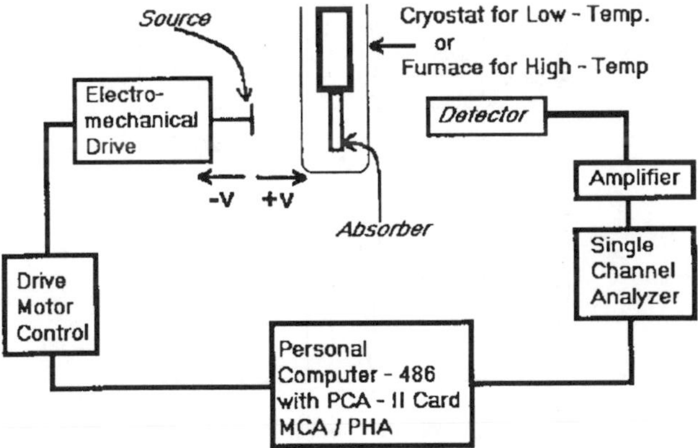

Fig. 1. Schematic arrangement of a typical Mössbauer spectrometer in which transmission of gamma-rays from a source through an absorber is studied as a function of a Doppler motion of the source.

[30]. An appropriately configured system can give trouble-free service 24 hours a day with excellent T-control ($\pm$ 0.2 K) and no instrumental line-broadening. Figure 2 taken from Ref. [30], reproduces Mössbauer spectrum of the inner two lines of α-Fe taken with a ^{57}Co source in Pd and the cryocooler off [Fig. 2(a)], and then on [Fig. 2(b)]. The observed linewidth of 0.231(3) mm/s measured with the cryo-cooler off revealed no change upon turning on the cryocooler and letting the system equilibrate at 300 K. In our case, we suspended the cold head from an angle aluminum framework anchored to a concrete beam in the laboratory ceiling. The interface assembly was independently supported from a heavy table (500 lbs) with the help of a trapezoidal network of angle

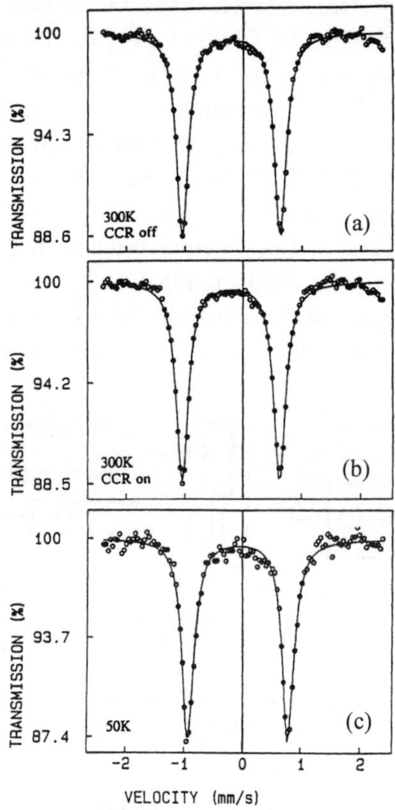

Fig. 2. Inner 2-lines of α-Fe taken with (a) He-cryocooler off, (b) He-cryocooler on and $T = 300$ K and (c) He-cryocooler on and $T = 50$ K, revealing no observable line broadening due to expander motion of the cryocooler. Figure taken from Ref. [30].

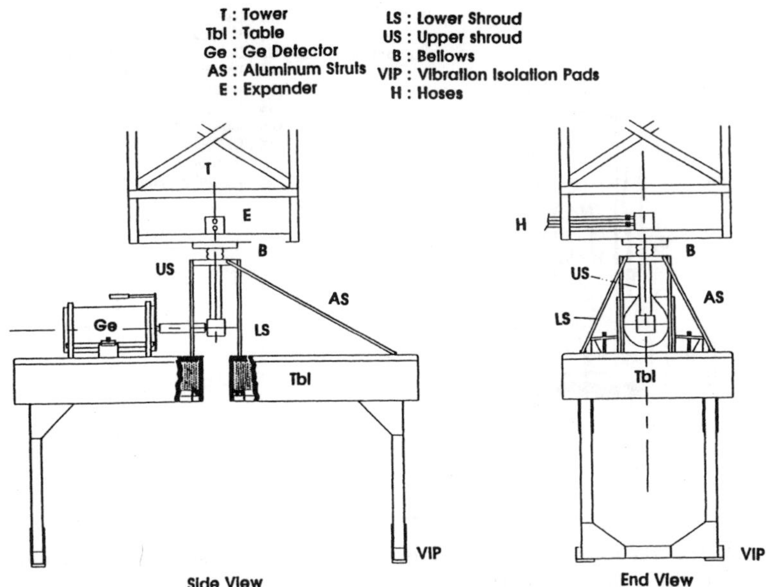

Fig. 3. Details of He cryocooler mount showing cold finger supported by a metal framework mounted to the concrete ceiling. The interface is independently supported from a heavy table with angle Al stock. Provision of a hole in the table is made to service the interface, particularly mounting temperature sensors and replacing heaters. See Ref. [30].

aluminum struts (see Fig. 3). To service the cold head, we incorporated a hole in the table center as illustrated in Fig. 3.

To minimize spurious residual motion between a source and an absorber, the transducer assembly was mechanically fastened to the Aluminum vacuum shroud of the He cryo-cooler directly with a pair of C-clamps on one end, and with the weight of the transducer assembly supported by a pair of legs resting lightly on the heavy table at the other end (Fig. 4). To facilitate sample changes, the Aluminum shroud of the DMX-20 interface was split into two parts. The upper part of the shroud was mechanically fastened to the transducer assembly (Fig. 4) which consisted of a dove-tail track on which the transducer could be moved back and forth. The lower half of the shroud was held in place with an O-ring vacuum seal, and could be dropped to access the sample holder for changing samples (Fig. 4).

With a PCAII card [31], one can store, retrieve, analyze and accumulate Mössbauer spectra almost simultaneously. A PC coupled to a HP Color Pro

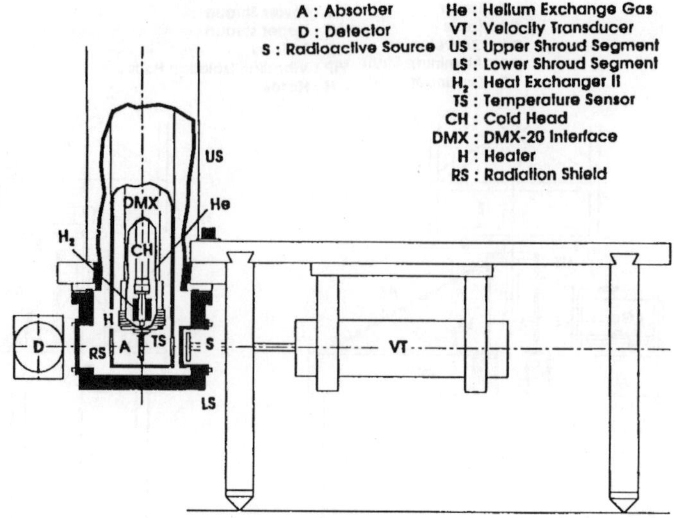

Fig. 4. Details of the mechanical coupling between transducer assembly and the upper part of the shroud (US). The transducer can be moved back and forth on a dove-tail track in the assembly. Sample interchange is achieved by dropping the lower half of the shroud (LS) colored black after breaking vacuum. See Ref. [30].

plotter can be used to plot the observed spectra along with the least squares fitted results in a matter of minutes.

3.2. *Measurements of Lamb-Mössbauer factors*

The development of emitter matrices [32] with respectable f-factors (> 0.5) at room temperature such as $CaSnO_3$ and Mg_3TeO_6 [33] for ^{119}Sn and ^{125}Te Mössbauer spectroscopy measurements represented an important step forward in the spectroscopy. Spectra of Te- and Sn-bearing glass samples (used as absorbers) mounted on a (**He** cryocooled) cold-finger can be systematically obtained as a function of temperature with an emitter held at room temperature outside the cryostat. Changes in the integrated area $A(T)$ under the resonance then derive solely from the T-dependence of the f-factor of the absorber $f_a(T)$. Before undertaking work on glasses, the full T-dependence of the integrated area for some standard crystalline compounds was obtained. By least-squares fitting the observed $A(T)$ variation, one can establish $f_a(T)$ using a standard harmonic vibrational density of states such as a Debye-like theory (see Eq. (13)).

In many cases, by working with thin-absorbers (effective nuclear thickness $T = n\sigma_o f_a < 3$), to avoid saturation effects, the observed integrated area $A(T)$ can be calibrated directly in terms of $f_a(T)$ in a fixed geometry. σ_o represents the maximum cross-section at resonance, n the number of resonant nuclei/cm^2 and f_a the recoil-free fraction of the absorber. The full T-dependence of $f_a(T)$ in several chalcogenide glass samples was studied and will be discussed in the next section.

4. ^{119}Sn Lamb-Mössbauer Factors and Nanophase Separation in GeSe$_2$ Glass

It is not often that Lamb-Mössbauer Factors (LM Factors) shed light directly on aspects of glass structure. The case of GeSe$_2$ is a notable exception in this regard. And in some sense this is a rather fortunate circumstance because the structure of GeSe$_2$ glass has been at the center of attention [34] in the

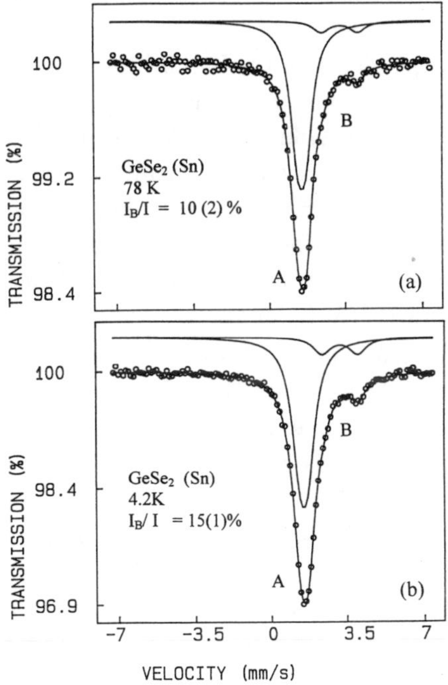

Fig. 5. ^{119}Sn spectrum of a Ge$_{0.98}$Sn$_{0.01}$Se$_2$ bulk glass sample taken at (a) 78 K and (b) at 4.2 K. Note the increase in the site-intensity ratio I_B/I upon cooling sample.

chalcogenides for the past two decades for interests that are both basic science and applications related. Thin-films of GeSe$_2$ have attracted interest because they display a rich variety of light-induced effects, such as optical bistability [35], giant photocontraction effects [36], reversible photomicrocrystallization [37], photodarkening [38, 39], and many others.

As discussed in Chapter 5(B), the ^{119}Sn Mössbauer spectrum of a (Ge$_{0.99}$ Sn$_{0.01}$)Se$_2$ glass at 4.2 K [see Fig. 5(b)] reveals [40] two features: a narrow single line (labelled A) and a doublet (labelled B). The narrow line represents Sn replacing Ge atoms that are tetrahedrally Ge(Se$_{1/2}$)$_4$ coordinated. The isomer shift and narrow line-width of the single line is consistent with the proposed assignment [40]. The doublet feature is identified with Sn replacing Ge in face-sharing (FS) ethane-like Ge$_2$(Se$_{1/2}$)$_6$ units and its integrated intensity as a function of x in Ge$_x$Se$_{1-x}$ glasses scales with the normalized scattering strength of the 180 cm^{-1} mode of FS tetrahedra [see Chapter 5(B)] in Raman scattering. The doublet is thought to result from the quadrupole interaction at Sn in such a unit, since Sn has one Ge and 3 Se nearest neighbors in a locally trigonal symmetry.

The complete T-dependence of the Lamb-Mössbauer factors $f(T)$ for both sites A and B in GeSe$_2$ glass appears in Fig. 6. These results were obtained [41] by recording spectra of a Ge$_{0.99}$Sn$_{0.01}$Se$_2$ glass as a function of T in the 10 K $< T < 200$ K range using a He cryocooler facility described in Sec. 3. In these experiments, an emitter of ^{119m}Sn in CaSnO$_3$, held at room temperature, was used to excite the resonance. The observed T-dependence of the LM-factors (Fig. 6) can be described in terms of an *effective Debye-temperature* θ_D of 130(5) K for site A and 100(5) K for site B.

The integrated intensity in a Mössbauer spectroscopy measurement depends on the product of the concentration of a species (N) and the probability of recoil-free absorption to occur at such a species (f). Consequently, the observed site B intensity ratio is given by

$$I_B/I = (N_B f_B)/(N_A f_A + N_B f_B) \tag{14}$$

where f_A and N_A designate the LM-factor and concentration of species A in the network. Since $f_A \simeq f_B$ at 4.2 K, from Eq. (14), it follows that $I_B/I = N_B/N = 0.15(1)$ where $N = N_A + N_B$. At higher temperatures, in general, $f_A > f_B$ and consequently the observed I_B/I ratio will be less than N_B/N. In particular, one can show that at a temperature $T(> 4.2$ K), the observed

I/I_B ratio is given by Eq. (15)

$$I/I_B(T) = 1 + [I/I_B(4.2) - 1]f_A(T)/f_B(T)$$

$$= 1 + 5.66 f_A(T)/f_B(T)$$

$$\text{or } I_B/I = 1/(1 + 5.66 f_A/f_B) \tag{15}$$

Figure 6 reveals that $f_A/f_B = 1.6$, at 78 K giving $I_B(78)/I(78) = 0.10$ from Eq. (15). This prediction is in excellent accord with the degree of broken chemical order observed $[(I_B(78)/I(78) = 0.10(1)]$ at 78 K in a GeSe$_2$ glass [Fig. 5(a)].

At room temperature, Fig. 6 shows that $f_A(300)/f_B(300) \simeq 6.3$. I_B/I calculated from Eq. (15) then yields 0.02. For that reason, one does not expect to observe the B-site in a spectrum of a GeSe$_2$ glass at 300 K.

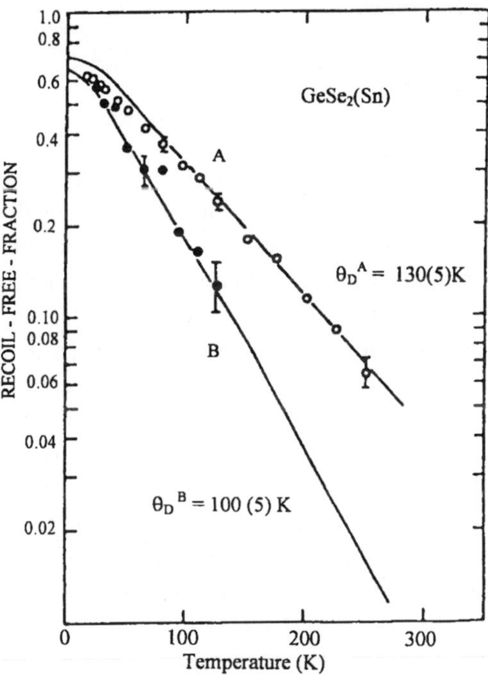

Fig. 6. Lamb-Mössbauer factors $f(T)$ of ^{119}Sn sites A and B in GeSe$_2$ glass. The sharper T-dependence of $f(T)$ of site B is suggestive of a lower θ_D as revealed by the fits. See Ref. [41].

The significantly lower LM-factor of site B in relation to site A is an *important result*. The atomic masses of Ge and Se of are within 8% equal to each other, and since sites A and B each possess four nearest neighbors (nns) with comparable chemical bond lengths (Sn–Se = 2.57 Å, Sn–Ge 2.63 Å) one would expect the LM-factor of sites A and B to be *nearly equal*, provided these local environments are formed in the *same polymerized network*.

The 23.8 keV γ-ray energy corresponds to a wavelength $\lambda = 0.083$ Å. It represents 3% of a typical covalent bond length a Sn atom forms either with a Se or a Ge nn. When the mean square displacement (msd), $\overline{u^2}$, of a Sn-atom becomes comparable to λ^2, one can expect the Lamb-Mössbauer factor [Eq. (8)],

$$f = e^{-\overline{u^2}/3\,\lambda^2} \tag{16}$$

to decrease from its maximum value of 1.0. From Fig. 6 one can calculate the msd of sites A and B at room temperature by extrapolating the results to $T \to 300$ K, and one obtains,

$$\overline{u_A^2}(300\text{K}) = 6.39 \times 10^{-2}\text{Å}^2$$

$$\overline{u_B^2}(300\text{K}) = 10.2 \times 10^{-2}\text{Å}^2 \tag{17}$$

The 60% larger msd of site B in relation to site A suggests that these sites are formed in separate molecular clusters. Specifically, the tetrahedral site A is thought to be formed in large (> 100 Å) Se-rich clusters in which all cations Ge(Sn) are tetrahedrally coordinated to Se nns in either corner-sharing or edge-sharing configurations. These clusters are dressed by Se–Se bonds on surfaces. The Se-rich clusters are large because their surface to volume ratios are small as measured by the low ($< 2\%$) concentration of homopolar bonds [see Chapter 5(B)]. On the other hand, the non-tetrahedral site B is thought to be formed in small (< 25 Å) Ge-rich chain-fragments consisting of ethane-like units (having Ge–Ge bonds). While the intracluster interactions are covalent in character, the intercluster ones are mediated by weaker van der Waals forces. The larger value of $\overline{u_B^2}$ in relation to $\overline{u_A^2}$ is thought to reflect the larger amplitude of vibration of the smaller-sized ethane-like cluster, as a whole, in relation to the larger Se-rich cluster.

Examples of small clusters (monomers) in chalcogenide glasses abound, such as S_8 rings in S-rich Ge–S or As–S glasses, As_4S_4 monomers in As–S glasses and P_4Se_3 monomers in P–Se glasses. And it is likely that ethane-like FS units first appearing near $x = 0.30$ in Ge_xSe_{1-x} glasses, also form part of a small molecular cluster, as suggested by the sharper T-dependence of the [119]Sn LM-factor at B-sites in GeSe$_2$ glass. Thus, the physical picture of the stoichiometric GeSe$_2$ glass suggested by these Mössbauer vibrational spectroscopy measurements consists of two nanophases (molecular clusters), large Se-rich and small Ge-rich clusters. Such a description was independently suggested by [129]I Mössbauer spectroscopy measurement on GeSe$_2$ glass and discussed in Chapter 5(B). Such a structural description in terms of clusters is at variance with the simplistic view [42] that the backbone of this glass consists of a completely polymerized network in which *homopolar bonds* are formed as *random defects* in an ordered-bond-network.

5. Rigidity Transition and Lamb-Mössbauer Factors in Glasses

In this section, we shall discuss Lamb-Mössbauer Factor results on several glass systems, including chalcogenides and oxide glasses, that provide striking evidence for the onset of rigidity through an abrupt rate of change of the mean-square-displacement (msd) as the connectivity of a network is progressively increased.

5.1. *Chalcogenides*

Reasonably complete Mössbauer spectroscopy results are now available on the two binary glass systems, Ge_xSe_{1-x} [41] and Ge_xTe_{1-x} [43] which will serve as the basis for our discussion below. In the former glass system, one probes the cation-centered Ge VDOS in Mössbauer spectroscopy measurements using [119]Sn as a dopant. In the latter glass system, on the other hand, one examines the anion-centered VDOS, using [125]Te as the Mössbauer probe.

5.1.1. Ge_xSe_{1-x}

Resonant absorption signals from a $(Ge_{0.99}Sn_{0.01})_xSe_{1-x}$ glass sample mounted onto a cold finger of a He cryocooler were systematically studied [41] as a function of T in the range 10 K $< T <$ 200 K. Spectra of glasses at $x < 0.32$ display a narrow single line resonance with an isomershift characteristic of tetrahedrally coordinated Sn as in $Sn(Se_{1/2})_4$ units. Figure 7 displays the observed T-dependence of the integrated area $A(T)$ at several glass compositions. From the observed high-T slopes, $dA(T)/dT$, one can uniquely define a Debye-temperature, and calibrate $A(T)$ in terms of the Lamb-Mössbauer factor $f(T)$. When this procedure is carried forward in $(Ge_{0.99}Sn_{0.01})_xSe_{1-x}$ glasses

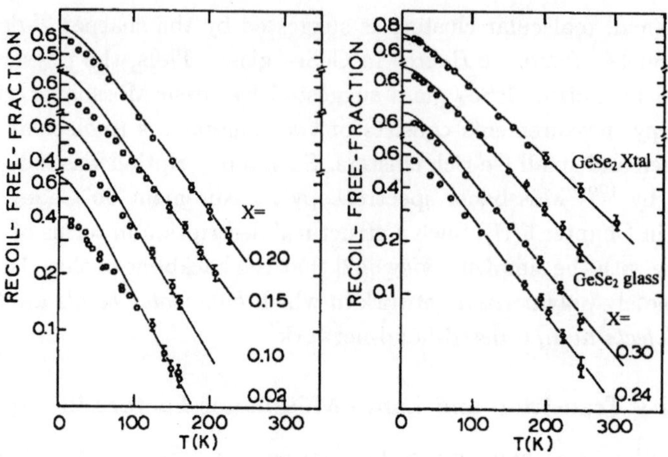

Fig. 7. T-dependence of $f(T)$ in $(Ge_{0.99}Sn_{0.01})_x Se_{1-x}$ glasses at indicated x. Note the qualitative softening of $f(T)$ at low-T (< 100 K) as $x \to 0$. See Ref. [41].

at several compositions x in the $0 < x < 1/3$ range, one finds that at $x = 1/3$, one obtains an excellent fit to the observed $A(T)$ over the studied T-range in terms of one θ_D. On the other hand, at glass compositions $x < 1/3$, such a procedure yields a *progressive softening* of the Lamb-Mössbauer factors at $T < 100$ K as $x \to 0$. For example, the pronounced softening in $f(T)$ observed at $x = 0.02$, when $T < 100$ K is ascribed to the presence of *cyclical modes* in a Se glass. The inelastic neutron scattering results on Se glass [20] display a mode centered at 5 meV which carries one-third of the spectral weight and is identified as the cyclical mode.

One can obtain the normalized first-inverse moment $(\overline{\omega_m \omega^{-1}})$ of the VDOS from the measured neutron $g(\omega)$ as a function of glass composition x in binary $Ge_x Se_{1-x}$ glasses. Here ω_m represents the maximum frequency in the VDOS taken from the neutron results as 42 meV. The neutron results for $(\overline{\omega_m \omega^{-1}})$ are compared to those deduced from the Lamb-Mössbauer factors in Fig. 8. One finds that the Mössbauer spectroscopy deduced $(\overline{\omega_m \omega^{-1}})$ moment decreases sharply with x to display a kink (change in slope) at $x = 0.20$ and then levels off to a value of about 4 as x increases to $1/3$. The kink at $x_c = 0.20$ is taken as *evidence for onset of rigidity* in $Ge_x Se_{1-x}$ glasses once x exceeds x_c. In the neutron scattering derived moment, on the other hand, the kink at $x = 0.20$, if present, is less obvious. In the Mössbauer experiments, one exclusively probes the *msd* of Sn in $Sn(Se_{1/2})_4$ local units. Such units

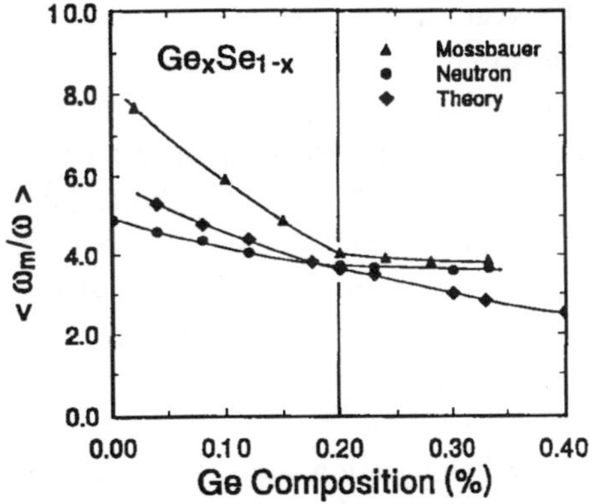

Fig. 8. A plot of the normalized first-inverse moment $\omega_m\overline{\omega^{-1}}$ as a function of Ge concentration x in Ge_xSe_{1-x} glasses. The filled triangle points describe the Lamb-Mössbauer results and reveal a kink at $x = 0.20$ identified with onset of rigidity. The neutron results were obtained by calculating the $\overline{\omega^{-1}}$ moment from the measured vibrational density of states. See Ref. [41] for additional details.

are formed at cross-links of entangled Se_n chains as $x \to 0$, while such units form part of chains of CS units branched by pairs of ES units as $x \to 1/3$. The significant difference in the compositional variation of $\omega_m\overline{\omega^{-1}}$ between the Mössbauer and neutron scattering measurements at $x < x_c$ is not only a curious result, but actually quite important in so far as Lamb-Mössbauer studies are concerned. The Sn-localized VDOS apparently select the lowest frequency cyclical modes of the network not only because of the higher mass of Sn (119 amu) in relation to that of Ge (72.59 amu) or Se (78.96 amu), but also because Sn possesses a larger covalent radius than Ge, and replaces Ge in the more flexible CS rather than ES units. On the other hand, the neutron $g(\omega)$ measures the global average of $\overline{\omega^{-1}}$ which is determined largely by the dynamics of Ge and Se atoms. In the $\overline{\omega^{-1}}$ moment, a larger weight accompanies the cyclical modes of low frequencies, and these considerations are responsible for the significant difference between the neutron and Mössbauer spectroscopy results. Remarkably at $x > x_c$, both types of measurements converge, suggesting that the strength of cyclical modes in the glasses decreases as the network is progressively cross-linked and the Sn dopant forced to mimic the vibrational behavior of Ge. These trends reinforce the view that softening of the Lamb-Mössbauer factor arises due to the presence of cyclical modes in these glasses. On the other hand, the Mössbauer and neutron scattering measurements for the second inverse moment, i.e. $\omega_m\overline{\omega^{-2}}$ reveal [41] that the rigidity transition is smeared. The smearing probably reflects the upshift ($\sim$ 5 meV) of the cyclical mode frequency rendering the second inverse moment insensitive to changes in mode scattering strength with network connectivity.

The onset of rigidity at $x = 0.20$ in the Ge_xSe_{1-x} glass system, suggested by the ^{119}Sn Lamb-Mössbauer factor results is in harmony with the variation of optical elasticities deduced from recent Raman scattering results [44] on these glasses. The Raman results are more detailed and reveal that although rigidity onsets at $x = 0.20$, the network does not become percolatively rigid until x increases to 0.23. Thus the transition as determined from the onset point at $x_c(1) = 0.20$ and completion point at $x_c(2) = 0.23$ in these optical measurements, possesses a finite width of $\Delta x = 0.03$.

5.1.2. $Ge_x Te_{1-x}$

The identification of Mg_3TeO_6 as a matrix for ^{125m}Te sources, which has a respectable Lamb-Mössbauer factor ($= 0.46$) at room temperature for the 35.5 keV gamma rays represented an important step [33] in the spectroscopy. Lamb-Mössbauer factors of Te-based compounds (crystals or glasses) can now

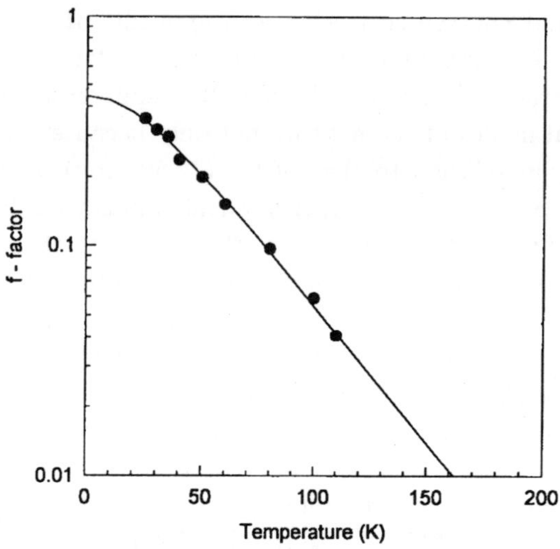

Fig. 9. T-dependence of ^{125}Te Lamb-Mössbauer factors in a $Ge_{0.20}Te_{0.80}$ glass along with a least squares fit to a Debye-like density of states.

be established with precision by systematically recording the resonance effect as a function of temperature in the 10 K $< T <$ 200 K range, using such a source kept at room temperature. Figure 9 provides the observed T-variation of the integrated area under the resonance for a $Ge_x Te_{1-x}$ binary glass at $x = 0.20$, fit to Debye-like vibrational states [43]. From such a plot one can extract $f(T \to 0)$ and also the high-T slope, $d\ln f/dT$, thus permitting measurement of the $\overline{\omega^{-1}}$ and $\overline{\omega^{-2}}$ moments of the Te centered VDOS. In Fig. 10, we provide a summary of the observed variation in these moments as a function of Ge content x. Bulk glasses can be synthesized by melt-quenching only in the narrow Ge concentration range $0.16 < x < 0.24$, while thin-films have been synthesized [45] by RF sputtering over much wider Ge concentrations. In Fig. 10, the open circles provide the variation of the $\overline{\omega^{-1}}(x)$ moment, while the filled circles give the variation of the $\overline{\omega^{-2}}(x)$ moment.

In the case of $Ge_x Te_{1-x}$ glasses, we thus find that both the $\overline{\omega^{-1}}(x)$ and $\overline{\omega^{-2}}(x)$ moments display a threshold behavior near $x = x_c = 0.22$. This concentration corresponds to a mean coordination of $\bar{r} = 2.44$ for the network backbone, if Ge and Te are taken to be 4- and 2-fold coordinated, respectively. These results constitute direct evidence for onset of rigidity in these glasses

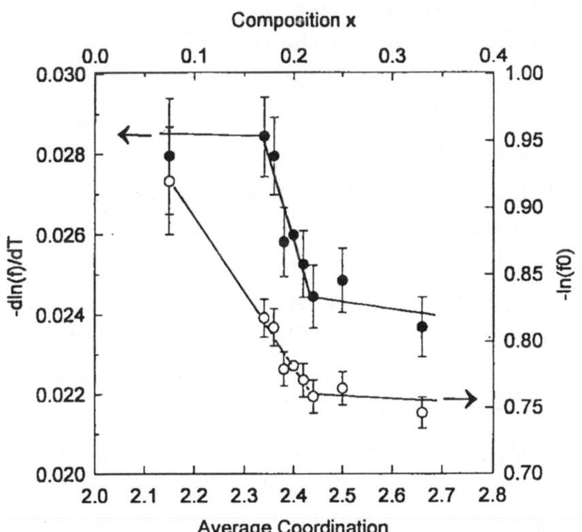

Fig. 10. First-inverse (right ordinate, open circles), and second-inverse (left ordinate, filled circles) moment of the VDOS in $Ge_x Te_{1-x}$ glasses plotted as a function of x or average coordination, providing a threshold behavior in both moments at $\bar{r} \simeq 2.45$.

at $\bar{r} > \bar{r}_c = 2.44$. Glass compositions at $\bar{r} < \bar{r}_c$ are considered to be floppy as reflected by the higher value of the *msd*, while compositions at $\bar{r} > \bar{r}_c$ are considered to be rigid as the *msd* of Te atoms in the network settles to a lower value. In these experiments, one probes the rigidity onset of the network by examining the *average vibrational behavior of Te atoms*. Such atoms not only bridge across different types of tetrahedra, such as corner-sharing, edge-sharing and face-sharing, but also form part of isolated Te_n-chains characteristic of the glasses or amorphous thin films. The intrinsically large natural width of the ^{125}Te resonance precludes resolving contributions from these different Te local environments. In each instance, Te atoms are believed to be 2-fold coordinated, with two Ge nearest-neighbors (nns) when acting as a bridge across tetrahedra, with two Te nns when forming part of a chain, and with one Ge and one Te nn when serving as a cross-link between a chain and a tetrahedron.

There are two broad issues raised by these Lamb-Mössbauer results on $Ge_x Te_{1-x}$ glasses. First, what significance can one attach to the rigidity transition manifesting in both the $\overline{\omega^{-1}}$ and $\overline{\omega^{-2}}$ moments of the VDOS. Second, in what manner does the network structure of the glasses evolve with Ge concentration across the rigidity transition and relate to the glass forming tendency in this binary in a global sense.

The observation of a threshold behavior in *both* the $\overline{\omega^{-1}}$ and $\overline{\omega^{-2}}$ moments of the VDOS in $Ge_x Te_{1-x}$ glasses at the rigidity transition is a striking result. It suggests that the frequency of cyclical modes in $Ge_x Te_{1-x}$ glasses is low and probably less than 3 meV. We noted earlier that ^{119}Sn Lamb-Mössbauer factors in $Ge_x Se_{1-x}$ glasses *revealed* the rigidity transition in the $\overline{\omega^{-1}}$ moment but that the transition in the $\overline{\omega^{-2}}$ moment was *smeared*. In relation to $Ge_x Se_{1-x}$ glasses, the mean cyclical mode frequency in $Ge_x Te_{1-x}$ can be expected to be downshifted not only because of the heavier mass of Te (127.6 amu) compared to that of Se (78.96), but also on account of a reduced bond-stretching force constant of the network (Ge–Te vs. Ge–Se). These constitute some of the considerations contributing to the observation of the stiffness transition in both the $\overline{\omega^{-1}}$ and $\overline{\omega^{-2}}$ moments in $Ge_x Te_{1-x}$ glasses.

The molecular structure of $Ge_x Te_{1-x}$ glasses has been the subject of investigations for the past three decades. Radial distribution functions in amorphous films were reported [46] by Bienenstock *et al.* to model Ge and Te local environments in such films. ^{125}Te Mössbauer spectroscopy measurements on such films were undertaken by Boolchand *et al.* [47] to examine the Te local environments. Raman and ^{119}Sn Mössbauer spectroscopy measurements on $(Ge_{0.99}Sn_{0.01})_xTe_{1-x}$ bulk glasses were recently undertaken [48] to elucidate the molecular structure of these glasses.

Raman experiments on Ge_xTe_{1-x} glasses used the 676.1 nm line from a Kr-ion laser. Measurements were performed as a function of decreasing laser power to suppress photomicrocrystallization of helical Te_{3n} chains from the amorphous Te network, and to identify vibrational features characteristic of the pristine glass. The principal modes in the glass consist of a feature at 125 cm^{-1} identified [48] with face-sharing $Ge_2(Te_{1/2})_6$ tetrahedra, a broad mode at about 165 cm^{-1} identified with distorted chains of amorphous Te [49], and a weak feature at 145 cm^{-1} identified with corner-sharing tetrahedra [48].

In samples of $(Ge_{0.98}Sn_{0.02})_xTe_{1-x}$ glasses probed by ^{119}Sn spectroscopy, one observes evidence for two local environments; a tetrahedrally coordinated

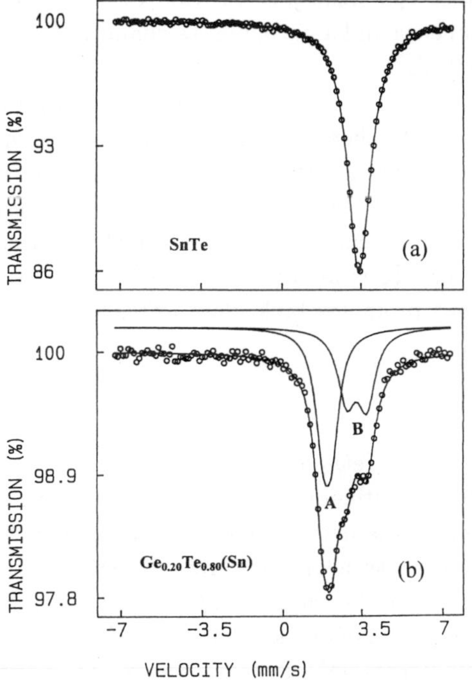

Fig. 11. ^{119}Sn spectrum of (a) crystalline SnTe (b) a $(Ge_{0.98}Sn_{0.02})_{0.20}Te_{0.80}$ glass. The A-site resonance in the glass is identified with tetrahedrally coordinated $Sn(Te_{1/2})_4$ while the narrow line resonance in c-SnTe to octahedrally coordinated $Sn(Te_{1/2})_6$.

local structure (A) giving rise to a narrow single line and a non-tetrahedrally coordinated one (B) giving rise to a doublet (Fig. 11). For comparison we also show the spectrum of cubic SnTe (rocksalt structure) which shows a single line as expected. The site integrated intensity ratio $I_B/I = 0.40$ at $x = 0.18$, and it is found to increase to 0.45 at $x = 0.20$ (Fig. 11) suggesting that B-sites represent signature of Ge–Ge bonds arising from face-sharing $Ge_2(Te_{1/2})_6$ local units. These Mössbauer spectroscopy results also show that the concentration ratio of Ge–Ge to Ge–Te bonds in a glass at $x = 0.20$, is 9.2%. This ratio is well below the maximum value of 24.5% expected for a strictly random covalent network based on bond-statistical estimates [42]. These results suggest that the backbone of a $Ge_{20}Te_{80}$ glass, displays some preference to form heteropolar Ge–Te bonds over homopolar Ge–Ge and Te–Te ones. The driving force to achieve chemical ordering is understandably weak because the Pauling electronegativities of Ge(2.01) and Te(2.1) are nearly the same. Both Raman and ^{119}Sn Mössbauer spectroscopies provide evidence of FS tetrahedra in addition to CS tetrahedra and amorphous Te-chains as the structural elements of glasses near $x = 0.20$.

The physical picture of the glass structure emerging from these measurements suggest a partially chemically ordered network structure of distorted Te_n chains characteristic of an amorphous Te phase that are cross-linked by Ge atoms with substantial Ge–Ge signatures. The existence of the rigidity transition near $x_c = 0.22$ can thus be reconciled in terms of a network in which Ge is 4-fold and Te is 2-fold coordinated. The onset of rigidity is an important factor that promotes bulk glass formation in this poorly ordered binary glass system.

5.2. *Oxide glasses*

SiO_2, GeO_2, B_2O_3, P_2O_5, TeO_2 and V_2O_5 represent some of the prototypical oxide glasses that form continuous networks of characteristic building blocks, and are generally known as network formers (NF). In such networks oxygen serves as a bridge across tetrahedra, or pyramids, or trigonal bipyramids. Such glass forming networks can be modified by alloying alkali oxides, or alkaline-earth oxides, or other oxides of divalent cations such as MgO, PbO, SnO which represent network modifiers (NM). For example, Na_2O alloying in SiO_2 or TeO_2 glass, is known to produce a pair of non-bridging oxygen (NBO) atoms per Na_2O molecule, with Na^+ ions compensating for the NBO O^- anions. Thus a network of tetrahedral $Si(O_{1/2})_4$ units or a network of trigonal bipyramidal

Te$(O_{1/2})_4$ units is progressively depolymerized as NBO are formed upon alloying Na_2O. Refractive index, thermal expansion, T_g, T_l, and even glass forming tendency, display systematic trends as the concentration of modifiers increase [49]. Historically, such investigations have served as the basis to tune oxide glasses for specific requirements including development of optical fibers.

The role of cross-linking or connectivity of a network on T_g was noted [51] in the early sixties for alkali-phosphate glasses. More recently, Kerner and Micoulaut [52] have developed a stochastic matrix description of network agglomeration. Changes in $T_g s$ for binary, ternary and quaternary glass systems have also been made using slope equations which explicitly include the coordination number of the modifier atoms. Remarkably, the model captures the essence of experimental trends on T_g-variation in chalcogenide and oxide glasses at low mean coordination number $\bar{r} \leq 2.40$. These results underscore the importance of *coordination number of the additive cations* in modifying the T_g of the backbone.

5.2.1. $(Na_2O)_x(TeO_2)_{1-x}$

Alkali-tellurate glasses attracted our interest because one can probe the vibrational behavior of Te cations in the telluria backbone using ^{125}Te Lamb-Mössbauer factors, as the backbone is progressively modified by alloying different proportions of Na_2O.

Figure 12 provides the T-dependence of the Lamb-Mössbauer factor in selected $(Na_2O)_x(TeO_2)_{1-x}$ glasses and crystals [11, 53]. These results were

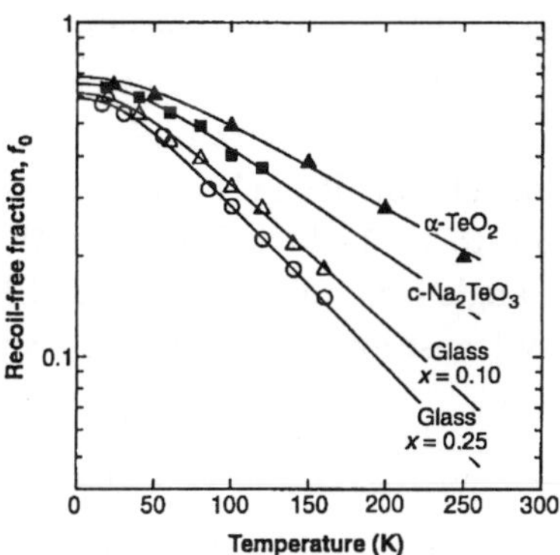

Fig. 12. ^{125}Te Lamb-Mössbauer factors $f(T)$ in $(Na_2O)_x(TeO_2)_{1-x}$ glasses at indicated compositions compared to those in c-Na$_2$TeO$_3$ and α-TeO$_2$.

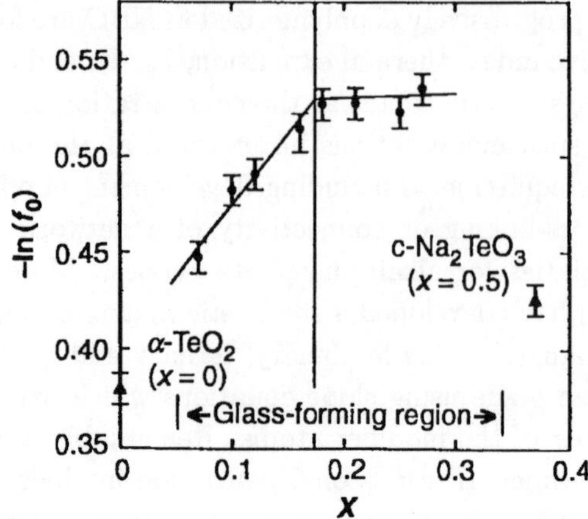

Fig. 13. First-inverse moment of the VDOS in $(Na_2O)_x(TeO_2)_{1-x}$ glasses (deduced from Lamb-Mössbauer factors) plotted as a function of x. Note the kink near $x = 0.18$ identified with onset of rigidity. See Ref. [11].

obtained by using an emitter of Mg_3TeO_6 kept at 300 K and with the glass sample cooled in a He-cryocooler. Figure 13 provides a semilog plot of the $T \to 0$ limit of the Lamb-Mössbauer factor, $-\ln f(0)$, or the zero point motion (zpm) plotted as a function of Na_2O-content x in $(Na_2O)_x(TeO_2)_{1-x}$ glasses. The zpm displays a progressive increase with x, and then saturates at $x > x_c = 0.18$. Figure 13 also reveals that the threshold behavior in zpm at x_c coincides with a minimum in the activation energy for enthalpy relaxation at T_g accessed from conventional DSC measurements [53].

The threshold behavior in the zpm at x_c constitutes direct evidence for the rigid to floppy transition in sodium tellurate glasses as can be seen from the following constraint count [11, 53]. If the Na, Te, and O coordination numbers in these glasses are taken to be 1, 4, and 2, respectively (see Fig. 14), and furthermore if the bond-angle constraint about the NBO atoms is considered to be broken, the rigidity transition is predicted to occur at

$$\bar{r}_c = 2.4 - 0.4 \left(\frac{n_1}{N} - \frac{m_2}{N} \right) \tag{18}$$

where n_1/N represents the concentration of 1-fold coordinated atoms, i.e. Na atoms/formula unit of $2x/3$ and m_2/N represents the concentration of 2-fold coordinated atoms with broken bond-angle constraint, i.e. the number of NBO

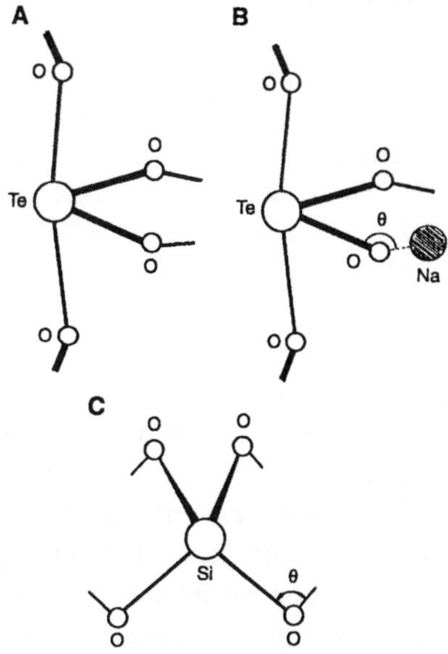

Fig. 14. (A) Pseudo trigonal bi-pyramid coordination of Te in crystalline TeO_2, (B) formation of NBO upon Na_2O-alloying in TeO_2, (C) tetrahedrally coordinated $Si(O_{1/2})_4$ unit in SiO_2 glass.

atoms/formula unit of $2x/3$. Since $n_1/N = m_2/N$, Eq. (19) reduces to

$$\bar{r}_c = \frac{2x_c + 4(1 - x_c) + 2(2 - x_c)}{3} = 2.4$$

or $x_c = 0.2$ \hfill (19)

The predicted value at $x_c = 0.20$ based on constraint counting is in close agreement [11, 53] with the observed value of $x_c = 0.18(1)$.

For a 1-fold coordinated atom only a bond-stretching force (constraint) need be considered [14]. Such a constraint requires only a central force, and can arise either from a covalent or an ionic interaction. The chalcohalides represent examples where the one-fold coordinated halogen atoms bond covalently with group IV or group V cations. In the present Na-tellurate glasses, on the other hand, the Na^+–O^- interaction is largely ionic, and the Coulomb interaction provides the underlying bond-stretching constraint.

The decrease in T_g upon alloying Na_2O with TeO_2 glass shows that the connectivity or the mean coordination $\bar{r}$ of the alloyed network decreases progressively [11, 53]. These thermal results unambiguously suggest that the mechanically effective coordination of Na is quite low and certainly much lower than the chemical coordination (5 or larger) inferred from the Na^{23} NMR shift [54–56] or EXAFS measurements [50] on oxide glasses in general. Indeed, to account for the observed $T_g(x)$ trends in $(Na_2O)_x(TeO_2)_{1-x}$ glasses, using the stochastic description developed by Kerner and Micoulaut [52], one requires a coordination number of 1 for Na, if Te and O are taken to have a coordination number of 4 and 2, respectively. Such a stochastic description has been remarkably successful in understanding T_g variation in the chalcogenide glasses where the *chemical* and *mechanical* coordinations coincide.

It is therefore natural to inquire why do the *chemical* and *mechanical* coordination differ qualitatively in the case of the alkali atoms in oxide glasses. There are several reasons. In the case of covalently bonded networks, the distinction between intact (first nn interactions) and broken (second nn interactions) constraints is straightforward, since the bond lengths describing first- and second nn interactions are quite different. The intramolecular interactions are covalent in character while the intermolecular interactions are van der Waals in character. Indeed, Si-centered NMR chemical shifts [57] and EXAFS measurements [50] on alkali-silicates are in harmony with results of Si coordination from diffraction measurements [58]. On the other hand, for the local environments of the ionically stabilized alkali atoms (Na^+) serving as network modifiers in the oxide glasses, the distinction between first nn interaction, i.e. Na^+ and O^- (NBO) and second nn interactions, i.e. Na^+ and neutral O (BO) is for all practical purposes lost since the underlying bond lengths are not all that different. Additionally, since ionic interactions are stabilized by increased coordination number, the number of second nn interactions far outweighs the solitary first nn interaction. Understandably, therefore, Na–NMR and EXAFS measurements yield the relatively high coordination number of 5 or even 6 for the alkali ions. Nevertheless, this does not imply that all these interactions are both chemically and mechanically *equivalent*. In fact, for glasses to form, only the constraint associated with Na^+–NBO is thought to be *intact*, while all others are *intrinsically broken*. In instances when the bonding of Na with its first nns is completely equivalent, both chemically and mechanically, as for example in NaCl, glass formation is totally precluded since the network is highly overconstrained.

Constraint counting algorithms have proved to be powerful in predicting and understanding glass formation. However, these algorithms must not be used indiscriminately. Caution needs to be exercised in identifying broken from intact constraints in specific systems and the case of alkali environment in oxide and chalcogenide glasses is certainly one to handle with discretion.

X-ray diffraction scans [11, 53] of several sodium tellurate melts, of varying alkali oxide content that received the same undercooling, are displayed in Fig. 15. These scans unequivocally show the existence of two compositional windows, one at $0.15 < x < 0.20$ and the other near $x = 0.27$, where the glass forming tendency is optimized. The first window is of topological

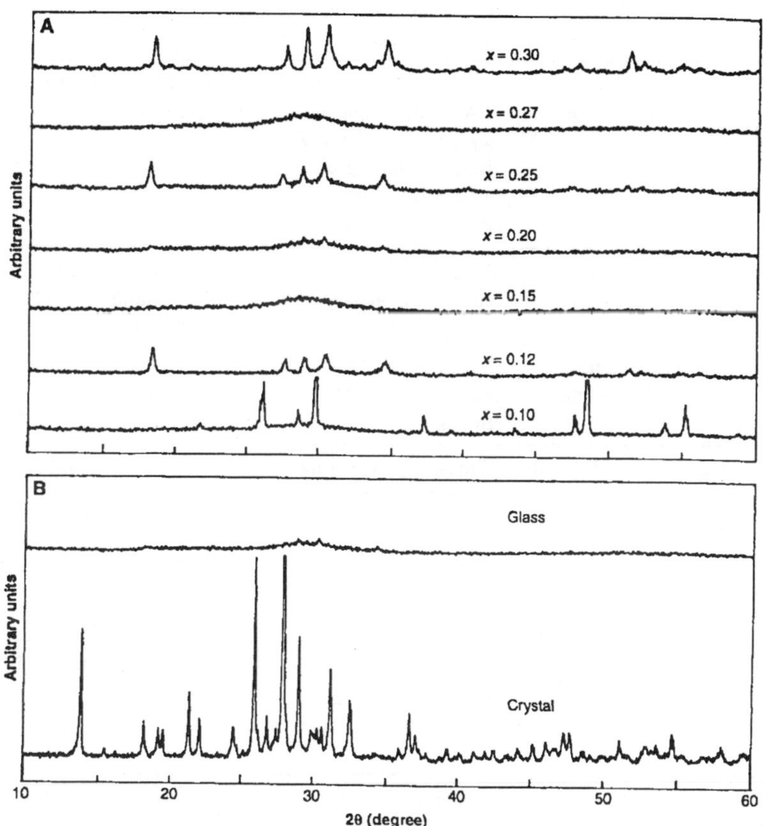

Fig. 15. (A) XRD scans of slowly cooled $(Na_2O)_x(TeO_2)_{1-x}$ melts that received the same undercooling. (B) XRD scan of c-$(Na_2O)(TeO_2)_4$.

origin, arising from the existence of the stiffness transition established from the Lamb-Mössbauer factors. The second window, on the other hand, is due to the existence of a eutectic in the quilibrium phase diagram which permit phase separation of melts resulting in large undercooling and glass formation.

Table 1. Experimental probes of stiffness transition in glasses.

Physical Property	Method	Result	Author (Yr) Ref
Elasticity Acoustic	Ultrasonic Moduli	Weak evidence for anomaly in C_{11} and C_{44} near $\bar{r} = 2.40$ in Ge–Se–Sb	Sreeram *et al.* (1991) Ref. [59]
Acoustic	Brillouin Scattering	Linear variation of LA mode frequency ω_{LA} with $\bar{r}$ at low P. Large softening of ω_{LA} with P near $\bar{r} = 2.46$; Ge–Se.	Finkler *et al.* (1999) Ref. [60]
Optical	Raman Scattering	Vibrational thresholds at $\bar{r} = 2.40$, 2.46; Ge–Se.	Feng *et al.* (1996) Ref. [13] Boolchand *et al.* (1999) Ref. [44]
	Raman Scattering	Vibrational thresholds at $\bar{r} = 2.40$, 2.54; Si–Se	Selvanathan *et al.* (1999) Ref. [70]
Kinetic Heat Flow	T-modulated DSC	Thermally reversing T_g in the $2.40 < \bar{r} < 2.46$ range; Ge–Se.	Feng *et al.* (1996) Ref. [13]
	T-modulated DSC	Thermally reversing T_g in the $2.40 < \bar{r} < 2.54$ range	Selvanathan *et al.* (1999) Ref. [70]
Activation Energy for Stress Relaxation	DMA	Shows broad min. near $\bar{r} = 2.40$; Ge–As–Se	Bohmer & Angell (1993) Ref. [61]
Activation Energy for Viscosity	DMA	Shows min. near $\bar{r} = 2.40$; Ge–Se	Tatsumisago *et al.* (1990) Ref. [8]

Table 1. (*Continued*)

Physical Property	Method	Result	Author (Yr) Ref
Thermal Expansion	DMA	Shows min. near $\bar{r} = 2.40$	Senapati & Varshneya (1995) Ref. [62]
Network Packing or Molar Volumes	Density-Archimedes Method	$V_M(\bar{r})$ show min. near $\bar{r} = 2.40$ in Ge_xSe_{1-x}	Feltz *et al.* (1983) Ref. [64] Feng *et al.* (1996) Ref. [13]
Floppy Modes	Neutron VDOS	5 meV mode in g–Se; Scattering strength of mode shows mild kink at $\bar{r} = 2.40$; Ge–Se	Kamitakahara *et al.* (1991) Ref. [20]
First- and Second-inverse Moments of VDOS	Lamb-Mössbauer Factor	^{119}Sn $\overline{\omega^{-1}}$ in Ge_xSe_{1-x} show a threshold at $\bar{r} = 2.40$	Boolchand *et al.* (1990) Ref. [41]
		^{125}Te $\overline{\omega^{-1}}$ in $(Na_2O)_x(TeO_2)_{1-x}$ show a threshold at $x = 0.18$	Zhang & Boolchand (1994) Ref. [11]
		^{125}Te in Ge_xTe_{1-x}, $\overline{\omega^{-1}}$ and $\overline{\omega^{-2}}$ show threshold at $\bar{r} = 2.42$	Enzweiler *et al.* Ref. [43]
Network Dimensionality and Morphology	Kohlrausch Fractional Exponent β	β increases with $\bar{r}$ and saturates at $\bar{r} \geq 2.40$; Ge–As–Se	Bohmer & Angell (1992) Ref. [64]
Insulator-Metal Transition Pressures (P_T)	Resistivity with Pressure	P_T shows kink at $\bar{r} = 2.40$; Ge–Se	Asokan *et al.* (1988) Ref. [65]
Network Morphology and Local Strain	^{129}I Mössbauer Spectroscopy	Site intensities display a local max. at $\bar{r} = 2.46$; Ge–Se	Bresser *et al.* (1986) Ref. [66]

6. Complimentary Probes of Rigidity Transition

Lamb-Mössbauer factors represent one of a number of experimental probes that have shed light on the *rigidity transition* in network glasses. In Table 1, we have compiled an overview of principal results on the subject. Broadly, two types of experiments have been undertaken in this field, one on *glassy melts* at $T \gtrsim T_g$ where thermal properties change remarkably with T and glass composition, and the second on glassy solids at $T \ll T_g$ where electrical transport, optical vibrations (Brillouin, Raman, IR), and inelastic neutron scattering in addition to Lamb-Mössbauer factors have revealed systematic changes with glass composition that bear on the onset of rigidity. In this section, we discuss selected results that provide new features of the rigidity transition in network glasses.

6.1. *Temperature modulated differential scanning calorimetry*

The intimate connection between the *rigidity transition* and *glass forming tendency* is elegantly illustrated by T-modulated DSC [67, 68]. This thermal method is a recent variant of conventional DSC and permits deconvoluting the heat flow at T_g into a kinetic or non-reversing component and a reversing component. The kinetic heat flow appears to be directly related to the *glass forming tendency* as we discuss below.

In a conventional DSC measurement, the signature of softening of a glass is an endothermic heat flow with respect to an inert reference sample, as the temperature of the glass and reference sample is swept linearly in time at a controlled rate. By programming a sinusoidal temperature variation over the linear T-ramp, it is possible to deconvolute the heat flow into two components, one that tracks the sinusoidal T-variation and is therefore called the *reversing heat flow* and the remainder that does not track the periodic T-variation and is called the *non-reversing heat flow* (ΔH_{nr}). Figure 16 provides MDSC scans [44, 69] of a GeSe$_2$ glass and Ge$_{0.23}$Se$_{0.77}$ glass illustrating the deconvolution of heat flow into the *non-reversing* and *reversing* components. Two noteworthy features become apparent from these scans, first the apparent glass transition temperature (T_g^{app}) deduced from the total *heat flow* is, in general, lower than the glass transition deduced from the *reversing heat flow*. Second, ΔH_{nr} is found to be miniscule for the $x = 0.23$ sample but it is an order of magnitude larger for the stoichiometric glass, GeSe$_2$. The shift, $T_g - T_g^{\mathrm{app}}$, is only 3°C for the $x = 0.23$ glass sample when ΔH_{nr} is nearly vanishing, but it is 12°C for GeSe$_2$ glass when ΔH_{nr} increases by almost an order of magnitude. The presence of a sizeable ΔH_{nr}-term in a glass will lower T_g^{app} in relation to T_g due to kinetic heat-flow effects which are intrinsically related to heat dissipated in a glass system as it reorganizes and network relaxation ensues.

Important insights into the physical origin of the ΔH_{nr}-term in chalcogenide glasses have emerged from the compositional trends illustrated in

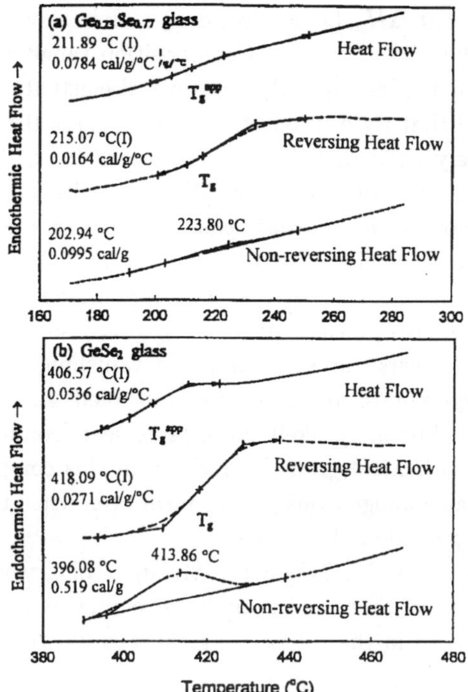

Fig. 16. MDSC scans of (a) Ge$_{0.23}$Se$_{0.77}$ and (b) GeSe$_2$ glasses revealing the deconvolution of the heat flow into the reversing and non-reversing components. Note the qualitative reduction in ΔH_{nr} for the optimally coordinated glass. See text for details.

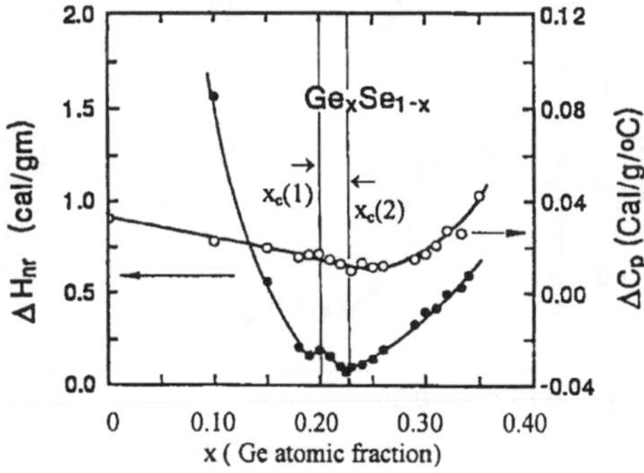

Fig. 17. Trends in ΔH_{nr} and ΔC_p in Ge$_x$Se$_{1-x}$ glasses revealing a deep and flat minimum in the non-reversing heat flow (ΔH_{nr}) in the transition region $x_c(1) < x < x_c(2)$.

Fig. 17. One finds that $\Delta H_{nr}(x)$ in Ge_xSe_{1-x} glasses shows a deep and nearly flat minimum in the $0.19 < x < 0.24$ composition range with steep walls on either side of this flat minimum. Raman optical elasticity results [13, 44] on these glasses reveal that rigidity onsets at $x = x_c(1) = 0.20$ and the network become percolatively rigid only when x increases to $x_c(2) = 0.23$. These Raman results in conjunction with the MDSC results suggest occurrence of a *transition region* or *an intermediate phase*, $x_c(1) < x < x_c(2)$, across which glass transitions are *thermally reversing*. The thermally reversing character of the glass transition in the $x_c(1) < x < x_c(2)$ compositional window suggests that configurational entropy changes between glassy solids and glassy liquids upon heating or vice-versa upon cooling are vanishing. When melts in the thermally reversing compositional window are cooled, glass formation instead of crystallization is thus more likely since large scale configurational changes are precluded on mechanical grounds. Indeed, in this compositional window, it is well known that homogeneous glasses form even when melts are cooled at a slow rate of $2°C$ per minute!

Recent investigations [70, 71] on the Si_xSe_{1-x} binary using Raman scattering and MDSC reveal that the thermally reversing compositional window is nearly twice as wide as in the Ge_xSe_{1-x} glasses as illustrated in Fig. 18. The

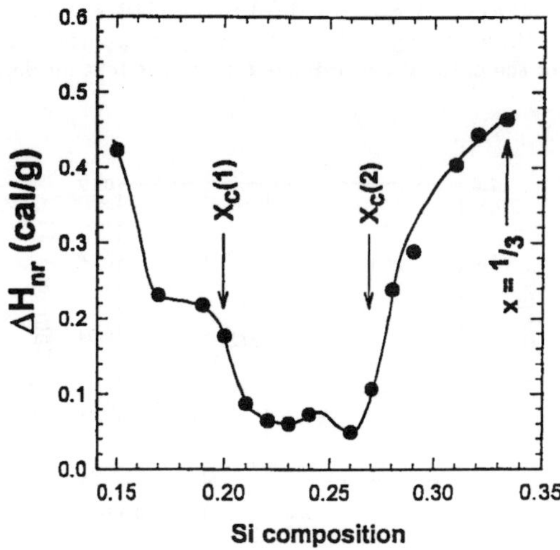

Fig. 18. Non-reversing heat-flow, $\Delta H_{nr}(x)$ in Si_xSe_{1-x} glasses showing the transition region $x_c(1) < x < x_c(2)$. See Ref. [70].

existence of a *transition region* between the *floppy* and the *rigid regions* represents an important recent discovery in the field.

6.2. *Raman scattering*

The power of Raman scattering as a probe of the rigidity transition in network glasses has emerged from recent results on two families of IV–VI glasses that have shown existence [70, 71] of a *transition region* separating the floppy- and rigid-regions. In the Ge–Se binary glass system, in the bond-stretching regime, one has observed that the frequency squared of CS modes (which provides a measure of optical elasticity) undergoes a functional change from being linear at $x < x_c(1) = 0.20$ to becoming superlinear at $x > x_c(2) = 0.23$ as shown in Fig. 19. The power-law describing the variation of optical elasticity with mean-coordination $\bar{r}$ of the network, i.e.

$$v^2 - v_c^2 = A(\bar{r} - \bar{r}_c)^p \tag{20}$$

is found to be $p = 1.33(5)$ and is good accord with the prediction ($p = 1.4$) of numerical simulations.

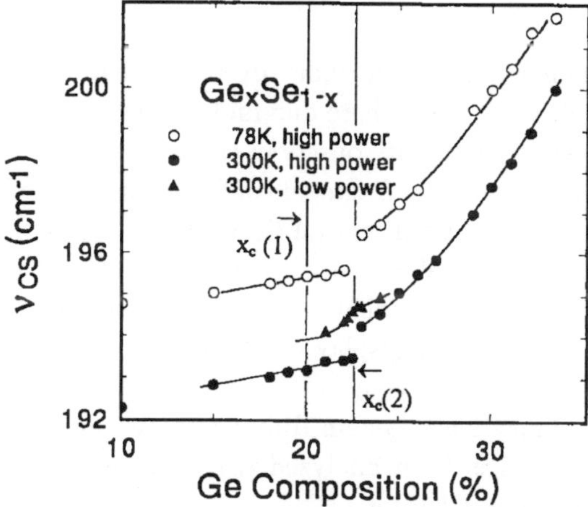

Fig. 19. Raman mode frequencies of corner-sharing Ge(Se$_{1/2}$)$_4$ tetrahedra in Ge$_x$Se$_{1-x}$ glasses studied at 750 μW laser power, using 647.1 nm line, show a linear behavior with x at $x \leq 0.23$, a jump at $x_c = 0.23$, and a power-law variation at higher x. At a low power of 82 μW, two kinks (change in slope) are observed, one at $x = x_c(1) = 0.20$ and a second at $x = x_c(2) = 0.23$. The presence of some light-induced structural changes of the pristine networks at 750 μW laser power cannot be excluded.

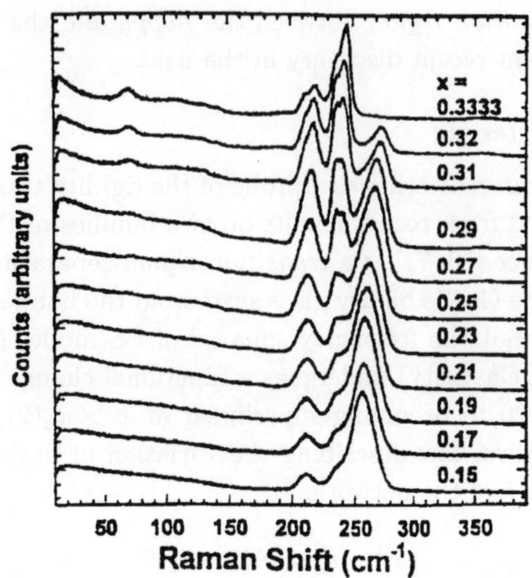

Fig. 20. Raman scattering in Si_xSe_{1-x} glasses taken with 647.1 nm line at low power, show splitting of the CS band (210 cm^{-1}) and ES band (235 cm^{-1}) and eventual disappearance of the Se$_n$-chain mode (255 cm^{-1}) as x approaches 1/3. See Ref. [70].

Parallel and somewhat more striking are the Raman results [70, 71] on the Si_xSe_{1-x} binary glass system, where one finds $x_c(1) = 0.20$ and $x_c(2) = 0.27$ (Fig. 20). It appears that rigidity onsets in chains of corner-sharing $Si(Se_{1/2})_4$ units at $x = x_c(1)$ separated by fragments of Se$_n$-chains. The network does not become percolatively rigid until $x = x_c(2)$, when the floppy Se$_n$ chain segment are cross-linked by additional Si. At $x_c(2)$ chains of ES $Si(Se_{1/2})_4$ tetrahedra also comprise the structural elements of the rigid backbone (Fig. 21) since the underlying optical elasticities display a power-law behavior (Fig. 22) with a power $p = 1.33(3)$.

In the Ge–Se binary glass system, Raman scattering in the low-frequency region (< 100 cm^{-1}) has been analyzed [74] and found to display a fractal behavior [see Murase in Chapter 6(C)]. A log-log plot of Raman scattering strength I against ω reveals three linear regions, with these regions in increasing ω identified with Debye phonons, bending fractons and stretching fractons. At $x > x_c(2)$ bending fractons are found to disappear. However, the fractal dimension of bending fractons was found to be larger than the fractal dimension of stretching fractons, a result that the authors of Ref. [74] recognized

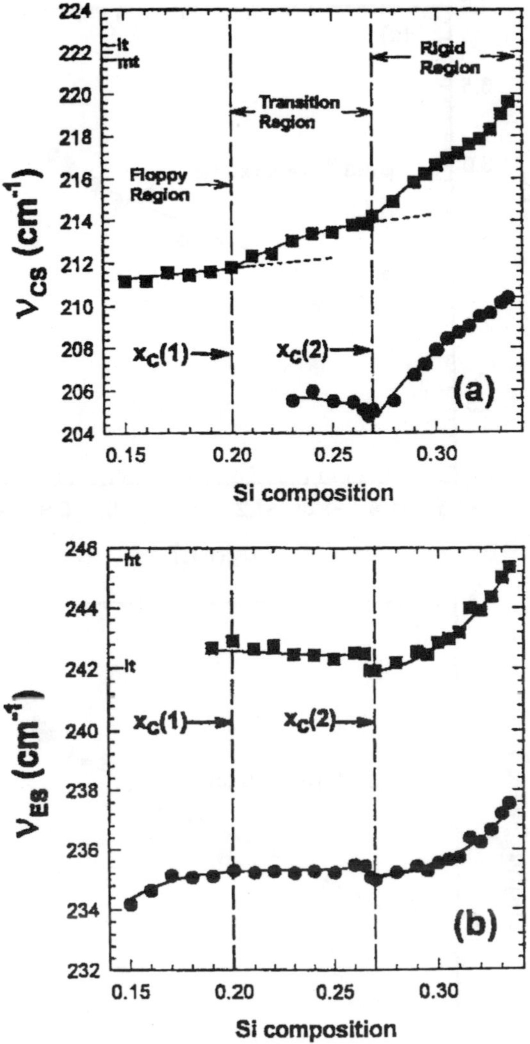

Fig. 21. Compositional trends in frequency variation of the Raman active (a) CS mode and (b) ES mode in Si$_x$Se$_{1-x}$ glasses showing existence of the transition region $x_c(1) < x < x_c(2)$ where kinks appear. Squares and circles in (a) represent mode frequencies of the majority and minority CS modes. Circles and squares in (b) represent the majority and minority ES modes. See Ref. [70].

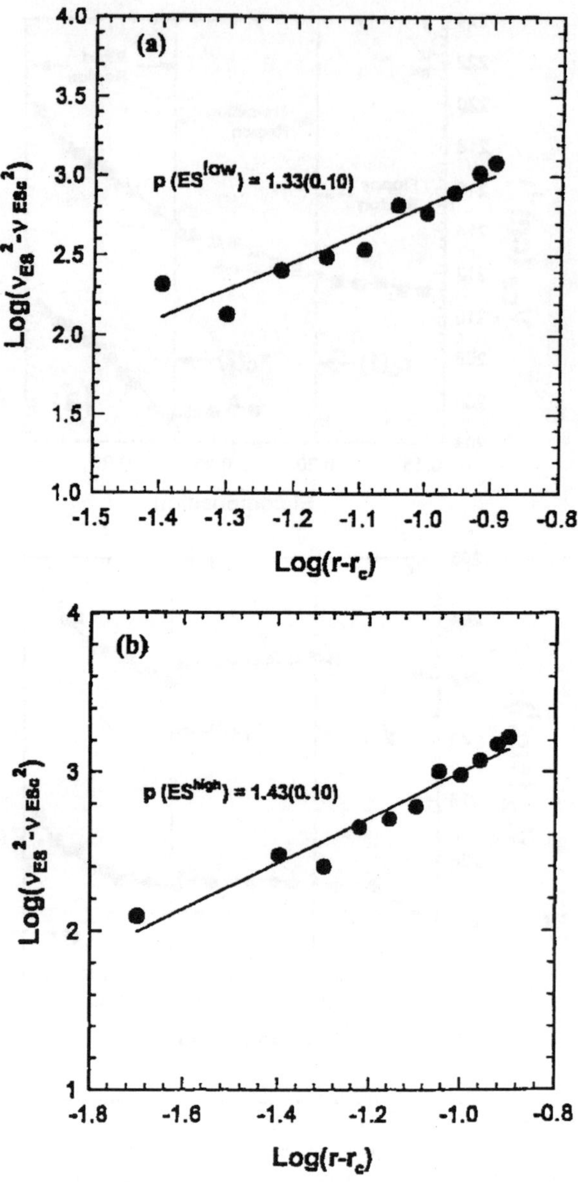

Fig. 22. Log-log plot of $v_{ES}^2 - v_c^2$ against $\bar{r} - \bar{r}_c$ for (a) the minority ES mode and (b) the majority ES mode revealing a power-law for the optical elasticity. See Ref. [70].

to be physically unacceptable. Phillips [75] has advanced an alternative identification of the three linear regions in terms of a domain-like morphological
structure of these glasses. Specifically, he has suggested, as before, that the
first linear region be identified with Debye phonons, but the second linear
region be identified with domain interface-bending excitations and the third
linear region with intradomain bending and stretching excitations. The disappearance of the second linear region at $x > x_c(2)$ when the network becomes
percolatively rigid can then be rationalized in terms of shear modes of domain
interface bending excitations vanishing. These Raman measurements at low
frequencies studied as a function of glass composition provide new insights into
the nature of localized or fracton-like excitations [76] in network glasses when
these become rigid.

6.3. *Mössbauer hyperfine structure*

The somewhat unique and special application of ^{129}I Mössbauer emission spectroscopy as a probe of the morphological structure of the stoichiometric glasses,
GeS_2 and $GeSe_2$, was discussed in Chapter 5(B) of this volume. These aspects
of the local probe stem from the *oversized nature* of the Te-chalcogen dopant
when studying S- or Se bearing glasses. In such systems, ^{129m}Te site occupancies of available Se or S sites in a glass network can be *qualitatively augmented*
by the distribution of *network strain* permitting one to distinguish networks
that possess a *homogeneous* versus a *heterogeneous morphology*. Available Se
sites in a host network will not be *randomly substituted* by the oversized ^{129m}Te
dopant if the network morphology is *heterogeneous* on a molecular scale, as
found in the case of g–$GeSe_2$ and g–GeS_2 [77, 78].

Historically, one of the earliest experiments to suggest a change in network
morphology of binary Ge-Se glasses near the composition $x \sim 0.23$, came from
^{129}I Mössbauer emission spectroscopy [66]. Mössbauer dopant site occupancies
at low $x(x < 0.15)$ are found to be *random* and these differ qualitatively from
those at high $x(x > 0.25)$ *which are selective*, with evidence of a rather striking
threshold behavior (local maximum) near $x = 0.23$ as shown in Fig. 23.

^{129}I spectra of bulk Ge_xSe_{1-x} glasses can be deconvoluted into two chemically inequivalent local environments [77], labelled as sites A and B (Fig. 24).
The former site is identified with an iodine species σ-bonded to a Ge nearest
neighbor (nn), while the latter site to an I σ-bonded to a Se nn. In a Se glass
only a B-type of local environment is expected while in a *chemically ordered*
$GeSe_2$ network only an A-type of local environment is expected as discussed

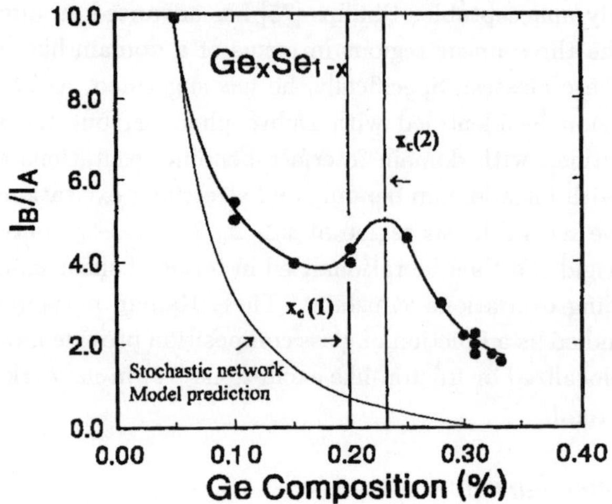

Fig. 23.   Te dopant site occupancies $I_B/I_A(x)$ in Ge$_x$Se$_{1-x}$ glasses deduced from ^{129}I Mössbauer spectroscopy, revealing a local maximum at $x = 0.23$ corresponding to $x_c(2)$. See Ref. [77].

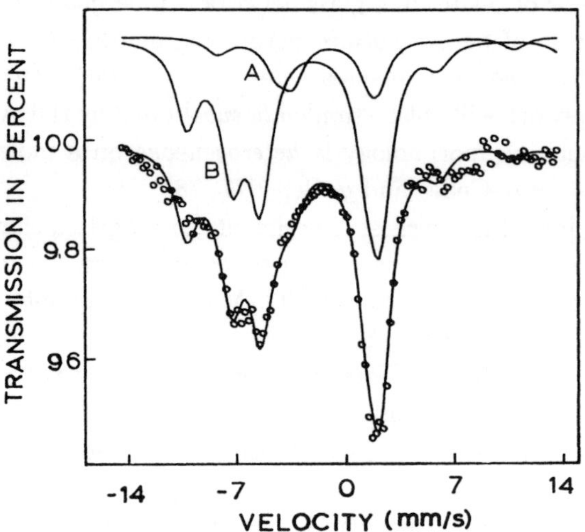

Fig. 24. Deconvolution of the observed ^{129}I spectrum of a Ge$_{20}$Se$_{80}$ glass in terms of two local environments site (A) and site (B). See Ref. [77].

elsewhere [77]. The Mössbauer site intensity ratio $I_B/I_A(x)$ variation deduced from the experiments appears in Fig. 23. There are two outstanding features of these $I_B/I_A(x)$ trends, *first* a monotonic reduction of $I_B/I_A(x)$ with increasing x in the $0.05 < x < 0.33$ range reflecting a baseline variation and *second*, a peak centered at $x = x_c(2) = 0.23(2)$ superimposed on this base line.

For a purely stochastic network description of these glasses in which cross-linking of Se_n chains by Ge atoms can proceed in a *random fashion*, the expected $I_B/I_A(x)$ variation is also shown in Fig. 23 as the thin continuous line. A stochastic network description of these glasses is clearly an appropriate description of these glasses at low $x(< 0.15)$ where the observed and calculated trends are qualitatively similar. Beginning at $x > 0.15$, the observed $I_B/I_A(x)$ variation differs *qualitatively* from the projected variation of a random network. Broadly, these results are the signature of network inhomogeneities i.e. ES rigid units nucleating in a floppy network of Se_n chains that are sparsely cross-linked by CS units. The oversized ^{129m}Te dopant is expelled from these rigid-regions, where A site occupancy would have occurred in ES units, thus qualitatively altering the random occupancy ratio of Se B to A sites in the network.

Although global network strain is a minimum at $x_c = 0.23$ as revealed by $\Delta H_{nr}(x)$ trends, locally the dispersion of the floppy- (Se_n chains) and rigid-fragments (ES- and CS-units) of the network backbone are the *finest* at the stiffness transition. This leads to the largest surface to volume ratio of the rigid fragments which is reflected in the *peak* of the Mössbauer site intensity ratio $I_B/I_A(x)$ near $x = 0.23(2)$. At $x > 0.23$, the rigid-fragments grow in extent, reducing the surface to volume ratio of the rigid regions and thus I_B/I_A. But even as x increases to $1/3$, I_B/I_A does not extrapolate to zero (as required by a chemically ordered continuous random network) and acquires a finite value (about 2) because rigid fragments do not *completely* fill the network backbone. The stoichiometric glass, $GeSe_2$, is only partially polymerized, and phase-separated into large Se-rich and small Ge-rich molecular clusters as demonstrated by these Mössbauer spectroscopy results [66] and those discussed earlier in Sec. 4, and as confirmed by Raman scattering measurements [13, 44].

In addition to the experimental results described above, a variety of thermal properties of glasses near and above T_g, such as thermal expansion [62], activation energy for viscosity [8], and activation energy of enthalpy relaxation [8] reveal a threshold behavior as a function of network connectivity at the rigidity transition. The reader is referred to a more extensive review of the subject elsewhere [44].

7. Conclusions and Future Work

Lamb-Mössbauer factors $f(T)$, provide an elegant means to probe site-specific low-frequency vibrational excitations including cyclical modes in network glasses. Experiments on chalcogenide and oxide glasses show that one can obtain both the first-inverse, $\overline{\omega^{-1}}$, and the second-inverse, $\overline{\omega^{-2}}$, moment of the VDOS. Changes in these moments as a function of mean coordination number ($\bar{r}$) display a kink near $\bar{r}_c = 2.40$ in Ge_xSe_{1-x} glasses, identified with the stiffness transition. Such measurements to date have been performed at coarse composition intervals. With the appearance of rather detailed Raman scattering and MDSC results in several glass systems, where rigidity transition occurs in two steps, there is a clear motivation to pursue the Mössbauer spectroscopy measurements at smaller composition steps and to independently ascertain the details of the rigidity transition in glasses.

In contrast to inelastic neutron scattering measurements that require a few cm^3 volume of a sample and provide the global VDOS in a glass, Lamb-Mössbauer measurements can be performed typically on several tens of a milligram of a bulk sample or even a thin-film and provide the local VDOS. The miniscule sample size requirements of Lamb-Mössbauer factor measurements can be a significant advantage in working with materials that do not form bulk glasses but can be rendered amorphous in a thin-film configuration by vapor deposition using either evaporation or sputtering or laser ablation.

In contrast to Raman scattering measurements on narrow gap glassy semiconductors which are not only poor light scatterers, but can readily also undergo photostructural changes or photocrystallization even at low (100 μW) exciting laser power, Lamb-Mössbauer factor measurements on such materials are immune from such problems. This is largely the case because low-energy gamma rays are only weakly absorbed by electronic processes in glassy materials of interest.

Finally, the onset of rigidity in glasses in two steps rather than a solitary step has evoked particular recent interest in the field [44, 70, 75]. The existence of a *transition region* or an *intermediate phase* separating the *floppy* from the *rigid* region in glasses may be intrinsically related to the presence of medium range structure, an issue that is likely to be examined in the future both experimentally and in numerical simulations. The existence of an *intermediate phase* in glasses may have parallels in other areas of condensed matter science such as the doping induced metal-insulator transition [79] in crystalline semiconductors and the underlying physical consequences pose new challenges and opportunities. It may actually bring to closure an intensive debate on the atomic scale structure of glasses that has persisted in the field for several decades.

Acknowledgments

I have benefited from several suggestions made by W.J. Bresser, B. Goodman, D.H. McDaniel and J.C. Phillips in preparing this review. This work is supported by the National Science Foundation grant DMR-97-01289.

References

[1] D.R. Uhlmann, *Glass Science and Technology 1*, eds. D.R. Uhlmann and N.J. Kreidl (Academic 1983) p. 1; also see D.R. Uhlmann, *J. Non Cryst. Solids* **7**, 337 (1972).

[2] F.H. Stillinger, *Science* **267**, 1935 (1995). R.A. LaViolette and F.H. Stillinger, *Phys. Rev.* **B35**, 5446 (1987).

[3] J.C. Phillips, *Rep. Prog. Phys.* **59**, 1133 (1996).

[4] J. Jackle, *Phil. Mag.* **B56**, 113 (1987).

[5] R. Kohlrausch, *Ann. Phys. Lpz.* **12**, 393 (1847).

[6] J.C. Phillips, *J. Non-Cryst. Solids* **34**, 153 (1979), ibid **43**, 37 (1981).

[7] B. Effey and R.L. Cappelletti, *Phys. Rev.* **B59**, 4119 (1999).

[8] M. Tatsumisago, B.L. Halfpap, J.L. Green, S.M. Lindsay and C.A. Angell, *Phys. Rev. Lett.* **64**, 1549 (1990).

[9] B. Uebbing and A.J. Sievers, *Phys. Rev. Lett.* **76**, 932 (1996) .

[10] S.P. Love, A.J. Sievers, B.L. Halfpap and S.M. Lindsay, *Phys. Rev. Lett.* **65**, 1792 (1990).

[11] M. Zhang and P. Boolchand, *Science* **266**, 1355 (1994).

[12] J.C. Phillips, *Solid State Commun.* **47**, 203 (1983).

[13] X.-W. Feng, W.J. Bresser and P. Boolchand, *Phys Rev. Lett.* **78**, 4422 (1997).

[14] P. Boolchand and M.F. Thorpe, *Phys. Rev.* **B50**, 10366 (1994).

[15] M. Mitkova and P. Boolchand, *J. Non-Cryst. Solids* **240**, 1 (1998).

[16] M.F. Thorpe, *J. Non-Cryst. Solids* **57**, 355 (1983).

[17] D.J. Jacobs and M.F. Thorpe, *Phys. Rev. Lett.* **75**, 4051 (1995).

[18] D. Selvanathan, M.S. Thesis, University of Cincinnati, 1998 (unpublished).

[19] D.G. Georgiev, M.I. Mitkova, P. Boolchand, *Bull. Am. Phys. Soc.* **44**, 1915 (1999).

[20] W.A. Kamitakahara, R.L. Cappelletti, P. Boolchand, B. Halfpap, F. Gompf, D.A. Neumann and H. Mutka, *Phys. Rev.* **B44**, 94 (1991).

[21] P. Debye, *Ann. Phys.* (New York) **43**, 49 (1914); I. Waller, PhD Thesis (1925) Uppsala.

[22] W.E. Lamb, Jr., *Phys. Rev.* **55**, 190 (1939); R.L. Mössbauer, *Z. Phys.* **151**, 124 (1958). See C. Kittel, *Quantum Theory of Solids*, John Wiley 1963 p. 386 for a discussion of Lamb-Mössbauer factors.

[23] F. Parak, J. Heidemeier and G.U. Nienhaus, *Hyperfine Interactions* **40**, 147 (1988).

[24] W. Doster, S. Cusack and W. Petry, *Nature* **337**, 754 (1989).

[25] C.A. Angell, *Relaxation in Complex Systems*, eds. K. Ngai and G.B. Wright (National Technical Information Service, US Department of Commerce, 1985) p. 1.

[26] B. Kolk, *Dynamical Properties of Solids*, ed. G.K. Horton and A.A. Maradudin (North Holland, 1984) p. 80.

[27] R.C. Zeller and R.O. Pohl, *Phys. Rev.* **B4**, 2029 (1971); W.A. Phillips, *Amorphous Solids: Low Temperature Properties* (Springer-Verlag, 1981).

[28] A.P. Sokolov, A. Kisliuk, D. Quitmann and E. Duval, *Phys. Rev.* **B48**, 7692 (1993). Also see B. Frick and D. Richter, *Science* **267**, 1939 (1995).

[29] M. Foret, B. Hehlen, G. Taillades, E. Courtens, R. Vacher, H. Casalta and B. Dorner, *Phys. Rev. Lett.* **81**, 2100 (1998); A. Boukenter and E. Duval, *Phil. Mag.* **B77**, 557 (1998); U. Buchenau, C. Schonfeld, D. Richter, T. Kanaya, K. Kaji and R. Wehrmann, *Phys. Rev. Lett.* **73**, 2344 (1994).

[30] P. Boolchand, G. Lemon, W.J. Bresser, R. Enzweiler, R. Harris, *Rev. Sci. Instrum.* **66**, 3050 (1995).

[31] PCAII, Oxford Instruments, Inc., P.O. Box 2560, Oak Ridge, TN 37831–2560, USA.

[32] For description of source matrices used in Mössbauer Spectroscopy, see for example N.N. Greenwood and T.C. Gibb, *Mössbauer Spectroscopy* (Chapman and Hall, 1971).

[33] W. Bresser, M. Zhang, L. Koudelka, J. Wells, P. Boolchand, G.J. Ehrhart and P. Miller, *Phys. Rev.* **B47**, 11663 (1993).

[34] S. Sugai, *Phys. Rev.* **B35**, 1345 (1987); also see P.M. Bridenbaugh, G.P. Espinosa, J.C. Phillips and J.P. Remeika, *Phys. Rev.* **B20**, 4140 (1979); K. Murase, Chapter 6(C), in this volume.

[35] J. Hajto, I. Janossy, W.K. Choi, *J. Non-Cryst. Solids* **114**, 304 (1989).

[36] K.L. Chopra, K.S. Harshwardhan, S. Rajagopalan, L.K. Malhotra, *Solid State Commun.* **40**, 387 (1981).

[37] J. Griffiths, G.P. Espinosa, J.P. Remeika and J.C. Phillips, *Phys. Rev.* **B25**, 1272 (1982).

[38] C.A. Spence and S.R. Elliott, *Phys. Rev.* **B39**, 5452 (1989).

[39] K. Tanaka, Y. Kasanuki and A. Odajima, *Thin Solid Films* **117**, 251 (1984).

[40] P. Boolchand, J. Grothaus, W.J. Bresser and P. Suranyi, *Phys. Rev.* **B25**, 2975 (1982).

[41] P. Boolchand, R.N. Enzweiler, R.L. Cappelletti, W.A. Kamitakahara, Y. Cai and M.F. Thorpe, *Solid State Ionics* **39**, 81 (1990).

[42] S.R. Elliott, Glasses and Amorphous Materials, ed. J. Zarzycki (VCH, 1991) p. 377. Also see S.R. Elliott, Physics of Amorphous Materials, 2nd Edn. (Longman Scientific and Technical, 1990, p. 127).

[43] R. Enzweiler, D. Selvanathan and P. Boolchand (unpublished).

[44] P. Boolchand, X.-W. Feng, D. Selvanathan and W.J. Bresser, *Rigidity Theory and Applications*, eds. M.F. Thorpe and P.M. Duxbury (Plenum Press, 1999) p. 279.

[45] J.P. deNeufville, *J. Non-Cryst. Solids* **8–10**, 85 (1972).

[46] F. Betts, A. Bienenstock and S.R. Ovshinsky, *J. Non-Cryst. Solids* **4**, 554 (1970).

[47] P. Boolchand, B.B. Triplett, S.S. Hanna, J.P. deNeufville, *Mössbauer Effect Methodology* **9**, 53 (1974), eds. J. Gruverman, C.W. Seidel, D.K. Dieterly (Plenum Press, NY).

[48] P. Boolchand, W.J. Bresser, D. Selvanathan and R.N. Enzweiler (unpublished).

[49] H. Rawson, *Glasses and Amorphous Materials*, ed. J. Zarzycki (VCH, 1991) p. 281.

[50] G.N. Greaves, *Glass Science and Technology* **4B**, eds. D.R. Uhlmann and N.J. Kreidl (Academic 1990) p. 25.

[51] N.H. Ray and C.J. Lewis, *J. Mat. Sci.* **7**, 47 (1972).

[52] R. Kerner and M. Micoulaut, *J. Non-Cryst. Solids* **210**, 298 (1997); also see M. Micoulaut, *Eur. Phys. J* **B1**, 277 (1998).

[53] M. Zhang, PhD Thesis, University of Cincinnati, 1995 (unpublished). Also see M. Zhang, W.J. Bresser, P. Boolchand, *J. Non-Cryst. Solids* **151**, 149 (1992).

[54] J.W. Zwanziger, S.L. Tagg and J.C. Huffman, *Science* **268**, 1510 (1995).

[55] S.L. Tagg, J.C. Huffman and J.W. Zwanziger, *Chem. Mater.* **6**, 1884 (1994); also see *Acta Chem. Scand.* **51**, 118 (1997).

[56] J.W. Zwanziger, J.C. McLaughlin and S.L. Tagg, *Phys. Rev.* **B56**, 5243 (1997).

[57] R. Dupree, D. Holland and M.G. Mortuza, *Nature* (London) **328**, 416 (1987).

[58] A.C. Wright, Chapter 5(A), this volume.

[59] A.N. Sreeram, A.K. Varshneya and D.R. Swiler, *J. Non-Cryst. Solids* **128**, 294 (1991).

[60] I. Finkler, H. Xia, R. Sooryakumar and P. Boolchand, *Bull. Am. Phys. Soc.* **44**, 1047 (1999).

[61] R. Bohmer and C.A. Angell, *Phys. Rev.* **B48**, 5857 (1993).

[62] U. Senapati and A.K. Varshneya, *J. Non-Cryst. Solids* **185**, 289 (1995).

[63] A. Feltz, H. Aust and A. Bleyer, *J. Non-Cryst. Solids* **55**, 179 (1983).

[64] R. Bohmer and C.A. Angell, *Phys. Rev.* **B45**, 10091 (1992).

[65] S. Asokan, M.V.N. Prasad, G. Parthasarathy and E.S.R. Gopal, *Phys. Rev. Lett.* **62**, 808 (1989).

[66] W.J. Bresser, P. Boolchand and P. Suranyi, *Phys. Rev. Lett.* **56**, 2493 (1986).

[67] B. Wunderlich, Y. Jin and A. Boller, *Thermochim. Acta* **238**, 227 (1994).

[68] T. Wagner, S.O. Kasap and K. Maeda, *J. Mat. Res.* **12**, 1892 (1997).

[69] X.-W. Feng, MS Thesis, University of Cincinnati, 1997 (unpublished).

[70] D. Selvanathan, W.J. Bresser, P. Boolchand and B. Goodman, *Solid State Commun.* **111**, 619 (1999).

[71] D. Selvanathan, W.J. Bresser and P. Boolchand (unpublished).

[72] D.S. Franzblau and J. Tersoff, *Phys. Rev. Lett.* **68**, 2172 (1992).

[73] H. He and M.F. Thorpe, *Phys. Rev. Lett.* **54**, 2107 (1985).

[74] M. Nakamura, O. Matsuda and K. Murase, *Phys. Rev.* **B57**, 10228 (1998).

[75] J.C. Phillips, *Rigidity Theory and Applications*, eds. M.F. Thorpe and P.M. Duxbury (Plenum Press, 1999) p. 155.

[76] S. Alexander and R. Orbach, *J. Phys. (France) Lett.* **43**, L625 (1982).

[77] W.J. Bresser, P. Boolchand, P. Suranyi and J.P. deNeufville, *Phys. Rev. Lett.* **46**, 1689 (1991).

[78] P. Boolchand, *Physical Properties of Amorphous Materials*, eds. D. Adler, B.B. Schwartz, M.C. Steele (Plenum Press, 1985) p. 221.

[79] J.C. Phillips, *Solid State Commun.* **109**, 301 (1999).

C.   RAMAN SCATTERING

K. MURASE

Department of Physics, Graduate School of Science, Osaka University,
1-1 Machikaneyama, Toyonaka City, Osaka 560-0043, Japan
murase@phys.sci.osaka-u.ac.jp

Contents

1. Introduction

Raman scattering, an inelastic scattering of light with typical frequencies between several and several thousands per cm, is reserved for scattering by phonons or by a number of other elementary excitations, such as magnons, plasmos, polaritons [1]. The selection rules for the first order Raman effect are

$$\omega_i = \omega_s \pm \Omega; \qquad \vec{k_i} = \vec{k_s} \pm \vec{K} \tag{1}$$

where $\omega_i, \vec{k_i}$ refer to frequency and wavevector of the incident photon; $\omega_s, \vec{k_s}$ refer to the scattered photon; and $\Omega, \vec{K}$ refer to the phonon created $(+)$ or destroyed $(-)$ in the Raman scattering event. The photon at $\omega_i - \Omega$ is called the *Stokes* process and that at $\omega_i + \Omega$ is the *anti-Stokes* process.

The intensity of such processes are temperature dependent through the statistical (Bose-Einstein) occupation number of the excitations n_K,

$$Stokes : I(\omega_i - \Omega) \propto n_K + 1$$

$$anti\text{-}Stokes : I(\omega_i + \Omega) \propto n_K$$

where $n_K = [\exp(\hbar\Omega/k_B T) - 1]^{-1}$. In thermal equilibrium, the temperature of a spot in a sample can principally be determined by a Raman microprobe from the ratio of *anti-Stokes* to *Stokes* intensities,

$$\frac{I(\omega_i + \Omega)}{I(\omega_i - \Omega)} = \frac{n_K}{n_K + 1} = \exp(-\hbar\Omega/k_B T) . \tag{2}$$

The so-called wave-vector conservation law in Eq. (1) applies to crystalline solids as it is a direct consequence of their periodicity. The magnitudes of $\vec{k_i}$ and $\vec{k_s}$ are:

$$k_{i,s} = \frac{2\pi n_{i,s}}{\lambda_{i,s}} \tag{3}$$

where $n_{i,s}$ is the refractive index and $\lambda_{i,s}$ the wavelength of the light in vacuum, around 500 nm. The typical values of $\vec{K}$ are determined by the size of the Brillouin zone $\cong 2\pi/a_0$, where a_0 is the lattice constant of nm order. Since $\lambda_{i,s}/n_{i,s} \gg a_0$ which makes $k_{i,s} \ll 2\pi/a_0$, only phonons very near the center of the Brillouin zone ($K \cong 0$) can participate in light scattering. The symmetry of the scattering system imposes a few more selection rules for crystals which can be derived by the methods of group theory. Since the photons are both *odd*, we find that the excitations involved in first order Raman scattering must be *even*. This selection rule is contrary to that found for infrared absorption (*odd* excitation). One can distinguish the excitations into the *odd* or the *even* with respect to the inversion.

In amorphous materials, $\vec{K}$ is not a good quantum number for phonons because losing the translational symmetry yields the break-down of the usual wave-vector conservation law. Raman scattering processes are allowed to occur from essentially all the vibrational modes of the material. Usually the intensity $I(\omega)$ of scattered light is proportional to the vibrational density of states $g(\omega)$:

$$I(\omega) = C(\omega)g(\omega)[n_K + 1]/\omega \tag{4}$$

for *Stokes* process. $C(\omega)$ describes the coupling of the vibrational modes of frequency $\omega/2\pi$ to the light.

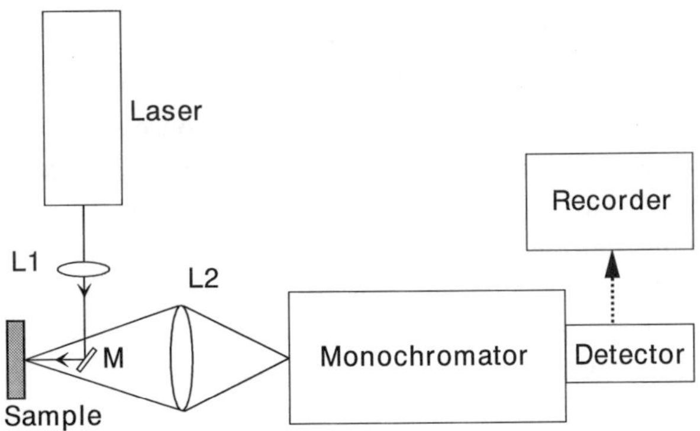

Fig. 1. Schematic diagram of a Raman measurement in the backscattering configuration. L1, L2: lens. M: plane mirror. The recorder is usually a computer.

A typical backscattering configuration is shown in Fig. 1. The incident laser light is focused on the sample. To avoid heating, using a cylindrical lens is a good choice. The scattering light is collected by a high aperture camera lens and focused on the entrance slit of a monochromator. For a fast measurement, the scattered light is generally analyzed with a double or triple grating polychromator and detected by a charge-coupled-device (CCD) detector. Spectra of both crystalline and amorphous materials are displayed in the following sections.

Another useful technique — the inelastic neutron scattering, which allows the observation of all the space in the Brillouin zone, is applied to investigate the glasses since one can also directly measure the vibrational density of state $g(\omega)$. Nevertheless, the inelastic neutron scattering technique is cumbersome when compared with Raman scattering. Generally an experimental setup of Raman measurement is set on a vibration isolation table and one can obtain a spectrum within one hour. The accuracy in the frequency determination and the resolution obtained with Raman spectroscopy is < 0.1 cm^{-1}. The good accuracy, high resolution and easy measurement of Raman scattering are advantageous to the investigation of glasses.

In this chapter, the Raman scattering study in the typical network glasses, Ge-(S,Se), is presented. Many interests of glasses, such as medium-range order, glass transition, light induced structural changes, low-frequency Raman spectra, fracton, fragility of glasses (see Chapter 1) and rigidity percolation [see Chapter 4, 6(c)] are discussed in detail.

2. Vibrational Spectra in Glassy GeSe$_2$

Raman spectra for the parallel (H,H) and perpendicular (H,V) polarization in the backscattering configuration, and infrared response spectra in the stoichiometric glass GeSe$_2$ are shown in Fig. 2(a) and (b) (after Ref. [2]). The A_1 band energy (203 cm^{-1}) is known primarily to be determined by the symmetric breathing motions of chalcogen atoms at GeSe$_{4/2}$ tetrahedra. The low energy side band at 180 cm^{-1} is due to the Ge–Ge bond. The presence of such homopolar bonds corresponds to partially broken chemical order in the stoichiometric glass in contrast with a perfect chemical order in crystalline GeSe$_2$. The number of Ge–Ge bonds must be equal to that of chalcogen–chalcogen bonds. It should be noted that the peak at 247 cm^{-1} has been identified as the chalcogen–chalcogen stretching mode (CS) and the peak at 145 cm^{-1} as the corresponding bending mode (CB) [3]. The higher energy side band at 220 cm^{-1} of the A_1 peak is called the companion A_1 line

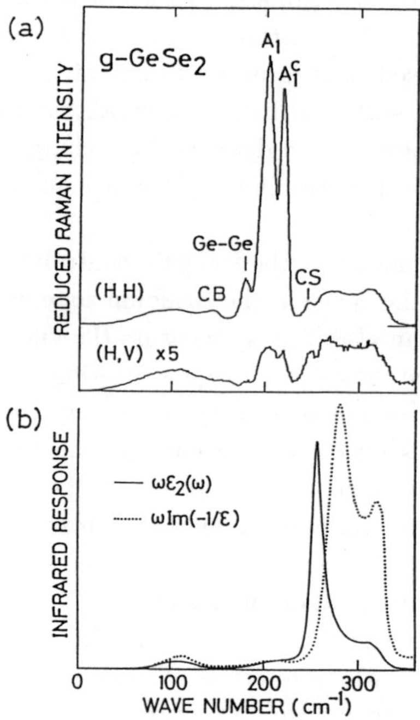

Fig. 2. Vibrational spectra of g–GeSe$_2$: (a) Raman spectra at 50 K with 6328 Å He-Ne laser, and (b) infrared response spectra at RT.

(A_1^C). Models for the A_1^C line had been subjected to controversy for many years [4–6]. The origin of the line will be clarified through Secs. 3 and 4.

In Fig. 2(b), the sharp band at 256 cm^{-1} and broad band peaking at 315 cm^{-1} in the transverse response spectra $\omega\varepsilon_2(\omega)$ have been attributed to the transverse F_2 modes (TO) of GeSe$_{4/2}$ clusters. The corresponding LO modes peaking at 280 cm^{-1} and 320 cm^{-1} are observed in the longitudinal response spectra $\omega\mathrm{Im}\,(-1/\varepsilon)$.

3. Vibrations in Crystalline GeSe$_2$

To know structural relation between crystalline and amorphous phases, Raman spectra of the well-known crystals were studied on the view point of the connectivity of the tetrahedra. The Raman spectra of two typical modifications of GeX$_2$, that is, high-temperature form HT–GeX$_2$ and low-temperature form LT–GeX$_2$ (X=S and Se) [7–9], were investigated through a simple model calculation using a combination of a valence-force-field (VFF) and a bond polarizability (BP) [10].

Common framework of HT– and LT–GeX$_2$ is a linear chain structure $(-Ge-X-Ge-X-)_n$ formed by repeats of the corner-sharing connection of the tetrahedra. The essential difference between the two modifications is that the

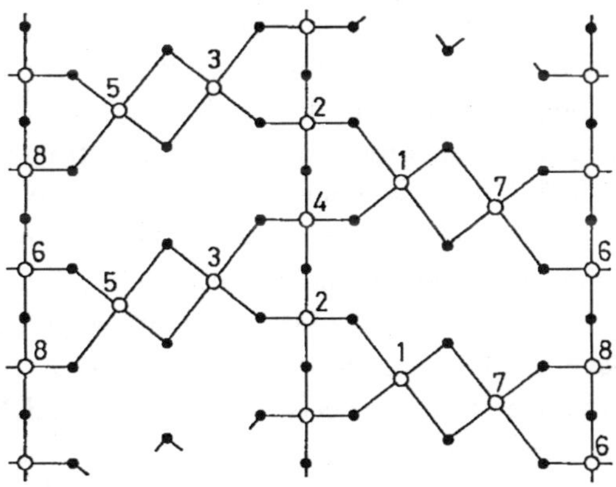

Fig. 3. Schematic illustration of network connection in a layer of the high-temperature form. Open circles show Ge atoms, and closed circles, chalcogen atoms. Numbers from 1 to 8 indicate the different GeX$_{4/2}$ tetrahedral sites in a unit cell in the layer. Linear chains (–2–4–2–4–) and (–6–8–6–8–) along with [100] direction are connected by bridges (1–7) and (3–5).

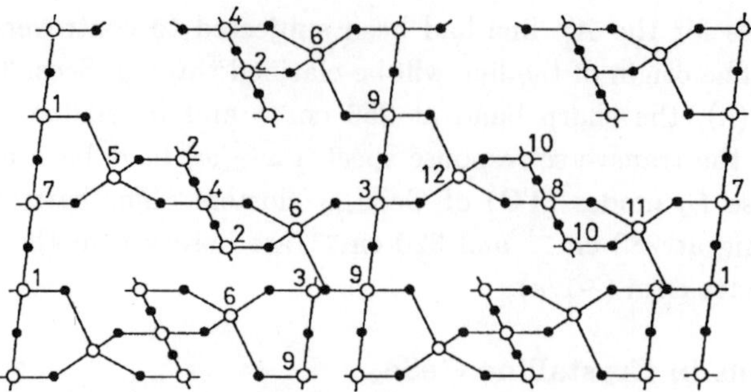

Fig. 4. Schematic illustration of network connection in the low-temperature form. Open circles show Ge atoms, and closed circles, chalcogen atoms. Numbers from 1 to 12 indicate the different $GeX_{4/2}$ tetrahedral sites in a unit cell. There are two kinds of linear chains (–1–7–1–7–), (–3–9–3–9–) along with [001] direction, and (–2–4–2–4–), (–8–10–8–10–) along with [101] direction. These chains are not included in the same plane, and they are connected mutually by corner-sharing tetrahedra, 5, 6, 11, and 12 in a 3-dimensional way.

high temperature form includes edge-sharing connections, while the low temperature form is composed of corner-sharing tetrahedra only [7–9]. The high temperature forms, HT–GeS_2 and HT–$GeSe_2$, are double layered crystals with the inversion symmetry (two dimensional, 2D). Figure 3 illustrates the schematic network connection in the mono-layer of HT–GeX_2. There are corner-sharing tetrahedral chains (–2–4–2–4–and–6–8–6–8–) parallel along the [100] direction, and the chains are connected like a ladder by bridges made of a pair of edge-sharing tetrahedra (1–7 and 3–5). On the other hand, in LT–GeS_2, the unit cell contains four corner-sharing tetrahedral chains: two, (–1–7–1–7–) and (–3–9–3–9–), are along the [001], and two, (–2–4–2–4–) and (–8–10–8–10–), are along the [101], as shown in the network in Fig. 4 [10]. The chains are connected by corner-sharing tetrahedra, 5, 6, 11, and 12, in a three dimensional structure (3D). The low temperature modification of $GeSe_2$ is considered to have a similar structure to the LT–GeS_2 structure from the resemblance between their Raman spectra [11], but the full X-ray analysis of the structure has not been done yet.

Crystalline samples were made by following ways. High temperature form of GeS_2 was crystallized from molten GeS_2 by slow cooling. Low temperature form of GeS_2 was made by a reaction of powdered GeS and S at 300°C. Two crystalline modifications of $GeSe_2$ were obtained from amorphous $GeSe_2$ films evaporated on glass substrates with about 600 nm thickness by annealing at 425°C for HT–GeSe$_2$, and at 325°C for LT–GeSe$_2$ (see Fig. 23). Raman spectra were measured with the 514.5 nm Ar ion laser line at room temperature. Polarization of the scattering light was not analyzed, but the component parallel to the incident polarization was dominant. The experimental spectra were compared with the spectra calculated in the (V,V) polarization condition. Since the structure of LT–GeSe$_2$ has not been well known, it was assumed that the structure of LT–GeSe$_2$ is geometrically similar to that of LT–GeS$_2$, and lattice constants of LT–GeSe$_2$ are 1.06 times of those of LT–GeS$_2$, from preliminary X-ray scattering measurement.

The Raman spectra were measured at room temperature, using several Ar-ion laser lines in the back scattering configuration. Figure 5(a) shows the Raman spectrum of a–GeSe$_2$ using 514.5 nm (2.41 eV) excitation. A broad

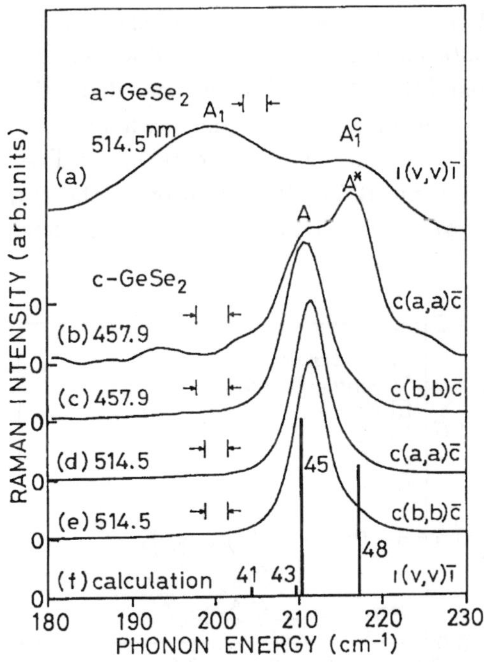

Fig. 5. Comparison of the Raman spectra for GeSe$_2$ in the amorphous state (a) and in the crystalline state (b)–(e). The vertical bars (f) show the calculated intensities for the crystalline Raman spectrum. The maximum intensities are internormalized. The excitation wavelengths, spectral resolutions (arrows), and polarization conditions are shown in the figure.

peak A_1 at 199 cm^{-1} [The spectral width is about 22 cm^{-1} full width at half maximum (FWHM)] and a companion peak A_1^C at 216 cm^{-1} (FWHM is about 12 cm^{-1}) are observed. From a Gaussian decomposition, the intensity ratio $A_1{:}A_1^C$ is about 3:1. On the other hand in the Raman spectrum of c–GeSe$_2$ (in the layered form hereafter, if not specified), only one intense peak, labeled A, is observed using 514.5 nm excitation at 211 cm^{-1} as shown in Figs. 5(d) and (e). It is quite natural to assume a correspondence between the A_1 peak and the A peak. The origin of the A_1^C peak, however, requires further discussion. In relation to this, it should be mentioned that the crystalline peak has an asymmetric tail on the high energy side [4] and it is worthy of note that in the highly-resolved spectra at low temperatures, a very weak side peak has been observed at about 216 cm^{-1}[12]. This side peak is tentatively labeled as A^*. The question is whether the A^* peak is intrinsic and whether it corresponds to the A_1^C peak.

To study the A^* peak in detail, the Raman spectra of c–GeSe$_2$ were measured at several excitation energies [13–15], and a resonant Raman effect was found, where the A and the A^* intensities changed depending on the excitation energy, especially in the $c(a,a)\bar{c}$ polarization configuration. The Raman peak energies did not change within the experimental error. The A^* peak increased remarkably with 457.9 nm (2.71 eV) excitation in the $c(a,a)\bar{c}$ configuration as shown in Fig. 5(b). The excitation energy dependence of the intensity

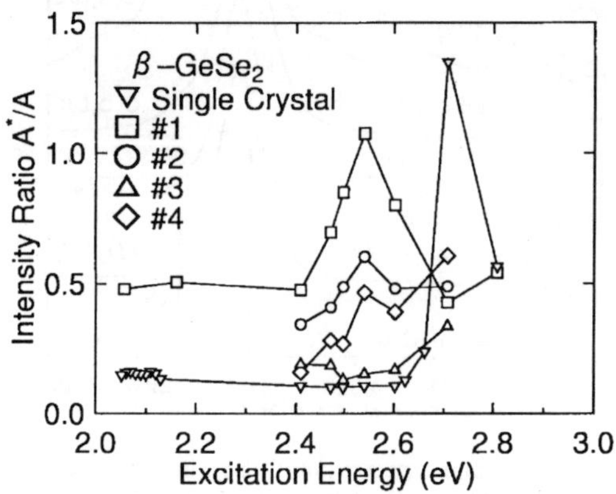

Fig. 6. Intensity ratio of the A^* band to the A band for the single crystals and the small crystals (#1–#4). The lines are guides to the eye. The error bars are smaller than the point symbols. The spectra are strongly influenced by the degree of disorder in the crystals.

ratio A^*/A is shown in Fig. 6. In the $c(a,a)\bar{c}$ configuration, the ratio (open triangles) shows a strong peak at about 2.7 eV, where a bulk exciton transition was observed as a step-like structure in the $E \parallel a$ absorption spectrum [16]. In contrast, there was no notable structure in the ratio for the $c(b,b)\bar{c}$ configuration or in the $E \parallel b$ absorption spectrum as shown in Fig. 5(d). Because the resonant enhancement of the ratio is closely related to the exciton, it has been concluded that the A^* peak is an intrinsic mode which can couple with the exciton. In the amorphous spectrum, the intensity ratio between A_1^C and A_1 bands did not change with excitation energies. It is supposed, however, that the structural origin of the A^* band in the crystal and the A_1^C band in the amorphous state would be closely related.

With usual excitation below or near the absorption edge, for example in Figs. 5(d) and (e), the A^* peak is very weak. On the other hand, in the amorphous spectrum, the A_1^C peak has nearly one third the intensity of the A_1 peak. To discuss such an intensity difference, the Raman selection rules in the crystal should be examined. The crystal symmetry of c–GeSe$_2$ is monoclinic P2$_1$/c [7] and consequently the A_g modes are Raman active in the $c(a,a)\bar{c}$ and $c(b,b)\bar{c}$ configurations, and the B_g modes in the $c(a,b)\bar{c}$ configuration. From experiments in these configurations, it was concluded that the A^* peak has the A_g symmetry as well as the A peak. The selection rules due to the crystal symmetry, however, do not explain the weakness of the A^* peak. The origin of the suppression of the A^* peak will be discussed later on.

In the calculation of the Raman spectrum of GeX$_2$, the vibrational levels as well as the atomic motions were calculated by the use of a VFF model [17], in which the interatomic potential U was described by the Keating potential,

$$U = \frac{1}{2}k_1 \sum_i \left(\frac{\delta r_i}{r_i}\right)^2 + \frac{1}{2}\sum_{i \neq j} k_x \left(\frac{\delta(\vec{r}_i \cdot \vec{r}_j)}{r_i r_j}\right)^2$$

where r_i is the bond-length of a bond i, δr_i is the displacement, and x indicates the central atom species (Ge or Se) which has neighboring bonds i and j. The values of the parameters adopted for the HT–GeSe$_2$ (c–GeSe$_2$) are $k_1 = 2.2$, $k_{\text{Ge}} = 0.22$, and $k_{\text{Se}} = 0.11$ ($\times 10^{-12}$ erg). At present interlayer interactions are neglected. To calculate the Raman intensities, an ellipsoidal bond polarizability is assigned to each Ge–Se bond [18] whose magnitude α was assumed to depend on the bond length:

$$\alpha_k = a_k + \delta a_k \left(\frac{\delta r}{r}\right)$$

where k indicates longitudinal (l) or transverse (t) to the bond direction. The following values were used for the parameters: $a_l - a_t = 1.0$, $\delta a_l = 2.0$, and $\delta a_t = 3.0$. In addition, to take the contribution of the Se lone-pair electrons into account, a similar ellipsoidal polarizability was assumed at a Se site, which was perpendicular to the Ge–Se–Ge bonding plane [10]. For this polarizability, $a_l - a_t = 2.0$, $\delta a_l = 1.0$, and $\delta a_t = 0.5$.

The calculated Raman intensities in c–$GeSe_2$ with a model using a VFF and BP are shown by the vertical bars in Fig. 5(f). The mode 45 corresponds to the A line and the mode 48 to the A^* line. The atomic motions in the mode 48 are illustrated in Fig. 7. The layer is composed of four basic $GeSe_{4/2}$ tetrahedra: C_1, C_2, E_1, and E_2. The motions of all the other tetrahedra in the crystal were obtained by the A_g symmetrical transformation of these four. In the mode 48, the edge-sharing tetrahedra E_1 and E_2 vibrate in a symmetric breathing motion, while the corner-sharing tetrahedra C_1 and C_2, which exist in a chain structure, move almost rotationally. On the contrary in the mode 45, which is not shown in the figure, C_1 and C_2 vibrate in a symmetric breathing motion, while E_1 and E_2 vibrate very little. Since both modes, 48 and 45, are due to in-phase breathing motions of the tetrahedra, it is reasonable that in the

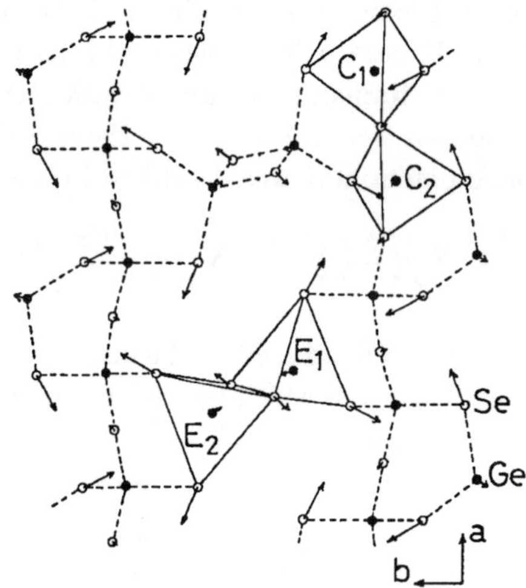

Fig. 7. A c–$GeSe_2$ layer with the atomic motions in the mode 48 indicated by arrows. The corner-sharing tetrahedra C_1 and C_2 make linear chains, which are interconnected by the edge-sharing tetrahedra E_1 and E_2.

calculated spectrum they have comparable intensities as shown in Fig. 5(f). For the explanation of the weakness of the A^* peak, it was imagined that the actual 48 mode intensity is suppressed by inter-layer interactions between adjacent edge-sharing tetrahedra drawing in their horns (the nearest Se atoms belonging to the adjacent layers). Such interaction is neglected in the present calculation. If the stacked structure of the crystal is disturbed by some defects, the A^* mode may appear with an intensity comparable to the A mode, because the cancellation will be released by a modification of the inter-layer interactions. Experiments on micro-crystalline $GeSe_2$ have indicated such phenomena [13, 19] as shown in Fig. 6.

In the crystal, the momentum selection rule for the Raman process should be considered: near Γ–point modes are Raman active. In the amorphous spectra, however, the momentum selection rule is completely broken. This breaking of the momentum selection rule will be an additional reason why the A^* peak becomes intense in the micro-crystals and in the amorphous sample. Figure 8(a) and (b) shows the calculated phonon dispersion curves as well as the VDOS for the crystal. It should be mentioned that the dispersion of the uppermost branch, which corresponds to the mode 48, is small. Vibrations in the mode 48 are quasi-localized at the edge-sharing tetrahedra E_1

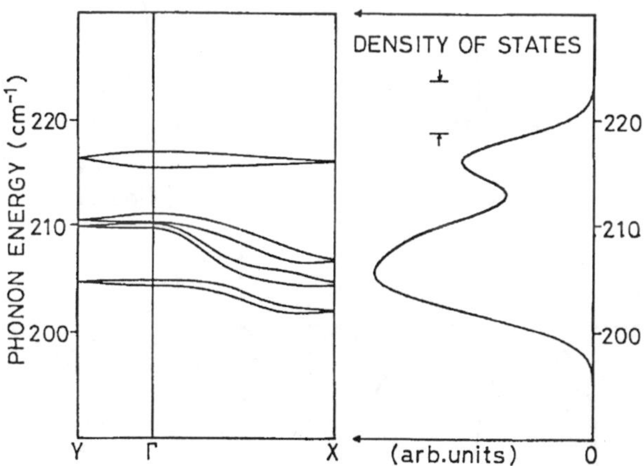

Fig. 8. The phonon dispersion curve (a) and the density of states (b) for the VFF model. Eight branches, the mode 41 through the mode 48, are shown. The density of states is calculated on the assumption of Gaussian peaks with a width of about 5 cm^{-1} FWHM. In the present model, phonons are dispersion-less in the Z direction since interlayer interactions are neglected.

and E_2 as shown in Fig. 7. In the VDOS this branch leads to a rather narrow peak at about 215 cm^{-1} as shown in Fig. 8(b). On the contrary, since the mode 45 is an extended mode in the chain structure (C_1 and C_2 in Fig. 7), it has a large dispersion and thus gives a broad peak at about 205 cm^{-1} merging with the other branches in Fig. 8. As the result, the total VDOS has a doublet peak which resembles the amorphous Raman spectra (the A_1^C and the A_1 peaks).

In a group theory, vibrations of a GeSe$_4$ tetrahedron are represented by 3 translational, 3 rotational, $2E$, $3F_2$, $1A_1$ and $3F_2$ modes. Figure 9(a) shows the A_1 vibration of an isolated GeSe$_4$ molecule. This mode is notable in the following points: (1) the central Ge atom does not move, (2) the movements of the four Se atoms are symmetric, and (3) it is the most intense Raman mode, as confirmed by the VFF–BP in Fig. 10(a). When two GeSe$_4$ tetrahedra are connected sharing one Se atom as shown in Fig. 9(b)(Ge$_2$Se$_7$), the vibrations of the two constituent tetrahedra are coupled. The coupling will be, however, rather weak, since the bonding angle around the shared Se atom is nearly 90°. Atomic motions of an in-phase coupled vibration of the A_1 tetrahedral modes

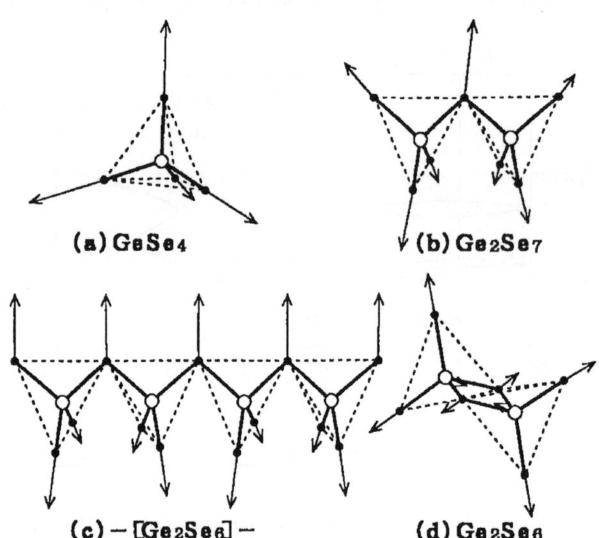

Fig. 9. Connected GeSe$_{4/2}$ tetrahedra (dotted lines) which vibrate in the in-phase A_1 mode. Open circles indicate Ge atoms; black dots, Se atoms; and heavy lines, covalent bonds. Atomic motions of eigenmodes calculated by VFF-BP are indicated by arrows.

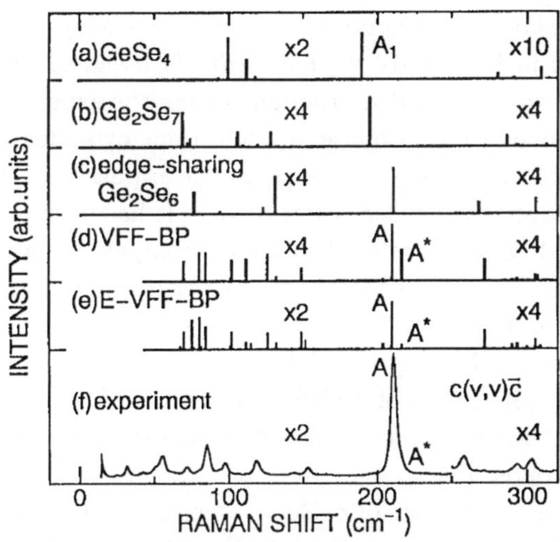

Fig. 10. Raman spectra of the connected $GeSe_4$ tetrahedra. The atomic structures for (a)–(c) are shown in Figs. 9(a), (b), and (d). Spectra calculated for c–$GeSe_2$ by VFF–BP (d) and by E–VFF–BP (e) are compared with the experimental spectra of c–$GeSe_2$ (f). Weak modes relating to E and F_2 vibrations (< 170 cm^{-1} or > 250 cm^{-1}) are multiplied by the factor indicated in the figure.

are shown in Fig. 9(b). This phonon has a vibrational energy close to the A_1 mode of the isolated $GeSe_4$ with a strong Raman intensity as shown in Fig. 10(b).

Another characteristic tetrahedral connection is an edge-sharing tetrahedron Ge_2Se_6 shown in Fig. 9(d). The bonding angles around the sharing Se atoms are also nearly 90°. The coupling of the two tetrahedra is, therefore, weak, although it is approximately twice as strong as that of the Ge_2Se_7 because of a double connection between the tetrahedra. Consequently, the phonon energy is a little higher than that of Ge_2Se_7 as shown in Fig. 10(c).

By repeating the connection in Fig. 9(b), a linear chain structure can be made as shown in Fig. 9(c), which is a main structural component of the crystalline $GeSe_2$ in the high temperature form (c–$GeSe_2$). In the unit cell of c–$GeSe_2$ in Fig. 3, there are four chains which are connected to each other by edge-sharing tetrahedra. The coupling between the A_1 modes of the chain and of the edge-sharing tetrahedra is so weak that the phonons relating the A_1 modes can be separated into extended chain modes and quasi-localized edge-sharing modes approximately [10].

Figure 10(d) shows the Raman intensities of c–$GeSe_2$ calculated by VFF–BP. As indicated in Fig. 10(d), we have already assigned two intense modes around 210 cm^{-1} to the A and A^* Raman lines in the experimental spectra (f) [15]. The A phonon is the in-phase A_1 vibrations extended along the $GeSe_4$ tetrahedral chain, while the A^* phonon is the in-phase A_1 vibrations quasi-localized at the edge-sharing $GeSe_4$ tetrahedra. It has been previously demonstrated that the A^* mode is related to the A_1 companion mode in the Raman spectra of amorphous $GeSe_2$ [15]. In the experimental spectra in Fig. 10(f), the A^* line located at 216 cm^{-1} is very weak, except for the resonant Raman spectra and polycrystalline spectra [19]. The peculiarity of the A^* mode intensity in the Raman spectra of c–$GeSe_2$ has remained an important question. To explain the A^* mode intensity, the following extended VFF–BP model(E–VFF–BP) was proposed.

Interactions between layers of c–$GeSe_2$ were ignored in the VFF–BP, since there are no covalent bonds (heavy lines in Fig. 11) between the layers. Natural extension of VFF–BP was obtained by taking the interactions between layers into account [25]. It was assumed that the Se–Se pairs, labeled A, B, and B'

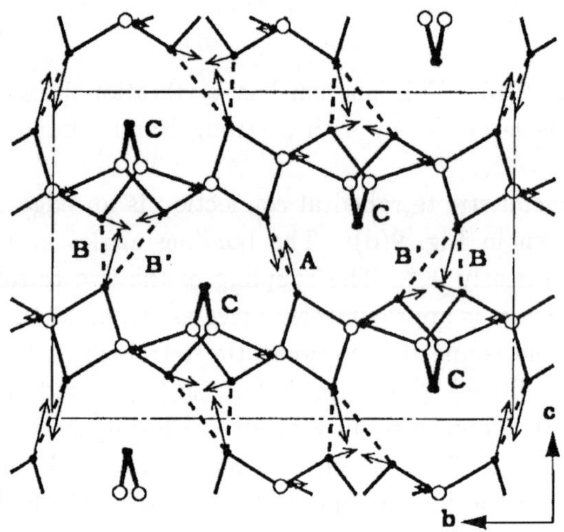

Fig. 11. Projection of layered structure of c–$GeSe_2$. Open circles indicate Ge atoms; black dots, Se atoms; and heavy lines, covalent bonds. Unit-cell is bounded by a dashed-and-dotted line. Arrows indicate the atomic motions of the A^* mode calculated by E-VFF-BP by taking Se–Se weak bonds, A, B, and B' (dashed lines), into account. The –Ge–Se–Ge–Se– chain (C) is normal to the plane of the figure.

in the figure (broken lines) are weak bonds, and a bond polarizability is assigned to them. The reason why these three bonds was chosen is that their Se–Se distance is about 4 Å (e.g. B' bond length is 3.5 Å) which is almost equivalent to the second nearest Se–Se distance of GeSe$_4$ tetrahedra, and that the lone-pair orbitals of the two Se atoms bound by these bonds are almost parallel (e.g. exactly parallel for bond A). The overlap integral of lone-pair orbitals will become large under the above two conditions. In the present calculation, the bond polarizability assigned to these weak bonds was twice as large as the ordinary bond polarizability: that is, anisotropy $a_l - a_t = 2.0$, and derivatives $\delta a_l = 4.0$, and $\delta a_t = 6.0$, while the force constant of these bonds is ignored for simplicity. As a result the calculated A^* intensity was notably reduced: thus the spectra in Fig. 10(e) agree well with the experimental spectra (f).

In summary, the relation between the phonons and the tetrahedral connectivity is studied by VFF–BP. In c–GeSe$_2$, the A phonon is extended along the tetrahedral chain, while the A^* phonon is quasi-localized at the edge-sharing tetrahedra. An extended VFF–BP model including interlayer Se–Se interactions is proposed to understand the Raman spectra of c–GeSe$_2$.

4. Glass Transition, Recrystallization and Melting in GeSe$_2$

The Raman spectra of GeSe$_2$ were extensively studied in the temperature range from 30°C to 730°C. The temperature range covers the glassy, super-cooled-liquid(SCL), crystalline and liquid states.

The GeSe$_2$ films ($\sim$1 μm) were prepared from melts in quartz cells with a narrow gap defined by two parallel plates [20]. A weak line-focused probing light (8 mW, 2.10 eV) was used in the Raman measurement to avoid an additional temperature rise and the effect of photo-induced crystallization. The temperature was controlled within ±1°C during the measurement by using an electric furnace with a small window, through which the incident and scattering light pass. The measurements were made in increasing temperature step mode from 30°C to 730°C.

Figure 12 shows Raman spectra of a GeSe$_2$ film at various temperatures. Three characteristic Raman bands named A_G, A_1 and A_1^C were observed around 200 cm^{-1}[21]. The A_1 and A_1^C bands are related to the breathing vibration mode of the corner-sharing tetrahedra (CST) and the edge-sharing tetrahedra(EST) respectively, and the A_G band is related to the vibration relating to the Ge–Ge bonds showing broken chemical order. At temperatures lower than 350°C, each spectrum was accumulated for 10 minutes at each

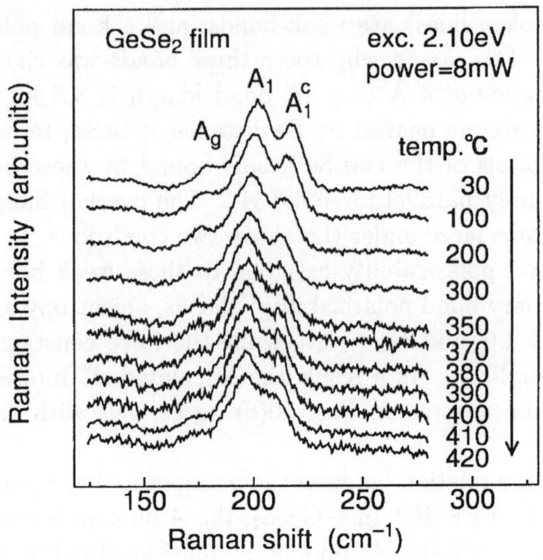

Fig. 12. Raman spectra of g–GeSe$_2$ film at various temperatures. The arrow indicates the sequence of measurement. Spectra around 200 cm^{-1} at 420°C are somewhat different compared with that at 30°C.

temperature. Because the thermal crystallization for usual amorphous GeSe$_2$ films prepared by vacuum evaporation occurs above 350°C [11], the measurement above 350°C was carefully made in order not to overlook any crystallization process. In the temperature range of 350 $\sim$ 420°C the accumulated time for a Raman measurement was about 2 hours at each temperature, except at 390°C and 410°C, where the spectra were monitored for over 16 hours. At each temperature, however, no time–dependent spectral change was observed. In all the spectra in Fig. 12, the Raman bands A_G, A_1 and A_1^C were distinguishable and they shift to the lower energy side with increasing temperatures. Comparing the two spectra at 30°C and at 420°C, the spectral shape around 200 cm^{-1} was significantly changed.

When the temperature was higher than 420°C, the Raman band of c–GeSe$_2$ (the A band in Fig. 13) began to appear with the passage of time. Figure 14 shows the time dependence of the scattering intensity of the A band at the crystallization temperature (T_c) 440°C, and 450°C. The intensity of the A band was obtained by four–Gaussian band fitting of the spectra including the three amorphous components (A_G, A_1, A_1^C) with a linear base line.

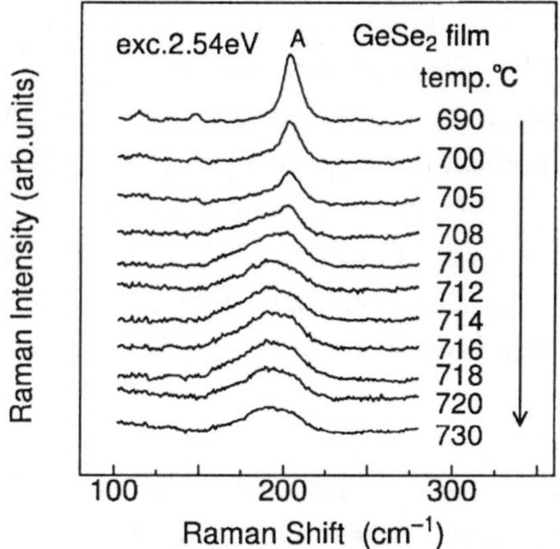

Fig. 13. Raman spectra of c– and l–GeSe$_2$ at various temperatures. The crystalline vibration A band disappears at 712°C, whose values is related to the melting temperature T_m. The Raman spectra of l–GeSe$_2$ are very similar to those measured at 420°C.

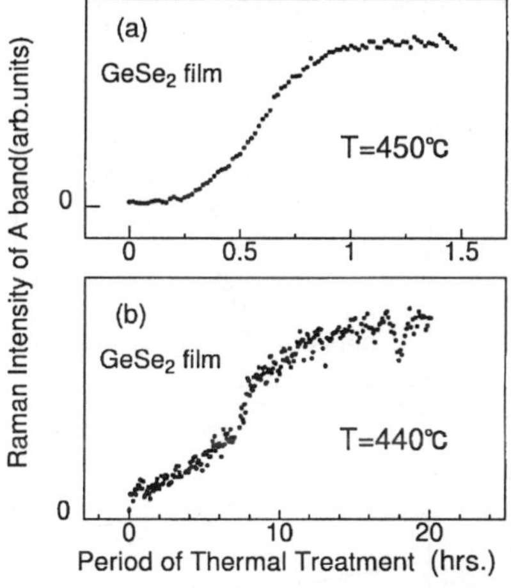

Fig. 14. Time dependence of the intensity of the crystalline Raman band A for the thermal crystallization process at 450°C (a) and 440°C (b).

The crystallization at 450°C proceeded much faster than at 440°C. Once the specimen turned into the crystalline phase, the vibration bands in crystalline Raman spectra shifted with temperature by a coefficient -0.01 cm^{-1}/K which was about the same value measured in bulk c–GeSe$_2$ [11].

Figure 13 shows the Raman spectra around the melting point. At 690°C, the characteristic c–GeSe$_2$ A band appeared clearly together with other crystalline vibrational bands on the lower energy side. The intensity of the A band became weaker with increasing temperature up to 712°C. When the temperature was higher than 712°C, the c–GeSe$_2$ A band disappeared in the Raman spectra and the specimen was molten. By these results, the melting point T_m was evaluated as 712°C in this specimen. Raman spectra of liquid (l–) GeSe$_2$ were accumulated for one hour at each temperature up to 730°C. The spectra around 200 cm^{-1} were similar to those of A_G, A_1 and A_1^C bands at 420°C.

Comparing the spectra of g–and l–GeSe$_2$ in Figs. 12 and 13, the spectral shapes around 200 cm^{-1} were found to be changed. By fitting the Raman spectra with three Gaussians corresponding to A_G, A_1 and A_1^C bands at various temperatures, the Raman intensity of each band normalized by the total intensity of the three Gaussians was obtained. The relative intensity of the A_G band kept a value of several percent in the temperature range of 30 $\sim$ 730°C, while that of A_1 and A_1^C bands had notable changes as shown in Fig. 15(a). The glass-transition temperature T_g was determined to be 392°C by using

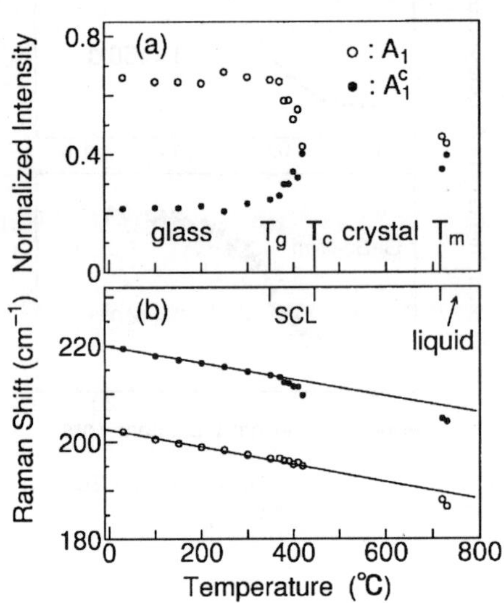

Fig. 15. Temperature dependence of phonon vibration A_1 and A_1^C bands in g–, SCL and l–GeSe$_2$; (a) intensity of A_1 and A_1^C bands normalized by the whole intensity of A_G, A_1 and A_1^C bands, and (b) peak position of A_1 and A_1^C bands. Lines are drawn as guides for the eyes.

differential thermal analysis for $GeSe_2$ bulk glass [22]. The intensity of the A_1 and A_1^C bands began to change drastically at a temperature around 370°C. The onset temperature is naturally related to the glass transition. So the glassy, SCL, crystalline and liquid phases are separated by characteristic temperatures: T_g (370°C), T_c(440°C) and T_m(712°C).

In the glassy phase, the intensity ratio between the A_1^C and A_1 bands (A_1^C:A_1) was kept at about 1:3 as shown in Fig. 15(a). In the SCL phase, the ratio A_1^C:A_1 approached 1:1. It should be noted that the ratio A_1^C:A_1 in the liquid phase at 720°C and 730°C was similar to that in the SCL at 420°C. From the temperature dependence of peak positions of the A_1^C and A_1 bands in the glassy phase, it was found that both A_1 and A_1^C bands in g–$GeSe_2$ shift with the same temperature coefficient -0.017 cm^{-1}/K, while the coefficient of the A peak in c–$GeSe_2$ was -0.01 cm^{-1}/K. This shows a large anharmonicity in the non-crystalline states. The peak positions of the A_1 and A_1^C bands in SCL and liquid $GeSe_2$ were near the extrapolated line from the temperature dependence of peak positions of g–$GeSe_2$ as shown in Fig. 15(b). The results also support our view based on the A_1 and A_1^C bands: there is a similar medium range structure(MRS) in g– and l–$GeSe_2$.

It has been shown that the A_1 and A_1^C bands are related to the corner sharing tetrahedra (CST) and the edge sharing tetrahedra (EST), respectively (Fig. 3) [21]. In the neutron diffraction experiment, the numbers of CST and EST in g– and l–$GeSe_2$ have been estimated to be comparable with each other [23, 24]. In the meanwhile, it is hasty to conclude that the numbers of CST and EST for g– and l–$GeSe_2$ are quite different in parallel with the fact that the ratio A_1^C:A_1 is changed between glass and liquid.

Before considering the above problem, we should discuss the Raman spectra of c–$GeSe_2$ which has a layered form. In the Raman spectra, the characteristic bands A and A^* exist around 200 cm^{-1}. The origin of the A and A^* bands has been ascribed to the breathing vibration of CST and EST, respectively [21]. In the usual non-resonant Raman spectra of the single crystal, the A^* band is unexpectedly weaker than the A band [21]. A model calculation using a combination of a valence-force-field and a bond polarizability model has revealed that the peculiar weakness of A^* band is due to the interlayer interactions in the two neighboring layers as described in the previous section and in Ref. [25]. The intensity ratio A^*:A in various kinds of disordered $GeSe_2$ crystals is much larger than that in the single crystal under the non-resonant condition as shown in Fig. 6 [21]. Then, it is probable that with increasing disorder, the interlayer interaction becomes weak so that the intensity ratio of A^*:A is enlarged.

As discussed above, the medium range structure (MRS) should be maintained both in g–, SCL–, and l–GeSe$_2$. Thus the rapid decrease of viscosity η with increasing temperature above T_g is mainly caused by a relaxation of the interlayer constraint between two successively stacked layer-like fragments. The relative flow of the layer-like fragments results in a decrease of interlayer interactions. The decrease of interactions should cause an increase of the ratio A_1^C:A_1 from 1:3 to 1:1 similar to the increase of the ratio A^*:A in crystals with increasing disorder.

In summary, both in g– and l–GeSe$_2$, a topologically crystalline layer-like fragment is observed. The fact that the intensity ratio of A_1^C band to A_1 band changes quickly from 1:3 to 1:1 in super-cooled liquid is related to a drastic decrease of the interaction between two neighboring layer-like fragments.

5. Chalcogen-Rich Glasses

The chalcogen-rich glasses $Ge_{1-x}Se_x$ and $Ge_{1-x}S_x$ are known to be micro-heterogeneous, which may have different kinds of low dimensional molecular clusters [26]. It is very important to investigate interactions among clusters [27] as well as stability, structure, strain, spatial extent and dimensionality of the clusters to understand the physics of the glass and specific effects, such as photo-doping and switching, which are applied to important devices. The morphology of the medium range order(MRO) has been signified by vibrational studies [26, 28], up to this time, as follows: (1) a cluster composed of Ge(Se, or S)$_{4/2}$ molecular units whose atomic connections may be similar to fragments of crystalline Ge(Se, or S)$_2$ with some relaxation and modification, (2) a selenium chain and Se$_8$ fragment, (3) selenium chains with a chain crossing germanium atom, (4) a sulfur chain and a S$_8$ ring. In the infra-red response spectra, Ge(Se, or S)$_{4/2}$ clusters play a leading part, each of which is surrounded by chalcogen clusters in a wide composition range of the glass [28]. Raman spectra give us complementary information about clusters in the glasses. For the Ge(Se, or S)$_{4/2}$ cluster, a tentative, but heuristic and influential model has been proposed, which is called an outrigger raft [4]. Figure 16 shows composition dependence of unpolarized Raman spectra at 50 K for $Ge_{1-x}Se_x$. The 203 cm^{-1} peak at $x = 2/3$ has been ascribed to the symmetric breathing A_1 mode of the GeSe$_4$ tetrahedron. The high energy side band (A_1^C) at 219 cm^{-1} at $x = 2/3$ is called the companion A_1 line which has similar polarization

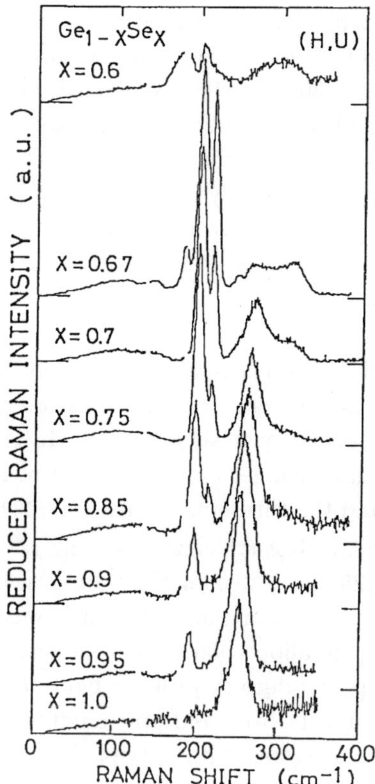

Fig. 16. Raman spectra of g–Ge$_{1-x}$Se$_x$ at 50 K with 6328 Å He-Ne laser light.

character as the A_1 band (see Sec. 3). The 180 cm^{-1} mode observed at $x = 2/3$ has been identified by a A_G mode of an ethane like molecule Ge$_2$Se$_6$.

It has been concluded that the GeSe$_{4/2}$ clusters in g–GeSe$_2$ are highly stressed even without external pressure. The stress will be relaxed with increasing selenium composition, as is supported by composition and pressure dependences of the A_1 band [29]. The composition behavior of the A_1^C peak position is quite similar to the A_1 peak (Fig. 16).

6. Structural Transformation in Glassy and Liquid Ge$_x$Se$_{1-x}$

The typical covalent network glass system Ge$_x$Se$_{1-x}$ has received a great deal of interest, in terms of the changes in the structural dimensionality of the glass with increasing Ge content. For the composition range $0.00 \leq x \leq 0.33$, the coordination numbers of Ge and Se are 4 and 2, respectively. Based on a mean-field constraint theory [27, 30], the rigidity percolation threshold occurs at an average coordination number, $\langle r \rangle$, of 2.4 (see Chapters 2 and 4), corresponding to $x = 0.20$ for $Ge_x Se_{1-x}$. The character of a network glass undergoes a qualitative change, from being easily deformable at $\langle r \rangle < 2.4$ to being rigid at $\langle r \rangle > 2.4$. Experiments have shown evidence of threshold behavior in the electronic [31] (see Sec. 7), structural [32], vibrational [2, 33, 34] and thermal [34–36] properties around $\langle r \rangle = 2.4$.

The Raman spectra for $Ge_x Se_{1-x}$ glasses and melts $(0 \leq x \leq 0.33)$ are extensively investigated at the glass-transition, recrystallization and melting temperatures. The glass-transition temperatures are also determined, using differential scanning calorimetry measurements. In the super cooled liquid (SCL) state, the structural changes and the rate of recrystallization are especially investigated around the threshold value, $\langle r \rangle = 2.4$.

Thin films of $Ge_x Se_{1-x}$ glass (thickness ~ 1 μm; $x = 0.00, 0.04, 0.06, 0.07, 0.10, 0.15, 0.18, 0.20, 0.25$ and 0.33), for the Raman scattering measurements, were prepared by quenching from the melt which was held in an evacuated fused-silica cell [37, 38] at about $900°$C, except the Se sample, which was held at $450°$C. A low power-density probing light $(h\nu = 2.41$ eV$)$ was used for the Raman measurements, to avoid any unintentional temperature rise. The scattered light was analyzed with a triple grating monochromator and detected by a multi-channel detector. The measurements were carried out sequentially at increasing temperatures, from $20°$C to $850°$C, which covers the glassy $(g-)$, SCL, crystalline $(c-)$ and liquid $(l-)$ states. In general, the spectrum was accumulated for 10 minutes at each sample temperature which was fixed within $\pm 1°$C and the rate of increase of the temperature between two successive measurements was 2–$5°$C/min. The glass-transition temperatures, $T_g(x)$, for ~ 20 mg of each glass were also measured at a heating rate of 10 K/min in sealed aluminum pans, using a Perkin–Elmer model DSC-7 differential scanning calorimeter.

The differential scanning calorimetry (DSC) measurements (Fig. 17) show that T_g increases monotonically with x for $x < 0.33$ but decreases for $x > 0.33$. This composition dependence is in good agreement with the published DSC [39] and modulated DSC [34] results.

Figures 18(a)–(f) show typical Stokes Raman spectra for various compositions. All the spectra are divided by $(n(\omega, T) + 1)$, where $n(\omega, T)$ is the Bose

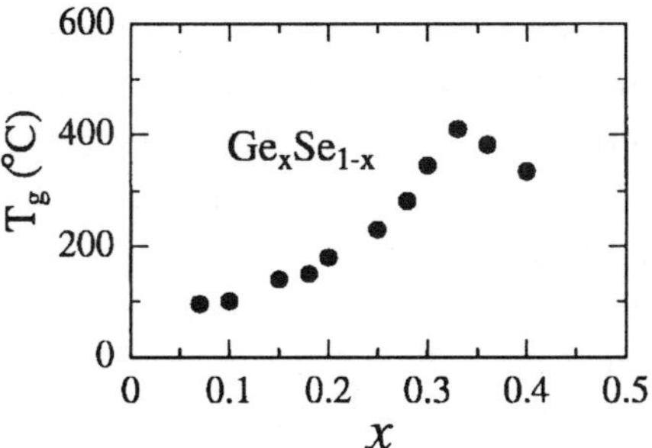

Fig. 17. T_g for Ge_xSe_{1-x} bulk glasses from DSC measurements. The random error in determining the T_g is smaller than the dot size.

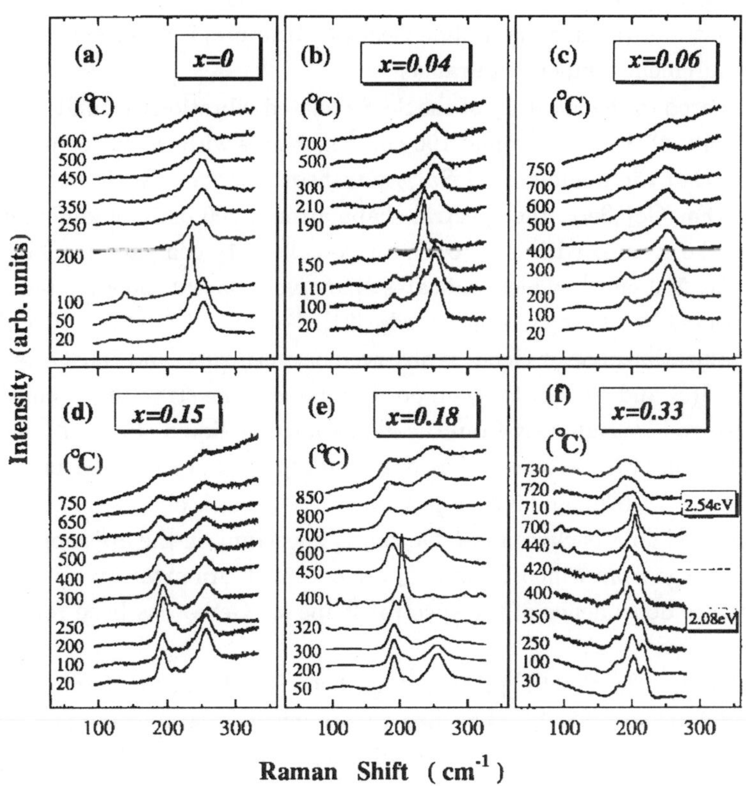

Fig. 18. Typical Stokes Raman spectra for Ge_xSe_{1-x} in the glassy, liquid or/and crystalline state.

factor, and are normalized to the maximum intensity. The spectrum at the successive higher temperature is plotted with the base line raised by the same value. The base line of some spectra in high temperature region, i.e. (d) at 750°C, going up with increasing Raman shift is affected by a comparable distribution for the scattering of the silica cell. For each composition, all the spectra were obtained in $\sim$ 10 hours. For the Se sample, a sharp peak at 235 cm^{-1}, characteristic of trigonal crystalline (tc–) Se, clearly appears at 100°C and disappears with increasing temperatures above 200°C, while, for Ge_4Se_{96}, the sharp tc–Se peak is observed in the temperature range 110–190°C. In this section we use the recrystallization temperature, T_c, and the melting point, T_m, as the temperatures at which the crystalline peak appears and disappears, respectively, regardless of whether the sample is stoichiometic ($x = 0, 0.33$). In the case of Ge_4Se_{96}, T_c is $\sim$ 110°C and T_m is $\sim$ 190°C. A peak at 190 cm^{-1}, which relates to the non-crystalline $GeSe_{4/2}$ structure, is observed throughout the measurement temperature range.

Compared to the intensity of the Se–Se related vibrations around 250 cm^{-1}, the intensity of the peak around 190 cm^{-1} increases with the Ge content. This indicates that the number of $GeSe_{4/2}$ tetrahedra increases with increasing Ge content. For $Ge_{15}Se_{85}$, the spectral shape around 200 cm^{-1} is different from that for the glasses with $x \leq 0.07$. A so-called A_1 companion mode (A_1^C) appears at 210 cm^{-1}, which indicates that two adjacent $GeSe_{4/2}$ tetrahedra are connected by sharing a common *edge* [21]. For $x = 0.06, 0.07, 0.10$ and 0.15, no crystallization of Se or $GeSe_2$ is observed between 20°C and 750°C, as shown in Figs. 18(c) and (d). In this composition region, the structure is considered to be changed continuously from the glassy into the liquid state, although the crystallization of tc–Se or c–$GeSe_2$ has been reported in the literature [26].

For $x = 0.18, 0.20$, and 0.25, a peak due to c–$GeSe_2$ is observed at 210 cm^{-1}, superimposed on a background l–Ge_xSe_{1-x} spectrum, in the appropriate temperature range from which T_c and T_m are obtained. An extensive study of the temperature dependence of the spectral shape for $GeSe_2$ has been reported in Sec. 4.

In Ge_xSe_{1-x} ($x \leq 0.07$) glasses, it has been found that the dominant structure is $(Se)_n$ chains [38, 40]. The glassy state involves parallel and non-parallel Se–Se chain regions, whose relative vibrational Raman bands are 235 cm^{-1} and 255 cm^{-1}, respectively. The decrease in viscosity above the glass-transition temperature causes large configurational fluctuations in the region of the parallel $(Se)_n$ chains, thus broadens the corresponding vibrational Raman band at 235 cm^{-1}, according to the comparison of the temperature dependence of the line shapes between the 235 cm^{-1} and 255 cm^{-1} bands.

For $0.10 \leq x \leq 0.25$, the Raman spectra in the range $150 \sim 300$ cm^{-1} are fitted by four Gaussian peaks plus a straight line. The peak around 255 cm^{-1} relate to the $(Se)_n$ chain structure, while the two peaks around 200 cm^{-1} comprise the A_1 and A_1^C GeSe$_{4/2}$ tetrahedral vibrational bands. The intensity ratio $I_{GeSe_{4/2}}/I_{Se}$, from the peak fits, is shown in Fig. 19, which also includes the values of T_g obtained from the DSC measurements.

For $0.18 \leq x \leq 0.25$, the intensity ratio, $I_{GeSe_{4/2}}/I_{Se}$, for the liquid state is the same as that below the value of T_g obtained from the DSC measurements. These results suggest that, in our experimental temperature range, the structures of the glassy and liquid states are mainly composed of a similar arrangement of $(Se)_n$ chains and GeSe$_{4/2}$ tetrahedra. For these samples,

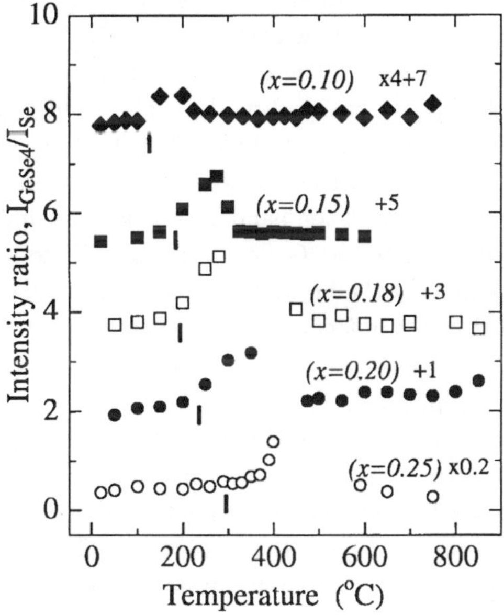

Fig. 19. The intensity ratio, $I_{GeSe_{4/2}}/I_{Se}$, for the glassy and liquid samples, obtained from fitting the experimental data with error bars smaller than the marks. For $x = 0.18$, 0.20 and 0.25, the blanks are due to the appearance of the spectrum for c–GeSe$_2$. The vertical bars indicate T_g from the DSC measurements.

$I_{GeSe_{4/2}}/I_{Se}$ increases with increasing temperature above T_g, prior to crystallization. The possibility of resonant Raman scattering is eliminated by the result that no excitation energy dependence has been found in the temperature range where the spectra begin to change [38]. The relative populations of the $(Se)_n$ chains and $GeSe_{4/2}$ tetrahedra are the same in the liquid and glassy states and so it is concluded that the increase in the intensity ratio above T_g relates to a nucleation process, following which the spectrum of crystalline $GeSe_2$ appears, superimposed on that of the liquid. In Fig. 18, with increasing Ge composition above 0.07, c–$GeSe_2$ –like fragments, which invole *corner*– and *edge*–sharred tetrahedra [21], enlarge their contribution to the Raman scattering. In the SCL state, the viscosity decreases very rapidly with increasing temperatures, which increases the rate of the diffusive motions of the particles, so that the structure varies on a time scale comparable with that of the experiment. The c–$GeSe_2$ –like fragments easily move and coalesce with each other, to form the crystalline embryo where the local electronic states are assumed to be similar to those of the crystal. This phase separation makes the Raman scattering cross section contributed by the crystalline embryo part and the rest of l–Ge_xSe_{1-x}. An expective widening of the Ge_xSe_{1-x} band-gap, based on the appearance of the crystalline embryo, should affect the contributions of the Se–Se and $GeSe_{4/2}$ vibrational modes. The band-gap of l–Se is much lower than the probing light energy, 2.41 eV, while, that of l–$GeSe_2$ is comparable. In general, the Raman intensity decreases rapidly when the probing light energy slightly increases around the band-gap. The widening of the Ge_xSe_{1-x} band-gap is more sensitive to the $GeSe_{4/2}$ vibrational modes in this respect. So that, the enlargement of the crystalline embryo for $GeSe_2$ leads to a large increase in the contribution of $GeSe_{4/2}$ to the Raman scattering cross section.

For $Ge_{10}Se_{90}$ and $Ge_{15}Se_{85}$, the intensity ratio, $I_{GeSe_{4/2}}/I_{Se}$, increases with temperature above T_g and, at high temperatures, returns to the initial value for the glassy state. As discussed above for $0.18 \leq x \leq 0.25$, the increase in $I_{GeSe_{4/2}}/I_{Se}$ should also relate to c–$GeSe_2$ –like fragments forming crystalline embryo. The decrease at high temperatures suggests that the embryo start to dissolve. At temperatures higher than 210°C, for $Ge_{10}Se_{90}$, and 260°C, for $Ge_{15}Se_{85}$, both of the samples are in the *normal* liquid state. Here, the critical temperature is denoted as the melting point, T_m, which identifies the samples as being in either the SCL or the *normal* liquid state, based on their structure.

From the temperature dependence of the Raman spectrum, the glass-transition temperature, T_g, is determined as the temperature above which the

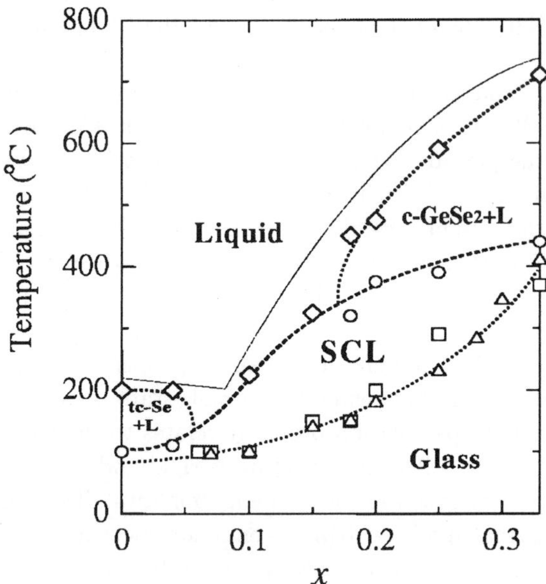

Fig. 20. The value of T_g is determined from Raman scattering (squares) and DSC measurements (triangle). The circles and diamonds, respectively, denote T_c and T_m from the Raman scattering measurements. The error bars are smaller than the marks. The broken lines are drawn as a guide to the eye. The liquidus line (solid) is from Ref. [26].

relaxation time for the structural changes is comparable with the experimental time periods. The phase diagram in Fig. 20 summarizes our results for T_g, T_c and T_m.

The values of T_g determined from the Raman scattering measurements are in general agreement with those from the calorimetric measurements. The present values for T_m agree with the results of Ref. [26], where the liquidus line indicates the primary crystallization temperature for c–$GeSe_2$ or tc–Se. In the equilibrium phase diagram, the eutectic point is at $x = 0.08$. In the present phase diagram, for $0.04 < x < 0.18$, neither tc–Se nor c–$GeSe_2$ appears in the measurement temperature range. The tendency for crystallization, as a function of composition, is clearly indicated by the phase diagram.

For $0.04 < x \leq 0.07$, the experimental results suggest that the relaxation time for the $(Se)_n$ chains forming the long-range order is longer than the experimental time period. The population of randomly distributed $GeSe_{4/2}$ tetrahedra, whose cross-linked structure is rather stable at the experimental temperature, is large enough to easily prevent the $(Se)_n$ chains from crystallizing.

For $Ge_{10}Se_{90}$ and $Ge_{15}Se_{85}$, although the $GeSe_2$ crystalline embryo are formed in the SCL, the time for the c–$GeSe_2$ growth is much longer than the measurement time period. With increases in x to 0.18, the glass crystallizes as easily as that in the Ge-rich region ($0.18 \leq x < 0.33$) [37]. We interpret the result that the rate of crystallization shows an abrupt change, when x increases from 0.15 to 0.18, as an evidence for a rigidity threshold at an average coordination number, $\langle r \rangle$, of 2.4. Previous authors [34–36] have studied the dependence of properties such as thermal expansion, molar volume, and heat capacity on $\langle r \rangle$. Distinct minima in the above properties at $\langle r \rangle = 2.4$ suggest that the structural rearrangements in the SCL state, at $\langle r \rangle = 2.4$, are minimized. As the rigidity percolation probably relates to the $GeSe_{4/2}$ tetrahedra structure, thus, for $\langle r \rangle > 2.4$, the structure is rigid and tends to become more continuous which may increase the tendency for crystallization. It should be noted that the normal and inverse photo-emission spectra also change abruptly at $x = 0.18 \sim 0.20$ as discribed in Sec. 7. According to the constraint theory, the properties of glasses should be optimized at $\langle r \rangle = 2.4$, which implies a maximized glass forming tendency. In this work, the rate of crystallization shows an distinct change at $x = 0.18$ ($\langle r \rangle = 2.36$), very close to $x = 0.20$ ($\langle r \rangle = 2.4$). The poor crystallization tendency for $x < 0.18$ is an evidence for the rigidity threshold which marks the *best* glass.

In summary, the structure of liquid and glassy Ge_xSe_{1-x} ($0.00 \leq x \leq 0.33$) and the structural changes occurring at the glass-transition have been studied by Raman scattering combined with differential scanning calorimetry measurements. For $x = 0.10$–0.25 range, a c–$GeSe_2$ –like crystalline embryo is formed in the super cooled liquid, while for the $x < 0.18$ the time for crystallization becomes extremely long compared to the experimental time period. The change in the tendency for crystallization is considered as evidence for the rigidity threshold at $\langle r \rangle = 2.4$ ($x = 2.0$) predicted by the mean-field constraint theory [27, 30].

7. Electronic Structure in Ge_xSe_{1-x}

Figure 21 shows a series of valence-band ultraviolet phoemission spectroscopy (UPS) and conduction-band inverse-photoemission spectroscopy (IPES) spectra of g–Ge_xSe_{1-x} with x from 0 to 0.33. The UPS and IPES spectra for $x = 0$ (g–Se) exhibit structures at -6.4 and -5.4, -2.7, 3.1 and 7.4 eV due to the Se $4p$ bonding, $4p$ lone pair(LP), $4p$ antibonding and $4d$ and/or $5s$ states,

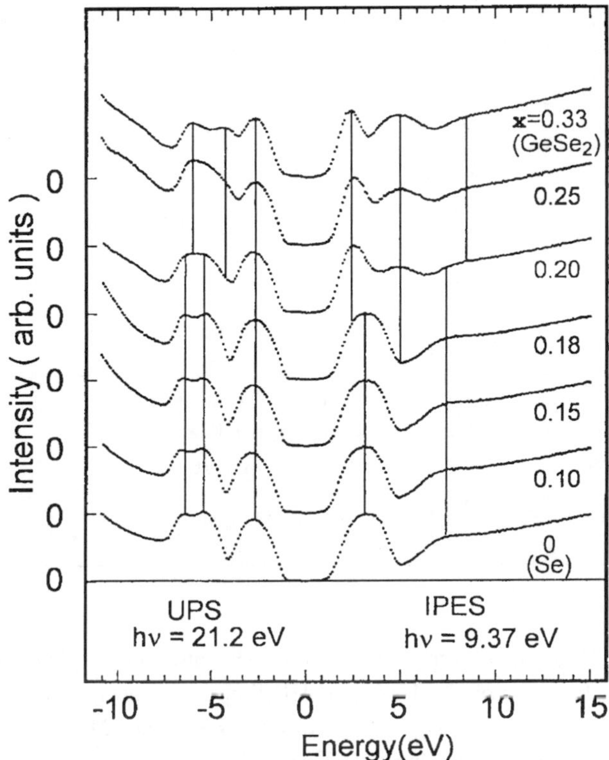

Fig. 21. A series of the valence-band UPS and conduction-band IPES spectra of g–$Ge_x Se_{1-x}$ with x from 0 to 0.33. Intensities of the UPS and IPES spectra are tentatively normalized at -2.7 eV and 3.1 ($x = 0$, 0.10, 0.15 and 0.18) and 2.4 eV ($x = 0.20$, 0.25, and 0.33), respectively. Vertical bars indicate the positions of structures. Energies are referred to the Fermi level [31].

respectively [41], while those for $x = 0.33$ (g–$GeSe_2$) represent peaks at -6.0 and -4.3, -2.7, 2.4 and 4.9, and 8.4 eV originating mainly from the Ge sp^3–Se p bonding, Se p LP, Ge sp^3–Se p antibonding states, and $4d$ and /or $5s$ states of Ge and Se atoms, respectively [42], as summarized in Table 1. With increasing x from 0 to 0.18, the UPS and IPES spectra do not show any noticeable change with respect to their spectral shape and energy positions of structures. At $x = 0.2$, however, one can recognize a slight blurring of the structures at -6.4 and -5.4 eV and increasing components around -6.0 and -4.3 eV for the UPS spectrum. A remarkable change is also observed for the

Table 1. Energy positions of peak structures in the UPS and IPES spectra of g–Se and g–GeSe$_2$, and electronic states which contribute predominantly to the peak structures [31].

	Energy (eV)	Electronic states
	8.4	$4d$ and/or $5s$ states of Ge and Se atoms
	4.9 ⎫	
	2.4 ⎭	Ge sp^3–Se $4p$ antibonding states
g–GeSe$_2$	−2.7	Se $4p$ lone pair states
	−4.3 ⎫	
	−6.0 ⎭	Ge sp^3–Se $4p$ bonding states
	7.4	Se $4d$ and/or $5s$ states
	3.1	Se $4p$ antibonding states
g–Se	−2.7	Se $4p$ lone pair states
	−5.4 ⎫	
	−6.4 ⎭	Se $4p$ bonding states

shape of the IPES spectrum as well as for the number of peaks. For further increase of x, features of the UPS and IPES spectra reach those of g–GeSe$_2$.

The UPS spectra for $x = 0.20$ and 0.33 in Fig. 21 are in qualitative agreement with spectra obtained by X-ray photoemission spectroscopy (XPS) [43, 44]. From modifications of the Ge 4s and Se 4s bands with x, these XPS studies emphasize that the 4(Ge):2(Se) coordination is manifested [43, 44] and chemically ordered structural units are favored [43].

The composition dependence of the UPS and IPES spectra in Fig. 21 will be related to the Phillips–Thorpe rigidity percolation theory [30, 45, 46]. In g–Ge$_x$Se$_{1-x}$ with $x < 0.2$, there exist Se chains and Ge(Se$_{1/2}$)$_4$ tetrahedral units, where Ge atoms act as cross-links between Se chains. For x above 0.2, the Ge(Se$_{1/2}$)$_4$ units grow to form fragments of the two-dimensional form of GeSe$_2$.

Such a transition from one kind of network to another shows up clearly in the UPS and IPES spectra near $x = 0.2$. One notices also that the present results correlate well with those of Raman scattering [3], Mössbauer experiments [47–49] and molar volumes [32]. The present UPS and IPES studies of g–Ge$_x$Se$_{1-x}$ provide a direct experimental finding of the structural transformation relating to the rigidity percolation threshold near $x = 0.2$.

8. Light Induced Crystallization

The glass is practically stable well below the glass transition temperature T_g. By light irradiation, the structure of chalcogenide glasses is easily changed as is evidenced by phenomena of photo-darkening, photo-bleaching, photo-crystallization, photo-doping etc. Light induced crystallization (LIC) is not only an intriguing photo-induced phenomena, but also is important for investigations of the structural relation between the glass and the crystal, as well as for technological applications. Light induced crystallization of $GeSe_2$ system has been studied by Raman scattering at first by Griffiths *et al.* [22, 50] and then by a few authors [11, 51–55]. Important difference of the light induced process from the thermal recrystallization of glass has been recently clarified [2, 11]. Light induced crystallization processes are investigated in details in slowly evaporated a–$GeSe_2$ films in comparison with the purely thermal annealing processes.

Glass-to-crystal change was observed in time resolved Raman spectra in every 1 minute with the excitation light (5145 Å) from an Ar ion laser, which was focused onto a film surface with a typical diameter 0.03–0.05 mm and the incident power 5-60 mW. The penetration depth of the light was comparable with the film thickness 600 nm. Experiments were made at several temperatures in a furnace (480 K), in the atmosphere (300 K), and in a cryostat (90 K and 30 K). The structure of glassy germanium chalcogenide films is known to be micro-heterogeneous [2]. At STP, a layered crystalline form (2D), which is composed of corner shared $GeSe_4/2$ chains and edge shared $GeSe_{4/2}$ bridges between the chains, is stable in c–$GeSe_2$ system, whereas there are two forms in c–GeS_2 [8, 9], i.e. a three dimensional form (3D) containing only the corner shared tetrahedra (low temperature phase) and the layered form (high temperature phase, 2D).

In light induced crystallization of a–$GeSe_2$, there are practically two types of crystallization [11]. In type A, only the 2D form grows as shown in Fig. 22(a). On the other hand, in type B, both of the 2D and the 3D form can grow [Fig. 22(b)]. The type A is more frequently observed than the type B. The 2D form is much more favorably produced in the light-induced process. By thermal annealing in the dark just above the glass transition temperature of the film $\sim 300°C$, the 3D form as compared with the 2D form, has a tendency to grow at high rate, while the 2D crystal growth becomes dominant at high temperature say $425°C$ (Fig. 23) [56]. This trend is consistent with the fact that the 3D form is the low temperature phase in GeS_2.

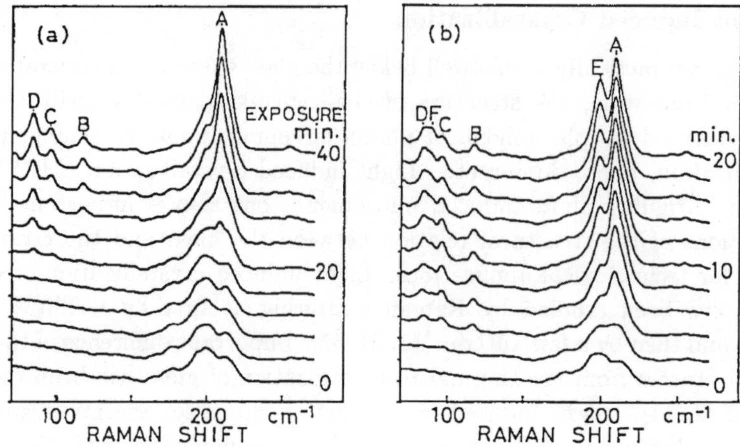

Fig. 22.   Time-resolved Raman spectra of a–GeSe$_2$; (a) type A and (b) type B for the crystallization are shown. Lines A, B, C, and D are due to the 2D form, and lines E and F are due to the 3D form. Exposure times are also shown.

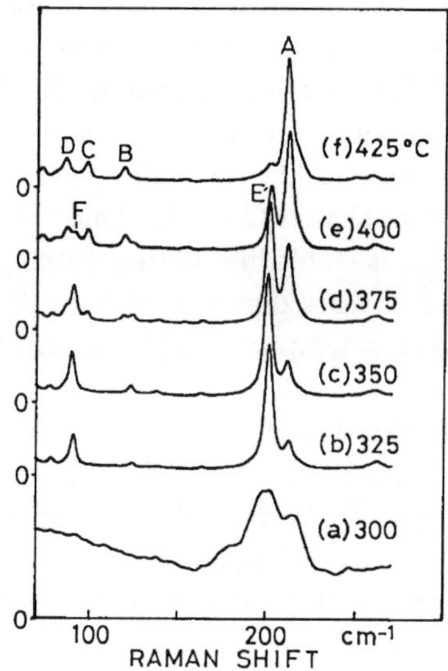

Fig. 23.   Raman spectra in GeSe$_2$ after thermal annealing of a–GeSe$_2$ in the dark for 18 hours in Ar gas. Annealing temperatures are shown in the figure. Lines A, B, C, and D are due to the 2D form and lines E and F are originated from the 3D form. Spectrum (a) is similar to that of as grown a–GeSe$_2$. The spectra are normalized at the maximum peaks.

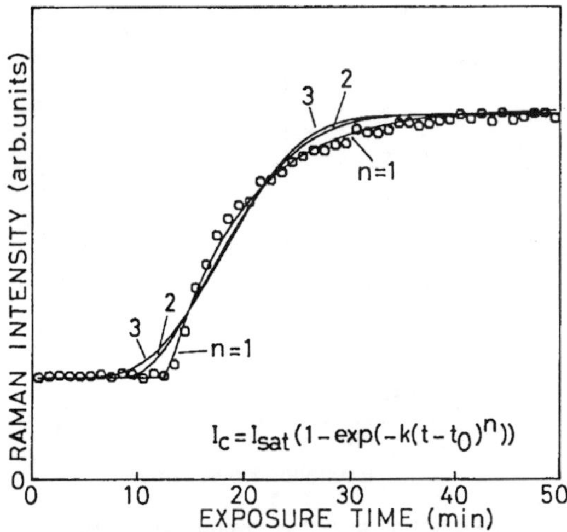

Fig. 24. Growth curve of the 2D crystal in light-induced crystallization type A by 22 mW incident laser power. The intensity at the A peak position are plotted. Solid lines illustrate the empirical curves; $I_0[1 - \exp(-k(t - t_0)^n)]$. The best fit is obtained for $n = 1$.

Also in $GeSe_2$, the 3D form will be stable at low temperatures below T_g. In the type A case, the intense peak grows at about 210 cm^{-1} which corresponds to the A mode of c–$GeSe_2$ (2D). The growth curve obeys nearly a $1 - \exp(-k(t - t_0))$ time dependence after a latent period t_0 as shown in Fig. 24. The temperature dependence of the Raman peak position of the laser induced microcrystals (LIMC) is almost the same as that of the bulk single crystal (c–$GeSe_2$) [56]. The latent period critically becomes very long with decreasing excitation power. There is no recognizable difference between the thresholds in the type A and B within experimental error. In order to investigate interplay between the thermal and photo-excitation effects in the light-induced crystallization, the threshold laser power was studied at several environmental temperatures. The latent period vs. excitation power at several environmental temperatures is shown in Fig. 25. The threshold is approximately 6 mW at 480 K, 14 mW at 300 K, 35 mW at 90 K and 60 mW at 30 K; it becomes increasingly high at lower environmental temperatures [56]. Local temperature at LIMC can be estimated by the Raman shift [54, 55]. It should be pointed out that the importance of the temperature shift of Raman spectra was overlooked in Refs. [22, 50–53]. Although the sample surface was cooled by the cooled gas

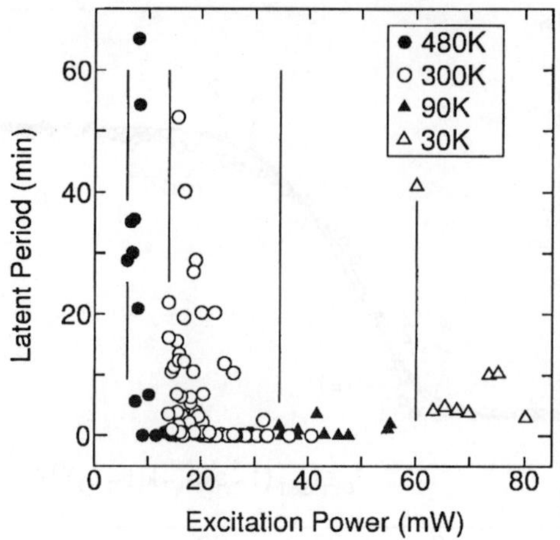

Fig. 25. Latent period vs. excitation power at several temperatures described in the inset. Threshold power level becomes increasingly high at lower environmental temperatures. Local temperatures at the threshold are $420 \sim 500$ K slowly depending on the environmental temperatures.

flow at 90 K and 30 K, the local temperature near the threshold was around 420 K, which is near the local temperature 450 K in the experiments at 300 K and much less than the glass transition temperature T_g (600 K). The thermal excitation near to 450 K may be necessary for the LIMC growth. It should be stressed, however, that crystallization does not occur by purely thermal annealing in the dark at least below 600 K. Both photo- and thermal-excitation are important in the laser-induced crystallization.

The distinction of the threshold power between the types A and B is not recognizable as mentioned above. It seems that both types appear rather independently of the laser power or environmental temperatures. This suggests that there might be precursors of the 2D and the 3D forms (MRS) in the amorphous films. In the evaporation process from the gaseous phase at high temperatures, the topological structure like the high temperature form (2D) will be kinetically frozen in rather than that like the 3D form, though the 3D form is stable at low temperatures. Thus the reason of the result that the type A appears much more frequently than the type B will be that the 2D like precursor is the majority in the grown glass. Since the temperature is sufficiently low in the laser-induced process, the critical nucleus of the 2D crystal growth may be produced from the 2D precursors without the mutual conversion between the 2D and the 3D precursors. On the other hand, in the thermodynamic process during the thermal annealing above T_g in the dark, the conversion might be efficient, and so, the equilibrium phase at annealing temperature will grow dominantly. The idea of the 2D precursor was supported by the previous optical study [16]. It should be noted that the rate of occurrence of the type A becomes comparable to that of the type B in amorphous films prepared by thermal annealing for 18 h at 300°C. This suggests that the medium range structure changes in thermodynamically favorable way at this temperature.

Next, one imagine what happens during the latent period. By the photoexcitation and relaxation processes, bond rearrangements will be induced, causing spatial extension of distorted topologically 2D clusters with average temperature raise up to 450 K. Then medium or mesoscopic range strain relaxation proceeds together with macroscopic strain relaxation around the light spot on the film. Cluster layers could rotate to grow the interlayer correlation with van der Waals interactions resulting in micro crystalline nuclei.

Microscopically, the light induced processes can be imagined in the following way. Electrons in the top of the lone pair band of Se or Se–Se sites or in Ge–Ge tail states will be excited by the band-gap tail photons, creating localized excitons. The excitons strongly interact with phonons with resulting large stokes shift by ~ 1 eV. Local heating may occur especially at the cluster edges, which favors the evolutional processes including bond rearrangement in the latent period. It is not probable that the steady state excitons density is high enough [52, 53] to cause directly the valence bond instability, since the exciton life time is too short at the local temperature 420 K. Overall structural change in the amorphous state during the latent period could not be detected in the Raman spectra, but appearance of the crystal line is clearly realized even in less than 1% of the saturated value. Once the crystalline nuclei are formed, crystals grow around there by the time evolution $1 - \exp(-k(t - t_0))$, which suggests that the total growth rate only linearly depends on the remaining amorphous volume at the irradiated portion. There should be many nuclei in the amorphous matrix around which crystals grow.

In summary, by laser light irradiation in a–$GeSe_2$, the crystallization proceeds even well below the glass transition temperature. In the light induced process in as-deposited a–$GeSe_2$, the 2D crystalline formation is much more favorable than 3D, rather independently of the laser power and temperatures. This suggests that the 2D like precursor is frozen in during the glass formation. The crystal growth curve followed a simple exponential curve after a latent period, which divergently increased in the vicinity of the threshold laser power. The threshold became increasingly high at lower environmental temperatures, but the local temperature near the threshold was around 450 K which was much less than T_g and irrespective of the environmental temperatures. The rate of light induced 3D formation increased in the amorphous film after thermal annealing just below T_g. This suggest that the medium range structure changes in thermodynamically favorable way, as is evidenced by the purely thermally induced 3D formation by annealing just above T_g. It should be noted that the crystallization corresponds to the thermotype or "light-type" of the medium range structure in the glass in a similar way as that "letters written by invisible ink are read out by heat or special light."

9. Low-Frequency Raman Spectra and Fracton

Low energy excitations in disordered materials presents universal features such as a plateau in their low temperature thermal conductivity and an appearance of a Boson peak in Raman and inelastic neutron scattering spectra. In this section, this problem has been tackled with Raman scattering measurements and the concept of fracton dynamics [58]. The extensive experimental investigations of Raman scattering from fractal structure have been carried out on aerogels [59, 60]. Besides studies on aerogels, low frequency Raman properties of disordered materials have been reported for polymers, such as PMMA [61] and glasses, such as borate glass [62] or amorphous arsenic [63]. Theoretical works for Raman scattering from fractal structure are also vigorously made [64, 65].

In glassy systems, the intensity of Raman scattering $I(\omega)$ at the frequency $\omega/2\pi$ is usually written in the form [66],

$$I(\omega, T) = C(\omega)g(\omega)[n(\omega, T) + 1]/\omega \tag{5}$$

where $n(\omega, T) + 1$ is the Bose factor for the Stokes scattering, $g(\omega)$ the vibrational density of states (VDOS), and $C(\omega)$ the coupling between the light and lattice vibration. The function $C(\omega)$ can be determined by comparison of inelastic neutron and Raman scattering spectra. It has been found that $C(\omega) \propto \omega$ for disordered material of various types [67]. This relation holds in the low frequency region above the Boson peak position. By taking into account a relation of $C(\omega) \propto \omega$, the Bose factor reduced Raman intensity $I_R(\omega)$ is proportional to VDOS in the low frequency region and is independent of temperature,

$$I_R(\omega) \equiv \frac{I(\omega, T)}{n(\omega, T) + 1} \propto g(\omega) \tag{6}$$

where a certain ω-dependence of VDOS, $g(\omega) \propto \omega^{\tilde{d}-1}$, refer to a fracton behavior [58]. The value of $\tilde{d}$, which is referred to as spectral dimensionality, is not an integer.

Figure 26 shows the $\log - \log$ plots of the reduced Raman spectra I_R of the $\mathrm{Ge}_x\mathrm{Se}_{1-x}$ glasses [68]. For the Se-rich samples ($x = 0.02, 0.15$), each spectrum is fitted by three kinds of straight lines and for the Ge-rich samples ($x = 0.33, 0.36$) by two kinds of straight lines. It is considered that the line #1 represents the Debye behavior ($g(\omega) \propto \omega^2$) and lines #2 $\sim$ #4 correspond to the fracton behavior. The slope of the line #1, however, is not an integer value. It may be due to the ω-dependence of $C(\omega)$ which is no longer proportional to ω in the very low frequency region below the Boson peak position.

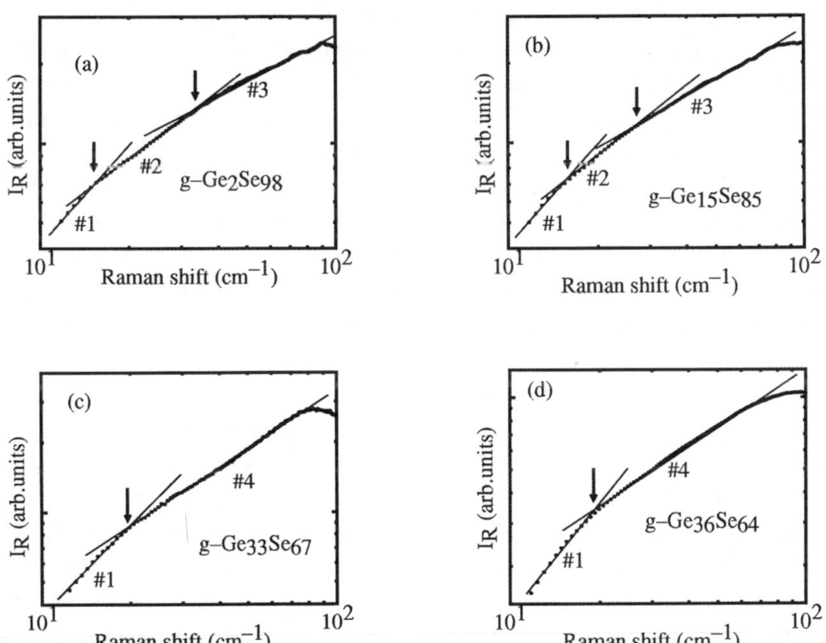

Fig. 26. The $\log - \log$ plots of reduced Raman spectra $I_R(\omega)$ of $\mathrm{Ge}_x\mathrm{Se}_{1-x}$ glasses in the low-frequency region. The spectrum of the Se-rich sample (a), (b) is fitted by three straight lines, while that of the Ge-rich sample (c), (d) is fitted by two straight lines.

It has been found that the log − log plot of the reduced Raman intensity is fitted by two or three straight lines depending on the composition ratio. Two kinds of low frequency properties will be discussed in terms of vector elasticity [57, 69, 70]. It has been pointed out that the vector nature of atomic displacement is essential for describing the fracton excitation in real materials. In the vector elasticity model, the potential energy is given by Refs. [57, 69]

$$V = \frac{1}{2}\alpha \sum_{\langle ij \rangle} g_{ij}[(\vec{u}_i - \vec{u}_j) \cdot \vec{r}_{ij}]^2 + \frac{1}{2}\beta \sum_{\langle ijk \rangle} g_{ij}g_{ik}(\delta\theta_{jik})^2 . \tag{7}$$

Here, $\vec{u}_i$ is the vector displacement of atom i, $\vec{r}_{ij}$ the directional unit vector between nearest neighbors $\langle ij \rangle$, $\delta\theta_{jik}$ the change in the angle between bond $\langle ij \rangle$ and $\langle ik \rangle$ due to the displacements of atoms; $g_{ij} = 1$ for the bonds that are occupied and $g_{ij} = 0$ for the bonds that are empty, and α and β are the bond stretching and bond bending force constants, respectively.

Two crossover lengths ξ and l_c independently are supposed to exist in the Ge–Se network glass as discussed in the followings. The length ξ is the *structural* crossover length. A characteristic length of interest is L, for example the wavelength of the excitation. When the relation of $L > \xi$ holds, the structure is regarded as homogeneous. A vibrational excitation whose wavelength equals to L belongs to extended modes (acoustic phonons). In the case of $L < \xi$, the structure is considered to be fractal, and a vibrational excitation of wavelength L is strongly localized (fracton). The length l_c is the *elastic* crossover length where the bond stretching motion is energetically favorable ($L \ll l_c$) or the bond bending motion becomes dominant ($L \gg l_c$). This crossover length l_c is determined from the ratio between the bond bending force constant β and the bond stretching force constant α as $l_c \propto (\beta/\alpha)^{1/2}$ [57, 69]. When $\xi < l_c$, the low frequency vibrational properties are dominated by acoustic phonon and stretching fracton as shown in Fig. 27(a). This situation corresponds to the case of the Ge-rich samples. On the other hand, in the case of $\xi > l_c$, bending fracton contributes to the low frequency vibrational properties besides acoustic phonon and stretching fracton as illustrated in Fig. 27(b). This relation causes the low frequency properties of the Se-rich samples. Accordingly, the origins of #3 and #4 straight lines are both due to stretching fracton modes.

As shown in Fig. 28, the spectral dimensionalities of bending fractons, $\tilde{d}_b$ (#2), have almost constant value, whereas those of stretching fractons, $\tilde{d}_s$

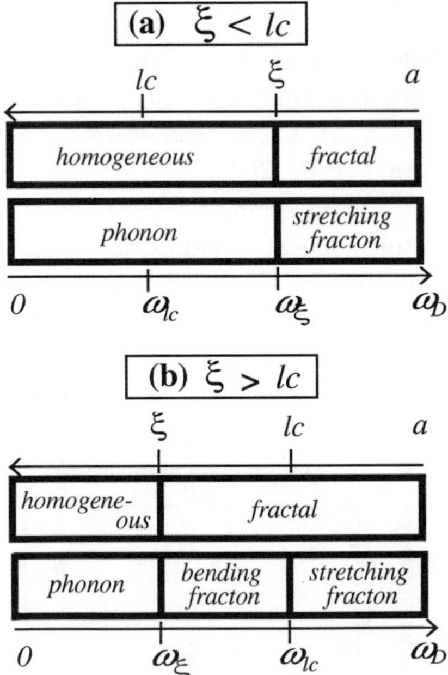

Fig. 27. Schematic diagram showing the correspondence between length scales and frequency (after Ref. [57]). (a) In the case of $\xi < l_c$, vibrational excitations in a low-frequency region are acoustic phonons and stretching fractons. (b) In the case of $\xi > l_c$, they are acoustic phonons, stretching fractons, and bending fractons.

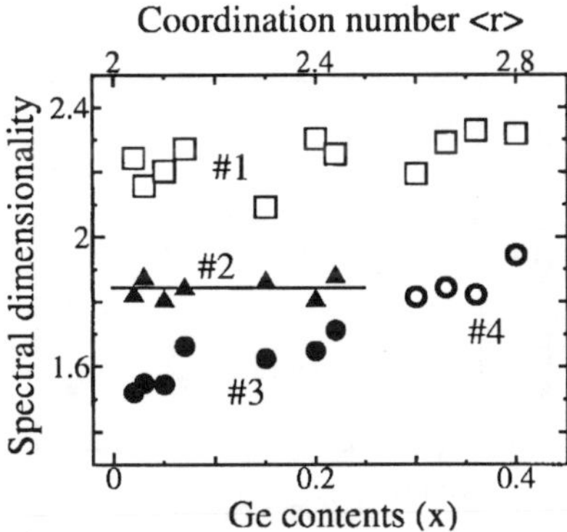

Fig. 28. The spectral dimensionalities obtained by the slope of the lines (#1–#4). The data of #1 are supposed to correspond to Debye behavior. The data of #2 are due to the bending fracton and those of #3 and #4 are due to the stretching fracton.

(#3 and #4), linearly increase with the germanium contents. In all cases, the relation $\tilde{d}_{b,s} < 2$ holds, which is a condition that excitations are localized. It is noteworthy that the relation of $\tilde{d}_b > \tilde{d}_s$ is realized at any composition. This relation contradicts a prediction by a scaling analysis for a percolating network, which results in $\tilde{d}_b < \tilde{d}_s$ [70].

In Ge_xSe_{1-x} glasses, Se-rich samples are 1D-like glasses mainly composed of Se chains and the stoichiometric composition $GeSe_2$ has 2D-like network structure. And it is reported that the GeS_2 glass has 1D-like network structure from the experiments on the sensitivity to pressure [71]. $Ge(S_xSe_{1-x})_2$ ternary glasses, whose network dimensionality can be controlled by varying

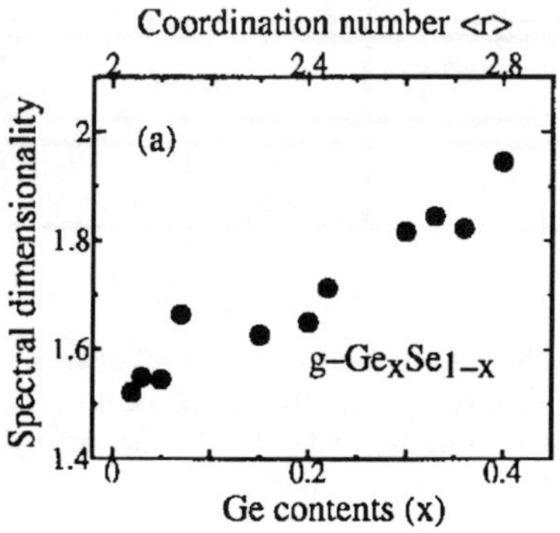

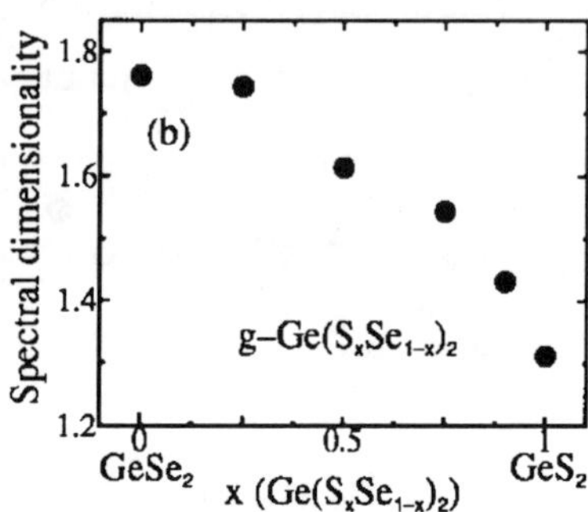

Fig. 29. The behavior of $\tilde{d}_s$ of (a) Ge_xSe_{1-x} glasses and (b) $Ge(S_xSe_{1-x})_2$ glasses. The value of $\tilde{d}_s$ of Ge_xSe_{1-x} glasses increases with Ge composition ratio and that of $Ge(S_xSe_{1-x})_2$ glasses decreases with S composition ratio.

the composition, is a suitable system to inquire further into the contradicted result, the relation between the $\tilde{d}_b$ and $\tilde{d}_s$. In $Ge(S_xSe_{1-x})_2$ glasses, two types of the low frequency lattice vibrations were observed which are attributed to the acoustic phonon and stretching fracton. The value of $\tilde{d}_s$ of $Ge(S_xSe_{1-x})_2$ glasses decreases with S composition ratio as shown in Fig. 29(b). These experimental results on spectral dimensionality of stretching motion suggest us the relation between spectral and actual dimensionality of Ge chalcogenide glasses.

Instead of using the critical exponents, it is fairly reasonable to adopt the facts deduced from the present experimental results. The diffusive volume, $V(t)$, is considered during the time t in the diffusion problem [72] and it is supposed that the $V(t)$ closely relates to the extensible range of localized vibrational modes. The probability of finding the particle at the origin at time t, $P(0,t)$ with $P(0,0) = 1$, relates to the VDOS through the Laplace transform as

$$P(0,t) = \int_0^\infty g(\varepsilon)\exp(-\varepsilon t)d\varepsilon \tag{8}$$

where $\varepsilon = \omega^2$. The $\tilde{d}_s$ is defined through $P(0,t)$ by $P(0,t) \sim t^{-\tilde{d}_s/2}$. Thus we have the relation:

$$V(t) \sim \frac{1}{P(0,t)} \sim t^{\tilde{d}_s/2}. \tag{9}$$

In uniform systems, it is naturally expected that the $V(t)$ has a following relation, $V(t) \sim t^{d/2}$. The diffusive volume $V(t)$ depends on the Euclidean dimension d.

Extending the relation of $V(t) \sim t^{d/2}$ in Euclidean space to $V(t) \sim t^{\tilde{d}_s/2}$ deduced from low-frequency Raman result of $g(\omega) \propto \omega^{\tilde{d}_s-1}$, the value of $\tilde{d}_s$ can be regarded as a parameter describing the degree of localization of lattice vibrations. The small value of $\tilde{d}_s$ corresponds to the "strong" localization and the lattice vibrations are confined to "narrow" network structure (low-dimensionality). By a same way, the large value of $\tilde{d}_s$ represents the "weak" localization and the lattice vibrations are extended to "wide" network structure (high-dimensionality). The spectral dimensionality of stretching motion $\tilde{d}_s$ can be accepted as an indicator of network dimensionality of disordered materials. It is concluded that the localization degree of lattice vibration depends on the network dimensionality of the glass, which is indicated by the width of the main structure of the glass. This argument is greatly suggestive of the Anderson localization of electrons in random potentials. It is emphasized again that the present arguments are *not* in need of the critical exponents discussed in Ref. [72].

In Fig. 28, the spectral dimensionality of bending fractons extends to $\langle r \rangle = 2.44$ which gives a similar result based on Raman investigations [34]. It has been claimed that the rigidity threshold in Ge–Se glasses occurs at $\langle r \rangle = 2.46$. While, Wang *et al* [40, 73] and Taniguchi *et al* [31] reported that the rigidity threshold occurs at $\langle r \rangle = 2.40$ or less based on their experimental result. Wang *et al.* investigated the temperature dependence of the structural changes in Ge–Se glasses by Raman scattering. Taniguchi *et al.* investigated the valence-band UPS and conduction-band IPES spectra of Ge–Se glasses.

Recently, Phillips discussed the relation between β/α ratio and the constraint theory of network glass [74] to explain the Uebbing and Sievers' data [33] for the relaxation rate γ of the H_2O symmetrical stretching mode. For $\langle r \rangle > 2.4$, β bending constraints has been replaced by α stretching constraints, which causes an increase in the concentration of α localized mode. In the present experimental data in Fig. 28, the contribution of bending motion seems to vanish in overconstrained glasses, which seems to reconsile his arguments.

In summary, the low frequency Raman scattering observed for germanium selenide glasses has been discussed in terms of fracton dynamics and the notion of rigidity percolation. The excited vibrational modes of Ge_xSe_{1-x} glasses in the low frequency region are acoustic phonon and fracton. Especially, the contribution of bending-fracton to the low frequency region appears only in the underconstrained glasses. These phenomena have been discussed with using the concept of the vector elasticity.

10. Rigidity Percolation and Fragility

Studying the relationship between the rigidity percolation and the fragility, a degree depending on the departure from the Arrhenius behavior of the average of the relaxation time, is an advantageous approach to the physical nature of glassy system [35]. In this section, a relative degree of fragility of the glassy Ge_xSe_{1-x} is given by an investigation of the line shape of low-frequency Raman spectra (3–100 cm^{-1}), and the correlation between the fragility and the rigidity percolation is also discussed [85].

Figure 30 shows the reduced Stokes-Raman intensity, $I(\omega) = I(\omega)_{\text{exp}}./\omega[n(\omega)+1]$, of the glassy Ge_xSe_{1-x} at selected composition from 3 to 100 cm^{-1}. The temperature factor $\omega[n(\omega) + 1]$ is $\omega[1 - \exp(-\hbar\omega/kT)]^{-1}$. In glassy materials, the low-frequency Raman scattering arose from two kinds of contributions: [66] quasielastic scattering, which is usually ascribed to some kind of

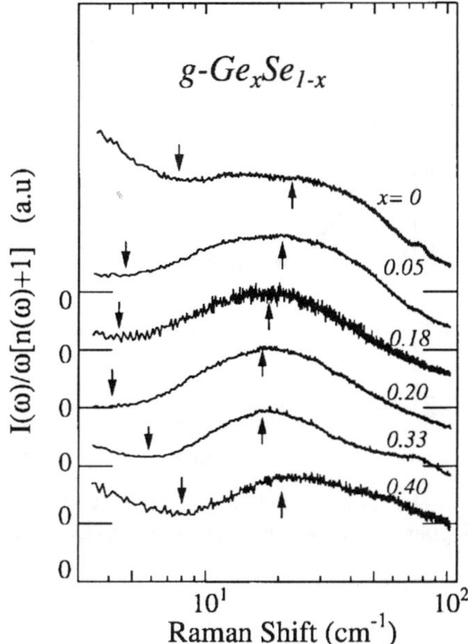

Fig. 30. Depolarized low-freqencey Raman scattering spectra for different composition of Ge$_x$Se$_{1-x}$. The spectra normalized to the intensity at the boson peak maximum (pointed by up arrows) around 20 cm^{-1}. Down arrows point the minimum discussed in the text.

relaxational motion, and the so-called *boson peak*, a vibrational motion. The boson peak which is a universal characteristic of glasses, as shown in Fig. 30 at about 20 cm^{-1}, is suggested to be a reflection of the peculiarities of the vibrational density states and not of the light-to-vibrations coupling coefficient determined by a comparision of Raman- and neutron-scattering spectra [75]. It has been recently shown [76] that the quasielastic scattering, dominating at frequencies lower than the boson peak, has a relationship to the β process analyzed within a mode coupling theory. In Fig. 30 the relaxational contribution decreases with ω, while the vibrational contribution increases with ω up to the maximum of the boson peak. Thus, a minimum, below the boson peak, appears due to the overlap of these two contributions. Sokolov *et al* [77] recently used the ratio, I_{min}/I_{max} (defined as the ratio between the intensities at the minimum and that at the boson peak maximum), at the glass-transition temperature to describe the fragility (i.e. departure from the Arrhenius viscosity or relaxation time behavior). They reported that the ratio, I_{min}/I_{max}, increased with the fragility basing on an analysis of a number of glass formers.

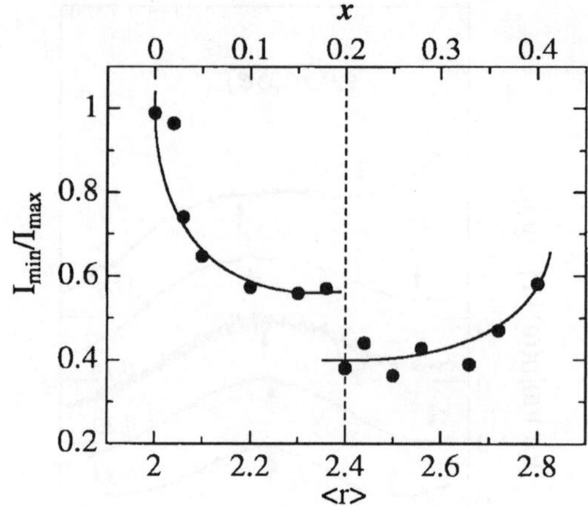

Fig. 31. $\langle r \rangle$ dependence of the ratio defined as the intensity at the minimum to that at the boson peak maximum. Upper axis shows the corresponding Ge content. The lines are drawn as a guide to the eye.

Figures 31 shows the ratio, $I_{\min}/I_{\max}$, changing with the average coordination number, $\langle r \rangle$. One can sort these changes into three parts according to the ranges of $\langle r \rangle$: (1) in the $\langle r \rangle$ range of 2.0–2.2, the ratio decreases rapidly with the $\langle r \rangle$ increasing; (2) in the $\langle r \rangle$ range of 2.2–2.6, there is a rapid drop at $\langle r \rangle = 2.4$ which is the rigidity percolation threshold value; (3) in the $\langle r \rangle$ range of 2.6–2.8, the ratio increases slightly. The notable feature of the ratio, $I_{\min}/I_{\max}$, holding a minimum at $\langle r \rangle = 2.4$–2.6 is in agreement with the result [35] that: the fragility, studied basing on the viscosity and on the changes in thermodynamic properties (i.e. excess heat capacity) at T_g, showed the minimum at the rigidity percolation threshold, $\langle r \rangle = 2.4$. Although the mechanism of light scattering at these frequencies is still controversial and this work were made at room temperature, it seems reasonable to suppose that the present ratio, $I_{\min}/I_{\max}$, qualitatively reflects the fragility of the Ge_xSe_{1-x} glasses.

In the equilibrium phase diagram [26], the eutectic point being at $x = 0.08$ indicates that the $GeSe_{4/2}$ tetrahedra will mainly determine the structural properties of the glasses in the range of $x > 0.08$. The rapid decrease of the fragility in the $\langle r \rangle = 2.0$–2.2, coresponding to the Ge content $x = 0$–0.10, is related to an enlargement of population of the randomly distributed $GeSe_{4/2}$ tetrahedra whose structure is more stronger than that of the Se–Se chains.

It has been discussed that the bulk elastic constants in Ge_xSe_{1-x} glasses displaying a threshold behavior close to the $\langle r \rangle = 2.67$ ($x = 0.33$) is a representation of chemical effect [49]. It is possible that in the Ge-rich region with the Ge–Ge bond increasing the structural disorder will also increase. The slight increase of the fragility in the $\langle r \rangle > 2.6$ seems to reflect some kinds of structural disorder increases.

In an undercoordinated network glass ($\langle r \rangle < 2.4$), the mean-field constraint theory predicts the existence of the so-called *floppy*-mode, a low-frequencey or zero-frequencey vibrational excitation [27, 30]. The inelastic neutron scattering measurements [78] directly evidence the floppy modes existing below the $\langle r \rangle = 2.4$ in the Ge_xSe_{1-x} glasses. The reported floppy modes should influence the Raman spectra particularly in the low-freqencey region through the mask of light-to-vibrations coupling coefficient, $C(\omega)$. As shown in Fig. 31, the rapid drop at $\langle r \rangle = 2.4$ points out a discontinuity of the fragility at the rigidity percolation threshold where the glass changes from *floppy* to *rigid* structures. The decrease of the low-frequency intensity ratio, $I_{\min}/I_{\max}$, can have two explanations: either a strong boson peak or a weak relaxational distribution which reduces the intensity at the minimum between the Rayleigh line and the boson peak, or both. The mechanism of light scattering at these frequencies is still controversial. Nevertheless, the behavior of the intensity ratio, $I_{\min}/I_{\max}$, at the perocolation threshold, $\langle r \rangle = 2.4$, is reasonable to be regarded as an important evidence of the *floppy*-mode existing in the undercoordinated and disappearing in the overcoordinated network glasses, Ge_xSe_{1-x}.

11. Concluding Remarks

In this chapter, structural features in typical network glasses of chalcogenides, mainly Ge–Se system, have been introduced in relation to rigidity percolation, fractal dimension, medium range structure, network dimensionality, fragile/strong characters and light-induced crystallization.

Recently it has been found by neutron scattering in a wide range of energy–momentum space of vitreous silica that an acoustic phonon dispersion in quasi Brillouin zones coexists with a localized mode, so called Boson peak in the low energy region [79]. Existence of three types of low energy modes due to acoustic excitation has been predicted by a simple mode in network-forming glasses; these are weakly, strongly and mesoscopically localized modes [80]. It has been discussed that the strongly localized mode is responsible for the boson peak observed in Raman and neutron scattering, the low temperature thermal conductivity plateau and specific heat [81–83]. It should be noted that low energy excitations in water has been analyzed by modified simple model analysis [84]. It should be stressed that riches of universal phenomena and individual characters of glasses and liquid in wide temperature range are increasingly attractive for us together with recent advance of new experimental techniques and computer physics in future.

Acknowledgments

The author wishes to thank T. Fukunaga, K. Inoue, O. Matsuda, Y. Wang and M. Nakamura for their collaboration and discussions. He is grateful to Prof. Taniguchi group in Hiroshima University for their cooperation and discussions in the electronic structures in chalcogenide glasses.

References

[1] M. Cardona, *SPIE Raman and Luminescence Spectroscopy in Technology* **822**, 2 (1987).

[2] K. Murase and T. Fukunaga, *Mat. Res. Soc. Symp. Proc.* **61**, 101 (1986).

[3] K. Murase, T. Fukunaga, K. Yakushiji, T. Yoshimi and I. Yunoki, *J. Non-Cryst. Solids* **59 & 60**, 883 (1983).

[4] P.M. Bridenbaugh, G.P. Espinosa, J.E. Griffiths, J.C. Phillips and J.P. Remeika, *Phys. Rev.* **B20**, 4140 (1979).

[5] R.J. Nemanich, F.L. Galeener, J.C. Mikkelsen, Jr, G.A.N. Connell, G. Etherington, A.C. Wright and R.N. Sinclair, *Physica* **B117 & 118**, 959 (1983).

[6] G. Lucovsky, C.K. Wong and W.B. Pollard, *J. Non-Cryst. Solids* **59 & 60**, 839 (1983).

[7] V.G. Dittmar and H. Schäfer, *Acta Cryst.* **B32**, 2726 (1976).

[8] V.G. Dittmar and H. Schäfer, *Acta Cryst.* **B31**, 2060 (1975).

[9] V.G. Dittmar and H. Schäfer, *Acta Cryst.* **B32**, 1188 (1976).

[10] K. Inoue, O. Matsuda and K. Murase, *Solid State Commun.* **79**, 905 (1991).

[11] K. Inoue, K. Kawamoto and K. Murase, *J. Non-Cryst. Solids* **95 & 96**, 517 (1987).

[12] Z.V. Popović and H.J. Stolz, *Phys. Stat. Sol. (b)* **108**, 153 (1981).

[13] O. Matsuda, K. Inoue and K. Murase, *Proc. 20th Int. Conf. Phys. Semicond., Thessaloniki, Greece*, eds. E.M. Anastassakis and J.D. Joannopoulos (World Scientific, Singapore, 1990) p. 2135.

[14] O. Matsuda, K. Inoue and K. Murase, *Solid State Commun.* **75**, 303 (1990).

[15] K. Inoue, O. Matsuda and K. Murase, *J. Non-Cryst. Solids* **150**, 197 (1992).

[16] K. Inoue, T. Katayama, K. Kawamoto and K. Murase, *Phys. Rev.* **B35**, 7496 (1987).

[17] P.N. Keating, *Phys. Rev.* **145**, 637 (1966).

[18] D.A. Long, *Proc. Roy. Soc.* **A217**, 203 (1953).

[19] O. Matsuda, K. Inoue, T. Nakane and K. Murase, *J. Non-Cryst. Solids* **150**, 202 (1992).

[20] S. Hashimoto and M. Itoh, *Jpn. J. Appl. Phys.* **95 & 96**, 726 (1987).

[21] K. Murase, K. Inoue and O. Matsuda, *Current Topics in Amorphous Materials: Science and Technology*, eds. Y. Sakurai, Y. Hamakawa, T. Masumoto, K. Shirae and K. Suzuki (Elsevier, Amsterdam, 1993) p. 47.

[22] J.E. Griffiths, G.P. Espinosa, J.P. Remeika and J.C. Phillips, *Phys. Rev.* **B25**, 1272 (1982).

[23] S. Susman, K.J. Volin, D.G. Montague and D.L. Price, *J. Non-Cryst. Solids* **125**, 168 (1990).

[24] I.T. Penfold and P.S. Salmon, *Phys. Rev. Lett.* **67**, 97 (1991).

[25] K. Inoue, O. Matsuda and K. Murase, *Physica* **B219 & 220**, 520 (1996).

[26] P. Tronc, M. Bensoussan and A. Brenac, *Phys. Rev.* **B8**, 5947 (1973).

[27] J.C. Phillips, *J. Non-Cryst. Solids* **43**, 37 (1981).

[28] T. Fukunaga, Y. Tanaka and K. Murase, *Solid State Commun.* **42**, 513 (1982).

[29] K. Murase, T. Fukunaga, Y. Tanaka, K. Yakushiji and I. Yunoki, *Physica* **B117 & 118**, 962 (1983).

[30] M.F. Thorpe, *J. Non-Cryst. Solids* **57**, 355 (1983).

[31] M. Taniguchi, T. Kouchi, I. Ono, S. Hosokawa, M. Nakatake, H. Namatame and K. Murase, *J. Electron Spectroscopy and Related Phenomena* **78**, 507 (1996).

[32] A. Feltz, H. Aust and A. Blayer, *J. Non-Cryst. Solids* **55**, 179 (1983).

[33] B. Uebbing and A.J. Sievers, *Phys. Rev. Lett.* **76**, 932 (1996).

[34] X. Feng, W.J. Bresser and P. Boolchand, *Phys. Rev. Lett.* **78**, 4422 (1997).

[35] M. Tatsumisago, B.L. Halfpap, J.L. Green, S.M. Lindsay and C.A. Angell, *Phys. Rev. Lett.* **64**, 1549 (1990).

[36] U. Senapati and A.K. Varshneya, *J. Non-Cryst. Solids* **185**, 289 (1995).

[37] Y. Wang, M. Nakamura, O. Matsuda, K. Inoue and K. Murase, *J. Non-Cryst. Solids* **198–200**, 753 (1996).

[38] Y. Wang, O. Matsuda, K. Inoue and K. Murase, *Proc. 23rd Int. Conf. Phys. Semicond.*, Berlin, eds. W. Scheffler and A. Zimmermann (World Scientific, Singapore, 1996) p. 153.

[39] D.J. Sarrach and J.P. deNeufville, *J. Non-Cryst. Solids* **22**, 245 (1976).

[40] Y. Wang, O. Matsuda, K. Inoue and K. Murase, *Progress of Theoretical Physics Supplement* **126**, 191 (1997).

[41] J.D. Joanopoulos, M. Schlüter and M.L. Cohen, *Phys. Rev.* **B11**, 2186 (1975).

[42] S. Hosokawa, K. Nishihara, Y. Hari, M. Taniguchi, O. Matsuda and K. Murase, *Phys. Rev.* **B47**, 15509 (1993).

[43] M.L. Theye, A. Gheorghiu, C. Senemaud, M.F. Kotkata and K.M. Kandil, *Phil. Mag.* **B69**, 209 (1994).

[44] E. Bergignat, G. Hollinger, H. Chermette, P. Pertosa, D. Lohez, M. Lannoo and M. Bensousan, *Phys. Rev.* **B37**, 4506 (1988).

[45] J.C. Phillips, *J. Non-Cryst. Solids* **34**, 153 (1979).

[46] J.C. Phillips and M.F. Thorpe, *Solid State Commun.* **53**, 699 (1985).

[47] M. Stevens, J. Grothaus, P. Boolchand and J.G. Hernandez, *Phys. Rev.* **B31**, 981 (1985).

[48] W. Bresser, P. Boolchand and P. Suranyi, *Phys. Rev. Lett.* **56**, 2493 (1986).

[49] P. Boolchand, W. Bresser, M. Zhang, Y. Wu, J. Wells and R.N. Enzweiler, *J. Non-Cryst. Solids* **182**, 143 (1995).

[50] J.E. Griffiths, G.P. Espinosa, J.P. Remeika and J.C. Phillips, *Solid State Commun.* **40**, 1077 (1981).

[51] E. Haro, Z.S. Xu, J.F. Morhange, M. Balkanski, G.P. Espinosa and J.C. Phillips, *Phys. Rev.* **B32**, 969 (1985).

[52] S. Sugai, *Phys. Rev. Lett.* **57**, 456 (1986).

[53] S. Sugai, *Phys. Rev.* **B35**, 1345 (1987).

[54] K. Murase, K. Inoue and K. Kawamoto, *Proc. 10th Int. Conf. on Raman Spectroscopy*, Eugene, USA, eds. W.L. Peticolas and B. Hudson (University of Oregon, Eugene, Oregon, USA, 1986) p. 7.19.

[55] K. Murase and K. Inoue, *Disordered Semiconductors*, eds. M.A. Kastner, G.A. Thomas and S.R. Ovshinsky (Plenum Press, New York, 1987) p. 297.

[56] K. Inoue, O. Matsuda and K. Murase, *Proc. 19th Int. Conf. Phys. Semicond.*, Warsaw, Poland, ed. W. Zawadzki (Institute of Physics, Polish Academy of Sciences, Poland, 1988) p. 1665.

[57] K. Yakubo, K. Takasugi and T. Nakayama, *J. Phys. Soc. Japan* **59**, 1909 (1990).

[58] S. Alexander and R. Orbach, *J. Phys. (France) Lett.* **43**, L625 (1982).

[59] E. Courtens, R. Vacher, J. Pelous and T. Woigner, *Europhys. Lett.* **6**, 245 (1988).

[60] Y. Tsujimi, E. Courtens, J. Pelous and R. Vacher, *Phys. Rev. Lett.* **60**, 2757 (1988).

[61] A. Mermet, E. Duval, S. Etienne and C. G'Sell, *J. Non-Cryst. Solids* **196**, 227 (1996).

[62] A. Fontana, F. Rocca, M.P. Fontana, B. Rosi and A.J. Dianoux, *Phys. Rev.* **B41**, 3778 (1990).

[63] P.P. Lottici, *Phys. Stat. Sol. (b)* **146**, K81 (1988).

[64] E. Duval, N. Garcia, A. Boukenter and J. Serughetti, *J. Chem. Phys.* **99**, 2040 (1993).

[65] S. Alexander, *Phys. Rev.* **B40**, 7953 (1989).

[66] J. Jäckle, *Amorphous Solids*, ed. W.A. Phillips (Springer-Verlag, Berlin, 1981) Chapter 8, p. 135.

[67] T. Achibat, A. Boukenter and E. Duval, *J. Chem. Phys.* **99**, 2046 (1993).

[68] M. Nakamura, O. Matsuda and K. Murase, *Phys. Rev.* **B57**, 10228 (1998).

[69] S. Feng, *Phys. Rev.* **B32**, 5793 (1985).

[70] Y. Kantor and I. Webman, *Phys. Rev. Lett.* **52**, 1891 (1984).

[71] B.A. Weinstein, R. Zallen, M.L. Slade and J.J.C. Mikkelson, *Phys. Rev.* **B25**, 781 (1982).

[72] T. Nakayama, K. Yakubo and R. Orbach, *Rev. Mod. Phys.* **66**, 381 (1994).

[73] Y. Wang, O. Matsuda, K. Inoue, O. Yamamuro, T. Matsuo and K. Murase, *J. Non-Cryst. Solids* **232–234**, 702 (1998).

[74] J.C. Phillips, *Phys. Rev.* **B54**, R6807 (1996).

[75] A.P. Sokolov, A. Kisliuk, D. Quitmann and E. Duval, *Phys. Rev.* **B48**, 7692 (1993).

[76] G. Li, W.M. Du, X.K. Chen, H.Z. Cummins and N.J. Tao, *Phys. Rev.* **A45**, 3867 (1992).

[77] A.P. Sokolov, E. Rössler, A. Kisliuk and D. Quitmann, *Phys. Rev. Lett.* **71**, 2062 (1993).

[78] W.A. Kamitakahara, R.L. Cappelletti, B.L. Halfpap, P. Boolchand, F. Gompf, D.A. Neumann and H. Mutka, *Phys. Rev.* **B44**, 94 (1991).

[79] M. Arai, Y. Inamura, T. Otomo, N. Kitamura, S.M. Bennington and A.C. Hannon, *Physica* **B263–264**, 268 (1999).

[80] T. Nakayama, *Physica* **B263–264**, 243 (1999).

[81] T. Nakayama and L. Orbach, *Physica* **B263–264**, 261 (1999).

[82] M. Yamaguchi, T. Nakayama and T. Yagi, *Physica* **B263–264**, 258 (1999).

[83] Y. Inamura, M. Aari, O. Yamamuro, T. Matsuo, N. Kitamura, T. Otomo and S.M. Bennington, *Physica* **B263–264**, 299 (1999).

[84] T. Nakayama, *Phys. Rev. Lett.* **80**, 1244 (1998).

[85] Y. Wang, M. Nakamura, O. Matsuda and K. Murase, *Physica* **B263–264**, 313 (1999).

D. LOW FREQUENCY VIBRATIONAL EXCITATIONS IN GLASSES BY BRILLOUIN AND RAMAN SCATTERING

CLAIRE LEVELUT

*Laboratoire des Verres, Université Montpellier II, Place Eugène Bataillon,
case 69, 34095 Montpellier Cedex 5, France*
claire@ldv.univ-montp2.fr

Contents

1. Introduction

Understanding the properties of glasses has remained a fundamental goal during the last decades. In the seventies, the low temperature properties of glasses and disordered materials were very intensively studied [1]. Interest in the glass transition has been renewed by the emergence of new models such as mode coupling theory in the eighties [2]. Nowadays, one of the most debated questions is the connection between the properties in the liquid phase near the glass transition and in the glass in the frozen phase.

The low-frequency dynamics of glasses and glass transition has received considerable attention in recent years. A great effort has been directed towards the frequency range from a few tens of GHz to a few THz. This frequency range is expected to reflect the cooperative motions of atoms over correlation lengths in the nanometer range. Many studies deal with a quasi-universal characteristic of disordered structure: a strong broad peak which arises in inelastic neutron scattering or low-frequency Raman scattering, around 20–50 cm^{-1} $\equiv$ 0.6–1.5 THz $\equiv$ 2–6 meV. This peak is called boson peak because its intensity varies with the temperature like the Bose population factor. This peak is generally assumed to be mostly due to vibrational contributions. It represents the main contribution of low frequency Raman spectra below the glass transition temperature T_g.

Above the glass transition, the dominant feature of low frequency Raman spectra is a strong quasi-elastic line below 10 cm^{-1} whose intensity increases with temperature faster than the Bose factor. This quasi-elastic contribution is generally assigned to a relaxational process. Many studies are devoted to the relaxational properties in the neighborhood of the glass transition. At the glass transition, the viscosity undergoes a drastic increase. As the viscosity reflects the underlying movements of the molecules in the system, the glass transition is associated to a relaxational process usually referred to as a structural or primary relaxation (α) which follows the changes of the viscosity. This long time relaxation is generally considered as cooperative and the relaxation time τ is well described by the Vogel–Fulcher–Tamman–Hesse equation. In many glass forming liquids, a secondary process, usually called Johari–Goldstein β relaxation [3], whose characteristic time exhibits an Arrhenius behavior has been observed below the glass transition. Mode coupling theory is one of the most discussed theories devoted to the relaxational processes related to the glass transition, in the liquid phase [2]. This theory concentrates on the dynamics in the mesoscopic range, which is specially interesting because of its implications on the microscopic mechanisms. It predicts [2, 4] a two-step relaxation scenario and the existence of a fast β-process in the picosecond range. The α region of the mode coupling theory is identical with the conventional α relaxation. The β-process corresponds to the minimum in the susceptibility spectrum between the α-process and the microscopic motions and its characteristic time exhibits power law divergences when approaching the temperature of the decoupling on both sides. One of the most important prediction of the mode coupling theory is the existence of a critical temperature T_c located above the calorimetric glass transition temperature T_g. At this temperature T_c, the transport properties change from those typical for a strongly coupled liquid to those characteristic of a glass corresponding to a viscoelastic behavior, where the motions are cooperative. Below T_c, there is a decoupling of various relaxation processes whereas the coupling of all the density modes leads to a single universal time scale (time-temperature superposition principle) above T_c. Many recent works involve inelastic neutron and light scattering techniques which are well suited for the study of the dynamics in this range.

The relaxational excitations have been studied mainly in "fragile" (in Angell's classification) [5] glasses (i.e. glasses with weak intermolecular interactions) where a quantitative agreement of the predictions of the mode coupling theory is fulfilled. It has been recently established that these vibrational and relaxational properties are connected. Several recent models imply that these relaxational and vibrational contributions cannot be treated separately [6, 7].

Several authors tried to connect the vibrational properties reflected by the boson peak with the first sharp diffraction peak (FSDP) [8, 9]. The FSDP can be observed in X-ray or neutron scattering diffraction spectra and is a characteristic of the disorder in the nanometer range [10, 11, 12]. On the other hand, another much discussed subject is the possibility of a common origin for the dynamics in the frequency range of the boson peak (about 1 THz) and the quasi-elastic line (about 0.3 THz) on the one hand and the Brillouin lines (at lower frequencies, around 10 to 30 GHz) on the other hand. Several studies examined the connection of the variation with the temperature of the position of the boson peak with that of the velocity of sound [13, 14, 15]. Other authors tried to connect the attenuation of sound, as measured by Brillouin scattering, to the quasi-elastic intensity [15, 16, 17].

2. Inelastic Light Scattering

2.1. *General*

Electric field of incident light induces an oscillating polarization of molecules of the scattering material. Charges within molecules oscillate and an induced dipole moment appears; molecules act as a secondary source and the accelerated electrons scatter light. If the medium is perfectly homogeneous, the scattered fields have the same amplitude in the whole material and interfere destructively. However, scattering is caused by spatial or temporal fluctuations of the dielectric constant ε or the dielectric susceptibility of the medium [18, 19]. Light scattering can then give information about any physical phenomenon responsible for fluctuations of the dielectric properties. The frequency of the scattered light is connected to the dynamics of the fluctuations. The wave vector $\vec{q}$ of the fluctuations is defined by the wave vector conservation law, $\vec{q} = \vec{k_i} - \vec{k_s}$, where $\vec{k_i}$ and $\vec{k_s}$ are the wave vectors of incident and scattered light respectively. However, the scattering process changes very little the wavelength of the light, so that $|\vec{k_s}| \simeq |\vec{k_i}|$. Then,

$$|\vec{q}| = 2n|\vec{k_i}| \sin\left(\frac{\theta}{2}\right) \tag{1}$$

where θ is the scattering angle and n the refractive index of the scattering medium. This relation reflects the Bragg's condition. Using visible light, the wavevector is about 1 to 3×10^{-3} Å^{-1}.

Static fluctuations of the dielectric constant ε (impurities, defects, frozen density or concentration fluctuations ...) scatter light of the same frequency ν_0 as the monochromatic incident light. Such inhomogeneities cause the *elastic Rayleigh scattering*.

Non-propagating time dependent fluctuations can occur due for example to heat diffusion, diffusive fluctuations of concentration or density, rotational motions for anisotropic molecules or aggregates. The duration of the emission of scattered light is finite and its amplitude depends on time. The spectrum of the scattered light is then made of one or several broadened lorentzian components centered on the frequency ν_0. This effect is called *quasi-elastic Rayleigh scattering*. In glasses, such non-propagating fluctuations are caused by entropy fluctuations of the disordered structure which are frozen-in at the glass transition temperature. Since these large fluctuations are nearly static, they give rise to a very intense narrow central peak.

Inelastic scattering is caused by propagating dynamic fluctuations. Due to the conservation laws, this scattering occurs with a change of wavelength. By analogy with crystals, the excitation which gives rise to the inelastic scattering can be qualified either as an acoustic phonon if the neighboring atoms vibrate in phase or as optical vibrations if the neighboring atoms have relatives motions. The former cause Brillouin scattering and the latter Raman scattering [20]. Propagating density fluctuations, or sound waves, gives rise to Brillouin scattering. Propagating fluctuations in the molecular polarizabilities caused by vibrations and rotations of molecules or vibrations and librations of the lattice are the origin of Raman scattering. Moreover, due to the disorder, all the acoustic modes can contribute to incoherent scattering which occurs in the frequency range between the Brillouin lines and Raman optical modes, this is the so-called low frequency Raman scattering.

2.2. *Brillouin scattering*

Brillouin light scattering is an inelastic light scattering process corresponding to frequency shifts (difference between the frequency of the incident light and the frequency of the scattered light) in the GHz range. It corresponds to coherent scattering associated with propagating vibrational motions. It is due to the interaction of light with long wavelength elastic waves (phonons) due to propagating thermal motions in condensed materials [21, 22], or to random thermal density fluctuations in liquids or gas. The wavelength of those vibrations is about 2000 Å, much larger than interatomic motions. Therefore, Brillouin scattering probes the scattering medium as a continuous one.

The first theoretical prediction is due to Brillouin who decomposed the thermal motions of molecules under a basis of elastic waves [23]. The light is scattered by the maximum density planes created by the elastic waves. The elastic waves propagate in the scattering medium, and thus the Bragg diffracted light suffers a Doppler frequency shift, proportional to the velocity V of propagation of the sound waves (sound velocity) at the frequency of the Brillouin doublet. The associated frequency change is given by

$$\Delta\nu = \nu - \nu_0 = \pm 2\nu_0 (V/c) n \sin(\theta/2) \tag{2}$$

where c is the velocity of light, $\Delta\nu$ is in the GHz range. In viscous liquids or in solids, in which mechanical shear waves can propagate, the Brillouin spectrum also contains a doublet of transverse modes, which propagates besides the longitudinal ones.

The spectrum of the scattered light reflects the dynamics of the elastic excitations, which is determined through the viscoelastic properties of the scattering medium. Thus, Brillouin scattering gives access to the elastic properties at small wavevectors and at high frequencies (GHz): velocity of sound in the hypersonic regime, attenuation of hypersounds; compressibility or shear modulus in the GHz range can be deduced from the values of longitudinal and transverse velocities.

The longitudinal Brillouin scattering (corresponding to vibrations parallel to the wavevector) has been explained by adiabatic pressure fluctuations propagating as a longitudinal sound wave [24, 25]. It corresponds to light scattered with the same polarization as the incident light. The transverse Brillouin scattering (vibrations perpendicular to the wavevector) is related to the existence of anisotropic fluctuations associated with shear deformation [26, 27]. It contributes to the depolarized scattered light, i.e. light scattered with a change of polarization with respect to the incident light.

The hypersonic shear modulus G can be deduced from the transverse velocity V_T using $G = \rho V_T^2$ where ρ is the density of the scattering material; the longitudinal modulus $M = K + 4/3G$, K being the compressibility modulus, is obtained from the longitudinal velocity V_L using a similar expression ($M = \rho V_L^2$).

The spectral width of the Brillouin peaks is determined by the life-time of the fluctuations responsible for the scattering and thus by the attenuation of the fluctuations. In supercooled liquids, the lifetime of the density fluctuations is related to the compressional and shear viscosities in liquids. In the glass phase, the line width is related to the sound attenuation. As in the ultrasonic frequency range [28], anharmonicity, scattering by the disorder itself, interaction with localized defects through relaxational processes or resonant interaction are possible origins for sound attenuation.

At very low temperature, the sound wave damping can be described by resonant interaction with defects described by two-level-systems and by relaxational absorption by those two level systems [29]. The term due to scattering by disorder is temperature independent and thus contributes mainly at low temperatures. At higher temperatures, thermally activated relaxation processes become possible and give rise to a broad absorption peak or a plateau. Anharmonicity, which is the main cause of attenuation of sound in crystals, is expected to contribute at temperatures where the attenuation in crystals is significant (above about 20 K).

2.3. *Low-frequency Raman scattering*

In disordered materials, Raman scattering corresponds to the inelastic scattering due to the interaction with optical modes and incoherent diffusion of high frequency acoustical modes. It corresponds to higher frequency shifts than Brillouin scattering: from about 10^{11} to 10^{14} Hz. Low-frequency Raman scattering corresponds to the frequency range from 100 GHz to about 3 THz. Since the selection rules that govern the Raman scattering in crystals breakdown in glasses because of the disorder, all the modes contribute to the scattering process. Thus, the Raman spectra become a continuum instead of a discrete series of line.

By selecting the polarization of the scattered light, it is possible to measure two distinct quantities: if one refers to a horizontal scattering plane, and using the usual notation V and H for vertical and horizontal polarization, one can measure I_{VV}, I_{VH}, I_{HV} and I_{HH} where the first index refers to the incident light and the second to the scattered light. It can be shown that, in terms of isotropic and anisotropic intensities:

$$I_{VV}(\nu) = I_{HH}(\nu) = I_{\mathrm{iso}}(\nu) + \frac{4}{3}I_{\mathrm{aniso}}(\nu) \qquad (3)$$

and

$$I_{VH}(\nu) = I_{HV}(\nu) = I_{\mathrm{aniso}}(\nu) \qquad (4)$$

where $I_{\mathrm{iso}}(\nu)$ and $I_{\mathrm{aniso}}(\nu)$ are respectively the self-correlation function of the isotropic scalar part of the polarizability tensor and of the anisotropic part. If one consider a molecule characterized by reorientational and vibrational modes of motions, then the isotropic spectrum depends only on vibrational motions whereas the anisotropic part depends both on vibrational and reorientational motions [19]. The reorientational motions probed by Raman scattering are the independent rotations of single molecules. Indeed, as Raman scattering is incoherent, it measures single particle motions. Another physical quantity of interest is the depolarization ratio, $\rho = I_{VH}/I_{VV}$, which gives indications about the physical mechanism responsible for the scattering. The scattering mechanism can be either direct or through dipole-induced dipole interaction.

Shuker and Gammon explained the Raman intensity in disordered systems by first order disordered-induced scattering: the disorder of the structure causes fluctuations of the elastic constants and then leads to an attenuation of the vibrational modes [30]. Vibrational modes with long wavelength (longer than a correlation length R_c) travel in the scattering medium as in a continuous one, except that they are attenuated by disorder; each vibrational mode corresponds to a fixed wavevector value k, determined by the selection rules. The vibrational modes of wavelength shorter than the correlation length are strongly attenuated in space, so that they correspond to a broadened band in k-space, instead of a narrow peak as expected from the selection rules. Shuker and Gammon made the hypothesis that all the modes originating from the same band contributes to the spectra with the same weight, corresponding to the coupling constant for those modes. Then, they obtained an expression for the first-order scattered intensity:

$$I(\nu) = \sum_b C_b^{ij} \frac{[1 + n(\nu)]}{\nu} g_b(\nu) \tag{5}$$

where the summation is taken over all the bands which contribute to the scattering. C_b^{ij} and $g_b(\nu)$ are respectively the coupling constant and the density of states for modes of band b. This expression is often written in a simplified manner, using an effective coupling constant $C(\nu) = \sum_b C_b^{ij} g_b(\nu)/g(\nu)$, where $g(\nu)$ is the total density of states, or assuming that the coupling constant is identical for all the bands. Thus, the Raman intensity is proportional to the density of vibrational modes $g(\nu)$, through a coupling constant $C(\nu)$ which represents the efficiency of the interaction between the light and the phonons [30].

2.4. *Experimental aspect*

Light scattering techniques do not require any physical contact with the sample, nor application of any external force (unlike ultrasonic techniques). These techniques are particularly suited to the study of transparent materials such as many glasses. They are efficient in liquids as well as in solids so that they can be useful to follow changes in physical properties related to the glass transition.

From the experimental point of view, the analysis of Brillouin scattering, with frequency shifts of a few GHz requires the dispersive power of the Fabry–Pérot interferometers, often used in a multipass mode to increase the contrast (i.e. the ratio of the maximum transmission of the apparatus to the minimal one). A very widespread and very efficient kind of spectrometer follows Sandercock's design and uses two plane Fabry–Pérot etalons with very slightly different thicknesses in a multipass configuration [31]. Other spectrometers make use of the higher resolution of spherical Fabry–Pérot interferometer [32]. An example of typical Brillouin spectrum in soda-lime glass taken at 90° scattering showing Stokes and anti-Stokes transverse lines at about ±14 GHz and the longitudinal ones at about ±24 GHz is shown in Fig. 1(a). The spectrum is registered using a two-pass plane Fabry–Pérot interferometer. Spectrum on the same sample, also in right-angle scattering geometry, using the high resolution instrument described by Vacher and Boyer [32] is shown in Fig. 1(b). One can note that the Brillouin line is much larger than the elastic residuals whereas in Fig. 1(a) elastic and inelastic line have the same width — determined by the apparatus function. This high resolution set-up is suitable for the measurements of the attenuation.

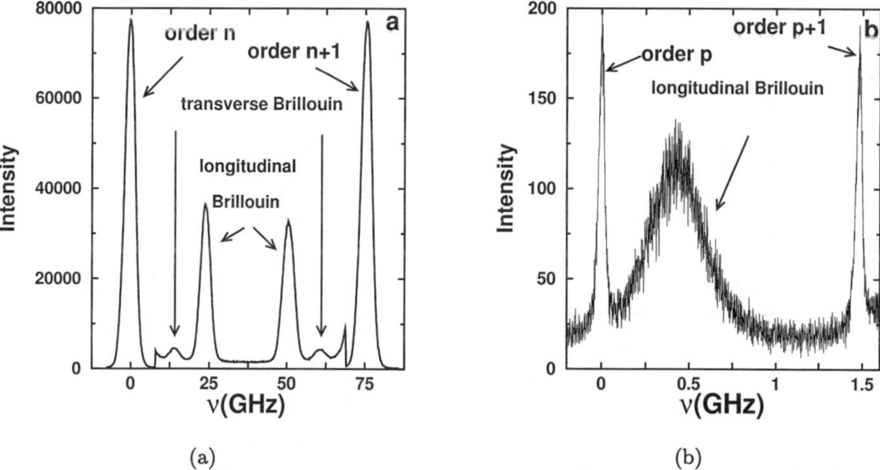

(a) (b)

Fig. 1. (a) Example of Brillouin spectrum in soda-lime glass recorded in right-angle geometry using a two-pass plane Fabry–Pérot interferometer. (b) Example of spectrum on the same glass using the high resolution instrument described by Vacher and Boyer [32]. Because a two-pass plane Fabry–Pérot is used as a filter whose maximum transmission is adjusted to the frequency of the Brillouin line to study, one observes only one Brillouin peak. In this figure as well as in the following figures, the intensities are in arbitrary units.

Low-frequency Raman scattering can be measured using either Fabry–Pérot interferometers with very small thicknesses or conventional grating Raman spectrometers. A typical example of spectra in an oxide glass (BaF4, from Schott) as a function of the temperature, across the glass transition temperature is shown in Fig. 2(a). As usual, we plotted the so-called reduced intensity:

$$I_R(\nu) = \frac{I(\nu)}{\nu(n(\nu,T)+1)} \quad \text{for Stokes Raman intensity} \tag{6}$$

$$I_R(\nu) = \frac{I(\nu)}{\nu n(\nu,T)} \quad \text{for anti-Stokes Raman intensity} \tag{7}$$

where $n(\nu,T)$ is the Bose population factor. This kind of plot removes the trivial temperature dependence for the boson peak and makes the peak more prominent. The boson peak appears at about 50 cm^{-1}. The quasi-elastic line is noticeable below 10 cm^{-1} and above 300 K. In Fig. 2(b), inelastic neutron scattering spectra in the same frequency range are represented at several temperatures for a polymer (polyurethane) which is a fragile glass in Angell's classification. The spectra exhibit the same feature as in oxide glasses.

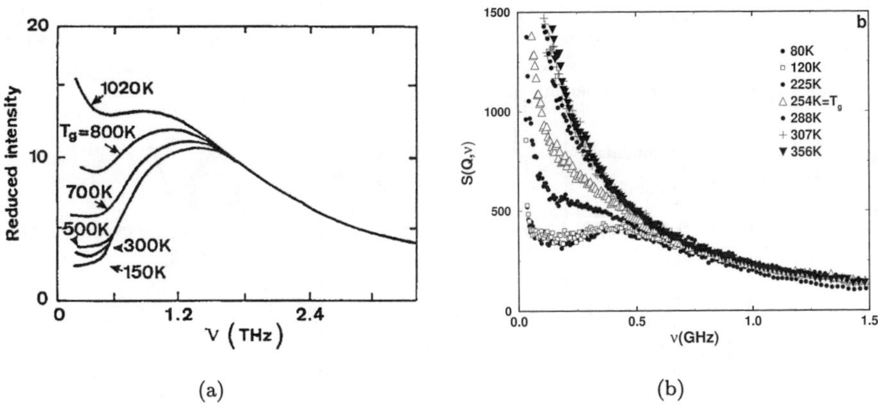

Fig. 2. (a) Low-frequency Raman scattering (reduced intensity) as a function of temperature in BaF4 (an oxide glass provided by Schott). One can note the boson peak at 1.5 THz and the strong increase of the quasi-elastic signal above room temperature. (b) Example of inelastic neutron scattering spectra on a fragile glass (polymer) showing the same feature. In this frequency range, the inelastic neutron scattering spectra are very similar to low frequency Raman data. The boson peak occurs at lower frequency (about 0.5 THz) than in the strong glass. At the glass transition temperature (800 K for BaF4 and 250 K for the polymer), the quasi-elastic contribution to the scattering is higher in the fragile glass.

However, it can be noticed that the ratio of the relaxational contribution over the boson peak intensity at the glass transition temperature (250 K) is higher than in the oxide glass at its glass transition at 800 K.

3. Models for the Hypersonic Properties of Glasses

Typical behavior for sound attenuation and sound velocity as a function of the temperature can be found for example in several articles of Pelous, Vacher *et al* [33–35]. At extremely low temperature, the attenuation of sound decreases with increasing temperature. Ultrasonic measurements have shown that, in this temperature range, the attenuation decreases with increasing intensity [36]. At slightly higher temperatures, the attenuation increases with temperature as T^3 and is independent of the frequency. Then it exhibits a small shoulder at low temperatures — below 20 K — whose position shifts to higher temperature with increasing frequency, as well as a large peak at "intermediate" temperatures (50 to 200 K) whose position hardly depends on frequency. The attenuation varies as ν^2 on the high-frequency side of this peak. Then it presents a very steep rise at the glass transition temperature, indicating the coupling of the acoustic waves with vibrations of molecules and structural rearrangements of the material.

At very low temperatures (below about 1.5 K), the sound velocity presents a maximum with a logarithmic variation on the low-frequency side of the peak [37]. In most glasses, at higher temperatures the sound velocity decreases with increasing temperature. This effect corresponds to a softening of the glass. Above the glass transition, the decrease becomes very steep. In tetrahedral glasses, such as SiO_2, GeO_2, BeF_2 and $Zn(PO_3)_2$, the sound velocity exhibits a minimum and then increases linearly with the temperature, corresponding to a hardening of acoustic modes with increasing temperature. This behavior is not completely understood and will be discussed in Sec.3.2.

3.1. *Models for the attenuation*

3.1.1. *Resonant absorption by two-level systems*

The behavior at extremely low temperature (below 4 K) can be accounted for by resonant interaction with defects described by two level systems [38]. The phonon is absorbed and re-emitted after recombination by two-level systems whose upper level is empty. This process has been described by Jackle *et al.* using a formal equivalence between the Hamiltonian of a two-level system and that of a spin $S = 1/2$ [39]. The contribution to the attenuation of resonant absorption by two-level systems depends on ν^2 and varies as $1/T$ if the acoustic intensity is weak.

3.1.2. *Relaxational absorption by two-level systems*

At higher temperatures, between 4 and 20 K, the upper level of the two-level systems is populated and the resonant absorption process is saturated. Then, the two-level systems contribute to the absorption through another mechanism called relaxational absorption. The sound wave traveling among the two-level systems disturbs their equilibrium and the systems relax to equilibrium by emitting and absorbing thermal phonons. The relaxation is phonon-assisted: two-level systems with splitting energy around $3\,kT$ couple to thermal phonons. This channel can explain the small plateau or shoulder below 20 K. An attenuation independent on frequency and varying like T^3 is predicted in the low-temperature side of this absorption peak. The high-temperature limit is not considered in details because it is masked by the strong thermally activated absorption peak around 50 K. A good agreement between the model and the experimental data has been found in silica [38, 40].

3.1.3. *Scattering by disorder*

Some authors introduce another contribution to the sound attenuation due to the disorder itself [41, 42]. The disorder induce density fluctuations or fluctuations of the elastic properties related to frozen-in strain-field which scatter the acoustic waves. This contributions does not depend on the temperature and is rather weak at hypersonic frequencies. It corresponds to the residual attenuation which can be measured at very low temperatures. This contribution increases with frequency as ν^4, so it dominates at high frequency, as will be discussed in Sec. 4.2.2.

3.1.4. *Interaction with thermal phonons — anharmonicity*

The anharmonic processes have been described in crystals in terms of interaction of the acoustic phonon with the thermal phonons, which induces changes in the velocity and in the attenuation of the acoustic phonon. Two distinct regimes must be taken into account for the description: $\omega\tau \gg 1$ and $\omega\tau \ll 1$, where $\omega = 2\pi\nu$ is the pulsation of the acoustic phonon and τ the lifetime of the thermal phonon. In glasses, the same mechanism is assumed to contribute to the attenuation of sound. Then, the mean free path of thermal phonons is nearly temperature independent above 20 K and of the order of the interatomic distance. Therefore, τ is about 10^{-13} s and $2\pi\nu\tau \ll 1$ for any temperature above 20 K. Below 20 K, $2\pi\nu\tau \gg 1$, the interaction can be described as an inelastic collision between the two phonons, considered as particles [43]. The probability for such collision is proportional to the energies of the acoustic phonon ($h\nu$) and of the thermal phonons (kT) and to the number of thermal phonon. This latter is proportional to T^3 for transverse phonons and increases even faster for longitudinal ones. The attenuation of phonons is then proportional to ν and T^n, with $n = 4$ for the transverse attenuation and $n = 6$ or 7 for longitudinal modes. In the regime corresponding to $2\pi\nu \ll 1$, the wavelength of the acoustic phonon is larger than the mean free path of the thermal phonon and the scattering material is seen as a homogeneous medium. The sound waves modulate the frequency ν_{th} of the thermal phonons by an amount $\Delta\nu_{\text{th}}$ determined by the anharmonicity of the lattice. The anharmonicity of the lattice is characterized by the Grüneisen parameter which is a measure of how the frequency ν_i of a phonon is altered by a small change of volume V on the sample (or on the unit cell for a crystal): $\gamma = -\partial\log\nu_i/\partial\log V$. Such constant can be calculated for one particular mode ν_i, for the Debye Frequency or as an average over several modes [44]. Due to this modulation, the distribution of the thermal phonons is out of equilibrium. It returns to equilibrium after a characteristic time τ and is responsible for absorption of energy from the acoustic phonon causing attenuation of hypersounds. The mean free path of the acoustic phonons is then proportional to T and ν^2 [45]. The experimental data above 20 K exhibit a ν^2 behavior, but quantitative agreement between experiment and theoretical calculation would need an unreasonably large value of the Grüneisen parameter [46].

3.1.5. *Thermally activated relaxational processes*

At temperatures higher than 20 K, though the model which ascribes the attenuation to interactions with thermal phonons predicts a frequency and temperature dependence in agreement with observations, anharmonicity is not sufficient to account for the strong attenuation observed in glasses. At those temperatures, another significant contribution can arise from the occurrence of a large distribution of thermally activated relaxational processes with relaxation times τ. The induced attenuation α is then given by:

$$\alpha = \frac{D^2}{4\rho V^3 kT} \int P(v) \frac{\nu^2\tau(v)dv}{1 + (2\pi\nu)^2\tau^2} \tag{8}$$

where $P(v)$ is the distribution of barriers v for the thermally activated relaxational processes and D is a deformation potential which expresses the energy shift of the relaxing states [39]. Therefore, at low temperature, $2\pi\nu\tau \gg 1$ and $\alpha \propto 1/\tau$, whereas, in the high temperature limit, $2\pi\nu\tau \ll 1$ and $\alpha \propto \nu^2$. Good agreement with experimental data for attenuations in the ultrasonic range can be achieved in the framework of this model [40, 47].

The strong absorption peak or plateau at intermediate temperature can be described by thermally activated relaxations. The ν^2-dependence of the absorption above this peak is due to the superposition of the high frequency side of this peak and anharmonicity [47, 48].

3.2. *Temperature dependence of the velocity*

The low temperature peak in the velocity can be explained by the resonant absorption by two-level systems, using the same model as for the attenuation. The decrease of the sound velocity with increasing temperature in glasses can be understood by the same relaxational processes which give rise to the absorption behavior. The change in velocity can be calculated from (8) using the Kramers-Kronig relation, thus

$$\frac{\Delta V}{V} = \frac{-D^2}{8\rho V^2 kT} \int P(v) \frac{dv}{1 + (2\pi\nu)^2\tau^2}. \tag{9}$$

This model provides a good description of the experimental behavior in many glasses [38]. However, it cannot describe the anomaly of elastic properties in tetrahedral glasses such as silica. In silica, this behavior was related to the anomalously small thermal expansion coefficient by Anderson and Dienes [49]. A more general explanation was proposed in the framework of the models relating elastic properties to density fluctuations. Kul'bitskaya *et al.* estimated the elastic constants as the difference between an average value corresponding to a homogeneous material and a quantity proportional to the amplitude of frozen-in fluctuations [50]. These fluctuations which are present in the glass are more important if the glass has been frozen at higher temperature. Once frozen in, the fluctuations, responsible for strains in the glass, cause the elastic modulus to decrease when the temperature is raised. If the variation of the average elastic property with temperature is low enough, the overall temperature dependence is dominated by the term arising from the fluctuations and the effective elastic constant can increase with increasing temperature.

4. The Boson Peak

4.1. *General presentation*

One of the anomalies of physical properties of glasses compared to the corresponding crystals, is the existence of harmonic modes in excess of the Debye contribution in crystals in the 1–3 THz frequency range. This excess of modes can be observed by low frequency Raman scattering [51], infrared absorption [52], specific heat measurements or inelastic neutron scattering [53]. It is a quasi-universal characteristic of glasses and disordered solids. It is not seen in the Raman spectra of corresponding crystalline phases. It occurs in all kind of glasses (molecular glasses, polymers, oxide glasses, metallic glasses). However, the presence of a broad peak in the same frequency region has also been reported in DNA, in some crystallized metallic glasses [54] and in crystalline cristobalite [55]. This peak is the dominant feature of the low-frequency Raman scattering spectra of glasses below the glass transition. In most glasses, this peak occurs around 0.6–1.5 THz. It appears in a wide range of wave vectors (ranging from about 10^{-3} Å^{-1} for light scattering to a few Å^{-1} for neutron scattering). In the light scattering spectrum it arises both in the VH and the VV polarizations. The physical origin of the boson peak is an object of debate: it can be ascribed for example to extra vibrational modes connected to the localization of phonons strongly scattered by defects at the frequency of the boson peak or to a different kind of excitation, called "soft" modes, coupled to the phonons.

4.2. *Models for the boson peak*

For the reduced intensity at low frequencies (such as $\hbar\omega \ll kT$), Shuker and Gammon's Eq. (5) gives:

$$I_{\text{red}}(\nu) = \sum_b C_b^{ij} \frac{g_b(\nu)}{\nu^2} \tag{10}$$

where $g_b(\nu)$ is the vibrational density of modes for the band b and C_b^{ij} the coupling constant describing the interaction of light with the modes of band b. This equation suggests two possible explanations for the boson peak: a maximum in $g(\nu)/\nu^2$, or a peak partially due to a coupling constant strongly dependent on the frequency.

4.2.1. *Martin and Brenig model*

Martin and Brenig [56] make the hypothesis that the density of vibrational modes is of Debye-type ($g(\nu) = \nu^2$, as in crystals) and the coupling constant expresses the disorder of both mechanical and electrical origins. The latter is due to the random character of the coupling between atomic motions which results in irregular bonding interatomic distances. The disorder results in spatial fluctuations of both elastic and photoelastic constants. The authors introduce a correlation length 2σ for the fluctuations, assuming that all the fluctuations have the same correlation function, and that this correlation function corresponds to a lorentzian profile. The correlation length is a measure of the extent of the short range atomic order; it represents the length over which the phonon can propagate without damping. Thus the coupling constant can be expressed as a function of this correlation length using the following expression:

$$C(\nu) \propto \left(\frac{\nu}{V_i}\right)^2 \exp(-(2\pi\nu\sigma/V_i)^2) \tag{11}$$

where V_i is the speed of sound in the amorphous materials for the polarization i (longitudinal or transverse velocity). The reduced intensity for the VV polarization is thus given by:

$$I_{\mathrm{red}}(\nu) \propto \nu^2 \left\{ A \exp(-(2\pi\nu\sigma/V_l)^2) + B \left(\frac{V_l}{V_t}\right)^5 \exp(-(2\pi\nu\sigma/V_t)^2) \right\} \tag{12}$$

where the coefficient A and B are comparable. The second term is dominant and the reduced intensity has a maximum for a position close to V_t/σ. The position of the boson peak depends on the sound velocity and on the correlation length. However, this model does not provide an accurate description of the spectral shape of the Raman signal above the position of the boson peak (the predicted intensity decreases faster than the experimental one).

4.2.2. *Maximum in the density of vibrational modes*

Nevertheless, this first interpretation of the boson peak as due to a maximum in the coupling constant was eliminated by the presence of a broad peak in inelastic neutron scattering in the same frequency range (the inelastic neutron scattering is proportional to the vibrational density of states) [57, 58]. An excess in the vibrational modes with respect to Debye behavior ($g(\nu) \simeq \nu^2$ in crystals) can be observed in specific heat measurements around 5 to 10 K, as a maximum in C_p/T^3 [59]. This maximum is due to the same excess of vibrational states as the boson peak in Raman or inelastic scattering. Several models have been proposed for the physical interpretation of the maximum in $g(\nu)/\nu^2$ responsible for this peak.

Some models assign this maximum to vibrations of structural units in particular materials. For example, in vitreous silica, Buchenau proposes a model identifying this excess of modes as a coupled rotation of SiO_4 tetrahedra [60]. Other authors explain these low frequency modes as vibrations associated with the lamellar structure in chalcogenide glasses [61]. However, the universality of the boson peak requires a more general explanation. Several interpretations relate the peak to the disorder itself or to a characteristic of the organization at medium range scale.

Some authors associate the boson peak with a characteristic frequency of clusters with size in the 10–20 Å range [62, 63, 64]. Duval *et al.* suggest a structural model in which the amorphous materials are formed of weakly interacting aggregates [62]. The aggregates are in general not microcrystals but disordered structures, called "blobs" in which the binding forces are stronger then between neighboring blobs. The Raman scattering spectra exhibit two regimes: a low frequency regime where the vibrations, of wavelength larger than the blob size, are delocalized over the glass, and a high frequency regime where they are localized inside the blobs. The maximum in the Raman reduced intensity occurs from localized vibrations inside the blobs. Pócsik also relates the boson peak to a log-normal distribution of cluster sizes; again the clusters are entities corresponding to strongly bound atoms and the correlations between the clusters are weaker [64]. Pang explains the boson peak by a log-normal distribution of clusters with individual vibrational density of states $g_L(\nu) = \nu^{D-1}$, where D is the fractal dimension of the clusters [63].

Other models attribute the excess of modes to a localization of phonons by the disorder. The localization can occurs through a mechanism of Rayleigh-like scattering by density fluctuations by resonant coupling with localized low energy excitations [65–69]. The amorphous materials act as a continuous medium for phonons with wavelength larger than the characteristic size of the density fluctuations, a Rayleigh-like scattering mechanism occurs and induces a mean free path for the phonons varying as $1/\nu^4$ (the attenuation increases as ν^4). Thus, because the attenuation increases strongly with frequency, the inverse mean free path of the phonon may become of the same order as the wave vector, and the condition of localization in the Ioffe–Regel definition is achieved:

$$|\vec{q}| \cdot \ell \simeq 1 \qquad (13)$$

The boson peak arises because, at the boson peak frequency, this condition is fulfilled. Akkermans and Maynard [65] showed that the localization of phonons is associated with an increase of the vibrational density of states. This increase corresponds to a breakdown of the dispersion relation, which implies that a full band, instead of a single frequency as in a crystal, is associated to one value of the wavevector. In the Ioffe–Regel regime, the attenuation of phonons is proportional to the frequency ν. Similarly, Elliot also associates the peak to a localization of phonons resulting from elastic scattering by density fluctuations domains [66]. The size R of the domains can also be identified as a correlation length over which short and medium range order are maintained. R can be either defined as the distance at which peaks in the radial distribution function are no longer discernible, or the point at which the atom-voids partial radial distribution function has no more peaks. The amorphous medium is homogeneous and isotropic beyond R. R is also related to the decay in amplitude of the fluctuations. In Klinger's model the acoustic phonons are strongly scattered due to resonant interactions with vibrational excitations in soft configurations [69]. The localization condition occurs at a frequency such that the smallest wavelength for extended scattered acoustic phonons is of the same scale as the characteristic size of the spatial correlations for fluctuations of the short range order parameters i.e. the effective size of the microscopic inhomogeneities forming the amorphous structure. A recent paper [70] shows that the Ioffe–Regel crossover to localized modes occurs at a frequency of about 1 THz in vitreous silica, corresponding to a wavevector of crossover of about 1 nm^{-1}.

4.3. *Connection with medium-range order*

Most of the models introduce a connection between the boson peak and a structural characteristic length ℓ of the glass [56, 71].

In Martin and Brenig's model, $\ell = 2\sigma$ is a correlation length for the fluctuations. In the clusters models [62, 63], this characteristic length may be identified with the cluster size and ν is a typical vibrational frequency of the clusters [62]. A characteristic size of "clusters" responsible for the occurrence of the boson peak can be derived from the position of the maximum of this peak, by analogy with glasses containing nanocrystallites, where it has been shown that the Raman spectrum contains peaks of frequency proportional to the inverse size of the crystallites: $\nu_{\max} = A\frac{V_i}{\ell}$, where ℓ is the size of nanocrystals and A a constant. For longitudinal modes, V_i is the longitudinal sound velocity and $A = 0.7$ whereas for torsional modes, V_i is the transverse velocity and $A = 0.85$ [58, 72]. In the models where the BP occurs due to localization of the phonons by the disorder [66, 69], ℓ has the meaning of a localization length in the Ioffe–Regel definition. In Elliot's model, ℓ can be identified with a correlation length over which short-range order and medium-range order are maintained [66]. The localization condition is fulfilled when Eq. (13) is achieved with $\ell = 2R$, where R is the size of fluctuation density domains. And, if one assumes a linear dispersion law ($2\pi = qV$, where V is the sound velocity) then $\nu_{\max} = V/2R$. To summarize, in all those models,

$$\nu_{\max} = \frac{AV}{\ell} \tag{14}$$

where $\nu_{\max}$ is the frequency of the maximum of the BP, V is the sound velocity and A is a constant, depending on the model, but generally close to one.

Using velocities V around 3000 ms^{-1} and frequencies of the boson peak around 1 to 1.5 THz, ℓ ranges from 1 to 2 nm. This scale corresponds to medium range order which also relates to another quasi-universal feature, the so-called First Sharp Diffraction Peak [73]. Several authors have tried to establish a correlation between ℓ and the lengths deduced from the position Q_1 or the width ΔQ of the FSDP.

This connection is underlying in some of the models for the boson peak: in Elliot's model, the correlation length R can be connected to the nearest neighbor atom-void separation D and to the nearest neighbor distance d by $R = pD = pd$ where p is a constant of order 3 or 4. Elliot's model for the FSDP relates the peak to the occurrence of low-density regions at the "cluster" boundaries; it predicts that the position of the FSDP is given by $Q_1 = 3\pi/2d$ [74]. Thus, a correlation between the position of the boson peak on one hand and of the first sharp diffraction peak on the other hand is expected through the equation: $\nu_{\max} = V/2pd = VQ_1/3\pi p$. Klinger also relates the scale length introduced for the interpretation of the boson peak, i.e. the characteristic size of the spatial correlations for fluctuations of the short range order parameters, to the length introduced in Elliot's model for the FSDP.

In two articles Sokolov and Novikov tried to establish experimentally such correlation by comparison of Raman and X-ray diffraction data [8, 9]. Novikov *et al.* observed a correlation between changes in the intensities of the boson peak and of the FSDP as well as between their positions [8]. A linear relationship is deduced between the position Q_1 of the FSDP and the "normalized" position of the boson peak $\nu_{\max}/V$. However, Sokolov *et al.* found that the meaning of this correlation was not clear and the length scale defined by $d_1 = 2\pi/Q_1$ is about 5 times smaller than the inverse of the normalized position. They infer that the position of the FSDP reflects some repetitive characteristic distance between structural units whereas a correlation length such as the one connected to the boson peak would be related to the width of the peak. An estimated correlation length connected to the FSDP can be deduced using a simplified version of Scherrer's formula for microcrystallites [75]: $d_2 = 2\pi/\Delta Q$, where the numerical coefficient depending on the form of the microcrystallites has be taken equal to 1; ΔQ is the half width at half maximum of the FSDP. A rather good correlation for several glasses and one polymer was found, which can be expressed as a linear approximation:

$$d_2 = 2\pi/\Delta Q \approx (0.90 \pm 0.11)\frac{V}{\nu_{\mathrm{max}}} \tag{15}$$

The authors also mentioned that the first correlation was fulfilled because in most glasses $\Delta Q \approx (0.2 - 0.3)Q_1$. Other authors also suggest that the correlation between d_1 and ℓ is not convincing because it has to be interpreted as two distinct correlations concerning two different families of glasses [76]: in a $Q_1 - (\nu_{\mathrm{max}}/V)$ plot, all the points concerning oxide glasses on one hand and chalcogenide glasses on the other hand are aligned, but with different slopes for the two families. This observation corresponds to the fact that oxide and chalcogenide glasses have distinct first-neighbor distances r_1. Therefore $Q_1 r_1$ and ν_{max}/V exhibit a good correlation for every kind of glass.

Several experimental studies try to confirm or refute such a connection. In particular, investigations were carried out in series of samples with small changes in composition or heat treatment, in order to get rid of effects of the chemical composition on short range order, which will necessarily modify the medium range order. For several samples of vitreous silica of different origins and different OH impurities content [16, 77], measurements of the FSDP show very little change within the experimental uncertainty (about 10%). This indicates that, in those strong glass formers, the medium range order is very similar for glasses formed at different temperatures in the transformation range. The variation of OH content does not change ℓ more than 10% (ℓ increases with increasing OH content). In contrast, noticeable changes are induced by fast-neutron irradiation: the irradiation produces a broadening of the FSDP, together with a shift to higher q-values. Thus d_2 decreases by about 20% and d_1 by about 5%. The characteristic length ℓ also decreases by 20% under the effect of irradiation, as already observed by Konstantinov [78], indicating that the width of the FSDP and the BP are well correlated for silica samples.

The existence of such a correlation was contradicted by Börjesson *et al* [79, 80]. The authors investigated one system (B_2O_3) in which they introduce changes by adding several network modifiers and dopant salts. They determined the position of the FSDP by neutron diffraction, the position of the boson peak by Raman scattering and the sound velocity by Brillouin scattering. No correlation was found between the variations of the FSDP and that of the boson peak. The medium-range order of the Ag_2O and Li_2O modified glasses is nearly unchanged whereas strong modifications of the boson peak are observed. The effect of dopant salts is just the reverse: these strongly affect the FSDP but only slightly modify the boson peak.

In a similar investigation carried out on a series of "barium crown dense" glasses (Sovirel) at various fictive temperature T_f, each sample was kept at a temperature T_f in the transformation range during a time sufficiently long to allow its physical properties to stabilize at their equilibrium value for that temperature, after which the sample was rapidly quenched to room temperature. The measurements of the FSDP demonstrate that the position and width of the FSDP are independent of T_f [81]. However, the boson peak is significantly affected by changes in T_f: its position decreases with increasing T_f, as does the sound velocity. The correlation length ℓ deduced from the ratio of the sound velocity and the position of the boson peak increases with increasing temperature. Therefore, there is no correlation between the FSDP and the boson peak for this glass with respect to changes in the fictive temperature.

Another study [82] concerns several oxide glasses: LaSF7 and BaF4 from Schott, a soda-lime glass and two phosphosilicates glasses. A good correlation was found between d_1 and ℓ.

4.4. *Connection with hypersonic velocity*

In most of the models, the sound velocity is introduced in the expression of the maximum of the boson peak. Several studies where the authors investigated the influence of quenching rate on one sample of fixed chemical composition showed that the changes in ν_{max} reflect the changes in the sound velocity [83, 84]. This shows that the correlation length ℓ does not depend on the quenching rate. It is therefore a thermodynamical property, depending on the composition of the glass [8]. However, in the "barium crown dense" glass of Sovirel [81], the correlation length ℓ increases with increasing fictive temperature.

It has often been mentioned that the position of the boson peak varies proportionally to the longitudinal sound velocity [13, 85]. An example of very similar temperature dependences of both quantities in an oxide glass can be observed in Fig. 3(a). An interesting question is to know whether the boson peak should be better correlated to the transverse or to the longitudinal

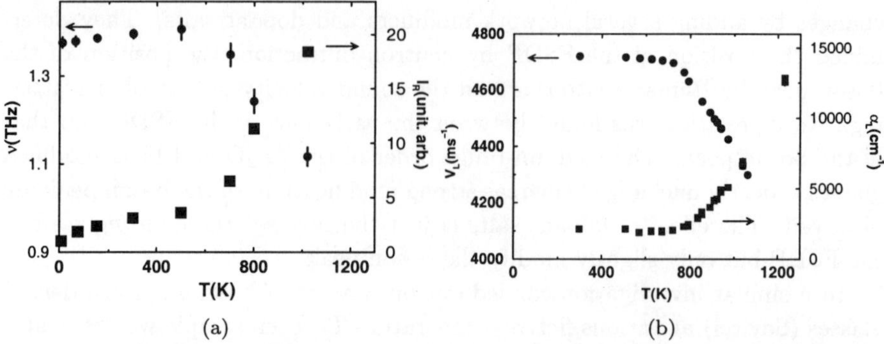

Fig. 3. Comparison in BaF4 glass between the temperature dependence of the boson peak and the quasi-elastic reduced intensity (a) on the one hand and that of the longitudinal sound velocity and the hypersound attenuation on the other hand (b).

velocity. It occurs in both the VV and the VH Raman spectra, so that it is expected to be at least partially due to depolarized modes which probably originate from transverse acoustic modes. Moreover, we expect that the maximum of the vibrational density of state is dominated by transverse modes by analogy with crystals where it is related to the Debye velocity which reflects mainly the variations of the transverse modes. Therefore, the variation of the boson peak is expected to be better correlated with the transverse velocity. Nevertheless, it was shown [86] that there is a linear relation between the longitudinal and transverse velocities for a wide variety of glasses at room temperature. Moreover, the velocity curves as a function of temperature are often parallel [16]. In amorphous orthoterphenyl, a partial correlation between the boson peak and the transverse velocity was found on the basis of their simultaneous disappearance at about $1.3\ T_g$ and qualitatively similar variations [59]. Another investigation carried out on B_2O_3 demonstrates a very close similarity of behavior between the position of the boson peak, the transverse and the longitudinal modes [13]. One can note that transverse modes are observable at more than 800 K above the glass transition temperature, which is equal to 740 K. The correlation length deduced from $\ell \propto V_a/\nu_{\max}(V_a = (V_L + 2V_T)/3$ is the average sound velocity, V_T and V_L are the transverse and longitudinal velocities respectively) decreases less than 2% in a wide temperature range from room temperature to 1300 K.

An experimental study carried out as a function of the temperature in a industrial optical glass (BaF4 provided by Schott) shows that the variation of $\nu_{\max}$ is similar to that of the transverse and longitudinal velocities: all quantities are nearly constant up to a temperature close to T_g, and then decreases sharply [15]. However, the authors show that the quantitative correlation between the variation with temperature of $\nu_{\max}$ and that of the sound velocities is strongly dependent upon the model used for the extraction of vibrational modes. In particular, a correct description of the quasi-elastic intensity and possibly, as will be presented in the following, of a correlation between relaxational modes responsible for this quasi-elastic line and vibrational modes which arises in the boson peak is a fundamental prerequisite for the study of such correlation.

Nevertheless, most studies make use of the apparent position of the BP. Indeed, few works present the models including relaxations and vibrations or a comparison of several models to determine the position of the maximum of the boson peak $\nu_{\max}$. An example of such comparative study is provided by Krüger *et al* [14]. The authors show that the variations of the position of the boson peak maximum are similar if they are determined using independent contributions to describe the boson peak and the quasi-elastic contributions from the one hand and using a model which assume coupling between relaxational modes responsible for the quasi-elastic intensity and vibrational modes at the origin of the boson peak on the other hand [7]. The only difference was the temperature at which $\nu_{\max}$ is no longer constant: this temperature is equal to T_y if one uses Gochiyaev's coupling model instead of being higher than T_g with the superposition model. Using both models, the correlation length deduced from the ratio of the sound velocity V and of the position $\nu_{\max}$ of the boson peak does not vary significantly with the temperature. This indicates a good correlation between the variation of V and $\nu_{\max}$.

5. Quasi-Elastic Scattering

The presence of an excess scattering at low frequencies in the Raman spectra has been reported for the first time by Winterling in silica [87]. He showed that at 5 cm^{-1} $\equiv$ 0.15 THz, and at room temperature, the measured intensity is one hundred time higher than the purely elastic intensity, and that the vibrational contribution expected in the Debye approximation represents only a small fraction of the measured intensity so that an excess scattering is evidenced. Fleury and Lyons extended this observation in a silica core fiber at lower frequencies and low temperatures (down to 1.6 K) [88]. In order to reach very low frequencies, they used a low-frequency technique based on an iodine cell for the reabsorption of the incident light [89]. It was then possible to investigate frequencies as low as 1 GHz. The quasi-elastic contribution exhibits a very strong temperature dependence, its intensity increases faster than the Bose factor. At high temperature, above T_g, the spectra are dominated by the quasi-elastic contribution. It is often assigned to relaxational contributions. These contributions are more important in fragile glasses than in strong ones, even below T_g [57]. However, the precise mechanism which account for this relaxation is the subject of a controversy.

The relaxation observed in the quasi-elastic Raman intensity has sometimes been identified with the fast β relaxation of the MCT, varying slowly with the temperature [90]. Other authors proposed a different approach where this so-called fast process is due to an increase of the density of states at low energy when the temperature is raised toward the glass transition [91, 92]. This β_{fast} process is then different from the β_{slow} of the MCT.

Several hints show that the boson peak and the quasi-elastic intensity are correlated. The boson peak appears in a frequency range (1THz) close to that of the β relaxation of the MCT. There is a continuous change as the temperature is raised from an almost purely vibrational behavior to a relaxational behavior above the glass transition temperature. The depolarization ratio, though having very distinct values in different glasses (from 0.25 to 0.75), is nearly constant over the frequency range of both features. Therefore, several models have been developed to account for both contributions, assuming a common origin for vibrational and relaxational modes [7, 57]. Those two contributions can also be explained in the framework of models which attempt to explain all the anomalies of glasses [93, 94].

5.1. *Models accounting for both the boson peak and the quasi-elastic scattering*

5.1.1. *Soft-potential model*

The soft potential model [68, 93] has been introduced as an extension of the two level systems model which provides a good description of the low temperature properties of glasses on the basis of a nearly constant density of two level systems [95, 96]. However, there is a deviation from this model at higher temperatures, due to the existence of other kinds of low-energy excitations. The soft potential model introduces "weak" forces for structural entities and "soft" modes caused by the local strain induced in glasses during their formation. Three kinds of soft modes can be described by a potential of the following form:

$$V(x) = \epsilon_0 \left[\eta \left(\frac{x}{a}\right)^2 + \xi \left(\frac{x}{a}\right)^3 + \left(\frac{x}{a}\right)^4 \right] \qquad (16)$$

x is a generalized coordinate associated with the low frequency modes, a is a characteristic length of the order of an interatomic distance, ϵ_0 is a characteristic energy of the order of a chemical bond, η and ξ are dimensionless parameters depending on the considered mode. Soft modes corresponds to $|\eta| \ll 1$ and $|\xi| \ll 1$. The three kinds of low frequency excitations are: tunneling two-level systems, thermally activated modes in double-well potentials and quasi-localized oscillators in harmonic or quasi-harmonic single-well potentials. In this approach, the boson peak is due to a localization of phonons by interactions with soft modes of spatial extension comparable to the wavelength of phonons. The quasi-elastic intensity is due to coupling of soft modes with propagating phonons: the displacement due to the soft modes induces a modulation of the dipole moment proportional to the polarizability tensor.

5.1.2. *Vibration-relaxation*

The "vibration-relaxation" model developed by Buchenau for the interpretation of neutron scattering data in polymers [57, 97] suggests a transformation from more or less damped vibrations at high frequencies to low-barrier relaxations at low frequencies. This model is a simplified classical version of the soft potential model for glasses. It predicts that high frequency (above the boson peak) modes are vibrational ones: two thirds coming from eigenmodes corresponding to positive eigenvalues which stay vibrational at lower frequencies, and one third corresponding to negative eigenvalues. The modes with negative eigenvalues move in a double-well potential and can turn into relaxational contribution at low frequency by jumping over the barrier. In this model, the boson peak arise from localized vibrations ("soft modes") coexisting and interacting with sound waves.

5.1.3. *Gochiyaev's coupling model*

Gochiyaev's model introduces a strong coupling between relaxational and vibrational contributions. The authors assume that the vibrational excitations seen as the boson peak at $T \ll T_g$ are strongly coupled to some relaxing variable which does not itself contribute to the scattering but induces a broadening of each frequency of the boson peak [7]. The coupling introduces two parameters: the damping rate γ related to a relaxation time τ by $\gamma = 1/\tau$ and the coupling strength δ. The adjustable parameters are τ, δ, the position of the maximum of the vibrational contribution $\nu_{\max}(T)$ and an amplitude factor I_0^{bp}. The Raman reduced intensity at a given temperature T is than described by the following expression:

$$I_R(\nu,T) = \frac{1}{\nu}\int \frac{I_R^0(\nu',T)\nu'^2\frac{\delta^2\gamma}{\nu^2+\gamma^2}d\nu'}{\left(\nu'^2-\nu^2-\frac{\delta^2\gamma^2}{\nu^2+\gamma^2}\right)^2+\left(\frac{\delta^2\nu\gamma}{\nu^2+\gamma^2}\right)^2} \qquad (17)$$

where $I_R^0(\nu',T)$ is the vibrational contribution deduced from a measurement at low enough temperature to assume that the relaxational processes are frozen out for this temperature. The softening of vibrational modes as the temperature increases can be accounted for by rescaling the lowest temperature spectra (10 K for example) along the frequency axis by the ratio of the adjustable variable $\nu_{\max}(T)$ and the position of the boson peak at low temperature.

As the low temperature spectrum $I_R^0(\nu',T)$ contains mainly vibrational contributions, the depolarization ratio is the same for the vibrational and quasi-elastic contribution.

Several authors proposed minor modifications of this model. Petscherezin *et al.* introduce the structural relaxation at low frequency in order to avoid divergence of the integral (17) [98]. Kojima introduced a model for the relaxation process which predicts the growth of the quasi-elastic intensity as a function of the free volume [99]. Novikov assumed that the phonon density fluctuations coupled to the boson peak due to vibration anharmonicity [100]. The strength of the relaxation process is thus proportional to the square of Grüneisen constants.

5.2. *Theodorakopoulos and Jäckle's model for the quasi-elastic scattering*

In order to account for the quasi-elastic intensity, Theodorakopoulos and Jäckle introduced a coupling of light with localized structural defects with two configuration states [101]. Their basic assumption is that the two states are states of different polarizability. Light scattering is caused by time-dependent fluctuations of the polarizability. It can be expressed by the polarizability autocorrelation function. These defects can be excited either by tunneling at low temperature or by thermal activation at higher temperature. For defects which can relax via thermal activation, this autocorrelation function is proportional to $(\Delta\alpha)^2/(1-i2\pi\nu\tau)$, where $\Delta\alpha$ is the polarizability difference between the two possible states of the defect and $\tau = \tau_0 \exp -v_j/kT$ is the inverse of the rate of thermally activated transition for a potential barrier v_j. Assuming a large distribution of the relaxation times associated to the defects, it yields, for the relaxational part of the intensity:

$$\frac{I^{\text{rel}}(\nu)}{\nu[n(\nu)+1]} \propto \frac{(\Delta\alpha)^2}{2\pi kT} \int P(v)\frac{\tau(v)dv}{1+4\pi^2\nu^2\tau^2(v)} \tag{18}$$

where $P(v)$ is the distribution of energy barriers and $\tau(v)$ the distribution of relaxation times.

5.3. *Connection with hypersonic attenuation*

In the framework of the two-level systems model or the soft potential model, a proportionality relationship between the attenuation $\alpha(\nu,T)$ and the Raman reduced intensity I_R has been predicted if the origins of the two physical properties are related. The relationship affects either the relaxational [101] or the total [102] reduced intensity. Thus, if the infrared or hypersonic attenuation are due to the same kind of modes as the Raman intensity, the intensity takes the following form:

$$I(\nu,T) = \frac{1+n(\nu)}{\nu}\alpha(\nu,T) \tag{19}$$

where $\alpha(\nu,T)$ is the hypersonic or infrared attenuation. This correlation was first proposed by Theodorakopoulos and Jäckle [51, 101] in the case of silica where defects which can have two distinct states of polarizability are responsible for a quasi-elastic intensity due to the relaxation of these defects as well as the hypersonic attenuation in the GHz range: then, the relaxational Raman I^{rel} scattering excess and the hypersonic $\alpha^{\text{rel}}(\nu,T)$ of relaxational origin are proportional. In the framework of the soft potential model, Gurevich *et al* [68, 102] have established this connection and discussed the temperature and frequency dependence of these quantities. Sokolov predicts such correlation by assuming that the quasi-elastic contributions arises from the scattering of light on damped vibrations which have a quasi-elastic part in their response function; the vibrations at the boson peak frequency are damped by the same process as the acoustic waves [103].

An attempt to calculate the hypersonic attenuation using a model of relaxation of defects and parameters extracted from ultrasonic measurements failed because of additional contributions to the attenuation due to anharmonicity or to the existence of defects at hypersonic frequencies inefficient at ultrasonic frequencies. Very few experimental data for the sound wave attenuation in the GHz range are available because they require measurements of the hypersonic attenuation α by Brillouin scattering using a high resolution spectrometer and α can be extracted only at the frequency of the Brillouin shift. A comparison of Brillouin attenuation and Raman quasi-elastic intensity was tested in boron oxide glass as a function of temperature (from 20 K to the glass transition temperature $T_g = 540$ K): a good connection was found below 300 K only and could not be extended to glasses of different composition (B_2O_3–xLi_2O) [17]. A rather good correlation was found between the ultrasonic attenuation and the elastic scattering excess [104]. Buchenau *et al.* evidenced similarities between the temperature dependence of the Raman intensity (or the inelastic neutron scattering data) and the ultrasonic or infrared attenuations for temperatures below 300 K in silica [53]. A more recent study establishes that the temperature and frequency dependence of the quasi-elastic scattering agrees qualitatively with the predictions of the soft potential model for the temperature and frequency dependence of the hypersonic attenuation [105]. These results support the hypothesis of a common origin on the basis of classical activated relaxation processes for the attenuation. A qualitative agreement for the variation of both quantities was obtained for commercial borosilicate (Sovirel) "barium-crown dense" glasses with different fictive temperatures [81]. A recent attempt to check if the connection is fulfilled for several samples was successfully performed at room temperature in silica samples [16]. The reduced Raman intensity and the hypersonic attenuation have clearly correlated variations for various silica samples. A correlation was clearly evidenced for a BaF4 (Schott) strong glass as a function of the temperature in the 300–1000 K range [15] [see Fig. 3(b)].

5.4. *Evidence of two-level systems*

Many attempts to observe two-level systems by direct interaction with the inelastic scattered light (in the frequency range of the quasi-elastic line of low-frequency Raman) have been unsuccessful [88, 106]. Recently, an anomalous temperature dependence of the scattered light has been observed below 10 K in LaSF7 glass (a heavy metal oxide glass from Schott) [107]. This glass exhibits an increase of the scattered intensity with increasing temperature below 10 K (see Fig. 4). This effect has been explained in terms of resonant interactions with tunneling states in the framework of Theodorakopoulos and Jäckle's model [101]. However, it was possible to confirm such an effect only on one

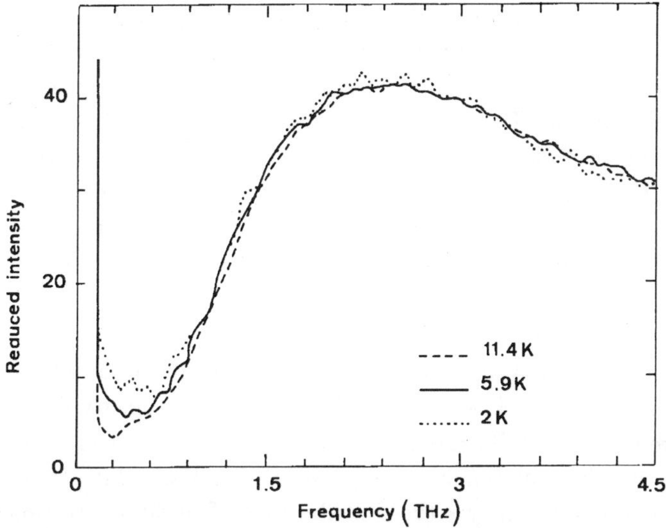

Fig. 4. Reduced Raman intensity for LaSF7 glass from Schott at low temperature, showing anomalous behavior: the scattered light at 4 K is more intense than the 10 K signal [82].

other particular heavy metal oxide glass [108]. Therefore the anomalous behavior seems to be strongly correlated to the presence of polar ions or electronic extrinsic defects in the glass.

6. Conclusions

The field of low frequency dynamics in glasses, though extensively studied, is full of open questions. In order to find a unifying view of glass properties, many papers try to relate the dynamics to the structure or several dynamics properties between them. Several models predicts a connection between the first sharp diffraction peak, which is a universal property of glasses, characteristic of the medium range order, to the boson peak, a characteristic of the dynamics of disordered systems. Even if this correlation has been observed by comparison of a great number of samples with different compositions: in a polymer such as polymethylmetacrylate, in amorphous germanium and selenium, in several chalcogenide glasses (As_2Se_3, As_2S_3, GeS_2, $SiSe_2$, $GeSe_2$), in AgI, $Ag2O + nB_2O_3$, in several silica samples, in GeO_2 with 5.4% of Na_2O, in six borosilicate commercial glasses from Schott: (SF1, F1, BaSF64, BaF4, LaSF9, LaSF7), in a soda-lime glass and in phosphosilicate glass with different fictive temperatures, it was contradicted by several works where the influence of small modifications of a given composition without changing the short range order was studied (addition of small amounts of modifiers Li_2O, Ag_2O or dopant salts LiCl or AgI to B_2O_3, changes of fictive temperatures in borosilicate glass from Sovirel-B90-30). This means that this correlation is not universal. However, one should keep in mind that in multicomponent glasses, the FSDP occurs from the superposition of several contribution (peaks at this position arise in several partial distribution function) and thus its apparent (average) position may not be the proper parameter to relate to the sound velocity.

The connection between the position of the boson peak, which represents the dynamics in the THz range and the hypersound velocity which relates to the dynamics in the GHz range, is qualitatively observed in a great number of glasses, but the achievement of a quantitative relationship strongly depends on the model used to analyze the contributions. The same assertion is true for the connection of the temperature dependence of the attenuation and of the quasi-elastic intensity as well as their modifications to minor changes in composition, which indicates that the same "defects" or "relaxational processes" are responsible for both the attenuation in the 10 GHz range and the Raman quasi-elastic scattering in the 100 GHz range.

However, due to the complexity of the glassy material and to the number of mechanism at the origin of the scattering, none of these correlations seems completely universal. The origin of the attenuation of sound waves and its relation to localization of modes is a much debated question. The question is whether the boson peak arises from localized modes in excess due to the disorder or from propagating modes. New experimental techniques such as inelastic X-ray scattering or neutron Brillouin scattering may bring new elements to the debate. Very recently, the existence of propagating phonons with wavelength in the nanometer range, observed by inelastic X-ray scattering has been reported [109] and contradicted [70, 110]. Recent approaches by numerical simulations will probably also help to solve this question.

Acknowledgments

I would like to thank Professor Jacques Pelous for his critical reading and many suggestions about this manuscript. I am also grateful to J.-L. Prat, D. Cavaillé, F. Terki and Y. Scheyer for their experimental data.

References

[1] *Amorphous Solids: Low Temperature Properties*, ed. W.A. Philips (Springer-Verlag, Berlin, 1981).

[2] W. Götze and L. Sjögren, *Rep. Progr. Phys.* **55**, 241 (1992).

[3] G.P. Johari and J. Goldstein, *J. Chem. Phys.* **53**, 2372 (1970).

[4] W. Götze and L. Sjögren, *Transp. Theor. Stat.* **24**, 801 (1994).

[5] C.A. Angell, *J. Phys. Chem. Solids* **49**, 863 (1988).

[6] U. Buchenau, C. Schönfield, D. Richter, T. Kanaya, K. Kaji and R. Wehrmann, *Phys. Rev. Lett.* **73**, 2344 (1994).

[7] V.Z. Gochiyaev, V.K. Malinovsky, V.N. Novikov and A.P. Sokolov, *Philos. Mag.* **B63**, 777 (1991).

[8] V.N. Novikov and A.P. Sokolov, *Solid State Commun.* **77**, 243 (1991).

[9] A.P. Sokolov, A. Kisliuk, M. Soltwisch and D. Quitmann, *Phys. Rev. Lett.* **69**, 1540 (1992).

[10] C. Moss and D.L. Price, Physics of Disordered Materials, eds. D. Adler, H. Fritzsche and S.R. Ovshinsky (Plenum Press, New York, 1985), p. 77.

[11] S. Susman, K.J. Volin, D.G. Montague and D.L. Price, *Phys. Rev.* **B43**, 11076 (1991).

[12] P.H. Gaskell and D.J. Wallis, *Phys. Rev. Lett.* **76**, 66 (1996).

[13] A.K. Hassan, L.M. Torell and L. Börjesson, *J. Phys. IV* **C2**, 265 (1992).

[14] M. Krüger, M. Soltwisch, I. Petscherizin and D. Quitmann *J. Chem. Phys.* **96**, 7352 (1992).

[15] F. Terki, C. Levelut, J.-L. Prat, M. Boissier and J. Pelous, *J. Phys. Cond. Matt.* **9**, 3955 (1997).

[16] F. Terki, C. Levelut, M. Boissier and J. Pelous, *Phys. Rev.* **B53**, 2411 (1996).

[17] J. Lörosch, M. Couzi, J. Pelous, R. Vacher and A. Levasseur *J. Non-Cryst. Solids* **69**, 1 (1984).

[18] A. Einstein, *Ann. Phys.* **33**, 1275 (1910).

[19] B.J. Berne and R. Pecora, *Dynamic Light Scattering* (Wiley, New York, 1976).

[20] P. Brüesch, *Phonons: Theory and Experiments II* (Springer Verlag, Berlin, 1986).

[21] J.G. Dil, *Rep. Prog. Phys.* **45**, 285 (1982).

[22] S. Candau, *Ann. Phys.* **4**, 21 (1969).

[23] L. Brillouin, *Ann. Phys.* **17**, 88 (1922).

[24] L. Landau and G. Placzek, *Z. Phys. Sovvjetunion* **5**, 172 (1922).

[25] R.D. Mountain, *Rev. Mod. Phys.* **38**, 205 (1966).

[26] J. Frenkel, *Kinetic Theory of Liquids* (Dover Publications, New York 1955).

[27] S.M. Rytov, *Soviet Physics JETP* **6**, 30 (1958).

[28] S. Hunklinger and W. Arnold, Ultrasonic Properties of Glasses at Low Temperatures, *Physical Acoustics, Principles and Methods*, Vol. XII, eds. Mason and Thurston (Academic Press, New York, 1976) p. 155; S. Hunklinger and M.V. Shickfus, in Ref. [1], p. 53.

[29] J. Pelous and R. Vacher, *J. Phys.* **38**, 1153 (1977).

[30] R. Shuker and W. Gammon, *Phys. Rev.* **B25**, 222 (1970).

[31] J.R. Sandercock, *J. Phys.* **E9**, 566 (1976).

[32] R. Vacher, Thesis, Montpellier (1972); L. Boyer, Thesis, Montpellier (1972); R. Vacher, M.V. Shickfus, S. Huncklinger, *Rev. Sci. Inst.* **51**, 288 (1980); R. Vacher, R. Vialla, D. Cavaillé, to be published.

[33] J. Pelous, R. Vacher and J. Phalippou, *J. Non-Cryst. Solids* **30**, 385 (1979).

[34] M. Schmidt, R. Vacher, J. Pelous and S. Hunklinger, *J. Phys Colloq.* **43**, C9-501 (1982).

[35] R. Vacher, J. Pelous and E. Courtens, *Phys. Rev.* **B56**, R481 (1997).

[36] S. Hunklinger, W. Arnold, S. Stein, R. Nava and K. Dransfeld, *Phys. Lett.* **A42**, 253 (1972).

[37] J. Pelous and R. Vacher, *Solid State Commun.* **19**, 627 (1976).

[38] R. Vacher, H. Sussner and S. Hunklinger, *Phys. Rev.* **B21**, 5850 (1980).

[39] J. Jäckle, L. Piché, W. Arnold and S. Hunklinger, *J. Non-Cryst. Solids* **20**, 365 (1976).

[40] D. Tielburger, R. Merz, R. Ehrenfels and S. Hunklinger, *Phys. Rev.* **B45**, 2750 (1992).

[41] J. Jäckle, *Proc. IV Int. Conf. on Physics of Non-Crystalline Solids*, ed. G.H. Frischat (Trans. Tech., 1977) p. 568.

[42] D.P. Jones, N. Thomas and W.A. Phillips, *Philos. Mag.* **B34**, 5684 (1986).

[43] L. Landau and G. Rumer, *Phys. Z. Sovvjetunion* **11**, 18 (1937).

[44] G. Grimvall, *Thermophysical Properties of Materials*, Series of monographs on Selected Topics in Solid State Physics, ed. E.P. Wohlfarth (North-Holland Physics Publishing, Amsterdam, 1986).

[45] A. Akhieser, *J. Phys. (USSR)* **1**, 277 (1939).

[46] A.S. Pine, *Phys. Rev.* **185**, 1187 (1969).

[47] R. Vacher, J. Pelous, F. Plicque and A. Zarembowitch, *J. Non-Cryst. Solids* **45**, 397 (1981).

[48] J.P. Bonnet, R. Vacher, J. Pelous and C. Laermans, *Phys. Rev.* **B45**, 557 (1992).

[49] O.L. Anderson and G.J. Dienes, *Non-Crystalline Solids*, *Proc. Conf. on Non-Crystalline Solids*, Alfred, New York, September 1958, ed. V.D. Frechette (John Wiley & Sons, New York, 1959).

[50] M.N. Kul'bitskaya, V.P. Romanov and V.A. Shutilov, *Sov. Phys. Acoust.* **19**, 399 (1974).

[51] J. Jäckle, Low-frequency Raman Scattering in Glasses, *Amorphous Solids: Low Temperature Properties*, ed. W.A. Phillips (Springer, Berlin, 1981).

[52] U. Strom and P.C. Taylor, *Phys. Rev.* **B16**, 5512 (1977).

[53] U. Buchenau, H.M. Zhou, N. Nücker, K.S. Gilroy and W.A. Phillips, *Phys. Rev. Lett.* **60**, 1318 (1988).

[54] J. Colmenero, *Physica* **A234–236**, 1236 (1997).

[55] M.T. Dove, M.J. Harris, A.C. Hannon, J.M. Parker, I.P. Swainsom and M. Gambhir, *Phys. Rev. Lett.* **78**, 1070 (1997).

[56] A. Martin and W. Brenig, *Phys. Status Solidi* **B64**, 163 (1974).

[57] U. Buchenau, M. Prager, N. Nücker, A.J. Dianoux, N. Ahmad and W.A. Phillips, *Phys. Rev.* **B34**, 5665 (1986).

[58] V.K. Malinovski, V.N. Novikov, A.P. Sokolov and V.G. Dodonov, *Solid State Commun.* **65**, 681 (1988).

[59] R.B. Stephens, *Phys. Rev.* **B8**, 2896 (1973).

[60] U. Buchenau, N. Nücker, A.J. Dianoux, *Phys. Rev. Lett.* **53**, 2316 (1984).

[61] R.J. Nemanich, *Phys. Rev.* **B16**, 1655 (1977).

[62] E. Duval, A. Boukenter and T. Achibat, *J. Phys.: Condens. Matter* **2**, 10227 (1990).

[63] T. Pang, *Phys. Rev.* **B45**, 2490 (1992).

[64] I. Pócsik and M. Koós, *Solid State Commun.* **74**, 1253 (1990).

[65] E. Akkermans and R. Maynard, *Phys. Rev.* **B32**, 7850 (1985).

[66] S.R. Elliot, *Europhys. Lett.* **19**, 201 (1992).

[67] W. Schirmacher and M. Wagener, Effective Medium Theory for Phonons in Glasses: The Mean Free Path, *Phonons 89*, eds. S. Hunklinger, W. Ludwig and G. Weiss (World Scientific, Singapore, 1993) p. 531.

[68] D.A. Parshin, *Phys. Solid State* **36**, 991 (1994).

[69] M.I. Klinger, *Phys. Lett.* **A170**, 222 (1992).

[70] M. Foret, E. Courtens, R. Vacher and J.-B. Suck, *Phys. Rev. Lett.* **77**, 3831 (1996).

[71] V.K. Malinovski and A.P. Sokolov, *Solid State Commun.* **57**, 757 (1986).

[72] E. Duval, A. Boukenter and B. Champagnon, *Phys. Rev. Lett* **56**, 2052 (1986).

[73] D.L. Price, S.C. Moss, R. Reijers, M.L. Saboungi and S. Susman, *J. Phys. Cond. Matt.* **21**, L1069 (1989).

[74] S.R. Elliott, *Phys. Rev. Lett.* **76**, 711 (1991).

[75] A. Guinier, *Théorie et technique de la radiocristallographie* (Dunod, Paris, 1956).

[76] D.L. Price, S.C. Moss, R. Reijers, M.L. Saboungi and S. Susman, *J. Phys.* **C21**, 1069 (1988).

[77] C. Levelut, F. Terki, Y. Scheyer and J. Pelous, Vibrational Dynamics of Glasses, *Amorphous Insulators and Semiconductors, NATO ASI Series, 3. High Technology*, Vol. 23, eds. M. Thorpe and M. Mitkova (Kluwer Academic Publishers, Dordrecht, 1997) p. 385.

[78] A.V. Konstantinov, L.V. Maksimov, A.R. Silin and O. V. Yanush, *J. Non-Cryst. Solids* **123**, 286 (1990).

[79] L. Börjesson, A.K. Hassan, J. Swenson, L.M. Torell and A. Fontana, *Phys. Rev. Lett.* **70**, 127 (1993).

[80] A.K. Hassan, L. Börjesson and L.M. Torell, *J. Non-Cryst. Solids* **172–174**, 154 (1994).

[81] C. Levelut, N. Gaimes, F. Terki, G. Cohen-Solal, J. Pelous and R. Vacher, *Phys. Rev.* **B51**, 8606 (1995).

[82] J.L. Prat, PhD Thesis of University Montpellier II (France), 1996.

[83] V.K. Malinovski, V.N. Novikov and A.P. Sokolov, *Phys. Lett.* **A13**, 19 (1987).

[84] V.Z. Gochiyaev and A.P. Sokolov, *Sov. Phys. Solid State* **31**, 557 (1989).

[85] W. Steffen, B. Zimmer, A. Patkowski, G. Meier, E.W. Fischer *J. Non-Cryst. Solids* **172–174**, 37 (1994).

[86] J.F. Berret and M. Meißner, *Z. Phys. B — Cond. Matt.* **72**, 1987.

[87] G. Winterling, *Phys. Rev.* **B12**, 2432 (1975).

[88] K.B. Lyons, P.A. Fleury, R.H. Stolen and M.A. Bösch, *Phys. Rev.* **B26**, 7123 (1982).

[89] P.A. Fleury and P.A. Fleury, *Phys. Rev. Lett.* **36**, 1188 (1976).

[90] G. Li, M. Du, M., X.K. Chen, H.Z. Cummins and J.T. Tao, *Phys. Rev.* **A45**, 3867 (1992).

[91] U. Buchenau, M. Galperin, V.L. Gurevich and H.R. Schober, *Phys. Rev.* **B43**, 5039 (1991).

[92] R. Zorn, D. Richter, B. Frick and B.Farago, *Physica* **A201**, 52 (1993).

[93] V.G. Karpov, M.I. Klinger, F.N. Ignat'ev, *Sov. Phys. JETP* **57**, 439 (1983).

[94] U. Buchenau, Y.M. Galperin, V.L. Gurevich, D.A. Parshin, M.A. Ramos and H.R. Schober, *Phys. Rev.* **B49**, 940 (1994).

[95] P.W. Anderson, B.I. Halperin, C.M. Varma, *Phil. Mag.* **25**, 1 (1992).

[96] W.A. Phillips, *J. Low-Temp. Phys.* **7**, 351 (1972).

[97] U. Buchenau, *Philos. Mag.* **B71**, 793 (1995).

[98] J. Petscherizin, S. Loheider, M. Soltwisch and D. Quitmann, Structure and Dynamics in the Glass ↔ Liquid Transformation, *Proc. Workshop on Non-equilibrium Phenomenon*, Pise, Italy, September 1995, eds. M. Giordano, D. Leporini and M.P. Tosi (World Scientific, Singapore, 1996) p. 315.

[99] S. Kojima and V.N. Novikov, *Phys. Rev.* **B54**, 222 (1996).

[100] V.N. Novikov, *Phys. Rev.* **B58**, 8367 (1998); V.N. Novikov, *Phys. Rev.* **B55**, R14685 (1997).

[101] N. Theodorakopoulos and J. Jäckle, *Phys. Rev.* **B14**, 2637 (1976).

[102] V.L. Gurevich, D.A. Parshin, J. Pelous and H.R. Schober, *Phys. Rev.* **B48**, 16318 (1993).

[103] A.P. Sokolov, V.N. Novikov and B. Strube, *Europhys. Lett.* **38**, 49 (1997).

[104] J.-F. Berret, J. Pelous, R.Vacher, A.K. Raychaudhuri and M. Schmidt, *J. Non-Cryst. Solids* **87**, 70 (1986).

[105] A. Brodin, A. Fontana, L. Börjesson, G. Carini and L.M. Torell, *Phys. Rev. Lett.* **73**, 2067 (1994).

[106] R.H. Stolen and M.A. Bösch, *Phys. Rev.* **B48**, 805 (1982).

[107] J.-L. Prat, F. Terki and J. Pelous, *Phys. Rev. Lett.* **7**, 755 (1996).

[108] F. Terki, J.-L. Prat and J. Pelous, *Phil. Mag.*, **B77**, 373 (1998).

[109] C. Masciovecchio, G. Ruocco, F. Sette, M. Krisch, R. Verbeni, U. Bergmann and M. Soltwisch, *Phys. Rev. Lett.* **76**, 3356 (1996).

[110] M. Foret, B. Hehlen, G. Taillades, E. Courtens, R. Vacher, H. Cosolta and B. Dorner, *Phys. Rev. Lett.* **81**, 2100 (1998).

7

Tunneling Systems in Crystalline and Amorphous Solids

SIEGFRIED HUNKLINGER and CHRISTIAN ENSS

Institut für Angewandte Physik, Universität Heidelberg,
Albert-Ueberle-Str. 3-5, D-69120 Heidelberg, Germany
hunklinger@urz.uni-heidelberg.de

Contents

1. Introduction

The idea of atomic tunneling can be traced back to Hund's publication [1] on
the optical spectrum of ammonia in 1927. In NH_3-molecules there are two
equivalent positions for the nitrogen atom, namely on either side of the plane
formed by the three hydrogen atoms. These two positions are separated by a
small energy barrier of about 0.3 eV. Transitions from one potential minimum
to the other are possible by tunneling through the barrier. In this way, the
degeneracy is lifted and a small splitting of the ground state of about 24 GHz
occurs. It is accessible to absorption measurements at microwave frequencies
and is the physical basis of the ammonia maser invented in 1955.

Tunneling of atoms in solids was first discussed by Pauling [2] in 1930
when he considered the hindered rotation of molecules in crystals. In 1962,
it was shown by Känzig [3] that oxygen impurities embedded in alkali halide
crystals are able to tunnel between adjacent positions. Intense experimental
and theoretical studies at low temperatures led to a satisfactory understanding
of the phenomena at small defect concentrations but at higher defect densities
many questions remained unanswered. A new chapter was opened in 1971 when
Zeller and Pohl [4] found that at low temperatures glasses exhibit thermal
properties fundamentally different from those of their crystal counterparts.
Shortly afterwards it was shown that also in this case it is the tunneling motion
on atomic scales that gives rise to those hitherto unexpected properties.

Quantum tunneling of atoms and molecules is a ubiquitous phenomenon
in disordered solids and turned out to be the origin of a large variety of

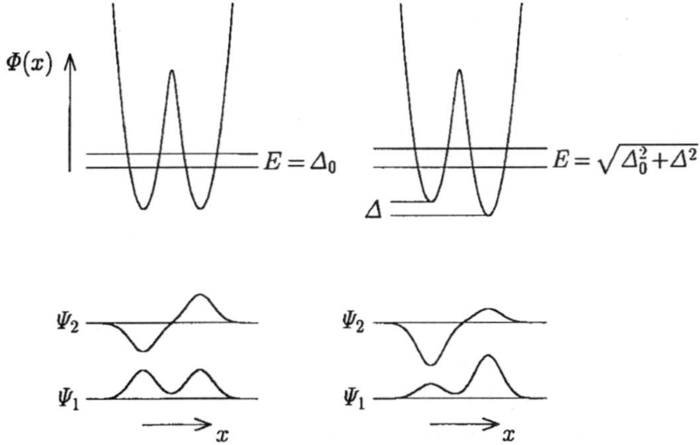

Fig. 1. Schematic representation of the potential $\Phi(x)$ and the wave functions Ψ_1 and Ψ_2 of a particle in a double-well potential. On the right-hand side the potential wells are shifted by the asymmetry energy Δ with respect to each other.

fascinating phenomena. Therefore, it was a matter of great interest during the last 40 years and still causes many experimental and theoretical surprises.

The tunneling motion of atoms, molecules, or small groups of atoms between potential minima with nearly the same depth gives rise to low energy excitations which determine the dynamical properties of most solids at low temperatures. In the simplest case such systems can be described as particles with an effective mass m moving back and forth in a double-well potential formed by smoothly joined identical wells of two harmonic oscillators. Such potentials are plotted in Fig. 1 together with the corresponding wave functions Ψ_1 and Ψ_2 of the two eigenstates. Let us start the discussion with the symmetric double-well potential shown on the left-hand side with two wells of identical depth. In this case neither of the wells is energetically preferred and the particle is found in both wells with equal probability. The energy difference E between the eigenstates is given by the exchange energy which we will call "tunnel splitting" Δ_0 in this article. Using elementary quantum mechanics the splitting can be calculated; in good approximation it is given by

$$\Delta_0 = 2\hbar\omega_0 \sqrt{\frac{2V}{\pi\hbar\omega_0}} \exp\left(\frac{-d\sqrt{2mV}}{2\hbar}\right) \qquad (1)$$

where $\hbar\omega_0$ is the vibrational energy of the particle moving in one well, d the distance of the two wells, and V the barrier height. Since the motion of the tunneling particles is not necessarily translational, d and m are, in a more general sense, quantities in configuration space. For the sake of simplicity the expression (1) is often replaced by

$$\Delta_0 \approx \hbar\omega_0\, e^{-\lambda} \tag{2}$$

Roughly speaking, the tunnel splitting Δ_0 is given by the vibrational energy $\hbar\omega_0$ of the particle multiplied by the probability $\exp(-\lambda)$ for tunneling. The so-called tunneling parameter $\lambda = d\sqrt{2mV}/2\hbar$ is a measure of the overlap of the wave functions of the particle localized in single wells.

A more general situation is shown on the right-hand side of Fig. 1, where the depths of the two wells differ by the asymmetry energy Δ. The wave functions have lost their symmetry and the particle preferentially dwells in the lower minimum. The energy difference between the two lowest eigenstates is now given by

$$E = \sqrt{\Delta_0^2 + \Delta^2} \tag{3}$$

Since the tunnel splitting Δ_0 is usually rather small, the thermal energy $k_{\mathrm{B}}T$ becomes comparable to Δ_0 only at low temperatures. Under this condition tunneling systems (TSs) have a decisive impact on the properties of solids. In particular, they dominate the thermal, elastic, and dielectric properties of disordered crystals and amorphous solids at low temperatures.

In Sec. 2, we will discuss the properties of TSs arising from substitutional defects in crystals. We will show that with rising defect concentration a transition from coherent to incoherent tunneling occurs, driven by the interaction between the TSs. In addition, we will discuss the properties of defect pairs and TSs in crystals with large random strains. Section 3 is devoted to TSs in structural glasses. We will deal with the elastic, dielectric, and thermal properties of this class of solids and we will show that also in these materials the interaction between TSs causes interesting and unusual phenomena.

2. Tunneling Systems in Crystals

In the simplest case, TSs in crystals are caused by substitutional atoms which do not properly fit into the lattice of the host crystal. Because they differ in shape and size from the atoms of the host material they experience a multi-valley potential landscape. The number of equivalent wells and their orientation within the crystal depends on both, the impurity atom and the host. At low temperatures, the thermal energy is not sufficient for the impurities to overcome the potential barriers between the wells via thermally activated processes, and tunneling through the barriers can be observed. As we will see the behavior of TSs strongly depends on their density: at very low concentrations, the average distance between the impurity atoms is so large that interaction between them can be neglected. In this case, the substitutional atoms, which we also call defects, can be treated as an ensemble of identical and independent TSs. With rising concentration, the interaction between them becomes increasingly important and leads to new collective phenomena which are fundamentally different from those observed for isolated TSs. Of course it is impossible to give in this article a complete review of the tremendous amount of experimental and theoretical work on the properties of TSs published during the last decades. Therefore, we will discuss only selected examples, which we believe to be typical. In particular, we will restrict our discussion to some substitutional defects in alkali halides which can be considered as simple model systems. For detailed reviews on the early work on TSs in crystals we refer to Refs. [5] and [6]. An overview including very recent results is given in Ref. [7].

2.1. *Isolated tunneling systems*

In cubic materials like alkali halide crystals, three different types of multivalley potentials are compatible with octahedral symmetry, namely systems with potential minima in $\langle 100 \rangle$-, $\langle 110 \rangle$- or $\langle 111 \rangle$-direction. As an exemplary case, we will discuss TSs originating from Li^+-ions in KCl. It is an especially simple and intensively studied model system in which the considerably smaller lithium ions substitute potassium ions of the host. Instead of occupying the centrosymmetrical lattice site, Li^+-ions prefer to dwell in an off-center position in $\langle 111 \rangle$-direction. In our further discussion, we will name this type of defect "$\langle 111 \rangle$-TS". As an illustration the ions in the (001)-plane of a KCl crystal are drawn in Fig. 2 together with a substitutional Li^+-ion. For the small Li^+-ion there exist eight off-center positions, namely four positions above and four below the plane of drawing. The eight potential minima are located at the corners of a cubic cage with side length $d = 1.4$ Å. The position of these minima is given by $r = (d/2)(\alpha, \beta, \gamma)$ with the values α, β, $\gamma = \pm 1$. We will denote the corresponding localized wave function by $|\alpha\beta\gamma\rangle$. Tunneling of ions between the corners of this cube lifts the degeneracy of the ground state.

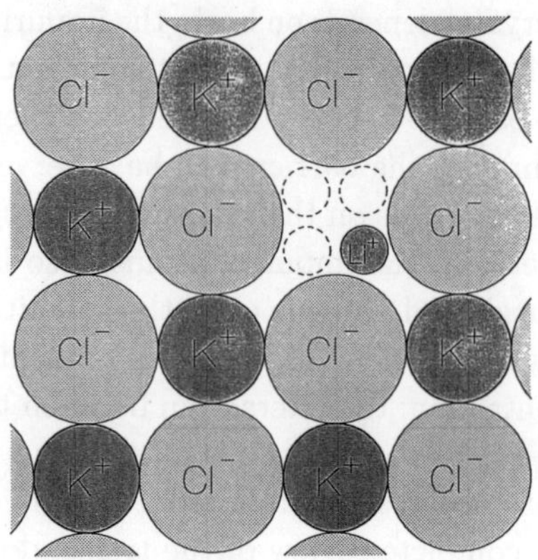

Fig. 2. Schematic representation of a (001)-plane of a KCl crystal with an embedded Li^+-ion. The eight off-center positions are located above and below the plane of drawing.

As the most important tunneling path is along the edges of the cage we neglect tunneling along faces and space diagonals (see e.g. Ref. [8]). The eigenstates of the TSs are constructed by a linear superposition of the localized states $|\alpha\beta\gamma\rangle$ with appropriate phase factors. In this way one obtains

$$|j\rangle = \frac{1}{\sqrt{8}} \sum_{\alpha,\beta,\gamma} \exp\left(i\frac{\pi}{d}\, j \cdot r\right) |\alpha\beta\gamma\rangle \tag{4}$$

where the components of j have the values 0 or 1. The corresponding eigenvalues E_j read as follows:

$$E_{111} = \hbar\omega_0/2 + 3\,|\zeta|$$
$$E_{110} = E_{101} = E_{011} = \hbar\omega_0/2 + |\zeta|$$
$$E_{100} = E_{010} = E_{001} = \hbar\omega_0/2 - |\zeta|$$
$$E_{000} = \hbar\omega_0/2 - 3\,|\zeta| \tag{5}$$

where $\hbar\omega_0/2$ represents the ground state energy and ζ the transition matrix element for tunneling along the edges of the cubic cage. The resulting four levels are equidistant and exhibit the degeneracies 1–3–3–1. The spacing $\Delta_0 = 2\zeta$ between the levels will be called tunnel splitting because of the similarity to the situation shown in Fig. 1. The level scheme, together with the

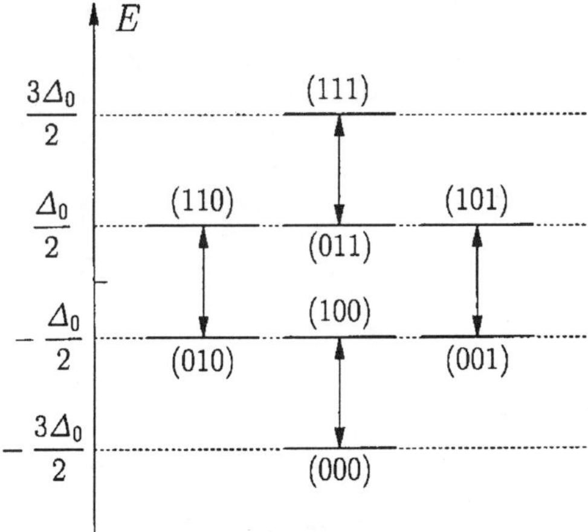

Fig. 3. Level scheme of a $\langle 111 \rangle$-TS in a cubic crystal. The allowed dipolar transitions in a microwave field pointing in $\langle 100 \rangle$-direction are indicated by arrows. The origin of the scale reflects the ground state energy $\hbar\omega_0/2$.

arrows indicating the allowed transitions in electric fields in $\langle 100 \rangle$-direction, is shown in Fig. 3.

Since the tunnel splitting depends on the mass of the tunneling particle, isotope effects are expected. Indeed, for ^{6}Li and ^{7}Li in KCl a ratio of $^6\Delta_0/^7\Delta_0 \approx 1.5$ has been reported (see e.g. Refs. [9] and [10]). However, Eq. (1) agrees only quantitatively with this experimental result although the parameters ω_0, V and d are known. It seems that the main reason for this discrepancy is not the simplification we have made by approximating the three-dimensional problem by a one-dimensional model, but that it is the assumption that the defect ions are moving in a static potential. The motion of the defect ions induces a rearrangement of the environment leading to an effective mass of the tunneling entity being different from that of the bare defect ions (see e.g. Ref. [11]). This mechanism is known as "phonon dressing" and has been studied theoretically in detail for lithium ions tunneling in KCl [12].

2.1.1. *Specific heat*

The existence of TSs is clearly demonstrated by their contribution to the specific heat. The internal energy (and thus the specific heat) of systems with the level scheme given by Eq. (5) and drawn in Fig. 3 follows directly from the partition function $Z = [1 + \exp(-\Delta_0/k_B T)]$ [3]. For independent TSs

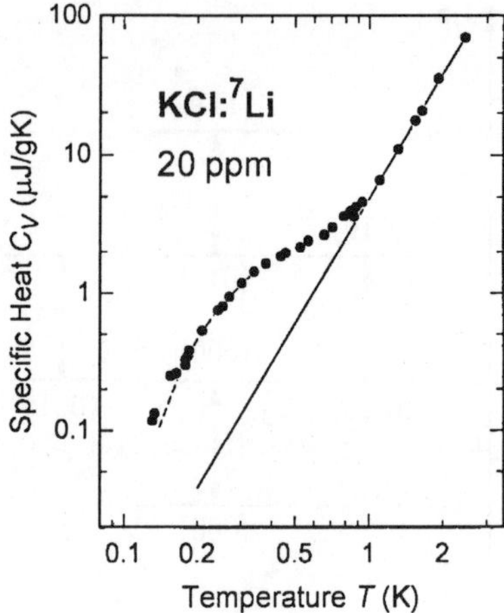

Fig. 4. Specific heat of KCl doped with 20 ppm ^{7}Li. The Debye specific heat of the pure crystal is shown by the straight line. From Ref. [9].

with the density n one obtains

$$C_{\mathrm{TS}} = 3nk_{\mathrm{B}} \left(\frac{\Delta_0}{2k_{\mathrm{B}}T} \right)^2 \mathrm{sech}^2 \left(\frac{\Delta_0}{2k_{\mathrm{B}}T} \right) \tag{6}$$

Except for the factor 3, the expression for $\langle 111 \rangle$-TSs is identical with the one for two-level systems which exhibit the so-called Schottky anomaly. A characteristic maximum is found at $T \approx 0.42 \, \Delta_0/k_{\mathrm{B}}$ reflecting directly the magnitude of the tunnel splitting. In Fig. 4, experimental data for a KCl crystal containing 20 ppm lithium[a] are plotted [9]. The solid line represents the Debye specific heat which originates from the phonons of the host crystal and which dominates above 1 K. Well below that temperature the contribution of the Li-TSs exceeds that of the phonons by far. Figure 5 shows the specific heat of two crystals doped with a comparable amount of ^{6}Li and ^{7}Li, after subtraction of the lattice contribution [9]. Note that different temperature scales have been used to make the data coincide despite the difference in tunnel splitting of the two isotopes. The solid line was calculated using Eq. (6). It agrees reasonably well with the measured specific heat of both samples. The deviations above 1 K and below 0.2 K are probably of experimental origin. The tunnel

---

[a]Natural lithium consists of a mixture of 92.6% ^{7}Li and 7.4% ^{6}Li. In the following, we will label this mixture simply by ^{7}Li.

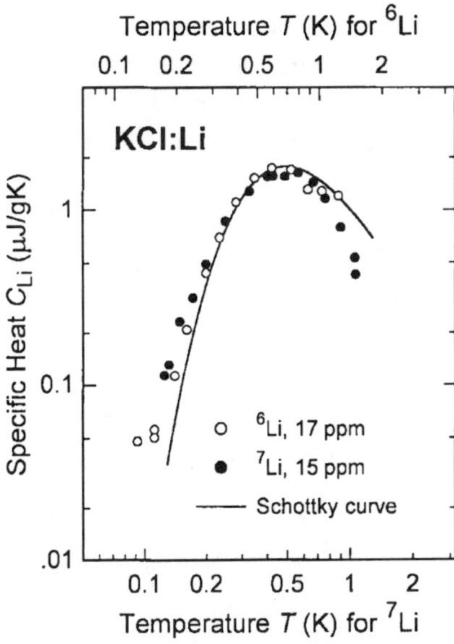

Fig. 5. Specific heat caused by ^{6}Li- and ^{7}Li-TSs. Note that two different temperature scales are used for the data of the two isotopes. The theoretical curve based on Eq. (6) is shown by a solid line. From Ref. [9].

splittings derived from these data are $^6\Delta_0/k_B = 1.6$ K and $^7\Delta_0/k_B = 1.1$ K for ^{6}Li and ^{7}Li, respectively.

2.1.2. *Dielectric susceptibility*

Without external field the wave functions corresponding to the four energy levels of $\langle 111 \rangle$-TSs have cubic symmetry. Therefore, no permanent electric dipole moment can exist. However, application of a small DC electric field induces a dipole moment which increases linearly with field strength. In strong fields, the defects become localized in field direction leading to a constant dipole moment. $\langle 111 \rangle$-TSs exhibit their largest moment if the strong DC field is pointing in $\langle 111 \rangle$-direction. Then the defect ions will reside in one corner of the cage and the resulting dipole moment p_{111} is simply given by the distance between the lattice site and the corresponding off-center position, i.e. $p_{111} = ed\sqrt{3}/2$. In the following, we will drop the subscript 111, i.e. we use $p \equiv p_{111}$. An electric field in $\langle 100 \rangle$-direction will drive the ion into the four corners of the cage along that direction leading to the dipole moment $p_{100} = p/\sqrt{3} = ed/2$.

For the discussion of the dielectric properties, the field dependence of the energy levels of the TSs and the selection rules must be known. Using the notation of Eq. (4), dipolar transitions are allowed, if $|j - j'| = \Delta j = 1$. Since in the following examples the applied AC electric field is pointing in $\langle 100 \rangle$-direction the conditions $\Delta j_x = 1$, and $\Delta j_y = \Delta j_z = 0$ must hold. The allowed dipolar transitions have already been shown by arrows in the level scheme of Fig. 3. Interestingly, only transitions between specific level pairs are possible. Under this experimental condition $\langle 111 \rangle$-TSs act like two-level systems.

The variation of the energy levels with the applied field is described by the following expressions [9]:

$$E_{111} = -E_{000} = \Delta_0 + \sqrt{\frac{1}{4}\Delta_0^2 + \frac{1}{3}p^2 F^2}$$

$$E_{011} = -E_{100} = \Delta_0 - \sqrt{\frac{1}{4}\Delta_0^2 + \frac{1}{3}p^2 F^2} \tag{7}$$

$$E_{101} = E_{110} = -E_{010} = -E_{001} = \sqrt{\frac{1}{4}\Delta_0^2 + \frac{1}{3}p^2 F^2}$$

The energy levels vary quadratically with the electric field at small field strengths but linearly at high fields. This behavior reflects the fact that the effective dipole moment first increases linearly with the field and finally becomes constant.

From the field dependence of the eigenvalues the partition function, and hence the static dielectric susceptibility χ'_{iso} (the subscript indicates that the TSs are considered to be isolated, i.e. non-interacting) can easily be calculated. For electric fields along the $\langle 100 \rangle$-direction one finds for isolated $\langle 111 \rangle$-TSs

$$\chi'_{\text{iso}} = \frac{1}{V k_{\text{B}} T} \left.\frac{\partial^2 \ln Z}{\partial F^2}\right|_{F=0} = \frac{2}{3} \frac{np^2}{\epsilon_0 \Delta_0} \tanh\left(\frac{\Delta_0}{2 k_{\text{B}} T}\right) \tag{8}$$

where ϵ_0 stands for the permittivity of the vacuum. It is worth mentioning that this expression is the same as for two-level systems. This result becomes plausible if we reconsider that the selection rules allow only transitions between two energetically neighboring, equally spaced levels. As expected, the susceptibility of isolated TSs increases proportional to the number density n of the defects. At high temperatures $T > \Delta_0/k_{\text{B}}$, Eq. (8) reflects the classical Langevin-Debye susceptibility.

In Fig. 6, the temperature dependence of the dielectric susceptibility of two KCl crystals is shown. One crystal was doped with 4 ppm ^{7}Li, the other

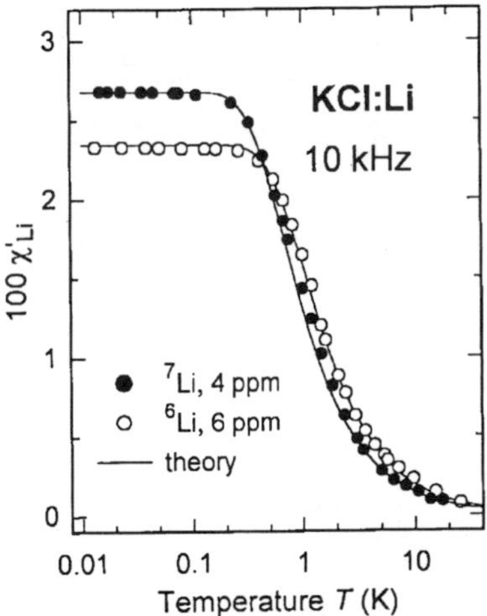

Fig. 6. Susceptibility of ^{6}Li- and ^{7}Li-TSs in KCl measured at 10 kHz. The solid lines represent the theoretical curves obtained from Eq. (8). From Ref. [13].

one with 6 ppm ^{6}Li [13]. Since p and Δ_0 are known quantities, the solid lines following from Eq. (8) without free parameter demonstrate the excellent agreement between theory and experiment. In particular, the $1/\Delta_0$-dependence of the susceptibility can be seen very clearly. Although the ^{6}Li-concentration is higher than the ^{7}Li-concentration, the latter sample exhibits a larger susceptibility. Moreover, the susceptibility of the ^{6}Li-sample shows its point of inflection at a higher temperature due to the fact that $^6\Delta_0 > {}^7\Delta_0$.

2.1.3. *Sound velocity*

Defects do not only give rise to elastic distortions or local strain fields but also couple to sound waves and thermal phonons. Because of the elastic anisotropy of crystals the coupling strength depends on the symmetry of the defect, the polarization of the sound wave and its direction of propagation. For example, TSs in KCl originating from lithium ions do not couple to longitudinal waves propagating along directions of high symmetry like the [110]-direction. The coupling of shear waves running in that direction depends on their polarization. As can be seen from Fig. 7, the [1$\bar{1}$0]-polarized wave is not influenced by the TSs but a strong variation of the velocity v with temperature is observed if the wave is polarized in [001]-direction. From such measurements it was concluded that the Li-TSs in KCl are oriented along the $\langle 111 \rangle$-directions [14].

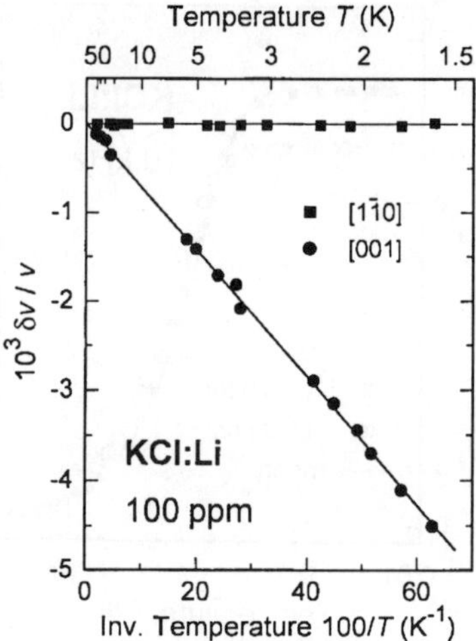

Fig. 7. Relative change of the velocity of shear waves propagating in [110]-direction as a function of inverse temperature. The measurements were carried out at 30 MHz with waves polarized in [1$\bar{1}$0]- and [001]-direction. From Ref. [14].

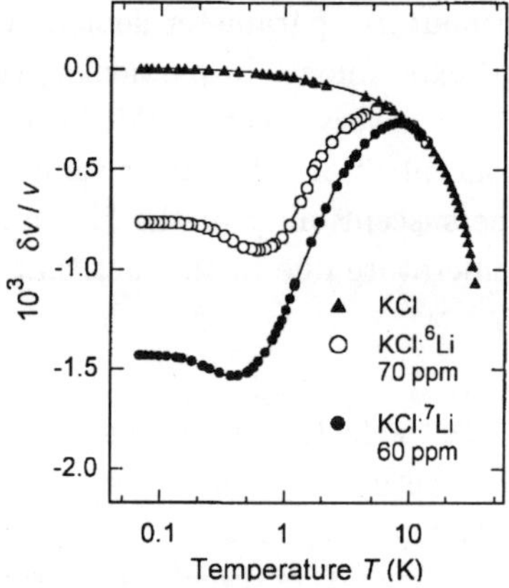

Fig. 8. Relative change of the velocity of torsional waves propagating in [100]-direction in pure and Li-doped KCl as a function of temperature. The measurements were carried out at 2 kHz. From Ref. [15].

Very recently the velocity of torsional waves in [100]-direction has been investigated in KCl:Li down to very low temperatures [15]. The relative change $\delta v/v$ of the velocity of a pure and a Li-doped KCl crystal is shown in Fig. 8.

In the pure crystal the velocity decreases monotonically with temperature, however, the observed variation is not proportional to T^4 as expected for perfect dielectric crystals. Instead, the data can be fitted by the relation $\delta v/v = -(aT + bT^4)$, where a and b are positive constants. The origin of the additional linear term is not known, but it seems likely that dislocations are responsible for this effect [16]. Apart from the sign, the velocity and the dielectric susceptibility of the doped crystals behave in a very similar way because both quantities follow directly from the theory of linear response. However, there is one significant difference: a minimum of the velocity is found just below 1 K whereas the corresponding maximum is missing in the dielectric susceptibility. In fact, the energy levels of the TSs respond differently to electric and elastic fields. A small torsion in the (100)-plane causes an energy shift which varies quadratically with strain only for four of the eight levels, whereas the other four levels experience a linear shift. As a consequence, the elastic susceptibility differs from the dielectric one where all levels show a quadratic field dependence. For the relative change of the sound velocity in the static limit, one obtains [15]:

$$\frac{\delta v}{v} = -\frac{2n\gamma^2}{\rho v^2}\left[\frac{1}{\Delta_0}\tanh\left(\frac{\Delta_0}{2k_\mathrm{B}T}\right) + \frac{1}{2k_\mathrm{B}T}\operatorname{sech}^2\left(\frac{\Delta_0}{2k_\mathrm{B}T}\right)\right] \tag{9}$$

The first term resembles expression (8) for the dielectric susceptibility: the dipole moment in Eq. (8) is replaced by the deformation potential γ reflecting the coupling between TSs and the strain field. In addition, the mass density ρ and the sound velocity v enter the prefactor. In the more general case where the sound wave propagates along an arbitrary direction, Eq. (9) becomes more complicated because of the tensorial nature of the deformation potential and of the strain field. The second term gives rise to the minimum in the temperature dependence of the velocity of sound. It reflects the relaxation of the linearly coupled levels under their modulation in the periodic strain field. A discussion of the frequency dependence of this contribution is given in Ref. [7]. As in case of the dielectric susceptibility, the isotope effect expected from Eq. (9) is in good agreement with the experimental data. However, a perfect quantitative agreement between experimental and theoretical curves is not obtained since at defect concentrations around 60 ppm the influence of interaction between the TSs cannot be neglected. We will discuss the consequences of this interaction in the following two subsections.

2.2. *Tunneling of defect pairs*

In general, point defects are randomly distributed on the lattice sites of the host crystal. With increasing concentration, the mean distance between the defects decreases and the interaction between them becomes gradually stronger. For the substitutional defects discussed here, there exist two types of interaction, namely via electric fields and via elastic strain fields. In KCl doped with lithium the electric interaction exceeds the elastic coupling by an order of magnitude. Hence only the electric dipole-dipole interaction is relevant. For the interaction energy J of a pair of TSs we have

$$J(\boldsymbol{p},\boldsymbol{p}') = \frac{1}{4\pi\epsilon_0\epsilon_{\mathrm{KCl}}} \left[\frac{\boldsymbol{p}\cdot\boldsymbol{p}'}{r^3} - \frac{3(\boldsymbol{r}\cdot\boldsymbol{p})(\boldsymbol{r}\cdot\boldsymbol{p}')}{r^5} \right] \tag{10}$$

where $\boldsymbol{r}$ is the vector connecting the two defects, $\boldsymbol{p}$ and $\boldsymbol{p}'$ are their dipole moments, and ϵ_{KCl} is the dielectric constant of the host. The separation between the interacting TSs can be considered to vary continuously if they are separated by many lattice constants. The fact that only discrete values of the separation are possible in a crystalline lattice becomes important at small distances because the relatively large interaction energy can assume only distinct values.

2.2.1. *Pairs of strongly coupled defects*

If two defects are close enough, they will form a pair with properties remarkably different from those of isolated defect states. As discussed in Sec. 2.1, the Li^+-ions occupy the eight corners of their cage so that the dipole moment of the defect can be oriented in eight different directions. The coupling of two defects thus leads to 64 configurations. The potential energy associated with these states is easily calculated using Eq. (10). The energy scale and the degree of degeneracy of the different levels depends on the relative position of the two defects. We focus our attention on near neighbors and denote them by NN1, NN2, ... with increasing distance between the individual Li^+-ions. In Fig. 9, the resulting classical level scheme for the four nearest neighbor pairs is shown. In addition, we want to mention that the classical reorientation of pairs of OH^--defects in KCl has been observed in dielectric absorption measurements [17].

Because of the large differences in the energy of the levels, only the degenerate ground state is of importance for the understanding of the properties of pairs at very low temperatures. In the further discussion, E_0 denotes the

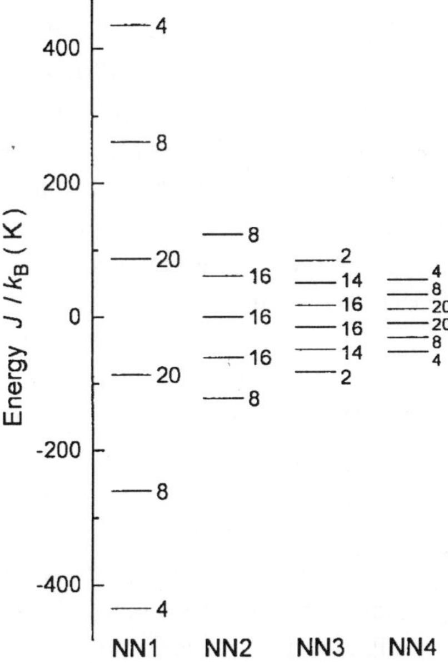

Fig. 9. Classical level scheme of pairs of electrostatically interacting Li$^+$-ions. The numbers next to the levels indicate the degree of degeneracy. From Ref. [18].

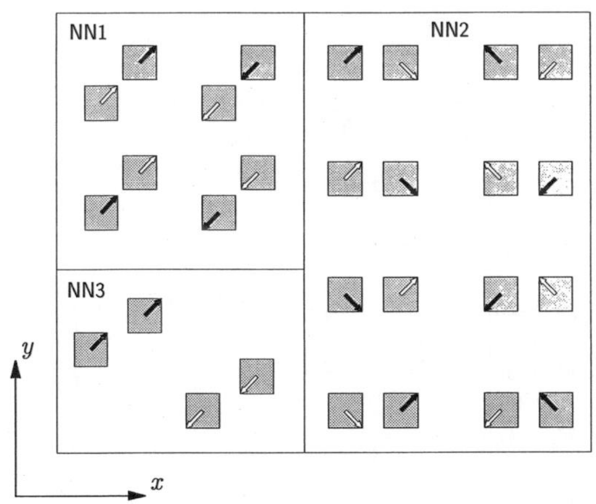

Fig. 10. Schematic representation of the spatial configurations of the NN1-, NN2-, and NN3-pairs that possess the lowest potential energy. The squares indicate the space accessible to the Li$^+$-ions in KCl. The positive and negative z-components of the dipole moments are indicated by full and open arrows, respectively. From Ref. [18].

potential energy J of the lowest lying levels of the different pairs. The possible spatial configurations for NN1-, NN2-, and NN3-pairs are shown in Fig. 10. For NN1-pairs, four configurations exist, for NN2 there are eight, and for NN3 just two configurations. The interaction energy E_0/k_B decreases for these three types from 430 K to 120 K and 80 K. Our further considerations are based on the idea that the two coupled Li^+-ions perform a coherent tunneling motion. Of course the probability of tunneling of these new entities will be strongly reduced compared with the tunneling probability of single ions. An additional restriction occurs for the motion of the coupled Li^+-ions: during the tunneling process the relative orientation of the two dipoles has to remain unchanged since otherwise the potential energy J would rise tremendously. Therefore tunneling along the edges of the cage is only possible for certain pair configurations, e.g. for NN2-pairs but not for NN1- and NN3-configurations.

As mentioned above, the classical ground state of the NN2-pairs is eight-fold degenerate but its degeneracy is lifted by the tunneling motion. A tedious calculation [18] leads to the level scheme of the NN2-pairs shown in Fig. 11 which is very similar to that of isolated TSs. In this figure, arrows indicated the transitions which are allowed in rf-fields applied along $\langle 100 \rangle$-directions. Since only transitions between specific doublets are possible we may consider the defect pairs as effective two-level systems. The main difference to isolated $\langle 100 \rangle$-TSs lies in the new, much smaller level splitting η

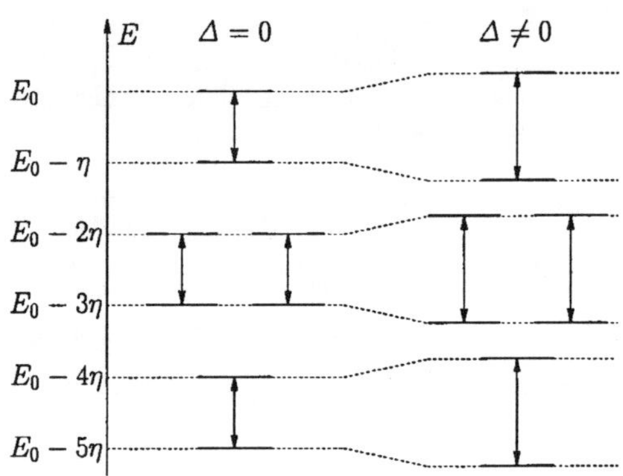

Fig. 11. Ground state-splitting of a Li^+-ion pair. Left side: Level scheme for an undisturbed pair. Right side: Level scheme for a pair with asymmetry energy $\Delta \ll \eta$. The arrows indicate the allowed dipole transitions. From Ref. [18].

given by

$$\eta = \frac{\Delta_0^2}{J} \tag{11}$$

In this formula, the tunnel splitting Δ_0 of isolated TSs enters. This indicates that the transition from one configuration to another occurs by a simultaneous tunneling of the two ions along the edges of the cage. A glance at Fig. 10 tells us that this is only correct for pairs of the type NN2. In all other cases, only tunneling along other paths is possible and Δ_0 has to be replaced by the appropriate tunnel splitting which is always considerably smaller [10]. Therefore, among all configurations up to NN8, the NN2-pairs exhibit the largest value of η although J is smaller for NN3 to NN7. For the tunnel splitting of pairs formed by ^{7}Li$^+$-ions we deduce the values $^7\eta/k_B = 0.001$, 9.3, 0.007, and 0.01 mK for NN1-, NN2-, NN3-, and NN4-pairs, respectively.

In the discussion of the KCl:Li system we have assumed so far that all potential minima have the same depth. In this case tunnel splitting and energy splitting E are identical, i.e. $E = \Delta_0$. From experiments it is known that this assumption is only valid for isolated defects for which Δ_0 is relatively large. For pair states the situation is not as simple because $\eta \ll \Delta_0$. Internal strain fields or residual impurities cause a shift of the potential wells relative to each other by an asymmetry energy Δ which will be distributed and may be comparable or even larger than η. The asymmetry energy gives rise to a shift of the unperturbed eigenvalues of pair states as shown in Fig. 11 on the right-hand side. In this case, the energy splitting E of the ground state is given by

$$E = \sqrt{\eta^2 + \Delta^2} \tag{12}$$

The asymmetry energy does not change the selection rules and the pairs respond to AC electric fields like two-level systems.

Very recently the coherent tunneling of pairs has been studied using a technique which is also known from NMR-measurements, namely by so-called dielectric rotary echoes [19]. In such experiments a microwave electric field is applied to the sample inducing transitions between the levels if its frequency $\omega/2\pi$ meets the resonance condition $E = \hbar\omega$. At very low temperature, the interaction of the TSs with the heat bath can be neglected and the pairs act like undisturbed quantum mechanical two-level systems. Under this condition, the occupation of the lowest two levels oscillates with the Rabi frequency

$$\Omega_R = \frac{1}{\hbar} \frac{\eta}{E} p_P \cdot F_0 \tag{13}$$

Here F_0 is the amplitude of the electric field and p_p the dipole moment of the pairs. Directly related with the varying occupation number is a macroscopic, oscillating electric polarization that can be detected experimentally. In real crystals random strain fields are present which cause a distribution of the asymmetry energy Δ and therefore also of the energy splitting E. As a consequence, pairs with different values of the Rabi frequency exist. Because of the superposition of the individual contributions the macroscopic polarization becomes smeared out and vanishes rapidly after switching on the microwave field. In the experiments discussed here [19], the decaying Rabi oscillations — called "free induction decay" in NMR-experiments — could not be detected since the amplifiers became saturated by the switching transient of the driving field. Therefore the phase of the microwave field was inverted after the time $t = t_p$, causing a reversal of the time evolution of the polarization. Consequently, Rabi oscillations appear again and can be detected as a rotary echo at $t = 2t_p$. In the experiments discussed here, frequencies between 700 and 1200 MHz — corresponding to approximately 50 mK — were applied and Rabi frequencies of the order of 1 MHz were observed.

Such experiments have been carried out at different frequencies on KCl doped with ^{6}Li, ^{7}Li, and a mixture of both isotopes. Because $E = \hbar\omega$ is fulfilled in experiments based on resonant interaction, a linear relationship

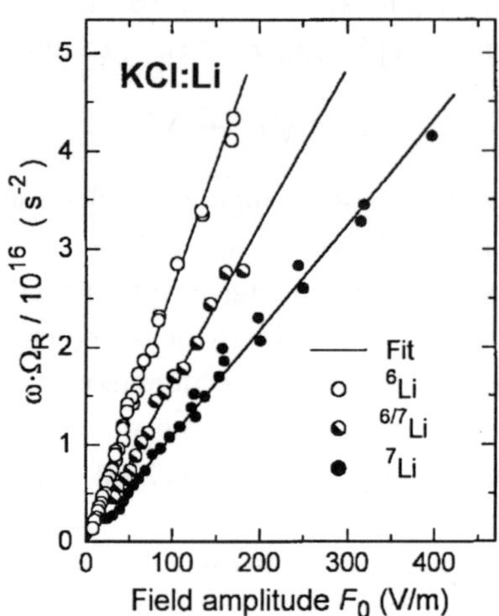

Fig. 12. Product $\omega \cdot \Omega_R$ as a function of the field strength F_0 for samples containing different kinds of pairs. From Ref. [18].

between $\omega \cdot \Omega_R$ and the field amplitude F_0 is expected from Eq. (13). As shown in Fig. 12, three clearly distinct straight lines were found for the three samples. Since the dipole moment p_p of NN2-pairs is known, the value of the tunnel splitting η follows directly from the slopes of the experimental curves. For $^6\eta$, $^{6/7}\eta$, and $^7\eta$ the values (expressed in η/k_B) 20.5, 12.9, and 8.5 mK were found which compare favorably with those deduced from Eq. (11), namely 22, 14.4, and 9.3 mK.

2.2.2. *Pair-model*

By means of rotary echoes only NN2-pairs are detectable. All other pairs exhibit a tunnel splitting which is either too small or too large for this experimental technique. However, because of the random distribution of the Li^+-ions a broad spectrum of pairs must be present in crystals. In fact pairs of the type NN $\geq$ 8 with their relatively large tunnel splitting were observed in measurements of the specific heat at moderately high Li-concentrations. In Fig. 13, the specific heat of KCl:Li is plotted for three different concentrations of lithium. Contributions to the heat capacity are found at temperatures well below the maximum of the Schottky anomaly which cannot be understood on the basis of isolated TSs.

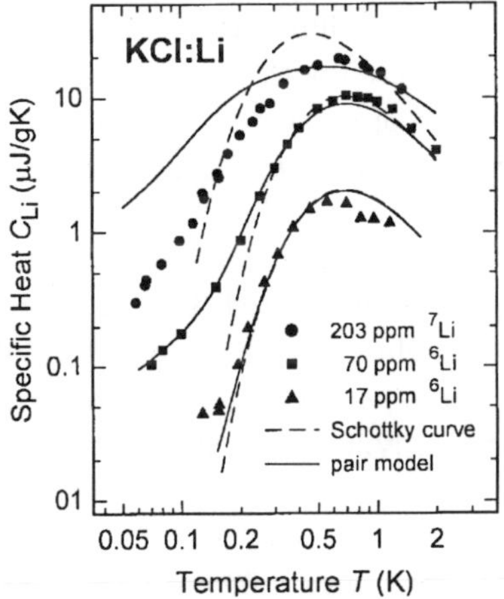

Fig. 13. Specific heat of KCl:Li after subtracting the contribution of the host lattice. The prediction of the pair model is shown by solid lines, the classical Schottky anomaly by dashed lines. Data from Refs. [9] and [20].

A qualitative explanation of this phenomenon is simple but the quantitative description is rather sophisticated. The level splitting of the pairs depends on the interaction energy J which decreases with rising separation of the TSs forming the pairs. The problem lies in finding the appropriate distribution function of their energy splitting. This problem is addressed in the pair-model (see e.g. Refs. [21–23]) which is based on three simplifications. Firstly, it is assumed that the two interacting eight-level systems can be replaced by two two-level systems. Secondly, the distance between the interacting TSs is treated as a continuous variable. Of course this assumption is not valid for strongly coupled ions. According to Eqs. (11) and (12), however, their tunnel splitting is so small that they do not influence the specific heat above 50 mK. Finally, the asymmetry energy Δ is ignored since it is expected to be small compared to $k_B T$ in this temperature range.

The theoretical curves drawn in Fig. 13 demonstrate that this model predicts the specific heat up to about 100 ppm Li correctly but fails at still higher concentrations for reasons we will discuss in the following subsection. Moreover, measurements of the dynamical properties of Li-doped crystals indicate that already at moderate concentrations a description based on the pair-model is incomplete [24].

2.3. *Incoherent tunneling*

The schematic drawings in Fig. 14 illustrate that the interaction among the TSs becomes increasingly complicated with rising defect density. The square grid represents the regular lattice of a crystal, the dots the impurities. A "radius of interaction" is drawn for which $J = \Delta_0$ holds. As long as the shaded areas do not overlap we may consider defects as "quasi-isolated". Overlapping areas indicate coupled TSs. In the example for low concentrations on the left-hand side of the figure only isolated TSs and one pair are present. At higher concentrations larger clusters come into play resulting in a complicated many-body problem. The experimental results shown in the following two figures demonstrate that the dynamics of the systems changes with increasing concentration.

In Fig. 15, the temperature dependence of the dielectric susceptibility is plotted for KCl samples doped with different amounts of lithium [13]. Three characteristic features are found. Most striking is the non-linear concentration dependence of the magnitude of the susceptibility. Below 200 mK the susceptibility of the 1100 ppm sample is even smaller than for the 210 ppm

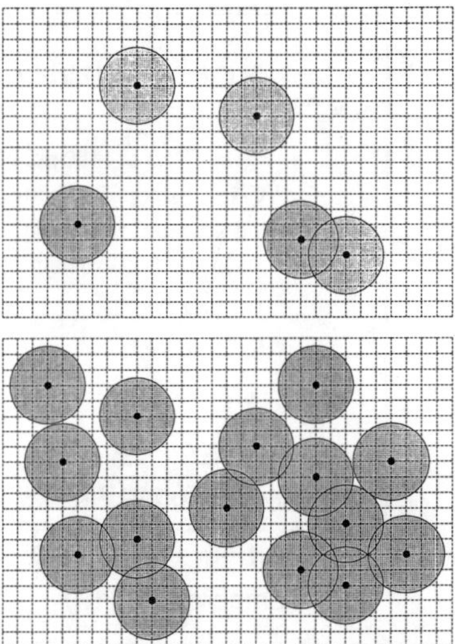

Fig. 14. Schematic representation of the interaction between TSs. The square grid represents the crystal lattice, the dots, the TSs. The dark areas indicate the regions within which the dipolar forces are "noticeable". In the lower panel, the concentration of the TSs has been increased. From Ref. [7].

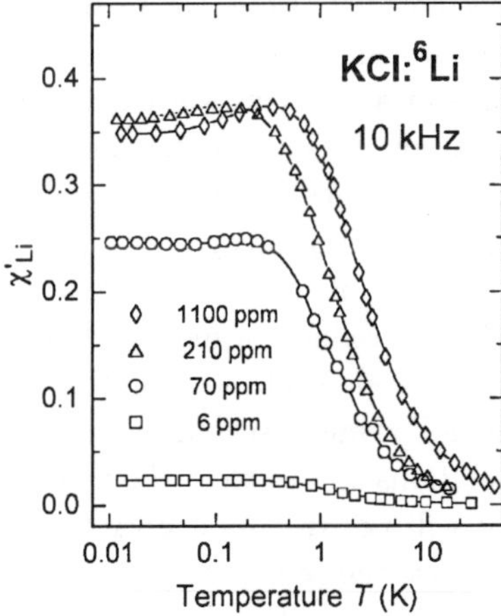

Fig. 15. Real part χ' of the susceptibility of KCl:Li-samples containing different amounts of lithium. The measurement was done at 10 kHz. The data of the sample containing 6 ppm lithium are also shown in Fig. 6. From Ref. [13].

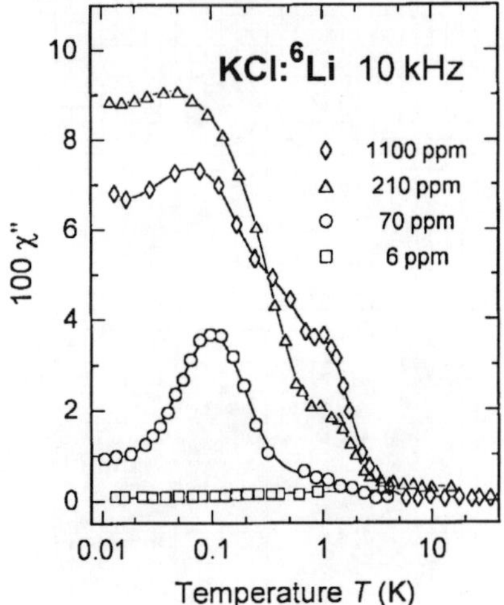

Fig. 16. Dielectric loss χ'' of KCl:Li-samples with different concentrations of lithium. The measurements were performed at 10 kHz. From Ref. [13].

sample! Secondly, with increasing concentration the steep decrease of χ' shifts to higher temperatures. Finally, at temperatures around some hundred mK a shallow maximum is observed indicating the occurrence of relaxation processes. The last phenomenon becomes more obvious in the imaginary part χ'' of the dielectric susceptibility shown in Fig. 16 [13]. The loss cannot be caused by resonant interaction as the measurement was carried out at 10 kHz so that $\hbar\omega$ was much smaller than Δ_0 or η. Therefore, relaxation must occur although such a process does not take place in samples containing only isolated TSs which do not exhibit an asymmetry energy.

This complex theoretical problem has been treated only recently [25]. The calculation is based on Mori's reduction method [26] which permits a derivation of a continued fraction representation for the relevant two-times correlation functions. In the following a scheme was developed to decouple the static and dynamic correlations and to treat the memory function in a Markov approximation. Here we will not discuss the rather sophisticated theoretical considerations but we will use the results to explain the experimental data. In the theory the dimensionless parameter

$$\mu = \frac{1}{3}\frac{np^2}{\epsilon\epsilon_0}\frac{1}{\Delta_0} \tag{14}$$

plays an important role. It reflects the ratio between the mean value $\bar{J} = np^2/3\epsilon\epsilon_0$ of the interaction energy and the tunnel splitting Δ_0.

For isolated TSs the phase of the superimposed localized wave functions is preserved, i.e. the tunneling motion between the potential minima is coherent. Coupling to neighboring TSs disturbs the free tunneling motion and thus the phase of the tunneling particle. With increasing value of μ, the number of interacting TSs rises and the perturbation becomes increasingly important. Finally, when $\mu > 1$ is reached the tunneling motion is virtually incoherent. At the same time the interaction of the TSs with external fields changes its character. Since the strength of the resonant interaction is determined by the time average of the overlap integral of the wave function it becomes more and more suppressed and relaxation phenomena set in. For $\mu > 1$ the perturbed TSs do in general no longer relax via interaction with phonons of the heat bath but by direct interaction with other TSs. We will come back to this point at the end of this subsection. For the dielectric susceptibility, the theory predicts in Ref. [7]

$$\chi'_{\text{res}} = \frac{2p^2}{3\epsilon_0 V} \sum_i (1 - R_i) \frac{1}{E_i} \tanh\left(\frac{E_i}{2k_{\text{B}}T}\right) \tag{15}$$

The summation has to be carried out over all TSs i in the sample with the volume V. The important quantity in this formula is R_i, the so-called relaxation amplitude. It is a measure of the extent to which the system i contributes to relaxation. At low concentrations the TSs interact only resonantly with the applied field and R_i vanishes. At high defect concentration, R_i approaches unity, i.e. the resonant process is completely suppressed and the response to electric fields is only possible via relaxation. Furthermore, the energy splitting E_i of the individual TSs depends on their local environment and is fluctuating in time. Obviously it differs from the energy splitting of isolated defects. Its mean value increases with defect concentration and is given by

$$\bar{E}_i = \Delta_0\sqrt{1 + \mu^2} \tag{16}$$

For the relaxation part of the susceptibility one finds

$$\chi_{\text{rel}} = \frac{p^2}{3\epsilon_0 V} \frac{1}{k_{\text{B}}T} \sum_i R_i \frac{1}{1 + i\omega\tau_i} \tag{17}$$

where τ_i is the relaxation time of the individual TSs. For a random distribution of TSs the mean value $\bar{R}$ of the relaxation amplitude can be approximated by $\bar{R} = \mu[\pi/2 - \arctan(\mu)]$. This means that the amplitude increases steeply in the range $0.1 < \mu < 1$ and approaches unity for $\mu > 5$. There exists qualitative agreement between theory and experiment, in particular with respect to the three characteristic experimental features stated above. A quantitative comparison is not possible so far because the relaxation spectrum entering Eq. (17) is not known and can only be calculated using very coarse approximations [25].

The dielectric loss has been measured at different temperatures in a wide frequency range [27, 28]. These measurements show that a broad distribution of relaxation times exists and that its width increases with defect concentration. Phenomenologically, it can be described by the inverse Havriliak–Nagami ansatz $\epsilon'' \propto \text{Im}[1 + (i\omega\tau_{\text{a}})^{-\alpha}]^{-\beta}$, where the parameters α and β reflect the symmetric and asymmetric broadening of the spectrum, respectively. In Fig. 17, the temperature dependence of the mean rate τ_{a}^{-1} is shown for KCl crystals containing different amounts of ^{6}Li and ^{7}Li [28]. Surprisingly, it turned out that the heavier TSs consisting of ^{7}Li$^+$-ions relax faster than the lighter TSs.

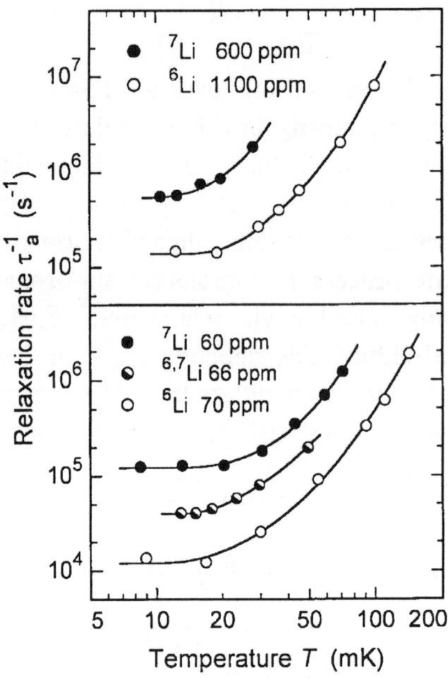

Fig. 17. Relaxation rate τ_{a}^{-1} of ^{6}Li- and ^{7}Li-TSs in KCl. Note that the crystal labelled ^{6,7}Li 66 ppm was doped with an equal amount of both isotopes. The lines serve as guides for the eye. From Ref. [28].

Clearly, this observation indicates that relaxation does not occur via absorption or emission of thermal phonons since this process would lead to the opposite relation. The result rather demonstrates that at high concentrations relaxation takes place via collective motions within the ensemble of TSs. The surprising dependence on the mass of the TSs is also expected from the theoretical treatment of the incoherent tunneling phenomenon, where $\tau \propto \mu^{-4} \propto \Delta_0^4$ has been predicted for $\mu < 1$ [7].

There exists an interesting possibility of a quantitative comparison between theory and experiment. On the one hand for $T \to 0$ the relaxational contribution χ'_{rel} to the dielectric susceptibility in Eq. (17) vanishes even at a high concentration of TSs since the relaxation times become extremely long; on the other hand the expression (15) for χ'_{res} can be approximated by Ref. [7]

$$\chi'_{\mathrm{res}} = \frac{2np^2}{3\epsilon_0 \Delta_0} \frac{(\sqrt{1+\mu^2}-\mu)^2}{\sqrt{1+\mu^2}} \tanh\left(\frac{\Delta_0\sqrt{1+\mu^2}}{2k_{\mathrm{B}}T}\right) \tag{18}$$

Here we have used analytic approximations for the mean values $\bar{R}$ and $\bar{E}$ of the relaxation amplitude and the energy, respectively. Combining Eqs. (8) and (18) we obtain for $T \to 0$ the simple relation

$$\frac{\chi'}{\chi'_{\mathrm{iso}}} = \frac{(\sqrt{1+\mu^2}-\mu)^2}{\sqrt{1+\mu^2}} \tag{19}$$

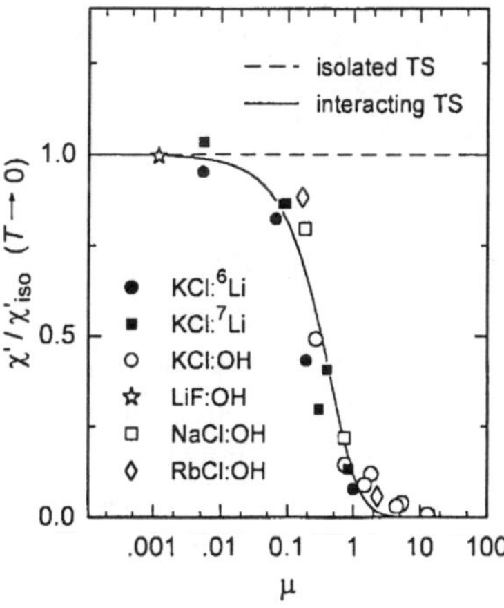

Fig. 18. Ratio $\chi'/\chi'_{\mathrm{iso}}$ of crystals with different concentrations of TSs as a function of μ. The theoretical curve (19) for interacting TSs is shown as solid line. From Ref. [29].

In Fig. 18, this ratio is plotted as a function of μ together with experimental points not only from KCl:Li but also from measurements on alkali halides containing OH$^-$-ions. Although the agreement is impressive it should be noted that Eq. (19) is expected to hold only if the elastic interaction between the TSs can be neglected, i.e. if the TSs are coupled by electrical dipolar forces.

2.4. *Host crystals with large random strains*

During the last 15 years great attention was paid to TSs in host crystals with strong random strain fields. Under this condition, TSs exhibit properties remarkably different from those we have discussed until now. Their dynamical behavior at low temperatures resembles that of TSs of structural glasses which we will consider in Sec. 3. Thus there was hope to find crystalline systems that are able to serve as model systems for glasses and could be used as guides towards a microscopic understanding of the TSs of amorphous materials. For long time most of the research was focused on the mixed crystal $(KBr)_{1-x}\,(KCN)_x$, in which the CN$^-$-ions form rotational TSs. One advantage of this system is that such crystals can be grown with arbitrary CN$^-$-concentration. Because of the random occupation of lattice sites by CN-molecules and their strong interaction the crystal freezes into a phase with randomly oriented CN-molecules for the concentration range $0.05 < x < 0.55$. This phase is termed "orientational glass state". In a variety of low-temperature experiments, it has been confirmed that a close similarity exists in the behavior of structural glasses and $(KBr)_{1-x}\,(KCN)_x$ in its orientational glass state (see e.g. Ref. [30]). Moreover, using Monte Carlo simulations of randomly located elastic dipoles, a macroscopic model has been developed which successfully describes the glass-like low temperature properties of $(KBr)_{1-x}\,(KCN)_x$ [31].

For the study of certain problems, however, systems of the type just discussed have the disadvantage that the TSs themselves cause the strain fields. In particular the question cannot be answered whether the observed broad distribution of the tunnel splitting Δ_0 and the asymmetry energy Δ is due to frozen-in strain fields, or due to the dynamic strain interaction between tunneling particles. Recently, it has been shown that these contributions can be separated in crystals like $(KBr)_{1-x}\,(KCl)_x$ doped with small amounts of CN [32]. There, the mismatch in ionic radii between bromine and chlorine causes strong random strain fields. Thus TSs made of CN$^-$-ions substituting either Cl$^-$ or Br$^-$, experience a randomly modulated potential landscape. As a consequence, one finds broad distributions of the characteristic parameters of the

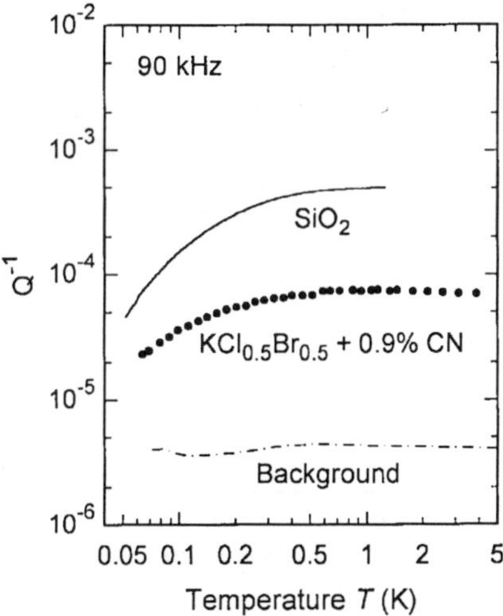

Fig. 19. Comparison between the internal friction Q^{-1} of $(KBr)_{0.45}$ $(KCl)_{0.55}$ containing 0.9% CN and of a–SiO$_2$, measured at 90 kHz with a torsional oscillator. The dashed-dotted line indicates the background loss in this experiment. From Ref. [32].

CN$^-$-TSs. In Fig. 19 we show the result of an internal friction measurement on $(KBr)_{0.45}$ $(KCl)_{0.55}$ doped with 0.9% CN. Clearly, the temperature dependence of the attenuation is very similar to that of vitreous silica which is also shown in this figure. A discussion of the acoustic properties of amorphous materials will be given in Sec. 3.2.

3. Tunneling Systems in Amorphous Solids

At low temperatures, the thermal, elastic, and dielectric behavior of amorphous solids differs significantly from that of crystalline solids. Prominent examples are the almost linear temperature dependence of the specific heat C_V [4] and the surprising temperature variation δv of the velocity of sound [33].

As an example, the specific heat of vitreous silica and, for comparison, of crystalline quartz are plotted in Fig. 20. At very low temperatures, the specific heat of the glass exceeds that of the crystal by orders of magnitude. It should be pointed out that the data for quartz have been extrapolated to very low temperatures. In measurements, a deviation from the expected T^3-dependence is observed below 0.4 K [4]. It is caused by defects and impurities which give rise to TSs similar to those discussed in Sec. 2. There is another

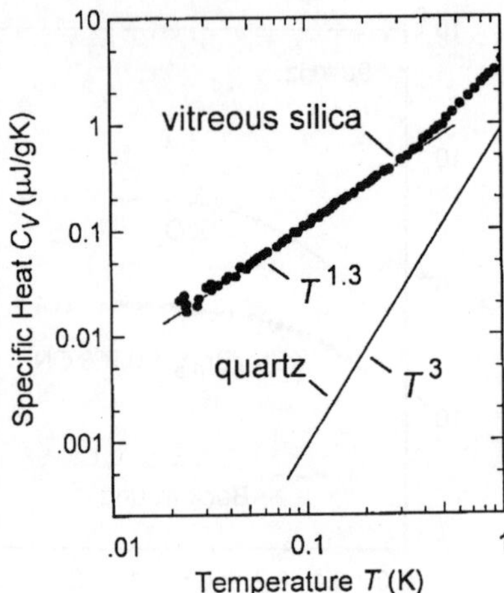

Fig. 20. Specific heat of vitreous silica and crystalline quartz as a function of temperature. From Refs. [4] and [34].

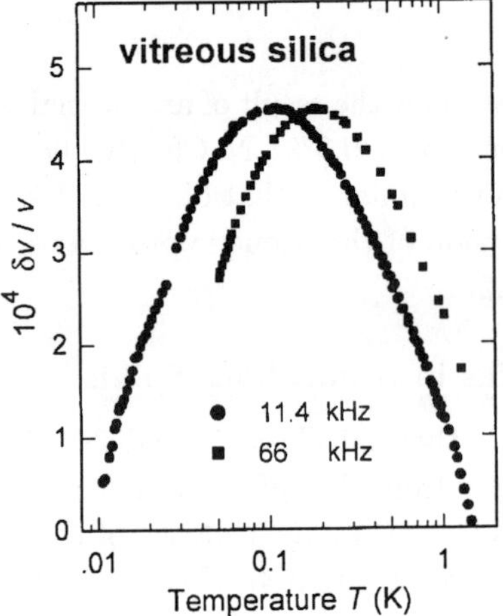

Fig. 21. Relative variation of the velocity of sound of vitreous silica measured at 11.4 kHz and 66 kHz. From Refs. [35] and [36].

remarkable aspect: the magnitude of the specific heat of glasses depends only weakly on their chemical composition and their impurity contents. Thus the number of TSs present in glasses is an almost universal quantity.

The temperature variation of the sound velocity of vitreous silica at two different frequencies is shown in Fig. 21 [35, 36]. The velocity first rises logarithmically, passes a maximum and then decreases again with temperature. The variation δv of the velocity is caused by the response of the TSs to strain fields and is not observed in crystals. In pure crystals the velocity is expected to vary with the internal energy and therefore δv should be proportional to T^4, i.e. below 1 K the velocity is expected to be virtually constant.

3.1. *Tunneling model*

In spite of intense efforts, these and other "anomalies" can at present only be described on the basis of phenomenological models. We restrict ourselves to the originally introduced Tunneling Model (TM) [37, 38] which is often called "standard" TM. It is based on two simple assumptions: first it is assumed that in glasses a relatively small number of TSs exists which couple to strain fields, i.e. they interact with phonons. The microscopic nature of these systems is not well understood since the absence of long-range order prevents a proper characterization of the atomic potential landscape. As confirmed by simulations [39, 40] the TSs do not correspond to well-defined atoms or molecules but rather involve collective structural rearrangements of local configurations. The coordinate used in Fig. 1 to characterize the double-well potential of TSs in glasses is therefore not a function of a simple translation but represents a collective coordinate involving the displacement of several atoms. Under such circumstances it is quite natural to assume that the relevant parameters of the TSs are distributed. Therefore, distribution functions are introduced on the basis of the second main assumption of the TM: asymmetry energy Δ and tunneling parameter λ [given by Eq. (2)] are assumed to be independent of each other and to be uniformly distributed, i.e. their distribution function P is given by

$$P(\Delta, \lambda)\, d\Delta\, d\lambda = \bar{P}\, d\Delta\, d\lambda \qquad (20)$$

where $\bar{P}$ is a constant, characteristic for the amorphous solid under consideration. An alternative model, the so-called Soft Potential Model [41, 42] involves a description of the TSs in terms of a fourth order potential with appropriate distributions of the parameters. Depending on the individual values of these parameters and, in particular, on the relative magnitude of the quadratic term, the resulting potential may exhibit either one or two minima. For the description of the properties of glasses at very low temperatures only the resulting double-well potentials are important and the Soft Potential Model can essentially be reduced to the TM. At higher temperatures, potentials with only one minimum become significant. These "quasi-harmonic potentials" arise from quartic potentials with a vanishing or even positive quadratic term and determine the behavior of glasses. Thus the Soft Potential Model is able to make predictions for experiments in a temperature range where tunneling is no longer important.

As mentioned before the specific heat of glasses hardly depends on their chemical nature; this means that also the magnitude of the spectral density $P(\Delta, \lambda)$ expressed by $\bar{P}$ is approximately the same for almost all amorphous solids. A more careful inspection shows, however, that $\overline{P}$ is smaller in amorphous solids with an overconstrained network like a–Ge [43]. Surprisingly, very recently it has been demonstrated by measurements of the acoustic loss (which will be discussed in the following subsection), that a–Si containing 1% hydrogen and prepared by "hot-wire-assisted chemical vapor deposition" exhibits virtually no TSs at all [44].

From an experimental point of view the energy splitting E of the TSs and the ratio $r = (\Delta_0/E)^2$ rather than Δ and λ are the more relevant quantities. Using these parameters Eq. (20) may be replaced by

$$P(E,r)\,dE\,dr = \frac{\bar{P}}{2r\sqrt{1-r}}\,dE\,dr \tag{21}$$

This distribution function tends to infinity at the boundaries of the allowed interval, i.e. for $r \to 0$ and $r \to 1$. An integration over r yields the density of states:

$$n(E) = \int_{r_{\min}}^{1} \frac{\bar{P}\,dr}{2r\sqrt{1-r}} = \bar{P}\ln r_{\min} \tag{22}$$

A lower limit $r_{\min}$ has been introduced as otherwise the integral and hence the number of TSs would diverge. This implies a lower limit $\Delta_0^{\min}$ for the tunnel splitting for which a value of approximately $\Delta_0^{\min}/k_{\mathrm{B}} \approx 1$ mK has been proposed [45]. This value seems to be rather small but it means that the tunneling parameter is roughly limited to values $\lambda < 12$. From a physical point of view such a limitation is difficult to understand and we will return to this question at the end of this section. It should be pointed out that the quantities discussed here are those which are experimentally accessible. This means that dressing effects [46] (see Sec. 2.1) are already included and are not considered separately.

The knowledge of the density of states allows a calculation of the internal energy and of the specific heat. Ignoring the weak logarithmic energy dependence of $n(E)$ we obtain the famous linear temperature variation of the specific heat, i.e. $C_V \propto T$. However, further subtle details are connected with the behavior of the specific heat of glasses which are not easily explained. Firstly, there is a small deviation from strict linearity. As shown in Fig. 20 the experimental points rather follow a $T^{1.3}$-dependence. Secondly, the measured specific heat seems to depend logarithmically on the duration of the measurement [47]. Several explanations have been proposed but recent calculations show that a treatment of the heat diffusion based on the coupled Boltzmann equations for TSs and phonons leads to a consistent description of the experimental results within the TM [48].

3.2. *Elastic properties of glasses*

Acoustic experiments have been a very important tool in the investigation of the low temperature properties of amorphous solids. In fact, they gave the first evidence of the two-level character of the low energy excitations observed in specific heat measurements. While TSs in crystals, at least at small concentrations, exhibit well-defined parameters, the parameters of TSs in amorphous solids are distributed according to Eq. (20). Therefore the average values of the relevant quantities have to be calculated in order to allow a comparison between experiments and theory. We do not write down the lengthy expressions here because we do not want to discuss them in detail (see e.g. Refs. [49, 50]).

3.2.1. *Internal friction*

As discussed in Sec. 2.3, TSs can give rise to resonant and relaxation effects. Carrying out the averaging procedure for the elastic compliance constant $s = s' + is''$ one finds for the internal friction $Q^{-1} \equiv \rho v^2 s''$ at ultrasonic frequencies

$$Q_{\text{res}}^{-1} = \frac{\pi C}{\sqrt{1 + I/I_{\text{c}}}} \tanh\left(\frac{\hbar\omega}{2k_{\text{B}}T}\right) \tag{23}$$

if the resonant interaction is considered only. The constant $C = \bar{P}\gamma^2/\rho v^2$ reflects the strength of the coupling between the sound wave and the TSs, where γ stands for the deformation potential as in Sec. 2.1.3. The last factor expresses the population difference of the two levels meaning that the contribution of the resonant absorption becomes negligible at higher temperatures. If the acoustic

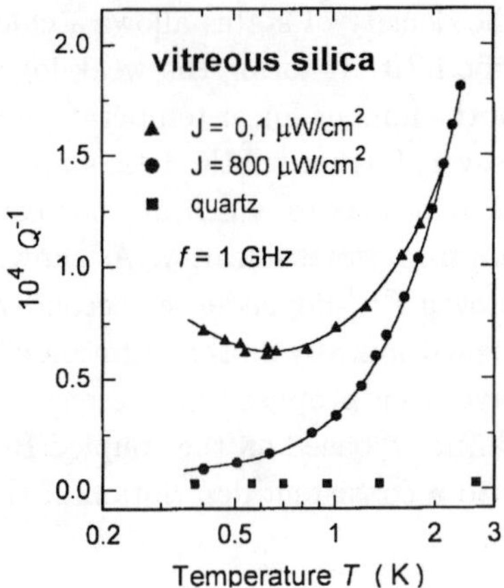

Fig. 22. Temperature variation of the internal friction of vitreous silica and of crystalline quartz measured at about 1 GHz. At the lowest temperature the magnitude of the internal friction in the glass strongly depends on the acoustic intensity. From Ref. [51].

intensity I is well below its critical value I_c the absorption process is linear. With rising intensity the upper level becomes increasingly populated due to the absorption mechanism itself. This effect is called "saturation" and causes a reduction of the resonant acoustic loss. In Fig. 22, data on the temperature variation of the internal friction of vitreous silica are presented at a frequency of about 1 GHz and at two different intensities [51]. The saturation effect and the increase of the absorption with decreasing temperature at low intensities can be seen clearly. As shown in Fig. 23, a corresponding behavior is found in measurements of the dielectric loss at 10 GHz [52]. The similarity between the two figures demonstrates the intimate relation between acoustic and dielectric properties.

In principle the density of states of the TSs could be determined by measuring the resonant absorption as a function of frequency at sufficiently low temperatures. Because of the wide distribution of the energies of the TSs a very wide span in frequency would be necessary. Since acoustic experiments of this type are virtually impossible and dielectric measurements almost as difficult, other methods have to be used to obtain information about the distribution functions.

The rise of the absorption at higher temperatures is caused by the relaxation process. Figure 24 shows data on the internal friction of vitreous silica taken

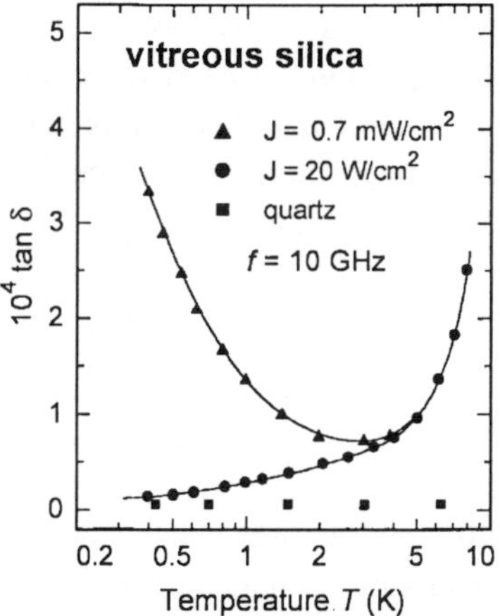

Fig. 23. Temperature variation of the dielectric loss $\tan \delta$ in vitreous silica and crystalline quartz measured at about 10 GHz. At low temperatures the loss in vitreous silica strongly depends on the microwave intensity. From Ref. [52].

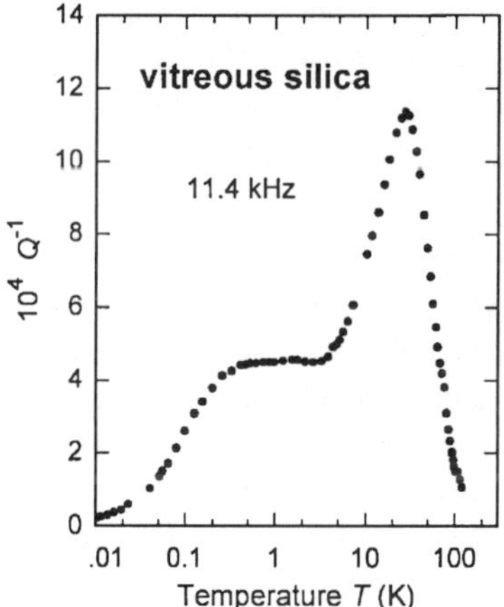

Fig. 24. Internal friction of vitreous silica at 11.4 kHz. The strong peak at higher temperatures is attributed to thermally activated processes, which are not discussed here. From Ref. [35].

at about 10 kHz in a wide temperature range [35]. At such low frequencies, the absorption due to the resonant process can be neglected and background losses can be kept small. Therefore measurements in this frequency range are especially suitable to extract information on the distribution of the characteristic parameters of the TSs. Before we express the internal friction due to this process in an equation we need to consider the relaxation mechanism. In general it is assumed that below 1 K the TSs of glasses relax via absorption or emission of a single thermal phonon. The relaxation rate for this so-called "one-phonon process" is given by

$$\tau^{-1} = A\, r E^3 \coth \frac{E}{2k_{\mathrm{B}}T} \tag{24}$$

where A is a constant containing the deformation potential and the sound velocity of the longitudinal and transverse phonon branches. As the parameter r varies in the range $0 \leq r \leq 1$ there exists a wide distribution of relaxation rates for a given energy splitting even at a fixed temperature. From ultrasonic experiments it is known that TSs with an energy splitting $E/k_{\mathrm{B}} = 1$ K exhibit a minimum relaxation time $\tau_{\min}(r = 1) \approx 1$ ns. On the other hand for TSs with the same energy splitting, relaxation times $\tau > 10^4$ s have been reported from so-called heat release experiments [53, 54]. In this type of measurement the sample is kept at a fixed temperature for a long time to establish thermal equilibrium. Then the temperature is lowered so rapidly that a considerable part of the TSs does not reach thermal equilibrium during cooling. Because of the wide distribution of relaxation times these TSs will successively relax and deliver energy to the phonon path resulting in a continuous release of heat. From Eqs. (21) and (24) this heat release is expected to decay inversely proportional to the observation time t. In some experiments deviations from this law were observed and have been interpreted as an indication for deviations from the distribution function Eq. (21) (see e.g. Ref. [55]). Recently it has been pointed out that these deviations from the $1/t$-law are likely to be caused by the diffusion of the released energy to the heat sink and agreement with the prediction of the TM is obtained if this process is taken into account correctly [56].

Like in crystals, the relaxational response can be expressed by a superposition of Debye relaxators describing the behavior of the individual TSs. For the contribution δs of the TSs to the elastic compliance constant s one finds:

$$\delta s = \frac{\gamma^2}{\rho^2 v^4} \int \int P(E, r) \frac{\Delta^2}{E^2} \frac{d}{dE}\left(\tanh \frac{E}{2k_{\mathrm{B}}T}\right) \frac{1}{1 - \mathrm{i}\omega\tau}\, dE\, dr \tag{25}$$

Beside the fact that the distribution of the parameters E and r has to be taken into account it should be mentioned that the main contribution stems from TSs with an energy splitting $E \approx k_B T$. This is understandable as only in this case sound waves cause a noticeable change of the occupation of the energy levels. In general, Eq. (25) can only be solved numerically but analytic solutions can be found for two limiting cases, namely for $\omega \tau_{min} \gg 1$ and for $\omega \tau_{min} \ll 1$.

Let us first consider the acoustic loss, i.e. the imaginary part of the integral. At low temperatures the relation $\omega \tau_{min} \gg 1$ holds meaning that even the fastest TSs are unable to attain thermal equilibrium within a period of the applied field. Only the contribution of the fastest relaxing systems have to be considered and

$$Q^{-1} = \frac{\pi^4 AC k_B^3 T^3}{12\omega} \tag{26}$$

is found, provided that the relaxation time is given by Eq. (24). The temperature and frequency dependence reflects the fact that the imaginary part $[\omega\tau/(1 + \omega^2\tau^2)]$ of the Debye factor can be approximated by $(\omega\tau_{min})^{-1}$ and that $\tau_{min} \propto E^3 \propto T^{-3}$ for the relevant TSs.

The relaxation time τ_{min} of the fastest systems decreases rapidly with increasing temperature and finally $\omega\tau_{min} \ll 1$. In this regime, the main contribution to the integral (25) comes from systems for which $\omega\tau \approx 1$. The calculation leads to the relation

$$Q^{-1} = \frac{\pi C}{2} \tag{27}$$

meaning that in this regime Q^{-1} is expected to be independent of frequency and temperature. It should be pointed out that this so-called plateau is a direct consequence of the uniform distribution of the parameters Δ and λ.

At first glance the measurement shown in Fig. 24, and also many other results reported in literature, appear to be in agreement with the predictions of the TM. But it is extremely difficult to prove the existence of the expected T^3-increase since inevitable background losses cover small values of the internal friction. It seems that the rise at low temperatures is always less steep and can be approximated by a power law with an exponent closer to unity than to three [35, 57]. We will discuss this important observation in Sec. 3.3.

At higher temperatures, the motion of the tunneling particle is disturbed and a transition from coherent to incoherent tunneling is expected to occur. In amorphous solids, the partial loss of phase coherence of the wave function is caused by perturbation through the thermal motion of the environment and not by the motion of neighboring TSs as in crystals. It has been shown that the onset of this phenomenon results in an increase of the absorption starting at temperatures between 5 K and 10 K in agreement with observation [58].

3.2.2. *Velocity changes*

Since real and imaginary part of the susceptibilities are intimately related, acoustic and dielectric absorption are accompanied by a change δv of the velocity of sound or a change $\delta \epsilon'$ of the dielectric constant. For $\hbar \omega \ll k_B T$, the resonant process leads to a logarithmic increase with temperature of the form

$$\left. \frac{\delta v}{v} \right|_{\text{res}} = C \ln \frac{T}{T_0} \tag{28}$$

where T_0 is an arbitrary reference temperature. This rise was first reported in Ref. [33] and is characteristic of glasses. The validity of this equation has been verified for a large number of amorphous solids (see also Fig. 21). Recently, it has been observed in low frequency measurements that the slope, i.e. the constant C, depends on the amplitude of the applied strain [35, 57]. In such experiments, the applied elastic field is generally rather high leading to a strong modulation of the energy splitting of the TSs and resulting in a time average of the splitting that exceeds the static value. As a consequence, a strong non-linear response of the sample is observed (see e.g. Ref. [59]).

The contribution of the relaxation process to the velocity change is negligible as long as $\omega \tau_{\text{min}} \gg 1$. At higher temperatures, when $\omega \tau_{\text{min}}$ exceeds unity, the contribution of the relaxation process becomes dominant and the velocity starts to decrease. In this regime, the TM predicts a logarithmic decrease with temperature with a slope depending on the relaxation mechanism. As long as one-phonon processes (24) are dominant, the relation

$$\left. \frac{\delta v}{v} \right|_{\text{rel}} = -\frac{3}{2} C \ln \frac{T}{T_0'} \tag{29}$$

should hold, where T_0' is again a reference temperature. The total change of the velocity is a sum of the two contributions (28) and (29). In this way, not only the occurrence of a maximum in the velocity is easily explained but also its shift to higher temperatures with increasing frequency. Roughly speaking the maximum temperature T_{max} indicates that $\omega \tau_{\text{min}} \approx 1$. As mentioned,

$\tau_{\min} \propto T^{-3}$ for the relevant TSs and therefore the relation $T_{\max} \propto \omega^{1/3}$ should hold, in good agreement with observation.

Above 1 K the situation is more complicated since higher order phonon processes set in and lead to a decrease of the velocity which is more pronounced. However, as pointed out before, above a few Kelvin the tunneling motion becomes incoherent. As a consequence, the velocity decreases linearly with temperature [58]. At still higher temperatures, tunneling recedes in importance and thermally activated processes dominate the relaxation behavior.

As can be seen from Fig. 21 and as is known from many other acoustic (and dielectric) measurements at low frequencies, the ratio of the slopes given by Eq. (28) and by the sum of the contributions of Eqs. (28) and (29) for temperatures below and above the maximum, respectively, is not $2/(-1)$, but rather $1/(-1)$ [35, 57]. An explanation for this obvious discrepancy has been given very recently and will be discussed in Sec. 3.3.

3.2.3. *Thermal conductivity*

As mentioned at the begining of this subsection, measurements of the thermal properties gave first evidence of the anomalous behavior of amorphous solids at low temperatures [4]. Figure 25 shows the thermal conductivity Λ of vitreous silica as a function of temperature. It increases proportional to T^2 up to about 3 K, levels off above that temperature and starts to rise again around 20 K. There is no doubt that at low temperatures heat is transported by phonons as in crystals but phonon propagation is hindered by the TSs. Using the kinetic formula

$$\Lambda = \frac{1}{3} C_{\mathrm{D}} v l \tag{30}$$

where C_{D} represents the T^3-dependent Debye specific heat of the phonons, the thermal conductivity can be described in a simple way. In order to obtain the correct temperature dependence the mean free path l must be proportional to the inverse temperature, i.e. $l \propto T^{-1}$ must hold. The mean free path $l = Qv/\omega$ of thermal phonons is easily obtained from Eq. (23) by putting $I = 0$. Since furthermore the main contribution to the heat transport arises from phonons with an energy $\hbar\omega \approx 2k_{\mathrm{B}}T$, we may use $\tanh(\hbar\omega/2k_{\mathrm{B}}T) \approx 1$ and find immediately the correct relationship. It is noteworthy that the magnitude of the conductivity can correctly be deduced from measurements of the temperature variation of longitudinal and transverse sound waves [60].

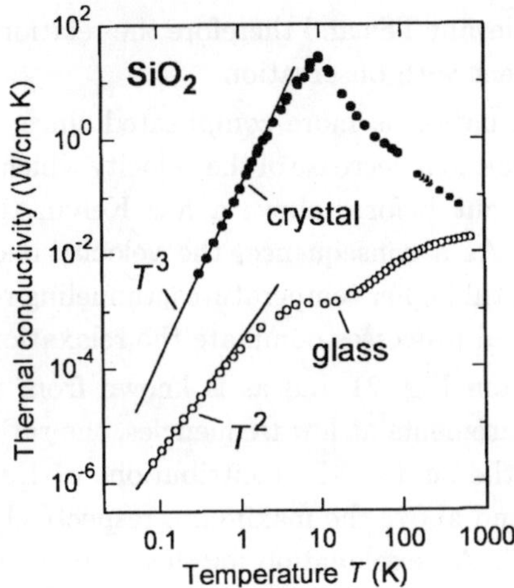

Fig. 25.   Thermal conductivity of vitreous silica and crystalline quartz as a function of temperature. From Ref. [4].

The conductivity in the plateau region between 1 K and 10 K is reduced by additional scattering processes. In the Soft Potential Model this scattering is attributed in a natural way to particles moving in quasi-harmonic potentials which have the same microscopic origin as the TSs [42, 61]. It should be noted that several alternative explanations for the heat transport at higher temperatures are under discussion.

3.3. *Interaction between tunneling systems*

So far we have discussed the interaction of TSs of glasses with elastic or electric fields. Many experimental results indicate that also the interaction between the TSs themselves plays an important role and is an essential ingredient to the understanding of amorphous solids especially at very low temperatures. As we have seen the TSs carry elastic and electrical dipole moments. They give rise to local fields and may thus be the origin of interaction. Let us first estimate whether the elastic or the electric interaction is predominant. Neglecting the dependence on the orientation of the two dipole moments i and j we may write for the interaction energy J between the two TSs:

$$J_{\rm a} = \frac{\gamma_i \gamma_j}{\varrho v^2} \frac{\Delta_i}{E_i} \frac{\Delta_j}{E_j} \frac{1}{r_{ij}^3} \quad \text{and} \quad J_{\rm e} = \frac{p_i p_j}{4\pi\epsilon\epsilon_0} \frac{\Delta_i}{E_i} \frac{\Delta_j}{E_j} \frac{1}{r_{ij}^3} \tag{31}$$

Here the subscripts a and e stand for acoustic (or elastic) and electric, respectively, and r_{ij} is the distance between the interacting TSs. Inserting typical numbers like $\gamma = 1$ eV and $p = 1$ Debye $= 3.2 \times 10^{-30}$ Asm, we obtain the ratio $J_a/J_e \approx 20$. This means that we have to consider only the elastic interaction in our further discussion.

There are several phenomena reflecting the interaction between TSs. We will discuss here briefly hole burning and echo experiments, and deal with the memory effect observed in dielectric measurements at very low temperatures. Finally, we will show that many effects which are only poorly described by the TM can be understood if due to the interaction a transition from coherent to incoherent tunneling occurs at very low temperatures.

3.3.1. *Hole burning*

As already mentioned and expressed by Eq. (23), the magnitude of the resonant absorption depends on the difference in the population of the two levels. At high acoustic (or electric) intensities the two levels become equally populated and the absorption is "saturated". This phenomenon can be expressed in a different way: the intense sound pulse burns a "hole" into the occupation number at the energy $E = \hbar\omega$ with a width $\Delta\omega$. A second pulse following immediately afterwards will see that the two levels at the resonant energy are almost equally populated and will therefore be subjected to a reduced attenuation. If the probing pulse has a slightly different frequency and is weak enough not to cause saturation by its own, its attenuation will reflect the occupation of the levels of the TSs at the corresponding frequency. Figure 26 shows the result of such a measurement carried out in a borosilicate glass [62]. (For experimental reasons the frequency of the probing pulse was kept constant and that of the saturating pulse was varied. This change does not influence the conclusions drawn from this experiment.) A surprisingly broad hole was observed which is much wider than expected from lifetime broadening or the pulse spectrum. Two important observations were made: the measured line width depends linearly on temperature and becomes larger with increasing pulse length of the saturating pulse.

This remarkable effect is ascribed to temporal fluctuations of the energy splitting of the TSs caused by the following mechanism: in thermal equilibrium TSs with an energy splitting smaller or comparable to $k_B T$ are continuously excited and de-excited. These transitions are accompanied by changes of the local strain field, or, in other words, by reorientations of the elastic dipole

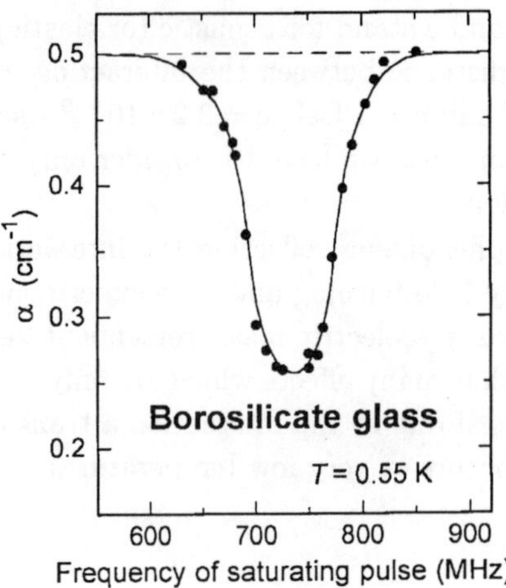

Fig. 26.   Resonant attenuation $\alpha = Q^{-1}v/\omega$ of a weak probing pulse at 735 MHz as a function of the frequency of the intense saturating pulse. The horizontal line indicates the attenuation of the probing pulse without saturating pulse. From Ref. [62].

moment. Since the energy splitting of TSs depends on strain, the resonating TSs in the hole burning experiment will have a fluctuating energy splitting. Therefore TSs with an energy splitting close to, but not exactly at the probing pulse frequency will temporarily be in resonance with the sound wave at finite temperatures although at $T = 0$ an energy difference would exist. As a consequence TSs can be excited by the resonant process within an energy range that is given by the width of the temporary excursions of the level splitting and a rather broad hole is registered by the probing pulse. The excursion of the energy splitting of a given TS will depend on time since the contributions of the neighboring TSs add statistically. Thus the line width will grow until all TSs have undergone a thermal transition, or, to be more precise, until half of the systems have changed their state an odd number of times. We do not carry out the proper average over the interacting neighbors j and over all existing TSs i for a estimate of the energy excursion δE_∞ in its stationary state but make the following simplifications: like in the TM we assume $\gamma_i \approx \gamma_j \equiv \gamma$, replace r_{ij}^{-3} in Eq. (31) by the concentration $\bar{P}k_\mathrm{B}T$ of thermally excited TSs, and put $\langle \Delta_i/E_i \rangle = \langle \Delta_j/E_j \rangle \approx 1$. In this way we obtain the simple expression in Ref. [63]

$$\delta E_\infty = \frac{\gamma^2 \bar{P}k_\mathrm{B}T}{\rho v^2} \tag{32}$$

This result explains the linear increase of the hole width with temperature, which obviously reflects the rising number of thermally excited TSs. The underlying mechanism of this broadening is called "spectral diffusion" by analogy with the mechanism known from magnetic resonance experiments.

In glasses, however, the stationary state will never be reached since the probability for a thermal transition of a TS is given by $[1 - \exp(-t/\tau)]$ and very slowly relaxing TSs are present in glasses. For short times $t \ll \tau_{\min}$ the expression may be approximated by $t/\tau_{\min}$, this means that the energy excursion is given only by this fraction of δE_∞. For long times $t \gg \tau_{\min}$ the situation is more difficult to treat because of the slowly relaxing TSs. According to Eq. (24) the relaxation rate $\tau^{-1} \propto r$ and hence $P(\tau^{-1}) \propto P(r)$. A quick glance at the distribution function (21) tells us that it diverges for TSs with $\tau^{-1} \to 0$. This means that with increasing time, a growing number of TSs contributes to the energy excursion. In this regime the hole width is expected to increase logarithmically with time [64]. These two results are put together in the following equation:

$$
\delta E =
\begin{cases}
\delta E_\infty \dfrac{t}{\tau_{\min}} & \text{for } t \ll \tau_{\min} \\[2ex]
\delta E_\infty \ln \dfrac{t}{\tau_{\min}} & \text{for } t \gg \tau_{\min}
\end{cases}
\tag{33}
$$

The long time limit is in agreement with the acoustic experiments mentioned above but unfortunately such experiments are only capable of covering a rather limited range in time and temperature so that a precise test of this theoretical prediction is not possible.

During the last years spectral diffusion has been studied extensively by optical means. We touch upon this powerful technique here and refer to a recent review of this field [65]. In most cases, hole burning experiments were carried out on organic glasses doped with dye molecules. An example for such a system is polymethylmethacrylate containing phthalocyanine. The optical absorption of the chromophore is inhomogeneously broadened by the random structure of the host. In most systems (but not in all) illumination causes photochemical changes of the dye. Thus a narrow, homogeneous hole can be burnt into the absorption band by a laser. At low temperatures, this hole is stable and changes can be monitored even after weeks by measuring the absorption with a weak light source. Spectral diffusion enters the problem in the following way: TSs close to the chromophore cause temporally fluctuating strain fields. Since the exact position of the absorption line of an individual dye molecule depends on strain, it will jitter as the energy splitting of the interacting TS does in the acoustic case. The theoretical description of this phenomenon will therefore be analogous to that given above.

Many interesting modifications of the described experiment have been worked out and measurements have been extended to very low temperatures. Here we only mention a recent result on the time-dependence of the hole width which apparently cannot be understood on the basis of the TM. In these experiments, the broadening of the burnt hole was measured at 0.5 K and 1 K for times up to 10^6 s [66]. Strong deviations from the logarithmic law (33) were observed for $t > 10^4$ s, the holes were broadening faster than expected. Since the logarithmic law is one of the consequences of the distribution function (22) a phenomenological description of the observation can easily be found by an appropriate change of this distribution. Of course more interesting is the question of the origin of such a modification. It has been argued that the distribution function is modified by the interaction between the TSs, but the final answer is still to be found.

As already pointed out, optical hole burning is a very elegant and efficient technique. However, one has to keep in mind that the dye molecules used in the experiments as local probes are rather big in size. They alter the structure of the host material in their environment. While in acoustic experiments "native" TSs are studied, the TSs contributing to spectral diffusion in optical hole burning might have properties slightly different from those of the undisturbed amorphous host.

3.3.2. *Coherent echoes*

With decreasing temperature both the relaxation time τ and the phase memory time τ_2, also known as "transverse" relaxation time, are continously increasing. Therefore, at sufficiently low temperatures coherent echoes can be generated like in NMR-experiments and similar to the rotary echoes discussed in Sec. 2.2. As an example, we consider here measurements of the decay time of spontaneous echoes in vitreous silica. The echo decay is also governed by the spectral diffusion processes just discussed in conjunction with the hole burning experiment. In fact the phase memory time is directly related to the hole width via $\tau_2 = \hbar/\delta E$, and thus reflects the interaction between the TSs. Figure 27 shows recent measurements of the decay of the amplitude as a

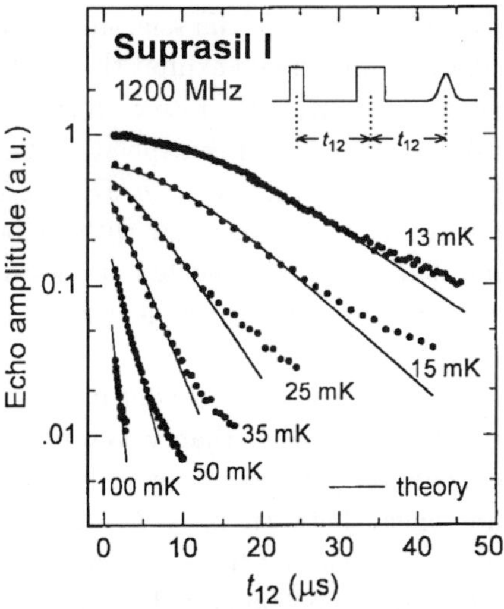

Fig. 27. Decay of the amplitude of the spontaneous echo in vitreous silica as a function of the pulse separation t_{12} at 1200 MHz and different temperatures. The curves are displaced vertically for clarity. The lines are fits using the model of spectral diffusion. The insert schematically shows the sequence of the exciting pulses and the echo. From Ref. [67].

function of pulse separation time t_{12} at different temperatures [67]. In the same figure the applied pulse sequence and the observed echo at time $2t_{12}$ are drawn schematically.

Roughly speaking, the dephasing or phase memory time τ_2 is the time for which the spread in phase is of the order of unity. This means that it is determined by the equation

$$\frac{1}{\hbar} \int_0^{\tau_2} \delta E(t) dt = 1 \tag{34}$$

Depending on the ratio $2t_{12}/\tau_{\min}$, two regimes can be distinguished. According to Eq. (33) the broadening is given by $\delta E_\infty \, t/\tau_{\min}$ in the short time limit resulting in a dephasing of the contribution of the individual TSs to the polarization which increases quadratically with time. From Eq. (34) the relation $\tau_2^2 = \hbar \tau_{\min}/2\,\delta E_\infty$ follows for the phase memory time. For the amplitude A of the echo, we expect therefore a non-exponential decay of the form

$$A(2\tau_{12}) = A(0) \exp[-(2t_{12}/\tau_2)^2] \tag{35}$$

This type of echo decay is often termed "Gaussian-like". Since $\tau_{\min}^{-1} \propto E^3 \propto T^3$, and $\delta E_\infty \propto T$ the dephasing time should exhibit the proportionality $\tau_2 \propto T^{-2}$.

In the long time regime $t_{12} \gg \tau_{\min}$, the rate with which the neighboring TSs undergo thermal transitions does not enter explicitly. We may put $\delta E \approx \delta E_\infty$ in Eq. (34) and hence obtain the result $\tau_2 = \hbar/\delta E_\infty \propto T^{-1}$. In this regime, the echo decay is expected to be exponential with τ_2 being proportional to temperature. It seems that we have neglected the logarithmic correction of Eq. (33) but a careful treatment of the problem leads in fact to the simple behavior just mentioned [64].

In several dielectric and acoustic echo experiments on vitreous silica, an exponential decay was observed rather than a Gaussian [68, 69]. Moreover, the temperature dependence of τ_2 derived from different experiments appeared to be inconsistent and ranged between $\tau_2 \propto T^{-1}$ and $\tau_2 \propto T^{-2}$. In recent experiments [67] it has been shown that the contradiction is only apparent since a hitherto unexpected frequency dependence of τ_2 exists and the measurements of Refs. [68, 69] had been performed at different frequencies.

In Fig. 27, numerical fits of the decay of the echo amplitude are shown together with the data. Bearing in mind that there are only two free parameters the agreement is satisfactory. The theory not only correctly describes the time evolution of the echo but also the temperature dependence. There are, however, significant discrepancies at long time intervals t_{12} for which no clear explanation exists. The most surprising result is the fact that the cross-over from the Gaussian-like to the exponential decay is observed within the experimental time-frame. From the fit of these data a value for the relaxation time $\tau_{\min}$ is found which is 250 times smaller than the value deduced from acoustic experiments like those shown in Fig. 24. This means that at very low temperature thermal relaxation of the TSs is much faster than expected from the one-phonon process. Furthermore, the frequency dependence of the echo decay explains why the results of earlier experiments were in apparent contradiction. However, the frequency dependence and the short relaxation time are not understandable within the framework of spectral diffusion.

3.3.3. *Incoherent tunneling*

In Sec. 2.3, we have seen that at high concentrations TSs interact between each other resulting in an incoherent tunneling motion. In this case, TSs hardly contribute to the elastic or electric properties through the resonant process, although this is the dominant process in crystals if only isolated defects are present. In the discussion of this phenomenon, the central parameter was the factor $\mu = \bar{J}/\Delta_0$ which is defined as the ratio of the mean interaction energy $\bar{J}$ and the tunnel splitting Δ_0. For TSs interacting via electrical dipole moments, the parameter μ is given by Eq. (13). Recently, it has been argued that this concept should also be applicable to glasses [70]. As TSs in these solids show a wide distribution of their parameters, the quantity μ does not exhibit a well-defined value but depends on temperature. As mentioned before, the main contribution to acoustic or electric properties stems from TSs with an energy splitting $E \approx k_{\mathrm{B}}T$. According to Eq. (21), the average value $\bar{\Delta}_0$ of the tunnel splitting is proportional to E and therefore we may state that $\mu = \bar{J}/\bar{\Delta}_0 \propto T^{-1}$. Thus, on cooling, interaction becomes increasingly important for the relevant TSs and a continuous transition from coherent to incoherent tunneling should occur. Assuming a mean interaction energy of the order of $\bar{J}/k_{\mathrm{B}} \approx 1$ mK several observations can be explained which hitherto were not understood.

First, we want to mention that incoherent tunneling will enhance the dephasing of TSs in the two-pulse echo experiments and hence leads to a reduction of τ_2. Roughly speaking, the phase of an incoherently tunneling TS is fluctuating due to the interaction with the coupled TSs. On the one hand, this process resembles that taking place in spectral diffusion, but on the other hand it is basically different because no thermal transitions are necessary in the neighboring TSs. Eventually incoherent tunneling leads to the loss of phase coherence of the ensemble of excited TSs and has therefore the same effect on the evolution of the macroscopic polarization as spectral diffusion. It also gives an explanation of the frequency dependence of τ_2. Since $\hbar\omega \approx \Delta_0$ in echo experiments and $\mu \propto 1/\bar{\Delta}_0$ coherent tunneling is more pronounced in TSs with higher energies, i.e. in TSs which are excited in experiments at higher frequencies.

The phenomenon just described also gives an explanation of the weak temperature dependence of the internal friction at very low temperatures mentioned in Sec. 3.2. Incoherent tunneling opens an additional relaxation channel which reduces the relaxation time τ. In this way, it leads to an enhancement of the internal friction at the lowest temperatures and thus causes a variation with temperature weaker than T^3, the dependence predicted by the TM in Eq. (26). It should be mentioned that there exists an alternative explanation which is based on the formation of intimately coupled pairs of TSs (see also Sec. 3.3.4). In turn the relaxation of these pairs induces an additional relaxation of the ensemble of TSs with a rate proportional to T [71].

The most important support for the idea of incoherent tunneling comes from the temperature variation of the sound velocity and the dielectric constant. Below 100 mK these quantities are expected to vary less with temperature than predicted by the original TM as with decreasing temperature, incoherent tunneling develops more and more, and the resonant effect becomes weaker. At the same time the contribution of the relaxation process will be enhanced due to the additional relaxation channel, causing a further flattening of the temperature variation of the two quantities. Thus both effects modify the ratio of the slopes in the direction observed in experiment. A comparison between the prediction of the standard TM and the modification just discussed as shown in Fig. 28. Measurements of the temperature variation of the dielectric constant of vitreous silica are shown down to temperatures below 1 mK [72]. Clearly, the variation with temperature levels off below a few Millikelvin. Assuming a mean value of the interaction energy of $\bar{J}/k_B \approx 1$ mK almost

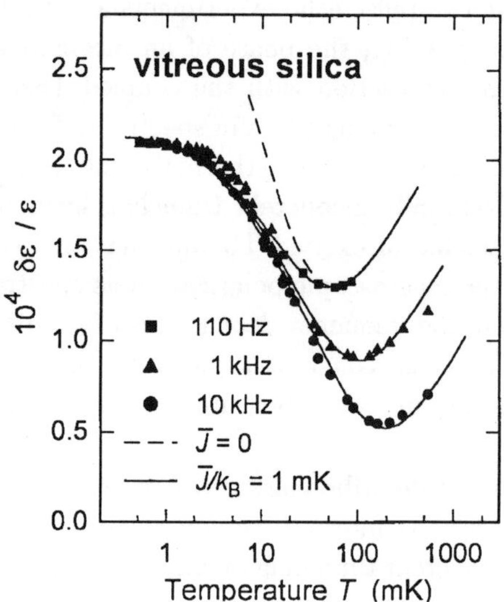

Fig. 28. Variation of the dielectric constant of vitreous silica with temperature. The measurements were carried out at frequencies between 110 Hz and 10 kHz and temperatures down to less than 1 mK. The dashed line represents a calculation with the TM, the solid lines are calculated taking into account incoherent tunneling processes. The points show experimental data from Ref. [72].

all relevant TSs will tunnel incoherently at those temperatures and the contribution of the resonant process will have vanished. The full lines represent the fitting curves obtained under the assumptions just discussed. The dashed line depicts the prediction of the original TM which clearly disagrees with the experimental data. It should be pointed out that a flattening of the curves at temperatures below 10 mK could also be explained by a cut-off of the density of states. However, arguments for the existence of such a high value of the cut-off energy are difficult to find. It should be added that this so-called saturation of the dielectric constant at ultra-low temperatures has not been observed in all amorphous substances. This is easily understood because the mean interaction energy $\bar{J}$ will depend on the material under investigation.

3.3.4. *Memory effects*

Finally, we want to discuss briefly the surprising memory effects discovered recently in dielectric experiments at very low temperatures [73, 74]. As shown in Fig. 29, the dielectric constant ϵ' (which is proportional to the measured capacitance C) determined at audio frequencies jumps rapidly upward after the sudden application of a DC electric field, and then begins to decay with a rate which is logarithmic in the time elapsed since the field was applied.

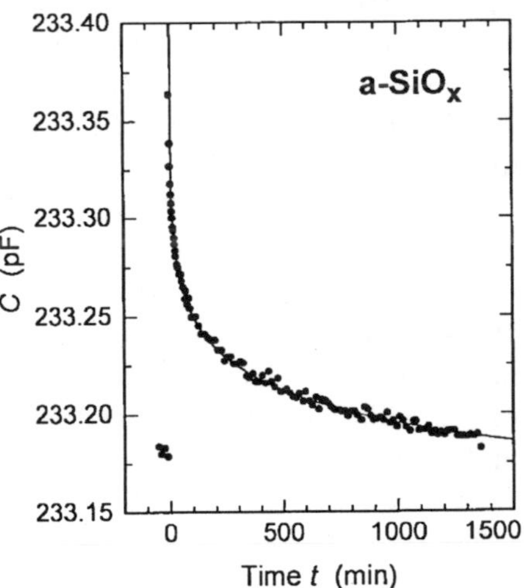

Fig. 29. Response of a capacitor filled with SiO$_x$ to a sudden application of a large DC electric field at 50 mK. From Ref. [74].

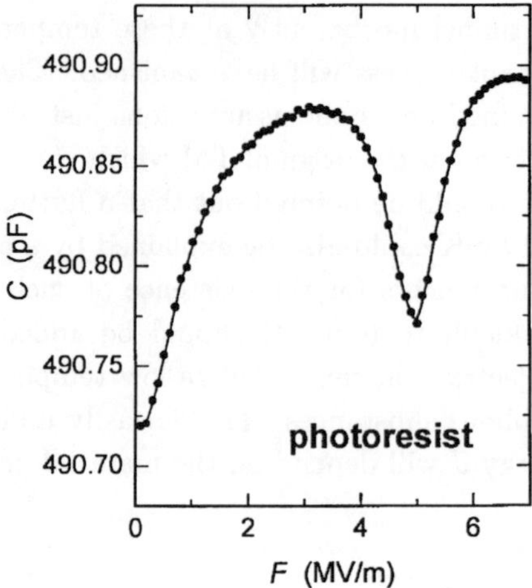

Fig. 30. Capacitance of a photoresist film as a function of the applied DC electric field after keeping the film under a field of 5 MV/m for several minutes. From Ref. [74].

Please note the time scale: the decrease was monitored during the course of a whole day. Another interesting observation was made by sweeping the DC field slowly. It was found that the dielectric constant ϵ' has a minimum value at the field at which the sample was cooled down from high temperatures. In Fig. 30, the response of a thin film of a photoresist (which is amorphous) is shown which had been cooled down without an electric field. Clearly, a minimum was found at the field $F = 0$. However, if the field was kept at a different value, e.g. at 5 MV/m, for several minutes, a new local minimum developed the depth of which was growing logarithmically with time.

Obviously, the jump of the dielectric constant ϵ' shown in Fig. 29 reflects the shift of the ensemble of TSs out of its equilibrium value at $F = 0$. The decay is associated with the formation of a new hole at the applied DC electric field. In further measurements it was shown that the effect is more pronounced at lower temperatures. In addition, a weak dependence on the frequency of measurement was found at very low temperatures whereas the frequency dependence disappeared at temperatures above the minimum of the dielectric constant (see e.g. Fig. 28).

Two theoretical approaches have been worked out to explain these phenomena [75, 76]. The theory derived in Ref. [76] leads to a better agreement with the experiment and will be briefly considered here without going into details.

The theory is based on the assumption that the interaction between the TSs leads to the formation of pairs if the interaction energy exceeds the thermal energy, i.e. if $J > k_\mathrm{B}T$. As shown in Ref. [76], the interaction causes a change of the distribution function (21).

Most important to the explanation of the memory effect is the fact that the difference $\delta P' = P' - P$ between the new distribution function P' and the distribution function P of the TM depends on energy and temperature, namely like $\delta P' \propto -\ln(E + k_\mathrm{B}T)$. Because of the weakness of the interaction this change is expected to be rather small. An estimate leads to $\delta P'/P \approx 0.01$.

The application of an DC electric field shifts the position of the energy levels. Depending on the relative orientation of dipole moment and field, the energy splitting becomes larger or smaller. In the latter case the levels may even exchange their relative position if the field is high enough. These TSs are of particular interest. For further discussion, the TSs are divided into two categories: slowly relaxing (non-adiabatic) TSs and fast relaxing (adiabatic) TSs. Switching on the field will leave the non-adiabatic TSs in their quantum state. Thus by application of a strong field these TSs become inverted. In contrast, the fast relaxing systems will remain in the ground or in the excited state, respectively.

It can be shown that the non-adibatic TSs contribute most to the observed change of ϵ'. They give rise to an increase of the density of states just after switching by the amount $\delta P'' = P'' - P' \propto \ln[(E + k_\mathrm{B}T + 2pF_\mathrm{DC})/(E + k_\mathrm{B}T)]$. This means that the application of the DC electric field is equivalent to a rise of temperature. The change is followed by a relaxation to equilibrium. This idea has been worked out quantitatively with respect to the real and imaginary part of the dielectric function ϵ and the relaxation time as well, and quantitative agreement with the experimental results has been obtained.

4. Concluding Remarks

The investigation of TSs has a long tradition in low temperature physics; nevertheless many open questions still remain. Isolated TSs in model systems have been studied in great detail and are well described by theory. Although the understanding of their structure and their properties is based on optical measurements [76] in many cases we did not discuss these experiments for the sake of brevity but only used their findings. However, the consequences of the interaction between the TSs — though studied for a long time — are still not fully understood. For TSs interacting via electrical dipole moments, an extensive theory has been worked out recently which leads to good agreement with experimental results. The situation is less clear if the TSs interact elastically because of the complexity of the problem.

Here we have focused the discussion on lithium-doped KCl which often serves as a simple model system. It has been shown that in this material, with increasing defect concentration, a transition occurs from the coherent tunneling of unperturbed TSs to the incoherent tunneling of interacting TSs. Due to this interaction, the dynamic response of the TSs has almost exclusively relaxational character at high defect concentrations. At moderate defect concentrations, however, a particularly interesting situation occurs where the coherent tunneling of pairs of strongly coupled Li-TSs can be observed. For brevity we only have mentioned the interesting theoretical and experimental results for TSs in crystals with large random strains like KBr:KCN, although these materials have very often been considered as crystalline models for glasses.

TSs in amorphous materials can presently only be described on the basis of phenomenological models. Much theoretical and experimental work has been put into the determination of the relevant distribution functions. Despite intense efforts, the fundamental question of the atomic nature of the TSs in amorphous solids still remains and has not been addressed in this article at all. No generally accepted answers can be given at present. Attempts have been made to attribute the TSs to material-specific local atomic configurations. If such an approach leads to a good description of the experimental results the universality of the observed phenomena is difficult to explain. At present, it seems that computer simulations of simple glasses are the most promising attempts to come to an understanding of the structure of the TSs [40]. Like in crystals, another open problem remains, namely the interaction between TSs. There is no doubt that TSs also interact in glasses and that this interaction is responsible for many phenomena observed especially at very low temperatures. Among the experiments allowing to study this interaction we have discussed here hole burning, coherent echoes and the recently discovered memory effects which provide information on the strength of the elastic and the electric coupling among the TSs. However, the consequences of this interaction are by no means understood. Some theoretical models even postulate that interaction is responsible for the creation of the experimentally observed TSs and for the distribution of their parameters [78].

We want to add that very recently surprising observations have been made in studies of the dielectric properties of multi-component glasses at ultra-low temperatures: in contrast to expectation, below 100 mK the dielectric response becomes sensitive to weak magnetic fields. Dielectric constant and dielectric loss show an oscillatory behavior with increasing magnetic field; below 6 mK even a phase transition within the ensemble of TSs was observed [79].

Acknowledgments

We would like to thank J. Claßen, F. Luty, M. von Schickfus, R. Weis, and A. Würger for many comments and valuable discussions and the Deutsche Forschungsgemeinschaft for financial support.

References

[1] F. Hund, *Z. Physik* **43** (1927) 805.
[2] L. Pauling *Phys. Rev.* **36** (1930) 430.
[3] W. Känzig, *J. Phys. Chem. Solids* **23** (1962) 479.
[4] R.C. Zeller and R.O. Pohl, *Phys. Rev.* **B4**, 2029 (1971).
[5] V. Narayanamurti and R.O. Pohl, *Rev. Mod. Phys.* **42**, 201 (1970).
[6] F. Bridges, *Crit. Rev. Solid State Sci.* **5**, 1 (1975).
[7] A. Würger, *From Coherent Tunneling to Relaxation*, Tracts in Modern Physics **135** (Springer, Berlin, 1997).
[8] M. Gomez, S.P. Bowen and J.A. Krumhansl, *Phys. Rev.* **153**, 1009 (1967).
[9] J.P. Harrison, P.P. Peressini and R.O. Pohl, *Phys. Rev.* **171**, 1037 (1968).
[10] X. Wang and F. Bridges, *Phys. Rev.* **B46**, 5122 (1992).
[11] H.B. Shore and L.M. Sander, *Phys. Rev.* **B12**, 1546 (1975).
[12] J.P. Sethna, *Phys. Rev.* **B24**, 698 (1981); **B25**, 5050 (1982).
[13] C. Enss, M. Gaukler, S. Hunklinger, M. Tornow, R. Weis and A. Würger, *Phys. Rev.* **B53**, 12094 (1996).
[14] N.E. Byer and H.S. Sack, *J. Phys. Chem. Solids* **29**, 677 (1968).
[15] G. Weiss, M. Hübner and C. Enss, *Physica* **B263–264**, 388 (1999).
[16] G.A. Alers, *Physical Acoustics*, ed. W.P. Mason (Academic Press, New York, 1966) **IVA**, p. 277.
[17] D. Moy, R.C. Potter and A.C. Anderson, *J. Low Temp. Phys.* **52**, 115 (1983).
[18] R. Weis, C. Enss, A. Würger and F. Lüty, *Ann. Physik* **6**, 263 (1997).
[19] R. Weis, C. Enss, B. Leinböck, G. Weiss and S. Hunklinger, *Phys. Rev. Lett.* **75**, 2220 (1995).
[20] J.N. Dobbs and A.C. Anderson, *Phys. Rev.* **B33**, 4172 (1986).
[21] M.E. Baur and W.R. Salzman, *Phys. Rev.* **178**, 1440 (1969).
[22] M.W. Klein, *Phys. Rev.* **B35**, 1397 (1987).
[23] O. Terzidis and A. Würger, *Z. Physik* **B94**, 341 (1994).
[24] R. Weis and C. Enss, *Czech. J. Phys.* **46**, 3287 (1996).

[25] A. Würger, *Z. Phys.* **B94**, 173 (1994); **B98**, 561 (1995).

[26] H. Mori, *Progr. Theor. Phys.* **33**, 423 (1965); **34**, 399 (1965).

[27] A.T. Fiory, *Phys. Rev.* **B4**, 514 (1971).

[28] C. Enss, M. Gaukler, M. Nullmeier, R. Weis and A. Würger, *Phys. Rev. Lett.* **78**, 370 (1997).

[29] A. Würger, R. Weis, M. Gaukler and C. Enss, *Europhys. Lett.* **33**, 533 (1996).

[30] J.J. De Yoreo, W. Knaak, M. Meissner, and R.O. Pohl, *Phys. Rev.* **B34**, 8828 (1986).

[31] E.R. Grannan, M. Randeria and J.P. Sethna, *Phys. Rev.* **B41**, 7784 (1990).

[32] S.K. Watson, *Phys. Rev. Lett.* **75**, 1965 (1995).

[33] L. Piché, R. Maynard, S. Hunklinger and J. Jäckle, *Phys. Rev. Lett.* **32**, 1426 (1974).

[34] J.C. Lasjaunias, A. Ravex, M. Vandorpe and S. Hunklinger, *Solid State Comm.* **17**, 1045 (1975).

[35] J. Classen, C. Enss, C. Bechinger, G. Weiss and S. Hunklinger, *Ann. Physik* **3**, 315 (1994).

[36] J.E. Van Cleve, A.K. Raychaudhuri and R.O. Pohl, *Z. Phys.* **B93**, 479 (1994).

[37] W.A. Phillips, *J. Low Temp. Phys.* **7**, 351 (1972).

[38] P.W. Anderson, B.I. Halperin and C.M. Varma, *Phil. Mag.* **25**, 1 (1972).

[39] A. Heuer and R.J. Silbey, *Phys. Rev. Lett.* **70**, 3911 (1993).

[40] A. Heuer and R.J. Silbey, *Phys. Rev.* **B53**, 609 (1996).

[41] V.G. Karpov, M.I. Klinger and F.N. Ignatiev, *Sov. Phys. JETP* **57**, 439 (1983).

[42] D.A. Parshin, *Physica Scripta* **T49**, 180 (1993).

[43] J.Y. Duquesne and G. Bellessa, *Phil. Mag.* **B52**, 821 (1985).

[44] X. Liu, B.E. White, Jr., R.O. Pohl, E. Iwanizcko, K.M. Jones, A.H. Mahan, B.N. Nelson, R.S. Crandall and S. Veprek, *Phys. Rev. Lett.* **78**, 4418 (1997).

[45] M. Meißner and P. Strehlow, *Czech. J. Phys.* **46**, 2233 (1996).

[46] A. Würger, *Phys. Rev.* **B57**, 347 (1998).

[47] J.L. Black, *Phys. Rev.* **B17**, 2740 (1978).

[48] P. Strehlow, *Glastech. Ber. Glass Sci. Technol.* **68**, C1, 67 (1995).

[49] S. Hunklinger and W. Arnold, *Physical Acoustics*, eds. W.P. Mason and R.N. Thurston (Academic Press, New York, 1976) **XII**, p. 155.

[50] S. Hunklinger and A.K. Raychaudhuri, *Progress in Low Temperature Physics*, ed. D.F. Brewer (Elsevier, Amsterdam, 1986) **IX**, p. 267.

[51] S. Hunklinger, W. Arnold and S. Stein, *Phys. Lett.* **A45**, 311 (1974).

[52] M. von Schickfus and S. Hunklinger, *Phys. Lett.* **A64**, 144 (1977).

[53] J. Zimmermann and G. Weber, *Phys. Rev. Lett.* **46**, 661 (1981).

[54] M. Schwark, F. Pobell, M. Kubota and R. Mueller, *J. Low. Temp. Phys.* **58**, 171 (1985).

[55] D.A. Parshin and S. Sahling, *Phys. Rev.* **B47**, 5677 (1993).

[56] P. Strehlow, *Quantenthermodynamik von Gläsern*, Habilitation Thesis, TU Berlin, 1997.

[57] P. Esquinazi, R. König and F. Pobell, *Z. Phys.* **B87** (1992) 305.

[58] S. Rau, C. Enss, S. Hunklinger, P. Neu and A. Würger, *Phys. Rev.* **B52**, 7179 (1995).

[59] Yu.M. Galperin, V.L. Gurevich and D.A. Parshin, *J. Physique Lett.* **45**, L747 (1984); J. Stockburger, M. Grifoni and M. Sassetti, *Phys. Rev.* **B51**, 2835 (1995).

[60] S. Hunklinger and L. Piché, *Solid State Comm.* **17**, 1189 (1975).

[61] L. Gil, M.A. Ramos, A. Bringer and U. Buchenau, *Phys. Rev. Lett.* **70**, 182 (1993).

[62] W. Arnold and S. Hunklinger, *Solid State Comm.* **17**, 833 (1975).

[63] W.A. Phillips, *Rep. Prog. Phys.* **50**, 1657 (1987).

[64] B.D. Laikhtman, *Phys. Rev.* **B31**, 490 (1985).

[65] P. Schellenberg and J. Friedrich, *Disorder Effects and Relaxation Processes*, eds. R. Richert and A. Blumen (Springer, Berlin, 1994) p. 407.

[66] H. Maier, B.M. Kharlamov and D. Haarer, *Phys. Rev. Lett.* **76**, 2085 (1996).

[67] C. Enss, R. Weis, S. Ludwig and S. Hunklinger, *Czech. J. Phys.* **46**, 2247 (1996).

[68] J.P. Graebner and B. Golding, *Phys. Rev.* **B19**, 964 (1979).

[69] L. Bernard, L. Piché, G. Schuhmacher and J. Joffrin, *J. Low Temp. Phys.* **45**, 411 (1979).

[70] C. Enss and S. Hunklinger, *Phys. Rev. Lett.* **79**, 2831 (1997).

[71] A.L. Burin and Y. Kagan, *JETP* **80**, 761 (1995).

[72] H. Nishiyama, H. Akimoto, Y. Okuda and H. Ishimoto, *J. Low Temp. Phys.* **89**, 727 (1992).

[73] S. Rogge, D. Natelson and D.D. Osheroff *Phys. Rev. Lett.* **76**, 3136 (1996).

[74] D.D. Osheroff, S. Rogge and D. Natelson, *Czech. J. Phys.* **46**, 2247 (1996).

[75] H.M. Carruzzo, E.R. Grannan and C.C. Yu, *Phys. Rev.* **B50**, 6685 (1994).

[76] A.L. Burin, *J. Low Temp. Phys.* **100**, 309 (1995).

[77] S. Kapphan and F. Luty, *J. Phys. Chem. Solids* **34**, (1973) 969.

[78] C.C. Yu and A.J. Leggett, *Comments Cond. Mat. Phys.* **14**, 231 (1988).

[79] P. Strehlow, C. Enss and S. Hunklinger, *Phys. Rev. Lett.* **80**, 5361 (1998); P. Strehlow, M. Wohlfahrt, A.G.M. Jansen, R. Haueisen, G. Weiss, C. Enss and S. Hunklinger **84** (2000).

8

Electronic Transport in Disordered Semiconductors

HARALD OVERHOF

Department of Physics, University Paderborn,
Warburgerstr. 100, D–33098 Paderborn, Germany
h.overhof@toranaga.uni-paderborn.de

PETER THOMAS

Department of Physics and Material Sciences Center,
Philipps Universität, D–35032 Marburg, Germany
peter.thomas@physik.uni-marburg.de

Contents

1. Introduction

The general physics and specifically the electronic transport in heavily disordered, glassy, and amorphous semiconductors is much less understood than the corresponding processes in lightly disordered crystalline semiconductors. This is due to the fact that in the latter material one can make use of an idealization, the perfect crystal, as a starting point for the description of the real crystal. For the perfect crystal, the atomic coordinates are known, as are the electron states which are extended Bloch states. Since Bloch states are eigenstates of the Hamiltonian for the perfect crystal, these states (in a single-particle approximation, which turns out to be a good approximation for surprisingly many problems in semiconductor physics) are not scattered by the *periodic* lattice.

For a real crystal, we have lattice imperfections like point defects, dopants e.g., more extended imperfections like dislocations, and extended imperfections like the deviations from the lattice periodicity due to lattice vibrations. The scattering by these lattice imperfections can be treated by first-order perturbation theory. The results of the calculations can be compared with the carrier densities and mobilties etc. obtained from a variety of different transport experiments. For many semiconductor systems experimental and theoretical data fit into a consistent picture. Thus, one can consider the electronic transport in lightly disordered crystals as well understood. Therefore, the transport experiments can be used as a diagnostic tool for the further investigation of these crystals.

In contrast, the heavily disordered amorphous and glassy semiconductors provide a much more rigorous obstacle to the researcher who undertakes an investigation of the electronic properties. To begin with, because of the disordered nature itself, there cannot exist a unique set of coordinates for the atomic sites. Instead, all structural information can be determined in a statistical sense at most: one can investigate average bond length and bond angles and in fortunate cases also the distribution of these quantities. This information, however, is much less than what is needed for a complete description of the structure of an amorphous semiconductor.

The lack of knowledge about the atomic structure causes a corresponding lack of knowledge of the eigenstates for electrons. One way to circumvent this difficulty is to study simple model Hamiltonians like the Anderson Hamiltonian instead. We shall learn in Sec. 2 from the Anderson Hamiltonian that the electronic single-particle states are completely different from their crystalline counterparts. Depending on the degree of disorder built into the Anderson Hamiltonian, one has delocalized states (which however are not Bloch-like) and localized states close to the band extremities (the tails). Both types of states may play the dominant role in transport, subject to the degree of disorder and to the temperature regime under consideration.

In this chapter, we shall concentrate on inorganic amorphous semiconductors. Furthermore, we shall treat electronic dc transport in thermal quasi-equilibrium exclusively: this excludes the discussion of the most interesting but more complex time- and frequency-dependent transport and also the treatment of high-field and photo-conductance. Although the latter property is the basis of the majority of the applications of amorphous semiconductors, we shall exclude it: photoconductivity cannot be discussed without a thorough investigation of the carrier recombination which is another wide, rather complex and still poorly understood field of amorphous semiconductor physics. Restricting our discussion to quasi-equilibrium transport, we can treat a comparatively simple system: there is no need to investigate in detail how the different recombination and excitation processes balance. It suffices to know that this results in the equilibrium concentration which follows from elemental statistical mechanics.

We shall base the discussion in this chapter on an Anderson Hamiltonian. In Sec. 3, we discuss the elements of transport theory deriving the formal Kubo and Kubo–Greenwood formulae. We then apply these formulae in Sec. 4 to the different transport modes that can be observed. We start with the transport near the so-called mobility edge, an energy regime which marks the transition from localized to delocalized states. We show how the interplay of localized states with delocalized states in the presence of electron-phonon interaction leads to a mixed mode of electronic transport. Although the transport mechanism is considerably more complex than the transport envisaged by Mott when proposing his "minimum metallic conductivity", the transport properties of both models are strikingly similar.

We also discuss the transport which is promoted by well-localized states deeper in the tails, i.e. hopping. We shall discuss a variety of situations for which hopping can occur including R-percolation, variable-range hopping, hopping in tails and small polaron hopping.

Section 5 attempts to give an overview over the experimental electronic transport properties of amorphous semiconductors. Although the field has been investigated for at least two decades, our knowledge concerning the underlying transport mechanism for most materials is still rather diffuse: there is no problem to find transport models that describes the data. In the contrary, one has several different mechanisms that would describe the experimental data perfectly, but how can one find out which mechanism is present?

To avoid this uncertainty as much as possible we shall focus our attention on hydrogenated amorphous silicon, a–Si:H. This system is most thoroughly studied, best understood, and still under intense scientific investigation to date.

In Sec. 6, we give a detailed interpretation of the transport phenomena in a–Si:H. We show that the model of transport near the mobility edge discussed in Subsec. 4.4 can successfully describe the observed transport phenomena if the influence of a long-ranged potential and also the statistical shift of the Fermi energy are taken into account properly.

2. Electrons in Disordered Solids

2.1. *Anderson localization*

Electronic eigenstates $\psi(\mathbf{r})$ in crystalline solids are spatially extended Bloch–states. An electron with given energy has equal density $|\psi(\mathbf{r})|^2$ in every unit cell throughout space. In the absence of scattering, transport under an applied external electric dc-field would be ballistic. Scattering at imperfections of the lattice, however, leads to a constant drift velocity and thus to a well defined conductivity. Theoretically, the conductivity is calculated within perturbation theory, starting from extended Bloch–states of the perfect lattice and treating the scattering events as a perturbation. One usually writes the conductivity σ in terms of the mobility μ and density n of electrons in the conduction band: $\sigma = ne\mu$, where e is the elementary charge. For vanishing temperature T the conductivity approaches a finite value in a normal metal.

In disordered solids, however, this approach fails. The disorder is so intense, that one has to incorporate the disorder into the electronic states from the beginning, avoiding perturbation theory. The wave functions then turn out to be no longer extended for all energy eigenstates. Instead, localized states occur for certain energies, i.e. $|\psi(\mathbf{r})|^2$ is finite only in a limited range of space.

The consequences of this localization for transport can best be understood if one uses Einstein's relation between the conductivity σ and the diffusion coefficient D: $D \propto \sigma$. If a particle (a wave-packet) composed from localized states is placed at a particular position in space, it will be unable to diffuse away. Consequently $D = 0$ and hence $\sigma = 0$, i.e. without further interactions (e.g. with phonons) electrons in localized states do not contribute to transport. In addition to localized states there may be a range of energies with extended (however, not of Bloch–type) eigenstates. Electrons in these states are able to diffuse away, i.e. to contribute to transport even in the absence of other interactions.

In order to obtain some insight into the action of strong disorder on electrons in solids, we study the simplified model due to Anderson [1]. We consider a single band composed from atomic-like levels $|i\rangle$ centered at sites i with single-site (atomic) energies ϵ_i and ignore the electron-electron interaction. It suffices to incorporate the disorder into the distribution (of width W) of the energies ϵ_i and place the sites on a regular lattice. For simplicity, we choose a simple cubic three-dimensional lattice with lattice constant a. Since there is some overlap between the wave functions residing at the individual sites, the states $|i\rangle$ are not eigenfunctions of the system. We further simplify the model by considering overlap only between nearest neighbors of the lattice. The overlap integral is then denoted by J. The Hamiltonian of this model system reads

$$H_0 = \sum_i \epsilon_i c_i^\dagger c_i + J \sum_{ij} c_i^\dagger c_j \tag{1}$$

where the double sum extends only over nearest neighbors. Although this Hamiltonian describes a rather idealized model, which is called "Andersonium", it can be argued that the results obtained for this system have a wide applicability even for real systems. In fact, the ordered version of H_0 (all $\epsilon_i = \epsilon_0$) describes a perfect crystal in the tight-binding scheme. Its eigenfunctions are Bloch–functions and its eigenvalues form a cosine-like band structure with total bandwidth $B = 12J$ in three dimensions. It is the simplest model for an s-like conduction band.

For finite W, due to the absence of translational symmetry, it is impossible to calculate the wave functions and the distribution of eigen-energies, the density of states (DOS), for Andersonium of macroscopic size. One either uses elaborate numerical methods for finite systems with suitable extrapolation schemes to obtain information about the DOS [2] and the localization properties of the system, or one tries to set up analytical, however approximate, non-perturbative microscopic theories [3, 4]. In this presentation, we choose the second approach, although for a complete picture a combined analytical and numerical study is necessary. It turns out that at the band extremities the states are always localized. If the ratio $\eta = W/J$ is small, then the states in the center of the band are delocalized. The localized eigenstates are separated from the delocalized ones by mobility edges [5, 6] E_C and E_C' (see Fig. 1). For increasing η the mobility edges move towards the band center, until for $\eta = \eta_{\text{crit}}$ the mobility meet, $E_C = E_C'$. For $\eta > \eta_{\text{crit}}$, all states are localized. This "Anderson transition" occurs if W/B is a number of the order of 1.

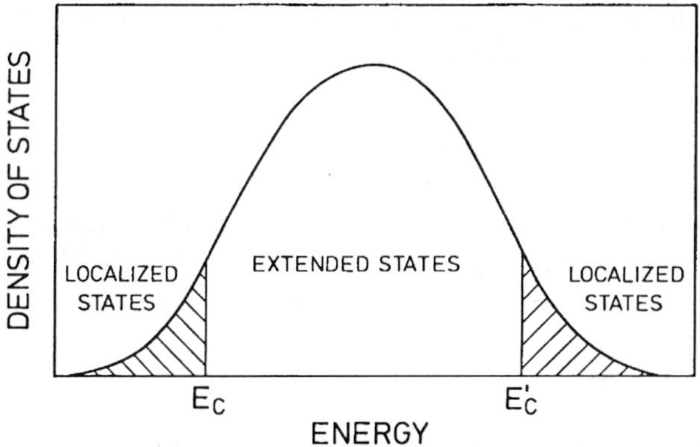

Fig. 1. Density of states distribution of Andersonium showing the mobility edges E_C and E'_C.

The degree of localization of energy eigenstates can be determined by explicitly calculating wave functions for disordered model systems, i.e. by diagonalizing the Hamiltonian for finite model systems. Although these systems are necessary quite small, by suitable scaling arguments one finds that for three dimensions the wave functions can be characterized by a decay like $\exp(-r/\xi_{\rm loc})$ with a localization length $\xi_{\rm loc}$. For states far below E_C, the localization length is small, comparable to the spatial extent of the basis states $|i\rangle$ (which in this regime are close to the eigenstates of the whole system which then have energies $E \simeq \epsilon_i$). It diverges if $E \to E_C$ from below.

Throughout this chapter, we shall treat the case that J is fixed and $\eta < \eta_{\rm crit}$, i.e. that E_C is located in the tails of the band. We fill, at temperature $T = 0$, the states in the band tail with electrons. The equilibrium distribution of electrons is box-shaped, all states below the Fermi level E_F are occupied, all others are empty. We are now interested in the transport properties of this system when a small external electric field is applied. The conductivity $\sigma(E_F)$ tells us whether, as a function of E_F, the system is an insulator or a metal.

For $E_F < E_C$, the system must be an insulator ($\sigma = 0$), since all electrons occupy localized states. For $E'_C > E_F > E_C$ the system must be a conductor with a finite conductivity. How does the conductivity vanish at E_C if E_F approaches E_C from above? Mott has argued that upon approaching E_C from above the mean free path becomes shorter and shorter, until at E_C it has the minimum value $l_{\min}$. Beyond this energy the conductivity is zero. Thus σ

approaches a finite value σ_M for $E_F \to E_C$. This value is called the minimum metallic conductivity. According to Mott [7] it is given by

$$\sigma_M = \frac{e^2}{\hbar} \frac{1}{l_{\text{min}}} C_3 \tag{2}$$

with $C_3 \simeq 0.03$. Later, however, both experimental [8] and theoretical work [2–4] showed that in fact σ vanishes like $[(E_F - E_C)/E_C]^\lambda$ with $\lambda \geq 1$ for situations where the electron-electron interaction can be ignored.

2.2. *Scaling theory*

As an example for the theoretical reasoning, we present a discussion of the conductivity using scaling arguments, following in Ref. [9]. The conductance $G(L)$ of a macroscopical cube of linear dimension L is considered. Let $g(L) = \hbar G(L)/e^2$ be the dimensionless conductance. In the ohmic regime, i.e. for sufficiently high conductance, $g(L)$ scales like $g(L) \propto L$. On the other hand, far down in the localized regime $g(L) \propto \exp(-L/\xi_{\text{loc}})$. Again, this behavior can be deduced from the diffusivity, which reflects the spatial extent of the energy eigenfunctions. Diffusion towards the electrodes is possible only to the extent that the wave functions connect both electrodes, which is exponentially decreasing with the exponent given above.

Now consider a microscopically small cube of linear extension l. However, l must be larger than that length scale where diffusion (i.e. scattering limited motion) can be defined. On that scale, we define a microscopic conductance $g_0 = g(l)$, which is always finite. Upon scaling up the length scale from l to $L \gg l$, depending on the value of g_0, we either arrive at the ohmic or at the localized regime. We define a scaling function

$$\beta(g) = \frac{d\ln(g)}{d\ln(L)} = \frac{L}{g}\frac{dg}{dL} \tag{3}$$

and assume that $\beta(g)$ is a single-valued, monotonous and continuous function of g. Its limiting behavior is $\beta(g) = 1$ for $g \to \infty$ and $\beta(g) = \ln(g) + \text{const}$ for $g \to 0$. At some value $g_0 = g_3$ the β-function must pass through zero, $\beta(g_3) = 0$ (see Fig. 2). For $g_0 > g_3$ $(\beta > 0)$ we arrive at the ohmic regime, while for $g_0 < g_3$ $(\beta < 0)$ we arrive at the localized regime. Since g_0 depends on E_F, $g_0(E_F = E_C) = g_3$ is the microscopic conductance at the mobility edge E_C. The corresponding conductivity $e^2 g_3/\hbar l$ is then the smallest value of the microscopic conductivity which upon scaling up leads to the ohmic

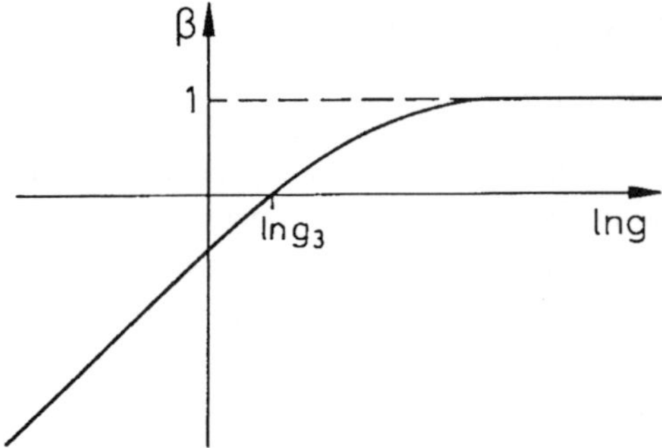

Fig. 2. The scaling function β for three-dimensional Andersonium.

behavior. It can, therefore, be identified with Mott's minimum metallic conductivity σ_M.

From the scaling function one can obtain information about the macroscopic conductivity in the vicinity of E_C. With a few technical approximations one arrives at (for details, see [9, 10])

$$\sigma = \frac{e^2}{\hbar} \frac{A g_3}{l} \left(\frac{E_F - E_C}{E_C} \right)^\lambda \tag{4}$$

with some constant λ and $A = (e/\lambda)^\lambda$. Microscopic theories [3, 4], although necessarily approximate suggest that the assumptions made for the scaling function are justified. In addition, numerical calculations have been performed (see, e.g. [2]) which also yield a continuous transition with an exponent λ which seems to be somewhat larger. In any case, the conductivity does not show a step at E_C but vanishes continuously. For our purposes, it suffices to use $\lambda = 1$, thus

$$\sigma = \sigma_M \frac{E_F - E_C}{E_C} \tag{5}$$

with $\sigma_M = (e^2/\hbar)(e g_3/l)$. Note that this relation is valid for $T = 0$ only.

In normal metals the conductivity is always finite. Here we find that it tends to zero as E_F approaches the mobility edge. This can be understood as a precursor of localization which sets in at E_C. When we consider finite temperatures and inelastic interactions and we will find that the conductivity stays finite for all energies, but still "remembers" in some way the transition at E_C.

3. Elements of Transport Theory

3.1. *The Kubo formula*

Unfortunately the scaling theory does not provide us with a numerical value of σ_M, nor is it applicable to transport processes further up in energy. Furthermore, in semiconductors the Fermi level lies outside the conduction band and conduction is only possible at finite temperature.

In order to describe transport in a semiconductor, we consider only the conduction band (i.e. an n-doped situation). The Fermi level is supposed to be several kT below the conduction band mobility edge E_C. We still ignore the electron-electron and the electron-phonon interaction.

Following in Refs. [11, 12], we start with the observation that the real part of the frequency dependent conductivity, $\sigma'(\omega)$, is related to the power dissipation density P by

$$P = \frac{1}{2}|\mathbf{F}|^2\sigma'(\omega) \tag{6}$$

if the electric field $\mathbf{F}$ is given by $\mathbf{F} = \mathrm{Re}(\mathbf{F}_0 e^{i\omega t})$. The interaction Hamiltonian $H' = eFx$, valid for a field in x-direction, leads to transitions from energy eigenstates $|n\rangle$ to energy eigenstate $|m\rangle$ with eigenvalues E_n and E_m, respectively. In first order perturbation theory the transition rate is

$$W_{nm} = \frac{2\pi e^2}{\hbar}\frac{|\mathbf{F}_0|^2}{2}|\langle n|x|m\rangle|^2(\delta(E_n - E_m - \hbar\omega) + \delta(E_n - E_m + \hbar\omega)) \tag{7}$$

The contribution of this transition to the power dissipation is given by

$$W_{nm}f_F(E_n)(1 - f_F(E_m))(E_m - E_n) \tag{8}$$

where f_F is the Fermi distribution. For a system of volume Ω, the real part of the conductivity then reads

$$\sigma'(\omega) = \frac{\pi e^{2\omega}}{\hbar\Omega}\sum_{nm}|\langle n|x|m\rangle|^2(f_F(E_m) - f_F(E_n))\delta(E_n - E_m - \hbar\omega) \tag{9}$$

This is one of the forms of the famous Kubo formula for the dc conductivity, written in terms of the energy eigenstates and eigenvalues. Of course these quantities are not known for a disordered system (except for small finite systems, which can be evaluated numerically).

3.2. *The Kubo–Greenwood formula*

With $[x, H] = (i\hbar/m)p_x$, we have $(E_m - E_n)\langle n|x|m\rangle = (i\hbar/m)\langle n|p_x|m\rangle$ which allows to write

$$\sigma'(\omega) = \frac{\pi e^2 \hbar}{\Omega m^2} \sum_{nm} |\langle n|p_x|m\rangle|^2 \frac{f_F(E_m) - f_F(E_n)}{\omega} \delta(E_n - E_m - \hbar\omega) \qquad (10)$$

which in the dc-limit $(\omega \to 0)$ takes the form

$$\sigma(\omega \to 0) = \sigma(T) = \int_{-\infty}^{\infty} \sigma(E) \left(-\frac{f_F(E)}{dE} \right) dE \qquad (11)$$

with the differential conductivity

$$\sigma(E) = \frac{\pi e^2 \hbar}{\Omega m^2} \sum_{nm} |\langle n|p_x|m\rangle|^2 \delta(E_n - E)\delta(E_m - E) \qquad (12)$$

Eq. (11) is called the Kubo–Greenwood formula. It contains the unknown differential conductivity $\sigma(E)$ which tells us the contribution of a state at energy E to transport.

Unfortunately, $\sigma(E)$ is written in a form that does not tell us immediately that $\sigma = 0$ if the states at E are localized. By rewriting this formula in terms of Green's functions [4], or by applying a correlation function approach [3] and after some approximations, one arrives at a self-consistent formulation which leads to the result discussed in connection with Eq. (5).

The Seebeck coefficient is given by the average energy relative to the Fermi level transported by the electrons [13], thus, with $\beta = (kT)^{-1}$ and $q = -e$ for electrons

$$S(T) = \frac{k}{q} \int_{-\infty}^{\infty} \beta(E - E_F)\frac{\sigma(E)}{\sigma(T)} \left(-\frac{d}{dE} f_F(E) \right) dE \qquad (13)$$

For non-degenerate statistics the term $-df_F(E)/dE$ can simply be replaced by the term $\beta \exp(-\beta(E - E_F))$.

We can now insert our previous result of Eq. (5) for the differential conductivity, and find

$$\sigma(T) = \sigma_M \frac{1}{\beta E_C} e^{-\beta(E_C - E_F)} \qquad (14)$$

and

$$S(T) = \frac{k}{q}(\beta(E_C - E_F) + 2) \qquad (15)$$

It is advantageous to combine these two transport quantities into a single function, the Q-function

$$Q = \ln(\sigma\Omega\text{cm}) + \frac{q}{k}S \tag{16}$$

which shows only a weak logarithmic temperature dependence (for 150 K $\leq$ $T \leq$ 350 K)

$$Q = 2 + \ln(\sigma_M\Omega\text{cm}/(\beta E_C)) \tag{17}$$

In our present model, we obtain an essentially activated conductivity with an activation energy $E_\sigma = E_C - E_F$, which is typical for semiconductors where transport is dominated by mobile carriers at a conduction band edge E_C. Also the Seebeck coefficient shows the behavior expected for a simple semiconductor model, with activation energy $E_S = E_\sigma$ and with a "heat of transport" term equal to 2. (The T-dependence of the prefactor of the conductivity and the heat of transport term of 2 instead of 1 are due to the linear increase of $\sigma(E)$ in our model, as opposed to the usual assumption of a constant mobility at E_C.) We also learn that the prefactor of the conductivity, and the value of Q, are essentially given by σ_M, which is the natural scale of the conductivity in a disordered system.

Hindley [14] and Friedman [15] have evaluated the Kubo–Greenwood formulae using a random phase model for the energy eigenstates. (Note that the assumption of random phases is by no means trivial, instead there are indications that correlation survives to a certain extent [16].) In this model, the differential conductivity can be cast into the form

$$\sigma(E) = kTeg(E)\mu(E) \tag{18}$$

where $g(E)$ is the DOS, and $\mu(E)$ the mobility, which turns out to be itself proportional to $g(E)$ in this model. For a DOS which is not too strongly dependent on E this result yields similar answers for σ and S as described above, provided the band edge is identified with E_C.

The derivation of the Kubo–Greenwood formula requires that at the transport path at E there are no inelastic processes. In a semiconductor, however, we have always scattering with phonons. The Kubo–Greenwood formula can then only be an approximation to a more realistic description. This approximation is the better, the smaller the energy transfer at the transport path is compared to kT. It fails for the description of some hopping transport processes, where energies of several kT are exchanged between the electrons and the phonons.

4. Transport Theory for Disordered Semiconductors

4.1. *The model*

As discussed in the Introduction to this chapter, there is no complete transport theory for a particular semiconducting glass or disordered semiconductor. Even the parameters entering the Hamiltonian for these systems are generally not known in detail. Instead, one treats simplified model systems, which contain the most relevant physical interactions, and hopes to arrive at answers which are applicable to the real systems. In this approach, it is necessary to use some empirical values for certain microscopic quantities, which in some cases can be taken from experiments. For amorphous semiconductors, e.g. there now exists a reasonably coherent picture of the DOS for the valence and conduction band tails, and within the pseudogap. A model is shown in Fig. 3. One also has reasonable numbers for the positions of the mobility edges. The theoretical problem then reduces to the calculation of transport coefficients for a model system which in turn gives rise to the phenomenologically known DOS and position of the mobility edge. This approach will be used in the following.

We will ignore the electron-electron interaction, although this interaction produces interesting physics if transport close to the Fermi level at very low temperatures is considered. The Hamiltonian of the model system is that of Andersonium, amended by phonons and a term describing the interaction

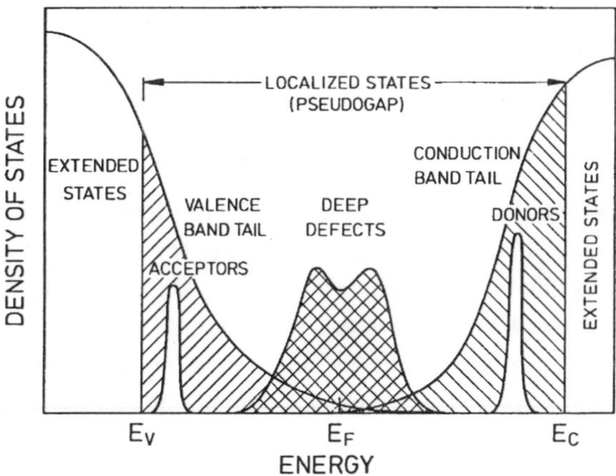

Fig. 3. Density of states distribution model for amorphous semiconductors [10].

between electrons and phonons:

$$H = H_e + H_p + H_{ep} \tag{19}$$

where the electronic Hamiltonian

$$H_e = \sum_i (\epsilon_i - E)c_i^\dagger c_i + J \sum_{ij} c_i^\dagger c_j \tag{20}$$

is known from Eq. (1). The ϵ_i will be distributed according to a certain distribution function such that $\langle \epsilon_i \rangle = 0$ and $\langle \epsilon_i \epsilon_j \rangle = V^2 \delta_{ij}$. V quantifies the degree of disorder and, depending on the distribution function, is equal to W up to a factor of order of unity.

$$H_p = \sum_{\mathbf{q}} \hbar \omega_{\mathbf{q}} \left(b_{\mathbf{q}}^\dagger b_{\mathbf{q}} + \frac{1}{2} \right) \tag{21}$$

describes the acoustical phonons. In our theory, the phonon system enters only as a bath, i.e. only frequency convolutions are involved and details of the phonon system are of no importance.

The electron-phonon interaction will be approximated by a a diagonal term ($\propto c_i^\dagger c_i$). The nondiagonal terms describing recoil are frequently neglected [28] since their contribution to the interaction term is exponentially smaller.

$$H_{ep} = \sum_{\mathbf{q}} \left(A_{i\mathbf{q}} b_{\mathbf{q}}^\dagger + A_{i-\mathbf{q}} b_{\mathbf{q}} \right) c_i^\dagger c_i \tag{22}$$

The interaction with the external field can be written in terms of the potential $V_j = e\mathbf{R}_j \cdot \mathbf{F}(t)$ at site j as

$$H' = -\sum_j V_j c_j^\dagger c_j \tag{23}$$

We shall not present the full development of the theory, which is described in detail elsewhere [10, 17–19]. Instead we discuss the results in a form familiar to solid state physicists. The theory aims at the calculation of the differential conductivity $\sigma(E)$ for a system with strong scattering at the disorder and scattering at the phonons. This differential conductivity is the dc-limit ($\omega \to 0$) of the real part of the frequency dependent conductivity $\sigma'(\omega)$ of the system, where the Fermi level is situated at E. Once $\sigma(E)$ is determined, it can be inserted into Eq. (11) and Eq. (13) to obtain the temperature dependent conductivity and the Seebeck coefficient.

4.2. *Localization theory*

We start by considering a phonon-free model at $T = 0$. This must lead us
to an expression for $\sigma(E)$ which shows Anderson localization. The approach
closely follows the mode-coupling theory of Götze [3].

In metal physics, the Drude formula is frequently used to describe the
motion of electrons under a small external field

$$\sigma'(\omega) = \frac{ne^2}{m} \frac{\tau^{-1}}{\omega^2 + \tau^{-2}} \tag{24}$$

where m is the electron mass and τ^{-1} the scattering rate. The Fourier transform of this quantity

$$\propto e^{-t/\tau} \tag{25}$$

shows that a given current fluctuation decays to zero with a relaxation time
τ, which is therefore also called the current relaxation time. $\sigma'(\omega)$ is the real
part of the complex conductivity $\sigma(z)$ [i.e. the Laplace transform of Eq. (25)]
for $z = \omega + i\varepsilon$, where $\varepsilon \to 0$, thus

$$\sigma(z) = \frac{ne^2}{m} \frac{i}{z + i\tau^{-1}} \tag{26}$$

This description is appropriate for conductors with sufficiently small scattering
for which scattering events are isolated and uncorrelated. In a strongly disordered system, however, scattering events have a finite temporal memory since
their action depends to some degree on previous scattering events as pointed
out by Götze [3]. For a dynamic system with memory, the decay of fluctuations is no longer purely exponential. Instead, the decay time τ has to be
replaced by a frequency dependent function, the memory kernel $M(z)$. Thus,
the conductivity has the form

$$\sigma(z) = \frac{ne^2}{m} \frac{i}{z + iM(z)} \tag{27}$$

The analytical properties of $M(z)$ for $z \to 0$ then determine the dc conductivity. We know, that for $E < E_C$, $\sigma'(E) = \sigma(z \to 0) = 0$ if we ignore the
phonons. This implies that in this case $M(z)$ must diverge in the localized
regime with $z \to 0$, while it tends to a finite value in the delocalized regime
$E > E_C$. The microscopic theory indeed produces this behavior [3, 10] with
$M(z) \propto z^{-1}$ for $z \to 0$ and for an energy E which is thus to be identified with

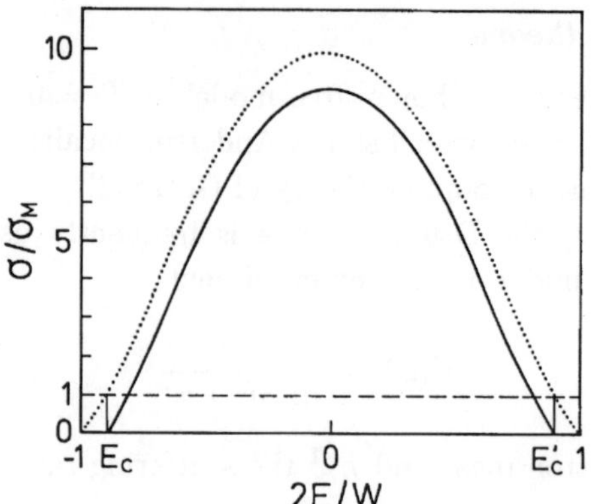

Fig. 4. Illustration of $\sigma(E)$ (solid line) for a model characterized by a box shaped band of width W. The dotted line is $\sigma_{00}(E)$, the dashed line is σ_M.

the mobility edge E_C. The result of this theory for $\sigma(E)$ can be written in an appealing compact form as

$$\sigma(E) = \sigma_{00}(E) - \sigma_M \tag{28}$$

for $E > E_C$ and zero otherwise. The mobility edge E_C is given by $\sigma_{00}(E_C) = \sigma_M$. σ_{00} is the lowest order approximation to the conductivity, e.g. the Drude result, which is non-zero everywhere in the band. One finds [10] $\sigma_M = 0.03e^2/(\hbar a)$ which surprisingly coincides with the value of Mott's minimum metallic conductivity (the present theory, however, does not yield a minimum metallic conductivity; in the region close to E_C it reproduces the result of the scaling theory instead). If the band edge is taken as the zero of the energy scale we can parameterize $\sigma(E)$ to a good approximation by (see Fig. 4)

$$\sigma(E) = \sigma_0 \frac{E - E_C}{E_C} \tag{29}$$

which substantiates the result of the scaling theory, Eq. (5).

Note that the prefactor $\sigma_0 = \sigma_M$ depends on one single microscopic parameter only, the lattice spacing a of Andersonium. Our model provides a tight-binding representation of the low-energy region of the conduction band of a disordered semiconductor. The structural units (sites) from which this portion of the band is composed in reality will have mutual separation of several Å, a number which should not depend much on sample preparation.

In tetrahedrally coordinated amorphous semiconductors, these units may be identified with distorted tetrahedra or small structures composed from these tetrahedra. Their mutual distance is then of the order of 5 Å. Inserting this value into Eq. (29) we obtain

$$\sigma_0 = 150\,\Omega^{-1}\text{cm}^{-1} \tag{30}$$

The memory effect mentioned above can alternatively be described as follows [24]:

Consider an electronic wave packet prepared to move into a given direction. The simultaneous multiple scattering of this wave produces back-scattered contributions to the amplitude, which have to be added up and squared in order to yield the scattered intensity. As a result, the back-scattered intensity is stronger than classical random walk theory would predict. For certain values of the physical parameters of the system this backscattering can be so effective, that the wave is completely reflected, becomes back-scattered in the reverse direction, etc. The wave is, therefore, oscillating around a given point in space, i.e. it is localized. This localization is due to a coherent wave phenomenon. In fact it also exists for other waves scattered in random media, e.g. for light and microwaves.

Looking at the eigenfunctions one realizes [20] that close to E_C they are large irregular objects with an envelope which are either, for $E < E_C$, slowly decaying in space according to the localization length ξ_{loc}, or are completely extended for $E > E_C$.

4.3. *Phonon-induced delocalization*

The interaction of the electrons with the phonon bath disturbs the memory. Consequently, the divergence $M(z) \propto z^{-1}$ for $z \to 0$ at E_C is lost and hence the dc conductivity becomes nonzero everywhere in the band. This is plausible for electrons in strongly localized states deep in the tails of the band, since phonons-induce transitions between these localized states (hopping) which otherwise do contribute to transport. On the other hand, in the regime just below E_C this effect is called [21] "phonon-induced delocalization". The phonon-induced delocalization occurs if the inelastic scattering length L_i due to the electron-phonon scattering is less than ξ_{loc}.

If we allow for a non-zero electron-phonon interaction, $\sigma(z)$ still has the form of Eq. (27)

$$\sigma(z) = \frac{ne^2}{m} \frac{i}{z + i(M_1(z) + M_2(z))} \tag{31}$$

where $M_2(z)$ is proportional to the electron-phonon coupling constant $|A_q|$ [2] and to T at sufficiently high temperatures. The kernel $M_1(z)$ reduces to $M(z)$ discussed above if $M_2 = 0$. These two kernels are intimately related to each other. For energies close to E_C the kernel M_1 predominantly determines the dc conductivity, however, it does no longer diverge due to the presence of the finite M_2. The hopping kernel M_2 is responsible for the finite values of the conductivity deep in the tail, while M_1 is negligible there.

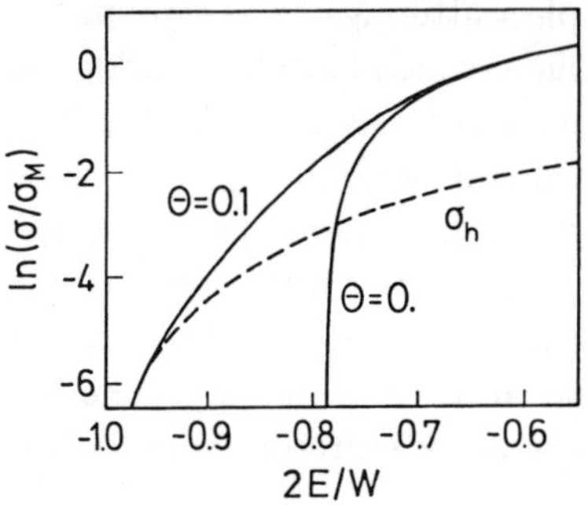

Fig. 5. Normalized differential conductivity $\sigma(E)/\sigma_M$ close to E_C ($2E_C/W = -0.82$) for a model characterized by a box shaped band of width W. The parameter Θ is proportional to the temperature and the electron-phonon coupling constant squared, $\Theta = 0$ is the localization limit (no inelastic interaction with phonons). σ_h is the pure hopping contribution to σ [10].

Both phonon-induced delocalization close do E_C and hopping far below E_C render the dc conductivity $\sigma(E)$ finite throughout the entire band. The detailed theory [10, 17–19] yields the result shown in Fig. 5. Close to E_C $\sigma(E)$ is enhanced due to phonon-induced delocalization and develops a tail, which interpolates between this regime and the hopping regime down in the tail of the DOS. Note, however, that the scale of $\sigma(E)$ for E close to E_C still is given by σ_M.

4.4. *Transport close to E_C*

The transport coefficients σ, S, and their combination Q, are obtained by inserting $\sigma(E)$ into the Kubo–Greenwood Eqs. (11) and (13), respectively. We first note that both the occupation factor df_F/dE and $\sigma(E)$ depend on energy. The energy dependencies, however, are opposite, such that at a certain energy, which we call the dominant transport energy E_t, the integrand in both equations has a sharp maximum. This maximum will shift with temperature, since both functions depend on temperature as well, again with opposite trends. While with increasing T $\sigma(E)$ develops a more prominent tail at lower E, the occupation factor decreases more slowly at high E. The temperature dependence of the transport energy resulting from the trade-off of the two trends can be parameterized in first order by

$$E_t(T) = E_t(0) + \gamma_t T \tag{32}$$

Model calculations [10, 19] give values up to $\gamma_t = \pm 1k$ depending on the detailed shape of the DOS close to E_C.

Using the transport energy E_t, both the dc conductivity and the Seebeck coefficient can be approximated in first order as

$$\sigma = \sigma_0 e^{-\beta(E_t - E_F)} \tag{33}$$

and

$$S = -\frac{k}{e}(\beta(E_t - E_F) + A) \tag{34}$$

with a certain heat-of-transport term A of order unity and σ_0 given by Eq. (30). Thus, the temperature dependence of both quantities is dominated by the same and relatively large $\beta(E_t - E_F)$ term. We have seen that this latter term is canceled if we combine conductivity and thermopower to the Q-function, which contains the fine details of the temperature dependencies only.

In fact, since the integrands of Eqs. (11) and (13) are slightly different, the temperature dependencies of $\ln \sigma(T)$ and $S(T)$ will not be exactly the same. Model calculations [19] show that the Q-function has the form

$$Q(T) = \ln(\sigma_0 \Omega \text{cm}) + A - \beta E_Q \tag{35}$$

where $E_Q = E_\sigma - E_S$ can be as large as 50 meV. Had we used the phonon-free result Eqs. (5) or (28) we would have obtained $E_Q = 0$. We thus find that the phonon-induced delocalization, leading to the tailing of $\sigma(E)$ close to E_C, already leads to a finite albeit small difference of the activation energies of σ and S. The mentioned tailing of $\sigma(E)$ also yields a deviation of the activation energy E_σ of the conductivity from $E_C - E_F$, which can be as large as $\pm$ 50 meV depending on details of the DOS [19]. We shall see below that the experimental data often require significantly larger values for E_Q. For these, we will have to present an additional mechanism.

For low temperatures eventually hopping within the localized tail states takes over, until finally hopping occurs in the states around E_F in the pseudogap. In the first case there exists, at least for exponential tails, another (hopping-) transport energy [22, 23], which will be discussed in Sec. 4.5.5.

4.5. *Hopping theory*

In order to present the principles of hopping theory (for a comprehensive review of hopping theory see, e.g. [25]) we can again start from the expression for the complex conductivity $\sigma(z)$ given in Eq. (31). However, we now turn to energies where the electronic states are strongly localized. In this regime, we may ignore the coherent kernel M_1 completely. The hopping kernel M_2 can be treated in the so-called binary approximation, i.e. only phonon-induced transitions between pairs of localized states are considered (this approach fails if transport in the presence of a magnetic field is considered, here at least three site processes have to be considered).

Starting from the general theory, we arrive at [10]

$$\sigma(z) = -\left\langle \frac{e^2}{6kT\Omega} \sum_{ij} R_{ij}^2 \left(\frac{z}{z + iCF^{-1}} C \right)_{ij} \right\rangle_{\text{config}} \tag{36}$$

where R_{ij} is the distance between sites i and j, and $\langle \ldots \ldots \rangle_{\text{config}}$ means configurational average over all realizations of the disorder. C_{ij} is the collision matrix

$$C_{ij} = -\Gamma_{ij}^0 + \sum_l \Gamma_{il}^0 \delta_{ij} \tag{37}$$

formed from the equilibrium hopping rates Γ_{ij}^0, and $F_i = f_i^0(1 - f_i^0)$, where $f_i^0 = f_F(\epsilon_i, E_F, T)$ is the occupation of state i in equilibrium.

The same result can be obtained if one starts directly from the rate equation for the non-equilibrium occupation f_i of site i in the presence of an external field

$$\frac{\partial}{\partial t} f_i(t) = \sum_j (f_j(1 - f_i)W_{ji} - f_i(1 - f_j)W_{ij}) \tag{38}$$

where W_{ij} is the field-dependent transition rate. After linearizing with respect to a weak external field and by taking the Laplace transform [26] one arrives at the result of Eq. (36) for the corresponding conductivity.

The major task has still to be worked out, namely the configurational average. This can be done using Green's function techniques [27], or by applying percolation theory or numerical simulations. For certain simple situations, there exist quite elegant approaches, which we will describe in the following.

4.5.1. *The hopping rates*

Quite generally hopping transport is discussed on the basis of a model originally developed for transport within impurity bands of crystalline semiconductors [28]. It applies to situations where the electron-phonon interaction is not very strong (i.e. no small polaron formation). It is also assumed that the electronic eigenstates are more or less compact objects (i.e. not highly extended and irregular like those just below E_C).

The Hamiltonian is assumed to have the standard from

$$H = H_e + H_p + H_{ep} \tag{39}$$

with H_p and H_{ep} as above but

$$H_e = H_0 + \sum_{ij} J_{ij} c_i^\dagger c_j \tag{40}$$

and

$$H_0 = \sum_i \epsilon_i c_i^\dagger c_i \tag{41}$$

where J_{ij} now connects all pairs of sites with values that strongly depend on the distances R_{ij}. The disorder is assumed to be strong compared with the typical nearest neighbor coupling $J_{ij} = J_0 < 0$, i.e. $\Delta E = \epsilon_i - \epsilon_j \gg |J_0|$. We may then concentrate on just a pair of states $|i\rangle$ and $|j\rangle$ with energies ϵ_i and ϵ_j and calculate the phonon-assisted transition rate in the binary limit. The quantum mechanical admixture of all the other sites to the eigenstates, centered at positions other than $\mathbf{R}_i$ and $\mathbf{R}_j$, is thereby neglected. These two particular states are not eigenstates as they are still coupled by J_{ij}. The 2×2-Hamiltonian of the two-site system can easily be diagonalized yielding an upper eigenstate $|u\rangle = r(|i\rangle - s|j\rangle)$ and a lower eigenstate $|l\rangle = r(|j\rangle + s|i\rangle)$. The eigenvalues are close to the single-site energies, $E_u \simeq \epsilon_i$ and $E_l \simeq \epsilon_j$ such that $s \simeq |J_0/\Delta E|$ and $r = (1 + s^2)^{-1/2}$. Since $\Delta E \gg J_{ij}$, electrons in states $|l\rangle$ and $|u\rangle$ are centered at position R_j and R_i, respectively. A phonon-induced transition described by the matrix element $\langle l|H_{ep}|u\rangle$ therefore moves the electron from R_j to R_i. As we consider single phonon processes here, the phonon must have the energy $\hbar\omega_\mathbf{q} = \Delta E$. It is available according to (approximately) a

Boltzmann function $\exp(-\beta\hbar\omega_{\mathbf{q}})$. If the hop is downwards in energy, phonons can always be emitted spontaneously, and this factor is to be replaced by unity. The matrix element contains the factor $J_0 \propto \exp(-R_{ij}/\xi_{\text{loc}})$ because it determines the degree of admixture of state $|j\rangle$ into the upper eigenstate $|u\rangle$. All other factors containing the details of the electron-phonon coupling are usually lumped together into an "attempt-to-escape" frequency ν_0. For some situations, it is possible to calculate ν_0 (see, e.g. [28, 29)]. However, in disordered semiconductors the wave functions are not known, nor is the electron-phonon coupling strength, so ν_0 remains a free parameter. Often one assumes $\nu_0 = 10^{12}/\text{sec}$, although values differing by several orders of magnitude can also be found in the literature.

The equilibrium rate then reads

$$W_{ij}^0 = \nu_0 e^{-2R_{ij}/\xi_{\text{loc}}} \cdot \begin{cases} e^{-\beta\Delta E} & \text{if} \quad \Delta E > 0 \\ 1 & \text{else} \end{cases} \tag{42}$$

The same form results if the incoherent kernel M_2 of Eq. (31) is evaluated in the binary limit for the present model system.

As it stands, Eq. (42) has been derived for processes where a single phonon is involved. This requires that $\Delta E < \hbar\omega_p$, where ω_p is a typical phonon frequency. Inspection of the electron-phonon coupling constant [28] shows that only those phonons couple effectively to the localized states, for which the phonon wavelengths exceed twice the localization length. The action of phonons with shorter wavelength (hence higher energy acoustical phonons) is averaged out over the extension of the electronic state. Therefore, strongly localized states (ξ_{loc} small) are more strongly coupled to phonons than weakly localized states. If the coupling is strong, more than a single phonon can be involved in the hop, even if T is low and ΔE is smaller than ω_p. The respective multi-phonon rates have been calculated on the basis of Holstein's molecular crystal model [30–32]. It turns out [33], that for not too strong coupling and at low T the rates remain nearly activated, with an activation energy given essentially by ΔE. One can therefore use Eq. (42) as a good approximation even if ΔE exceeds ω_p. Qualitatively different behavior of the rates, however, is found for strong coupling, when the carriers form small polarons. This case is discussed in Sec. 4.5.7.

We have calculated the equilibrium transition rate W_{ij}^0 above. The equilibrium occupation at the sites i and j is f_i^0 and f_j^0, respectively, such that the equilibrium hopping rate Γ_{ij}^0 describing Brownian motion reads

$$\Gamma_{ij}^0 = f_j^0(1 - f_i^0)W_{ij}^0 = \Gamma_{ji}^0 \tag{43}$$

Taking the Fermi level as the zero of the energy scale, the equilibrium hopping rate Γ_{ij}^0 according to Eq. (43) then reads for non-degenerate statistics and for energies large compared to kT

$$\Gamma_{ij}^0 = \nu_0 e^{-2R_{ij}/\xi_{\mathrm{loc}} - \beta(|\epsilon_i| + |\epsilon_j| + |\epsilon_i - \epsilon_j|)/2} \tag{44}$$

We are ultimately interested in transport when a small external dc field $\mathbf{F}$ is applied. The field changes both the transition rates and the occupation of the states which is reflected by changes of the local electrochemical potential. Linearizing the transition rate with respect to the applied field [26] we obtain

$$\Gamma_{ij}(\mathbf{F}) - \Gamma_{ji}(\mathbf{F}) = e\Gamma_{ij}^0\beta(V_i^{\mathrm{int}} - V_j^{\mathrm{int}}) \tag{45}$$

with V_i^{int} the sum of the applied potential and the electrochemical potential at site i.

4.5.2. *The equivalent network*

The result of Eq. (45) can be re-interpreted in a way which is most convenient for the evaluation of the configurational averages. We consider conductances

$$G_{ij} = e^2\beta\Gamma_{ij} \tag{46}$$

and realize that the linearized rate equations Eq. (38) reduce to Kirchhoff's equations [34], which give the conductivity of an equivalent electrical circuit [35] built from these conductances.

In the equivalent network every site i in our original system is considered as a node in a network that is connected to all other nodes j by conductances G_{ij}. It is important to note that the variation of these conductances in the system is extremely wide, since they depend exponentially both on spatial separation and energy. An estimate for the network conductivity can be obtained by the following percolation argument [35, 36] (see Fig. 6). We start with an empty network where all sites are unconnected. Then we identify the largest conductance and "solder" it into its place. We continue with the next largest, etc. During this process we first observe isolated conductances, which eventually form small isolated clusters. These clusters grow and tend to be interconnected. At a certain stage, the first cluster connecting both electrodes appears. This is the "percolation threshold", and the respective

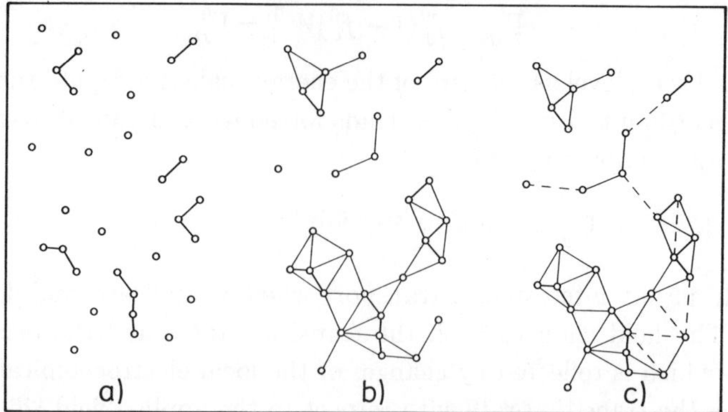

Fig. 6. A disordered bond-percolation network at different bond concentrations. (a) and (b) have isolated clusters only, (c) the first critical path appears.

current path is called the critical path. The conductance which establishes this critical path is called the critical conductance. Since all conductances put into the network previously are much larger than the critical one and already mutually interconnected, the latter essentially determines the conductivity of the critical path. Placing now all the rest of conductances, which are again much smaller than the critical one, will not change much the conductivity of the network.

For a given arrangements of sites, percolation theory [37] tells us the critical bond concentration p_c which establishes the critical path. The conductances $G_{ij} > G_c$ are then identified with the "bonds" of percolation theory. The critical conductance is then determined in such a way that the bond concentration equals p_c [38].

4.5.3. *R-percolation*

Depending on the physical situation two limiting cases of hopping transport theory have been formulated. If the localized states form a narrow band of width ΔE, such that $\beta \Delta E$ is small compared to R_{nm}/ξ_{loc} for nearest neighbors, then the critical path is determined entirely by the spatial arrangement of the sites (nearest-neighbor hopping). The "bonding criterion" for this "R-percolation" problem is $R_{nm} \leq R_c$ with $2R_c = -\xi_{\text{loc}} \ln(G_c)$. A given site can, therefore, have bonds to all sites within a sphere of radius R_c. On average a site has thus

$$p = \frac{4\pi}{3} R_c^3 n_{\text{loc}} \tag{47}$$

bonds, where n_{loc} is the density of sites. Since p must be equal to p_c in order to have a critical path we find

$$G_c = e^2 \beta e^{-2R_c/\xi_{\text{loc}} - \beta \delta E} \tag{48}$$

where δE is of the order of ΔE. The resulting conductivity has the form

$$\sigma = \sigma_0 e^{-\beta \delta E} . \tag{49}$$

If the Fermi level is below this band of current carrying localized states by an amount of E_L, then all values of G_{ij} are altered by the *same* factor $\exp(-\beta E_L)$ and therefore there is no change in the geometry of the critical path. The conductivity Eq. (49) will thus be modified by the same activation factor $\exp(-\beta E_L)$.

4.5.4. *Variable-range hopping*

We consider a situation where the Fermi level lies inside a wide band of localized states. The lower the temperature, the more will most of the hopping processes be confined to sites close to E_F in energy, according to Eq. (44). The critical path will thus depend on temperature. For simplicity, we consider a system where the DOS at the Fermi energy, $g(E_F)$, is energy-independent. Following Mott [39] we consider a site at E_F (which is still the zero of the energy scale). This site is connected by bonds to all sites located within a sphere of radius R_c and having energies less than a certain energy δE. On the average there will be p bonds emanating from this site, with

$$p = \frac{4\pi}{3} R_c^3 g(E_F) 2\delta E . \tag{50}$$

We equate p with p_c which relates R_c with δE. If we write the critical conductance as

$$G_c = e^2 \beta e^{-\zeta} \tag{51}$$

we find

$$\zeta = \frac{2R_c}{\xi_{\text{loc}}} + \frac{3p_c}{4\pi R_c^3 2g(E_F) kT} . \tag{52}$$

The largest conductance (minimal value $\zeta = \zeta_c$) is found for an optimal distance R_{opt} given by

$$\frac{d\zeta}{dR_c}\Big|_{R_{\text{opt}}} = 0 \tag{53}$$

which results in

$$R_{\text{opt}}^4 = \frac{9p_c\xi_{\text{loc}}}{16\pi g(E_F)kT} \tag{54}$$

and

$$\zeta_c = \left(\frac{256}{9\pi}\frac{p_c}{g(E_F)kT\xi_{\text{loc}}^3}\right)^{1/4} \tag{55}$$

or

$$G_c = e^2\beta e^{-(T_0/T)^{1/4}} \tag{56}$$

with

$$T_0 = \frac{256}{9\pi}\frac{p_c}{\xi_{\text{loc}}^3 g(E_F)k} \tag{57}$$

which is Mott's famous $T^{1/4}$-law for "Variable-Range-Hopping" at the Fermi energy. The conductivity is given by

$$\sigma = \sigma_{VRH}e^{-(T_0/T)^{1/4}} \tag{58}$$

where the prefactor σ_{VRH} depends on details of the wave functions and the electron-phonon coupling [29].

Deviations from this law are known to exist if the density of states depends on energy in a range of $\pm zkT$, where z can be as large as 10. Following Pollak [36], these situations can be treated numerically. It turns out that the prefactor σ_{VRH}, T_0 and the power of the temperature depend strongly on details of the density of states profile close to E_F [49]. Often, a non-exponential temperature dependence of the conductivity is used as an indication of variable-range hopping. The data are then plotted vs. $T^{-1/4}$, and the density of states is determined from T_0. This procedure can be, however, highly misleading. First, in a limited range of temperature it is usually hard to determine the exponent precisely. Second, Eq. (57) is applicable only if the density of states is indeed constant in a range of several kT around the Fermi level. If $g(E)$ has, e.g. a minimum with finite value at E_F, σ still shows a behavior close to Mott's law. The parameter T_0, however, which results from a fit of this expression to the data, is no longer given by Eq. (58).

The thermopower in the variable-range hopping regime has also been investigated theoretically [40, 41] and experimentally [42, 43].

4.5.5. *Hopping in band tails*

In amorphous semiconductors, the DOS in the tail region can well be described by an exponential law, $g(E) = N_0/\varepsilon_0 \exp(E/\varepsilon_0)$, where N_0 is the total number of localized tail states and ε_0 the tailing parameter. For low temperatures, the contribution of carriers near the mobility edge can be neglected and transport is due to hopping within the tail states. It has been realized [22–24] that a certain energy exists, called the hopping-transport energy E_t, which governs not only the equilibrium transport properties, but also transient phenomena like time-of-flight [45] and non-equilibrium phenomena like recombination and photoconductivity [46] in this regime. It is given by

$$E_t = 3\varepsilon_0 \ln \frac{3\varepsilon_0 (4\pi N_0/3)^{1/3}\xi_{\mathrm{loc}}}{2kT}.\tag{59}$$

Its significance lies in the fact that the fastest hop, starting at an energy deeper in the tail than E_t, $\epsilon_i < E_t$, ends in a state with energy in the vicinity of E_t, independent of ϵ_i. The width of the distribution of hopping rates around its maximum at E_t is given by $(6\varepsilon_0 kT)^{1/2}$. For hops starting with ϵ_i higher up in the tail than E_t, the fastest hop is to a nearest neighbor at an energy lower than ϵ_i. The hopping-transport energy thus plays the role of an effective, however temperature dependent, mobility edge [45]. This fact greatly simplifies the analysis of transport and relaxation phenomena in situations, where hopping in band tails is involved. It has been shown recently that a hopping-transport energy exists also for other forms of the tail-DOS, e.g. square root and Gaussian [47].

4.5.6. *Hopping of interacting electrons*

So far we have neglected the electron-electron Coulomb interaction completely. At low temperatures, however, this interaction must lead to a highly correlated system of electrons. Pollak [48], Srinivasan [50] and Efros and Shklovskii [51] were the first to point out that the one-particle density of states is depleted at the Fermi level (Coulomb gap). Consequently, the low-energy long-range hops are depleted as well. Efros and Shklovskii give an argument which shows that the density of states in three dimensions is given by $g(E) \propto (E - E_F)^2$. However, the precise form of $g(E)$ near E_F is still a matter of debate. Even more disputed is the influence of the Coulomb interaction on variable-range hopping transport at very low temperatures. Inserting the parabolic $g(E)$ into an interaction free variable-range hopping expression for σ yields a $T^{-1/2}$ behavior. Often experimental data seem to show such a temperature dependence, which is then taken as an indication of a Coulomb gap. However, it is still not completely understood how interactions modify the conductivity.

Certainly the assumption of binary, uncorrelated single-electron hops must be questioned. Both equilibrium [52] and dynamical properties [53] are affected by energy correlations due to the interaction. Instead of single-electron hops sequential hops and correlated many-electron hops are discussed in the literature [53]. They influence the temperature dependence of the conductivity at low temperatures.

With increasing temperature, electrons become more mobile and the Coulomb gap is smeared out. This starts to happen already at surprisingly low temperatures [54], until finally the temperature dependence expected for noninteracting electrons is recovered [54].

Another aspect of the Coulomb interaction has recently been discussed controversially [55, 56]: Besides changing the spectrum of eigenvalues it has been argued that the interaction may also lead to delocalization.

The notion of a "Coulomb-Glass" has been introduced for disordered systems with localized, interacting electrons [57]. This field is still very active, and exact results and statements for macroscopic systems are extremely hard to obtain.

4.5.7. *Small polaron hopping*

So far we have assumed that the electron-phonon coupling is sufficiently weak such that small polaron effects can be neglected. In some materials, however, it is known that carriers form small polarons or bi-polarons. A small (bi-) polaron is one (two) carrier(s) trapped in a potential well formed by the lattice distortion occurring around the carrier due to a strong short-range (e.g. deformation potential) electron-phonon interaction.

To illustrate the small polaron concept we first consider an electron placed into the conduction band of a perfect crystal. In the adiabatic limit the electron adjusts to the local changes of the surrounding atoms. On the other hand, due to the (short-range) electron-lattice interaction the atoms move in response to the presence of the electron. It is this self-consistent feedback process which eventually can lead to an electron confined to just one elementary cell, which is then called a small polaron. Relative to the free electronic states in the conduction band its energy is lowered by the polaronic binding energy E_b. This collapse of free carriers into strongly localized small polarons is favored by disorder [58]. We will concentrate on this situation here.

The motion of a small polaron is due to hopping [30]. At not too low temperatures the rates are activated [59] with an activation energy equal to E_b

if the disorder energies are much smaller than E_b. Otherwise disorder energies contribute to the activation energy. At low temperatures ($kT < \hbar\omega_p/3$) the rates become non-activated [31]. The prefactor is given by an average over the relevant phonon frequencies [59] in Holstein's molecular crystal model.

If we place a second electron into the site of a small polaron, we have to consider in addition the on-site Coulomb repulsion U. Such a situation is called a small bi-polaron. It is stable with respect to a decay into two small polarons if the energy of a small bi-polaron is smaller than that of two separated small polarons. One finds [60] that this occurs if $2E_b > U$ since the ground state energy of a small bi-polaron is $-4E_b + U$. Necessarily, the two carriers must then have anti-parallel spins, i.e. must form a singlet. Emin discussed various scenarios how carriers bound to bi-polarons are able to move and found that even if the density of bi-polarons exceeds that of polarons, the conductivity is dominated by single-polaron hops. The rates are then determined by at least the polaronic binding energy E_b and are otherwise similar to the rates discussed in Sec. 4.5.1. The existence of bi-polarons can be inferred from spin resonance measurements. A necessary criterium for the existence of small bi-polarons is that the number of spins is much less than that of charge carriers deduced from independent measurements, i.e. most carriers must be paired.

5. A Survey of Transport Experiments in Amorphous Semiconductors

From their transport properties, amorphous semiconductors can roughly be divided into two groups. In the first group, the dc conductivity at lower temperatures is no longer activated, but essentially follows Mott's law for variable-range hopping given by Eq. (58). The other group of materials does not show this type of transport. Instead, the dc conductivity appears to be activated down to the lowest temperatures, although the simple law Eq. (33) is not strictly obeyed in many cases.

5.1. *Variable-range hopping*

Tetrahedrally bonded amorphous semiconductors can be prepared by evaporation or sputtering in the absence of atomic hydrogen. Under these conditions the samples usually exhibit a rather large (of the order of 10^{18}–10^{20} cm^{-3}) density of deep states in the gap and, therefore, the introduction of dopants has little effect on the conductivity. A typical result is shown in Fig. 7 for amorphous Germanium [61]. At elevated temperatures the dc conductivity

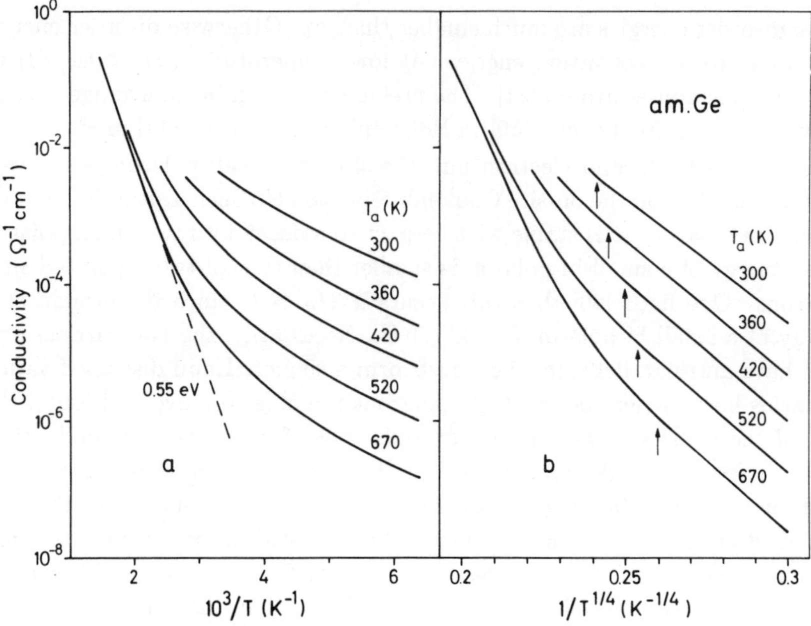

Fig. 7. Arrhenius plot (left) and $T^{-1/4}$ plot of the dc conductivity for evaporated a–Ge films after annealing at different temperatures (after Mell, 1974, [61]). Arrows indicate cross-over from activated to nonactivated behaviour.

appears to be activated like in Eq. (33). At about room temperature, already the conductivity changes to the typical $T^{-1/4}$ — law of Eq. (58) with a characteristic temperature T_0 of the order of 10^8 K, which is taken as evidence for variable-range hopping at these temperatures. In the variable-range hopping regime, the thermoelectric power usually is very small and only weakly temperature dependent [42, 62]. Such a thermopower can be calculated [40, 41] if one assumes a DOS distribution that is slightly energy-dependent.

The interpretation of the low-temperature transport as variable-range hopping faces two difficulties:

(1) the experimentally observed prefactor of the variable-range hopping conductivity is too large. It should be about six orders of magnitude smaller according to virtually all theories; and

(2) there is little if any variation of the observed characteristic temperature, T_0, which according to Eq. (57) should scale like the inverse of the DOS at the Fermi energy. From independent experiments (spin densities, e.g.)

one is convinced that the density of deep states is substantially (by more than one order of magnitude) altered upon annealing. Yet in the transport experiment (Fig. 7 is a good example), one generally observes no change of T_0 at all, whereas the prefactor, σ_{VRH}, is changed upon annealing by more than three orders of magnitude [61].

It has been shown [41, 63] that both difficulties can in principle be solved. One can carefully design a set of DOS distributions that depend on the energy in such a way as to yield a large spread of σ_0 values while T_0 is kept at a fixed value. In the same way, it is also possible to construct other DOS distributions that give rise to a $T^{1/2}$ law rather than to a $T^{1/4}$ law as is frequently observed in the experiments. While these solutions of the apparent discrepancy are available it is still unsatisfactory that there is no sequence of conductivity experiments in the variable-range hopping regime which show the behavior expected for a structureless DOS distribution, that is annealed uniformly: here we expect a significant variation of T_0 (by a factor of 100, say) without a major change of σ_{VRH}.

5.2. *Activated transport*

For many amorphous semiconductors the midgap DOS is sufficiently low to prevent the observation of variable-range hopping. Most of these semiconductors, hydrogenated amorphous silicon (a–Si:H) being a prominent example, can easily be doped and the doping effect is already observed at doping levels in the 10^{15} cm^{-3} range. Closer inspection of the temperature and doping dependence of the transport parameters shows several characteristic deviations from the expected activated transport given by Eq. (33).

5.2.1. *The Meyer–Neldel rule of the dc conductivity*

For hydrogenated a–Si:H at elevated temperatures the electronic transport takes place at the mobility edge. In the usual Arrhenius plot (log(conductivity) vs. inverse temperature) one would expect from Eq. (33) that the conductivity is given by a straight line. For a sequence of samples with different positions of E_F (determined by doping, e.g.) the slopes of the curves in this plot should give a measure of E_C–E_F or E_t–E_F with a common intercept, σ_0, at $1/T = 0$. The transport path is not considered to be affected by doping and, therefore, one would assume that the observed values for σ_0 agree to within a factor of 2, say.

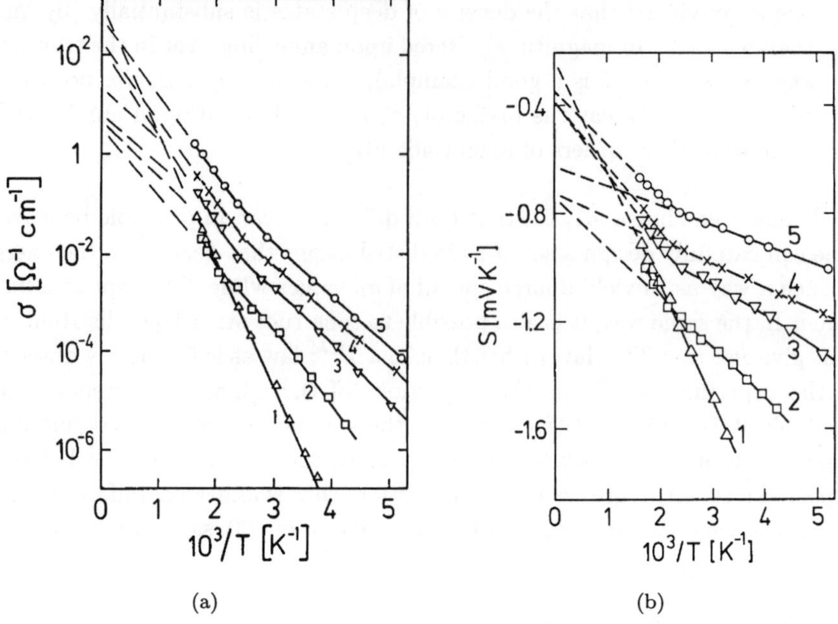

Fig. 8. Arrhenius plot of the dc conductivity (a) and the thermoelectric power (b) for n-type a–Si:H films doped by different amounts of PH_3 (after Beyer *et al.* 1977 [69]). Figures 1 to 5 indicate increasing doping [10].

The experimental data in Fig. 8(a) show significant deviations from our expectations: for undoped and for both n-type and p-type doped samples one generally observes conspicuous kinks with a change of slope at the kink which would correspond to a change of the activation energy by as much as 0.5 eV. In fact, we observe similar curves as in Fig. 7 except that in the latter case the kink is replaced by a more gradual transition from the high-temperature behavior to the low-temperature conductivity.

The presence of these kinks has been taken as evidence for a second transport channel [64, 65]: the high-temperature side was considered due to transport in extended states at the mobility edge while for the low-temperature regime hopping in tail states was assumed to be dominant. We consider this assumption to be unlikely, however, for the following reason:

If we force-fit the experimental data in the limited temperature range below room temperature, we obtain intercepts at $1/T = 0$ for the different curves that scatter by more than six orders of magnitude. It would be hard to explain a variation of this order of magnitude for hopping in tails, although the tail state distribution might be sensitive to the details of the preparation of the material. For transport close to E_C, however, such a variation is unacceptable.

We, therefore, conclude that the observed variation is not connected with the microscopic prefactor, σ_0 of the conductivity at E_C. Instead, it should either be due to some transport mechanism that is very sensitive on the details of the sample preparation (like hopping in some defect/dopant states band) or due to the temperature variation of some reference energy like the Fermi energy E_F. We shall show that the latter is in fact the case and, therefore, shall denote in the following intercepts and slopes by σ_0^* and E_σ^* to indicate that these are "quantities taken from an Arrhenius plot".

For a series of differently doped a–Si:H samples the plot of σ_0^* vs. E_σ^* is shown in Fig. 9(a). Apparently

$$\sigma_0^* = \text{const.} \cdot \exp(E_\sigma^*/E_{\mathrm{MNR}}) \tag{60}$$

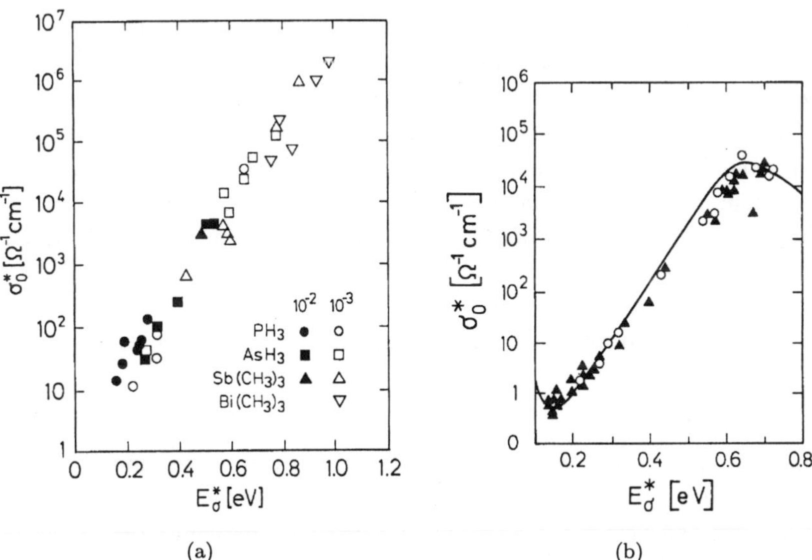

Fig. 9. (a) The Meyer–Neldel rule for a–Si:H doped with various dopants (from Carlson and Wronski, 1979) and (b) for a single a–Si:H sample after light soaking and annealing (from Irsigler [83]).

Such a law had been observed already in 1937 by Meyer and Neldel [66] in a conductivity study of baked oxide powder samples. A similar Meyer–Neldel rule can be observed in a different way [67]: by extended illumination additional deep states are produced (Staebler–Wronski-effect, [68]) which in part balance the doping of the sample. Combining the data for a single sample at different stages of light soaking and annealing, one obtains a perfect Meyer–Neldel rule shown in Fig. 9(b).

5.2.2. *Thermoelectric power*

The sign of the thermoelectric power reflects the charge of the predominant carrier [69]: undoped a–Si:H has a negative thermopower as is expected for n-type conduction. Doping of a–Si:H with phosphorus should lead to n-type conduction and in fact the thermopower keeps its negative sign. Doping by boron results in hole conduction and a positive sign of the thermopower as is observed at a sufficiently high doping level.

The Arrhenius plot of the thermoelectric power in Fig. 8(a) shows the same kinks as the corresponding plot of the dc conductivities for the same samples. The variations of the intercepts A^* mirror the variations of σ_0^* suggesting a common origin. If one compares the apparent activation energies E_σ^* with E_S^* one finds a systematic difference, $E_\sigma^* = E_S^* + \Delta E$ where ΔE is in the $0.05\ldots0.25$ eV range and is larger in the apparently more disordered samples.

The difference ΔE has been taken as further evidence for the presence of the two parallel transport channels [70]. In fact, in the temperature range where the predominant transport channel changes from transport in extended states to hopping at the bottom of the tails, one would expect that the apparent activation energies of the conductivity do not coincide with E_S^*. However, closer inspection of transport data taken from a large variety of samples suggests a different and more satisfactory explanation as we will see later.

5.2.3. *The Q-function*

We want to separate effects that arise from the transport channels from those that are caused by a possible temperature dependence of the reference energies like the Fermi energy. This can be done by combining experimental thermoelectric power and dc conductivity data measured simultaneously for the same sample [71]. Figure 10(a) shows the result of a combination of the data in Fig. 8(a) with those of Fig. 8(b). The resulting $Q(1/T)$ curves are free from any observable kinks. This would be entirely unexplained if we had considered that in the temperature range accessible to the experiments the dominant transport path changes. In this case, the Q-function in an Arrhenius plot would show an "inverted-S"-like shape and the intercept, Q_0, would sensitively depend on the particular values for the parameters used.

From the experimental data one instead finds the empirical law

$$Q(T) = Q_0^* - \frac{E_Q^*}{kT} \tag{61}$$

with $Q_0^* = 10 \pm 1$ and $E_Q^* = \Delta E = 0.05 \ldots 0.25$ eV. Note the remarkably small scatter of the Q_0^* intercepts as compared to the enormous variation of the σ_0^* and A^* values. The implication of this experimental fact is that the observed Meyer–Neldel rules for the dc conductivity do in fact not reflect the variation of the microscopic prefactor but must be related to the temperature dependence of the reference energies.

We show in Fig. 10(b) similar $Q(T)$ data for a variety of chalcogenide glasses including amorphous selenium and arsenic. Again the Q_0^* values agree

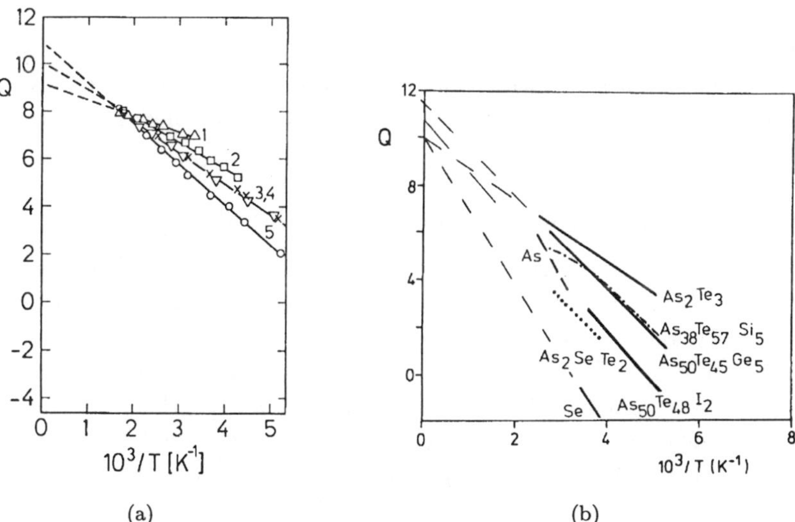

(a)   (b)

Fig. 10. (a) Arrhenius plot of the Q-function from the data of Fig. 8 for a–Si:H and (b) for some chalcogenide glasses: As$_2$SeTe$_2$ from Arnoldoussen *et al.* 1974 [72], As$_2$Te$_3$, As$_{50}$Te$_{45}$Ge$_5$ and As$_{50}$Te$_{48}$I$_2$ from Seager *et al.* 1973 [73], As$_{38}$Te$_{57}$Si$_5$ from Nagels *et al.* 1974 [65]. Data for a–Se from Jušča *et al.* 1974 [74] and a–As from Mytilineou, 1977 [75] are also included.

very closely with each other and with the values obtained for a–Si:H and also for a–Ge:H [76].

5.2.4. *Hall effect*

In crystalline semiconductors, the Hall coefficient is widely used in order to determine the carrier density. There are many cases where the carrier density thus obtained matches the number expected from the doping level. By comparison with the dc conductivity, the carrier mobility (Hall mobility) can be determined. In all amorphous semiconductors, the Hall effect suffers from a sign anomaly [69, 78]: the sign of the Hall effect is opposite to that of the thermopower. Since there is no theory that gives a satisfactory explanation for this sign anomaly, the Hall effect can hardly be used for the determination of carrier densities or mobilities. Simply neglecting the anomaly, one finds a Hall mobility that appears to be activated with an activation energy $E_H^* \sim 1/3 E_Q^*$.

We would like to stress that the sign anomaly of the Hall effect appears to be the most unfortunate obstacle towards a satisfactory understanding of the experimental transport data in all disordered semiconductors: Since we do not know the carrier mobilities we cannot discriminate the different transport processes.

It seems that the sign anomaly is not connected simply with the disorder present in the samples, but with some specific property of the amorphous state: for recrystallized a–Si:H samples [77] the sign of the Hall effect is normal as is the case for μc–Si:H [88].

6. Interpretation of the Transport Properties of a–Si:H

In this section, we give a detailed interpretation of the transport properties of a–Si:H, a material that is investigated more completely than any other amorphous semiconductor. We restrict our attention to the electronic transport in quasi-equilibrium for which there is no trace of hopping conduction. The material, however, has localized tail states and, therefore, there are transport processes which involve hopping in tail states. These include low-temperature photoconductivity, relaxation of optical excitation in transient luminescence experiments, and also high-field dc conductivity.

We shall concentrate on the interpretation of two phenomena that do not seem to fit into the theoretical scheme presented in Sec. 4: the Meyer–Neldel rule, i.e. the enormous variation of the prefactor σ_0^* with E_σ^* and the variation of the slope of the Q-function, E_Q^*.

6.1. *The statistical shift of the Fermi energy and the Meyer–Neldel rule*

The variation of the apparent prefactor, σ_0^*, exceeds 3 orders of magnitude for *undoped* a–Si:H [82]. For doped a–Si:H the variation exceeds 5 orders of magnitude. Even for a single a–Si:H sample a variation of the observed σ_0^* values of some 3 orders of magnitude can be achieved experimentally *in a reversible manner* by light-soaking and annealing. In most cases, the variation of the prefactor σ_0^* with the apparent activation energy E_σ^* follows the empirical law

$$\ln(\sigma_0^* \Omega \mathrm{cm}) = B_{\mathrm{MNR}} + E_\sigma^* / E_{\mathrm{MNR}} \tag{62}$$

with $B_{\mathrm{MNR}} \sim 0$ and $E_{\mathrm{MNR}} \sim 0.05$ eV.

This Meyer–Neldel rule (MNR) has first been observed by Meyer and Neldel, 1937 [66] in their study of baked semiconductor powders and the implication could be that the structure of a–Si:H must be considered as similar to baked powders if it showed the same features of the conductivity. In contrast, our explanation for the MNR in a–Si:H does not involve any structural inhomogeneities. Instead, we refer to a mechanism that is well-known for crystals.

For all singly doped semiconductors (for simplicity we treat n-doped crystals but for p-type crystals the physics is *mutatis mutandis* the same) at lower temperatures the Fermi energy E_F is close to the donor level and the carrier density (and thereby the dc conductivity) shows an activation energy which is essentially given by the energetic distance from the conduction band edge (for the present discussion it suffices to identify E_t with E_C) to E_F

$$n(T) = N_{\mathrm{eff}}(T) \exp\left(-\frac{E_C - E_F}{kT}\right) \tag{63}$$

The effective density of states, $N_{\mathrm{eff}}(T)$, depends on the temperature as $T^{3/2}$. If we raise the temperature, we enter the "exhaustion regime" where all donor levels are empty. Since the density of carriers in the conduction band is now temperature independent, the increase of kT in the denominator of the exponential in Eq. (63) must be balanced by a concomitant increase of the numerator, i.e. a decrease of E_F. Thus E_F must experience a "statistical shift" which persists until the center of the gap is reached and intrinsic conduction sets in with the onset of electron-hole pair production thus stopping the movement of E_F.

In any solid the position of $E_F(T)$ is governed by the neutrality condition which requires that the total density of electrons in a sample, n_{tot}, does not change with temperature (except for some trivial change due to lattice expansion which is ignored here)

$$n_{\text{tot}} = \int_{-\infty}^{\infty} g(E) f_F(E, E_F, T)\, dE \tag{64}$$

Here we have explicitly denoted that the Fermi distribution function f_F depends on E, E_F, and also on T. If e.g. E_F is close to E_C at lower temperatures in a DOS distribution similar to that shown in Fig. 3, it will move towards midgap with rising temperatures in a similar way as for an n-type doped crystal. This statistical shift will be larger in general if the variation of the DOS distribution is large.

To demonstrate the effect, we evaluate the neutrality condition Eq. (64) for a simple DOS distribution model given by Fig. 11. The result is shown in Fig. 12 for the case that at $T = 0$ the Fermi energy is 0.2 eV below E_C. As T is raised the statistical shift slowly sets in (proportional to T^2) until at about room temperature the shift is approximately linear with T. In most transport

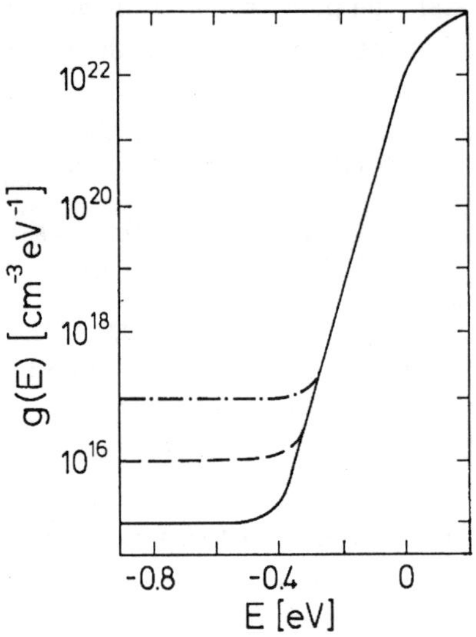

Fig. 11. Simplified model for the density of states distribution of a–Si:H. The exponential tail is augmented by a constant density of midgap states for which three different densities are assumed [10].

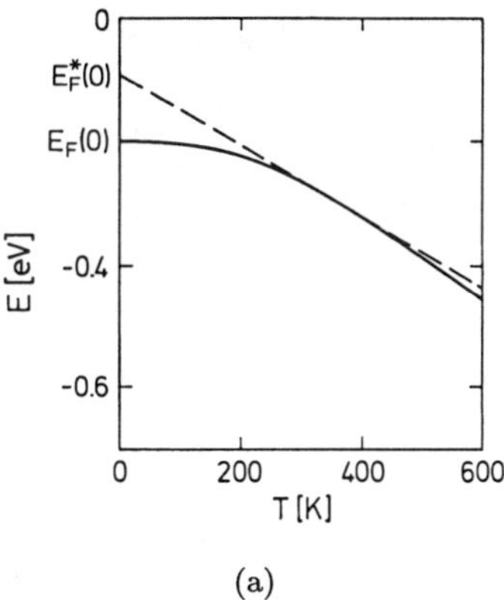
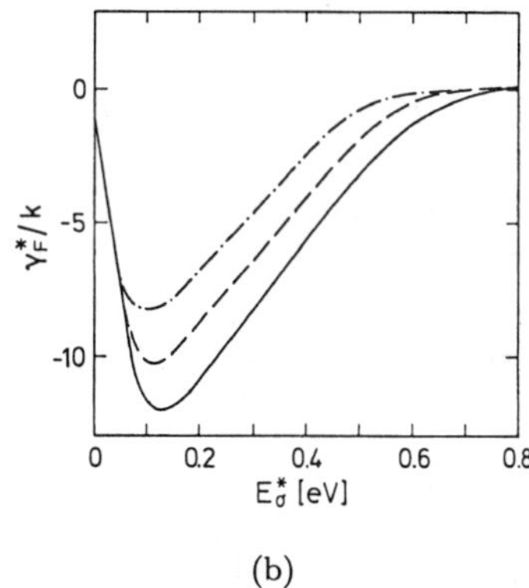

(a)

(b)

Fig. 12. (a) Statistical shift of the Fermi energy E_F calculated from the neutrality condition for the dos distribution models of Fig. 11. The linear approximation of $E_F(T)$ is also shown. (b) Mean statistical shift γ_F^* as a function of the apparent activation energy E_σ^* calculated for the dos distribution models of Fig. 11 [10].

experiments on a–Si:H the data are accumulated for temperatures above room temperature. We approximate $E_F(T)$ in Fig. 12 by its linear approximation

$$E_F(T) = E_F^*(0) + \gamma^* \cdot T \tag{65}$$

(both $E_F^*(0)$ and γ^* depend somewhat on the energy interval used for the linearization). Inserting Eq. (65) into Eq. (33) we obtain

$$\sigma(T) = \sigma_0^* \exp(-E_\sigma^*/kT) \tag{66}$$

with

$$\sigma_0^* = \sigma_0 \exp(\gamma_F^*/k) \quad \text{and} \quad E_\sigma^* = E_C - E_F^*(0) \tag{67}$$

Although we do not consider the DOS distribution model of Fig. 11 to be very realistic, it is instructive to evaluate the Meyer–Neldel rule for it. For simplicity, we keep the DOS distribution fixed but assume that by doping we can control the Fermi level position $E_F(T = 0)$. We evaluate for all positions of $E_F(T = 0)$ the neutrality condition Eq. (64). The resulting $E_F(T)$ is similar to that plotted in Fig. 12(a). From these plots the values for γ_F^* and E_σ^* are obtained in the 250 K $< T <$ 350 K temperature range. Results for several initial positions of $E_F(T = 0)$ are displayed in Fig. 12(b). There is a wide range of activation energies E_σ^* for which the shift parameter γ_F^* varies linearly with the apparent activation energy E_σ^*. For the three DOS models in Fig. 11

we derive the following Meyer–Neldel parameters from Fig. 12(b):

$$B_{\mathrm{MNR}} = \ln(\sigma_0 \Omega \text{ cm}) - 12 \pm 2 \text{ and } E_{\mathrm{MNR}} = 0.054 \text{ eV}$$

In the late 1970's, the DOS distribution models determined from field effect [84] where very popular. For these models the DOS at E_C was only about 3 orders of magnitude larger than the midgap DOS. Consequently the statistical shift derived from these DOS models was too small to account for the MNR observed in a–Si:H. We therefore proposed the use of a DOS distribution model [80] for which this variation was nearly 5 orders (which was considered rather unrealistic at that time) in order to explain the MNR. Such a variation is now generally accepted (see e.g. [85]). Unfortunately our previous attempts failed [80] to invert the method: we have tried to determine the statistical shift from experimental conductivity data and to find a DOS distribution from the requirement that it must give rise to the observed statistical shift. This procedure can be expected to yield satisfactory results only if the DOS distribution is temperature-independent. At the time this was the general believe of most experimentalists: state-of-the-art a–Si:H was claimed to be free from any hysteresis effects upon thermal cycling. Later it became apparent [79] that in contrast the DOS distribution of a–Si:H depends sensitively upon thermal history, illumination, etc. which prohibits a determination of the DOS distribution from the statistical shift alone.

Using realistic DOS distribution models we can easily reproduce the MNR found experimentally in the dc conductivity of doped a–Si:H . We can also easily account for the MNR rule found experimentally in light-soaking experiments [86, 83] as is shown by the full line in Fig. 9 (see e.g. [10] for details). In a similar way, the MNR observed in the field effect currents [87] is readily explained by the statistical shift of $E_F(T)$.

Note that a linear MNR is observed according to Fig. 9 only if E_σ^* exceeds 0.15 eV: the statistical shift is largest if $E_T(T = 0)$ is just in the low-energy onset of the tail states. If it is moved into the upper part of the tail, the statistical shift is progressively more impeded because the exponential tail state distribution must bend over close to E_C. Thus for $E_F(T = 0)$ approaching E_C the shift parameter γ_F^* goes to zero and we predict some "Anti-Meyer–Neldel rule" for the smallest values of E_σ^*. This has not been observed in a–Si:H, but is well documented for microcrystalline μc–Si:H [88].

For a–Si:H, the MNR in the field effect has recently [90] been observed in a novel thin film transistor structure which resulted in activation energies E_σ^*

as small as 0.1 eV. In these devices the turn-over from the usual Meyer–Neldel rule to the Anti-Meyer–Neldel rule was observed as predicted.

The statistical shift as the origin for the MNR has been doubted [89] because it was argued that the MNR relied on a particular shape of the DOS distribution while the MNR is observed quite generally. It is a fact, however, that one finds a MNR for quite general DOS distributions. Furthermore, any alternative explanation of the MNR in a–Si:H will have to cope with the difficulty that all the DOS distributions which have been proposed as realistic for a–Si:H must necessarily give rise to a large statistical shift, a feature that cannot be considered absent.

6.2. *The long-ranged random potential*

In order to investigate the origin for the variation of E_Q^* we consider an amorphous semiconductor that prior to doping has no charged centers. Donors are supposed to be built in at random positions in space. The donors are further assumed to be always ionized as their extra electrons have condensed into deep states in the gap. These latter states are thereby negatively charged and in addition the Fermi level is raised from its intrinsic position towards the conduction band edge. We are thus left with a system of equal numbers of positive charges and negative point charges, both distributed at random positions in space and both essentially immobile. We will further assume that also the deep defect states which are generated as a by-product of the doping process are not correlated in space with the donors. Otherwise, we would find the positive and negative charges predominantly paired as small dipoles that would hardly contribute to a long-ranged random potential. The assumption that all fixed charges are distributed at random therefore is vital for the rest of the analysis.

Branz and Silver [91] have recently shown that a short-ranged random potential can lead to a significant density of charged dangling bonds. They show that this density can be much larger than the density of neutral dangling bonds. This random distribution of charges would then also contribute to the long-ranged potential, in particular in undoped material.

Some 25 years ago, Shklovskii and Efros [92] have shown that in a 3D system a random distribution of point charges gives rise to a random potential and that the amplitude of this potential would diverge if there was no screening. The explanation is rather elementary: consider a 3D sample that is subdivided into cubes of length L. The sample contains a distribution of positive point charges with mean densities $\bar{n}_{(+)}$, neutral donors with a mean density $\bar{n}_{(0)}$, and negatively charged acceptors with a density $\bar{n}_A$ which shall be replaced by negative jellium. Since the point charges are distributed at random, a cube will contain on the average

$$N_{(+)}(L) = \bar{n}_{(+)} \cdot L^3 \tag{68}$$

positive charges with a mean deviation from the average that is given by

$$\Delta N_{(+)}(L) = \sqrt{N_{(+)}(L)} = \sqrt{\bar{n}_{(+)} \cdot L^3} \tag{69}$$

While the jellium balances the averaged density of positive charges, the fluctuations of the point charge density, $\Delta N_{(+)}(L)$, give rise to a random electrostatic potential. In the center of the adjacent cube and for a medium with dielectric permittivity ϵ_s, this potential has the value

$$\Delta V(L) = \frac{e^2}{4\pi\epsilon_0\epsilon_s} \frac{\Delta N_{(+)}}{L} = \frac{e^2}{4\pi\epsilon_0\epsilon_s} \sqrt{\bar{n}_{(+)} \cdot L} \tag{70}$$

This simple consideration shows that charged centers give rise to a random potential. For a 3D system, the mean amplitude of this potential is proportional to the square root of the length and thus diverges unless the potential is screened.

A more quantitative estimate of this random potential has been given by Jäckle [93]. He considers the density fluctuation $\Delta n_{(+)}(\mathbf{r})$ of a random distribution of charges. The long-ranged part of these density fluctuations will be obtained by coarse-graining $\Delta n_{(+)}(\mathbf{r})$. Since coarse-graining does not remove the divergence of the random potential, this divergence must be removed by screening, i.e. by a rearrangement of charge in response to the random potential.

This can be seen best if for the moment we discuss a partially compensated crystalline semiconductor where we may assume that there are no states in the gap apart from the donors (compensating acceptors are replaced by jellium). Potential fluctuations will be screened by the redistribution of the $\bar{n}_{(0)}$ electrons left in donor states due to incomplete compensation. A fluctuation of length l will be completely screened if the number of screening charges exceeds the coarse-grained density $\langle \Delta N_{(+)} \rangle_l$. Since the number of screening charges rises with l^3, much faster than $\langle \Delta N_{(+)} \rangle_l$, there will be a maximum length $l_{\max}$ for which the donor density fluctuation is screened uncompletely: fluctuations on a scale much smaller than $l_{\max}$ are not screened at all, fluctuations which have a larger length scale than $l_{\max}$ will be screened completely, while density fluctuation of size $l_{\max}$ give rise to the largest random potential.

If we use a Gaussian of variance σ as coarse-graining function

$$g_\sigma(\mathbf{r}) = \frac{1}{\sqrt{2\pi}\,\sigma} \exp\left(-\frac{r^2}{2\,\sigma^2}\right) \tag{71}$$

we obtain a random potential characterized by $\sigma_{\max}$ the mean square amplitude δ of which is given by ([93], see also [10])

$$\sigma_{\max} = \frac{1}{\sqrt{4\pi}} \frac{\bar{n}_{(+)}^{1/3}}{\bar{n}_{(0)}^{2/3}} \qquad \delta^2 = 4(\sqrt{2}-1)\left(\frac{e^2}{4\pi\epsilon_o\epsilon_s}\right)^2 \frac{\bar{n}_{(+)}^{4/3}}{\bar{n}_{(0)}^{2/3}} \tag{72}$$

If compensation is nearly perfect both the fluctuation length $l_{\max}$ (note that Gaussian coarse-graining function with $\sigma_{\max}$ gives rise to a random distribution with length $l_{\max} \sim 2.3\,\sigma_{\max}$) and the amplitude δ can grow until δ is comparable to the donor binding energy. In crystalline semiconductors, a further increase of the random potential beyond the donor binding energy is possible for perfect compensation only (with E_F near midgap) because otherwise electrons transferred into the conduction band would screen the potential very effectively.

For compensated a–Si:H, the compensation can be almost perfect because the minority dopant is preferentially accommodated in fourfold coordination. Furthermore, the same mechanism keeps the density of midgap states at least as low as for undoped material [94]. In this case, the Fermi energy is located near midgap and the random potential can grow until the band tails in the extrema of the random potential start to be occupied. The maximum length of the potential fluctuation is thus given by $l_{\max} \sim N_+^{-1}$ (see Howard and Street [95]). For a compensated sample doped nominally with 1 % phosphine and diborane, Howard and Street obtain $l_{\max} \simeq 125$ Å.

For singly doped amorphous a–Si:H, the donor states are also unoccupied. Electrons in the deep states around the Fermi energy E_F will be redistributed by the random potential and, therefore, a density of screening charges

$$n_{\mathrm{scr}} = g(E_F) \cdot \delta \tag{73}$$

limits the maximum fluctuation length of the random potential as well as its mean amplitude:

$$l_{\max} = 2.3 \left[\frac{\epsilon_0\epsilon_s}{e^2 \cdot g(E_F)}\right]^{1/2} \qquad \delta = 2 \cdot \left[\frac{e^2}{4\pi\epsilon_0\epsilon_s}\right]^{3/4} \frac{\bar{n}_{(+)}^{1/2}}{g(E_F)^{1/4}} \tag{74}$$

As a numerical example for highly doped a–Si:H, we take $g(E_F) = 10^{16}$ cm^{-3} eV^{-1}, $\epsilon_s = 12$ and $\bar{n}_{(+)} = 10^{18}$ cm^{-3} eV^{-1} which results in $l_{\max} = 4000$ Å and $\delta = 0.2$ eV. Within this fluctuation there are about $2 \cdot 10^5$ donors, hence a typical density fluctuation is caused by the excess of some 400 donors only.

The redistribution of electrons near E_F also contributes to screening in the case of compensated a–Si:H. For nominal doping levels below $\simeq 5 \cdot 10^{-3}$ the screening will be predominantly due to this mechanism. At this doping level, the potential fluctuations reach their maximum value and upon further doping the occupation of the tails limits the fluctuation length $l_{\max}$.

Thus far we have not considered screening by mobile charges in states at E_C as e.g. by carriers which have been excited thermally or optically or injected from the electrodes. These will be shepherded by the random potential and can contribute to the screening. For the numerical examples given above, we find that screening of the random potential for the 1% doped compensated sample requires $2.5 \cdot 10^{19}$ cm^{-3} mobile carriers whereas in the case of the singly doped sample $2 \cdot 10^{15}$ cm^{-3} would be sufficient. If the density of extra carriers is just sufficient to screen the random potential with length $l_{\max}$ then a random potential with smaller characteristic length will be left over and a higher density of screening charges would be required to screen these fluctuations.

For semiconductors with negative-U defects [96, 97] the situation is exactly analogous to that of a perfectly compensated semiconductor with the modification that in this case the amplitude of the random potential is limited by the absolute value of U.

Besides these electrostatic reasons for a long-ranged random potential one can also envisage random potentials that are due to structural and/or compositional inhomogeneities as are frequently reported for amorphous semiconductors. We do not want to exclude these poorly known random fields. Our discussion concentrated on the electrostatic field of the random distribution of charges in doped systems because here we can give a quantitative estimate.

6.3. *Transport in the presence of long-ranged random potentials*

For the calculation of the transport data, we shall assume that in the absence of the random potential the electronic transport would take place above a well-defined mobility edge at E_C with a constant prefactor σ_0 of the dc conductivity. We assume that the Fermi energy is located near midgap and use non-degenerate statistics throughout. The prefactor $\exp(-(E_C - E_F)/kT)$, however, will be left out in the calculations of the transport data except when explicitly needed. The influence of the random potential can be treated in a semiclassical manner: since $l_{\max}$ exceeds the inelastic scattering length L_i by more than one order of magnitude, the random potential will not contribute to scattering which is responsible for the existence of E_C. Instead, it modulates all energies by $V(\mathbf{r})$, in particular the energy E_C of the mobility edge. As a consequence the local density of carriers in extended states above E_C will be modulated by the random potential via the Fermi (Boltzmann) distribution function. Since this density will be constant over several L_i, it is possible to define a local conductivity $\sigma(\mathbf{r})$ which differs from the conductivity σ_0 by the Boltzmann occupation factor only

$$\sigma(\mathbf{r}) = \sigma_0 \, \exp(-V(\mathbf{r})/kT) \tag{75}$$

The evaluation of this model requires a numerical treatment which has been performed using two different numerical models:

6.3.1. *The resistor network model*

In this model, we divide our sample, a cube, into $N \cdot N \cdot N$ subcubes. Each subcube is considered to have a constant conductivity which is given by the local conductivity averaged over this subcube. The current through this array of subcubes can be obtained from the corresponding Kirchhoff equations [98]. The solution of Kirchhoff's equations requires the inversion of a matrix of rank N^3 which limits practical calculations to rather small values of N ($N \leq 10$) and therefore it is mandatory to minimize surface effects. This is achieved using periodic boundary conditions in the transverse directions for the random potential as well as for the current.

6.3.2. *The random walk model*

For the random walk model the sample is divided into $N \cdot N \cdot M$ subcubes with N of order 20–50 and M of order 200–1000. With these dimensions effects due to the limited sample size are negligible for N, $M \geq 20$ as can be shown by computations for different values of N and M. Carriers are generated in the interior of the sample and their movement is monitored assuming that the probability for the transit from subcube i to subcube j is given by

$$p_{ij} = \begin{cases} \exp(-\beta \Delta V_{ij}) & \Delta V_{i,j} \geq 0 \\ 1 & \Delta V_{i,j} \leq 0 \end{cases} \tag{76}$$

The random walk model has been used to calculate the dc conductivity and also the Peltier heat which through Onsager's relation gives the thermoelectric power. It can also be employed [10] to simulate time-of-flight (TOF) experiments for photo-generated carriers that are subject to a random potential. We have further applied the scheme [10] for the calculation of the dc current in external strong fields which may exceed the random field in which case one expects a strong field dependence of the conductivity.

6.3.3. *Results of the network calculations*

Several authors [95, 99–101] have assumed that the principal result of a long-ranged random potential would be a shift of the mobility edge toward higher energies. Such a shift is however expected for additional short-ranged disorder which adds to the disorder already present in an undoped amorphous semiconductor. The effect of a long-ranged random potential is quite different as is shown in Fig. 13. It rather gives rise to a percolative flow of the carriers along the valleys of the long-ranged potential. Here the conductance g of a resistor network model is plotted as a function of inverse temperature for different

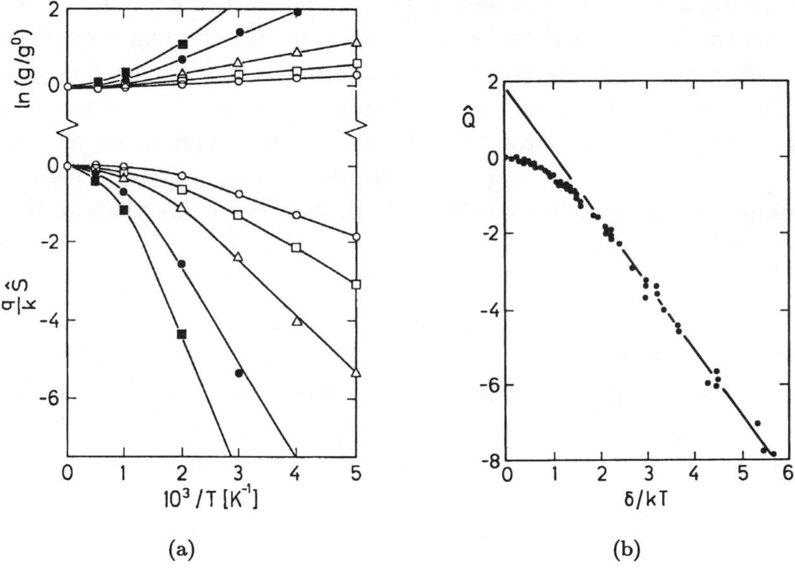

(a)　　　　　　　　　　　(b)

Fig. 13. (a) Calculated average network conductance g (upper part) and thermoelectric power (lower part) for a $10 \cdot 10 \cdot 10$ resistor model with different values for δ [98]. (b) $\hat{Q}$ function calculated from the results of Fig. 13(a) [10].

values of δ (the conductance is given in units of g_0, the conductance in the absence of a random potential). The fact that the conductance rises with rising inverse temperature shows that in a three-dimensional random potential the current flows predominantly through "valleys" where the highest intervalley "passes" that determine the sample conductance are still lower than the average potential (in a two-dimensional system the cols would coincide with the average value). Also shown in the figure are the corresponding results for the thermoelectric power $\hat{S} = q/(kT)\langle V \rangle$. We call the calculated quantity $\hat{S}$ because it contains the contribution from the potential energy only without the kinetic energy of the carriers which usually is contained in the heat of transport term, A. The configurational average denoted by $<>$ is an average weighted with the local current density. For lower temperatures both g and $\hat{S}$ appear to be activated, whereby the slopes for the thermoelectric power data are significantly larger that those for the corresponding conductances. This reflects the fact that the thermoelectric power essentially averages the potential energy of the carriers within the valleys whereas (at least at low temperatures) the conductance predominantly reflects the height of the passes. If we combine the logarithm of the sample conductance with the calculated thermoelectric power $\hat{S}$ we obtain the $\hat{Q}$ function shown in Fig. 5

$$\hat{Q} = \ln(g/g_0) + \frac{q}{k}\,\hat{S} \tag{77}$$

which again at low temperatures exhibits an activated behavior. At elevated temperatures the curves bend over with the effect that $\hat{Q}_0^*$, the extrapolated intercept at $1/T = 0$, is different from $\hat{Q}(1/T = 0)$. A discussion of the results obtained from the evaluation of the random walk model is not given here since the differences to the results derived from the resistor network shown above are not essential (see [10]).

6.4. *Discussion*

We shall start the discussion by a determination of the microscopic prefactor of the conductivity from published transport experiments. As we have shown in our discussion of the MNR, this requires the exact knowledge of the statistical shift. There are two cases in the literature for which $E_C(T) - E_F(T)$ can be estimated quite accurately: Dersch *et al.* 1983 [102] have studied a sample which by electron irradiation had an extremely high density of dangling bonds such that the statistical shift of E_F can be shown to be negligible. The other case is a compensated sample (Beyer *et al.* 1981 [103],) for which the Seebeck coefficient was approximately zero over a sizable temperature range which shows that the contribution of electrons and holes to the current is symmetrical and also that the Fermi level is situated very closely midgap. While the former experiment could be described by $\sigma_0^* \cong 2000\,\Omega^{-1}\text{cm}^{-1}$, the latter resulted in $\sigma_0^* \cong 4700\,\Omega^{-1}\,\text{cm}^{-1}$. Since in the latter experiment the contribution of both electrons and holes was very accurately the same, the factor of 2 difference to the former result is a confirmation of the interpretation. Having established that E_F in both experiments is at midgap we can estimate the shift of $E_\sigma^*(T)$ from the shift of the optical band edge ($\gamma^{\text{opt}} \sim 5.4k_B$) which leaves us with $\sigma_0 = 150\,\Omega^{-1}\,\text{cm}^{-1}$. This result should be valid for both electrons and holes in a–Si:H and we would estimate it to be correct within less than a factor of two.

Our theory developed in Sec. 4 provides a microscopic interpretation for this particular value of σ_0. There we found that σ_0 is determined by just one microscopic parameter, a. The value $\sigma_0 = 150\,\Omega^{-1}\,\text{cm}^{-1}$ is compatible with $a = 5$ Å. From the atomic structure of an amorphous semiconductor one may in fact anticipate such a value of being realistic. Furthermore, similar amorphous semiconductor systems like a–Ge:H and chalcogenides should not differ significantly, as far as the value of a is concerned.

Next, we discuss under which circumstances the effect of the random potential is observed in a–Si:H. We shall begin with the discussion of the Q-function. For undoped a–Si:H the non-zero slope E_Q^* (which is close to 0.05 eV for state-of-the-art material [104]) is explained by the transport near E_C, see Sec. 4.4. For lightly doped n-type samples the contribution of the random potential to the observed slope E_Q^* is small. This means that the extrapolated experimental values for Q_0^* are taken from that part of $\hat{Q}(1/T)$ [see Fig. 13(b)], that bends over on the high-temperature side. This provides a natural explanation for the fact that in Fig. 10(a) the extrapolated intercepts increase with increasing slopes. For the samples of Fig. 10(b) this effect is no longer important because the slopes E_Q^* are larger.

As an alternative explanation for the finite slope E_Q^* variable-range hopping in the exponential distribution of tail states has been proposed. Hopping will produce a similar Q-function [22] as a random long-ranged potential with a slope E_Q^* depending on the tail parameter T_c

$$E_Q^* = 3.5 \cdot kT_c \tag{78}$$

For larger values of E_Q^* as e.g. observed for moderately doped and for compensated samples [104] this relation would predict a broadening of the conduction band tail upon n-doping which is not observed except for very high doping levels.

Fascinating evidence for a very large random potential in compensated heavily doped a–Si:H has been presented by the Xerox group [95] from time-of-flight experiments: the observed decrease of the drift mobility upon raising of the doping level is interpreted as evidence that the random potential is increased with doping until the width δ is close to 0.5 eV. Further increase then merely leads to a reduction of $l_{\max}$ since the occupation of the tail states limits the further growth of δ. At this stage, the drift mobility of the holes (which for undoped samples is more than two orders of magnitude below that of electrons) becomes practically identical to that of the electrons. Howard and Street argue that tail broadening with doping which in the multiple trapping model would lead to a decrease of the mobility cannot explain the observed decrease of the drift mobility: the tail width observed in the optical absorption does not increase for moderate doping levels while the hole drift mobility is significantly reduced. We have shown [105] that the broadening of the optical absorption edge and the reduction of the observed drift mobility can be explained consistently with the large random potential amplitudes expected for highly doped and perfectly compensated a–Si:H samples.

From photoemission yield spectroscopy experiments [106, 107] it has been concluded that for compensated a–Si:H there is no broadening of the conduction band tail, i.e. the random potential is not observed. This result is most probably a particularity of the photoemission yield experiment itself: any electrostatic long-ranged potential modulates all energies including the vacuum level in exactly the same way. Random potentials therefore should be undetected unless the final kinetic energy of the photoelectrons is monitored.

Recently, Dyalsingh and Kakalios [108] have measured the activation energy of the short-circuit current of a sample with a temperature gradient and found it to be equal to E_σ^*. The authors take this an additional evidence for the presence of a random potential. A brief consideration shows, however, that this result is trivial for any Ohmic system: both dc current and thermocurrent are determined by the fraction of a voltage divided by some resistance. Experimentally, in both cases the resistivity of exactly the same piece of resitive material is examined and, therefore, not only the activation energies, but also the absolute values of the resistivities (or conductivities) must agree.

7. Summary

Although after more than 40 years of extensive research we still do not have a complete picture of the electronic properties of amorphous and glassy semiconductors, we have a fairly convincing case for the transport properties of a–Si:H which can be summarized by the following: the transport in quasi-equilibrium is dominated by states near E_C. The microscopic transport parameters of these states are well described by phonon-induced delocalization, Eqs. (30) and (33). We find a nearly universal value for the microscopic prefactor of the conductivity $\sigma_0 \cong 150\Omega^{-1}\mathrm{cm}^{-1}$. Note that the same prefactor applies for hole conduction in the valence band as well as for conduction by electrons in the conduction bands, as our theory leaves no space for a substantial difference between electrons and holes. The different band mobilities as e.g. derived from time-of-flight experiments can thus not be explained by the properties of carriers close to E_C and must be ascribed to the different tail DOS responsible for the multiple trapping process.

The observed apparent prefactor σ_0^* of the dc conductivity as derived by an Arrhenius plot is different from the microscopic prefactor because of the statistical shift of the Fermi energy. This statistical shift gives rise to an apparent prefactor following a Meyer–Neldel-rule for wide class of density of states distributions.

For undoped state-of-the-art a–Si:H the observed values for E_Q^* can be as small as 50 meV, a value compatible with the theory for phonon-induced delocalization. The fact that the observed values for E_Q^* are often much larger than 50 meV can be ascribed to the effect of a long-ranged random potential. For this potential, we have given a rather accurate estimate for the case that the potential is caused entirely by the residual inhomogeneity of a random distribution of donors. There will be other sources for a long-ranged potential in amorphous semiconductors like structural and growth inhomogeneities, inhomogeneities of the hydrogen distribution, etc. but for these the random potential amplitude and typical fluctuation length are hardly known.

A mystery remains: the sign of the Hall effect is anomalous and we have no real explanation for this. One might find some solace by noting that the normal Hall effect arises from the Lorentz force acting on free particles and that the carriers envisaged in our theoretical model are not at all free particles. But this excuse offers no answer for the question why the sign of the Hall effect is always anomalous for amorphous, but regular for crystalline semiconductors with arbitrary disorder.

Acknowledgments

We are grateful to Dr D. Emin and Prof. M. Pollak for a stimulating discussion about parts of this chapter.

References

[1] P.W. Anderson, *Phys. Rev.* **109**, 1492 (1958).

[2] B. Kramer and A. MacKinnon, *Reports on Progress in Physics* **56**, 1469 (1993).

[3] W. Götze, *J. Phys.* **C12**, 1279 (1979); W. Götze, *Phil. Mag.* **B43**, 219 (1981); D. Belitz, A. Gold and W. Götze, *Z. Phys.* **B44**, 273 (1981); D. Belitz and W. Götze, *Phys. Rev.* **B28**, 5445 (1983).

[4] D. Vollhardt and P. Wölfle, *Phys. Rev. Lett.* **45**, 482 (1980); *Phys. Rev.* **B22**, 4666 (1980).

[5] L. Bányai, *Physique de Semiconducteurs*, ed. M. Hulin (Dunod, Paris, 1964) p. 417.

[6] N.F. Mott, *Adv. Phys.* **16**, 49 (1967).

[7] N.F. Mott, *Phil. Mag.* **26**, 1015 (1972).

[8] G.A. Thomas, Y. Ootuka, S. Katsumoto, S. Kobayashi and W. Sasaki, *Phys. Rev.* **B25**, 4288 (1982).

[9] E. Abrahams, P.W. Anderson, D.C. Liciardello and T.V. Ramakrishnan, *Phys. Rev. Lett.* **42**, 673 (1979).

[10] H. Overhof and P. Thomas, *Electronic Transport in Hydrogenated Amorphous Semiconductors*, Springer Tracts in Modern Physics **114** (Springer, Berlin, Heidelberg, New York, 1989).

[11] N.F. Mott and E.A. Davis, *Electronic Processes in Non-Crystalline Materials* (Clarendon Press, Oxford, 1971 and 1979).

[12] P.N. Butcher, *Amorphous Solids and the Liquid State*, eds. N. Marsh, R.A. Street and M. Tosi (Plenum Press, New York, 1985) p. 311.

[13] M. Cutler and N.F. Mott, *Phys. Rev.* **181**, 1336 (1969).

[14] N.K. Hindley, *J. Non-Cryst. Solids* **5**, 17 and 31 (1970).

[15] L. Friedman, *J. Non-Cryst. Solids* **6**, 329 (1971).

[16] H. Overhof and K. Maschke, *J. Phys.: Condens. Matter* **1**, 431 (1989).

[17] H. Müller and P. Thomas, *Phys. Rev. Lett.* **51**, 702 (1983).

[18] H. Müller and P. Thomas, *J. Phys.* **C14**, 5337 (1984).

[19] P. Fenz, H. Müller, H. Overhof and P. Thomas, *J. Phys.* **C18**, 3191 (1985).

[20] H. Aoki, *Physica* **A114**, 538 (1982); *J. Phys.* **C16**, L 205 (1983).

[21] M. Kikuchi, *J. Non-Cryst. Solids* **59 & 60**, 25 (1983).

[22] M. Grünewald and P. Thomas, *Phys. Stat. Sol. (b)* **94**, 125 (1979); M. Grünewald, P. Thomas and D. Würtz, *Phys. Stat. Sol. (b)* **94**, K1 (1979).

[23] S.D. Baranovskii, T. Faber, F. Hensel and P. Thomas, *J. Non-Cryst. Solids* **198–200**, 222 (1996).

[24] G. Bergmann, *Phys. Rep.* **107**, 1 (1984).

[25] H. Böttger and V.V. Bryksin, *Hopping Conduction in Solids* (Akademie Verlag, Berlin, 1985).

[26] W. Brenig, G.H. Döhler and P. Wölfle, *Z. Phys.* **246**, 1 (1971); *Z. Phys.* **258**, 381 (1973).

[27] B. Movaghar, *J. de Physique* **42**, C4-73 (1981); B. Movaghar and W. Schirmacher, *J. Phys.* **C14**, 859 (1981); B. Movaghar, M. Grünewald, B. Pohlmann, D. Würtz and W. Schirmacher, *J. Stat. Phys.* **30**, 315 (1983).

[28] A. Miller and E. Abrahams, *Phys. Rev.* **120**, 745 (1960).

[29] D. Würtz and P. Thomas, *Phys. Stat. Sol. (b)* **88**, K73 (1978).

[30] T. Holstein, *Ann. Phys.* **8**, 343 (1959).

[31] D. Emin, *Phys. Rev. Lett.* **32**, 303 (1974); *Adv. Phys.* **24**, 305 (1975).

[32] S. Nagel and P. Thomas, *Proc. 7th Int. Conf. on Amorphous and Liquid Semicond.*, ed. W.E. Spear (Edinburgh, 1977) p. 266.

[33] P. Thomas and D. Würtz, *Phys. Stat. Sol. (b)* **86**, 541 (1978).

[34] L. Kirchhoff, *Annalen der Physik und Chemie* **64**, 497 (1845); *Annalen der Physik und Chemie* **72**, 497 (1847).

[35] V. Ambegaokar, B.I. Halperin and J.S. Langer, *Phys. Rev.* **B4**, 2612 (1971).

[36] M. Pollak, *J. Non-Cryst. Solids* **11**, 1 (1972).

[37] S. Kirkpatrick, *Rev. Mod. Phys.* **45**, 574 (1973).

[38] B.I. Shklovskii and A.L. Efros, *Electronic Properties of Doped Semiconductors*, Springer Ser. Solid-State Sciences **45** (Springer, Berlin, Heidelberg, New York, 1984).

[39] N.F. Mott, *Phil. Mag.* **19**, 835 (1969).

[40] I.P. Zvyagin, *Phys. Stat. Sol. (b)* **58**, 443 (1973).

[41] H. Overhof, *Phys. Stat. Sol. (b)* **67**, 709, (1975); *Festkörperprobleme XVI*, ed. J. Treusch (Vieweg, Braunschweig, 1976) p. 239.

[42] W. Beyer, PhD Thesis, Marburg, 1975 (unpublished); W. Beyer, J. Stuke and H. Wagner, *Phys. Stat. Sol. (a)* **30**, 231 (1975).

[43] M.-L. Theye, A. Gheorghiou, T. Rappenau and A. Lewis, *J. de Physique* **41**, 1173 (1980).

[44] F.R. Shapiro and D. Adler, *J. Non-Cryst. Solids* **74**, 189 (1985).

[45] D. Monroe, *Phys. Rev. Lett.* **54**, 146 (1985).

[46] B.I. Shklovskii, E.I. Levin, H. Fritzsche and S.D. Baranovskii, *Advances in Disordered Semiconductors* **3**, ed. H. Fritzsche (World Scientific, 1990) p. 161.

[47] S.D. Baranovskii, T. Faber, F. Hensel and P. Thomas, *J. Phys.: Cond. Matter* **9**, 2699 (1997).

[48] M. Pollak, *Discuss. Faraday Soc.* **50**, 11 (1970).

[49] H. Overhof and P. Thomas, *Electronic Phenomena in Non-Cryst. Semicond.*, ed. B.T. Kolomiets (Nauka, Leningrad, 1976) p. 107.

[50] G. Srinivasan, *Phys. Rev.* **B4**, 2581 (1971).

[51] A.L. Efros and B.I. Shklovskii, *J. Phys.* **C8**, L49 (1975).

[52] A. Möbius and M. Pollak, *Phys. Rev.* **B53**, 16197 (1996). A. Möbius and P. Thomas, *Phys. Rev.* **B55**, 7460 (1997).

[53] M. Pollak, *Phil. Mag.* **B65**, 657 (1992); *Phil. Mag.* **B42**, 781 (1980). M. Pollak and M. Mochena, *The Physics of Semiconductors*, eds. P. Jiang and H.Z. Zheng (World Scientific, 1992) p. 137; A. Perez–Garrido, M. Ortuno, E. Cuevas, J. Ruiz and M. Pollak, *Phys. Rev.* **B55**, R8630 (1997).

[54] I. Shlimak, M. Kaveh, R. Ussyshkin, V. Ginodman, S.D. Baranovskii, P. Thomas, H. Vaupel and R.W. Van der Heijden, *Phys. Rev. Lett.* **75**, 4764 (1995).

[55] D.L. Shepelyanski, *Phys. Rev. Lett.* **73**, 2607 (1994); Y. Imry, *Europhys. Lett.* **30**, 405 (1995).

[56] R.A. Römer and M. Schreiber, *Phys. Rev. Lett.* **78**, 515 (1997).

[57] J.H. Davies, P.A. Lee and T.M. Rice, *Phys. Rev.* **B29**, 4260 (1982); M. Grünewald, B. Pohlmann, L. Schweitzer and D. Würtz, *J. Phys.* **C15**, L1153 (1982).

[58] D. Emin and M.-N. Bussac, *Phys. Rev.* **B49**, 14290 (1994).

[59] D. Emin, *Phys. Rev.* **43**, 11720 (1991).

[60] D. Emin, *Phys. Rev.* **53**, 1260 (1996).

[61] H. Mell, *Amorphous and Liquid Semiconductors*, eds. J. Stuke and W. Brenig (Taylor and Francis, 1974) p. 203

[62] A. Lewis, *Phys. Rev.* **B13**, 2565 (1976).

[63] M. Ortuno and M. Pollak, *J. Non-Cryst. Solids* **59 & 60**, 53, (1983).

[64] P.G. LeComber, A. Madan and W.E. Spear, *J. Non-Cryst. Solids* **11**, 219, (1972).

[65] P. Nagels, R. Callaerts and M. Denayer, *Amorphous and Liquid Semiconductors*, eds. J. Stuke and W. Brenig (Taylor and Francis, 1974) p. 867.

[66] W. Meyer and H. Neldel, *Z. Techn. Phys.* **12**, 588 (1937).

[67] P. Irsigler, D. Wagner and D.J. Dunstan, *J. Phys.* **C16**, 6605 (1084).

[68] D.L. Staebler and C.R. Wronski, *Appl. Phys. Lett.* **31**, 292 (1977); *J. Appl. Phys.* **51**, 3262 (1980).

[69] W. Beyer, H. Mell and H. Overhof, *Amorphous and Liquid Semiconductors* ed. W.E. Spear (CICL Edinburgh, 1977) p. 328; W. Beyer and H. Mell, same volume, p. 333.

[70] F. Ghiassy, D.I. Jones and A.D. Steward, *Phil. Mag.* **B52**, 139 (1985).

[71] W. Beyer and H. Overhof, *Solid State Comm.* **31**, 1 (1979).

[72] D.A. Arnoldussen, C.A. Menezes, Y. Nagakawa and R.H. Bube, *Phys. Rev.* **B9**, 3377 (1974).

[73] C.H. Seager, D. Emin and R.K. Quinn, *Phys. Rev.* **B8**, 4746 (1973).

[74] G. Jušca, S. Vengris and J. Viščascas, *Amorphous and Liquid Semiconductors*, eds. J. Stuke and W. Brenig (Taylor and Francis, 1974) p. 363.

[75] E. Mytilineou and E.A. Davis, *Amorphous and Liquid Semiconductors*, ed. W.E. Spear (CICL Edinburgh, 1977) p. 328.

[76] D. Hauschildt, W. Fuhs and H. Mell, *Phys. Stat. Sol. (b)* **111**, 171 (1982).

[77] O.R. Reilly and W.E. Spear, *Phil. Mag.* **B38**, 295 (1978).

[78] P.G. LeComber, D.I. Jones and W.E. Spear, *Phil. Mag.* **36**, 1173 (1977).

[79] J. Kakalios and R.A. Street, *Phys. Rev.* **B34**, 6014 (1986).

[80] H. Overhof and W. Beyer, *J. Non-Cryst. Solids* **35 & 36**, 375 (1979).

[81] C E. Nebel, *J. Non-Cryst. Solids* **137 & 138**, 395 (1991).

[82] W.E. Spear, D. Allan, P.G. LeComber and A. Ghait, *Phil. Mag.* **B41**, 419 (1980).

[83] P. Irsigler, D. Wagner and D.J. Dunstan, *J. Phys.* **C16**, 6605 (1983).

[84] P.G. LeComber and W.E. Spear, *Amorphous Semiconductors*, ed. M.H. Brodsky, Topics in Applied Physics **36** (Springer, Heidelberg 1979) Chap. 9.

[85] R.A. Street, *Hydrogenated Amorphous Silicon* (Cambridge University Press, Cambridge, 1991).

[86] D. Wagner, P. Irsigler and D.J. Dunstan, *J. Non-Cryst. Solids* **59 & 60**, 413 (1983).

[87] R. Schumacher, P. Thomas, K. Weber, W. Fuhs, F. Djamdji, P.G. LeComber and R.E.I. Schropp, *Phil. Mag.* **B58**, 389 (1988).

[88] G. Willeke, *Amorphous and Microcrystalline Semiconductor Devices*, ed. J. Kanicki (Artech House, London 1991) p. 55.

[89] A. Yelon and B. Movaghar, *Phys. Rev. Letters* **65**, 618, (1990).

[90] M. Kondo, Y. Chida and A. Matsuda, *J. Non-Cryst. Solids* **198–200**, 178 (1996).

[91] H.M. Branz and M. Silver, *Phys. Rev.* **B42**, 7420 (1990).

[92] B.I. Shklovskii and A.L. Efros, *Fiz. Tekh. Poluprov.* **4**, 305 (1970) [English translation: *Soviet Physics Semic.* **4**, 247 (1970)]; *Z. Exp. Theor. Fiz.* **60**, 867 (1971) [English translation: *Soviet Physics JETP* **33**, 468 (1971)]; *Z. Exp. Theor: Fiz.* **62**, 1156 (1972) [English translation: *Soviet Physics JEPT* **35**, 610 (1972)].

[93] J. Jäckle, *Phil Mag.* **B41**, 681 (1980).

[94] R.A. Street, D.K. Biegelsen and J.C. Knight, *Phys. Rev.* **B24**, 969 (1981).

[95] J.A. Howard and R.A. Street, *Phys. Rev.* **B44**, 7935 (1991).

[96] P.W. Anderson, *Phys. Rev. Lett.* **34**, 953 (1975); *J. Physique (Paris)* **C4**, 339 (1976).

[97] R.A. Street and N.F. Mott, *Phys. Rev. Lett.* **35**, 1293 (1975); N.F. Mott, E.A. Davis and R.A. Street, *Phil. Mag.* **32**, 961 (1975).

[98] H. Overhof and W. Beyer, *Phil. Mag.* **B43**, 433 (1981).

[99] R.A. Street, J. Kakalios and M. Hack, *Phys. Rev.* **B38**, 5603 (1988).

[100] D.M. Goldie, W.E. Spear and E.Z. Liu, *Phil. Mag.* **B62**, 509 (1990).

[101] B.V. Roedern and A. Madan, *Phil. Mag.* **B63**, 293 (1991).

[102] H. Dersch, 1983, PhD Thesis, Marburg (unpublished).

[103] W. Beyer, H. Mell and H. Overhof, *J. Physique (Paris)* **42**, **C2**, 103 (1981).

[104] W. Beyer and H. Overhof, *Semiconductors and Semimetals*, eds. R.K. Willardson and A.C. Beer **21c** (Academic Press, New York, 1984) p. 257.

[105] H. Overhof, *Mat. Res. Soc. Symp. Proc.* **258**, 681 (1992).

[106] K. Winer, I. Hirabayashi and L. Ley, *Phys. Rev. Letters* **60**, 2697 (1988).

[107] S. Aljishi, S. Jin and L. Ley, *J. Non-Cryst. Solids* **137 & 138**, 387 (1991).

[108] H.M. Dyalsingh and J. Kakalios, *Phys. Rev.* **B54**, 7630 (1996).

9
Molecular Dynamics Simulations of Network Glasses

DAVID A. DRABOLD

*Department of Physics and Astronomy, Ohio University,
Athens, Ohio 45701-2979 USA*
drabold@ohio.edu

Contents

1. Introduction and Background

1.1. *History and use of MD*

The idea of *molecular dynamics* (MD) is a very old and fundamental notion of condensed matter physics. The essential idea is very simple: if we know the interaction potential between atoms and some initial conditions, then for a classical system we can use $\mathbf{F} = m\mathbf{a}$ to obtain the full time development of the atomic coordinates by numerically integrating the equations of motion. Even simple considerations of these ideas without using much detail leads to kinetic theory. Naturally, classical statistical mechanics, a cornerstone of science, originates substantially in the statistical consequences of the equations of motions and their integrals. In fact, a goal for workers on the foundations of statistical mechanics is to derive the laws of statistical mechanics from nothing more than the equations of motion: an example is Liouville's theorem.

For a classical system, the intuitive appeal of an MD simulation is the feeling that a complex experiment can be implemented on a computer. In principle, such a calculation proceeds without information loss: the coordinates and velocities of each atom are specified at any instant of time, and with this information comparison to any experiment imaginable should be possible. We will repeatedly stress in this paper just how far an MD simulation is from this ideal because of shortcomings of interatomic potentials, and length/time scales many orders of magnitude removed from experiment; nevertheless, my essential conclusion that MD is a very powerful tool in the theorist's arsenal for studies of disordered systems. This paper will describe when such calculations are appropriate, and otherwise.

Perhaps the first many-particle classical MD simulation was carried out by an astronomer, Erik Holmberg in 1941 [1], who was concerned with energy transfer between colliding galaxies. His idea was to exploit the identical radial $(1/r^2)$ dependence of the light intensity and the gravitational force, to enable simulations based upon an "optical analog" of the gravitational interaction. Holmberg constructed a remarkably ingenious apparatus in which two groups of 37 lamps each (each lamp representing a star, each group representing a galaxy) interacted optically, and the light intensity at each lamp ("star") was measured with photocells. Since the intensity was proportional to the interstellar forces, the time evolution of a galactic collision could then be simulated, and Holmberg was able to see spiral arms and learn something about capture probability. While there have been many MD simulations since, I wonder if any have been as clever as this one [2]!

Only with the advent of computers since the Second World War has it become possible to directly integrate the equations of motion for systems with large numbers of particles. Probably the best known early work is that of Alder and Wainwright [3] for hard spheres, and the simulations with the comparatively realistic Lennard–Jones potential due to Rahman [4] a few years later. The field has steadily grown in activity and importance since these pioneering studies and is now almost a third paradigm of physics, intermediate between experiment and "conventional" theory, heavily relying on and contributing to both.

The synergism between MD simulations and experiment merits a little more reflection. For amorphous and glassy systems MD is absolutely essential since it is a means to a *structural model* (more bluntly — "where the atoms are") which can never be completely inferred from experiments. The experiments nevertheless let us know whether the model is sensible: for example, an experimental static structure factor can be viewed as a "sum rule" any theoretical model must be in agreement with if it is to be seriously considered. The logic is therefore that *if* MD provides a model acceptably consistent with *all* the experiments we believe, *then* the model becomes worthy of additional study to investigate new phenomena of the glassy state, and make predictions beyond what is experimentally known or otherwise theoretically understood. It is in this mode of use that MD has come to revolutionize the study of disordered systems.

1.2. *The role of the potential*

In the field of network glasses, the MD method has actually evolved in three related, but rather philosophically different ways, each with its own range of validity. The complexity of the interatomic potential defines the boundary between the approaches.

(1) Use of MD with very simple potentials, which are not "realistic" for any physical system, but which include the "generic character" of a reasonable interaction (a repulsive core and some kind of weak attractive tail, possibly long range — e.g. power law decay). Prodigious numbers of atoms and relatively long simulations are possible for such potentials, and one can seek to grasp some characteristics of non-equilibrium processes like glass formation.

(2) Empirical interatomic potentials constructed to model a given material. Here, one uses more or less ingenious functional forms built to mimic the configurational (potential) energy of a given type of glass. "Empirical" means that the scheme requires input from an external source: experiment or better theory. The venerable, but still useful Stillinger–Weber [5]

potential for Si is the classic of this approach, and consists of the sum of a pair and three-body term; the later unapologetically built to favor tetrahedral bonding. More recent empirical models can mimic many particle correlations in more subtle and reliable ways [7, 8]. The fanciest empirical models are based on empirical electronic structure (tight-binding) Hamiltonians.

(3) *"Ab initio"* calculations of the interatomic interactions. Here, one starts from a more basic view that the complexity of the interatomic potential originates in the details of the electronic structure of the material. In this approach, one tries to accurately approximate the electronic eigenstates of the model without appeal to external information. As the reader will recall, the quantum many-body problem presents a formidable challenge, even with serious approximations. Acceptably accurate schemes do exist however, and have the key feature that they *do not depend upon fits to external information* (*e.g. experiments*): this is essential for studies of new materials very different than those for which adequate empirical potentials have been formulated, as for example amorphous GaN [9]. A serious *shortcoming* of the *ab initio* approach is that it is much slower than its competitors (1) and (2), and therefore must be used sensibly often in conjunction with (2) above to make optimal progress. Also, there are many "theoretical knobs" (approximations) in any *ab initio* simulation, and it takes experience to know what approximations are acceptable and which are not (this depends critically on the questions being addressed). Certainly, improperly conceived *ab initio* simulations can do worse than a proper empirical potential calculation.

1.3. *Scope of the method*

Thanks to a happy coincidence of continuing algorithmic breakthroughs and ever faster/cheaper computers, the frontier for MD is ever receding, and more and more systems are proving to be susceptible to MD modeling. There have been important and very different breakthroughs in both *ab initio* [10] and empirical methods [11] enabling computational cost to scale linearly with the number of atoms in the system. At the same time, accuracy continues to improve on both fronts. So while there are severe limits on what MD can accomplish, rapid progress is being made on most of the outstanding problems.

The MD methods are often used directly to simulate a glass (by simulated quenching from an equilibrated liquid). It is sometimes asserted that a MD approach to obtaining a topologically disordered model is intrinsically superior to other methods which are entirely based on the principle of "the end justifies the means", since the MD method provides an explicit trajectory. In fact, this argument by itself is not at all convincing, since the time scales (and sometimes the potential itself) in the MD simulation is qualitatively different from Nature. For a system like a–Si, a simulated quench from an equilibrated liquid always freezes in an unphysical concentration of electronically active defects, because the extreme quench rates of MD freeze in an unrealistic concentration of strained and over-coordinated conformations: remnants of the metallic-coordinated liquid Si phase (exceedingly different from the essentially tetrahedral character of the amorphous matrix). Thus, a key feature determining the realism of an MD quench model of a glass is the similarity (or lack thereof) between the liquid phase and the glass. It is undoubtedly the case that this is important in Nature too: good glass formers (from the melt) like SiO_2 and $GeSe_2$ are known to have liquid topology much like the glass [12]; whereas it is hard or impossible to make a–Si with a sensible defect density by a quench from liquid Si. Thus, while Angell's emphasis on the importance of the crystalline phases to glass formation [13] is well founded, one cannot neglect the topology of the liquid (relative to the glass and crystal) either.

1.3.1. *Use of a priori information*

Conventional "cook and quench" MD may be used to model complex systems, provided that one uses *a priori* information. For example, to properly model a glass which is known to exhibit strong chemical ordering (for example, $GeSe_2$ [12]), it is foolish to start with an initial configuration which is remote from this order. In other words, we cannot expect an MD procedure with its picosecond time scales to fully sort out the detailed chemical ordering given only the right number of each atomic species in the right volume at the first time step. Not only would substantial diffusion be required, but also "diffusion to the right place" — e.g. to where an atom would yield proper chemical order. Thus, for our work with $GeSe_2$ [12, 14, 15], we started with a periodic lattice with the chemical order built in (Ge surrounded by Se, Se by Ge); this structure then was "melted" so that the original lattice structure was forgotten *but* the chemical ordering initially in place by *fiat* was largely maintained (presumably because there is an energetic reward for heteropolar bonding). The net result was a model in pleasing agreement with most of the experimentally available information, and some new observations based upon the models.

1.3.2. *Appraising a model*

Obviously a critical element of MD simulations of complex materials involves criteria for whether a model is acceptable or otherwise. The mission, of course, is to "match" as many experiments as possible at the same time. It is cause for great concern if a model matches some experiments and not others which we have equal confidence in. Thus, global quantities like the radial distribution function (or static structure factor), the (total) electronic density of states or the (total) vibrational power spectrum offer global information averaged over of order 10^{22} atoms. Such spectra must be faithfully reproduced by a model. The electronic and vibrational states will typically exhibit "gaps", such that there are few or no states in the "gap" energy range. To the extent that such states do exist, they originate in some type of defective structure in the network, sufficiently different (chemically and/or topologically) to be pushed out of the bands into the gap. Agreement or discrepancy between gap features will usually imply *spatially local* information involving a modest number of atoms. Such isolated states (electron or vibrational) are usually localized (in the technical sense of Anderson localization [16]). Properly reproducing this local information is important for example, if one is interested in transport, and the defects are electron states near the Fermi level. Then the concentration, and position of the states will determine the entire transport process. To be specific, the literature is replete with models of a–Si with ridiculous concentrations of defects (10% or even more), with satisfactory radial distribution functions. Such a model is quite unacceptable, since the concentration of defect states in the gap is about 1 in 10^{3-4}.

2. MD Method

There is a considerable technical lore available for "classical" (based upon assumed interaction potentials) MD simulations. There are several books and review articles written on the "nuts and bolts". Here, I will survey the kinds of calculations that are available and point the interested reader to the appropriate literature for derivations and details. There is an unfortunate tendency, especially displayed by workers using *ab initio* methods, to neglect the vast amount of excellent work performed with empirical potentials, and many important generic findings about how best to implement simulations, how to deal with artifacts of simulations etc. My favorite source of information on these questions is the delightful book by Allen and Tildesley [17]. See also the recent work of Frenkel and Smit [18].

2.1. *Equations of motion*

As the name suggests, MD implies *time evolution*: the time evolution of Newton's equation $\mathbf{F} = m\mathbf{a}$; a coupled system of nonlinear ordinary differential equations (in general depending on $3N$ coordinates and requiring $3N$ velocities as initial conditions). The forces $\mathbf{F_i} = -\nabla_i \Phi(\mathbf{R_1}, \mathbf{R_2}, \ldots \mathbf{R_N})$ where Φ is a potential energy function (not necessarily expressed in any closed form, as from an *ab initio* calculation of the total energy Φ). There are many methods available for integrating the equations of motion, ranging from the simple first order Taylor method, which is nothing more than using $\delta\mathbf{R_i} = \mathbf{V_i}\delta t$, and $\delta\mathbf{V_i} = \mathbf{a_i}\delta t = \mathbf{F_i}\delta t/m$, and δt is a small time interval, the "time step", typically of order 1fs (1fs $= 10^{-15}$ sec). Low order methods like this are sometimes used, especially for cases where potential energies and forces are obtained from lookup tables or are otherwise not very smooth. *If* the force is a very smoothly changing function of interatomic coordinates, then it pays to use a *higher order method*, which extrapolates $\mathbf{R}$ and $\mathbf{V}$ to higher order in δt; this enables a longer time step, while still faithfully tracking the equations of motion. Allen and Tildesley [17] discuss a predictor-corrector method, which should be ideal for most applications. Some authors [19] particularly encourage the use of the Bulrisch–Storer method for the smooth case.

A simple but dependable means of gauging the reliability of a particular quadrature scheme is to check numerically the value of conserved quantities (like the total energy) in a microcanonical thermal simulation. If a time step is selected to be too long to properly track the equations of motion, the energy will not be constant.

2.2. *Energy minimization and equilibration*

To find a set of coordinates which minimize the total energy (e.g. such that the forces vanish), one can employ one of many strategies. For cases where *any* nearby energy minimum conformation is sufficient, a steepest descent quench may be sufficient. In this approach, atomic positions are iteratively changed to push the coordinates to the closest accessible energy minimum in configuration space, by changing coordinates $\mathbf{R_i}$ to iteratively follow the local downhill gradient, $\mathbf{F_i} = -\nabla_i\Phi$, until a set of coordinates $\mathbf{R_i}$ is obtained such that all $\mathbf{F_i} = 0$. A better method [19] for getting to a minimum more rapidly is the "conjugate gradient" scheme, which determines an optimal sequence of directions in which to displace the atoms to get to the minimum efficiently. The conjugate gradient method has the advantage that it requires the storage only of a few vectors of length $3N$, for N the number of atoms. A result of considerable *theoretical* interest is that the conjugate gradient scheme is guaranteed to find the exact minimum of a *quadratic form* potential in N dimensions in N steps. There is no such guarantee for a general potential, of course. An alternate group of "Newton" methods can in principle be applied for energy minimizations, but the storage requirements are more stringent (the Hessian matrix, which involves $O(N^2)$ elements must be saved), and in general the method is no more efficient than conjugate gradient anyway.

While the question of energy minimization is distinct from that of thermal MD, the two are closely related. In fact, the usual "MD" approach to forming a model of a glass (equilibrating a liquid and rapidly quenching it to $T = 0$) is usually accomplished by a velocity rescaling scheme. The simplest approach is due to Berendsen and coworkers [17, 20]. Suppose that one wishes to drive a system to some predetermined temperature T. Then one can simply rescale all the velocities according to the rule:

$$\chi = \sqrt{1 + \delta t(T/T_i - 1)/t_r} \tag{1}$$

where, T_i is the current instantaneous temperature, δt is the time step, and t_r is a relaxation time, which determines how rapidly the system is to be driven to the target temperature T. While the method is very simple, it works quite well for equilibrating a system to some desired temperature, or for quenching a liquid into a disordered solid. In general, one cannot make t_r too long; for quenches from a liquid to a glass, I have found that $t_r \approx 500 fs$ is both computationally tolerable and can produce good results for g–GeSe$_2$. It is obviously desirable to probe the effect of the cooling rate on the final structure obtained.

2.3. *Deeper or global minima*

It is in general a famous outstanding problem of computational physics to reliably determine a *global*, or at least deeper *local* minimum of the potential energy function for large systems. For our interests here, we can contrast the physical glass formation process which occurs sufficiently slowly in Nature to enable much annealing, and therefore a rather "deep" (but local!) minimum in the potential energy to a "shallow" minimum that comes from an extremely rapid simulated quench.

2.3.1. *Simulated annealing*

A method somewhat similar in spirit to the velocity rescaling scheme above is the simulated annealing method of Kirkpatrick *et al* [21]. Rather than using (velocity rescaled) MD to determine the atomic trajectories, a set of rules (the exact choice of which is investigator and problem dependent) is invoked according to which atoms can explore a larger volume of configuration space before finally getting trapped in some (probably local) minimum. A typical process is to accept new coordinates according to a Metropolis [22] (Monte Carlo) rule: (1) if a change lowers the total energy keep that change, (2) if the change increases the total energy, keep that change with probability $\propto \exp(-\Delta E/k_B T_A)$, where ΔE is the (positive) energy difference between the original coordinates and the "trial" coordinates, k_B is Boltzmann's constant and T_A is the "annealing temperature". T_A is analogous to a true annealing temperature in the sense that the higher one chooses T_A, the more freedom the system has to explore the configuration space.

2.3.2. *Genetic algorithms*

In a related vein, so called *genetic algorithms* have recently been applied to complex optimization problems in physics including simple problems involving disorder [23]. Ho [24] has recently applied the method to essentially random (in an appropriate volume) collections of 60 carbon atoms and found that the method can consistently find the icosahedral fullerene ("bucky ball") structure as the deepest minimum when used with a tight-binding Hamiltonian and some rules about how the "evolution" proceeds. While this result is very impressive, its relevance to glasses is not yet clear both because even for 60 atoms in free space huge amounts of computer time are needed, and also the determination of "rules" is substantially an art form.

2.3.3. *Activation-relaxation technique*

Away from the melting temperature, many materials explore a sequence of metastable states separated by energy barriers much larger than $k_B T$, the typical excitation energy scale at the atomic level. For long periods of time, the configuration vibrates around a given metastable state, then, due to rare energy fluctuations, it will find enough energy to jump over a barrier and move to a new metastable state. The dynamics of these materials is therefore dominated by the rates controlling the jumps from one metastable state to another.

It is possible to do better by directly focusing on the rare events. To a first approximation, these are fully determined by the activation energy, i.e. the energy needed to bring a configuration from a local configurational minimum to a nearby saddle point. A low temperature characterization of the dynamical properties of a disorder system can therefore be made by reducing the configuration energy landscape to a network of local minima connected by paths going through first order saddle points. Recently, Barkema and Mousseau have proposed such a procedure, the activation-relaxation technique (ART), which provides a local prescription for moving from one minimum to another one with a trajectory passing by a shared saddle-point (event) [25].

The advantage of ART is that it defines moves directly in the configurational energy landscape, which really controls the dynamics, instead of trying to map events into complicated real-space moves. ART is independent of the details of the interaction potential and the specificity of a given material and requires only a local and continuous description of the energy landscape.

An event in ART is defined as a move from a local energy minimum $\mathbf{M}^{(0)} \equiv (\mathbf{x}_1^{(0)}, \ldots, \mathbf{x}_N^{(0)})$ to another nearby minimum $\mathbf{M}^{(1)} \equiv (\mathbf{x}_1^{(1)}, \ldots, \mathbf{x}_N^{(1)})$ following a two-step process mimicking a physical activated processes:

(1) the **activation** during which a configuration is pushed from a local minimum to a nearby saddle-point; and

(2) the **relaxation** which brings the configuration from this saddle-point to a new local minimum.

The details of how the saddle point is reached, and other technical points concerning the implementation may be found in the literature [25].

ART has already been applied with success to a series of static problems in metallic glasses, a–Si [25] and a–GaAs [27, 28]. A slightly different version of the algorithm was also developed independently by Doye and Wales and applied to map the full energy landscape of a 13-atom Lennard–Jones cluster [29].

Because it necessitates only the calculation of the force, ART is fully scalable with the size of the systems studied. In general, about a 1000 force evaluations per event are necessary. This means that ART will be useful for activated processes with barriers significantly larger than the temperature of the material studied. In this case, though, it provides a unique tool for the description of rare events such as diffusion and relaxation mechanisms in disordered materials.

2.4. *Alternate dynamics*

Allen and Tildesly [17] discuss methods to perform Lagrangian constrained dynamics (or relaxations) of systems such that a particular symmetry is enforced through the simulation.

The simplest approach to MD simulation is to work in the microcanonical (constant NVE) ensemble. Other, perhaps more realistic ensembles are the canonical (NVT), isothermal/isobaric (NPT) and even the grand canonical (μVT). Allen and Tildesly [17] also give a complete discussion of these alternate ensembles. The most sophisticated approach to a canonical simulation [30] is to introduce a new degree of freedom to represent the "heat bath" and allow heat flow between the physical degrees of freedom (e.g. the atomic coordinates and velocities) and the fictitious heat bath. A "Nose' thermostat" faithfully produces a trajectory which samples from the canonical distribution function.

Virtually all *ab initio* MD simulations are carried out at constant volume (or some crude "adiabatic" volume shift from liquid volume to glass volume in the course of a simulated quench to form a glass). The state of the art for isobaric simulation is the work of Parrinello and Rahman, who also enable a changing box shape [31]. An exception to this rule is the work of Wentzkovitch and coworkers [32].

Vashishta [33] has fully implemented these more physical ensembles in simulations of glasses, and simulations of phase changes such as amorphization and densification, with impressive results which fully exploit the possibility of performing *very* large scale MD simulations with empirical potentials. This is a reminder that the most accurate method (an *ab initio* method) is not necessarily the right choice, in this case, if the choice of ensemble is important to the final structure obtained.

2.5. *Modeling infinite systems: Periodic boundary conditions*

To model a bulk topologically disordered system, it is usually not a reasonable approach to simply study a cluster, terminated by vacuum. For example, if one was studying a tetrahedral semiconductor like a–Si, this would produce *many* electronic states in the optical gap due to the dangling bonds at the surface. Since some of these states would be occupied in general, this would change the energetics, forces and topology of the system especially near the surface: an unphysical boundary. One way out is to passivate the dangling bonds with hydrogen, although this is not so easy in practice as it sounds for surface structures which are not near ideal dangling bonds.

Of course if one is interested in the surface of an amorphous material, the truncation to vacuum is physical (we describe a calculation on an amorphous C surface later in this article).

The usual choice to avoid truncation effects is to apply Born–von Karman (periodic) boundary conditions. All such models are then technically crystals, albeit with a large unit cell. An essential part of the analysis of any MD simulation is determining how the periodic boundary conditions manifest themselves in predictions of physical quantities. Vashishta [34] has studied this point, and there is plenty of discussion of size and boundary condition artifacts in the classical MD literature. We [14] have observed in a 63 atom model of g–GeSe$_2$ that the static structure factor is modulated by an unphysical "ringing" presumably connected with the relatively small box size used. When the calculation was repeated in a 216 atom cell (at the same density), the ringing virtually disappeared [15]. The "first sharp diffraction peak" (near $k = 1$ Å^{-1}) also agreed much more closely with experiment for the larger cell. Clearly, low $\mathbf{k}$ features are most affected by small box sizes. In a similar way, periodic boundary conditions enable meaningful studies of spatial correlations only up to half the box size.

More subtle problems also arise with periodic boundary conditions for calculations based on an electronic structure calculation (either tight-binding or *ab initio*); to compute the electronic contribution to the total energy and forces one needs to approximately integrate over the first Brillouin zone and sum over the occupied energy bands. It is usual to neglect this and pretend that there is no $\mathbf{k}$ dispersion in the energy bands and evaluate the energy bands only at $\mathbf{k} = 0$. The validity of this approximation, *particularly for forces* needs to be checked in small (roughly speaking less than 100 atom) cells. We have discussed this in more detail elsewhere [35]. Related issues arise in calculations of the vibrational states; we discuss this later.

3. The Interatomic Interactions

3.1. *Overview*

It is obvious that any attempt to perform MD simulations requires an interatomic potential in some form. In this article, I will concentrate on potentials useful for realistic simulations of particular materials, rather than "generic" potentials for studies of the glass transition or other network order-disorder transitions.

For glasses, the interatomic potential $\Phi(\mathbf{R}_1, \mathbf{R}_2 \ldots \mathbf{R}_N)$ arises from the chemical bonding (in the general case including important covalent and ionic contributions) between the various constituent atoms. The chemical bond is an intrinsically spatially non-local entity, the details of which depend critically on the local topology. This complexity makes simple partitions of Φ into two-body, three-body and higher multi-body interactions only approximately valid. In effect, one is attempting to model the subtle physics of chemical bonding in a disordered environment with some relatively simple assumed form. Depending on the problem this can meet with varied degrees of success.

The most reliable approach to modeling Φ is to acknowledge the origin of the complexity (multi-atom forces originating in complicated electronic effects), and start with an attempt to model the electronic structure of the system. From this, one can proceed to compute interatomic forces. As I will describe below, there is a mature and highly successful approach to approximately solving the electronic structure problem: the local density functional method; now implemented in a variety of forms. Such accuracy comes at a high price: even with clever algorithms and fast computers, these computations are limited usually to a few hundred atoms and a few picoseconds ($1\text{ps} = 10^{-12}$ sec) of time evolution, which is inadequate for many problems.

It is important to note that amorphous and glassy materials pose a formidable challenge to an assumed potential, because the disorder implies a wide range of bonding environments. For any empirical potential, including empirical tight-binding formulations, there is inevitably a "memory" of the database used to fit the potential in the first place. This means that the assumed potential will typically be reliable for structures topologically similar to what was included in the fitting database, but increasingly unreliable for topologies that are "new" to the potential. In the interatomic potential lore, the ability of a potential to properly describe a broad range of local bonding environments is called *transferability*. One can expect that an assumed functional form which includes the underlying chemistry/physics of the interatomic potential will be more transferable than a generic ansatz which fits large numbers of free parameters with little or no *a priori* information about the electronic structure included.

In the following subsections, I will review some of the more popular approaches to modeling Φ.

3.2. *Empirical classical potentials*

A venerable potential which has been heavily used (and misused) as a tool to investigate disordered phases of Si is the Stillinger–Weber (SW) potential [5], which consists of a two-body part, plus a three-body term given by:

$$\Phi_3(R_i, R_j, R_k) = C f(R_{ij}) f(R_{ik})[\cos(\theta_{ijk}) + 1/3]^2 \qquad (2)$$

where θ_{ijk} is the angle at atom i subtended by atoms j and k. Here, $C f(R_{ij})$ $f(R_{ik})$ is positive definite, and quite obviously, the sensible underlying idea is to favor tetrahedral bonding (the triplet term vanishes for $\theta_{ijk} = 109.47°$, the tetrahedral angle). In fact, there is such an overwhelming propensity for tetrahedral (sp^3) bonding in *solid* phases of Si, that this very simple potential is a respectable "zeroth order" approximation to the interatomic interactions. It is worth pointing out that liquid Si is presumably poorly described by this functional form (where the topology is highly non-tetrahedral); it is also apparent that non-tetrahedral network defects are poorly described. As an indication of how much care is needed when considering the appropriateness of a potential to a given problem, I have seen published studies of Si surface reconstructions and adatom adsorption energetics which use the SW potential, even though it gives the wrong surface reconstructions! Despite this warning, I emphasize that the SW potential has been a very valuable tool in the hands of researchers that understand what it can and cannot accomplish [6].

Where empirical potentials are concerned, we have performed explicit comparisons between several proposed empirical potentials and an *ab initio* calculation for a–Si [35]. This work illustrates the high degree of mutual inconsistency between the various potentials themselves and more sophisticated methods.

It is worth noting that some systems are much harder to model than others. For example, carbon is especially difficult since it can form almost degenerate tetrahedral sp^3 bonds (as for diamond), and trigonal sp^2 bonds (as for graphite). Thus, in this extreme case, it is very difficult for a carbon potential to properly differentiate different topologies which are possible in amorphous carbon. Silicon is easier, but still quite difficult to construct a *generally applicable* potential form. For either C or Si, it is not difficult to construct a potential which is accurate for a single topology, and small variations about that ideal. In some ways SiO_2 is easier to handle, since the basic unit of the glass (the crystals, or even the *liquid* [36]) is the $Si(O_{1/2})_4$ tetrahedron, rather than *atoms*, which are all too "flexible" in their bonding attributes. The ionicity of silica forbids "wrong" (homopolar) bonds, which is another key constraint which can be built into a reliable SiO_2 potential; no such information is available for elemental Si or C. Vashishta [37] has proposed a successful silica potential which is very useful, particularly since *ab initio* plane wave calculations are almost non-competitive because of tremendous demands on computer power to properly represent oxygen bonding.

3.3. *Potentials from electronic structure*

A very sensible and relatively successful approach to modeling the interatomic potential is to take an electronic structure viewpoint, and then determine approximate potentials based on the foundation of some electronic structure scheme (for example, a tight-binding Hamiltonian), and then develop functional forms extracted from the approximate version of the electronic structure calculation. An excellent review on this type of approach is that of Anders Carlsson [38]. I begin with a brief tutorial on the tight-binding total-energy method, with no approximations (beside the choice of tight-binding Hamiltionian).

3.3.1. *The tight-binding method*

In the "empirical tight-binding (ETB) approximation", we imagine that the electronic eigenstates can be represented by a linear combination of atomic orbitals: $|\psi_i\rangle = \sum_\mu a_\mu^i |\mu\rangle$ where μ is a site-orbital index and i indexes the band or state. This method enables the calculation of an approximate one-body Hamiltonian matrix, whose eigenvalues are taken to approximate the allowed electronic energies and the eigenvectors are the states. In the usual implementation of ETB calculations, the basis is taken to be orthonormal: $\langle\mu|\nu\rangle = \delta_{\mu\nu}$. Also, most ETB Hamiltonians include interactions only with near neighbors and include only two-center contributions. The sum of the occupied eigenvalues is the (attractive) electronic contribution to the total energy. To compute the system (ions + electrons) energy a repulsive interaction must be added to the electronic part. This is obtained from some fitting procedure. ETB is the simplest approach enabling an estimate of the many-body forces characteristic of covalently bonded materials. I outline here how an ETB calculation is implemented, since it is simple and is illustrative of many of the concepts of electronic structure calculations in amorphous solids.

(1) We begin by considering a supercell model (large unit cell with periodic boundary conditions) of an amorphous solid with N atoms and atomic co-ordinates $\{\mathbf{R}_i\}_{i=1}^N$ and three lattice vectors specifying the periodic boundary conditions. We can view the $\{\mathbf{R}_i\}_{i=1}^N$ as specifying a set of "basis vectors" for a crystal with a very large and topologically complex unit cell. Such a large unit cell possesses a band structure, as does any periodic system, but since the cell is supposed to represent an *amorphous* system, it must be large to have credibility. We will suppose that the **k** dispersion

is negligible because of the large cell size. In calculations *this point must be checked.* The significance of this is discussed further in Sec. 4.2.1. We limit our discussion here to the $\mathbf{k} = \mathbf{0}$ point of the Brillouin zone, valid for a large enough model.

(2) Next, we set up the Hamiltonian matrix $H_{\mu\nu} = \langle \mu | \hat{H} | \nu \rangle$ where $\hat{H}$ is the Hamiltonian operator and $|\mu\rangle$ are a set of orbitals (typically s, p_x, p_y, and p_z for a column IV material) centered on each atom. As the matrix elements in this representation depend in detail on the network topology, it is convenient to work with "molecular coordinates" specifying the interatomic hopping. These are V_{ss}, $V_{sp-\sigma}$, $V_{pp-\sigma}$, $V_{pp-\pi}$, in the usual chemistry nomenclature for an sp^3 model. Explicit forms for the distance dependence of these interactions (see for example Harrison [39]) plus simple rules [40] connecting the molecular and $|\mu\rangle$ representation enables the calculation of $\hat{H}$ in the $|\mu\rangle$ representation. The H matrix eigenvalue problem then reads $H|\psi_i\rangle = \epsilon_i |\psi_i\rangle$, the usual orthogonal eigenvalue problem, where the electronic eigenvalues are supposed to be approximated by ϵ_i. The electronic or "band structure" energy is given by

$$E_{BS} = 2 \sum_{i\ occ} \epsilon_i = 2 \sum_{i\ occ} \langle \psi_i | \hat{H} | \psi_i \rangle \tag{3}$$

For electronic state density calculations or questions of the spectral signature of a defect, an exact diagonalization of H is sufficient.

(3) The calculation of forces is easy when we possess the exact eigenvalues and eigenvectors; then the Hellmann–Feynman theorem can be employed in the form:

$$\mathbf{F}_\alpha^{BS} = -\partial E_{BS} / \partial \mathbf{R}_\alpha$$

$$= -2\partial \sum_{i\ occ} \langle \psi_i | H | \psi_i \rangle / \partial \mathbf{R}_\alpha$$

$$= 2 \sum_{i\ occ} \langle \psi_i | - \partial H / \partial \mathbf{R}_\alpha | \psi_i \rangle \tag{4}$$

Here, the sum on i is restricted to occupied states. Since total energies, forces, charge densities and other ground state properties depend upon the *occupied* eigenfunctions and eigenvalues, it is natural at this point to introduce the single-particle density operator $\hat{\rho} = 2\sum_{i\ occ} |\psi_i\rangle\langle\psi_i|$, which is just the projector onto the occupied subspace at zero temperature. In a particular representation $|\mu\rangle$, this is just the usual density matrix:

$$\rho_{\mu\nu} = 2 \sum_{i\ occ} \langle\mu|\psi_i\rangle\langle\psi_i|\nu\rangle \tag{5}$$

giving the usual expression for the electronic energy, $E = \text{Tr}(\rho H)$, and the forces are:

$$\mathbf{F}_\alpha = -\sum_{\mu\nu} \rho_{\mu\nu}\partial H_{\nu\mu}/\partial\mathbf{R}_\alpha \tag{6}$$

Note that this form assumes that the overlap matrix remains exactly the unit matrix through any atomic motions, which is suggestive of why forces in particular are sensitive to the assumption of orthogonality. It is also important to remind the reader that Eq. (6) prescribes only the (purely attractive) electronic part of the force; *an additional empirical repulsive pair potential must be added to the electronic energy to specify a system energy*; the derivative of this term contributes likewise to the forces, and a dynamical simulation is a complicated balancing act between the electronic and repulsive terms. Atomic trajectories are obtained by integrating the (classical) equations of motion for the atoms using one of many schemes for performing numerical integrations as we discussed above.

A useful local quantity that is very easy to extract in a tight binding calculation is the local electron charge projected onto a site (or orbital). This quantity is not uniquely defined, but a suitable definition (the "Mulliken" charge [41]):

$$q_A = Z_A - \sum_{\mu\epsilon A} \rho_{\mu\mu} \tag{7}$$

where Z_A is the charge of the atomic nucleus A; the sum is over basis functions centered on atom A.

As to the question of specific ETB Hamiltonians "on the market", there are several, and their transferability is variable. In particular, some are designed for producing good band structures [42], but are at the same time quite inappropriate for total energies and forces. As with the empirical potential methods, these Hamiltonians are most reliable for conformations "near" what the Hamiltonian was fit to in the first place. The entire underlying idea of an orthogonal TB model is the notion that there exist some underlying set of generalized Wannier functions with the property that these are mutually orthonormal in real-space. This is a justified point of view, which, however is valid only for a particular topology — if a different structure is considered, another set of Wannier functions emerge, and the justification for an orthogonal basis fails (unless the Hamiltonian is refit to the new structure). To be fair, it is probable failure is not serious until significant distortions are introduced; such that the is sometimes the case in amorphous insulators. The most widely used ETB Hamiltonians for force calculations are those of Goodwin–Skinner–Pettifor [43] type. The original [43] was for Si; this was adapted by Xu *et al* [44] to carbon systems. The difficulties with the orthogonal Hamiltonians seem to emerge most dramatically in *force* calculations [a universal feature of simulations is that the forces are more sensitive to approximations than quantities like the energy] [45, 46].

More consistently reliable results can be obtained from a variety of non-orthogonal Hamiltonians. Menon [47] has proposed an empirical, nonorthogonal TB Hamiltonian which has been applied widely to clusters and recently has been generalized to bulk systems. Also in this category is a Hamiltonian whose form is motivated by density functional theory, explicitly using squeezed atomic orbitals as basis functions [48]. This Hamiltonian has been tested extensively on total energies and equilibrium structures of simple structures [49], and neglects three-center integrals. Both of these methods seem to provide reasonably faithful descriptions of the chemistry while paying only a slightly higher price than the orthogonal Hamiltonian (the only computational difference is that the "generalized eigenvalue problem": $H|\psi\rangle = \epsilon S|\psi\rangle$, where H, S are the Hamiltonian and basis positive-definite overlap matrix, respectively, and ϵ is the energy eigenvalue). The equations of this section must be modified to accommodate a non-orthogonal basis. One key factor is that "Pulay corrections" [50] must be added, since the derivative of the matrix elements is no longer the matrix element of the derivative in Eq. (4).

3.3.2. *Approximate methods based on tight-binding*

The tight-binding prescription for total energies and forces is useful in its own right, as we saw above. It still is a demanding procedure to implement without additional approximations for large model systems, since the method requires the diagonalization of a matrix of dimension $N_{\text{atoms}} \times N_{\text{basis}}$. Some discussion of alternatives to diagonalization are discussed in Sec. 3.5 below.

A major industry of materials theory has involved extracting simpler approximate interatomic potentials from tight-binding models. Carlsson [38] has written a very useful review of this field and shows that even some potentials like the "embedded atom method" (applicable for certain metals) are most easily understood from tight-binding theory, and low-order moment expansions of the density of states (appropriate partial integrals of which provide the Fermi level and band energy). This work provides convincing evidence that an accurate description of the interatomic potential and forces *must* be based upon an electronic structure framework of some kind; well considered approximations can then be made with that starting point. Merely compelling a function with huge numbers of parameters to conform to a large number of (perhaps slightly discrepant) experimental parameters is unlikely to ever produce accurate results.

Many examples of successful empirical potentials and their origin is reviewed [38]; some selected recent developments include the work of Ercolessi [51] for Al, and that of Kaxiras and coworkers for Si [52]. Vashishta [37] has also proposed empirical potentials for binary systems.

3.4. *First principles*

As we discussed above, the complexity of electronic structure and force calculations arises from the many-body nature of the interactions between the electrons. Currently, it would seem that direct attacks on the many-electron problem is too difficult to have direct impact on amorphous systems, requiring as they do a large number of atoms to provide a model worth investigating. Thus, all the successful electronic structure calculations salient to amorphous insulators have involved some kind of mapping of the many-body problem into an effective one-electron problem. Historically, the Hartree and Hartree–Fock approaches were the first success in this direction; descendants of these methods are widely used today, particularly in quantum chemistry. The ETB work captures some of the many-body effects, albeit in an approximate fashion, and the exact connection between the ETB model and the real many-body problem is obscure.

Of course, the true many-body Hamiltonian treats both the ions and electrons on a quantum mechanical basis. Because of the large mass difference between the electrons and nuclei, it is standard to decouple the nuclear and electronic degrees of freedom with the adiabatic or Born–Oppenheimer approximation [53], in which the electrons are assumed to respond instantly to motions of the ions (the electrons are taken to be in their ground state for all instantaneous ionic conformations). Moreover, the nuclei are treated as classical particles which move in a potential determined by the electrons in their ground state (computed for the given ionic coordinates). For most studies of amorphous solids, this is a reliable approximation, and I am unaware of any such calculation which has not started with the Born–Oppenheimer approximation.

The key breakthrough for good approximate solution of the ground state many-electron problem came in the sixties: the density functional method of Hohenberg, Kohn and Sham [54]. In a nutshell, these workers showed that the electron charge density $\rho(\mathbf{r})$ determines the *ground state* energy (and related quantities, such as forces) for an interacting many-electron system *exactly*. Then, it was shown that the ground state energy satisfied a variational principle; if one possessed the exact energy functional, and could consider all possible charge densities as input to the functional, that density which minimized the energy would be the physical charge density and E would be the exact ground state. Moreover, the variational optimum density is unique.

Besides the fact that density functional methods provided (within some approximations) explicit computational recipes which work surprisingly well, it also gave researchers a solid and simple way to think about the extremely complex problem of a large number of interacting fermions. Clearly, density functional methods share much in philosophy with simpler "charge density based" theories of total energies such as the Gordon–Kim [55] scheme (applicable only for closed-shell systems — non-bonded interactions) and of course the Thomas–Fermi theories [56]. The density functional methods underlie practically every first principles calculation in glasses or amorphous materials. It is an unfortunate fact that $E[\rho]$ is not exactly known. Some rather good and relatively simple approximations *are* available, such as the "local density approximation", (LDA) (which is loosely an approximation that the charge density is slowly varying in the solid or molecule). The LDA is more reliable than its derivation would suggest [57], and is often very reliable for structural and vibrational calculations. For well understood reasons, it is poor at estimating energy gaps (the band gap is typically underestimated by 50%–100%). It is an area of intense development to improve the LDA, and there have been some successes [58]. Self-energy corrections [59] largely fix the gap problem, but are extremely computationally demanding. There are even promising developments toward fixing the excitation energies in a density functional framework [60].

I will not develop the well-known density functional equations in this article. I will instead summarize the "mathematical strategy" needed for approximately implementing density functional theory. One would guess from the dependence of E on ρ that the object subject to variation would be ρ. However, it turns out that in order to obtain a good estimate for the electronic kinetic energy, it is better to work with a set of $n_{\mathrm{elec}}/2$ orbitals, $|\chi_i\rangle$ (we have made the assumption that the orbitals are all doubly occupied). It is necessary that these $|\chi_i\rangle$ should form an orthonormal set $\langle\chi_i|\chi_j\rangle = \delta_{ij}$, with the connection $\rho = 2\sum_{i\ occ}\chi_i^*\chi_i$. This then specifies E as a functional of the $|\chi_i\rangle$ subject to the orthogonality constraint. To the extent that the basis set upon which the $|\chi_i\rangle$ is expanded on is *complete*, one can expect to get a reliable estimate of the energy and other ground state properties.

Consistent with chemical intuition, a minimal basis set of s and p functions is primarily responsible for the bonding and ground state electronic properties. The pseudopotentials are introduced so that only the really relevant electrons, namely the valence orbitals are *explicitly* treated in the calculation. Thus the valence electrons in the LDA solid are interacting with each other and with the ions via exotic potentials mimicking the nucleus dressed with the core electrons. This is an excellent approximation for most applications in the area of amorphous insulators. Pseudopotentials themselves have developed a considerable lore, and the classic modern work in the area is the paper of Bachelet, Hamann and Schlüter [61]. The choice of pseudopotential is governed partly by the basis used: for plane-waves special "soft" pseudopotentials are needed [62, 63]; plane-wave calculations also require a special "separable" form, enabling a factorization that saves considerable computational effort [64]. Real space local basis functions are immune to the need for "softness".

3.4.1. *Local basis*: *"ab initio tight binding"*

Because of formal similarity to the ETB method of Sec. 2.2, I begin by describing a spatially local basis approach to construct the density functional orbitals $|\chi_i\rangle$. This is a program that was developed by Sankey [65], and his coworkers, and has been extensively applied to studies of amorphous materials.

The essential approximations are: (1) the LDA; (2) nonlocal (angular momentum dependent), norm-conserving pseudopotentials; (3) a minimal basis set of valence orbitals per site (this has recently been extended by Yang [66] to include d states in the basis). These basis functions are slightly excited (confined) pseudoatomic orbitals (PAO) (a PAO is a valence eigenfunction for a given atom type in free space, calculated self consistently within the LDA using pseudopotentials to eliminate the core: in the case of Si, the PAO's can be thought of as approximate atomic $3s$ and $3p$ levels); (4) A linearized (non self consistent) version of density functional theory, the "Harris functional" [67] is used which enables calculations without self consistent iterations. This approximation is best discussed by Foulkes and Haydock [68] and is quite remarkably good except in the most ionic glassy or amorphous systems like SiO_2. For additional discussion, see the work of Smith [69].

The use of PAO's is appealing, because the chemistry is naturally built into the basis functions. The payoff is that for materials like Si and C, a minimal basis of four orbitals per site can be used, so that in a model with N atoms, the overlap and Hamiltonian matrix has dimension $n_b N$, for $n_b = 4$ or 10, the number of basis functions per atom (*sp* or *spd*) whereas $n_b \approx 100\text{--}1000$ for many plane-wave calculations (depending in detail on how carefully such a plane-wave calculation is implemented, and how "hard" the pseudopotential is, which determines the number of plane-waves needed). Also the eigenvectors are expressed naturally in terms of the local PAO's, which is helpful in interpreting the physical meaning of the results. Finally, since the method is implemented entirely in *real-space*, no artificial periodicity is ever imposed, making the approach much more amenable to surface and cluster calculations, where desired.

Sankey and coworkers constructed this method to optimize performance for *ab initio* molecular dynamics simulations; the details are well beyond the scope of this paper. The method is able to handle systems with up to ≈ 500 atoms with exact diagonalization on a workstation, and has been extended recently to include quantum order N methods. The ability to perform exact diagonalizations (as opposed to "iterative minimization" schemes) is valuable for systems with states in the gap and particularly metallic systems. The limitation of non self consistency has been lifted recently by Demkov *et al* [70] and Ordejón *et al* [71].

While this approach is relatively easy to understand, it is vastly more complex than the ETB method above. In particular, to make the scheme efficient, all the two and three-center matrix elements have to be tabled, so no integrals have to be computed during an electronic structure calculation or MD simulation; rather many calls are made to one and two dimensional interpolators. The calculations of forces, particularly with respect to the exchange correlation interaction, is tedious. Also, it is no longer correct to use the Hellmann–Feynman result (the derivative of the matrix elements is not the matrix element of the derivative, since the overlapping basis functions move with the atoms).

The formal expressions for matrix elements, forces etc. are substantially more complex than for plane-wave pseudopotential methods of the next section. I refer the reader particularly to Refs. [65, 70, 71] for the details of the theory. This scheme has found considerable use in amorphous insulators.

A very recently developed method which appears to be very promising is the "SIESTA" code of Ordejón, Soler and Artacho [72]. This is a fully self-consistent implementation of Kohn–Sham LDA with a rich local basis (double zeta with polarization functions), post-LDA corrections (generalized gradient approximations) and local spin density (applicable to non-spin polarized systems).

3.4.2. *Plane-waves: Car–Parrinello methods*

The first *ab initio* molecular dynamics simulation was performed in 1985 by Car and Parrinello (CP). I begin with a comment on what "CP" means. Often a "CP simulation" refers to a molecular dynamics calculation using a plane wave basis and an iterative minimization scheme to solve the electronic structure problem (the self consistent LDA equations). More properly, "CP" refers only to a method for coupling the approximation of stationary states of giant basis eigenvalue problems with associated ionic dynamics, even for a basis other than plane-waves.

In the next paragraphs, I will discuss the basic CP equations of LDA on a plane-wave basis [73]. To keep to the essentials, I will write the expressions only for the Γ point of the Brillouin zone. As usual, the LDA equations take the form (for LDA Hamiltonian $\mathcal{H} = p^2/2m + V_{\text{eff}}$):

$$\mathcal{H}|\chi_i\rangle = \epsilon_i|\chi_i\rangle \tag{8}$$

where

$$|\chi_i\rangle = \sum_{\substack{\vec{G} \neq \vec{0}}} A^i_{\vec{G}}\left|\vec{G}\right\rangle \tag{9}$$

where $|\mathbf{G}\rangle = 1/\sqrt{\Omega}\exp(i\mathbf{G}\cdot\mathbf{x})$, Ω is the cell volume, and $\mathbf{G}$ labels reciprocal lattice vectors (*a plane-wave basis restricts our studies to periodic systems*), so that the usual apparatus of reciprocal lattices is appropriate. The sum is cut off for $G = G_c$ large enough ("wavelength" small enough) that the smallest salient features in the problem are adequately described. An obvious advantage of this representation is that the kinetic energy operator is diagonal and therefore trivial. Matrix elements of V_{eff} are obtained by fast Fourier transforms, the existence of which makes the plane-wave approach tractable. The *problem* with Eq. (8) is that for systems large enough to interest us, the Hamiltonian matrix is too large to diagonalize with "classical" methods.

The key contribution of CP was to note that the diagonalization (Eq. 8) is unnecessary; it is adequate instead to consider the LDA expression for the total energy (a "partial trace" over the occupied electronic subspace) and vary the χ_i:

$$E(\{\chi_i\}) = \sum_{i\ occ} \langle\chi_i|\mathcal{H}|\chi_i\rangle \tag{10}$$

where two important facts must be emphasized: (1) the sum is over the *occupied* subspace and (2) $\langle \chi_i | \chi_j \rangle = \delta_{ij}$. Thus, using Eq. (9), E can be viewed as a function of an extremely large number of parameters A_G^i ($n_{pw} \times n_e/2$) for n_{pw} the number of reciprocal lattice vectors used and n_e the number of electrons (double occupancy assumed). The idea is that if a set of mutually orthogonal $|\chi_i\rangle$ can be found, adequate in number to accommodate the electrons in the problem, and minimizing the functional (Eq. 10), then the E so obtained is the LDA ground state energy, to the extent that the basis is complete (the plane wave cutoff is sufficiently large).

To perform the minimization, CP introduces a fictitious Lagrangian [74] which includes both the electronic and ionic degrees of freedom:

$$\mathcal{L} = \sum_i \mu/2 \langle \dot{\chi}_i | \dot{\chi}_i \rangle + \sum_\alpha 1/2 M_\alpha \dot{\mathbf{R}}_\alpha^2 - E(\{\chi_i\}, \{\mathbf{R}_\alpha\}) \qquad (11)$$

In this equation, $E(\{\chi_i, \mathbf{R}_\alpha\})$ is the same as that given in Eq. (10) (the expectation value for the electronic energy, viewed as a functional of the orbitals χ_i), and the quadratic terms involving time derivatives (dots) are for the electrons (χ_i) and ions ($\mathbf{R}_\alpha$). Parameter μ is a fictitious mass assigned to the electron (it has nothing to do with the real mass, and is adjusted to make the calculation proceed efficiently). Note at this point that this method *does not* provide any information about electron dynamics in the usual quantum mechanical sense of time evolution. Rather it is just a trick to find the stationary states of the LDA Hamiltonian (approximate the occupied eigenstates from eigenvalue problem Eq. 8). Thus, the "dynamics" for the electrons has only the utility of helping to solve the eigenvalue problem, and in fact if the fictitious kinetic energy becomes significant compared to the physical ionic kinetic energy, the dynamics for the ions will be suspect. In the absence of the "fictitious electronic kinetic energy", this Lagrangian would just generate the Newton equations of motion for the ions $M_\alpha \ddot{\mathbf{R}}_\alpha = -\partial E/\partial \mathbf{R}_\alpha$, as one would expect. The additional term adds an extra equation however: $\mu | \ddot{\chi}_i \rangle = -\delta E/\delta \langle \chi_i | + \sum_k \Lambda_{ik} | \chi_k \rangle$. Here, Λ_{ik} is a matrix of Lagrange multipliers required to maintain orthogonality $\langle \chi_i | \chi_j \rangle = \delta_{ij}$. In practical methods, the Λ_{ik} are obtained by compelling the eigenvectors $|\chi_i\rangle$ to be orthonormal at each time step, and computing Λ_{ik} from $\Lambda_{ik} = \langle k | \mathcal{H} | i \rangle - \mu \langle \dot{\chi}_k | \dot{\chi}_i \rangle$. Then the ionic and fictitious electronic "forces" are completely specified, and the simulation can proceed. Note that there are effectively two time scales in the dynamics so generated: one is determined by the artificial "mass" μ of the electrons; the physically salient time scale is set by the ionic masses M_α.

This method generates unreliable dynamics for the ions if significant energy is transferred from the ionic (physical) degrees of freedom to the fictitious electronic degrees of freedom. That this must happen for long times is apparent from the equipartition theorem (Eq. (11) is a quadratic form, and asymptotically energy will reach equipartition at least if the ionic coordinates are near an energy minimum). It develops that the rate of transfer from the ions to the unphysical electron kinetic energy depends critically on the energy gap (rapid transfer for small gap), which makes the CP method difficult to apply to metals [75].

3.5. *Efficient ab initio methods for large systems*

The area of network glasses and associated dynamics is a challenge not only to transferability (which requires first principles methods, at least in cases like a–C or a–Si), but requires models with *many* atoms. Depending on the precise question being addressed it can be necessary to have thousands of atoms. This is beyond the reach of plane wave based density functional methods, and is quite difficult even for the local basis methods (by contrast, simulations with millions of particles are possible for empirical potentials and using special linear scaling methods there [11]).

Several groups have been working on first principles total energy and even MD codes which have CPU and memory demand scaling *linearly* with the system size. These are dependent on two ideas: the existence of a real-space localized representation for the electron states, and a way to avoid explicitly keeping electron orbitals mutually orthogonal.

3.5.1. *The need for locality of electron states in real space*

The condition that a system should possess a localized real-space representation is connected closely to the idea of the "Wannier functions" in crystals. It has long been known [76] that an alternative to the space-filling Bloch representation (with band label n and quasicontinuous vector $\mathbf{k}$) is the Wannier function with labels $(n, \mathbf{R})$, where $\mathbf{R}$ specifies a lattice vector near where the electron is localized. It has been shown that idealized Wannier functions can exhibit exponential decay for insulators [76]. In this representation, the overlap between Wannier functions becomes exponentially small (and therefore negligible) for sufficiently separated Wannier centers. This also justifies use of a similar truncation in the single particle density matrix [77, 78], which can similarly be truncated for (local) basis functions sufficiently mutually remote. This ability to "cut off" the tails of the Wannier orbitals (with negligible harm done to the accuracy) is the key to efficient electronic calculations for insulators. In effect, it makes quantum mechanics a local theory (although there are fundamental bounds on *how* local it can be made). In a complicated way, the possibility for the real space local representation requires the existence of a significant optical gap; in a metal the density matrix (or Wannier orbitals) decay as a power law at zero temperature. A representation-independent measure of this spatial locality is the decay of the off-diagonal elements of the real-space single-particle density matrix $\rho(\mathbf{x}, \mathbf{x}')$; it seems clear that the Wannier orbitals cannot be *more* localized than the range of ρ. It is also possible [10, 79] to compute "generalized" Wannier functions for non-crystalline systems. Such generalized functions usually differ from the classical Wannier functions also by involving more than one band. Recently, "visualizations" of these functions have been published for diamond and discussed in detail for amorphous diamond [79]. The applicability of quantum order-N methods to disordered insulators is a considerable boon for our community, and will enable studies that are currently unthinkable in metallic systems. Ordejón and Drabold [80] performed local basis *ab initio* order-N MD on a 512 atom model of amorphous diamond [81], and Stephan and Drabold have recently applied *ab initio* methods to an a–C model with 4096 atoms [81], and showed [79] using projection methods [82, 83] that the local representation works in a disordered insulator as well as for crystals.

3.5.2. *Avoiding explicit orthogonalization*

The second requirement for a linear scaling algorithm is to avoid explicitly orthogonalizing the electronic orbitals at each step through a variational minimization scheme. The most obvious (but notoriously unstable) Gram–Schmidt orthogonalization [19] process has cpu demand scaling as N^3. This problem has been solved by Vanderbilt for calculations in which the density matrix is the object to be variationally determined [77] and independently by Ordejón *et al* [10] and Mauri *et al* [84] for a "generalized Wannier orbital" based approach. I refer the interested reader to the original references.

4. Connecting Simulation to Experiment

4.1. *Structure*

Diffraction measurements (typically X-rays or neutrons) provide information about the distribution of radial distances between atomic pairs. In particular, neutron diffraction experiments can measure the static structure factor:

$$S(Q) = 1 + \frac{1}{N\langle \overline{b^2} \rangle} \sum_i \sum_{j,j \neq i} b_i b_j \frac{\sin(Qr_{ij})}{Qr_{ij}} \tag{12}$$

The summations are over the atoms of the supercell, r_{ij} being the magnitude of the difference between the positions of atom i and atom j. The $b's$ are the scattering lengths for whatever atoms are present in the system. It is often more convenient to work with the pair-distribution function $g(r)$, related to $S(Q)$ according to Ref. [85]:

$$g(r) = 1 + \frac{1}{8\pi^3 n_0} \int_0^\infty [S(Q) - 1](\sin(Qr)/Qr)4\pi Q^2 dQ \tag{13}$$

An essential requirement for any physically credible model is agreement on static structural measurements, especially from X-ray or neutron diffraction studies. It cannot be too strongly emphasized however that while sufficient agreement on radial distribution function $g(r)$ and closely related static structure factor $S(q)$, is *necessary*, it is *not sufficient*. That is, the literature is replete with calculations yielding models with "good" radial distribution functions, which are wholly unrealistic. Some especially revealing work along these lines is by Putszai [86] for a–Si, who has shown that essentially perfect agreement for $S(q)$ can be obtained for models with a bond angle distribution at least twice as broad as what occurs in nature. For a complete discussion, see Wright's contribution in this volume.

Other *local* probes like NMR, NQR, Mössbauer and certain electronic and vibrational experiments are also critically important, and discussed by experts elsewhere in this volume.

4.2. *Network dynamics*

Inelastic neutron scattering offers the most direct probe of lattice dynamics experimentally available. For a proper discussion, see the article of Cappelletti in this volume. It can provide an accurate measure of the vibrational power spectrum, something easy to compute for a given structural model and interatomic potential (first principles or empirical). Other spectroscopies (such as Raman [87]) also probe the vibrational density of states, albeit, modulated by matrix elements that are hard to compute very accurately. It is common to compare the position of peaks in the Raman spectrum to peaks in the vibrational density of states, though this can be misleading if a particular sought after feature is not Raman active (symmetry forbidden). It would be rare to find a truly symmetry-forbidden feature in a glass, because of the lack of order. Certainly however, some modes are much more Raman active than others because of *local* order which certainly is present in all glasses.

4.2.1. *Computing the harmonic modes*

In the harmonic approximation, one models small oscillations of atomic positions about a local or global minimum of the potential energy function. In crystals with translational invariance, it is customary to label vibrational modes with band (branch) and $\mathbf{k}$ indexes, where it is sufficient to consider only those $\mathbf{k}$ in the first Brillouin zone. For a perfect crystal, all the vibrational modes are extended in real space.

In disordered systems, the lattice vibrational excitations are not so easily categorized, since $\mathbf{k}$ is no longer a suitable label. One can proceed to think about the vibrational modes in either of two ways: (1) The simplest approach is to view the harmonic modes as the normal excitations of a very large molecule (the model under consideration). It can be very instructive to visualize the time development of the atomic positions *executing a single vibrational eigenmode* [88]. (2) If periodic boundary conditions are used, as is usually the case in modeling a glassy or amorphous material, then there is in fact a Brillouin zone, and technically the apparatus of crystalline solids should apply. From this point of view, the $\mathbf{k} \to 0$ (acoustic) modes are not properly reproduced by a $\Gamma(\mathbf{k} = \mathbf{0})$ point calculation, and the smallest frequencies in the density of states to be accurately represented by a Γ point calculations should be $\omega_{\min} \approx 2\pi c/a$, where c is the speed of sound and a is the smallest dimension parameter of the supercell. One could of course compute the dynamical matrix as a function of $\mathbf{k}$ and compute the dispersion curves, but this is of doubtful value, since the dispersion so computed originates from an unphysical (but convenient) periodicity. For a sufficiently large cell and ω, it is appropriate to limit oneself to $\mathbf{k} = \mathbf{0}$ modes.

The dynamical matrix D is the key ingredient to a calculation of the normal vibrational modes, and is easily obtained from a system *well* relaxed to equilibrium for which interatomic forces are available. In particular, $D_{\alpha I,\beta J}$ is defined as $D_{\alpha I,\beta J} = (M_I M_J)^{-1/2} \partial^2 E/\partial u_{\alpha I} \partial u_{\beta J}$, where E is the total energy of the system, α and β are Cartesian coordinates on atoms I and J, respectively, and the M's are ionic masses, is computed by finite differences: atom I is displaced an amount $\Delta x_{\alpha I}$ in the α direction, and the forces on all the atoms $F_{\beta J}$ are computed. Then, $D_{\alpha I,\beta J} \approx (M_I M_J)^{-1/2} F_{\beta J}/\Delta x_{\alpha I}$, since the force on an atom is the derivative of the total energy with respect to the displacement of that atom: $F_{\beta J} = -\partial E/\partial x_{\beta J}$. In practical implementations, one usually has to symmetrize the dynamical matrix (it will not be *exactly* Hermitian unless the potential is exactly harmonic) [89].

If one is interested in a very large system, the dynamical matrix can be accurately approximated as a sparse matrix (a matrix with the vast majority of elements being zero), if the forces on atoms closer than a certain cutoff R_f from the displaced atom are computed, the rest of the forces being set to zero. This approximation takes into account the fact that the forces on each atom depend only on its local environment, and not on the details of the structure in distant regions. This also reflects the fact that the dynamical matrix elements between distant atoms decreases rapidly with distance. Our approximation can therefore be put in terms of imposing a cutoff on the dynamical matrix:

$$D_{\alpha I,\beta J} = 0 \quad \text{if} \quad |\mathbf{R}_I - \mathbf{R}_J| > R_f \tag{14}$$

so that the resulting matrix is sparse. Its calculation and storage are both $O(N)$ [90]. With the dynamical matrix in hand one can exactly diagonalize D for small systems or use the Maxent spectral technique [91, 92] for large systems. In addition, a Lanczos method can be used for computing a limited number of vibrational eigenvectors in a user specified energy range [93]. Of course this method also works for empirical potentials in which case the germain approximations are that D have finite range and one of the spectral techniques described in this paper.

It would be interesting to try these ideas on extremely large models of amorphous semiconductors to study the floppy-rigid transition [94, 95] for a model with Keating [96] springs. The density of states would be easy to compute $O(N)$. Such an approach would not be as efficient or elegant as *The Pebble Game* [97], but it would also be more "realistic", with relatively realistic interatomic interactions.

4.2.2. *Dynamical autocorrelation functions*

Sometimes it is inconvenient or even impossible to implement a calculation of the dynamical matrix because the harmonic approximation is not valid (as for a liquid). One can still compute the vibrational power spectrum directly from the trajectory given by a thermal MD simulation. Of course this method can also be applied to a system in the solid state, though in practice, it should be inferior to the dynamical matrix calculation since it gives less information (only the vibrational power spectrum), and in practice requires at least as many times steps of MD as the $3N$ steps needed to compute the full dynamical matrix. The dynamical approach also requires a numerical Fourier transform, which requires a long time series (simulation) to produce good estimates for the vibrational spectrum for $\omega \to 0$.

In this approach, the vibrational density of states (VDOS) $g(\omega)$ of a model can be obtained from the Fourier transform of the velocity autocorrelation function $G(t)$:

$$G(t) = \frac{\langle \mathbf{V}(0) \cdot \mathbf{V}(t) \rangle}{\langle \mathbf{V}(0) \cdot \mathbf{V}(0) \rangle} \tag{15}$$

$$\langle \mathbf{V}(0) \cdot \mathbf{V}(t) \rangle = \frac{1}{3N} \sum_i^N \sum_j^3 V_{ij}(0) V_{ij}(t) \tag{16}$$

$$g(\omega) = \frac{2}{\pi} \int_0^T G(t) \cos(\omega t) W(t)\, dt \tag{17}$$

The values of N and T are the number of atoms and the total simulation time respectively. The summations in Eq. (16) are over the total number of atoms and the three components of each atom's velocity. We use a standard Blackman window [19] $W(t)$.

$$W(t) = 0.42 + 0.5 \cos\left(\frac{\pi t}{T}\right) + 0.08 \cos\left(\frac{2\pi t}{T}\right) \tag{18}$$

The spectral resolution of the Fourier transform depends on how much time series data can be generated and a rule of thumb is to simply check whether the spectral function is too broad to address the question of interest. Our experience is that the Blackman window gives the best results over the entire spectral range of our VDOS. Note that periodic boundary conditions introduce artifacts in $G(t)$ for times longer than $\tau \approx a/c$, where c is the speed of sound and a is the box size [17]. For completeness, we should also mention that the diffusion constant can in principle be computed from velocity autocorrelations via the value of g at zero frequency:

$$D = (1/3) \int_0^\infty \langle \mathbf{V}(0) \cdot \mathbf{V}(t) \rangle W(t)\, dt \tag{19}$$

A much more sophisticated and potentially more powerful approach to spectral estimation (e.g. to finding the "frequency content" of a time series) is to use the methods of Bayesian parameter estimation [98]. Here, one can actually compute the probability that a given model (say a harmonic series with p distinct frequencies) describes a given data set [time series $G(t)$]. I have applied this approach to carbon clusters [99] and it is easy to see that a shorter MD simulation is needed to get an estimate of the normal mode frequencies than one would obtain by Fourier transform with windows. For additional details I refer the reader to the Bayesian literature [98]. While this approach is clearly useful for small molecules [99], the possibilities for this approach are uninvestigated to date in glasses.

4.2.3. *Dynamical structure factor*

An additional advantage to the dynamical matrix calculation (which gives the normal modes eigenvectors) is that one can easily compute the dynamical structure factor $S(\mathbf{Q}, E)$ (See R.L. Cappelletti, Chapter 6A in this volume). I refer the interested reader to the discussion given there.

4.3. *Electronic structure*

A very important and sensitive probe of structure is the nature of the electronic states and the electronic density of states in disordered systems. Since electron states near the gap are always significantly localized due to disorder, one can link an electronic signature (defect state) with a finite (usually small) collection of atoms upon which the state is localized (where the state's charge is located). In this way, electronic information can directly imply structural information. The nature of the electronic states in the vicinity of the optical gap is of key interest, since these states are responsible for conduction and are undoubtedly of key importance to light-induced defect creation and more exotic forms of light-induced effects.

Now it must be admitted that the experimental determination of the electronic density of states can be a challenge. In particular, since any imaginable experiment involves *transitions* in electronic states one naturally measures something like a *joint* density of states involving a convolution of the valence and conduction band tails (note however the impressive "total yield photoelectron spectroscopy" experiments of Aljishi and coworkers [100], which are able to sort out the valence and conduction tails separately). In principle, transport measurements depend upon the electronic density of states (and, unfortunately, on many other things too); so that it is extraordinarily difficult to infer the density of states given information about transport (for some progress on this interesting problem see the article of Adriaenssens [101]).

A key point which is only now coming to be appreciated [102, 103], is that informative calculations of anything but the most localized (midgap) states requires models which are extremely large by usual standards. This is because the band*tail* states are not compactly exponentially localized as midgap defect states are (for example in a–Si). A proper (atomistic) theory of transport within linear response in disordered systems is under development, and one (of several) essential features is to have models large enough to properly describe the states (e.g. without much effect from finite size artifacts and periodic boundary conditions) responsible for the conduction.

4.3.1. *Density of states*

It is clear that to get accurate eigenstates, either electronic or vibrational, one needs *large* — in many cases several *thousand* atom models. The most natural way to compute the density of states of a large Hermitian matrix (either the electronic Hamiltonian in a selected representation or the dynamical matrix for lattice vibrations) is to exactly diagonalize it with standard iterative methods (usually Householder tridiagonalization followed by a "QR" algorithm) [104]. This approach is extremely numerically reliable and it provides all the eigenvalues and eigenvectors to machine precision. The problem is that the cpu demand for this scales like the dimension of the matrix *cubed*. Practically speaking this means that a 2000×2000 matrix is quite tractable with these full diagonalization methods, but $10\,000 \times 10\,000$ is both a cpu and memory challenge.

Fortunately there are efficient alternatives to full diagonalization if one seeks only the density of states. These schemes can have cpu and memory cost that scales *linearly* with the dimension of the matrix. One simple and effective approach is the maximum entropy approach [91, 105], and the closely related recursion methods [106]. These methods apply for local basis representations of the Hamiltonian or the dynamical matrix, and fully exploit sparseness (when properly implemented, all matrix-vector operations are constructed to involve *only* the non-zero elements of the matrix. For a detailed discussion of methods to extract the DOS in "order N" fashion, I refer the reader to the literature [91, 105].

For the calculation of eigenvalues and eigenvectors "in a gap" (again, either in the electronic or vibrational DOS), e.g. where the spectrum is not very dense, Lanczos or inverse iteration methods are very efficient. We have recently used a Lanczos method [107] to compute 500 states around the optical gap in a

4096 atom model of a–Si, which enabled some new insight into the Anderson transition [16] in a real amorphous material (rather than for the Anderson model with its diagonal-only disorder). Knowledge of such states is critical to the optical and transport properties of disordered materials.

4.3.2. *Thermal modulation of the electron states* [108]

Since our focus in this article is *dynamics*, we also consider here the time evolution of electron energy eigenvalues and associated eigenvectors due to their modulation by lattice vibrations, e.g. we study the electron-phonon interaction in a quasi-classical approximation in which the atomic positions are modified from their ideal $T = 0$ positions by thermal disorder (superposed on the existing structural or topological disorder); this thermal motion induces a time dependence in the energy eigenstates and eigenvalues with experimentally observable consequences.

To illustrate these ideas, I consider the case of amorphous Si, for which we are motivated by the temperature-dependent photoemission measurements of Aljishi *et al* [100]. We have studied the modulation of the electronic energy levels due to thermal motion, as characterized by LDA simulations at different moderate temperatures in a WWW 216 atom cell [108]. In Fig. 1, we illustrate the time evolution of the band-tail and deep trap eigenvalues. Several points

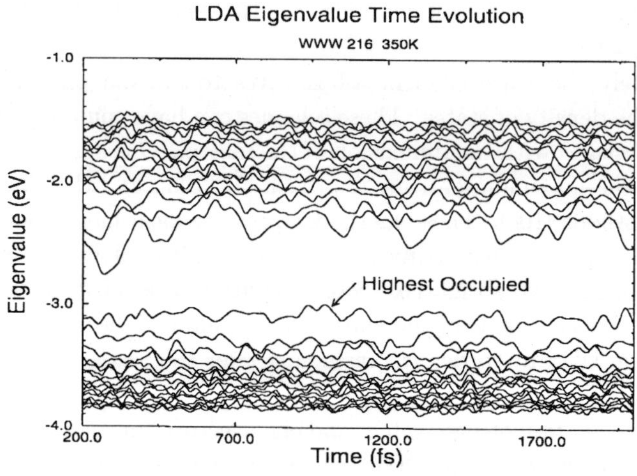

Fig. 1. Thermal fluctuation of LDA eigenvalues in the gap. The dynamics are for the 216 atom WWW amorphous Si cell at 350 K.

bear comment:

(1) The thermal modulation of the gap states is very large (the lowest energy unoccupied state fluctuates by *several tenths of a volt*, an effect *much* larger than thermal energies), and the RMS fluctuation of twelve states in the vicinity of the gap (out of 864 states total) is larger than 0.03 eV. Naturally, the structure of the LDA eigenvectors conjugate to these eigenvalues is changing too. This variation reflects the large electron-phonon coupling between the lattice vibrations and a localized (defect or band tail) electronic state.

(2) If one computes the inverse participation ratio (IPR) for the states illustrated in Fig. 1, there is considerable correlation between the RMS time fluctuation (a dynamical property) and the IPR (a static property) [more local, more fluctuation] [12].

(3) For the conductivity, the thermal modulation of the electronic states is critical, most particularly for the states near the middle of the gap, as they fluctuate most strongly (and are also the key states for transport in the intrinsic material).

(4) The conduction tail states fluctuate more than the valence tail states; this immediately suggests that electron transport should exhibit a more pronounced temperature dependence than holes in the valence tail. This is seen in drift mobility measurements [109].

(5) By making the (remarkably crude) identification of the RMS fluctuation of the highest occupied electronic eigenvalue with the (temperature dependent part) of band-tailing, we obtained semi-quantitative agreement [108] with the temperature dependent band tail experiments of Aljishi *et al* [100]. It must be admitted that much more work needs to be done along these lines, but I believe that it is a promising beginning.

(6) Spectral analysis of the time series for each eigenvalue shows the influence of recognizable phonons. Highly localized states originate in a locally strained part of the network, which usually also induces high frequency local vibrational modes [110], the modulation of which has particular strength in the vicinity of the defect. These effects do not depend upon system size, at least for the most localized states. These thermal effects are rather well approximated by local basis LDA methods, as detailed tests on the phonon spectrum of c–Si, a–Si:H and a more complex glass like GeSe$_2$ reveal Ref. [15].

(7) Large fluctuations in the *eigenvectors* of localized states are observed.

The dynamics of the electronic properties as described in this way are an essential precursor to modeling the finite temperature transport properties of a–Si (or other disordered systems), and point the way to a theory more connected to microscopic models and consequently to a detailed understanding of the connection between transport and microstructure.

4.4. *Transport*

One of the most important probes of electronic, topological *and* vibrational structure is transport. In particular, interpreting experimental measurements of transport properties, particularly in the presence of disorder, is a challenging problem, precisely because the conductivity depends critically on *all* of these physical properties.

An emerging area of theory work in glasses and amorphous semiconductors is the formulation of a microscopic, indeed *atomistic* scale approach to modeling conduction. The idea is to explicitly connect the conductivity to electronic wave functions, doping (the position of the Fermi level), etc. This is based on a famous result from statistical mechanics, the "Kubo formula" [111, 112] for the electrical conductivity (the name is actually generic and refers to a variety of related correlation functions which give the linear response of a system to a weak external perturbation). The Kubo formula for the AC electrical conductivity can be expressed in many different but equivalent forms; one convenient form is:

$$\sigma(\omega) = \frac{2\pi e^2}{3m^2 \Omega \omega} \sum_{i,k} [f_i - f_k] |\mathbf{M}_{i,k}|^2 \delta(E_k - E_i - \hbar\omega) \qquad (20)$$

where $\mathbf{M}_{i,k} = -i\hbar \langle \psi_i | \nabla | \psi_k \rangle$ is the momentum operator matrix element between eigenstates ψ_i and ψ_k, Ω is the cell volume, and E_i is the energy of state i. f_i is the Fermi–Dirac distribution function evaluated at energy E_i. This result can also be obtained [113] from a Fermi Golden rule [114] argument. Equation (20) describes conductivity in terms of quantum transitions of electrons near the Fermi level. By inspection of the Kubo formula, it is clear that the localization of the states, overlap (and momentum matrix elements), and proximity in energy to each other and to the Fermi level, are primary determinants of the DC conductivity.

The Kubo formula is quite correct and also deceptively innocent in its appearance. In particular it is nearly always the case for amorphous semiconductors that one is interested in the $\omega \to 0$ limit of σ, and this is tricky to evaluate for any finite system, since for any finite model the density of states is *discrete*, and therefore $\sigma(\omega = 0)$ is strictly zero! In fact, there are two key broadening mechanisms to the energy levels which must be considered in some fashion:

(1) for finite temperatures, the electron-phonon interaction modulates the position of the energy eigenvalues, especially those near the optical gap (since these are usually well localized); this can have a dramatic effect on some eigenvalues [108]; causing them to fluctuate several *tenths* of an eV about their reference $(T = 0)$ energy;
(2) in a supercell calculation there *is* a Brillouin zone associated with the supercell periodic boundary conditions, and thus there is a k-space broadening which scales inversely with supercell volume.

This latter effect is evidently important even for $T = 0$ and gives a properly continuous spectrum for the electronic eigenvalues. In Fig. 2, we show an early example of a theoretical calculation of the DC conductivity in amorphous Si [103]. There is a developing literature on the proper ways to evaluate the Kubo formula and related results from linear response theory [115].

To properly include thermal effects (e.g. the effects of thermal disorder superposed upon topological disorder) a necessary first step is to thermally average the expression 20 over a sufficiently long MD trajectory carried out at an appropriate temperature. This work is in progress [116] and it is too early to say how effective this will be in predicting the *temperature dependence* of the conductivity, for example. One point that is already apparent to us however is that a "real MD" trajectory is needed to gain understanding of the (vibration-induced) time evolution of the electronic stationary states; an attempt to just make small *random* displacements of the atomic positions from their $T = $ conformation leads to modulations of the electronic energies much weaker than observed in experiments which directly probe the density of states [116].

It is possible too that the most effective approach to understanding transport will involve a hybrid of atomistic electronic structure calculations in conjunction with a Monte Carlo or integral equation theory [101] of transport which requires input such as the scattering cross section of a given electronic defect states, information about the localization of the trapping state (and therefore information about expected lifetimes in that state), etc.

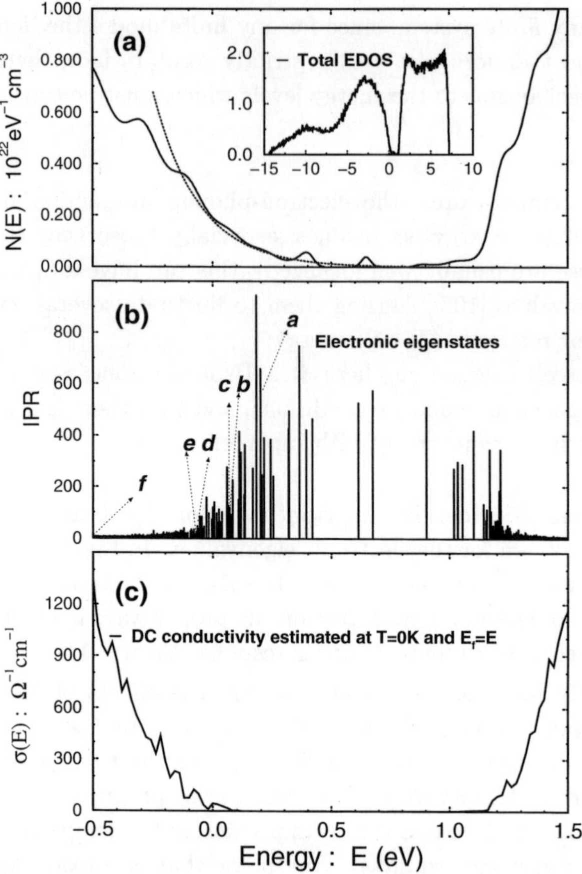

Fig. 2. The density of electron states, localization and estimated DC conductivity (as a function of doping — the position of the Fermi level) for 4096 atom model of *a*–Si.

5. Applications

In this section, I will review a few applications of the MD method to amorphous or glassy materials, and again, provide pointers to published work.

5.1. *g–GeSe$_2$*

We have recently been modeling glassy chalcogenides, especially *g*–GeSe$_2$ [12, 14, 15], using the methods of local basis *ab initio* MD. To summarize the principal results:

(1) *ab initio* MD with supercells involving 216 or more atoms provides a rather faithful structural model of g–GeSe$_2$, including the first sharp diffraction peak (FSDP). In Fig. 1 of Ref. [12] we show the total static structure factor for the liquid and glass. In the paper [12] we also elucidate the partial distribution functions, and make a connection between the rings and the FSDP;

(2) by explicitly computing the dynamical matrix (Sec. 4.2.1), and studying the vibrational eigenmodes, we substantially justified the proposal of Sugai [117] that the A_{1c} mode is predominantly a tetrahedral breathing mode associated with corner-sharing tetrahedra, albeit with some mixing from network defects [15];

(3) the electronic structure of the liquid [12] has the interesting property that the electronic eigenvalues *cross* in the course of a thermal simulation, and by study of the electronic states conjugate to the crossing eigenvalues, one can see bond-forming and bond-breaking events. It is conjectured that these events are connected with the remarkable light-induced effects observed in the glass [118]. The time evolution of the LDA eigenvalues is given in the literature [12], and also discussed by Cappelletti in his contribution in this volume. There is a wealth of information in the "spaghetti" of this figure, especially pertaining to transport and the electron-phonon interaction and its energy dependence.

5.2. g–$Ge_x Se_{1-x}$ glasses

A particularly interesting recent micro-Raman study was by Boolchand's group [119]. A large collection of GeS and GeSe glasses with different mean coordinations were fabricated and a rather spectacular and abrupt shift in the frequency of the A_1 (tetrahedral breathing) mode was observed for mean coordination 2.46, very near the stiffness threshold predicted by Phillips [94] and Thorpe [95]. Simulations are underway to probe the microscopic origin of this important experiment.

5.3. *Amorphous carbon surface*

As an indicator of another interesting frontier, only beginning to be explored, I describe our work on an amorphous diamond *surface*. There has been surprisingly little work done on the structure of surfaces for which the bulk material is topologically disordered. To the extent that the surface is always the "interface

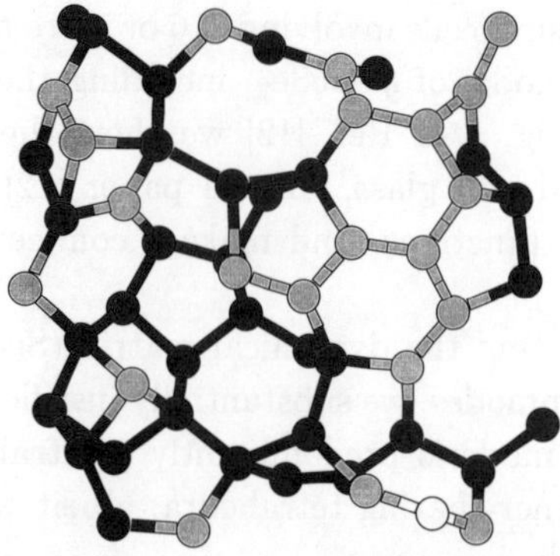

Fig. 3. Topology of relaxed amorphous diamond surface from *ab initio* MD. Note the strong tendency to graphitization and ring formation at the surface. The lighter atoms are three-fold, to emphasize the graphite-like ring formation.

to the world" this seems like an interesting area for theory and experimental work. The first work we know of on this problem was on a–Si [120].

As a first example, we started with a 216 atom supercell model of amorphous diamond [81], which we have studied previously [121]. This material is (by construction) entirely four-fold coordinated in the bulk, and is topologically very similar to amorphous Si. For this calculation, we simply maintained the periodic boundary conditions in only two dimensions (thus creating a slab geometry). This structure was then thoroughly relaxed, and naturally there were major rearrangements (the "amorphous analog of surface reconstruction") near the surface, where the local bonding environment had changed dramatically. In Fig. 3, we illustrate the surface. Note the strong tendency to graphitization (hexagonal rings), even though the bulk structure is entirely sp^3 (tetrahedral), and disordered. A similar graphitization effect is known in diamond; we establish here that it happens even for an amorphous system. Beside the intrinsic interest of an "amorphous surface", any microscopic approach to growth requires information about surface reactivity and microstructure, which this type of calculation provides directly.

6. Where to Get Codes to Get Started

For many of the computational methods discussed in this paper, it is a considerable task to develop codes which implement the calculations from the beginning. Because of this, I would like to point out that some of the necessary codes are available for free from generous individuals who took the trouble to make "user friendly" versions of codes they developed in their research, or specifically for pedagogic purposes. In order to make these hyperlinks conveniently accessible, and also to keep them reasonable up to date, I am establishing a WWW page:

http : //www.phy.ohiou.edu/~drabold/software.html

which will have links to the software I describe below and other useful codes I encounter. Of course this service is provided with the stern injunction: *caveat emptor*; I do not vouch for these codes, although naturally I will not put up links if I doubt their quality. It must also be understood, of course, that some of these codes are very sophisticated and require *much* background work before the researcher new to the area can make much progress.

Acknowledgments

This work was supported in part by the National Science Foundation under grants DMR 96-18789, DMR 96-04921 and DMR 93-22412. I have benefited greatly from discussions and collaboration with Prof. R.L. Cappelletti. I thank Prof. Normand Mousseau for providing the text on his activation relaxation technique. Peter Fedders, Otto Sankey, Pablo Ordejón, Richard Martin, Uwe Stephan, Mark Cobb and Jianjun Dong have contributed to many parts of the work I discuss here, and I thank them for all their insights and help.

References

[1] E. Holmberg, *Astrophys. J.* **94**, 385 (1941).

[2] I thank Prof. Tom Statler of our department for pointing this work out.

[3] B.J. Alder and T.E. Wainwright, *J. Chem. Phys.* **27**, 1208 (1957).

[4] A. Rahman, *Phys. Rev.* **A136**, 405 (1964).

[5] F. Stillinger and T.A. Weber, *Phys. Rev.* **B31**, 5202 (1985).

[6] An example of important pioneering work implemented with the Stillinger–Weber potential is the "inherent structure" studies of Stillnger and coworkers; see for example F. Stillinger, *Science* **267**, 1935 (1995), and references therein.

[7] M. Baskes, *Phys. Rev. Lett.* **59**, 2666 (1988).

[8] J. Tersoff, *Phys. Rev.* **B37**, 6991 (1988).

[9] P. Stumm and D.A. Drabold, *Phys. Rev. Lett.* **79**, 677 (1997).

[10] P. Ordejón *et al.*, *Phys. Rev.* **B51**, 1456 (1995). References to other order N schemes may also be found in this paper; P. Ordejón, *Comp. Mat. Sci.* **12**, 157 (1998).

[11] L. Greengard and V. Rokhlin, *J. Comp. Phys.* **73**, 523 (1987).

[12] M. Cobb and D.A. Drabold, *Phys. Rev.* **B56**, 3054 (1997).

[13] A. Angell, Amorphous Insulators and Semiconductors, eds. M.F. Thorpe and M.I. Mitkova (Kluwer, Dordrecht, 1997). p. 1.

[14] R.L. Cappelletti *et al.*, *Phys. Rev.* **B52**, 9133 (1995).

[15] M. Cobb *et al.*, *Phys. Rev.* **B54**, 12162 (1996).

[16] P.W. Anderson, *Phys. Rev.* **109**, 1492 (1958).

[17] M.P. Allen and D.J. Tildesley, Computer Simulation of Liquids (Oxford, Clarendon, 1987).

[18] D. Frenkel and B. Smit, Understanding Molecular Simulation: From Algorithms to Applications (Academic Press, New York, 1996).

[19] W.H. Press *et al.*, Numerical Recipes, The Art of Scientific Computing (Cambridge University Press, Cambridge, 1986).

[20] H.J.C. Berendsen *et al.*, *J. Chem. Phys.* **81**, 3684 (1984).

[21] S. Kirkpatrick *et al.*, *Science* **220**, 671 (1983); D.H. Vanderbilt and S.G. Louie, *J. Comp. Phys.* **56**, 259 (1984).

[22] N. Metropolis *et al.*, *J. Chem. Phys.* **21**, 1087 (1953). See also more recent references like Allen and Tildesley.

[23] P. Suttion and S. Boyden, *Am. J. Phys.* **62**, 549 (1994).

[24] D.M. Deaven and K.M. Ho, *Phys. Rev. Lett.* **75**, 288 (1995), and references therein.

[25] G.T. Barkema and N. Mousseau, *Phys. Rev. Lett.* **77**, 4358 (1996); ibid **81**, 1865 (1998).

[26] N. Mousseau and G.T. Barkema, *Phys. Rev.* **E57**, February 1998 (tentative date); N. Mousseau and G.T. Barkema, *Comput. Sci. Eng.* **1**, 74 (1999).

[27] N. Mousseau and L.J. Lewis, *Phys. Rev. Lett.* **78**, 1484 (1997).

[28] N. Mousseau and L.J. Lewis, *Phys. Rev.* **B56**, 9461 (1997).

[29] J.P.K. Doye and D.J. Wales, *Z. Phys.* **D40**, 194 (1997).

[30] S. Nosé, *Mol. Phys.* **52**, 255 (1984).

[31] M. Parrinello and A. Rahman, *J. App. Phys.* **52**, 7182 (1981).

[32] R.M. Wentzkovitch, J.L. Martins and G.D. Price, *Phys. Rev. Lett.* **70**, 3947 (1993).

[33] See for example, W. Jin, *et al.*, *Phys. Rev. Lett.* **71**, 3146 (1993).

[34] A. Nakano *et al.*, *J. Non-Cryst. Solids* **171**, 157 (1994).

[35] D.A. Drabold *et al.*, *Phys. Rev.* **B42** (5345); ibid (5135).

[36] J. Sarnthein *et al.*, *Phys. Rev. Lett.* **74**, 4682 (1995).

[37] See for example, P. Vashishta, Amorphous Insulators and Semiconductors, eds. M.F. Thorpe and M.I. Mitkova (Kluwer, Dordrecht, 1997) p. 151.

[38] A.E. Carlsson, Solid State Physics, Advances in Research and Applications, eds. H. Ehrenreich and D. Turnbull (Academic Press, New York, 1990) **43**, p. 1.

[39] W. Harrison, Electronic Structure: The Physics of The Chemical Bond (Freeman, San Francisco, 1980).

[40] J.C. Slater and G.F. Koster, *Phys. Rev.* **94**, 1498 (1954).

[41] A. Szabo and N. Ostlund, Modern Quantum Chemistry (Dover, Mineola, 1996).

[42] P. Vogl *et al.*, *J. Phys. Chem. Solids* **44**, 365 (1983).

[43] L. Goodwin *et al.*, *Europhys. Lett.* **9**, 701 (1989); see also J. Mercer and M.Y. Chou, *Phys. Rev.* **B47**, 9366.

[44] C.H. Xu *et al.*, *J. Phys. Cond. Matt.* **4**, 6047 (1992).

[45] D.A. Drabold *et al.*, *Phys. Rev. Lett.* **72**, 2666 (1994).

[46] C.Z. Wang and K.M. Ho, *Phys. Rev. Lett.* **72**, 2667 (1994).

[47] M. Menon *et al.*, *Phys. Rev.* **47**, 12754 (1993).

[48] G. Seifert and R. Jones, *Z. Phys.* **D20**, 77 (1991).

[49] D. Porezag *et al.*, *Phys. Rev.* **B51**, 12947 (1995).

[50] P. Pulay, *Theor. Chim. Acta* **50**, 299 (1979).

[51] F. Ercolessi *et al.*, *Europhys. Lett.* **26**, 583 (1994).

[52] E. Kaxiras, *et al.*, *Phys. Rev.* **B56**, 8542 (1997).

[53] M. Born and K. Huang, Dynamical Theory of Crystal Lattices (Oxford University Press, Clarendon, 1954).

[54] P.C. Hohenberg and W. Kohn, *Phys. Rev.* **B136**, 864 (1964); W. Kohn and L.J. Sham, *Phys. Rev.* **A140**, 1133 (1965).

[55] R.G. Gordon and Y.S. Kim, *J. Chem. Phys.* **56**, 3122 (1972).

[56] For a complete discussion of all the Thomas–Fermi theories, see R. Parr and W. Yang, Density-Functional Theory of Atoms and Molecules (Oxford, Clarendon, 1989). See also, L. Wang and M.P. Teter, *Phys. Rev.* **B45**, 13196 (1992).

[57] See Parr and Yang, *opt. cit.*

[58] J. Perdew, *Int. J. Quant. Chem.* **57**, 309 (1996).

[59] M.S. Hybertsen and S.G. Louie, *Phys. Rev.* **B34**, 5390 (1984).

[60] M. Stadele *et al.*, *Phys. Rev. Lett.* **79**, 2089 (1997).

[61] G.B. Bachelet *et al.*, *Phys. Rev.* **B26**, 4199 (1982).

[62] D.H. Vanderbilt, *Phys. Rev.* **B41**, 7892 (1990).

[63] N. Troullier and J.L. Martins, *Phys. Rev.* **B41**, 1754 (1992).

[64] L. Kleinman and D.M. Bylander, *Phys. Rev.* **B48**, 1425 (1982).

[65] O.F. Sankey and D.J. Niklewski, *Phys. Rev.* **B40**, 3979 (1989).

[66] S. Yang, PhD Thesis, University of Illinois 1996 (unpublished).

[67] J. Harris, *Phys. Rev.* **B31**, 1770 (1985).

[68] W.M.C. Foulkes and R. Haydock, *Phys. Rev.* **B39**, 12520 (1989).

[69] J.R. Smith *et al.*, *Phys. Rev. Lett.* **74**, 3084 (1995).

[70] A.A. Demkov *et al.*, *Phys. Rev.* **B52**, 1618 (1995).

[71] P. Ordejón *et al.*, *Phys. Rev.* **B53**, R10441 (1996).

[72] D. Sanchez-Portal, P. Ordejón, E. Artacho and J.M. Soler, *Int. J. Quant. Chem.* **65**, 453 (1997).

[73] M. Payne *et al.*, *Rev. Mod. Phys.* **64**, 1045 (1992).

[74] R. Car and M. Parrinello, *Phys. Rev. Lett.* **55**, 2471 (1985).

[75] G. Pastore *et al.*, *Phys. Rev.* **B44**, 6334 (1991).

[76] See, for example, W. Kohn, *Phys. Rev.* **115**, 809 (1959).

[77] X.-P. Li *et al.*, *Phys. Rev.* **B47**, 10891 (1993).

[78] N.H. March, W.H. Young and S. Sampanthar, The Many-Body Problem in Quantum Mechanics (Dover, New York, 1995). p. 10.

[79] U. Stephan and D.A. Drabold, *Phys. Rev.* **B57**, 6391 (1998); U. Stephan, D.A. Drabold and R.M. Martin, *Phys. Rev.* **B58**, 13472 (1998).

[80] P. Ordejón and D.A. Drabold (unpublished).

[81] B.R. Djordjevic *et al.*, *Phys. Rev.* **B52**, 5685 (1995).

[82] O.F. Sankey *et al.*, *Phys. Rev.* **B50**, 1376 (1994).

[83] S. Goedecker and L. Colombo, *Phys. Rev. Lett.* **73**, 122 (1994).

[84] F. Mauri *et al.*, *Phys. Rev.* **B47**, 9973 (1993).

[85] N. Cusack, The Physics of Structurally Disordered Materials (Adam Hilger, Bristol, 1987).

[86] See, for example, L. Pusztai and S. Kugler, *J. Non-Cryst. Solids* **164–166**, 147 (1993).

[87] P. Yu and M. Cardona, Fundamentals of Semiconductors (Springer, Berlin, 1996).

[88] M. Cobb, PhD. Thesis, Ohio University, 1997 (unpublished).

[89] O.F. Sankey, 1990 (unpublished).

[90] P. Ordejón *et al.*, *Phys. Rev. Lett.* **75**, 1324 (1995).

[91] D.A. Drabold and O.F. Sankey, *Phys. Rev. Lett.* **70**, 3631 (1993).

[92] D.A. Drabold *et al.*, *Solid State Commun.* **96**, 833 (1995).

[93] C. Lanczos, Applied Analysis (Prentice Hall, New York, 1956).

[94] J.C. Phillips, *J. Non-Cryst. Solids* **34**, 153 (1979).

[95] M.F. Thorpe, *J. Non-Cryst. Solids* **57**, 355 (1983).

[96] P.N. Keating, *Phys. Rev.* **145**, 637 (1966).

[97] D.J. Jacobs and M.F. Thorpe, *Phys. Rev. Lett.* **75**, 4051 (1995).

[98] G.L. Bretthorst, Bayesian Spectrum Analysis and Parameter Estimation (Springer, New York, 1988).

[99] D.A. Drabold *et al.*, *Phys. Rev.* **B43**, 5132 (1991).

[100] S. Aljishi *et al.*, *Phys. Rev. Lett.* **64**, 2811 (1990), and references therein.

[101] G. Adriaennsens, Amorphous Insulators and Semiconductors, eds. M.F. Thorpe and M.I. Mitkova (Kluwer, Dordrecht, 1997). p. 437.

[102] J. Dong and D.A. Drabold, *Phys. Rev.* **B54**, 10284 (1996).

[103] Jianjun Dong and D.A. Drabold, *Phys. Rev. Lett.* **80**, 1928 (1998).

[104] G.H. Golub and C.F. Van Loan, Matrix Computations (Johns Hopkins University Press, Baltimore, 1983).

[105] R.N. Silver and H. Röder, *Phys. Rev.* **E56**, 4822 (1997).

[106] R. Haydock, Solid State Physics: Advances in Research and Applications, eds. F. Seitz, H. Ehrenreich and D. Turnbull (Academic Press, New York, 1980).

[107] M. Jones and M. Patrick, "lanz"; available through netlib.

[108] D.A. Drabold and P.A. Fedders, *Phys. Rev.* **B60**, R721 (1999).

[109] T. Tiedje, Semiconductors and Semimetals **C21** (Academic, Orlando, 1984).

[110] P.A. Fedders and D.A. Drabold, *Phys. Rev.* **B53**, 3841 (1996).

[111] R. Kubo, M. Toda and N. Hashitsume, Statistical Physics II, Nonequilibrium Statistical Mechanics (Springer, Berlin, 1985).

[112] D. Thouless, *Phys. Rept.* **13**, 94 (1974).

[113] N.F. Mott and E.A. Davis, Electronic Processes in Non-Crystalline Materials, 2nd Edn. (Oxford University Press, Clarendon, 1979).

[114] L. Schiff, Quantum Mechanics, 3rd Edn. (McGraw Hill, New York, 1968).

[115] T. Iitaka *et al.*, *Phys. Rev.* **E56**, 1222 (1997).

[116] Jianjun Dong and D.A. Drabold (unpublished).

[117] S. Sugai, *Phys. Rev.* **B35**, 1345 (1987).

[118] S.R. Elliot, Physics of Amorphous Materials, 2nd Edn. (Longman, London, 1991).

[119] X.W. Feng, *et al.*, *Phys. Rev. Lett.* **78**, 4422 (1997).

[120] K. Kilian *et al.*, *Phys. Rev.* **B48**, 17393 (1993).

[121] P. Stumm *et al.*, *Phys. Rev.* **B49**, 16415 (1994).

10

Light-Induced Structural Changes in Glasses

H. FRITZSCHE

*Energy Conversion Devices, Inc.,
1675 W. Maple Road, Troy MI 48084
Phone 800-5280617; 248-2802697
sybillef@aol.com*

Contents

1. Introduction

Photo-induced structural and chemical changes generally occur in materials having low average coordination and steric hindrance of their atoms or relatively large internal free volume. These factors are also important for glass formation. Light-induced changes in material properties, photochromism, photopolymerization and photodegradation are, therefore, observed in many organic and inorganic glasses. Many of these materials exhibit another important feature promoting light-induced changes of their properties — a strong localization of the photo-excited electron-hole pairs. This localization can be the result of tight-binding character of the material, or of the disorder and lack of long-range periodicity of their structure. The localization of the electron-hole pair probably has two effects on the probability for local changes in atom configurations. It concentrates the recombination energy of the electron-hole pair in a very small volume element of the material and it changes the charge state and, hence, the valency of atoms involved in the charge localization just before recombination takes place.

Photo-induced changes in local atom configurations can produce defects and the cumulative effect of many local changes can alter the structure of a material in a major way. From these changes one can learn a great deal about the nature of defects, the recombination processes, and the relation between the structure and the properties of various materials.

This article selects four groups from the large number of materials that exhibit light-induced changes: chalcogenide glasses, photochromic oxide glasses, a prototype amorphous semiconductor, and a tight-binding metal oxide. We exclude from our discussion interesting light-induced effects which result from electronic disequilibration or space charge effects of charge migration from illuminated to dark regions if they do not involve structural changes. The article attempts to discuss our present understanding of the observed phenomena in a few typical materials. References to related materials and the wider scope of research can be found in the literature cited.

2. Chalcogenide Glasses

These materials offer a wealth of phenomena which are related to light-induced structural changes. This field got started by Ovshinsky who realized that the steric flexibility of chalcogenide glasses permits controlled changes in their local structure [1]. These, in turn, result in metastable and reversible alterations of the opto-electronic and chemical properties [2]. These features formed the basis for various electronic switching and memory devices [3], optical memories [4] and imaging applications [5]. Ovshinsky and his collaborators pointed out the important role played by the lone-pair electrons in altering the local bonding configurations in chalcogenide glasses after excitation by light, electric interactions, or by energetic particles [6]. Moreover, local light-induced structural changes result from recombination events which occur in a very small volume in these glasses because of the strong localization of the excited charge carriers [7]. These concepts and phenomena will be discussed in the following sections.

2.1. *Recombination and structural changes*

Even though glasses lack the translational symmetry of crystals, they maintain short-range order by fulfilling the chemical requirement that the valency of their constituent atoms determines the number of covalent bonds to nearest neighbors. Hence, Group IV elements (Si, Ge) are 4-fold; Group V elements, the pnictogens N, P, As and Sb, are 3-fold; and Group VI elements, the chalcogens O, S, Se and Te, are 2-fold coordinated.

2.1.1. *Valence alternation defects*

Atoms which do not obey this rule of short-range order are called coordination defects, first proposed by Street and Mott [8]. Since one missing covalent bond costs more than 2 eV energy, the lowest energy defects come in pairs — one an over-coordinated atom and the second defect an under-coordinated atom — such that the average number of covalent bonds remains the same [9, 10]. Creation of a defect pair can be written as

$$A_m + B_n \rightarrow A_{m+1}^+ + B_{n-1}^- \tag{1}$$

where the subscript indicates the covalent coordination and the superscript the charge of the atom. Charge transfer from the over- to the under-coordinated atom renders both diamagnetic and spin-paired and further lowers the energy (~ 0.7 eV) needed to create a defect pair. Since the valency of an atom depends on its outer shell, the charge transfer has changed the atom's valency and this, in turn, agrees with their new coordination number. The two defects on the right hand side of Eq. (1) are called valence alternation (VAP) defects [9, 10]. There are about 5×10^{15}–10^{17} cm^{-3} such native defects in chalcogenides, the concentration found in equilibrium at the glass transition temperature and frozen in by cooling to room temperature. Entropy considerations favor a random distribution of defects [10]. When a pair gets close and becomes an intimate valence alternation pair (IVAP), there is a good chance for self-annihilation; i.e. the inverse of the reaction described by Eq. (1) [11].

2.1.2. *Bands and localization*

The chalcogenide glasses derive their name from the dominant presence of Group VI chalcogen elements. These, as already mentioned, are 2-fold coordinated. Thus, two of their four outer p-electrons are not used in bonding. These lone-pair orbitals form the highest filled band, which is the valence band [12]. The unfilled conduction band consists of antibonding orbitals of the covalent bonds. Because of strong overlap with neighbors, the wave functions for the conduction and valence band states extend throughout the glass. The lack of atomic periodicity, however, results in tails of localized states which extend from the bands into the gap. Electrons and holes excited into extended band states localize quickly into the band tail states before recombining. This localization of the charge carriers is a general feature of amorphous and vitreous semiconductors which might justify the very local charge picture which will be used to explain a variety of structural changes [13].

2.1.3. *Interconversion of defects*

The charged VAP defects of Eq. (1) can lose a charge by photo-excitation into a band, or they can trap photo-excited electrons and holes and promote recombination after trapping another charge. When a VAP defect pair reverses its charge state by trapping sequentially two carriers of the appropriate charge, the valency has changed by 2 at each defect, forcing a local change in coordination number

$$A^+_{m+1} + B^-_{n-1} \to A^-_{m-1} + B^+_{n+1} \tag{2}$$

such that defect A is now under-coordinated and negative while B is over-coordinated and positive. Such local bonding changes are aided by the steric freedom of the low-coordination atoms of chalcogenide glasses. The interconversion property of VAPs gives rise to the negative value of their correlation energy [8, 9].

2.1.4. *Creation of metastable defects*

As mentioned in Sec. 2.1.2, the photo-excited electron-hole pair gets quickly trapped and localized in tail states. The trapped charges change the local

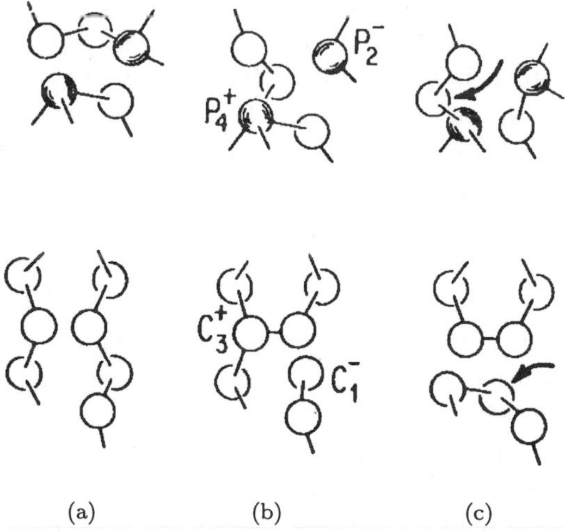

Fig. 1. Examples of elementary steps of photostructural changes. White balls are chalcogens and shaded balls are pnictide atoms. (a) Initial bonding configuration, (b) transient self-trapped exciton after photon excitation, (c) one of several new bonding configurations after recombination with motion of atom indicated by arrow.

valency and induce a bonding change which results in an IVAP as illustrated by Fig. 1(b). Such close IVAPs tend to annihilate with recombination of the electron-hole pair. It is possible, however, that the IVAP separates sufficiently by two or more bond rearrangements to become metastable [14]. Evidence for a photo-induced increase in metastable VAP defect concentration was found by Biegelsen and Street [15]. They made them neutral and, hence, paramagnetic by light-induced electron and hole trapping at low temperatures and measured that number by electron spin resonance. The photo-induced defect density is in the range 10^{18}–10^{20} cm^{-3} depending on the material. Their atomic structure is often different from that of native defects [15] because native defects are created at thermodynamic equilibrium near T_g, favoring the defects having lowest energy.

2.1.5. *Bond switching and diffusion*

The self-trapped exciton pictured in Fig. 1(b) is transient and non-radiative recombination occurs with dissipation of the rather large recombination energy via a local deformation associated with its formation and subsequent annihilation [11]. While many of the recombination processes will restore the original bonding configuration of Fig. 1(a), this does not necessarily occur and conversion to other configurations is possible [16–19], two of which are illustrated in Fig. 1(c). These depict examples of elemental steps of photostructural change. Although all the atoms in Figs. 1(a) and (c) are neutral and have their normal coordination number, the local structure has changed and bond switching has occurred. The number of homopolar and heteropolar bonds may also change in these processes as shown by Raman spectroscopy [17] and by XAFS [18].

Figure 1 demonstrates another important fact. Atoms can be displaced by more than an interatomic distance [19, 20] as illustrated by the black arrow in Fig. 1(c) which extends from the old to the new atom position. This light-induced diffusion tends to relieve strain and equalize density variations as will be shown later. Other ways leading to changes in bond configurations without involving transient self-trapped excitons have been described by Ovshinsky [6, 7], Tanaka [20], and by Krecmer *et al* [21].

Under illumination, a chalcogenide glass is in a dynamic state: local bonding changes and light-induced diffusion continue to occur even when an overall dynamic equilibrium and saturation of some macroscopic changes are established as a result of competing thermal and light-induced relaxation processes. A change in light intensity or polarization and temperature yields a new dynamic balance. Some relaxation is expected to occur when the light is turned off and a metastable structural state of energy higher than that of the annealed state gets frozen in.

2.1.6. *Additives inhibiting photostructural changes*

Certain additives to bulk chalcogenide glasses, such as Cu, Ni, Bi and Pb, greatly inhibit the occurrence of photo-induced changes [22]. It has been suggested that the bonding states with these impurities form the top of the valence band, replacing the lone-pair orbitals of the chalcogens and, thereby, diminishing the role of the lone-pair orbitals by promoting changes of the bond configurations [23]. Since only a few percent of impurities are often sufficient to inhibit the photo-induced changes [22], we suggest instead that the impurities are incorporated such that they act as efficient recombination centers, thus decreasing greatly the recombination probability via transient excitons which initiate photo-induced changes. Some of the additives act as inhibitors only in bulk glasses, but not in evaporated films, which suggests differing bonding environments for the additive atoms.

2.2. *Morphological changes*

2.2.1. *Photo-induced fluidity*

Hisakuni and Tanaka discovered that the glass transition can be induced by light in a number of chalcogenide glasses [24]. Figure 2 shows the elongation of a 50 μm thick and 0.2 mm wide As_2S_3 flake under uniaxial stress of $4.4 \times 10^7 dyn/cm^2$ in dark and during illumination by the focused beam of a 10mW He–Ne laser. Since the absorption depth of bandgap light is small compared to the thickness of their sample, Hisakuni *et al* [24] used more penetrating subgap light. The energy $h\nu = 2$ eV of the He–Ne laser is sufficient, however, to create transient self-trapped excitons in As_2S_3 whose bandgap is 2.36 eV. The effect increases linearly with light intensity. The light-induced viscosity is lowered further by decreasing the temperature (see Fig. 3) which is convincing evidence for the athermal nature of this phenomenon. Most likely, it can be observed even at helium temperatures because the recombination energies far exceed thermal energies. These observations confirm the light-induced flow reported by Feinleib *et al* [4]. That light can also relieve strain under pressure was shown earlier by Tanaka [25] who found that As_2S_3 flakes under hydrostatic pressure contract further with illumination.

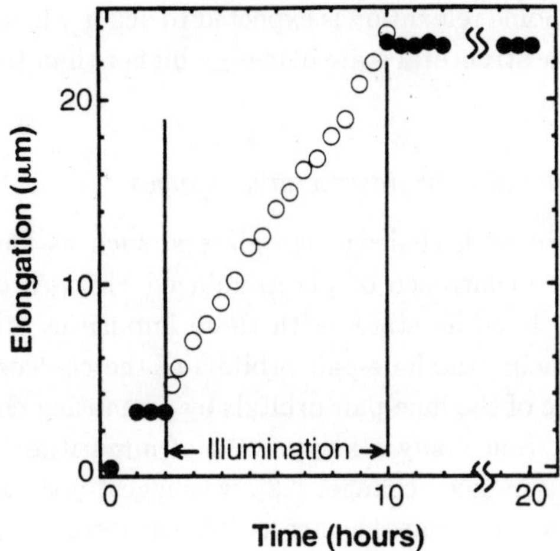

Fig. 2. Elongation of an As_2S_3 flake in the dark (solid circles) and under illumination (open circles). In this experiment, an As_2S_3 flake with a dimension of about 50 μm by 0.2 mm by 2 mm was pulled along the longest side length under a stress of 4.4×10^7 dyn/cm^2, and the elongation was measured *in situ* with an optical lever technique. After Ref. [24].

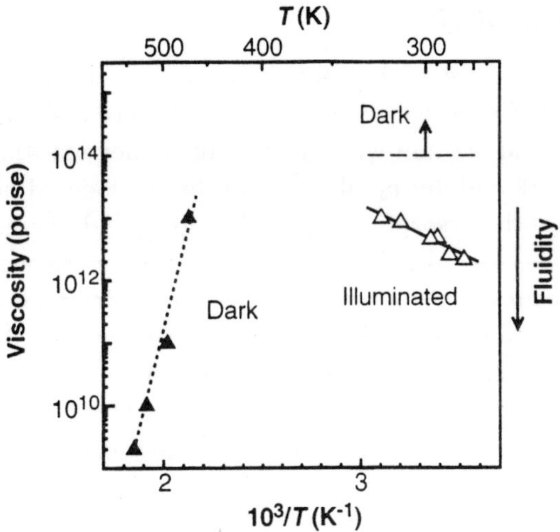

Fig. 3. Temperature dependence of the viscosity in As_2S_3 in the dark (greater than the dashed line) and under illumination (open triangles). The results were obtained from experiments of Fig. 2 at several temperatures. After Ref. [24].

As explained in Sec. 2.1.5, one can visualize the illuminated state of the glass as a dynamic state with constantly-occurring changes in local bonding and atomic motions which tend to relieve the local strain promoting flow under stress [4, 26]. It will be clear from the later Sec. 2.4.2 that one might expect the photo-induced fluidity to be greater with light polarized with the electric vector parallel to the stress axis than with perpendicular polarization.

Recently it was shown [27] that the viscosity of glasses in the dark and at ordinary temperatures is much too small to account for the alleged sagging of stained glass windows in 12th Century cathedrals during the past 800 years. Perhaps light reduces the viscosity sufficiently to make this possible.

2.2.2. *Phase changes and phase separation*

It is clear from the foregoing that light-induced fluidity and diffusion may result in photocrystallization and phase separation. Phase-change optical memory devices [28] utilize the optically-induced reversible change between crystalline and amorphous phases first demonstrated by Ovshinsky [29] who used the quasi stability of the modified eutectic composition $Te_{81}Ge_{15}S_2Sb_2$ which is very similar to the compositions being used commercially today [28, 30]. Soon thereafter, Herd *et al* [31] observed reversible amorphization and crystallization in an evaporated Se film sandwiched between two carbon layers and exposed to an electron beam at $T = 120$ K. Even though heating plays an important role in some of the laser-pulse induced phase changes, it was shown that these processes proceed also without heating [32–35].

Berkes *et al* [36] observed photodecomposition of evaporated As_2Se_3 films in air which results in micron-size crystallites of As_2O_3 on the surface. They confirmed what mineralogists have known since the end of last century — orpiment (As_2S_3) and realgar (As_4S_4) are unstable on exposure to light. This shows that photodecomposition is not restricted to the amorphous phase.

2.2.3. *Photopolymerization and giant densification*

Chalcogenide glasses and amorphous films can exist in different structural states depending on the quench rate from the melt or, for films, on details of the deposition process. Considerable confusion originates when these structural differences are not considered. For example, freshly evaporated binary or ternary films contain distinct molecular entities which are incompletely cross-linked remnants of bond-saturated molecular units that exist in the depositing vapor such as As_4 and As_4S_3 [37–39]. Moreover, voided structures with columnar growth and density deficits of up to 20% often result from self-shadowing effects when the deposition direction deviates strongly from normal [40].

These molecular and/or voided structures densify and polymerize upon exposure to light or other electron-hole producing radiation. The final structure is similar to that of an annealed glass, but not identical to it as will be explained in the next section. These irreversible photostructural changes affect most physical and chemical properties such as density, chemical dissolution, Raman spectra and the optical absorption edge. It is most likely that these major morphological and structural changes are the cumulative effect of many elemental steps of local photostructural changes as explained in Sec. 2.1.5 and illustrated in Fig. 1. Bond switching promotes cross-linking and the motion of atoms tends to equalize the density [19].

2.3. *Reversible isotropic changes*

Many photo-induced changes can be reversed by annealing the material close to its glass transition temperatures T_g. A necessary condition for reversibility is that the original material be annealed. The voided or molecular structure of a freshly-evaporated film, in contrast, is changed irreversibly. The condition of an annealed starting state is, however, not sufficient; we have to assume, in addition, that no large-scale chemical modifications occur.

Let us explore the possible structure of the light-saturated state which is reached after extended light exposure [19]. The state will depend on temperature and light intensity and, possibly, also on photon energy when saturation is caused by the dynamic balance of forward and reverse processes. It will depend on the mode of excitation, light, X-rays or electrons. It must be a state of higher energy than the annealed state and it must be a dynamic state because recombination events leading to local structural changes continue as long as the light is on. Evidence for the dynamic state is the light-induced fluidity and the reorientation of the axis of optical anisotropies discussed in Sec. 2.4.

It has been suggested [19, 41] that the light-saturated state has less medium-range order than the annealed state. This suggestion is based on (1) the observation that the first sharp diffraction peak decreases with light exposure [42], (2) the relation of the first sharp diffraction peak with medium-range order [43–45], and (3) computer simulations which show that a covalently-bonded random network can lower its energy by adopting a structure with medium-range order such as increased dihedral angle correlations [46]. In addition, the light-saturated state has more VAPs or IVAPs and more homopolar bonds than the annealed state.

Models offered to explain the light-induced phenomena discussed in the following sections fall into two groups: those that invoke changes involving most of the atoms and, therefore, structural changes in the whole material [16, 19, 41, 47] and others that are based essentially only on changes of the defects or homopolar bond concentrations [17, 20, 48, 49].

2.3.1. *Photodarkening*

As shown in Fig. 4, upon light exposure, chalcogenide glasses exhibit a redshift ΔE of their optical absorption edge. Annealing near the glass transition temperature T_g restores the original absorption. Since the absorption increases at any photon energy ($h\nu$), the reversible effect is called photodarkening. Also, the refractive index for photon energies below the optical gap increases as demanded by the Kramers–Kronig relations. The quantum efficiency for inducing photodarkening is nearly constant for light above the gap energy and in the exponential absorption tail below the gap energy [50], in contrast to earlier reports [51] which claimed that the efficiency dropped sharply for $h\nu < E_g$. Nevertheless, the saturation value of the redshift ΔE depends on $h\nu$ and light intensity. Figure 5 suggests that there is a relation between ΔE and

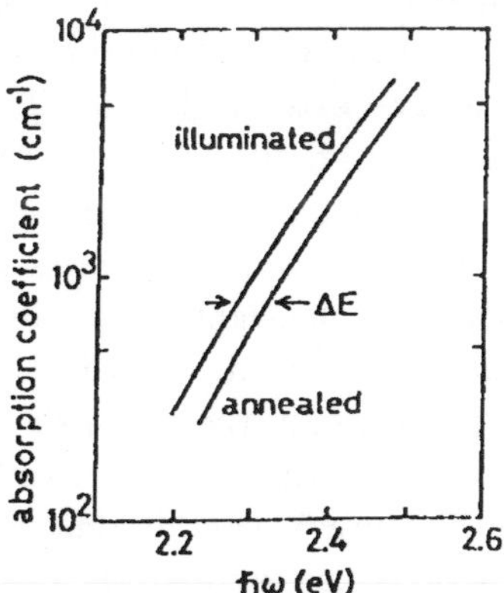

Fig. 4. Optical absorption of As_2S_3 annealed near the glass transition temperature and after illuminating it with bandgap light until saturation. The redshift ΔE of photodarkening is shown. After Ref. [20].

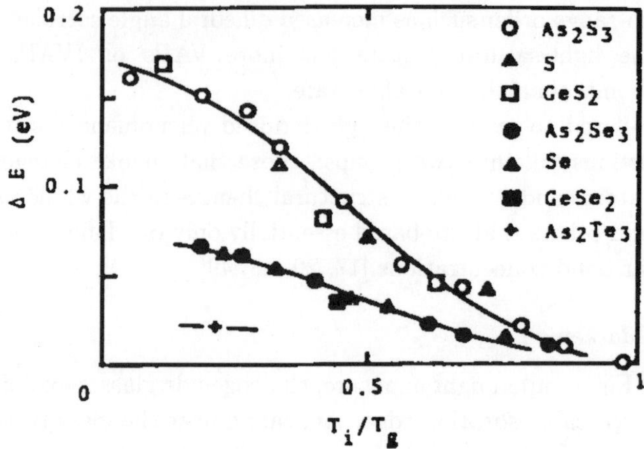

Fig. 5. Saturated reversible shift ΔE of photodarkening as a function of exposure temperature T_i normalized by the glass transition temperature T_g for different glass compositions. After Ref. [52].

the chalcogen element in the glass, sulphur yielding the largest and tellurium the smallest value of ΔE [52]. Figure 5 shows, moreover, that the redshift ΔE is largest at low illumination temperatures T_i and that it vanishes near T_g.

Annealing of photodarkening induced at 80 K begins already near 150 K and is characterized by a distribution 0.5–1.5 eV of activation energies [53]. The annealing effect is enhanced under illumination as shown in Fig. 6 [53].

This is a short summary of the vast literature on this phenomenon which has been reviewed and discussed extensively [16, 19, 20, 38, 41, 52, 54]. The relation of ΔE with the chalcogen element (Fig. 5) and the fact that the valence band is formed by the lone-pair orbitals [12] of the chalcogens led early to the suggestion that bond changes bring chalcogen atoms into new positions where the lone-pair interactions with neighbors are enhanced [6, 7, 19, 20]. The resulting broadening of the valence band will reduce the bandgap. The light-induced bond changes affect not only the lone-pair interactions and, hence, the valence band, but also the lengths and angles of the covalent bonds and, thus, the conduction band. Figure 5 suggests that the former dominate. An increase in lone-pair interactions also results from a decrease in medium-range order. The suggestion that the light-saturated state is a dynamic balance between light-induced creation of higher energy local bonding arrangements and light-induced as well as thermal relaxations agrees also with the observations

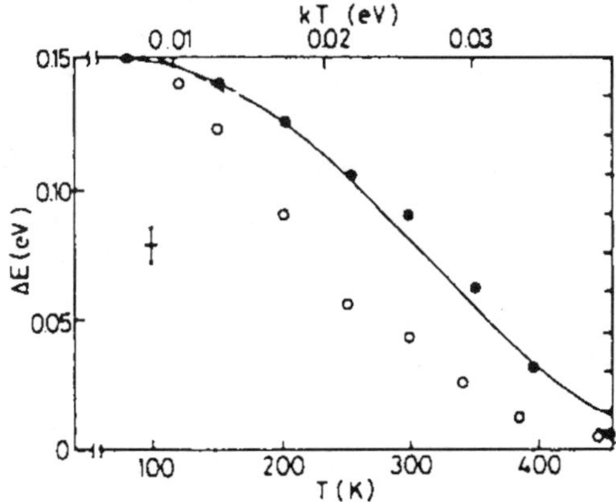

Fig. 6. Photodarkening shift ΔE for As$_2$S$_3$ induced by bandgap light at 80 K as a function of annealing in dark (solid symbols) and under simultaneous illumination (open symbols). After Ref. [53].

(Fig. 5) that ΔE is smaller for high T_i and that annealing is enhanced by illumination (Fig. 6). Such dynamic balance implies that a change in light intensity or photon energy at fixed temperature will yield a change in the light-saturated state and, hence, in absorption. These phenomena have been observed [55, 56].

An increase in defect density and homopolar bonds has been observed, but their concentration remains in the percent range or less and should affect photodarkening and related effects in only minor ways. In Sec. 2.1.3, we discussed interconversion of defects which can be induced by midgap light which promotes electrons or holes from defects into band states. Those processes must also be negligible for photodarkening because Tanaka and Hisakuni [50] observed that its quantum efficiency is essentially the same for bandgap and subgap Urbach tail light, but is practically zero for light absorbed in the defect band.

Moreover, the increase in concentration of homopolar bonds, such as As–As in As-rich glasses has been suggested as origin for photodarkening. However, many of these glasses exhibit an increase in gap energy relative to the stoichiometric glass instead of the expected decrease. Furthermore, elemental sulphur exhibits photodarkening (see Fig. 5) although all bonds are homopolar. From this, we conclude that a light-induced increase in the number of homopolar bonds is not the cause of photodarkening.

2.3.2. *Structural changes*

Although redshifts $\Delta E \sim 0.15$ eV suggest changes of the whole bonding network, one prefers to have direct evidence. One is the reversible photo-expansion of about 0.4–0.8 percent accompanying photodarkening [57, 58]. In a recent review, Tanaka [59] contrasts this with the 1–3% volume contraction of vitreous SiO_2 exposed to energetic neutrons or electrons. The other piece of evidence is a decrease and shift of the first sharp diffraction peak (FSDP) with light exposure [42], as shown in Fig. 7. This FSDP is characteristic of glasses [60] and is associated with medium-range order on the 7–12 Å length scale. An intensity decrease signifies a decrease in medium-range order which involves the whole bonding network. Further evidence is the optomechanical effect discussed in Sec. 2.4.2.

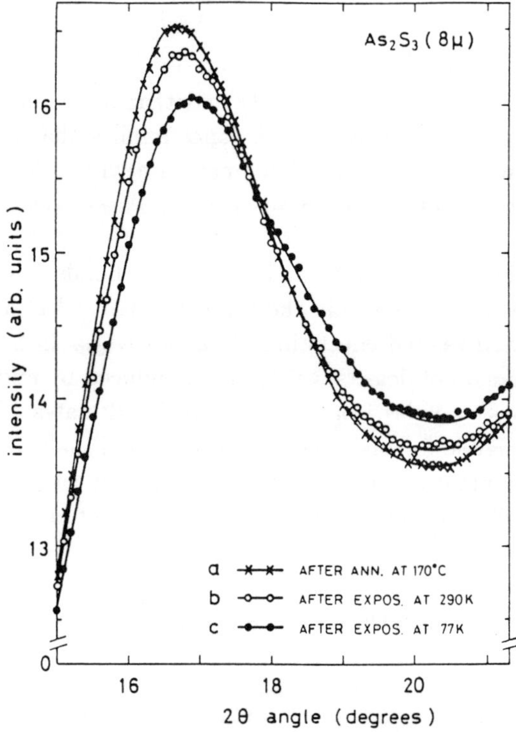

Fig. 7. First sharp diffraction peak of As$_2$S$_3$ after annealing and illumination at 299 K and 78 K. The light-induced changes are reversible and reproducible. After Ref. [42].

2.4. *Photo-induced anisotropies*

Optically isotropic materials such as chalcogenide glasses can become optically anisotropic because they consist of and contain entities which are optically anisotropic [61]. The original macroscopic isotropy originates from the random orientations of the microscopic anisotropic entities. A recombination event which leads to a structural change of a microscopic anisotropic entity will change the orientation or nature of this anisotropy. This constantly happens everywhere in the material during illumination without, however, necessarily producing a macroscopic anisotropy. For this to happen, it is necessary that the recombining electron-hole pair is excited in the same microscopic anisotropic entity which undergoes the structural change. This means essentially that macroscopic anisotropies result from geminate recombinations of electron-hole pairs which do not diffuse out of the microscopic entity in which they were created by absorbed photons [61]. The lack of electron-hole pair diffusion and the geminate nature distinguish the recombination events leading to anisotropies from all the other events which yield isotropic (or scalar) photo-induced changes. This important difference is the cause for the fact that the dependencies on temperature, light intensity and photon energy, among other parameters, are different for the anisotropic (vectoral) and isotropic (scalar) photo-induced effects. What are these microscopic optically anisotropic entities in chalcogenide glasses?

Following the discussion of Sec. 2.1, we first mention IVAPs whose dipole moment will be changed and reoriented if they are involved in a photostructural bonding change [62]. Close IVAPs, however, are unstable against self-annihilation and somewhat more distant IVAPs are more difficult to reorient in one recombination event. As a second possibility, we note that interconversion of distant VAPs (Sec. 2.1.3) produces local bonding changes and, hence, a change in local anisotropy [63] at the two sites, but the distance might be too large for most defects to preserve the polarization memory of the absorbed photon. On the other hand, if a VAP defect is included in the microvolume in which absorption of the polarized photon as well as recombination take place, then this can contribute to a macroscopic anisotropy. However, because of the low covalent coordination of the chalcogens, any other microvolume of the material is also optically anisotropic [61] and altered by a photostructural change (Sec. 2.1.5) as illustrated, for example, in Fig. 8. Hence, all optical transitions, i.e. interband, Urbach tail, and defect transitions, are polarization dependent. Each elemental step of photostructural change alters the local

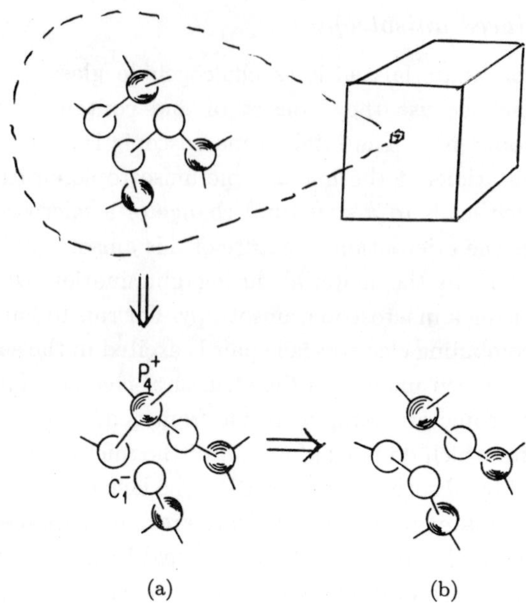

(a)   (b)

Fig. 8. On a nanometer scale, each microvolume is optically anisotropic. Absorption of favorably polarized light creates a self-trapped exciton (a). The local bonding structure is changed after recombination (b) with a chance of creating a microvolume with a changed anisotropy. C stands for chalcogen and P for pnictogen.

anisotropy. However, a macroscopic anisotropy can result only from non-radiative recombinations of electron-hole pairs that have not diffused away. Which of the various optical excitations fulfills this requirement can be determined by measuring the quantum efficiency for a given anisotropic effect as a function of photon energy.

2.4.1. *Photo-induced dichroism and birefringence*

Zhdanov *et al* [64] discovered that exposure to linearly polarized light produces an anisotropy of the dielectric tensor which results in dichroism and birefringence. An example is shown in Fig. 9(a) taken from the work of Krecmer *et al* [21]. Here the dichroism given by the parameter

$$D = \frac{2(I_{11} - I_1)}{I_{11} + I_\perp} \tag{3}$$

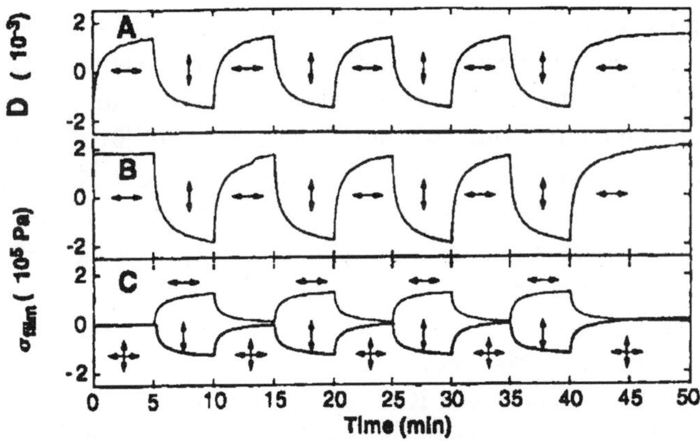

Fig. 9. Photo-induced dichroism D and stress σ in AsSe caused by linearly polarized light. (A) and (B) Orthogonal polarizations were altered. (C) Unpolarized and polarized light (in two orthogonal directions) were altered. After Ref. [21].

is produced with a He–Ne laser in a 2 μm thick film of AsSe. I_{11} and $I_\perp$ are the transmitted intensities of a weak probe beam polarized parallel and perpendicular to an arbitrary y-direction while traversing the sample in the z-direction. The dichroism D is observed to become positive for exposing light polarized parallel to the y-direction ($\leftrightarrow$) and negative for light polarized perpendicular to it [$\updownarrow$ in Fig. 9(a)]. It is most remarkable that the axis of the optical anisotropy can be turned essentially without fatigue by turning the polarization direction of the exposing light. Isotropy is restored by annealing. The kinetics and magnitude of the dichroism may [65] or may not [66, 67] be the same in the annealed and the photodarkened state depending on the magnitude of the structural and density difference between the two. The results shown here taken in the photodarkened state illustrate again the dynamic nature of the light-saturated condition where photostructural changes continue to occur although changes in the macroscopic properties, such as photodarkening, have come to saturation.

The sign of D and that of birefringence, the absorption coefficients $\alpha_\perp > \alpha_{11}$ and refractive indices $n_\perp > n_{11}$ are the same for all chalcogenide materials studied. They indicate that anisotropic microvolumes, which preferentially absorb light polarized parallel to that of the exposing light, become photostructurally changed microvolumes whose bonding arrangements are such that they preferentially absorb light polarized in the perpendicular directions.

Since the polarization direction of the exposing light determines which microvolumes absorb the most and undergo structural changes, the oscillator strength will be redistributed and the axis of anisotropy can be reoriented by changing the polarization direction of the exposing light [61].

The anisotropic matrix elements governing optical excitations within a microvolume will depend on the photon energy. Therefore, if the photon energy of the probe light differs greatly from that of the exposing light, it is possible that the photo-induced anisotropy disappears at a certain energy of the probe light or reverses its sign [68]. While dichroism depends on the anisotropic absorption coefficient at the photon energy of the probe, the refractive index which governs birefringence is a Kramers–Kronig weighted average of excitations over all energies larger than the probe energy. The reflectivity near the bandgap energy is solely governed by the refractive index while the extinction coefficient need be considered only at relatively high photon energies. A more complete picture of the physical processes would require measuring the induced anisotropy of the complex dielectric tensor as a function of photon energy.

These arguments can clearly be extended to anisotropies involving circularly polarized light because the small coordination number of chalcogenide glasses provides a local structure with sizable helicity [61, 69]. Moreover, since electromagnetic waves are anisotropic because of their transverse nature, even unpolarized light exposure will produce an optical anisotropy with an optical axis in the direction of propagation [61]. This prediction has been confirmed by experiment [70, 71].

Some chalcogenide glass compositions might not exhibit photodarkening because they might not develop a pronounced medium-range order in their relaxed annealed state and, hence, might not experience any appreciable increase in the lone-pair interactions in the light-saturated state. These same glasses may, nevertheless, show photo-induced optical anisotropies due to changes in local bonding structure.

2.4.2. *Anisotropic optomechanical effect*

The panels (b) and (c) of Fig. 9 show the discovery of a reversible anisotropic volume change accompanying the dichroism induced by polarized light. Krecmer *et al* [21] deposited a thin AsSe film on a micro-cantilever which is part of an atomic force microscope. The deflection of the cantilever caused by volume change of the film was recorded. Figure 9(b) shows the change in stress in the film as it is exposed to a polarized He–Ne laser. The y-direction mentioned in the previous section is along the axis of the cantilever and positive stress σ means contraction of the film. Panel (c) of Fig. 9 shows a superposition of two sequences. In the upper one light polarized along the cantilever axis is alternated with unpolarized light and in the lower one the axis of polarized light is turned by 90°. This experiment shows that the anisotropies produced extend to other material properties besides the optical tensor. The magnitude of the effect suggests that the anisotropic microvolumes of the whole material are involved and not only IVAP defects. This is also the interpretation of the authors [21]. These new results imply that the elastic properties, sound propagation and probably many other material parameters of chalcogenide glasses become anisotropic with light exposure.

2.4.3. *Light scattering*

Many experiments studying photo-induced anisotropies are conducted by sending a laser beam of inducing polarized light through the sample and measuring, with a polarized probe beam of greatly reduced intensity, the transmitted light intensity and its polarization. Inhomogeneities and density fluctuations will cause scattering which, in addition to absorption, decreases the transmitted light intensity [72]. Scattering can be measured separately in directions away from the beam axis [73].

Inhomogeneities are self-enhancing as localities, which photodarken, absorb more photons from the inducing beam and darken further. The increase in refractive index associated with photodarkening produces light scattering inhomogeneities. With polarized inducing light, all these phenomena carry isotropic as well as anisotropic components. The scattered light, in turn, produces photo-induced changes and inhomogeneities while passing through the material. Concurrently, the inducing light beam experiences self-focusing due to the increased refractive index caused by photodarkening in the laser beam channel. It is clear, therefore, that the recombination rates leading to photo-structural changes depend on time and location even when the intensity of the inducing light remains constant.

These complex and interrelated processes may explain why one observes such a variety of interesting and often puzzling phenomena in these experiments.

2.4.4. *Anisotropic crystallization*

Lyubin *et al* [74, 75] discovered recently that the properties of photodarkening and dichroism induced in $Se_{10}Ag_{15}I_{15}$ films by a 2.75 W/cm^2 polarized He–Ne laser beam were very unusual and large as well as irreversible. They found the formation of oriented Se, Ag_2Se and AgI microcrystals producing the large dichroism whose sign depended on the polarization of the inducing light. One wonders whether the orientation of crystallites is a strain-relief process of the strains caused by the optomechanical effect (Sec. 2.4.2).

A different mechanism was proposed recently to explain anisotropic laser crystallization of a–Se [76, 77]. These authors suggest that nearest-neighbor IVAPs reorient without bond breaking by excitation of polarized light [78] and then form oriented nuclei for crystal growth. On the other hand, one should keep in mind the low probability [10] for the presence of IVAPs, particularly in a–Se, and their tendency to self-annihilate [11].

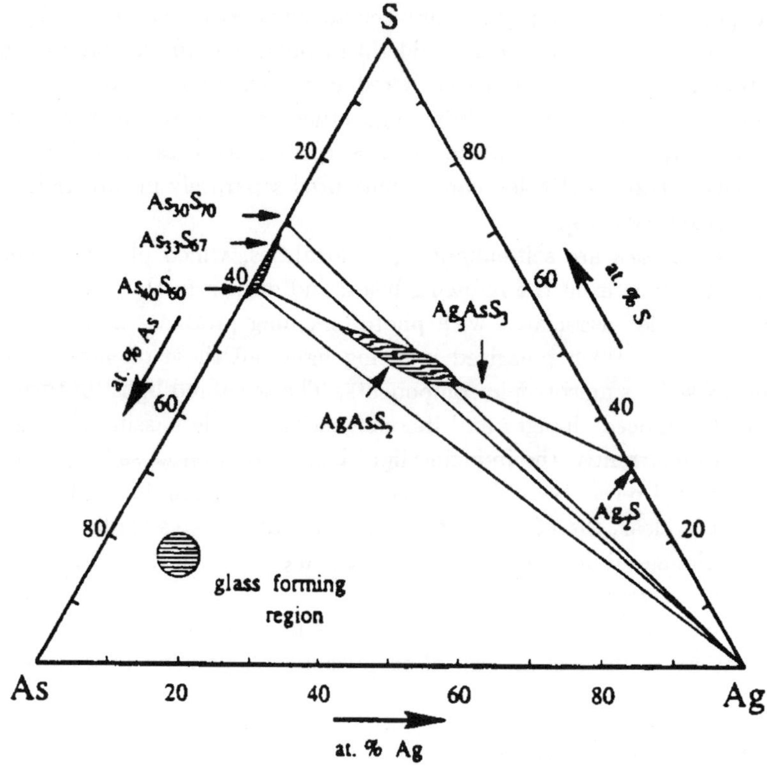

Fig. 10. Phase diagram of the Ag–As–S system showing the regions of glass formation and the location of known crystalline compounds. After Ref. [82].

2.5. *Photo-induced chemical modification*

During the past 30 years since the discovery [79] of photo-induced metal dissolution in chalcogenide glasses, a large number of experiments have been conducted to elucidate this and related photo-induced chemical modification [80, 81]. The topic has been the subject of several reviews which give a guide to the literature. Only the major features and problems can be highlighted here. Of all the metal-chalcogenide systems studied, we restrict our discussion to the As–S–Ag ternary shown in Fig. 10 and predominantly to the $Ag/As_{33}S_{67}$ system because this is one of the few systems which remains a homogeneous glass during the chemical modifications [80, 82]. Even after choosing the simplest model system, one finds that the available experimental data from different laboratories are sometimes difficult to compare because the photocarrier generation rates in the photosensitive regions are often not published and the photodissolution rates are influenced at the beginning by photopolymerization processes (see Sec. 2.2.3) occurring in freshly-deposited chalcogenide films.

It should be noted that the size of a glassforming region depends on the experimental conditions and increases with the quench rate. The data in Fig. 10 was taken from the work of Kawamoto *et al* [83]. A much larger region of glass formation was found by Kawaguchi *et al* [84].

2.5.1. *Photodissolution of Ag*

When a layer of Ag is deposited onto or below a chalcogenide film, Ag rapidly dissolves in the film when the sandwich is exposed to bandgap light. The boundary of the alloyed region is quite sharp and progresses into $As_{33}S_{67}$ as shown in Fig. 11 [85]. The non-linear region, which is alternately described by a $t^{1/2}$ or a [1-exp(-at)] time dependence, corresponds to the period during which the Ag source is being consumed [86]. During this period, a well-defined reaction product having a composition close to $Ag_{40}As_{20}S_{40}$ advances with a step-like profile. Once the Ag film is depleted, the reaction product becomes the source of Ag and photodissolution continues with a linear time dependence as shown in Fig. 11 [87]. The final Ag concentration depends on the thickness of the remaining unreacted chalcogenide [87]. It would be interesting to know the lower limit of this concentration. The photodissolution rate increases nearly linearly with light intensity and is thermally activated with $E_A \sim 0.1$–0.13 eV. The dependence of E_A on light intensity is not known. The dissolution rate shown in Fig. 11 is about 0.1 nm/s in the linear region.

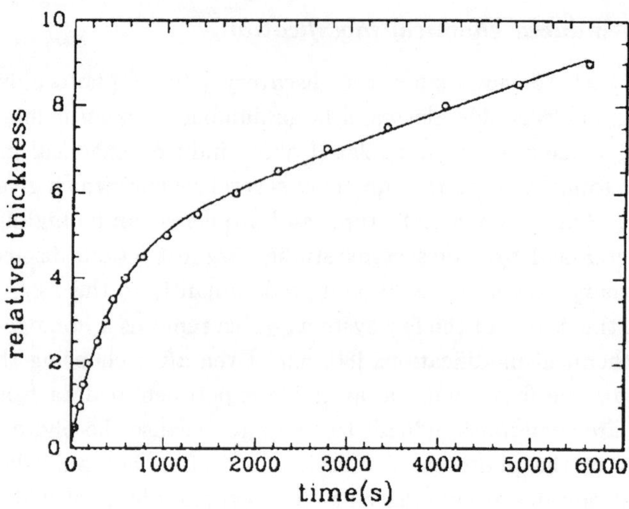

Fig. 11. Increase of Ag-alloy thickness as a function of illumination time. After Ref. [85].

An important insight into the dissolution mechanism is gained by studying the lateral progression of the alloy front into a chalcogenide film which extends sideways beyond the Ag-chalcogenide sandwich, first explored by Matsuda *et al* [88]. Here, Owen *et al* [80] conducted a crucial experiment — one part of the sample was deposited on glass while another part had an underlayer of Au which does not react with the chalcogenide film. Under illumination, the photodissolution front hardly moved laterally on the glass, but on the Au-covered substrate it progressed with a velocity of 1 μm per min as shown in Fig. 12. The large intercept at $t = 0$ is probably an artifact caused by the finite width of the probe light used to determine the front's position by optical transmission. This experiment established that only illumination of the alloy front is important. Illumination of the metal or of the rest of the alloyed or unalloyed chalcogenide film has little effect on the dissolution rate. The alloy front moves under illumination until the source of Ag is depleted.

The conducting (Au) underlayer promotes fast lateral photodissolution. It assures equipotential in the Ag-alloy and chalcogenide film. The driving force of the Ag motion to the moving front must, therefore, be the Ag chemical potential difference. The Ag-chalcogenide alloy is a fast ion conductor with a conductivity in excess of 10^{-6} ohm^{-1} cm^{-1} which increases with Ag content [89]. Assuming 20% Ag content, the dissolution rate of 1 μm per min shown

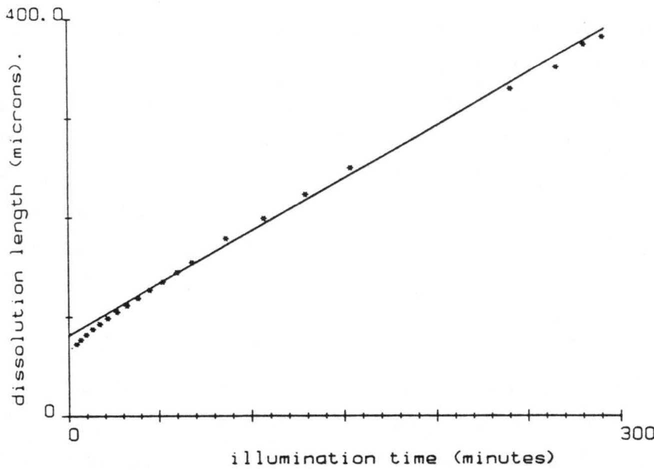

Fig. 12. Lateral progression of photo-dissolution front on metallized substrate. After Ref. [80].

in Fig. 12 for a 30 nm thick As_2S_5 film corresponds to a flux of 1.6×10^{16} Ag/cm^2s or an Ag^+ current density of $2.5\ mA/cm^2$.

Dale *et al* [82] compared the activation energies of the various processes involved to determine the rate-limiting step. Since the activation energy $E_A \approx$ 0.15 eV (depending on light intensity) for photodissolution is considerably smaller than those for Ag self-diffusion (0.55 eV) and for Ag^+ conduction (0.55 eV) in the Ag-chalcogenide alloy, they concluded that motion of Ag in the alloy cannot be rate limiting because one would then expect 0.55 eV to be the activation energy for photodissolution. The smaller value of about 0.15 eV must, therefore, be associated with the Ag-alloying process at the interface with the Ag-free chalcogenide film. The nearly flat Ag profile in the alloy and the sharp Ag concentration gradient at the front suggest also that the bottleneck is at this interface. Moreover, that is the place where illumination keeps the process going. In contrast, thermal dissolution of Ag is activated by a much larger energy of 1.4 eV.

We suggest that photodissolution is driven by the gradient in Ag chemical potential but it is made possible by the photostructural changes which, adjacent to the interface, provide constantly altering bonding arrangements and motions of atoms as described in Sec. 2.5. The dynamic state which also yields photo-induced fluidity lowers the activation energy for Ag incorporation.

Once the Ag alloy is formed, it becomes part of the fast ion conductor supplying more Ag to the new interface.

We now discuss the question why photodissolution is slower and follows approximately a $t^{1/2}$ time dependence when the fast ion-conductor, i.e. the Ag-alloy, is not at equal potential. Note that the rate of Fig. 11 is more than a factor 1000 slower than that of Fig. 12. Unfortunately, the photocarrier generation rates in these two samples were not given; however, we know that lateral photodiffusion on an insulating substrate with the same light as in Fig. 12 did not noticeably occur [80]. If Ag transport is via Ag^+ ions, then current continuity demands an equal and opposite current carried by the dominant electronic charge carriers (presumably holes) in the Ag-chalcogenide alloy. These may be dark as well as photocarriers flowing in the (Nernst) electric field set up by the advancing dissolution front and directed toward the Ag metal [90]. The higher (photo) conductivity of selenide glasses, compared to sulphur glasses, might explain the larger dissolution rate of the former in addition to the basic thermodynamic arguments [91]. Experiments of lateral photodissolution on conducting and insulating substrates can distinguish the photochemical from the photo-electronic component of this phenomenon. Quantitative comparisons require information about photocarrier generation rates, their effect on the activation energy of photodissolution and about the electronic and ionic conduction of the Ag-alloys.

We now discuss the important question of the structure of the Ag–As–S alloys produced by photodissolution. It is believed that these alloys have similar structure to bulk-quenched glasses of the same composition [92]. The latter have been studied by neutron diffraction and isotopic substitution. One finds that Ag is threefold coordinated by chalcogens in the ion-conducting glasses of high Ag content in both the Ag–As–S and Ag–As–Se systems [92]. The As atoms remain threefold coordinated by the chalcogens. A similar situation holds in vitreous $Ag_4Ge_3Se_9$ [93]. The Ge atoms are fourfold and the Ag atoms threefold coordinated by Se. This increased demand for chalcogen coordination can only be satisfied if many of the chalcogens increase their coordination from two to three as Ag is added.

In order to describe the Ag-chalcogen bonds in greater detail, we follow the arguments of Kastner [94], who predicted fourfold coordination of metal atoms by chalcogenide atoms provided that d-states are not involved and no steric hindrances interfere. In Kastner's scheme, one of the Ag-chalcogen bonds is a normal covalent bond because Ag is monovalent, while the others are coordinate bonds. Bonds are called coordinate when both bonding electrons are supplied by one of the atoms. The other atom supplies empty orbitals. Chalcogenides can form coordinate bonds to metal atoms with empty orbitals by supplying their lone-pair electrons. The chalcogen becomes threefold coordinated in the process. The covalent and coordinate bonds are similar enough that neutron diffraction cannot tell the difference. The large chemical shift of the Se $3d$ core level in Ag-photodissolved Ge_xSe_{1-x} films observed by XPS [95] and electron-energy-loss spectroscopy [96] is evidence that lone-pair electrons are involved in bonding of Ag to the chalcogen atoms. Why Ag is found to be threefold and not fourfold coordinated as predicted by Kastner [94] is not clear.

Ag, threefold coordinated by chalcogens, forms a neutral entity even though there is a shift of negative charge as a result of ionicity from Ag to the chalcogen in the normal covalent bond. The Ag^+ ions which obviously are present in the fast-ion conducting Ag-As-chalcogen glasses must be differently bonded and the compensating negative charge must be dissociated from the Ag^+ ion and bound elsewhere in the structure. We suggest that the Ag^+ ion forms three coordinate bonds with neighboring chalcogens and that the negative charge is located at a twofold coordinated As atom or a onefold coordinated chalcogen. This follows the rules of valence alternation (Sec. 2.1.1) which associate the addition of an electron to a decrease in valency and, thus, a decrease in coordination of pnictide and chalcogen atoms. The charged bonding configuration represents a higher energy state, but the dissociation of the charges is favored by enhanced entropy effects [94].

The fact that Ag^+ is bonded by coordinate bonds only and can move without breaking a normal covalent bond may partly explain its high mobility. Upon leaving a site, the coordinate bond electrons revert back to lone-pair electrons while the reverse happens at the new site, lone pair electrons are ready to form coordinate bonds. It would be interesting to test whether this is the preferred bonding scheme and to measure the concentration of Ag^+. The ion Cu^+ might be similarly bonded in Cu–As–S and Cu–As–Se, even though it has a smaller diffusivity than Ag^+. Even though the charged species are diamagnetic, they can be made paramagnetic by capture of a photo-excited charge carrier at low temperatures and detected by light-induced electron-spin-resonance.

2.5.2. *Reversible chemical modification of Ag-chalcogenides*

The Tanaka group [97] discovered that Ag migrates in Ag-chalcogenide glasses to a region which is exposed to light or a 10–50 keV electron beam from the surrounding dark region. The increase in Ag concentration is about 8%. The Ag-chalcogenide glass can be made either by quenching from the melt or by photodissolution of Ag into a chalcogenide film (see Sec. 2.5.1). The photo-induced Ag motion can be reversed by changing the illumination region. Larger changes in Ag concentration are accompanied by a volume expansion [98].

2.5.3. *Photo-induced surface deposition of Ag*

Maruno *et al* [99] and Kawaguchi *et al* [84, 100] discovered that light induces a migration of Ag to the surface of silver-rich Ag-chalcogenide films. Annealing causes the metal to redissolve back into the glass. This suggests that the Ag content is above the solubility limit at room temperature and that light induces bonding changes and diffusion (see Sec. 2.1.5). The reverse process of annealing indicates that the Ag solubility increases with temperature. The formation of Ag_2S crystallites, which happens at higher anneal temperatures and is thermodynamically expected [91], stops the photo-induced surface deposition process. See also Mitkova's chapter in this book.

3. Photochromic Glasses

Photochromism — a change of color resulting from exposure to light — occurs in many organic and inorganic materials [101]. Complete reversibility, as found in chalcogenide glasses for instance (see Sec. 2.3.1), is much rarer. Here we discuss photochromism in alkali metal- or aluminum-borosilicate glasses which contain copper sensitized silver halide crystallites. These glasses, often used for ophthalmic lenses, darken when exposed to UV light and clear again when irradiation is stopped. We devote a section to these glasses even though the phenomena occur in the crystallites, because the photochromic mechanism has been thoroughly investigated and, therefore, might serve as a benchmark for photo-induced structural changes.

Proper heat treatment of these glasses which contain less than 1 wt.% of silver halide and copper-oxide sensitizer produces a colloidal segregation of about 5×10^{15} cm^{-3} silver halide crystallites which range in size between 50 and 120 Å [101, 102]. We write AgCl (Cu) for the silver halide because the mechanism is the same for AgBr (Cu) or solid solutions of the two halides.

The darkening produced by the UV light is a neutral grey extending through the near UV and the entire visible spectrum. This is caused by a near monolayer of Ag partially covering the crystallite surface. Light absorbed in the darkened spectral region yields accelerated bleaching.

Flash photolysis revealed a very fast darkening followed by a slow darkening whose rate is inversely proportional to some fractional power of the AgCl crystallite diameter [103]. This slow component is associated with the motion of Ag^+ to the surface, and along the surface, forming the silver aggregate.

Electron paramagnetic resonance studies were crucial in revealing the following details of the darkening [102] and bleaching [104] mechanisms. Cu^+ is known to substitute for Ag^+ in the crystal. The holes of the photo-excited electron-hole pairs are quickly captured forming divalent Cu^{++} while the electrons drift to surface states. The hole capture releases sufficient energy to kick a neighboring Ag^+ ion into an interstitial site, thus creating a nearest neighbor silver vacancy V_{Ag} in the [110] direction. It turns out that this (Cu^{++} V_{Ag}) complex is the only stable defect configuration while the light is on. This is so because the effective negative charge of the silver vacancy makes this complex overall neutral so that it has a very small cross-section for a recombining electron capture. The creation of the $Cu^{++}V_{Ag}$ complex, which has been observed to occur also with other metal dopants [105, 106], is the primary photostructural step. The interstitial Ag^+ ion drifts to the crystallite surface and becomes neutralized by the photo-electron while aggregating. The low thermal activation energy of the Ag^+ diffusion indicates that it moves by a collinear interstitialcy jump which means that the interstitial Ag^+ displaces a lattice Ag^+ which then becomes an interstitial in a collinear direction [102]. The rapid darkening component is attributed to optical transitions involving the Cu^{++} V_{Ag} defect complex [103].

The mobile Ag^+ interstitials are essentially all created during the trapping process of the holes at Cu^+ dopants. There are about 90–100 Cu^+ in a 100 Å AgCl crystallite [101]. Hence, at most, 90–100 Ag^+ reach the surface. Why do they aggregate? The process is the same as the formation of the latent image in a photographic emulsion [107]. There are very few surface states for electrons. The first interstitial Ag^+ reaching the surface and getting neutralized there becomes a low-lying electron trap. After capturing an electron, another interstitial Ag^+ is attracted, neutralized, and ready to trap the next electron and so forth. The scarcity of competing electron traps at the surface causes only one silver aggregate to grow.

While in a photographic emulsion the silver aggregate of about 10 Ag atoms on a much larger AgCl crystal is quite stable and is waiting to catalyze the reducing process of the developer, the 90 Ag atom aggregate on the 100 Å size AgCl crystallite of the photochromic glass disappears after the UV light is turned off. Why? The darkening process cannot be simply reversed because there is no incentive for the Ag aggregate to disperse in the form of Ag interstitials. The paramagnetic resonance studies provide the answer [104].

During illumination, the $Cu^{++}V_{Ag}$ defect is the only stable one because of its small electron capture cross-section. After the light is turned off, two processes unfold. The nearest-neighbor vacancy of some Cu^{++} V_{Ag} moves to a second nearest position, becoming a Cu^{++} $Cl\text{-}V_{Ag}$ complex. Further dissociation of V_{Ag} from Cu^{++} seems to be hindered by the Coulomb attraction between the effective negative charge of V_{Ag} and Cu^{++}. This problem is overcome, however, by transferring the hole from Cu^{++} to a lattice Ag^+ next to the vacancy, forming $Ag^{++}V_{Ag}$ and leaving behind Cu^+. This neutral defect complex $Ag^{++}V_{Ag}$ diffuses now to the surface and annihilates with Ag^o of the silver aggregate, thereby reversing the darkening process. The activation energy 0.3 eV of the thermal bleaching process indeed agrees with that of vacancy motion.

A noteworthy feature of the Cu^+ sensitizing center is that hole capture leads to a structural change, in this case a vacancy formation, which compensates the hole charge and, thus, greatly inhibits recombination. In addition to the processes yielding silver deposits, there are, of course, many recombinations via self-trapped holes and electron capture which lower the efficiency of the process. It should be mentioned that the unusually high dielectric constant of silver halides, which arises from the large quadrupolar deformability of the silver ions, decreases Coulombic energies such that the interstitial Ag^+ can move as described without capturing a photo-electron. Finally, a connection between this section and chalcogenides can be found by noting that silver sulphide (Ag_2S) is an indispensable sensitizing agent attached to the silver halide crystals in photographic emulsions [107].

4. Hydrogenated Amorphous Silicon

Amorphous Si is tetrahedrally coordinated and should, therefore, resist light-induced changes of its structure. However, because of its high coordination, the structure is very overstrained and relieves this strain by forming a high concentration (10^{20} cm^{-3}) of broken (or dangling) bonds. These coordination defects have electronic states in the gap center which act as efficient electron-hole recombination centers. By hydrogenating this amorphous silicon (a–Si:H), the defect density is reduced to a tolerable level (5×10^{15} cm^{-3}) and the material can be doped n-type and p-type and becomes useful for numerous applications [108, 109]. However, in 1977 Staebler and Wronski discovered [110] that exposure to bandgap light decreases the photoconductivity in doped and undoped a–Si:H as shown in Fig. 13. The original state can be restored by annealing near 180°C. Electron-spin-resonance experiments demonstrated that the cause for the degradation is the photo-induced creation of recombination centers. The efficiency is relatively low as more than

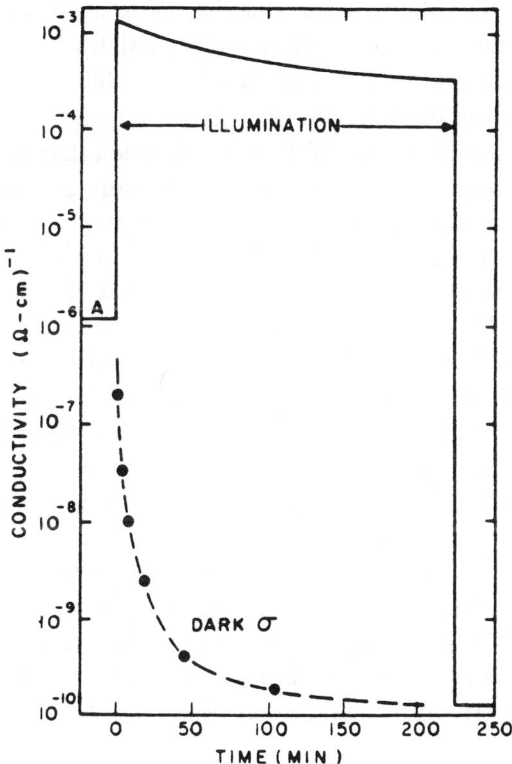

Fig. 13. Decrease of photoconductivity (solid curve) and of dark conductivity (dashed curve) of a–SiH with time of exposure to 0.2 W/cm^2 filtered tungsten light. The dark conductivity decreases only when the Fermi level is not near the gap center in the annealed state. After Ref. [110].

10^{26} cm^{-3} recombination events are needed to reach the saturation value of about 10^{17} cm^{-3} defects for continuous light and about 10^{18} cm^{-3} for short-pulse laser irradiation.

The Staebler–Wronski effect (SWE) is one of several metastabilities of a–Si:H and its related alloys. Other metastabilities are produced by rapid quenching from $T > T_E$, the dopant and defect equilibration temperature [111], or by carrier accumulation in field-effect transistors [112]. These other metastabilities are caused by changing equilibrium conditions while the SWE is produced by electron-hole-producing excitations under non-equilibrium conditions. Nevertheless, the annealing kinetics of all these metastabilities show remarkable similarities [113], and the underlying mechanism is presumed to be closely related to hydrogen motion which is the basis for understanding the equilibration kinetics and the significance of T_E with the hydrogen glass model [114]. T_E is the temperature above which hydrogen diffusion is sufficiently fast to promote defect and dopant reactions.

One attempt to explain the SWE is the suggestion that defects are created by non-radiative recombination of electron-hole pairs at spatially correlated bandtail states that correspond to the bonding and anti-bonding states of weak Si–Si bonds [115]. The recombination energy of about 1.4 eV, released locally, allows a bonded hydrogen to insert itself such that the weak bond cannot reform. This bond-breaking model has run into some difficulties recently as experiments failed to observe the expected close spatial correlation between dangling bond defects and between these and hydrogen [116].

It is possible that the expected spatial correlation of the original defect pair gets destroyed by subsequent recombination events which cause bond switching, which lead both to light-induced annealing and to randomization of defect sites [117]. These, in turn, might cause a change in the overall structure from the annealed, bond-relaxed state to a less-relaxed structure which in some ways is similar to a quenched state. Evidence for such photo-induced structural changes will be much more subtle than in chalcogenide glasses because of the steric hindrance of the tetrahedrally bonded network. Some evidence for photo-induced structural changes has been quoted [117], others are being disputed [118].

Recently, Branz [119] proposed a new mechanism for the SWE in which spatially uncorrelated dangling bond defects are created. His model suggests that non-radiative electron-hole pair recombinations break Si–H bonds. A Si dangling bond is left behind and the H-atom inserts itself into another Si–Si bond creating Si–H — Si which can move by switching to other Si–Si bonds. When two of these mobile Si–H — Si entities meet, they form a complex in which all covalent bonds are satisfied. The dangling Si bonds left behind are the spatially uncorrelated metastable SWE defects. Since one finds that the SWE needs no thermal activation and occurs at very low temperatures while H-diffusion is activated by about 1.4 eV, one must assume for Branz' mechanism that the diffusion of Si–H — Si is predominantly light-induced at low temperatures. That needs experimental verification.

The elimination of the SWE is of great importance for the long-term stability of a–Si:H based solar cells. Some progress has been made by improving the microstructure of the material by means of preparation conditions which allow the removal of weak bonds and the growth of films having a high density, a low microvoid fraction [120] and a low hydrogen concentration [121, 122]. A high density, more ordered structure would tolerate recombination events without damage and defect creation more easily than a strained, disordered, and voided structure.

5. Amorphous Indium Oxide

Stoichiometric indium oxide (In_2O_3) is an insulator with a gap of $Eg = 3.5$ eV. Non-stoichiometric In_2O_{3-x} is an n-type semiconductor with a maximum carrier concentration of the order of 10^{21} cm^{-3} arising from the excess In. The high conductivity and transparency of these films, or of tin-doped indium oxide (ITO), make them useful as transparent electrodes for solar cells and flat panel displays. They can be deposited amorphous or microcrystalline depending on the substrate temperature, and amorphous films can be crystallized by heating above 200°C.

These films, both amorphous and crystalline, undergo drastic changes in their electronic and optical properties upon exposure to electron-hole pair producing ultraviolet (UV) light [123–126]. A semi-insulating film becomes conducting upon exposure to UV light as shown in Fig. 14. The conducting state remains as long as the sample is kept in vacuum or in an inert atmosphere. An oxidizing atmosphere restores the non-conducting state. To test whether these changes are caused by light-induced effusion of oxygen and subsequent reoxidation, the UV exposure was carried out at 78 K and 4.2 K where oxygen diffusion would not occur [127]. It was found that the photo-induced conducting state could be established with comparable efficiency at these low temperatures. Since no oxygen was able to leave the film in these

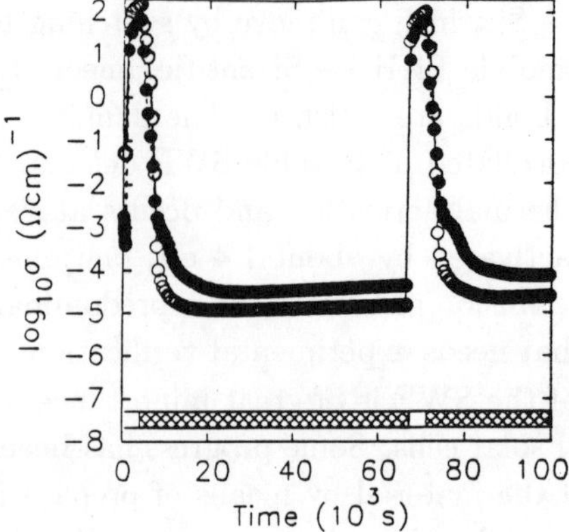

Fig. 14. Conductivity of 75 nm thick amorphous In_2O_{x-3} sputter deposited at 300 K in oxygen partial pressure of 0.5 m Torr (open symbols) and 1 m Torr (solid symbols). The conductivity increases during UV exposure in vacuum or inert gas (open bar) and decreases in an oxidizing atmosphere (shaded bar). After Ref. [125].

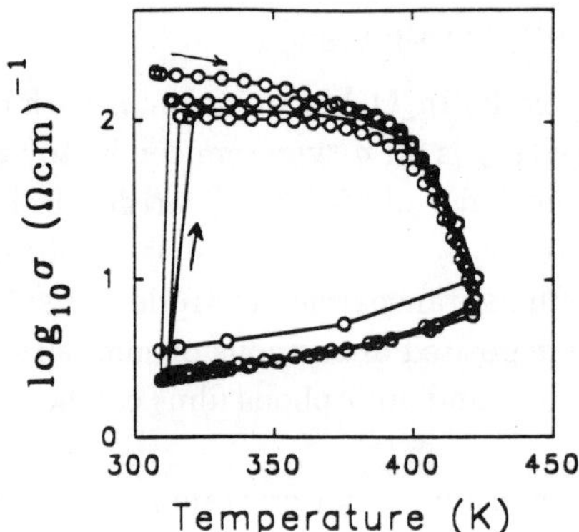

Fig. 15. Temperature cycling to 420 K in vacuum decreases conductivity of amorphous In_2O_{3-x} film, while UV light exposure for 40 min at 310 K increases the conductivity. Three cycles are shown. After Ref. [125].

experiments, it is conceivable that UV light had produced a metastable conducting state which could be returned to the non-conducting ground state by annealing. This was found to be the case with anneal temperatures near 420 K. Figure 15 shows four cycles of UV light exposure at 300 K and subsequent annealing at 420 K. Higher annealing temperatures would probably have yielded an even smaller conductivity. Unfortunately, this could not be tested because, at higher temperatures, thermal effusion of oxygen starts which increases the conductivity.

The Fermi level shifts from the gap into the conduction band as the material becomes conductive. This produces a shift of the optical absorption edge to higher energies (Burstein shift) because higher photon energies are needed to reach unoccupied states. This shift is associated with a decrease in refractive index. As a result, one finds a rather large photorefractive effect which can be used for holographic recording [128].

Our comments about the origin of these light-induced phenomena are very speculative at present. An analysis of the non-stoichiometry of In_2O_{3-x} revealed that only about 10% of the excess In (or O-deficiency) yields a conduction electron [129]. This suggests that not all In atoms remain in the In^{3+} oxidation state, but that many have the In^+ state of the In_2O suboxide with an atomic environment which differs from that of In^{3+}. Without knowing that environment, it is difficult to be sure; but it is not unreasonable to assume that the In^+ region has an overall negative charge acting as a hole trap. Since In^{2+} is not a stable valence state, In^+ might prefer to trap two holes:

$$In^+ + 2e^- + 2h^+ \rightarrow In^{3+} + 2e^- \tag{4}$$

The new In^{3+} center can be stabilized by an appropriate change in local oxygen environment. This change will tend to make the In^{3+} center appear neutral, offering a very small capture cross-section to the two electrons of Eq. (4). This is similar to hole trapping in silver halides (see Sec. 3), except that two holes are needed in Eq. (4). The light-induced conducting state is metastable. We might not need to exclude trapping of one hole and forming the uncommon In^{2+} oxidation state since we consider here a metastable non-equilibrium atom configuration. Reaction (4) can be reversed by annealing. The non-conducting state can also be re-established by an oxidizing atmosphere. The major difficulty with the latter process is to understand the repeatability of the photoreduction and oxidation cycle when one-half of the cycle involves the addition of oxygen.

Photo-induced reactions confined to intergrain surfaces have been considered as an alternative explanation. If the Debye screening length is of the order of the grain size, the conductivity and absorption properties of the whole material could be affected as experiments indicate [130]. A study of photo-induced reduction and oxidation of In_2O_{3-x} films as a function of crystallite size should be valuable for deciding whether the effects occur in the bulk or at grain surfaces.

Acknowledgments

I wish to thank S.R. Ovshinsky and other colleagues at ECD for their helpful discussions, as well as A.E. Owen, S.K. Deb and D.L. Price for their valuable comments.

References

[1] S.R. Ovshinsky, *J. Appl. Photographic Eng.* **3**, 35 (1977), and references therein.

[2] S.R. Ovshinsky and H. Fritzsche, *Metallurgical Trans.* **2**, 641 (1971); *IEEE Trans. on Electron Devices* **ED20**, 91 (1973).

[3] S.R. Ovshinsky, *Phys. Rev. Lett.* **21**, 1450 (1968).

[4] J. Feinleib, S. Iwasa, S.C. Moss, J.P. deNeufville and S.R. Ovshinsky, *J. Non-Cryst. Solids* **8–10**, 909 (1972).

[5] S.R. Ovshinsky and P. Klose, *Int. Symp. Soc. for Inf. Display, Proc. SID* **13**, 188 (1972).

[6] S.R. Ovshinsky and K. Sapru, *Proc. 5th Int. Conf. on Amorphous and Liquid Semicond.*, eds. J. Stuke and W. Brenig (Taylor and Francis, UK, 1974) p. 447.

[7] S.R. Ovshinsky, *Phys. Rev. Lett.* **36**, 1469 (1976); *Proc. Int. Conf. on Structure and Excitations of Amorphous Solids*, eds. G. Lucovsky and F.L. Galeener, *AIP Conf. Proc.* **31**, 178 (1976).

[8] R.A. Street and N.F. Mott, *Phys. Rev. Lett.* **35**, 1293 (1975).

[9] M. Kastner, D. Adler and H. Fritzsche, *Phys. Rev. Lett.* **37**, 1504 (1976).

[10] M. Kastner and H. Fritzsche, *Phil. Mag.* **B37**, 199 (1978).

[11] R.A. Street, *Phys. Rev.* **B17**, 3984 (1978).

[12] M. Kastner, *Phys. Rev. Lett.* **28**, 355 (1972).

[13] N.F. Mott and E.A. Davis, *Electronic Processes in Non-crystalline Materials* (Clarendon Press, Oxford, 1979).

[14] K. Shimakawa, S. Inami and S.R. Elliott, *Phys. Rev.* **B42**, 11857 (1990).

[15] D.K. Biegelsen and R.A. Street, *Phys. Rev. Lett.* **44**, 803 (1980).

[16] S.R. Elliott, *J. Non-Cryst. Solids* **81**, 71 (1986).

[17] M. Frumar, A.P. Firth and A.E. Owen, *Phil. Mag.* **B50**, 463 (1984).

[18] C.Y. Yang, M.A. Paesler and D.E. Sayers, *Phys. Rev.* **B36**, 9160 (1987).

[19] H. Fritzsche, *Phil. Mag.* **B68**, 561 (1993).

[20] Ke. Tanaka, *Japan J. Appl. Phys.* **25**, 779 (1986); *Reviews of Solid State Sciences* **4**, 641 (1990).

[21] P. Krecmer, A.M. Moulin, R.J. Stephenson, T. Rayment, M.E. Well and S.R. Elliott, *Science* **277**, 1799 (1997).

[22] J.Z. Liu and P.C. Taylor, *Phys. Rev. Lett.* **59**, 1938 (1987); ibid *Phys. Rev.* **B41**, 3163 (1990).

[23] V. Lyubin, T. Tada, M. Klebanov, N.N. Smirnova, A.V. Kolobov and K. Tanaka, *Materials Letters* **30**, 79 (1997).

[24] H. Hisakuni and Ke. Tanaka, *Science* **270**, 975 (1995).

[25] Ke. Tanaka, *Phys. Rev.* **B30**, 4549 (1984).

[26] H. Fritzsche, *Solid State Commun.* **99**, 153 (1996).

[27] E.D. Zanotto, *Am. J. Phys.* **66**, 392 (1998). See also J.M. Pasachoff, *Am. J. Phys.* **66**, 1021 (1998).

[28] J. Gonzalez-Hernandez, B.S. Chao, D. Strand, S.R. Ovshinsky, D. Pawlik and P. Gasiorowski, *Appl. Phys. Commun.* **11**, 55T (1992); S.R. Ovshinsky, see chapter in this book.

[29] S.R. Ovshinsky, *Proc. 5th Annu. Nat. Conf. Industrial Research* (Industrial Research, Inc., 1969) p. 86; J. Feinleib, J.P. deNeufville, S.C. Moss and S.R. Ovshinsky, *Appl. Phys. Lett.* **18**, 254 (1971).

[30] T. Ohta, N. Akahira, S. Ohara and I. Satoh, *Optoelectronics* **10**, 361 (1995).

[31] S.R. Herd and P. Chandhari, *J. Appl. Phys.* **44**, 4102 (1973).

[32] J.E. Griffiths, G.P. Espinosa, J.P. Remeika and J.C. Phillips, *Phys. Rev.* **B25**, 1272 (1982).

[33] S.R. Elliott and A.V. Kolobov, *J. Non-Cryst. Solids* **128**, 216 (1991).

[34] M. Frumar, A.P. Firth and A.E. Owen, *J. Non-Cryst. Solids* **192 & 193**, 447 (1995).

[35] O. Matsuda, H. Oe, K. Inoue and K. Murase, *J. Non-Cryst. Solids* **192 & 193**, 524 (1995).

[36] J.S. Berkes, S.W. Ing and W.J. Hillegas, *J. Appl. Phys.* **42**, 4908 (1971).

[37] J.P. deNeufville, S.C. Moss and S.R. Ovshinsky, *J. Non-Cryst. Solids* **13**, 191 (1974).

[38] J.P. deNeufville, Optical Properties of Solids, New Developments, ed. B.O. Seraphin (North Holland, Amsterdam, 1976) Chap. 9.

[39] U. Strom and T.P. Martin, *Solid State Commun.* **29**, 527 (1979).

[40] S. Rajagopalan, K.S. Harshavardhan, L.K. Malhotra and K.L. Chopra, *J. Non-Cryst. Solids* **50**, 29 (1982).

[41] G. Pfeiffer, M.A. Paesler and S.C. Agarwal, *J. Non-Cryst. Solids* **130**, 111 (1991).

[42] Ke. Tanaka, *Appl. Phys. Lett.* **26**, 243 (1975).

[43] J.C. Phillips, *J. Non-Cryst. Solids* **34**, 153 (1979); **43**, 37 (1981).

[44] L.E. Busse, *Phys. Rev.* **B29**, 3639 (1984).

[45] S.R. Elliott, *Phys. Rev. Lett.* **67**, 711 (1991).

[46] G. Pfeiffer, C.J. Brabec, S.R. Jefferys and M.A. Paesler, *Phys. Rev.* **B39**, 12861 (1989).

[47] V.L. Averianov, A.V. Kolobov, B.T. Kolomiets and V.M. Lyubin, *J. Non-Cryst. Solids* **45**, 343 (1981).

[48] J.C. Phillips, *J. Non-Cryst. Solids* **51**, 301 (1982).

[49] S. Ducharme, J. Hautala and P.C. Taylor, *Phys. Rev.* **B41**, 12250 (1990).

[50] Ke. Tanaka and H. Hisakuni, *J. Non-Cryst. Solids* **198–200**, 714 (1996).

[51] Ka. Tanaka, *J. Non-Cryst. Solids* **35 & 36**, 1023 (1980).

[52] Ke. Tanaka, *J. Non-Cryst. Solids* **59 & 60**, 925 (1983).

[53] Ke. Tanaka and A. Odajima, *J. Non-Cryst. Solids* **46**, 259 (1981).

[54] K. Shimakawa, A. Kolobov and S.R. Elliott, *Advances in Physics* **44**, 475 (1995).

[55] Ke. Tanaka, *Solid State Commun.* **34**, 201 (1980).

[56] M. Frumar, A.P. Firth and A.E. Owen, *J. Non-Cryst. Solids* **59 & 60**, 921 (1983).

[57] H. Hamanaka, K. Tanaka, A. Matsuda and S. Iizima, *Solid State Commun.* **19**, 499 (1976).

[58] H. Hamanaka, K. Tanaka and S. Iizima, *Solid State Commun.* **23**, 63 (1977).

[59] Ke. Tanaka, *Phys. Rev.* **B57**, 5163 (1998).

[60] D.L. Price, *Current Opinion in Solid State and Material Science* **1**, 572 (1996).

[61] H. Fritzsche, *Phys. Rev.* **B52**, 15854 (1995).

[62] V. Lyubin, M. Klebanov, V. Tikhomirov and G. Adriaenssens, *J. Non-Cryst. Solids* **198–200**, 719 (1996).

[63] A.V. Kolobov, V. Lyubin, T. Yasuda and K. Tanaka, *Phys. Rev.* **B55**, 23 (1997).

[64] V.G. Zhdanov, B.T. Kolomiets, V.M. Lyubin and V.K. Malinovskii, *Phys. Stat. Solidi (a)* **52**, 621 (1979).

[65] V.M. Lyubin and V.K. Tikhomirov, *J. Non-Cryst. Solids* **114**, 133 (1989).

[66] V. Lyubin and M. Klebanov, *Phys. Rev.* **B53**, R11924 (1996).

[67] V. Lyubin, M. Klebanov, V. Tikhomirov and G. Adriaenssens, *J. Non-Cryst. Solids* **198–200**, 719 (1996).

[68] A.V. Kolobov, V. Lyubin, T. Yasuda, M. Klebanov and K. Tanaka, *Phys. Rev.* **B55**, 8788 (1997).

[69] V.M. Lyubin and V.K. Tikhomirov, *Sov. J. Glass Phys. Chem.* **18**, 374 (1992).

[70] V.K. Tikhomirov and S.R. Elliott, *Phys. Rev.* **B49**, 17476 (1994).

[71] H. Hisakuni, M. Notani and Ke. Tanaka, *Solid State Commun.* **95**, 461 (1995).

[72] V. Lyubin, M. Klebanov, S. Rosenwaks and V. Volterra, *J. Non-Cryst. Solids* **164–166**, 1165 (1993).

[73] V.K. Tikhomirov and S.R. Elliott, *J. Phys: Condens. Matter* **7**, 1737 (1995).

[74] V. Lyubin, M. Klebanov, M. Mitkova and T. Petkova, *Appl. Phys. Lett.* **71**, 2118 (1997).

[75] V.M. Lyubin, M. Klebanov, M. Mitkova and T. Petkova, *J. Non-Cryst. Solids* **227**, 739 (1998).

[76] V.K. Tikhomirov, P. Hertogen, C. Glorieux and G.J. Adriaenssens, *Phys. Stat. Solidi (a)* **162**, R1 (1997).

[77] V.K. Tikhomirov, P. Hertogen, G.J. Adriaenssens, C. Glorieux and R. Ottenburgs, *J. Non-Cryst. Solids* **227**, 732 (1998).

[78] V.K. Tikhomirov, G.J. Adriaenssens and S.R. Elliott, *Phys. Rev.* **B55**, R660 (1997).

[79] M.T. Kostyshin, E.V. Mikhailovskaya and P.F. Romanenko, *Sov. Phys. Solid State* **8**, 451 (1966).

[80] A.E. Owen, A.P. Firth and P.J.S. Ewen, *Phil. Mag.* **B52**, 347 (1985).

[81] A.V. Kolobov and S.R. Elliott, *Advances in Physics* **40**, 625 (1991).

[82] G. Dale, A.E. Owen and P.J.S. Ewen, Physics and Applications of Non-Crystalline Semiconductors in Opto-electronics, eds. A. Andriesh and M. Bertolotti (Kluwer Academic Publications, Dordrecht, 1997) p. 45.

[83] Y. Kawamoto, M. Agata and S. Tsuchihashi, *J. Ceram. Soc. Japan* **82**, 502 (1974).

[84] T. Kawaguchi and S. Maruno, *J. Appl. Phys.* **77**, 628 (1995).

[85] T. Wagner, M. Vlček, V. Smrčka, P.J.S. Ewen and A.E. Owen, *J. Non-Cryst. Solids* **164–166**, 1255 (1993).

[86] A.E. Owen, private communication (1998).

[87] T. Wagner, M. Vlček, K. Nejezchleb, M. Frumar, V. Zima, V. Peřina and P.J.S. Ewen, *J. Non-Cryst. Solids* **198–200**, 744 (1996).

[88] A. Matsuda and M. Kikuchi, *Suppl. Japan Soc. Appl. Phys.* **42**, 239 (1973).

[89] Y. Kawamoto and M. Nishida, *Phys. Chem. Glasses* **18**, 19 (1977).

[90] T. Wagner, M. Frumar and V. Šuškové, *J. Non-Cryst. Solids* **128**, 197 (1991).

[91] J.C. Phillips, *J. Non-Cryst. Solids* **64**, 81 (1984).

[92] I.T. Penfold and P.S. Salmon, *Phys. Rev. Lett.* **64**, 2164 (1990); C.J. Benmore and P.S. Salmon, *J. Non-Cryst. Solids* **156–158**, 720 (1993).

[93] R.J. Dejus, S. Susman, K.J. Volin, D.G. Montague and D.L. Price, *J. Non-Cryst. Solids* **143**, 162 (1992).

[94] M. Kastner, *Phil. Mag.* **B37**, 127 (1978).

[95] S. Zembutsu, *Appl. Phys. Lett.* **39**, 969 (1981).

[96] A.E. Meixner and C.H. Chen, *Phys. Rev.* **B27**, 7489 (1983).

[97] N. Yoshida, M. Itoh and Ke. Tanaka, *J. Non-Cryst. Solids* **198–200**, 749 (1996).

[98] Ke. Tanaka and N. Yoshida, *Solid State Phenomena* **55**, 153 (1997).

[99] S. Maruno, *J. Non-Cryst. Solids* **59 & 60**, 933 (1983).

[100] T. Kawaguchi and S. Maruno, *Japan J. Appl. Phys.* **33**, 6470 (1994).

[101] W.H. Armistead and S.D. Stookey, *Science* **144**, 150 (1964).

[102] D. Caurant, D. Gourier and M. Prassas, *J. Appl. Phys.* **71**, 1081 (1992).

[103] C.L. Marguardt and G. Gliemeroth, *J. Appl. Phys.* **50**, 4584 (1979).

[104] D. Caurant, D. Gourier, D. Vivien and M. Prassas, *J. Appl. Phys.* **73**, 1657 (1993).

[105] L. Cordone, S.L. Fornili and S. Micciancio, *Phys. Rev.* **188**, 1404 (1969).

[106] E. Laredo, W.B. Paul, L. Rowan and L. Slifkin, *J. Phys. C: Solid State Phys.* **16**, 1153 (1983).

[107] J.F. Hamilton, *Advances in Physics* **37**, 359 (1988).

[108] J.I. Pankove, Hydrogenated Amorphous Silicon, *Semiconductors and Semimetals* (Academic Press, Orlando, FL, 1984) Vol. 21.

[109] R.A. Street, Hydrogenated Amorphous Silicon (Cambridge University Press, Cambridge, UK, 1991).

[110] D.L. Staebler and C.R. Wronski, *Appl. Phys. Lett.* **31**, 292 (1977); *J. Appl. Phys.* **51**, 3262 (1980).

[111] R.A. Street, J. Kakalios, C.C. Tsai and T.M. Hayes, *Phys. Rev.* **B35**, 1316 (1987).

[112] W.B. Jackson and M.D. Moyer, *Phys. Rev.* **B36**, 6217 (1987).

[113] W.B. Jackson, J.M. Marshall and M.D. Moyer, *Phys. Rev.* **B39**, 1164 (1989).

[114] J. Kakalios, Semiconductors and Semimetals, ed. J.I. Pankove **34** (Academic Press, New York, 1991) p. 381.

[115] M. Stutzmann, W.B. Jackson and C.C. Tsai, *Phys. Rev.* **B32**, 23 (1985).

[116] S. Yamasaki and J. Isoya, *J. Non-Cryst. Solids* **164–166**, 169 (1993).

[117] H. Fritzsche, *Solid State Commun.* **94**, 953 (1995); *Mat. Res. Soc. Symp. Proc.* **467**, 19 (1997).

[118] Lin Jiang, Qi Wang, E.A. Schiff, S. Guha and J. Yang, *Appl. Phys. Lett.* **72**, 1060 (1998).

[119] H.M. Branz, *Solid State Commun.* **105** & **106**, 387 (1998); *Mat. Res. Soc. Symp. Proc.* (1998).

[120] Sugiyama, J.C. Yang and S. Guha, *Appl. Phys. Lett.* **70**, 378 (1997); *Mat. Res. Soc. Symp. Proc.* **467** (1997).

[121] H. Mahan, J. Carapella, B.P. Nelson, R.S. Crandall and I. Balberg, *J. Appl. Phys.* **69**, 6728 (1991).

[122] E.C. Molenbroek, A.H. Mahan, E.H. Johnson and A.C. Gallagher, *Mat. Res. Soc. Symp. Proc.* **336**, 43 (1994).

[123] B. Pashmakov, B. Claflin and H. Fritzsche, *Solid State Commun.* **86**, 619 (1993); *J. Non-Cryst. Solids* **164–166**, 441 (1993).

[124] H. Fritzsche, B. Pashmakov and B. Claflin, *Solar Energy Materials and Solar Cells* **32**, 383 (1994).

[125] B. Pashmakov, H. Fritzsche and B. Claflin, *Mat. Res. Soc. Symp. Proc.*, 1996.

[126] C. Xirouchaki, G. Kiriakidis, T.F. Pedersen and H. Fritzsche, *J. Appl. Phys.* **79**, 9349 (1996).

[127] P. Stradins, H. Fritzsche and B. Claflin, *Mat. Res. Soc. Symp. Proc.*, 1996.

[128] S. Mailis, L. Boutsikaris, N.A. Vainos, C. Xirouchaki, G. Vasiliou, N. Garawal, G. Kiriakidis and H. Fritzsche, *Appl. Phys. Lett.* **69**, 2459 (1996).

[129] J.R. Bellingham, A.P. MacKenzie and W.A. Phillips, *Appl. Phys. Lett.* **58**, 2506 (1991).

[130] B. Claflin and H. Fritzsche, *J. Electronic Materials* **25**, 1772 (1996).

11

Chalcohalide Glasses

JACQUES LUCAS

Université de Rennes I, Laboratoire des Verres et Céramiques,
Campus de Beaulieu, Avenue du Général Leclerc,
35042 Rennes Cédex, France
jacques.lucas@univ-rennes1.fr

Contents

1. Introduction

Chalcohalide glasses can be described as vitreous materials containing chalcogen and halogen atoms. They could be considered as materials intermediate between halide glasses and chalcogenide glasses, a kind of by-product of these two basic families of glasses. As a matter of fact, many of these compounds have to be considered not only as substitution materials but as glasses having their specific structural characteristics and properties.

The main motivations which have been at the source of the research and development programs on such materials are related to their potential optical or optoelectronic properties. Due to the high atomic weight of halogens such as Cl, Br, I or the chalcogens S, Se, Te, the chemical bonds with different kinds of metals or semi-metal lead to vibrational modes which are chararacterized by low phonon energies Wp.

Consequently, with Wp typically in the range 400 cm^{-1} to 50 cm^{-1}, these glasses are characterized by excellent infrared transmission and are then potential candidates for optics operating in the strategic atmospheric window lying from 8 to 12 μm. For this purpose, they compete with the pure chalcogen or pure halogen based glasses.

Recently, chalcohalides glasses have been considered as potential hosts for rare earths in active devices such as micro-lasers, or optical amplifiers. Indeed, as low phonon materials, they offer the benefit of very weak phonon relaxation during the fluorescence process. Furthermore, because of the presence of the electronegative halogen, they exhibit better transmission in the UV-visible region. The later advantage, compared to pure chalcogenide glasses, allows rare earth pumping and emission in a broader optical window.

On the other hand, there are several intrinsic disavantages associated with chalcohalide glasses, when considered as potential candidates for applications. Because of the presence of halogen atom in the structure, often the metal-halogen bond competes with M–OH bond, when the glass entered comes in contact with water or moist air. This results in a poor resistance to corrosion and a loss in good optical properties. Further, halogen atoms (X) have a tendency to directly react with elements such as Ge, As and also chalcogens to produce molecular species such a GeX_4, AsX_3. The consequence is a severe competition for crystallite formation and gas bubble nucleation in the glasses as inclusions. One of the ultimate way, to select and to qualify good chalcohalide compositions, is to use them as infrared optical fibers. Indeed, to be shaped into infrared transparent low loss optical fiber, the glasses need to exhibit strong resistance towards phase separation, crystallization and moisture corrosion during the preform fabrication and the fibering process. The field of chalcohalide glasses has been reviewed at the end of the eighties by Sanghera, Heo and Mackenzie [1–5] and then by Gan Fuxi [6]. These papers give an overview on the glassforming systems and a model of structural description from vibrational spectroscopy. The present chapter will contain the most significant results already presented in the previous reviews and will discuss new materials and recent approaches concerning glass structure and applications of chalcohalide glasses.

2. Glass Formation on Systems Containing Chalcogens and Halogens

If one excludes the element fluorine F which has a very special behaviour, due to its high electronegativity, it is clear that S and Cl, Se and Br, Te and I

Table 1. Electronegativity, normal covalent radii and ionic radii for the halogen Cl, Br, I compared to chalcogen S, Se, Te.

Element	Electronegativity coefficient	Normal covalent radii	Ionic radii r
Cl	3.0	0.99	$Cl^- = 1.81$
Br	2.8	1.14	$Br^- = 1.96$
I	2.4	1.33	$I^- = 2.20$
S	2.5	1.04	$S^{2-} = 1.84$
Se	2.4	1.17	$Se^{2-} = 1.98$
Te	2.1	1.37	$Te^{2-} = 2.21$

are neighbours in the periodic chart, and possess similar electronegativities. Also, the size of the neighbouring atoms S/Cl, Se/Br, Te/I when entering in a covalent bond, namely covalent radii, or when entering in a pure ionic bond, namely ionic radii, are very similar as indicated on Table 1, extracted from Wells [7].

The halogen atoms belong to the group 7 and possess an outer electronic shell $s^2px^2py^2pz^1$. They have one bonding electron or hole. The chalcogens which belong to the group 6, have the electronic structure $s^2px^2py^1pz^1$, which indicates the presence of two unpaired bonding electrons and two lone pairs electrons which usually have a very important steric effect on the bond angles around the chalcogen.

Except the case of the tellurium halide glasses, most of the chalcohalide glasses are Ga, Ge or As based. The Pauling electronegativities for Ge and As, which are respectively 1.7 and 2.0, indicate that the chemical bond in these compounds is essentially covalent. The halogens and chalcogens chemically bond with an intermediate situation between pure covalent and pure ionic. The following classification of chalcohalide glasses take into account the nature of the chemical bond and the topology of the structural framework.

2.1. *The tellurium halide glasses and derivatives*

2.1.1. *The binary TeX glasses, X = Cl, Br, I*

The simplest chalcohalide glasses are without doubt compounds formed by reaction of Te with Cl or Br or I [8]. The experimental conditions are

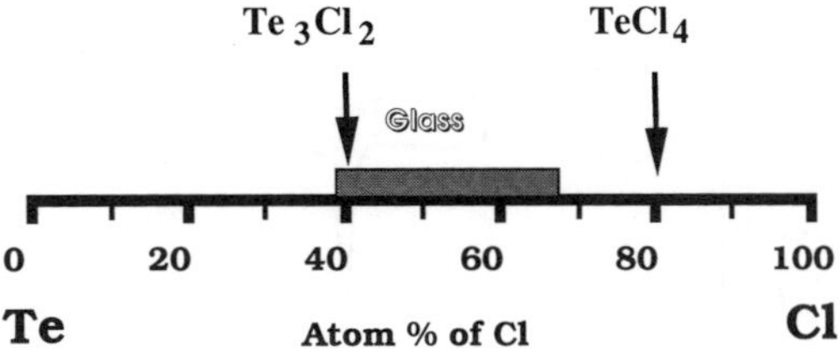

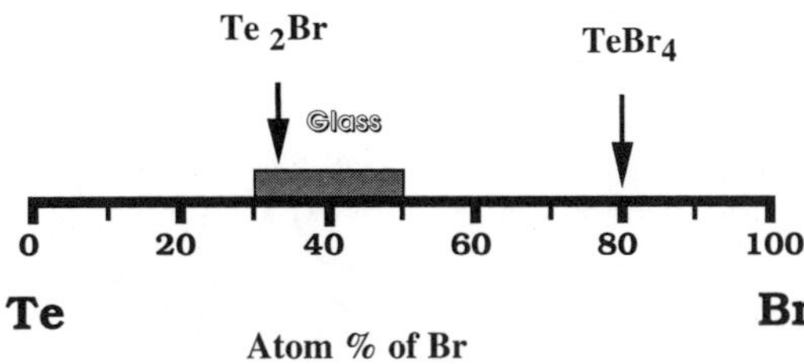

Fig. 1. Glassforming regions in the binary sytems Te/Cl and Te/Br.

described in Ref. [8], but basically glasses are obtained by reaction in a sealed pyrex tube of calculated quantities of Te powder with Cl$_2$ gas, liquid Br$_2$ or solid I$_2$. Usually, the sealed tube is heated in a rocking-furnace at 300°C, in order to homogenize the melt. When the sealed tubes are quenched by cooling down from 300°C to room temperature in air, the vitreous domain is located in the composition domain indicated on Fig. 1 for the Te/Cl and Te/Br system. In the case of Te–I system, glasses can be also obtained, but to prepare crystallite-free samples, fast quenching such as cold water quenching is needed. The composition which exhibits the easiest glassforming capability is closed to Te$_3$I$_2$.

All these glasses have been characterized by Differential Scanning Calorimetry, spectroscopical investigations as well as resistance to moisture corrosion [9–12].

Results of DSC analysis of three glasses belonging to the binary systems Te–Cl, Te–Br, Te–I are represented on Fig. 2. These materials are characterized by low glass transition temperatures T_g lying between $T_g = 44°$C for the Te$_3$I$_2$ glass to $T_g = 82°$C for Te$_3$Cl$_2$ glass. These observations are in clear agreement with the expected low-dimensionality of the glassy framework which

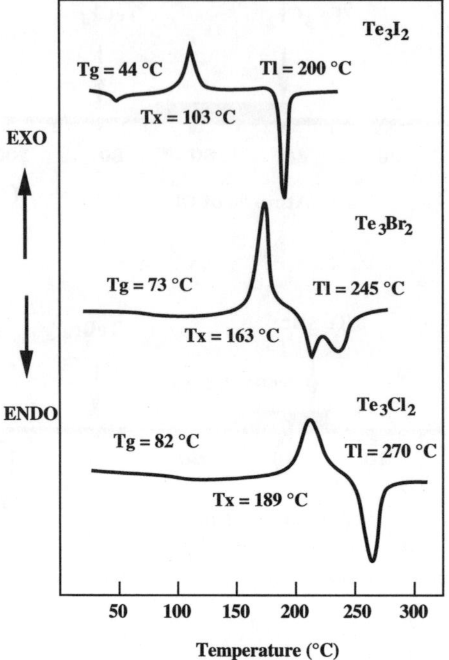

Fig. 2. D.S.C. analysis of three TeX glasses.

by reference to the parent crystalline material Te_3Cl_2 is well represented by a chain model [13]. The structural aspect and the crystal to glass competition will be discussed later on view of a very interesting structural investigation by Mössbauer spectroscopy.

2.1.2. *TeX glasses stabilized by S or Se*

It is well known that S or Se are characterized by chain or ring based structures and that the equilibrium between these two models is due to an easy rotation of the chalcogen–chalcogen bonds which have essentially a σ character. The degree of polymerization of the S_n or Se_n chain fragments as well as the distorted stereochemistry around the S or Se atoms due to the two lone pairs, introduce a spectacular variation of melt viscosities leading to easy glass formation.

The obvious similarity, between the chain-like structure of Te_3Cl_2 on one hand and S or Se on the other, facilitates formation of large domains of

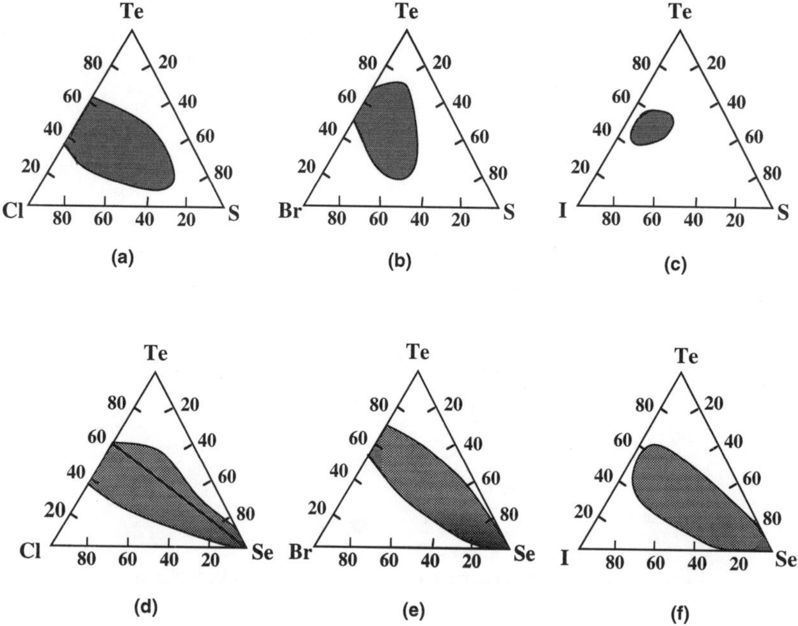

Fig. 3. Glassforming domains when TeX glasses are combined with S or Se.

miscibility in the liquid state as well as in the glassy state. This is illustrated by the very broad glassforming domains observed in the Te–X–S ternary with X = Cl, Br, I and S completely replaced by Se as portrayed in Fig. 3. Figure 4 represents a fragment of the Te_3Cl_2 chains structure, where Te atoms can be substituted by Se in keeping the same stereochemistry. In this model, the local lone pair effect introduces an helical variation of the bond angles and the possibility of easy rotation along the chains leading to an easy loss of periodicity.

Several remarks can be made concerning the Te–I–Se ternary system. Figure 5 shows the glassforming compositions in this ternary. The material compositions located in the middle of the diagram (zone B) are extremely stable against devitrification and reveal no crystallization peak in DSC scans [14]. At the edges of the glassy domain (shaded gray in Fig. 5), phase separation cannot be excluded specially in the Se rich region where ring formation such as Se_6, Se_8, Se_{12} is possible.

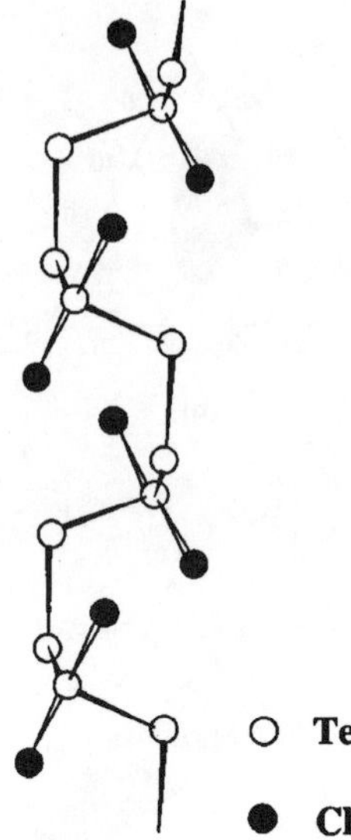

Fig. 4. The chain-structure of the Te_3Cl_2 crystalline form.

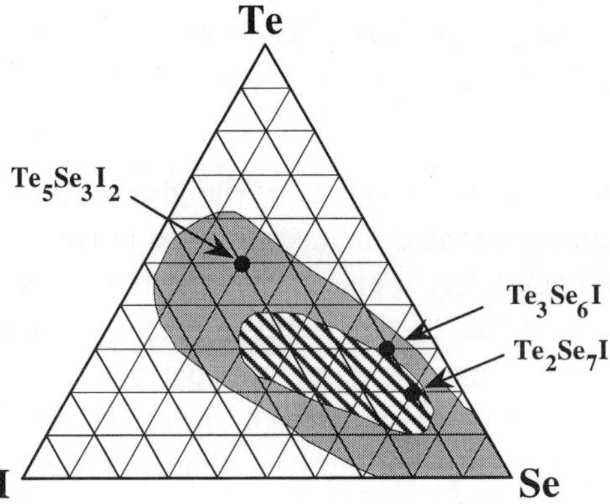

Fig. 5. The glassforming domains in the Te–Se–I systems. In the hatched zone, the glasses do not show any crystallization peak.

From a Mössbauer spectroscopy investigation on the glasses of the systems Te–Se, Br or Cl, Bresser *et al* [15] and Boolchand [16] have suggested a strong molecular origin of the glassforming tendency. Halogen atoms bond in the chains and are also responsible to fragment the chains by acting as chain-terminators. The floppiness of the chain-fragments then explains the glassforming tendency, and the low T_g, $T_g \approx 70°\text{--}90°\text{C}$, for these low dimensional glasses.

In a systematic investigation of the Se–Te–I glasses, Tichy [17] also remarks that the glassforming tendency is strongly enhanced by addition of Se to the Te–I system, in avoiding the formation of α or β Te–I crystallites. But they also suggested from infrared spectroscopy that the formation of Se_8 rings is highly probable in the Se rich region. From examination of the vibrational spectra of the same glasses, Wang and Chen [18] also reach the conclusion that these chain-like glasses can contain some Se_8 rings. These two remarks raise the problem of immiscibility and potential phase separation during glass processing or fiber-drawing.

2.1.3. *The TeXAs glasses*

From a purely technological point of view, the TeX glasses, discussed above suffer from several disavantages such as a low T_g, possible long-relaxation times for enthalpy changes at room temperature, high thermal expansion coefficient and sensitivity to thermal shocks. In other words, their thermomechanical properties do not conform to military applications. One of the routes to reach this objective is to improve the rigidity of the glassy framework by increasing the structural dimensionality. This can be realized by cross-linking the chains together by a trivalent atom, such as arsenic As [19].

In order to illustrate the improvement of the structural rigidity, several glasses have been prepared using as reference material the chain-like TeX glass Te_2Se_7I ($T_g = 58°\text{C}$) and replacing a part of the Se element in the chains by trivalent As. The target was to cross-link chains via the additional bond of the As atoms, as represented schematically in Fig. 6. Several $Te_2Se_{7-x}AsI$ glass compositions with $x = 1$ to 6 have been prepared and compared. The glass transition temperatures (T_g) change monotonically (Fig. 7) from $T_g = 58°\text{C}$ at $x = 0$ to $T_g = 152°\text{C}$ at $x = 6$, Te_2SeAs_6I. Most of these TeXAs glasses are very stable towards devitrification or moisture corrosion, and as discussed later some compositions, for instance the glass $Te_2Se_3As_4I$, have been selected for infrared fiber-drawing.

There are two interesting features about these TeXAs glasses, one is associated with their very good resistance to moisture or water corrosion, which begins at $T > 80°\text{C}$, and second that addition of As does not severely modify the infrared transmission. The infrared edge instead of being located at 20 μm

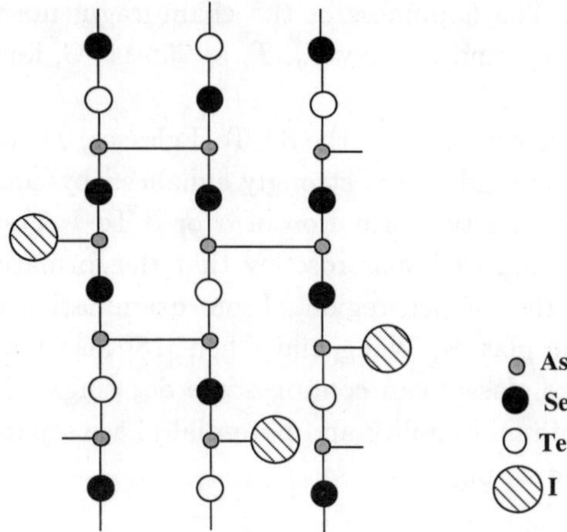

Fig. 6. The trivalent As atoms serve to cross-link TeX chains and increase the dimensionality of the TeXAs glass structure.

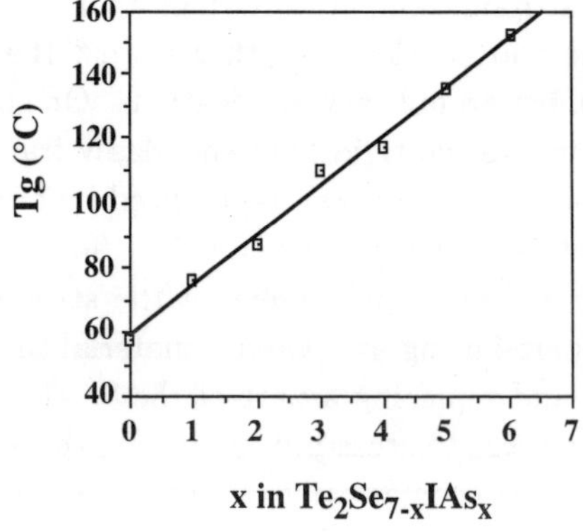

Fig. 7.   Evolution of the glass temperature T_g with the As content in Te$_2$Se$_{7-x}$As$_x$I glasses.

is moved to $\lambda = 18$ μm, but the low optical loss in the 8–12 μm region is held intact.

2.2. *Chalcohalide glasses related to the As_2S_3 prototype glass*

The very well-known glass As$_2$S$_3$ is certainly the most popular glass in the chalcogenide family. As$_2$Se$_3$ also exists but As$_2$Te$_3$ cannot be formed easily.

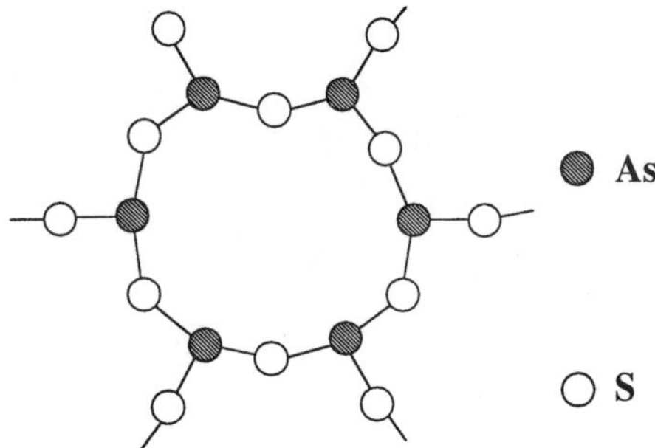

Fig. 8. The building block of the 2D glass As_2S_3. A lone-pair electron on As and two lone-pair electrons on S give rise to a locally distorted stereochemistry.

These glasses are considered as model networks of two-dimensional character in which the polymeric expansion of the chemical bond is built on the trivalency of As plus one lone pair and the divalency of S plus two lone pairs. The infinite aperiodic covalent polymer is then formed with distorted hexagonal rings as shown on Fig. 8.

The chemistry behind these chalcohalide glasses derived from arsenic consists essentially in decreasing the dimensionality of the framework by replacing connecting S or Se or Te by terminal Cl, Br or I. One of the theoretical limit of this chemistry is the total transformation of the 2D framework in a 1D chains-like structure. For instance, the composition $AsS_{2/2}X$ corresponds to the formation of the chains — S–As–S–As–S–As — with the halogen X regularly bonded to each As atoms.

2.2.1. *The arsenic-sulfur-halogen-based glasses*

Much information can be found in the review by Sanghera [1]. Perhaps the most interesting system is the As–S–I ternary which has been investigated by several authors during the sixtees. Solid I_2 is more easily handled than liquid Br_2 because it is less volatile. The iodine based compounds are better infrared transmission materials.

Investigations of the As–S–I ternary by Flaschen *et al* [20] have to be considered as the first comprehensive work on chalcohalide glasses. Since then, a significant amount of work has been published on the preparation, properties and structural aspects of the As–S–I glasses [21–24]. Figure 9 represents the glassforming domain which is remarkably large extending towards the sulfur

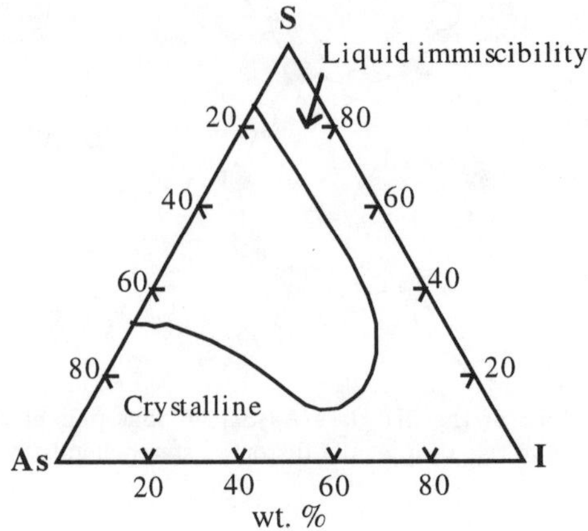

Fig. 9. Glassforming region in the As–S–I region taken from Ref. [20].

region but also towards the I region. For instance, the ideal 1D linear glass composition AsSI, just in the middle of the diagram, is well located in the vitreous domain. The most common way to prepare pure glasses is to heat the starting pure elements in an evacuated silica tube. Melts need to be homogenized in a rocking-furnace at 600°C before cooling down the tubes. Of course, the limits of the glass domain depends strongly on the quenching rate. The glassforming domain is limited on the S side by liquid immiscibility of S_n and on the arsenic side by crystallite formation.

As expected the evolution from a 2D glass towards a 1D glass is very well illustrated by a spectacular decrease of the glass temperature T_g. For instance, while the T_g of As_2S_3 is $T_g = 205°C$, it falls down continuously to $T_g = 70°C$ for the glass AsSI [25].

In addition to sulfur immiscibility and crystallite formation in the As region, this system is also prone to the nucleation of AsI_3 molecules in the iodine rich region as demonstrated by the vibrational spectroscopy work of Koudelka [23]. AsI_3 is indeed a solid with a melting point at 145°C but having a rather high tendency to sublimation.

Generally speaking, these glasses tend to be colored from deep red to yellow in the S-region and orange in the I-region. Glass formation has been also investigated in the As–S–Br system [26–30]. Except certain details, the results of these studies are reaching almost the same conclusion as for the iodine containing system. Also, the system As–S–Cl has received a very

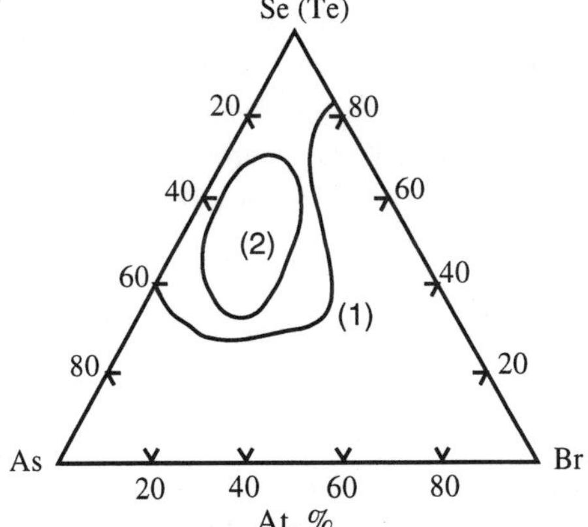

Fig. 10. Glassforming region in the As–Se–Br (1) or As–Te–Br systems (2) taken from Ref. [28].

limited interest due to the high volability of the materials produced in this system.

2.2.2. *The arsenic-selenium-halogen-based glasses*

For the same reasons, as discussed earlier, there are not reports in the literature on the As–Se–Cl system. However, in the As–Se–Br and As–Se–I ternaries up to 40 atomic % of halogen can be added to the As–Se compositions to obtain a glass. Figure 10 represents the glassforming region in the As (Se, Te) Br system from Turyanitsa *et al* [28]. The glassforming region is smaller than in the sulfur containing counterpart, but here also it extends towards the ideal chains-like glass [31] As–Se–Br and As–Se–I. Some observations can be made concerning the T_g values which drop down from $T_g = 195°C$ for As_2Se_3 to $T_g = 80°C$ for the glass in the middle of the diagram AsSeI. The general thermal stability of the iodine based glasses has been discussed by Petrovic [32].

2.2.3. *The arsenic-tellurium-halogen-based glasses*

First of all, binary glasses in the As/Te sytem are difficult to obtain under traditional melting and cooling techniques. Also no reliable glassforming results are available on the As–Te–Cl system.

However, several authors have noted that the addition of Br or I to the As–Te system improves the glassforming capability [26, 33]. However, the glass

AsTeI cannot be produced using conventional techniques due to its high tendency to crystallize. On the contrary, in the As–Te–Br system, glasses are formed along the binary line As_2Te_3–AsTeBr. In the iodine based system, the values of T_g range from 100°–150°C whereas in the As–Te–Br system, they range from $T_g = 80°C$ to $T_g = 120°C$. All these glasses are black or silver-black in color.

2.3. *Chalcohalide glasses derived from the Ge-chalcogen systems*

First of all, it must be mentioned that in the Ge–S and Ge–Se binary systems the glassforming capability is more pronounced close to the GeS_4 or $GeSe_4$ composition than in the vicinity of GeS_2 or $GeSe_2$.

The poor resistance to crystallization of the GeS_2 glass is due to the rigid connectivity of the GeS_4 tetrahedra by corner sharing but also edge sharing tetrahedra. On the other hand, in GeS_4 glass, the Ge atoms in a four-fold coordination are tridimensionally connected by $-S = S-$ bonds. Consequently, the $[Ge(S = S)_{4/2}]_n$ framework is much more flexible to bend and to rotate along the chemical bonds resulting in an easy loss of periodicity.

2.3.1. *The Ge–S–I glasses*

This system has been most extensively investigated [35]. The main reason is probably associated to the very large glassforming domain as noted by Dembovskii [34] and Heo [2]. It can be seen (from Fig. 11) that the vitreous domain extends very far from GeS_4 composition to the iodine-rich region, including for example the ideal linear glass $GeSI_2$ in which the tetrahedral building blocks GeS_2I_2 are 1D connected via the S atoms to form infinite chains $[GeS_{2/2}I_2]_n$. As expected, the reduced dimensionality results in more easy floppiness of the skeleton and a severe decrease of T_g: for instance $T_g = 428°C$ for GeS_2 and $T_g \approx 200°C$ for the glass containing 30 atomic % I_2. As noted by Seddon [36], the increase of iodine content is also correlated with the risk of S_8 ring formation. The composition $GeSI_2$ rich in iodine tends to be viscous liquid at room temperature.

A general examination of Ge–Se–I system indicates clearly that the easy glassforming area is roughly located between GeS_4, the stable 3D glass, and $GeSI_2$ the ideal 1D glass. In other words, in the region where the risk of phase separation is minimized.

This observation has been discussed in terms of percolation of rigidity by Boolchand and Thorpe [38], in modifying the general conditions enunciated

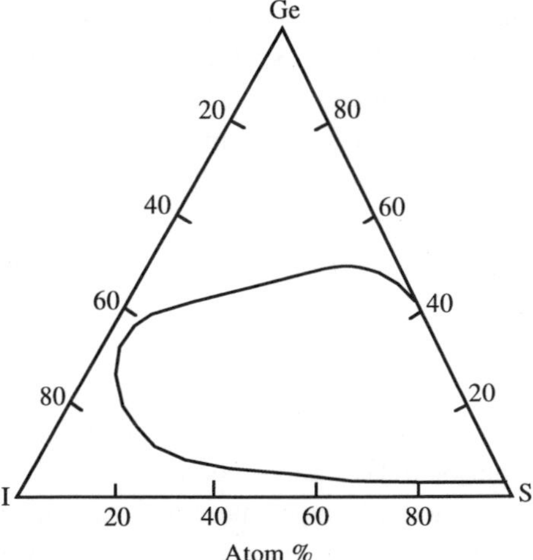

Fig. 11. Glassforming region in the Ge–S–I system taken from Ref. [34]. The vitreous area extends from GeS_4 towards the I rich region.

by J.C. Philips [39], due to the presence of one-fold coordinated atoms namely terminal iodine in the glass-framework.

2.3.2. *The Ge–S–X glasses* $(X = Cl, Br)$

No significant information is available in the literature on glasses belonging to the Ge–S–Cl system. However, Br containing glasses have been investigated [2, 33]. Starting materials used for the glass synthesis are Ge powder, S and liquid $GeBr_4$ which is preferred to Br_2 because of the high vapor pressure of liquid bromine. The same trend, as in the Ge–Se–I system, is observed here; the addition of Br has the effect of lowering the T_g values of the glasses and to lead sometimes to viscous liquid at room temperature. As an illustration for instance $T_g = 428°C$ for GeS_2 and $T_g = 225°C$ for the glass $Ge_{23}S_{97}Br_{30}$.

As discussed by Loheider [37], the compound $GSBr_2$ corresponding to the ideal chain-like type glass $GeS_{2/2}Br_2$ built up from the 1D connection of GeS_2Br_2 tetrahedra is a viscous liquid at room temperature. The glass transition of such glasses occurs at $T_g = -30°C$.

2.3.3. *The Ge(Se, Te)I glasses*

Glasses, in these systems, can be prepared by heating the elemental starting materials at around 800°C, within evacuated silica ampoules and subsequently quenching the ampoules in air. Under normal quenching conditions, glasses are difficult to obtain in the Ge/Te system while glass formation is easy in the Ge/Se binary system, specially closed to the composition $GeSe_4$. In both systems, addition of iodine facilitates vitrification [41–42].

As observed in the previous systems, addition of iodine tends to break down the connectivity of the Ge–Se network which is illustrated by a rapid decrease of the T_g values. In the Ge–Te–I system, the glass temperatures lie around $T_g \approx 125°C$.

As remarked by Dembovskii [40], there is an interesting region in the Ge–Se–I, specially close to the Se rich area. Indeed, there is a great similarities between the Se glass with $T_g = 32°C$ and the $GeSeI_2$ glass with $T_g = 37°C$. Both of them have the ideal 1D chain-like structure with the only difference that in the second glass, the building element is the tetrahedra $GeSe_2I_2$ instead of a single atom in the Se glass. Consequently, it is not surprising that a vitreous domain exists in the region between Se and $GeSeI_2$.

2.4. *Antimony-based chalcohalide glasses*

The pioneering work in these glasses have been performed almost entirely in the USSR. There are no stable glasses formed within the binary systems Sb–S or Sb–Se. On the contrary, chalcogenides such as Sb_2S_3 are known as crystalline materials having a high lattice energy.

However, addition of a third atom, such as Br or I, facilitates glass formation as mentioned by Turyanitsa and Kopperless [43, 44]. Many glasses have been prepared and characterized in the Sb–S–I and Sb–S–Br systems. The glassforming domain is rather small and is located along the Sb_2S_3–SbI_3 or Sb_2S_3–$SbBr_3$ binary line close to the Sb_2S_3 side at about 80 atomic percent. To prepare these glasses starting elements such as Sb, S, Sb_2S_3 and $SbBr_3$ are used in the appropriate amounts, then melted at about 600°C in evacuated silica tube. A small deviation out of the binary line leads to materials which need to be fastly quenched due to their tendency to crystallize. The glassforming region in the Sb–S–Br system seems to be located very close to an eutectic region [46, 47].

The Sb–Se–I system is quite different in the sense that the glassforming area is not along the Sb_2Se_3–SbI_3 line, but in a region located between this line at 80% Sb_2Se_3 and close to the Se region [45]. It is clear that the easy glassforming ability of the 1D Se chain plays an important role which facilitates the vitrification of the Sb–Se–I glasses.

It is interesting to note that the ideal 1D chain-like compositions SbSeI or SbSI in which the quasi-tetrahedra SbS_2ILP units are connected via S atoms, do not form glasses. LP means a lone pair and corresponds to the $5s^2$ non-bonding electrons of the antimony atom. On the contrary, SbSBr is just on the edge of the glassforming area; this solid indeed has a rather low melting point of 320°C, and Turyanitsa and Kopperless [43] observed that the melts are characterized by a high viscosity which of course indicate a high degree of polymerization in the melt which favors glass formation.

It must be noted that in the Sb–Se–I system, the T_g values are rather low, lying from $T_g = 60°C$ in the Se rich region to $T_g = 170°C$ for the glass [45] $Sb_{33}Se_{53}I_{14}$.

2.5. *Some others chalcogen-halogen-based glasses*

In this section, we discuss chalcohalide glass compositions, which have received a limited attention and for which more structural information is needed.

For instance, Kunze [47] and then Robinel [48] have investigated the Ge–S–Ag–I system and examined the optical and electrical properties of the GeS_2–Ag_2S-AgI glasses. About 50% of AgI can be combined with the GeS_2–Ag_2S pseudo-binary leading to glasses with a variation of T_g from 301°C to about 240°C. The motivation for this work was the influence of halogen addition on the Ag^+ ions mobility in the glass matrix.

The silicon-based chalcohalide system Si(Se, S)(Cl, I) has also been investigated by Dembovskii and Popova [40], but most of those glasses are severely corroded by moisture. Compositions, such as $SiSeI_2$ have T_g less than 0°C and $SiSCl_2$ is a viscous liquid at room temperature. Dembovskii and Popova proposed that these compounds are linear glasses related to the Se type materials.

Recently, Arif and Baidakov [49] have tried to identify the glassforming conditions in the Sb_2S_3–PbI_2 system. From chemical bonding considerations, which will not be discussed here, they predicted glass formation in the $(Sb_2S_3)_{1-x}(PbI_2)_x$ pseudo-binary at $0.3 < x < 0.7$. Experimentally, they were able to prepare glasses for $0.2 < x < 0.6$.

Mitkova and co-workers in several articles have discussed [50–52], the glassforming conditions in systems such as Se–Ag–X = Cl, Br, I and Te–Ag–I. In the tellurium-based system, a glassforming area is found in a region closed to the binary glasses Te–I with an addition of about 10% of Ag atoms. The role of silver is not well understood, except that it accelerates the crystallization process by formation of Ag–I crystallites. Indeed, in the molecular Te–I chains, it is difficult to understand where the electrons produced by the silver ionization could be trapped.

In the Se-silver halide systems, the glassy area, as expected, is located in the Se rich region. Here also, it is clear that AgCl or AgBr microcrystallites formation appears as the main phenomenon leading to devitrification. One cannot exclude the possibility that silver halide in small quantity exists as a nano-size dispersed phase in the Se vitreous matrix.

2.6. *Chalcohalide glasses containing modifiers cations*

A more marginal class of chalcohalide glasses, closer to the traditional image of glass formation with modifier cations are built from a giant aperiodic lattice, such as Al^{3+}, Ga^{3+}, Ge^{4+} covalently bonded to anions such as halogens or chalcogens. The giant polymer carries a negative charge which is compensated by large modifier cations, such as Cs^+, Ag^+, Rb^+, K^+ etc.... statistically distributed in the network.

2.6.1. *The cesium thiohalide glasses*

Berg and co-workers [53, 54], while investigating the solubility of sulfur and sulfides in chloro-aluminate melts, remarked the formation of a polymeric material of the composition $CsAlSCl_2$. The same phenomenon was observed when Al was replaced by gallium [54]. In both cases, the melts, when cooled at room temperature, formed fully transparent glasses very stable against crystallization. This chemistry was extended [55] to other thiohalide compositions and the only glasses which have been successfully prepared and characterized are $CsAlSBr_2$, and the corresponding Se glasses, $CsAlSeCl_2$ and $CsAlSeBr_2$. These glasses are easily prepared in evacuated silica tubes at 600°C, by reaction of the compounds CsCl, Al, S and $AlCl_3$ in appropriate amounts.

These glasses are highly hygroscopic and moisture sensitive. Therefore, they show a high resistance to devitrification. A $CsAlSCl_2$ glass exhibits a T_g at 193°C, but no crystallization exotherm peak. As discussed later, this vitreous material belongs to a large family of 1D chain-like glasses with an infinite connection of AlS_2Cl_2 tetrahedra connected by sharing S corners. The infinite $AlSCl_2{}^-$ anionic chains are counter-balanced by the large Cs^+ cations. These materials are obviously isotypic with the molecular $GeSI_2$ glass previously discussed.

2.6.2. *Thiohalide glasses based on the Ga_2S_3–GeS_2 system*

In a general and well-documented article, Tveryanovich and co-workers [56] have reviewed the work on the system GeS_2–Ga_2S_3–MCl mainly performed in Russia. This contribution translated in English, referred to previous articles published in Russian. The monovalent cation M represents M = Li, Na, K, Rb, Cs, Ag and Tl. With an ionic radius r(Cs) = 1.67 A, the cesium ion is the biggest cation existing, and it is well adapted to counter-balance an anionic framework made of Cl and S, which also occupies a significant part of the glass-skeleton volume. In this case, not surprisingly, in analogy to the glasses discussed above, the largest glassforming domain belongs to the Cs family GeS_2–Ga_2S_3–CsCl. The glassforming region, shown in Fig. 12 is most representative of the series. The vitreous domain extends continuously from the binary Ga_2S_3–GeS_2 system in the GeS_2 rich region to the Ga_2S_3–MCl side which is characterized by an eutectic region with low liquidus temperature around Tl = 500°C, and where glasses can be formed. However, more information is needed to qualify these glasses specially against the tendency to devitrification. For these glasses, the resistance to moisture corrosion decreases with the Cl content. From Raman spectroscopy [57] it is claimed that the glass-forming ability of this material is due to sulfur corner-sharing connection of $GeS_{4/2}$ and $GaS_{3/2}Cl$ tetrahedra.

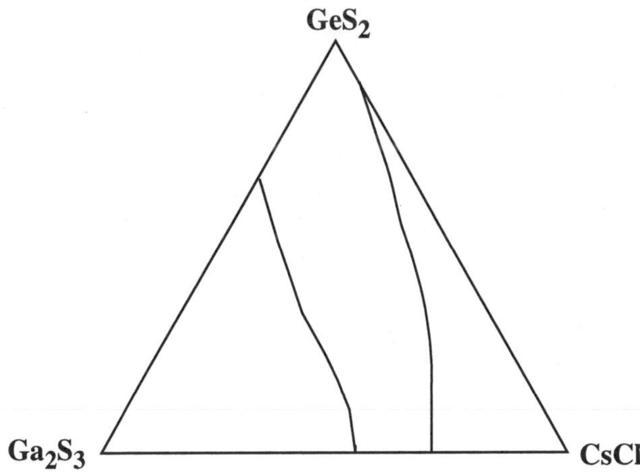

Fig. 12. Glassforming region in the Ga_2S_3–GeS_2–CsCl system taken from Ref. [56].

3. Structural Models for Chalcohalide Glasses

Since these materials belong to a rather new generation of glasses and also because they are often hygroscopic, the number of structural investigations are rather limited. Most of the results come from vibrational spectroscopy such as infrared or Raman scattering. From these very limited numbers of investigations, it is difficult to identify with certainty the observed vibrations. By nature, in these liquid-like solids, the polyhedra have all the possible orientations and are also coupled together by the fact that the lattice is a giant viscous polymer.

A powerful source of information on structure in chalcohalide glasses can be found in X-ray investigations of the crystalline-parent materials, Mössbauer spectroscopy used as local probe, glass transition temperatures and viscosity and relaxation measurements which illustrate the dynamics of structural relaxation.

3.1. *Vibrational spectroscopy on chalcohalide glasses*

Heo and Mackenzie [2, 3] have carried out a systematic investigation of the Ge–S–Br and Ge–S–I systems by infrared and Raman spectroscopy, in order to elucidate the role of the halogen in the Ge/S glasses, when S is substituted by Br or I. A significant number of vibrational modes occur in the 100 cm^{-1} to 600 cm^{-1} region, which represent signatures of various building blocks of the network. Table 2 contains respectively the frequency and assignment of the Raman bands for the Ge–S–Br and Ge–S–I glasses.

It is clear that the interpretation of the vibrational spectra suggest a very strong analogy between the two families of glasses Ge–S–Br and Ge–S–I. Heo and Mackenzie have proposed the following interpretation. When the halogens $X = Br$, I are added to the binary Ge–S glass, first, there is formation of Ge–X terminal bonds due to the S/X substitution in the GeS$_4$ tetrahedra. Second, the excess of sulfur is identified as forming S$_8$ rings or S$_n$ chains, which are dispersed as a solid solution in the network structure formed by GeS$_{4-x}$Br$_x$ groups connected through short bridging sulfur chains. The role of monovalent halogen is to decrease the network connectivity and to lead to a more open structure where the glass composition GeSI$_2$ appears as the ideal chain-like model.

The proposed model by Heo and Mackenzie, for the glass Ge$_{23}$S$_{47}$Br$_{30}$, selected as an example and represented in Fig. 13, is a good representation

Table 2. Frequency and assignment of the Raman bands in the two glass-families Ge–S–Br and Ge–S–I.

Frequency ν cm^{-1}	Assignments : sd : symmetrical deformation ss : symmetrical stretching as : asymmetrical stretching
130	Ge–Br$_3$ sd in GeBr$_3$S
158, 220	S$_8$
233	GeBr$_4$ ss
254	Ge–Br$_{ss}$ in GeBr$_2$S$_2$ and GeBr$_3$S
264	Ge–Br$_{ss}$ in GeBrS$_3$
288	Ge–Br$_{as}$ in GeBr$_3$S$_2$
300	Ge–Br$_{as}$ in GeBr$_3$S
342	Ge–S$_4$ ss
375	Ge–S$_4$ edge-shared
475	S$_8$ and S_n chains
100–108	GeI$_3$ sd in GeI$_3$S
150	S$_8$
201	GeI$_{ss}$ in GeI$_2$S$_2$ and GeI$_3$S
218	S$_8$
227	GeI$_{ss}$ in GeIS$_3$
239	GeI$_{as}$ in GeI$_2$S$_2$
253–261	Ge–I$_{as}$ in GeI$_3$S
343	GeS$_4$ ss
373	GeS$_4$ edge shared
432	S$_3$Ge–S–GeS$_3$
477	S$_8$ and S_n chains

of all the bonding and coordination possibilities such as S$_8$ rings, GeBr$_3$S, GeBr$_2$S$_2$, GeBrS$_3$, GeS$_4$ tetrahedra connected together to form the aperiodic vitreous matrix.

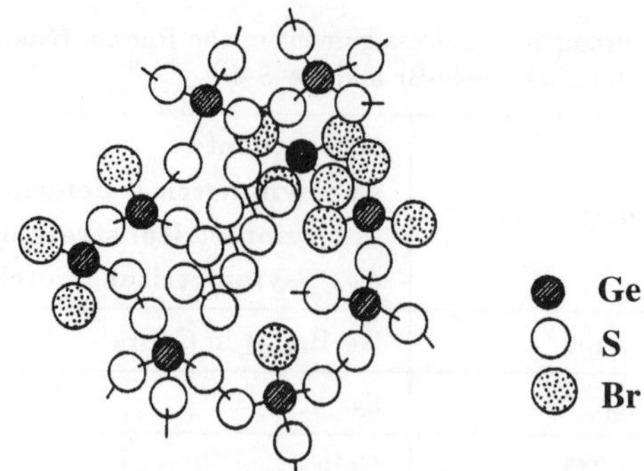

Fig. 13. A structural model for $Ge_{23}S_{47}Br_{30}$ glass.

Koudelka and Pisarcik [58] have studied the short-range order in Ge–S–I glasses by Raman scattering, and have reached the same conclusion that the dimensionality of the glass network decreases with the I content. They paid special attention to the $Ge_{30}S_{50}I_{20}$ glass composition located in the middle of the glassforming region, and which has an optimal average coordination number $\langle m \rangle = 2.4$, as predicted by the Philips and Thorpe theory [59]. Although, we now know that in the presence of 1-fold coordinated halogen atoms, the correct description of constraint theory is the one recently advanced by Boolchand and Thorpe [38]. The same authors [60–62] Koudelka and Pisarcik have also investigated the effect of S/I or S/Br substitutions on the glasses belonging to the As–S–Br or I system. The Raman studies suggest a bimodal character of the Raman spectra in As_2S_3–AsX_3 glass series X = Br, I. These anomalies are explained by a molecular phase separation due to the formation of individual AsX_3 molecules which are distributed inside the $AsS_{3/2}$ pyramids interconnected network.

Two other Raman contributions will be also mentioned here because they provide the only structural information on the Ga or Ge chalcohalide glasses containing modifiers cations such as Cs^+. Berg and co-workers [53, 54] in studying the Raman spectra of the $CsAlSCl_2$ glass have suggested that the strong peak at 325 cm^{-1} is due to the vibrational mode of AlS_2Cl_2 tetrahedra. They explained that the glass formation is due to the formation of an infinite linear polymeric network of $[AlS_{2/2}Cl_2]^-$ anions consisting of tetrahedra units connected by corner-sharing S atoms as indicated in Fig. 14. One can visualize the giant anionic polymer to be isotypic with the chain-like structure of the molecular glass $GeSI_2$.

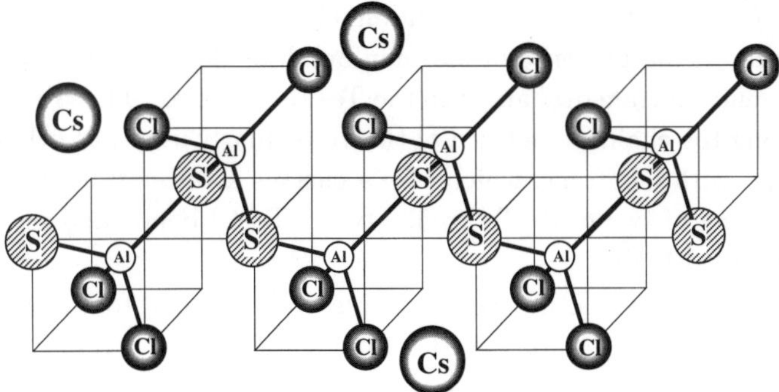

Fig. 14. The AlS_2Cl_2 tetrahedra sharing corners and forming the quasi-1D polymeric anion $AlSCl_2^-$ in the glass $CsAlSCl_2$.

Chalcohalides glasses based on gallium sulfide have been recently investigated by Tverjanovich [57] and co-workers. They conclude that the main reason for glass formation in the GeS_2–Ga_2S_3–MCl system, M = Ag, Tl, Br, Cs is the formation of complex anions, such as $GaS_{3/2}Cl$, formed by S connected tetrahedra in which Ga^{3+} occupies the center. These glasses are obviously not too far from the Berg glasses described previously.

3.2. *Other routes to obtain structural informations on chalcogenide glasses*

The earliest structural work on chalcohalide glasses using X-ray diffraction was by Hopkins and co-workers [63] on the As–S–I system. On the basis of the radial distribution functions showing a peak at 2.45 A, they proposed a structural model of twisted $As_2S_{3-x}I_x$ chains in which the iodine atoms are bonded to As. But they also suggested that the chains could have a finite length with some terminal $-S-As-I_2$ groups.

The case of the tellurium halide phases TeX or TeXAs glasses (X = Cl, Br, I) is of special interest because of the different approaches which have been used to give a reasonable structural model.

First, the structure of the crystalline parent material Te_3Cl_2 and Te_2Br have been investigated by Kniep and co-workers [13]. They proposed a polymeric chains structure as represented on Fig. 4. This is a rather fortunate situation to have a material having a so simple chemical formula and existing in both form: crystalline and vitreous.

To form the Te chains, two σ bonding electrons are used on each Te atom which results in the presence of two non-bonding electronic lone pairs, which give this pseudo-tetrahedral angle to the Te–Te–Te bond. This Te site is called A site. Along the chains, one third of the Te regularly used two other electrons for the bond with Cl atoms, which give a curious coordination around this Te atom consisting of two σ Te–Te bonds, two σ Te–Cl bonds and one lone pair; this site is called B site.

This chains-like structure is, of course, very similar to the Se structure and it is not surprising to observed a total solid solution, as discussed before between the TeX glasses and the selenium glass.

Wells [64] and co-workers and Boolchand [16], in a very interesting paper, have used a local probe approach to obtain more informations on the medium range structure in the TeX glass network. To do this, they remark that the Te_3X_2 (X = Cl, Br) chalcohalide glasses are ideally suited for Mössbauer effect studies, because one can probe the coordination of the Te cation elegantly using both ^{125}Te and ^{129}I spectroscopy.

Their conclusions indicate clearly that in the glassy state, the Te atoms are indeed in three kinds of coordination as indicated on Fig. 15. In addition to the coordinations, just discussed before, denoted A and B, they prove the existence of a third situation called A' in which Te atoms are connected to the chains on one side and to a halogen on the other side, indicating that some Cl and Br atoms play here the role of cutting the chains. From the quantitative estimation of the intra-chains Te sites and Te chains-end sites, it is possible to conclude that the average chains length is 1.5 nm. The fact that half of the halogen atoms play the role chain terminator have been also observed by Bresser [15] in glasses belonging to the ternary Te–Se–Cl or Te–Se–Br glasses. The low glass transition temperature $T_g \approx 70°$–$90°$C, the low average coordination number of the proposed chain fragment model suggest that the extensive glassforming tendency in this chalcohalide glasses derive from the floppiness of the chain-fragments.

This tendency of chain-fragmentation is probably inducing some phase separation, in the Te–Se–I glasses for instance, explaining Se_8 rings formation observed by Wang and Chen [18].

Structural relaxation [65] and viscosity measurements [66] performed on TeX and TeXAs glasses indicate that the chain conformation plays a key parameter to understand the annealing conditions of such low T_g glasses. Relaxation, near room temperature, has been investigated and it has been

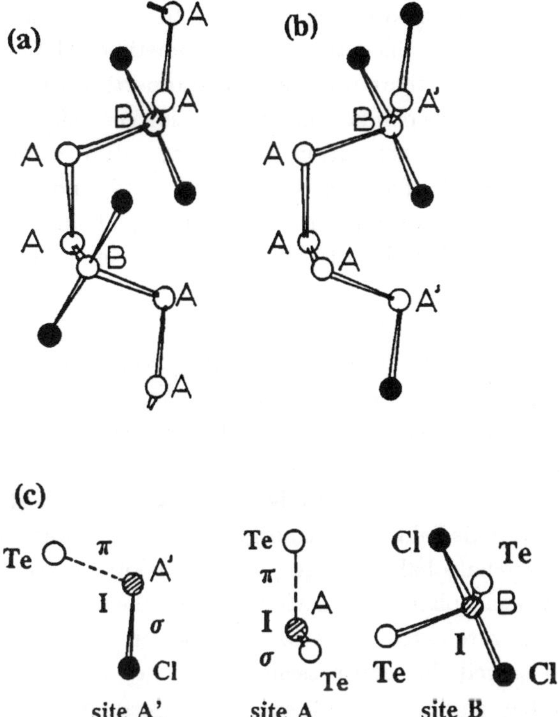

Fig. 15. The structural model for Te_3Cl_2 glass taken from Ref. [13]. (a) The infinite chain in the crystalline form, (b) the short segment with terminal Cl in the glassy form, (c) the three kinds of coordination for the Te atoms in the glass.

observed that the glass $Te_3I_3Se_4$, $T_g = 49°C$, relaxed to equilibrium within four days, the glass $Te_3Se_5Br_2$, $T_g = 71°C$ shows substantial relaxation but did not fully equilibrate over a period of months. The glass $As_4Se_3Te_2I$, $T_g = 118°C$, showed no relaxation in a two months period.

From all these informations, it can be concluded that the glass $\leftrightarrow$ crystal equilibrium in the case of the TeX glass is governed by a competition between long chain configuration, which favors the 1D ordering namely crystal formation while the glass formation derives from the floppiness of chain fragments.

4. Physical Properties and Applications of Chalcohalide Glasses

As already discussed, these are several intrinsic properties that have limited the number of investigations and tentative applications of the chalcohalide glasses.

Indeed, many compositions are very sensitive to moisture and their optical properties are severely modified by OH and sometimes H_2O contamination. Also in many cases, the thermo-mechanical properties are not suitable for applications. Low T_g corresponds to large expansion coefficient, poor resistance to thermal shock and risk of relaxation. Finally, by nature, chalcohalide glasses are composed of two kinds of anions from group 6 and 7, and in some situations the thermodynamics favors the formation of pure halide or pure chalcogenide compound leading to a big risk of phase separation in the glassy state. For instance Se_8, S_8 rings, AsI_3 crystal, GeI_4, $TeCl_4$ molecule as well as Cl_2, Br_2, I_2 vapor formation cannot be excluded in the chalcohalogenide chemistry.

Up to now, these glasses have almost only be considered for optical applications. As low phonon energy material, they offer an excellent transmission in the infrared region, as long as they are not contaminated by parasitic vibrational groups, such as OH^-, SH^- or SeH^-.

They also present some intrinsic advantages as optical host for rare earth doped devices [67]. This field is just emerging and more information is needed. But it is clear that chalcohalide glasses, with their low phonon vibrations, are characterized by weak phonon relaxation, namely heat dissipation in all the processes involving rare earth excitation for laser or optical amplification. In addition, it is observed that the presence of halogen in the matrix increases the band gap value making these glasses more transparent in the UV-visible region than the corresponding pure chalcogenides glasses.

The glass to crystal equilibrium has been also the subject of speculation specially under laser irradiation for potential application of TeX glass thin films in the optical storage technology.

4.1. *Laser irradiation and crystallization of TeX glasses*

A very accurate investigation on the crystallization of Te–Se–I thin films have been performed by Blatter and Ortiz [68]. They studied the thermally induced amorphous-to-crystalline transformation in RF sputtered Te–Se–I thin films with an approximatively constant Te/Se Ration = 3/4 and with different I concentrations in the range 0°–20%. It is observed that the crystals, which are formed during the heating process up to 160°C, were essentially halogen free, namely belonging to the Te/Se system.

Crystallization occured, via the formation and growth of cylindrities which are two-dimensional, radially aligned polycrystalline agregates and which exhibit a branch fibrillar morphology.

The major effect of iodine alloying is as follows:

(1) the halogen strongly stabilized the amorphous phase and the crystallization took place when there was substantial superheating at temperatures above 80°C after the halogen was depleted;
(2) the halogen restricted the cyclindrite formation to a thin surface layer and dictated the details of the morphology;
(3) in the I-containing films all the cylindrites appeared suddenly rather than being nucleate at a constant rate and their density was reduced by a factor of 100.

Adam and co-workers [69] have also investigated the effect of laser irradiation of Te–Se–Br glassy thin-films.

The glass composition used as the target for the sputtering deposition was $Te_{30}Se_{50}Br_{20}$. The composition of the As-deposited films has been analyzed by electron microprobe and was found to be $Te_{36}Se_{51}Br_{13}$ indicating a loss of Se and Br.

The target compositions does not exhibit crystallization peak while the thin film is characterized by exothermic crystallization near $Tc = 150°C$. The pulse laser irradiation from an argon laser at 514.5 nm induces in a thin-film an almost complete evaporation of bromine and a melting of the remaining alloy. When this Te/Se alloy crystallizes, it has been established that it starts with surface filamentary growth exhibiting fractal network formation.

4.2. *The optical properties of chalcohalide glasses*

4.2.1. *The infrared edge*

In a vitreous chalcohalide materials $ABXY$, where A and B represents the electropositive elements combined with X = halogen and Y = chalcogen, the position of the infrared edge is associated to the vibrational modes due to the bonds AX, AY, BX, BY. In these highly covalent materials, individual vibrational groups are also identified with their own modes such as the tetrahedra GeS_4, AlS_2Cl_2, etc. ...

Due to the high atomic weight of X and Y, the phonon energies Wp, associated to the fundamental vibrations, are located in the 200 to 700 cm^{-1}. This situation is very favorable for observing excellent infrared transparency extending towards the 12 to 20 μm region.

In these highly covalent compounds, in addition to the one phonon absorption, a weak but measurable two and sometimes three-phonons peak can be observed. This situation, of course, contraries results in IR absorption bands in some strategic regions, such as the 8–12 μm band, where the atmosphere is transparent.

In addition, to these intrinsic factors, there are some extrinsic reasons which can modify the infrared transparency. For instance, the contamination of the glasses by OH^-, SH^-, O^{2-} parasitic impurities can strongly modify the optical properties. The infrared transparency is much affected by these anomalies, specially in the region corresponding to the A–O, B–O, O–H, S–H vibrational region.

As discussed before, chalcohalide glasses are prone to moisture corrosion and in the literature information on good infrared optical chalcohalide glasses is rather scarce.

Furthermore, in order to be considered as serious candidates for infrared bulk optics, such as lenses, windows, an infrared glass needs also to be very resistant to crystallization, bubble formation, specially if large pieces have to be shaped. Consequently, the only data available for potential infrared optics are those concerning the TeX and TeXAs glasses.

Figure 16 represents the position of the phonon-edge in the infrared region for a TeX glass based on I and Se, $Te_3I_3Se_4$, a TeXAs glass $Te_3Se_3IAs_4$ compared to two other halogen free chalcogenide glasses. The transmission is in arbitrary unit and is for samples of millimeter size thickness. The TeX glass

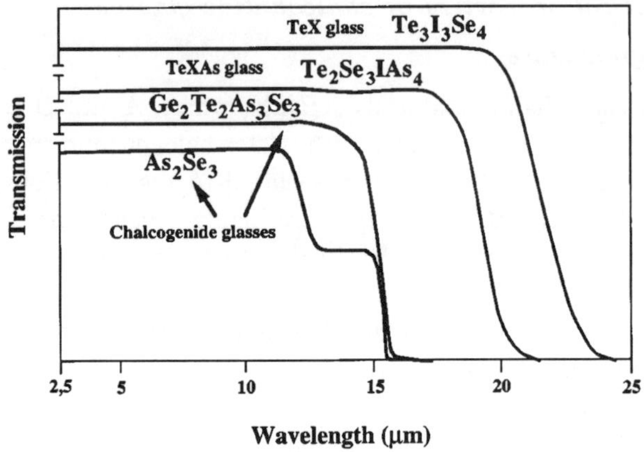

Fig. 16. The infrared edge for several chalcogenide or chalcohalide glasses. The thickness of the glass samples is about 2 mm.

transmits to about 20 μm which corresponds to a one phonon absorption with $Wp = 150$ cm^{-1}, as measured by Raman spectroscopy [18]. In the TeXAs glasses, due to the reinforcement of the chemical bond by arsenic, the infrared edge shifts towards the 17 μm region. It is interesting to note that, the spectra have been recorded on glass samples, maintained in normal moisture conditions and that, no parasitic absorption is observed all along the optical windows.

4.2.2. *The band gap edge*

The visible edge in the glasses is associated with electronic transition and is, of course, related to the bonding energy diagram for this chalcohalide materials.

The most remarkable feature is that these kind of glasses are not transparent in the visible region. They have a black color and exceptionally they transmit only a part of the visible spectrum, but never totally. Consequently, their colors change from black to red, orange and sometimes yellow.

This behaviour is a general characteristic of the chalcogen containing materials and specially in compounds where the non-bonding lone pair electrons are very active. The band gap energy in a covalent or iono-covalent material is due to the separation between the σ bonding orbitals and σ^* anti-bonding levels coming from the different bonds between A, B, X and Y as discussed before. Figure 17 gives a schematic representation of the energy band diagram.

In these chalcohalide compounds, a significant density of electrons also exist in non-bonding levels, namely the lone pair associated with the chalcogens and responsible for the very special local steric effect around Te for instance in

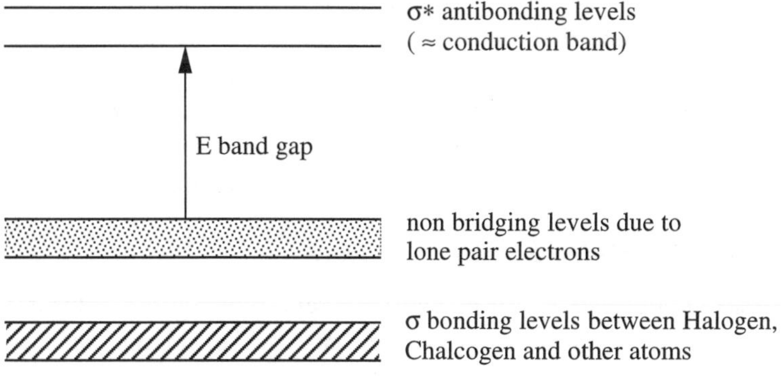

Fig. 17. The origin of the band gap E in a chalcohalide glass.

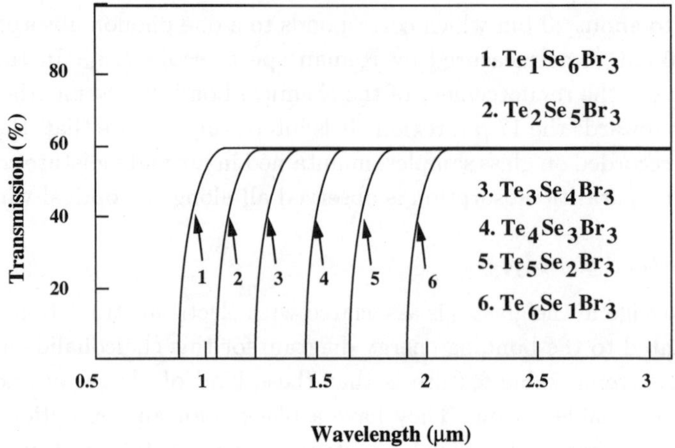

Fig. 18. Evolution of the band gap as a function of the Se/Te ratio in the indicated Te/Se/Br glasses.

TeX glasses. Consequently, the absorption mechanism occurs from a transition between the non-bonding and the anti-bonding levels giving rise to a smaller band gap value E close to 1 eV on average as represented in Fig. 17.

The value of E is very dependent on the chemical composition. For instance, in the homologous series of glasses $Te_{7-x}Se_xBr_3$, E varies from $E = 0.66$ eV, for Te_6SeBr_3 to $E = 1.39$ eV for $Te_1Se_6Br_3$. In Fig. 18, we show the evolution of E with composition for six glasses ($x = 1$ to 6). It is clear that the presence of the more metallic Te tends to decrease the band gap as expected.

It is also observed, that the presence of halogens and specially the most electronegative Cl tends to prevent the extension of the lone-pair electrons, and results in an increasing of E. On this point, the most interesting materials are the chalcohalide glasses belonging to the Ga_2S_3–GeS_2–MCl system, but more information is needed on the stability of these glasses.

4.3. *The chalcohalide glass optical fibers*

When chalcohalide glasses are enough resistant towards crystallization and moisture or oxygen contamination, they can be considered as reasonable candidates for infrared waveguides preparation [70, 71, 73]. With intrinsic absorption in the band gap region around 1 μm and in the phonon region far in the infrared, these kind of glasses are supposed to be ultratransparent in the middle of the optical window around 9–10 μm.

In practice, many extrinsic factors can affect the transparency, such as scattering centers due to crystallites, voids, phase separation and metallic particles. Probably, the most severe factors producing optical loss are associated with the presence of parasitic absorption bands, such as OH and SH, due to the contamination of the glass during preform fabrication process and fiber drawing.

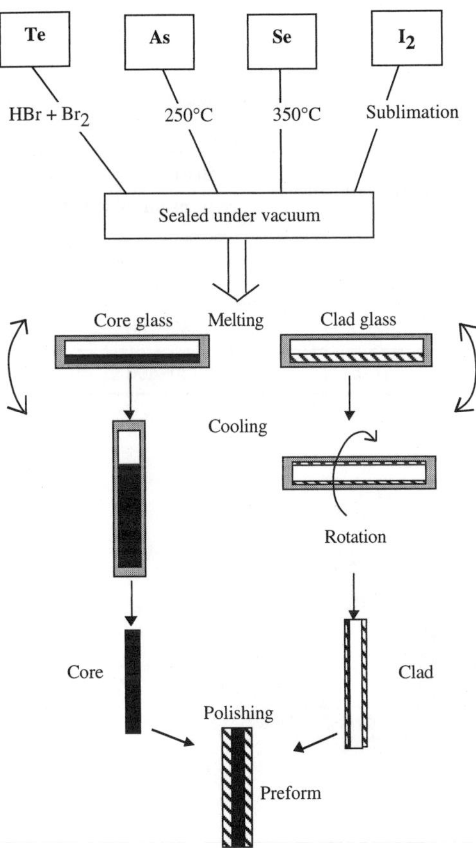

Fig. 19. The different steps in the fabrication of a preform before drawing of a TeX glass fiber. The purification of the starting elements is very critical as well as the quality of the interface between the core-glass rod and the clad-glass tube.

4.3.1. *Infrared fibers preparation*

The multistep operation necessary to obtain a good optical fiber are summarized on Fig. 19. In Table 3, we mention some glassy chalcohalide systems which have been optimized in order to meet the severe conditions necessary to obtain a loss low optical fiber. The preform preparation is a very delicate operation, where the core has to exhibit a higher refractive index than the cladding in order to achieve the guiding condition. The ultimate purification of the starting elements, as well as operation in clean room conditions are necessary to avoid any dust contamination during the introduction of the core glass rod into the cladding glass tube.

Table 3. Some chalcohalide glass compositions used for the preparation of infrared fibers and their main characteristics.

Glass Acronym	Composition	Softening Temperature (°C)	Transparency Domain (thickness 5 mm)
TeX	Te–Se–I	80–100	2–20 μm
TeXAs	Te–As–Se–I	130–160	2–18 μm
GASI	Ge–As–Se–I	320–400	1–16 μm

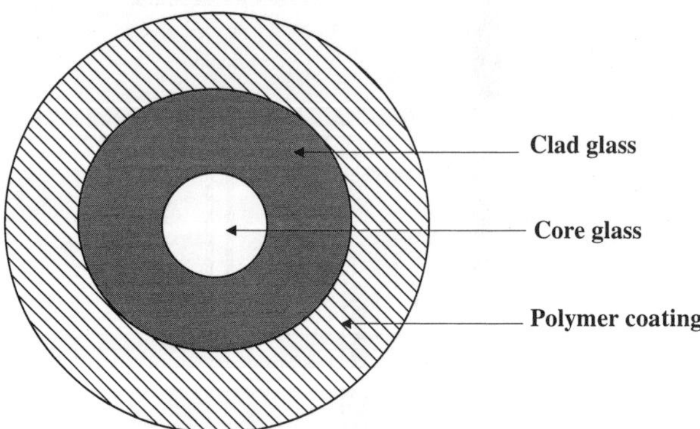

Fig. 20. The cross section of a TeX glass fiber. Infrared light propagates in the core glass which has a refractive index higher than of the clad glass. The polymer coating reinforces the mechanical properties of the fiber.

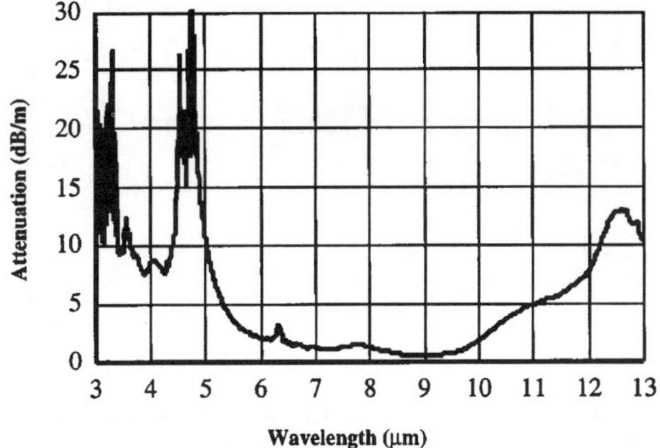

Fig. 21. Transmission spectra of a TeX glass fiber shows the low optical loss region from 6 to 11 μm.

After fibering in a drawing tower at a temperature close to the softening T_s, as indicated in Table 3, the fibers are coated on-line by a plastic jacket in order to make the fibers more resistant to mechanical damage. Figure 20 represents the section of a typical fiber showing the core glass where most of the light propagates, the clad glass having a slight difference in the glass composition and the polymer protection.

In Fig. 21, the typical transmission spectra for a fiber belonging to the TeX glass family is represented. The 3–5 μm region is contamined by H impurities which are very difficult to eliminate. The low loss region is, as expected, located in the 5 to 11 μm domain with an attenuation less than to 1 dB/m in the 9 μm band.

4.3.2. *The TeX glass fibers in some infrared devices*

In this section, we will briefly discuss three kinds of applications of these unique infrared waveguides.

4.3.2.1. TeX fibers for remote infrared spectroscopy of organic materials

This process takes advantage of two unique characteristics of the TeX glasses:
(a) the existence of a strong infrared evanescent wave propagating at the surface of the fiber;

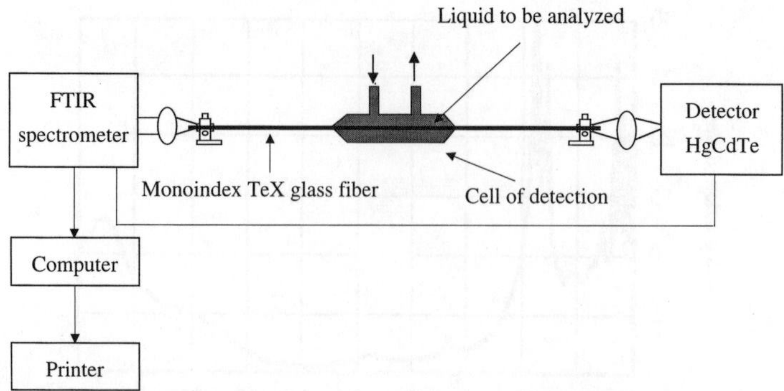

Fig. 22. Experimental set-up used for evanescent spectroscopy at the surface of a TeX glass fiber.

(a) a large optical window covering in particular the region where the organic species have their strong absorption band $5 \to 10$ μm. For instance the signature of glucose is around 10 μm while that of ethanol in the 9–9.5 μm region. In using the set-up showed in Fig. 22, it has been possible to follow *in situ* the fermentation process corresponding to the transformation of fruits juice into alcohol.

4.3.2.2. TeX glass fibers for temperature measurement

TeX fibers have found applications in a system aiming at carrying an infrared signal emitted by a thermal object towards an infrared MCT detector as shown in Fig. 23. Notice that the black body at room temperature has a maximum

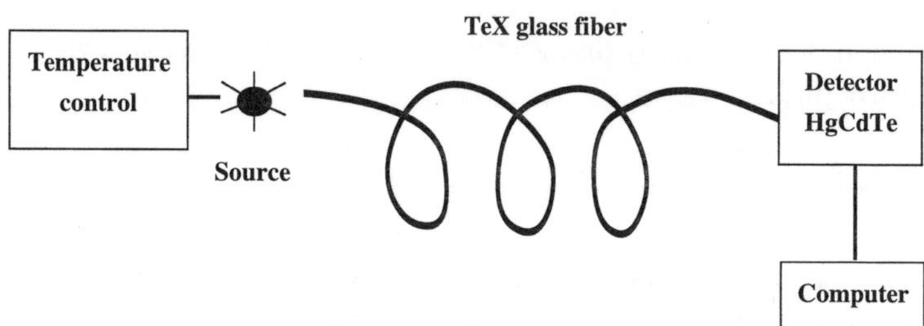

Fig. 23. Temperature control using a TeX glass fiber. The infrared emission of a source is transferred via the fiber towards an infrared detector.

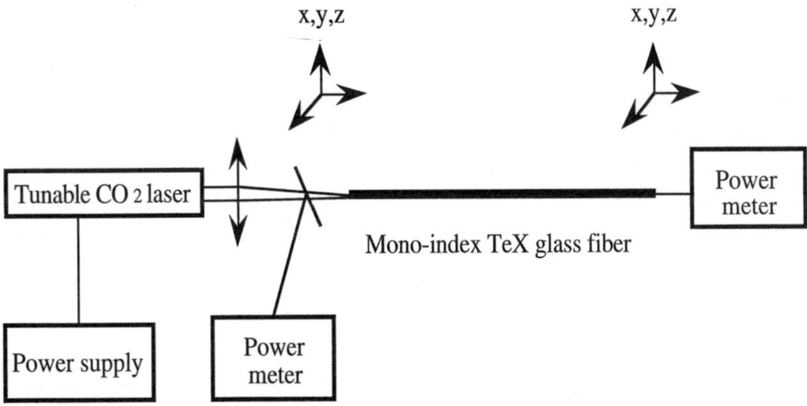

Fig. 24. Medium power transmission into a TeX glass fiber. The light emitted by a CO_2 laser is tuned at 9.3 μm which corresponds to the low loss region of the fiber.

of emissivity around 10 μm which corresponds to the low optical loss region of the fiber. With this set-up very precise, non contact measurement has been performed for instance to evaluate the human skin temperature.

4.3.2.3. The TeX glass fibers for CO_2 laser power transmission

As represented in Fig. 24, the light emitted by a CO_2 laser can be injected into a TeX glass fiber [74]. In order to match the low-loss optical region, the laser was tunned to emit in the 9.3 μm region. In using a powermeter to evaluate the transmitted energy, it was possible to verify that a maximum power of 3 watts was obtained on the target at the output of a 1 meter long fiber. Similar experiments have been performed in Japan by Inagawa [72].

5. Conclusions

Often, considered as substitution products of chalcogenide or halide materials, the chalcohalides glasses have now reached a status where they have to be considered as an original class of compounds. However, because they are, by nature, mixed-anion vitreous materials, they suffer from some risks of phase separation by formation of pure halide or pure chalcogenides based compounds. As a practical matter chalcohalide compositions have to be severely optimized in order to avoid bubble formation, microcrystallites and liquid immiscibility, factors which are of course very detrimental for optical transparency.

These observations explain, partly, why the use of chalcohalide glasses, in practical application, is still very limited. Among the intrinsic factors, which degrades the optical properties, is the high tendency for terminal halogen to react with water molecules rendering the material hygroscopic or moisture sensitive.

Concerning this last point, it is clear that hygroscopicity is less important in molecular materials, such as the TeX glasses than in glasses composed of giant anionic framework interacting with modifier cations such as the alkalis.

It is clear that good stable glasses can be expected in this specially difficult chemistry. If the problems mentioned above can be solved, this new class of glasses have a real potential as active hosts for rare earth dopants. Indeed, in this case, they offer, due to the presence of the electronegative halogen, a much better transparency in the visible region which correponds to the pumping region of the rare-earth lasers. Consequently, efficient micro-lasers, optical amplifiers and other optical devices could be developed in using this original optical matrix.

References

[1] J.S. Sanguera, J.Heo and J.D. Mackenzie, *J. Non-Cryst. Solids* **103**, 155 (1988).

[2] J. Heo and J.D. Mackenzie, *J. Non-Cryst. Solids* **111**, 29 (1989).

[3] J. Heo and J.D. Mackenzie, *J. Non-Cryst. Solids* **113**, 1 (1989).

[4] J. Heo and J.D. Mackenzie, *J. Non-Cryst. Solids* **113**, 246 (1989).

[5] J. Heo, J.S. Sanghera and J.D. Mackenzie, *Opt. Eng.* **30**, 4, 470 (1991).

[6] G. Fuxi, *J. Non-Cryst. Solids* **140**, 184 (1992).

[7] A.F. Wells, *Structural Inorganic Chemistry*(Oxford Science Publication, 5th Edn. Oxford, 1987).

[8] J. Lucas and X.H. Zhang, *J. Non-Cryst. Solids* **125**, 1 (1990).

[9] J. Lucas and X.H. Zhang, *Mat. Res. Bull.* **21**, 871 (1986).

[10] X.H. Zhang, G. Fonteneau and J. Lucas, *Mat. Res. Bull.* **23**, 59 (1988).

[11] X.H. Zhang, G. Fonteneau and J. Lucas, *J. Non-Cryst. Solids* **104**, 38 (1988).

[12] I. Chiaruttini, G. Fonteneau, X.H. Zhang and J. Lucas, *J. Non-Cryst. Solids* **111**, 77 (1989).

[13] R. Kniep, D. Mootz and A. Rabenau, *Angew. Chem. Int.* **12**, 499 (1973).

[14] J. Lucas, I. Chiaruttini, X.H. Zhang, H.L. Ma and G. Fonteneau, *Proc. SPIE* **1048**, *Infrared Fiber Optics*, 52 (1989).

[15] W.J. Bresser, J. Wells, M. Zhang and P. Boolchand, *Z. Naturforch.* **51a**, 373 (1996).

[16] P. Boolchand, *Z. Naturforch.* **51a**, 572 (1996).

[17] L. Tichy, V. Smrcka, H. Ticha and P. Tomiska, *J. Mat. Sci. Lett.* **9**, 681 (1990).

[18] Z. Wang and Q. Chen, *J. Non-Cryst. Solids* **184**, 177 (1995).

[19] X.H. Zhang, H.L. Ma, G. Fonteneau and J. Lucas, *J. Non-Cryst. Solids* **140**, 47 (1992).

[20] S.S. Flaschen, A.D. Pearson and W.R. Northover, *J. Appl. Phys.* **31**, 219 (1960).

[21] T.E. Hopkins, R.A. Parternak, E.S. Gould and J.R. Herndon, *J. Chem. Phys.* **66**, 733 (1962).

[22] O.V. Khiminets, P.P. Puga, V.V. Khiminets, I.I. Rosola and G.D. Puga, *Zh. Prikl. Specktr.* **28**, 700 (1978).

[23] L. Koudelka and M. Pisarcik, *Solid State. Commun.* **41**, 15 (1982) .

[24] L. Koudelka and M. Pisarcik, *J. Non-Cryst. Solids* **64**, 87 (1984).

[25] S.A. Demboski and A.P. Chernov, *Izv. Akad. Nauk. SSSR, Neorg. Mat.* **4**, 1229 (1968).

[26] A.D. Pearson, W.R. Northover, J.F. Dewald and W.F. Peck, *Advanced in Glass Technology* **2**, (Plenum Press, NY, 1962) p. 357.

[27] A.G. Fisher and A.S. Mason, *J. Opt. Soc. Am.* **52**, 721 (1962).

[28] I.D. Turyanitsa, V.V. Khiminets and O.V. Khiminets, *Fiz-Khim-Steckla* **1**, 170 (1975).

[29] A.T. Ward, *J. Phys. Chem.* **72**, 4133 (1968).

[30] L. Koudelka, J. Horak, M. Pisarcik and L. Sakal, *J. Non-Cryst. Solids* **31**, 339 (1979).

[31] V.A. Ananichev, L.A. Baidakov and N.I. Krylov, *Glass Physics and Chemistry* **21**, 6, 445 (1995).

[32] D.M. Petrovic, M.V. Dobosh, V.V. Khiminets, S.R. Kukic, A.F. Petrovic and S.M. Pogorelli, *J. Therm. Analys.* **36**, 2375 (1990).

[33] S.A. Dembovskii, *Phys. Chcm. Glasses* **10**, 73 (1969).

[34] S.A. Dembovskii, V.V. Kirilenko and Y.A. Buskaev, *Izv. Akad. Nauk. SSSR, Neorg. Mat.* **7**, 2, 328 (1971).

[35] S. Maneglier-Lacordaire and J. Rivet, *Ann. Chim.* **10**, 291 (1975).

[36] A.B. Seddon and M.A. Hemingway, *J. Thermal Analysis* **37**, 2189 (1991).

[37] S. Loheider, G. Vögler, I. Petscherizin, M. Soltwisch and D. Quitmann, *J. Chem. Phys.* **93**, 5436 (1990).

[38] P. Boolchand and M.F. Thorpe, *Phys. Review* **B50**, 10366 (1994).

[39] J.C. Philips, *J. Non-Cryst. Solids* **34**, 153 (1979); **43**, 37 (1981).

[40] S.A. Dembovskii and N.P. Popova, *Izv. Akad. Nauk. SSSR, Neorg. Mat.* **6**, 138 (1970).

[41] A. Feltz, H.J. Buttner, F.J. Lippmann and W. Maul, *J. Non-Cryst. Solids* **8–10**, 64 (1972).

[42] I.D. Turyanitsa and B.M. Koperless, *Izv. Akad. Nauk. SSSR, Neorg. Mat.* **9**, 707 (1973).

[43] I.D. Turyanitsa and B.M. Koperless, *Izv. Akad. Nauk. SSSR, Neorg. Mat.* 851 (1973).

[44] O.V. Khiminets, I.D. Turyanitsa, V.S. Gerasimenko and V.V. Khiminets, *Fiz. Khim. Stekla* **2**, 500 (1976).

[45] L.M. Belyaev, V.A. Lyakhovitskaya, G.B. Netsov and M.V. Mokhosdev, *Izv. Akad. Nauk. SSSR, Neorg. Mat.* **1**, 2178 (1965).

[46] T. Mori and H. Tamura, *J. Phys. Soc. Japan* **19**, 1247 (1964).

[47] P. Kunze, *Fast Ion Transport in Solids*, ed. W. Van God, (North-Holland, Amsterdam, 1973) 405.

[48] F. Robinel, B. Carette and M. Ribes, *J. Non-Cryst. Solids* **57**, 49 (1983).

[49] M. Arif and L.A. Baidakov, *Glass Phys. Chem.* **21**, 6, 443 (1995).

[50] M. Mitkova, T. Petkova and A. Yanakiev, *Mat. Chem. Phys.* **30**, 55 (1991).

[51] T. Petkova and M. Mitkova, *Mat. Chem. Phys.* **33**, 233 (1994).

[52] M. Mitkova and V. Boev, *Mat. Lett.* **20**, 195 (1994).

[53] R.W. Berg, S.Von Ninbush and N.J. Bjerrum, *Inorg. Chem.* **19**, 2688 (1980).

[54] R.W. Berg and N.J. Bjerrum, *Polyhedron* **2**, 179 (983).

[55] M. Le Toullec, P. Christensen, J. Lucas and R. Berg, *Mat. Res. Bull.* **22**, 1517 (1987).

[56] Y.S. Tveryanovich, E.G. Nedoshovenko, V.V. Aleksandrov, E.Y. Turkina, A.S. Tveryanovitch and I.A. Sokolov, *Glass Phys. Chem.* **22**, 1, 9 (1996).

[57] A. Tverjanovich, Y.S. Tveryanovitch and S. Loheider, *J. Non-Cryst. Solids* **208**, 49 (1996).

[58] L. Koudelka and M. Pisarcik, *J. Non-Cryst. Solids* **113**, 239 (1989).

[59] J.C. Phillips and M.F. Thorpe, *Solid State Comm.* **53**, 699 (1985).

[60] L. Koudelka and P. Pisarcik, *J. Non-Cryst. Solids* **64**, 87 (1984).

[61] L. Koudelka and M. Pisarcik, *Mater. Chem. Phys.* **9**, 571 (1983).

[62] L. Koudelka, M. Pisarcik, N.I. Krylovand and V.A. Ananichev, *J. Non-Cryst. Solids* **97** & **98**, 1271 (1987).

[63] T.E. Hopkins, R.A. Pasternak, E.S. Gould and J.R. Herndon, *J. Phys. Chem.* **66**, 733 (1962).

[64] J. Wells, W.J. Bresser, P. Boolchand and J. Lucas, *J. Non-Cryst. Solids* **195**, 170 (1996). M. Mitkova and P. Boolchand, *J. Non-Cryst. Solids* **240**, 1 (1998).

[65] H.L. Ma, X.H. Zhang, J. Lucas and C.T. Moynihan, *J. Non-Cryst. Solids* **140**, 209 (1992).

[66] H.L. Ma, X.H. Zhang, J. Lucas, H. Senapati, R. Bohmer and C.A. Angell, *J. Solid State Chem.* **96**, 181 (1992).

[67] D. Marchese, G. Kakarantzas, A. Jha, B.N. Samson and J. Wang, *J. Modern Optics* **43**, 963 (1996).

[68] A. Blatter and C. Ortiz, *J. Cryst. Growth* **139**, 120 (1994).

[69] J.L. Adam, C. Ortiz, J.R. Salem and X.H. Zhang, *J. Mat. Science* **26**, 2900 (1991).

[70] C. Blanchetière, K. Le Foulgoc, H.L. Ma, X.H. Zhang and J. Lucas, *J. Non-Cryst. Solids* **184**, 200 (1995).

[71] J. Lucas, X.H. Zhang, K. Le Foulgoc, G. Fonteneau and E. Fogret, *J. Non-Cryst Solids* **203**, 127 (1996).

[72] I. Inagawa, T. Yamagishi and T. Yamashita, *Japanese J. Appl. Phys.* **30**, 2846 (1991).

[73] X.H. Zhang, H.L. Ma, G. Fonteneau and J. Lucas, *J. Non-Cryst. Solids* **140**, 47 (1992).

[74] X.H. Zhang, H.L. Ma, C. Blanchetière, J. Lucas, P. Froissard and J.C. Favey, *Proc. SPIE* **2328**, 32 (1992).

12

Applications of Non-Crystalline Materials

<hr>

A. APPLICATIONS OF GLASSES, AMORPHOUS AND DISORDERED MATERIALS

<hr>

STANFORD R. OVSHINSKY

*Energy Conversion Devices, Inc., 1675 West Maple Road,
Troy, Michigan 48084, USA
ovonic@aol.com*

Contents

1. Introduction

The field of amorphous and disordered materials has made a significant impact on two of the largest and most fundamental areas of our global economy — information and energy.

The feature that permits amorphous and disordered materials to be used in devices for encoding, switching, transmission and storage of information, and also enables the development of new, pollution-free technology for the generation and storage of energy, is the ability to design local atomic order in these materials, free from the conventional constraints imposed by the periodicity of a crystalline lattice. Since 1955, we have focussed on developing the science and technology of amorphous and disordered materials that provide the basis for these products.

Amorphous and disordered semiconductor materials provide a matrix in which unique physical, electronic, chemical and, where desired, structural changes can take place. Following design rules, one can make materials, in thin-film form for example, which can be sensitive to light and provide the basis of the growing field of thin-film photovoltaics. In this case, the materials system is based on the tetrahedrally coordinated atoms such as silicon and germanium combined with hydrogen and/or fluorine. These materials systems are also useful for optical imaging arrays and for electrophotographic photoreceptors. One can design glassy thin-film semiconductor materials based on chalcogenide alloys in which the amorphous structure is stable to high field excitation and unique, high speed, reversible electronic switching takes place. These materials are the basis for the Ovonic threshold switch. In the Ovonic memory switch, which is also based on chalcogenide glass alloys, electric field excitation causes a reversible crystalline-to-amorphous phase change to take place. These materials form the basis for the very successful phase change erasable optical memory disk products.

The general principles involved in designing local order configurations to achieve particular optical, chemical, electronic, and structural properties in semiconductor materials can be extended to non-semiconductor materials, such as dielectrics — even to metals. In terms of dielectrics, utilizing my concepts of chemical modification, we have been able to make a wide bandgap material, normally nonconducting, with controlled and excellent conducting properties [1]. As to metals, multi-phase, compositionally complex metallic systems are not necessarily glassy, although they can have amorphous phases; similar types of atomic interaction mechanisms exist in these materials because of their compositional and structural disorder. These materials are now being used as advanced hydrogen storage materials and as electrodes in high energy storage capacity rechargeable batteries. In these metallic systems, there is a strong analogy between the design of *chemically reactive sites* through manipulation of local atomic interactions, such as d-orbital interactions affecting the density of states at the Fermi level, and the design of *the density of electronic states* in the bandgap of amorphous semiconductors.

Ovonic nickel metal-hydride (NiMH) batteries using multi-phase, compositionally complex electrodes have now replaced nickel cadmium as the battery of choice in consumer applications such as portable telephones, laptop computers, etc. What we mean by complex is not just that there are many elements, but that we build into a material a chemical, electronic and topological system which performs, in the same material, various functions such as catalysis, hydrogen diffusion paths, varying density of electrons, acceptor sites for hydrogen, etc. Such a material system in a battery must provide high energy density, power, long life and robustness [2–5]. Larger versions of these batteries have become the preferred choice to power electric and hybrid vehicles [6].

The essential difference between crystalline materials and amorphous and disordered materials is that, when one eliminates or reduces the effect of the periodicity as a controlling parameter, one can introduce local order optional bonding, generating various kinds of new bonding arrangements, configurations and conformations. The result is that one can design and atomically engineer many new materials not previously available "naturally" and can exhibit new mechanisms that can be the basis of many new products. We will describe some of the most important ones here.

I have previously stated [7–9] that the Rosetta Stone of amorphous materials is the understanding of the relationship between the normal structural bonding (NSB), which characterizes the great majority of atoms and is responsible for the cohesiveness of the amorphous solid, and the deviant electronic configurations (DECs) that control the transport properties and provide the active sites of the material. Tying these two together is the concept of the total interactive environment (TIE) [10] which takes into account the special nature of various local, chemical, topological, and electronic interactions which can be designed into disordered and amorphous solids.

We can write a new language of materials if we make use of this new alphabet. We need not be limited to the old dogmas of homogeneity and equilibrium chemistries. We have a new world of nonequilibrium chemistry and varying topological structures. *We can deliberately create combinations and geometries in which even the local short-range order can vary subtly or drastically from one part of the material to the other. Such new structural chemistry produces new electronic phenomena and new sites for chemical activity.*

Materials with unique submicron clusters and two-dimensional configurations with unusual bonding and orbital relationships intermediate between amorphous and crystalline [11], as well as those containing microcrystallites and microcrystalline inclusions and layers, are also part of the spectrum of atomically engineered materials discussed here. Such designed materials have far-reaching applications. We can synthesize and engineer materials where we "mismatch," that is, put various elements together in a manner not available in periodic materials and provide structures which do not have to deal with mismatching lattices. The use of multi-layers and compositional modulation provides additional design parameters [7]. In fact, for certain applications, heterogeneity becomes a welcome tool rather than a scare word.

We will first discuss the material characteristics and the physics of the devices that are utilized in the information field. In information storage, the Ovonic phase change memory [12, 13] has taken over the field from magneto-optics and is the basis for the new, high-capacity rewritable DVD optical memory disks.

The materials utilized for the Ovonic optical and electronic memory and the Ovonic threshold switch are chalcogenides — based on elements from Group VI of the Periodic Table which Kastner characterized as lone-pair materials [14]. These materials are generally two-fold coordinated and I, therefore, utilized polymeric concepts in their design and, particularly, utilized the lone pairs to devise new electronic mechanisms for switching and memory. I viewed these materials as being visco-elastic polymers whose stiffness and resistance to structural transformation depend upon the concept of connectiveness, C, which describes the average covalent coordination of the material. It increases from $C = 2$ to $C = 4$ as one moves from the Group VI elements through increasingly stronger, crosslinked alloy structures to the tetrahedral materials of Group IV. A high C results in a rigid matrix which is unlikely to deform or change its bonding appreciably [15]. Chalcogenide glasses having a low C offer structural flexibility. The small connectiveness, that is the undercoordination, provides the flexibility and mobility for bond switching and rearrangement. Such switching is connected with the excitation of the lone pairs [7, 16–19].

The uniquely fast switching effect in Ovonic chalcogenide threshold switching devices is used in two ways. The Ovonic Threshold Switch (OTS) is designed to be structurally stable to electronic excitation because of its many and strongly-bonded crosslinks, and the switching effect results in the rapid formation of an electrically-conducting channel which only exists during excitation. The Ovonic Memory Switch (OMS) responds to the threshold switching excitation by undergoing a reversible structural/phase change. Once switched, it remains in either the conducting or nonconducting state until reset. It is completely non-volatile. The OMS accomplishes this because it has fewer and weaker crosslinks designed into the chalcogenide alloy. It can be activated by non-electrical means, e.g. optically, which is the basis of the Ovonic optical phase change memory. Both the OTS and the OMS are radiation hard [20]. Both depend on coupling the electric field to the large number of lone-pair orbitals [21].

2. Information Technology

2.1. *History*

My work in amorphous materials began in 1955 as a result of my research activities in neurophysiology [22]. There was little understanding then of the physical basis of nerve cell and synaptic activity. I felt that it was important to understand the role of disorder and local order in surface activity of the nerve cells and built nerve cell and memory models based upon amorphous thin-film transition metal oxides. I developed reversible two- and three-terminal electronic switching devices [23]. These devices exhibited both volatile and non-volatile switching phenomena upon application of an electric field greater than a particular threshold value. The devices showed exceedingly high resistance in the Off state and high electrical conductivity in the On state. They also showed analog action.

The purpose of this activity was to develop computers that could learn. I had developed mechanisms for such a computer which would resemble neuronal and synaptic activity of the brain [22, 24–26]. I was particularly interested in the deep relationship between energy and information.

In 1960, Iris and I founded Energy Conversion Laboratories [later Energy Conversion Devices (ECD)] and, working with amorphous chalcogenide alloys, demonstrated the first threshold and memory switching devices. This 1960 work was reported in Physical Review Letters in a much-cited paper in 1968 [26, 27]. In July 1969, at a Gordon Conference, I described a laser-controlled optical memory that utilized the same structural phase that occurs in the electrical memory. This work was also reported in September 1969 [28]. I had previously shown that these materials also responded to energy pulses applied by electron beam, electronic strobe flash, etc. and that one could, in fact, produce a gray-scale response in these materials in which the structural phase change was proportional to the magnitude of the applied energy pulse. I demonstrated these phenomena using both electrical and optical excitations, and I showed that spot sizes less than 100 Å in size could be reversibly written and erased using an electron beam [21]. Optical information storage systems were also developed that used photonucleation and thermal crystallization not only in our inorganic materials, but also in organochalcogens which provide high sensitivity, non-reversible data storage [29]. These dry, non-silver, high resolution films exhibited photodiffusion phenomena and were capable of producing a recorded optical density of 1.0 with an exposure of 10^3 erg/cm^2. At this sensitivity, the amplification factor of the system was about 1000.

My collaborators and I published a large number of scientific and technical papers and gave many talks at international meetings describing the mechanism and operation of these optical memory devices, and the optical memory received some coverage nationally [30, 31].

I was issued a basic patent [32], which was filed in 1968 for this work. Various companies were licensed in this area, the first being IBM and later Matsushita. (I had introduced chalcogenide optical memory switching in Japan in the late 1960's starting with talks I gave at the Electrotechnical Laboratory, headed then by Makoto Kikuchi, who later became Director of Research and Member of the Board at Sony, with an attentive young scientist present, Kazunobu Tanaka, who then became quite active in the field.)

Since those early days, the Ovonic phase change optical memory has developed into a successful commercial data storage technology, now commonly referred to as the Phase Change Erasable (PCE) optical memory. ECD has generated many new patents in this area and is continuing to do so. Present licensees of this technology are Matsushita, Sony, Toray, TDK, Teijin, Asahi Chemical, Toshiba, Polaroid and Plasmon Data Systems who, among other manufacturers, are producing rewritable PD, CD–RW, and DVD optical data storage media.

Commercial optical data storage products using phase change materials have been on the market since 1983. The first product was a write-once video frame recorder introduced by Matsushita. The first computer peripheral drive, also using write-once media, was introduced in 1987 by IBM. The first drive that used phase change erasable media was introduced by Matsushita in 1991. The 650 Mbyte PD removable optical memory disk, which enjoyed substantial commercial success, was introduced at the end of 1995. CD–RW disks with the same 650 Mbyte capacity as the PD system were introduced in the beginning of 1997, and the next product, a very high capacity $\sim$ 2.6 Gbyte DVD rewritable disk, entered the market in the summer of 1998.

Products manufactured by the consumer electronics, communications and computer industries are beginning to converge into a single multiple-purpose digital device with entertainment, communications and computing functions. Rewritable DVD disks, with their very high data storage capacity, are the ideal removable storage media for these new generations of products.

2.2. *Phase change erasable optical memory*

2.2.1. *Basic physical process*

Phase change memories are based on thin-film alloys typically incorporating one or more elements from Column VI of the Periodic Table. These chalcogenide materials can exist in two or more distinct atomic structural states. An energy barrier, shown in Fig. 1, must be overcome before the structural state can be changed, thereby providing stability in each of the two states. Energy can be supplied to the material in various ways, including exposure to a laser beam or application of a current pulse. Laser exposure is used for recording, erasing, and rewriting in the case of an optical memory. If the laser energy applied exceeds a threshold value, the memory material will be excited to a state of high atomic mobility, in which it becomes possible for the chemical bonding to rapidly rearrange by slight movement of the individual atoms. In lone-pair materials (alloys containing chalcogen atoms such as selenium, tellurium, or sulfur) the Group VI atoms use two of their p-electrons to be divalently bonded, and the atomic rearrangement that

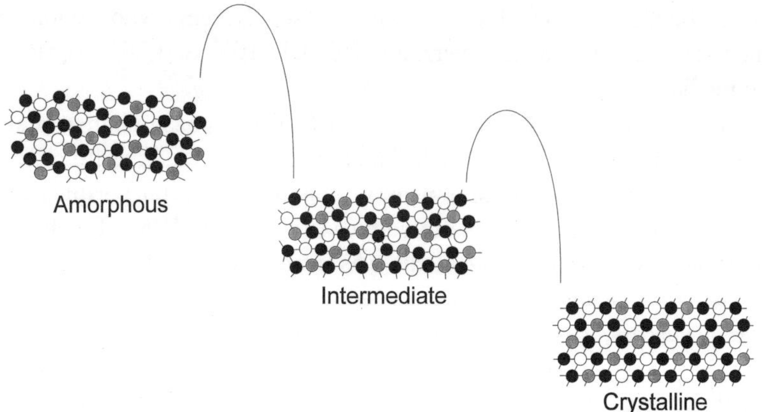

Fig. 1. Schematic representation of transformations between states with crystalline, intermediate and amorphous atomic structures. The energy barriers between these states can be traversed in either direction by the application of appropriate energy pulse profiles.

takes place during the phase change process may occur by simply shifting the other two non-bonding or weakly bonding "lone-pair" p-electrons to make new conformations, connections and configurations.

Selection of an appropriate composition for the memory alloys and the creation of a high mobility state during laser exposure are the underlying principles in direct-overwrite phase change erasable optical recording media [33, 34].

2.2.2. *Materials*

In order for phase change optical disks to be used in a high-speed optical data storage system, the optical memory material must be able to be transformed between the amorphous and crystalline structural states using laser pulses of extremely short duration. At the same time, both structures must be stable at room temperature.

Optical memory media using stoichiometric compounds show the necessary crystallization speeds, but they often exhibit a dramatic loss of crystallization speed when the composition deviates even slightly from stoichiometry [35]. Adding Sb to the GeTe binary system transforms it into a nonstoichiometric material and greatly increases the compositional range over which rapid crystallization can be achieved [34]. A ternary phase diagram of the Ge–Sb–Te system is shown in Fig. 2.

In the Ge–Sb–Te ternary materials system, compositions can be selected in which minute changes in bonding position of the atoms can cause profound changes in the physical properties of the material, including its optical absorptivity and reflectivity. The index of refraction changes by almost two units and the absorption index increases by one unit. These properties can be utilized in a multiple-layer optical stack to give good optical coupling of the incident energy into the phase change materials, and also good reflectivity contrast between the written and erased regions.

In the optical memory, use of materials which have the same composition in the amorphous and crystalline phases is also essential for long cycle life. Since no diffusion is involved in the structural phase change process, no phase segregation occurs and rewrite life is limited only by the integrity of the substrate. A polycarbonate substrate will begin to show degradation in its surface smoothness after about 100,000 re-writes, and this will contribute to a background noise level which will limit cycle life to about one million cycles. Optical disk substrates made with advanced plastics or glass, or those which use

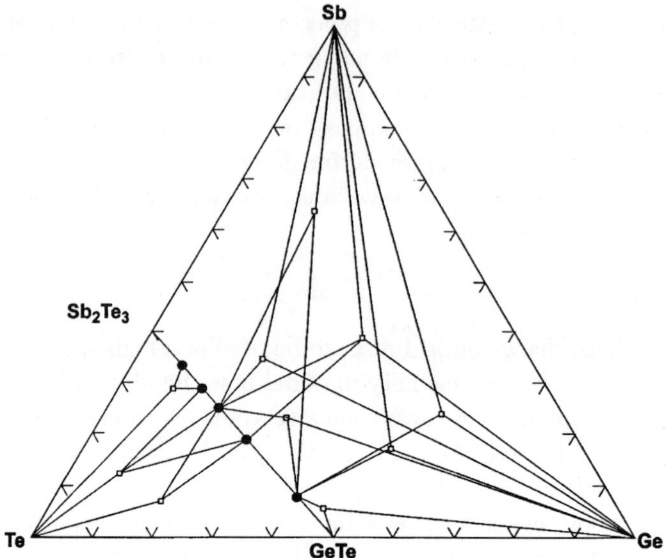

Fig. 2. Ge–Sb–Te ternary phase diagram. The open points represent compositions that have been prepared, and the lines show the compositions of the phases that were formed.

dielectric layers more effective in stabilizing the plastic surface, will have a much longer cycle life. The cycle life for the Ovonic electronic memory that uses similar chalcogenide alloys has been demonstrated to exceed 10^{13} cycles.

2.2.3. *Direct-overwrite recording*

Phase change disks have become the rewritable optical recording media of choice for cost reasons, and also because they are direct-overwrite media. This means that information can be written in a previously used location without need for an initial erase step. Magneto-optical (MO) media require an extra erase step. Although this requirement may be tolerable when writing small files to MO media, this extra step requires too much time for today's larger files. When recording the very large files needed for digital video, the need for a direct-overwrite process is absolute. This is one of the reasons that our phase change memory has replaced magneto-optics as the preferred technology.

Two material properties are required to provide direct-overwrite capability. First, the speed of the transition must be very fast. The structure of the current phase change erasable materials can easily be transformed in either direction by pulses of 50-nanosecond duration. Second, the energy delivered by the laser beam at both the amorphizing and crystallizing power levels must be equally absorbed by the phase change material when it is in either structural state. The refractive indices and the absorption coefficients of the phase change material in its two structural states inherently provide this capability, and appropriate design of the optical stack used to form the device provides the final tuning.

2.2.4. *Device configurations*

The device configurations used in products manufactured by our licensees are sophisticated, multi-layer optical designs. They incorporate principles we established for the protection of the phase change alloy from atmospheric contamination and chemical interaction with the protective layer itself. They also provide optimized optical coupling. The multi-layer, thin-film structure that is used in the PD media produced by Matsushita is shown in Fig. 3. High-yield, consistent manufacturing is, of course, a major consideration in the production of any product, and our licensees have done an outstanding job of developing a well-controlled process with good yield.

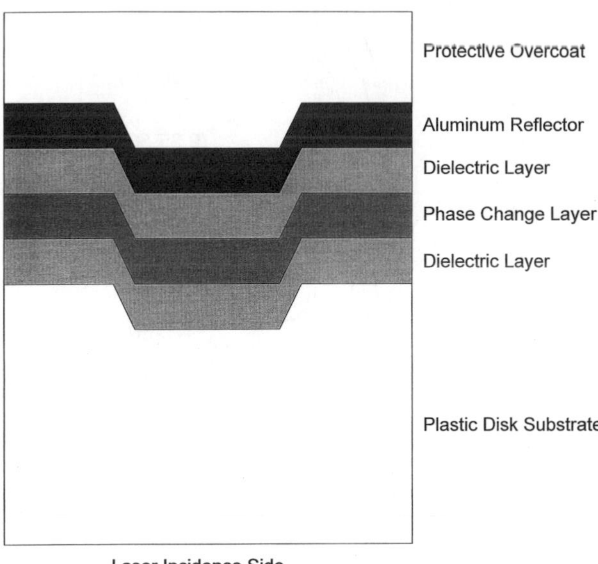

Fig. 3. Multi-layer device structure used in PD rewritable optical memory disks.

2.2.5. *Structural analysis*

Raman spectra obtained from amorphous Ge:Sb:Te films are shown in Fig. 4. These data and Raman studies of amorphous $Ge_x Te_{1-x}$ reported by others all support the random covalent network model for these materials, originally proposed by Betts, Bienenstock and Ovshinsky [36], in which Te and Ge atoms have two- and four-fold coordination, respectively. Increasingly substituting Sb for Ge in amorphous GeTe results in three important changes in the Raman spectra. The initial shift at low Sb concentrations is probably related to changes in the local environmental of Te–Te bonds. The gradual changes in the Raman spectra as the composition is changed from a–GeTe to a–SbTe strongly suggest a network model with a random distribution of homonuclear and

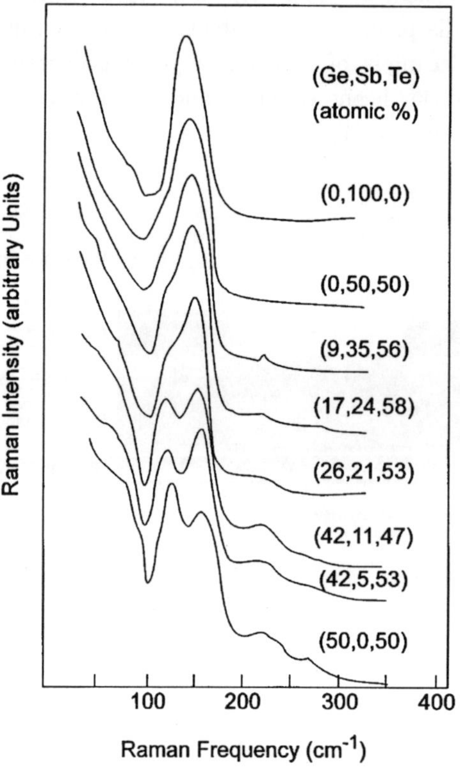

Fig. 4. Raman spectra of various amorphous Ge–Sb–Te alloys with compositions as indicated in the diagram.

heteronuclear covalent bonds. The Raman data show no evidence for sudden changes in chemical ordering or phase separation throughout the compositional range investigated. Sb likely participates in the structure with its preferred three-fold coordination in this network, which reduces the average coordination number of the alloy as the Sb level is increased at the expense of Ge.

2.2.6. *Thermal stability*

Crystallization occurs by nucleation and growth, and the kinetics of crystallization upon annealing of the phase change optical memory alloys should be described by the Johnson–Mehl–Avrami equation. This equation relates the volume fraction transformed into the crystalline state as a function of time, the nucleation rate and/or growth rate, and, in some instances, the growth morphology, and a rate constant which depends on both nucleation and growth rates.

Using the plots in Fig. 5, we estimate that the lifetime (at 50°C) of the amorphous phase is about 2×10^5 years for films with average Ge:Te:Sb

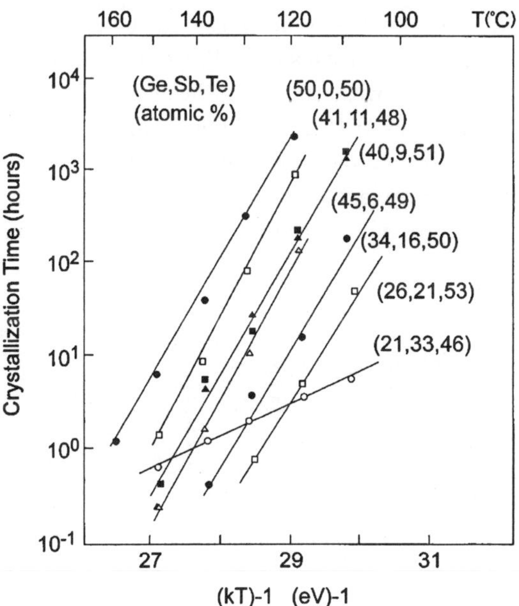

Fig. 5. The activation energy for thermal crystallization of three different alloys in the Ge–Sb–Te ternary materials system is shown in an Arrhenius plot.

composition (40:10:50) and about 200 years for films having a composition of (22:22:56).

2.2.7. *Phase transformation kinetics induced by a laser diode*

The kinetics of transformation between the amorphous and crystalline phases in an actual optical memory disk depend not only on intrinsic factors, such as composition-dependent crystallization time and thermal and optical properties of the media, but also on extrinsic factors such as disk structure and the energy profile applied by the semiconductor laser.

2.2.8. *Media performance*

PD systems are high-performance, optical data storage systems developed by Matsushita (Panasonic), ECD's lead optical memory licensee. The name PD (Phase-Change Dual) refers to the multi-function feature of the drive. Both phase change rewritable and CD-format disks, including CD-ROM, CD-I, Photo-CD, audio CD and others, can be read by the drive. This dual functionality is enabled by the similarity in the readback mechanism of the phase change and the CD media and is implemented in the optical head using a holographic element designed by Matsushita. The optical head is shown in Fig. 6. The media can be recorded with 10 mW of laser power, and a carrier-to-noise ratio of over 50dB is achieved (Fig. 7). The lifetime of the media, shown in Fig. 8, exceeds 105 cycles. As mentioned earlier, this lifetime is limited by the plastic substrate, not the phase change materials. The disk rotation rate in the PD drive is 2026 rpm, which results in linear track velocities of from 7 to

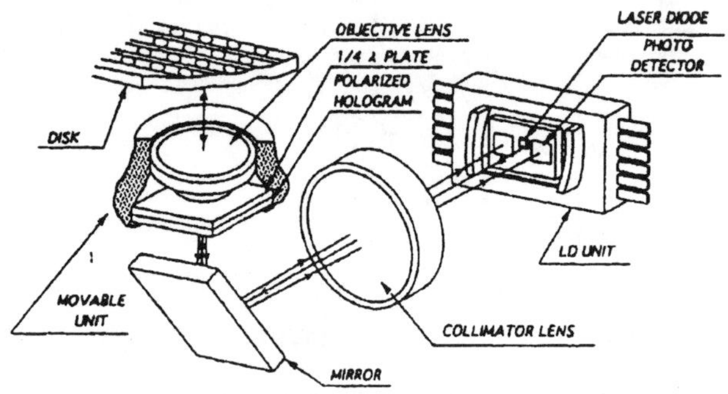

Fig. 6. Block diagram of the optical head for a PD rewritable optical memory drive.

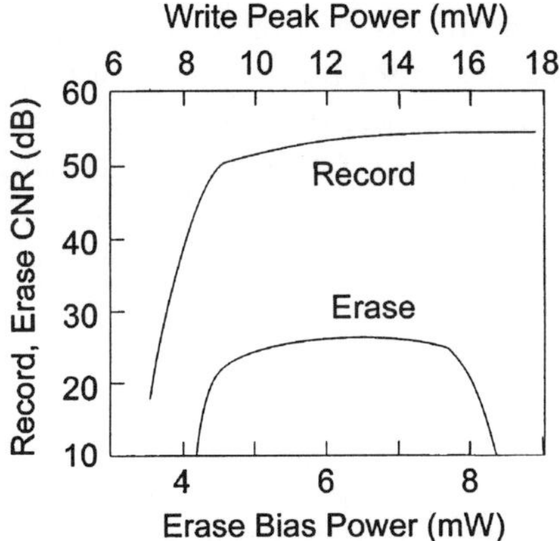

Fig. 7. The erase carrier-to-noise ratio as a function of erase bias power and write peak power for a PD optical memory disk.

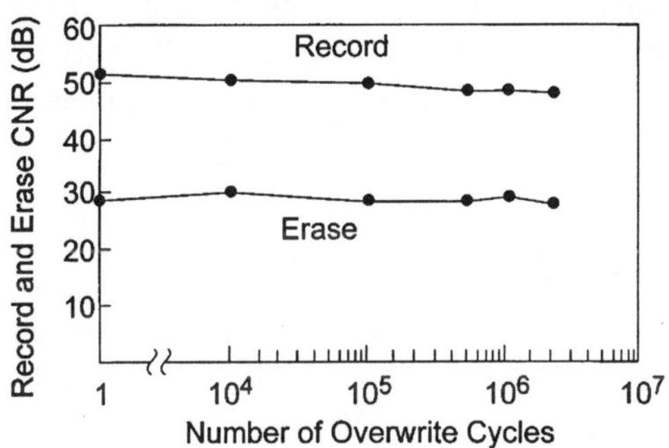

Fig. 8. Error bit count, erase C/N ratio, and reflectivity of PD media as a function of overwrite cycle number. These data show a direct-overwrite lifetime of phase change media in excess of one million cycles.

12 meters per second. The sustained data transfer rate at the optical head is 0.5 to 1.1 megabytes per second, depending on the disk diameter. The PD system uses a zoned CAV scheme to achieve high capacity of 650 megabytes on a 120 mm single-sided disk.

2.2.9. *Manufacturing*

The large differences in optical constants between the two structural states of the optical memory material lead to a major advantage of phase change optical disks in that the read contrast is very high. This high contrast can be used, in part, to relax manufacturing tolerances and lower product cost through improved production yields. Further, the read signal from phase change media is the same as for read-only CD, CD-ROM and DVD disks. This provides for inherent compatibility among these types of media. Further, phase change media will be able to take full advantage of advanced technologies like short-wavelength lasers, near-field optics and multi-level, gray-scale encoding to provide future increases in storage capacity.

ECD is now involved in a program to commercialize read only and phase change optical media in both WORM and rewritable configurations manufactured by a completely different process. This new, low-cost manufacturing method involves a continuous roll substrate which can be embossed with a disk format in a roll-to-roll process. The substrate material is placed in a vacuum chamber, formatted, and then layers of the dielectric, metallic and phase change materials are deposited as required — all in a continuous process. This general approach is intended to be used for production of pre-recorded DVD-ROM and rewritable DVD-RAM disks as well as other new disk formats. ECD is also involved in developing rewritable optical memory media with significantly increased storage capacity through the use of multi-level recording and near-field optics.

2.2.10. *Commercial applications*

Nearly every new computer sold today includes an optical memory disk drive. Most of these are CD (Compact Disk) readers, but CD-R (Recordable) and rewritable phase change drives like PD (Phase-Change Dual), shown in Fig. 9, and CD–RW (rewritable), shown in Fig. 10, are also important products on the market. Among all of the available technologies, phase change rewritable products are poised for the largest growth.

The optical disk market had been slow to become established, except for audio CD, principally because of the existence of several incompatible product formats. The market growth of audio CD, in contrast, has been extraordinary. Once this format was standardized by the industry leaders, the market grew with astonishing speed. CDs overtook audio LPs by first providing a quality advantage, and then later, as production processes matured and volumes expanded, by offering lower cost. The 650 megabyte digital storage capacity of these disks and the low-cost structure achieved by the success of audio CDs provided the opportunity for CD-ROM drives to enter and then become a dominant component in the PC industry.

Fig. 9. Matsushita PD optical memory drive and media.

Fig. 10. CD–RW media manufactures by Ricoh also use the phase change erasable optical memory technology.

DVD is building on the lead established by CD. The manufacturing costs attained in the CD industry are the starting point for DVD. The major difference between CD and DVD is that these new disks have a large-enough capacity, 4.7 gigabytes, to enable expansion to a third market: video. Matsushita introduced the first DVD-RAM products in mid 1998, and they project sales of over 10 million DVD-RAM phase change drives in the year 2000. If each drive is accompanied by the purchase of only five disks, the market would be 50 million disks, plus probably at least half that much more to service drives sold in earlier years. Phase change memory acceptance is developing rapidly.

2.3. *The Ovonic threshold switch*

Chalcogenide materials are distinguished from other alloys by their electronic structure wherein the structural integrity of the material is provided by the strong p-electron bonding orbitals, but the remaining lone pair p-orbitals are either free spin-up and spin-down configurations or have some weakly-bonded configurations based upon p-orbital interactions. This can result in a multi-elemental material with a large density of states in the mobility gap [37]. The overall bonding strength which controls the structural integrity of the material gives rise to the semiconductor bandgap which is sufficiently large as to make the material non-conducting. However, it is the density of states within the gap that controls the electronic properties of the material. Most chalcogenide alloys behave as intrinsic semiconductors with the Fermi level pinned at midgap. These materials can sustain a high electric field, exciting the lone pairs which then, because the material exhibits negative differential conductivity, permit a conducting plasma filament of electrons and holes to be formed. This filament expands and contracts in diameter in order to maintain a constant current density. The contacts to these chalcogenide devices are ohmic and the presence of a high electric field initiates current injection from the contacts [38–43]. The electron-hole "plasma" which results from current injection in the two-terminal device literally electrically short circuits its contacts, resulting in very little voltage drop across them. The switching of the Ovonic threshold switch (OTS) from a highly resistive state to a metallic-like conducting state at room temperature is so fast that it has only been equaled by the later-developed Josephson junction device which operates at liquid helium temperature. The operation of the OTS device is described below.

Figure 11 schematically shows the current-voltage characteristic of an OTS. OTS devices are fabricated from the material compositions shown in Fig. 12.

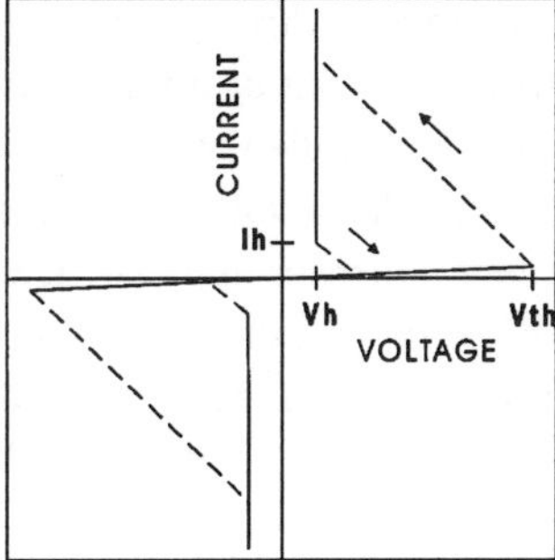

Fig. 11. Schematic current-voltage characteristic curve of the OTS. Vth is the threshold voltage, Ih is the minimum holding current, and Vh is the holding voltage.

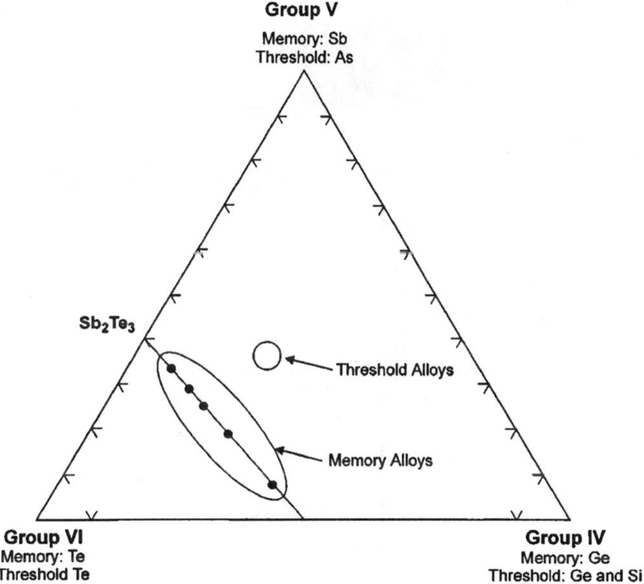

Fig. 12. Phase diagram of GeSiAsSTe alloys for OTS applications. Compositions for the threshold switch are shown in relation to alloy compositions used for memory device applications.

Comparing these compositions to the phase change optical memory ternary system shown in Fig. 2, one sees that the OTS composition uses additional elements that can cross-link and stabilize the structure against crystallization.

Figure 13 shows how the current through the OTS device grows while the voltage across the device stays constant at a fairly low value. For low currents, below the threshold voltage, the applied voltage across the device rises with increasing current until the threshold voltage is reached (condition A) which establishes the plasma filament. Further increase in current results in an actual decrease in voltage to the holding voltage (condition B). Still further increase in current results in essentially no increase in voltage (condition C) because of a rearrangement in the current distribution in the device (at the holding voltage). The current through a part of the device grows (until a saturated filament current density is reached) at the expense of current flowing in another part of the pore of the device. When the entire device cross-section is filled

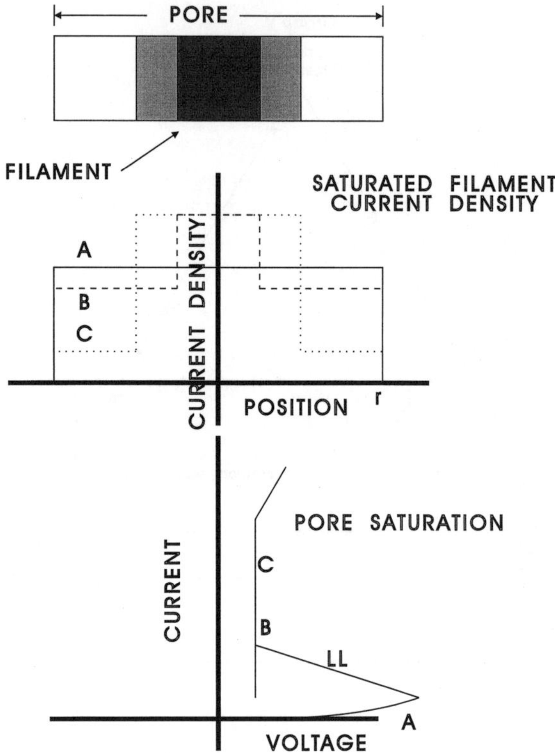

Fig. 13. Current density profile as a function of radial distance within the active pore of the OTS as a function of applied voltage. The different current profiles shown correspond to different points on the device current-voltage diagram shown.

with current at the saturated current density, the device begins to exhibit an ohmic current-voltage characteristic.

The OTS acts as a field-operated, exceedingly fast switch and it has many uses ranging from cross-point switching and logic to circuit protection. In operation, when the OTS threshold voltage is exceeded, the device switches into a highly conducting "on" state and remains there as long as the current flow through the switch is greater than a specific "holding" current [26, 44–46]. When the current through the switch drops below the holding current value, the device reverts back to its initial high impedance "off" state. Its operation is, therefore, comparable to a triac in the first and third quadrants of its I–V plot. The OTS differs from a triac, however, in its very high switching speed, low capacitance, and inherent simplicity [47]. These factors enable the OTS to be considered not only for computer circuitry, but for many different other applications as well, including their use as addressing elements in flat-panel displays. Interesting new phenomena are exhibited in the switch as one reduces the device cross section and not only two-terminal but three-terminal devices are made possible. The OTS has an indefinitely long lifetime — the same as a conventional crystalline silicon transistor.

Exceptionally high-current thin-film OTS devices have been used to demonstrate the operation of a typical EMP suppression circuit. In this demonstration, a fast, high amplitude 500 V transient pulse was superimposed onto a transmission line carrying a low-voltage simulated data signal. The OTS responded quickly by shorting the input circuit during the transient and, after the transient pulse terminated, the circuit resumed normal operation.

2.4. *The Ovonic electronic phase change memory*

The Ovonic electronic phase change memory is a unified semiconductor memory device that is a non-volatile replacement for DRAM, SRAM and Flash memory. It is both the newest type of non-volatile reprogrammable semiconductor memory, and a direct descendent of the oldest. This two-terminal, thin-film device was first developed in 1960–1961. In 1968, ECD published a description of what was called a read mostly memory (RMM) which employed proprietary amorphous semiconductor thin-film memory cells to create a 32-byte electrically erasable and programmable non-volatile solid state memory array. At that time, commercial semiconductor memory devices were either reprogrammable and volatile (RAM) or programmable once and non-volatile (ROM). The ECD RMM device, shown in Fig. 14, was, in effect, the first

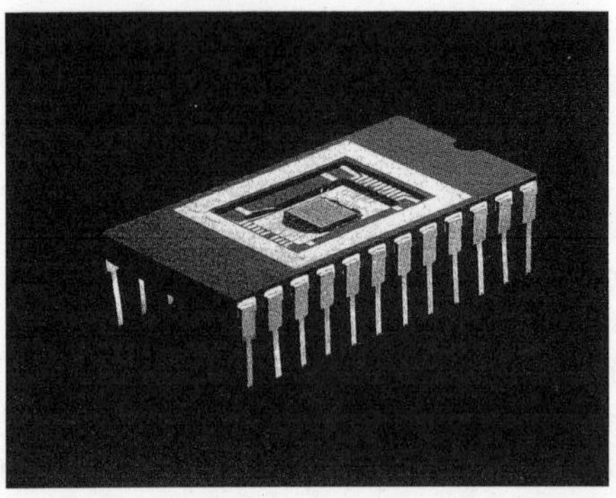

Fig. 14. Original ECD Read Mostly Memory (RMM) thin-film, two-terminal non-volatile semiconductor memory device.

EAROM (electrically alterable read only memory) or EEPROM (electrically erasable programmable read only memory) and the first Flash memory.

The RMM was fabricated as a thin-film amorphous semiconductor structure deposited on a crystalline silicon wafer which had been previously processed to provide an array of isolation diodes and interconnect lines. A later version of this pioneering technology, a 1024-bit ECD EEPROM [48], successfully opened the field for semiconducting memory and was well received. Gordon Moore of newly formed Intel Corporation, who was a co-author on one of our papers [49], then entered the memory field utilizing a conventional floating gate silicon approach. The ECD RMM was the only completely non-volatile, radiation hard device semiconductor memory.

Further development of the ECD amorphous semiconductor memory materials technology throughout the 1970's and 1980's was primarily directed toward their application to laser addressed optical memory disks, as described earlier. ECD's work has also provided a more profound understanding of the solid state physics and chemistry of amorphous semiconductor memory materials and this has led to significant progress in the development of new thin-film alloys and device structures which allow rapid, low power electronic switching of memory materials between the crystalline and amorphous structural state.

Recently developed, Ovonic Unified Memory[a] (OUM) devices utilize new proprietary materials and device design technology to bring the proven performance characteristics of our fast optical memory technology into the arena of

---

[a]Formerly referred to as "Ovonic Universal Memory".

non-volatile semiconductor electrical memories. In this way, the OUM achieves its remarkable characteristics of speed, reprogrammability and non-volatility. Compared to the original ECD RMM/EEPROM, the new OUM device offers six orders of magnitude faster programming time (less than 10 nanoseconds). OUM devices also have demonstrated stable, long-term data storage and an endurance in excess of 10^{13} programming operations with highly desirable low-voltage programming capability.

2.4.1. *Data storage mechanism*

The amorphous semiconductor memory materials which form the basis of Ovonic memory devices store information through changes in their atomic structures caused by the application of an energy pulse. These materials, which are multi-element chalcogenide alloys, can exist in a stable fashion in amorphous and crystalline structures, and also in a range of "intermediate" structural states as shown in Fig. 1. These different atomic structures have different characteristic physical properties, including different values of electrical conductivity. The structural state of the material in a memory cell can, therefore, be determined in a practical semiconductor memory array by a simple, non-destructive measurement of its two-terminal electrical resistance. The ability to be programmed to stable intermediate structures allows for storage of multiple bits of information in each memory cell location, to which we refer as the multi-state programming mode.

The memory cell in the OUM is programmed by application of an electrical pulse. The most important programming parameter is the current amplitude, although other characteristics, including pulse width and rise and fall time, describe an overall "pulse profile," which affects the programming characteristics in a secondary fashion. We use a load resistor in series with the memory cell element during the programming step and, for an appropriate choice of load resistance, we can program the OUM memory cell with less than two volts.

The intermediate structural states that can be established in the "active pore" of thin-film semiconductor material in the Ovonic memory cell give rise to a continuum of values of electrical conductivity between those characteristic of the completely amorphous structure and the completely crystalline structure. The choice of a practical number of usable intermediate states which can be reliably stored in a single memory cell depends on the precision and stability of the resistance sensing circuitry and the ability of the programming circuitry to set the resistance of the memory cell to the precisely desired value.

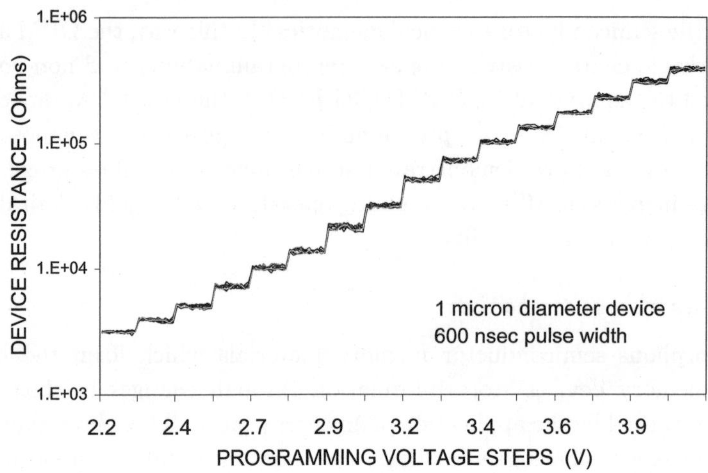

Fig. 15. Programmed resistance in an Ovonic unified memory cell as a function of programming voltage in 0.1 V steps.

Data in Fig. 15 show the programmed resistance of a memory cell as a function of the number of alternating erase/program events applied to the device. The different programmed-state resistance values plotted resulted from the application of programming pulses of discrete current amplitudes.

We have constructed thin-film, non-volatile memory storage cells which have been evaluated for both conventional binary and multi-level data storage. We use these test structures to investigate the performance of the new memory materials we have developed, and we have obtained the following memory cell performance characteristics.

The programming time for OUM memory cells is quite short. In fact, we have successfully programmed devices over the full range of achievable values of resistance using nanosecond pulses. Typically, we use pulse widths of about 40 to 200 nanoseconds for programming. Devices are more readily programmed to the low resistance state using 200 nanosecond pulses, and are programmed to the high resistance state using 40 nanosecond pulses.

Because it uses a simple resistance measurement, the read time of the OUM memory cell is determined in a practical device by some combination of RC time constant and the speed of the ancillary addressing and sensing circuitry. The inherent capacitance of our thin-film memory cell is quite low; therefore, the major source of capacitance is the metal conductor lines.

Table 1. Ovonic unified memory cell data.

	Binary Mode Devices	Multi-State Devices
Programming Endurance	10^{13} cycles without failure tests discontinued	10^8 cycles without failure tests discontinued
Data Retention	10 years at 100°C	Under Test
Biased Data Stability	Not affected by bias voltage below device switching threshold	Not affected by bias voltage below device switching threshold
Programming Voltages	Less than 2 V	Less than 5 V
Programming Currents	0.2 mA	1 mA to 4 mA
Programming Times	1 ns to 400 ns	1 ns to 400 ns
Low Resistance State	5 kΩ to 20 kΩ	1 kΩ to 20 kΩ
High Resistance State	20 kΩ to 1 MΩ	20 kΩ to 100 kΩ

Our present memory cells can be programmed with current pulses of about 0.1 mA to establish the lowest resistance state and about 0.2 mA to program the highest resistance state as shown in Table 1. Intermediate values of resistance are programmed at intermediate current levels. These current levels are used in programming pulses with widths of about 40 to 200 nanoseconds. Over this pulse width range, there is weak dependence of the required programming current on pulse width.

2.4.2. *Basic device operation*

OUM devices can be programmed to exhibit electrical resistance values that are directly related to the magnitude of programming current passed through the device. The electrical characteristics of the device during the programming pulse are essentially the same as those of a related thin-film chalcogenide alloy device, the Ovonic threshold switch. Both the threshold switch and the memory device show a non-ohmic, super linear, low-field resistance, although a linear approximation is good at electric fields below 10^4 V/cm. A memory device that has been programmed to a high resistance level will display a negative resistance switching phenomenon when a threshold voltage in the range of 0.5 to 1.5 volts is exceeded, and will enter a highly conducting "dynamic" state. A previously programmed low resistance does not exhibit a switchback when it enters the dynamic state. The voltage drop across the device in the dynamic state is referred to as the holding voltage and is relatively independent of the current conducted through the device. The holding voltage can be adjusted by changes in device thickness and alloy composition.

It is in the dynamic state that structural changes in the Ovonic memory material are initiated and, when the current pulse is removed, the final atomic structure is established. The conductivity of the thin-film chalcogenide alloy material depends strongly on its atomic structure and device resistances of about 1 KΩ for the low resistance state and about 100 KΩ for the high resistance state are routinely achieved. Higher off-state resistance can be achieved through materials design. It is also possible to achieve intermediate structural states and, hence, also possible to program a device to intermediate resistance levels. This can provide the basis for future increases in storage density via storage of multiple bits per memory cell and it can allow the use of OUM devices not only as electrically reconfigurable electrical interconnects in neural/synaptic networks, but also in other "learning" and encryption devices [50].

The programmed structural states are temporally stable and, therefore, information programmed into the memory is stored in a non-volatile fashion. The voltage required for switching and programming is low, the programming life is long, and the programming times can be quite short — on the order of nanoseconds. These characteristics, along with others such as the inherent radiation hardness of the device and the potential for small cell sizes and simplified processing, make the OUM technology very attractive for the development of a range of new non-volatile memory devices, including the Ovonic adaptive memory with its learning capability.

A schematic I–V curve of a memory device is shown in Fig. 16. The resistance of the device is determined by applying a low voltage and measuring the resultant current. Application of a higher voltage causes the device to enter a dynamic state where it can be programmed. The dynamic states of the initially-set and initially-reset devices are identical, which implies that the previous resistance state is completely erased. The final resistance state of the device, after the programming pulse is terminated, depends on the current level that passed through the device during the programming pulse.

2.4.3. *Commercial applications of Ovonic unified semiconductor memories*

Computers typically utilize a design architecture that incorporates a "memory hierarchy," including several types of memory devices. This is done to take

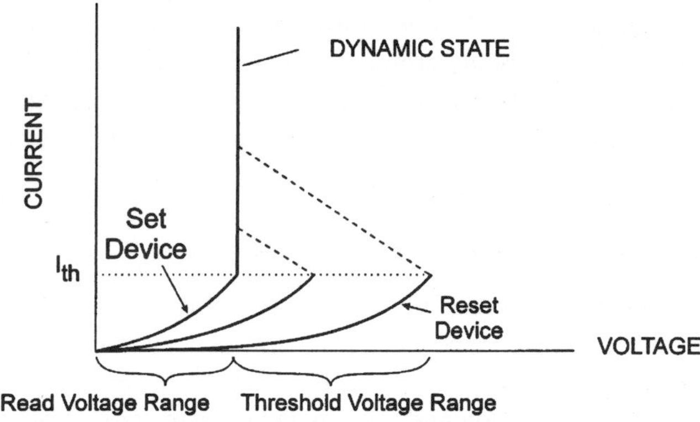

Fig. 16. Schematic current-voltage Characteristics of Ovonic unified memory.

advantage of the individual strengths of each of the types of memory devices and the particular combination of memory technology is chosen by computer manufacturers to provide the best price-to-performance ratio for their customers. SRAM, the fastest type of semiconductor memory, is used very sparingly, but provides a performance boost when used as a memory cache to transfer small blocks of information into and out of the microprocessor. DRAM, which is not quite as fast as SRAM, but significantly less expensive, is used, in turn, to hold larger amounts of information ready to be accessed by the SRAM cache. The information stored in DRAM comes not only from SRAM transfers, but also from rotating mass data storage devices — the fixed and floppy disk drives. Of all these types of memory, only the disk drives provide non-volatile storage; SRAM and DRAM loose all of their information when power is removed.

Ovonic memory technology has some important attributes which allow it to be used to replace all of these types of memory in a single device. The most important of these are speed comparable to SRAM, cost and endurance comparable to DRAM, and the non-volatility of magnetic disk media. This combination can allow computers to be built with a single plane of memory. This "collapse of the memory hierarchy" provides two important benefits. First, computer performance is increased because data transfer bottlenecks caused by moving information among levels of the hierarchy can be eliminated and processor time dedicated to managing data in the hierarchy is reduced or eliminated. Also, the stored code is executable in place. When the computer is first powered up, or when the operator desires to switch to another software application, the code is already stored in a medium that makes it completely available for use by the processor and no lengthy transfers from the disk drives will be necessary. Second, the architecture of the computer is simpler, which saves overall cost, size, and power requirements.

Ultra High Data Density Three-Dimensional Memories

In the mid 1980's, ECD successfully developed technology which allowed construction of the first three-dimensional all-thin-film memory devices. To do this, we developed a vapor deposited, thin-film diode device technology for memory cell electrical isolation and leveling layer technology. The required low temperature processes allow continued fabrication of additional layers without degradation of previously deposited, underlying layers. We demonstrated the feasibility of our approach by constructing a prototype 16 kbit, two-layer memory array, each layer of which consisted of a complete memory array, including lower metal contacts, semiconductor memory and diode isolation elements, insulator, and upper metal contacts.

There will be a need for high-density, low-cost memory as long as there are computers. The ability of a unified memory cell to be made smaller than 100 Å in diameter and also to be fabricated in three-dimensional arrays opens up a new vista for computer architecture.

Neural Network Computers

Continuing the journey since 1955, we are now in a position to have devices and circuitry to accomplish the goal of having neural networks with real learning ability. Neural networks function by storing data in a large number of points, or nodes, in an array. The nodes are each electrically connected to many other nodes, and each node has a particular "connectivity weight" assigned to it. It is the process of allowing the network to assign these weights and interconnections in response to outside stimuli that allows a neural network, in effect, to "learn." Once the network has learned its function, it can perform it with remarkable speed and tolerance for irregularities of the incoming data. The ability to interconnect the nodes, as provided by three-dimensional circuitry, and the ability of the nodes to be programmed to a continuum of intermediate connectivity states, as provided by the structural changes of Ovonic memories, make Ovonic memory technology ideally and naturally suited for truly synaptic neural network applications.

*Smart Cards and Other Tamper-Proof Secure Information
Storage Applications*

Ovonic memory cells have a unique capability to be programmed and interrogated in a novel, proprietary way that can allow them to be used in a range of secure memory storage applications. Unlike the conventional direct overwrite programming mode described previously, this "accumulation" mode uses the low current portion of the cell programming characteristic exclusively. In this regime, the programmed resistance of the memory cell drops discontinuously from the high resistance state to the low resistance state once the time integral of applied programming current exceeds a particular value. This behavior is a consequence of the nucleation and growth of microcrystalline regions within the volume of the memory cell pore. Once the microcrystalline regions form a percolation path that electrically connects the upper and lower contacts, the overall device resistance changes abruptly. The total integrated programming current required to form the percolation path is highly reproducible and the device can be set to the initial high resistance state in the same manner used for the normal programming mode.

Tamper-proof information storage is accomplished by subdividing the critical total integrated current into a series of current pulses. Information is recorded by applying the desired number of sub-critical pulses and the data is later read out by applying further sub-critical pulses until the conductivity transition is observed. Since the stored data is physically represented by an extremely small volume of crystalline material within the memory cell pore, forensic methods such as examination by electron microscope or X-ray diffraction will not be successful at determining the stored information, and electrical interrogation of the cell is destructive. Tamper-proof information storage that is invulnerable to forensic analysis is essential to a number of very important emerging data storage applications such as electronic cash.

2.5. *Amorphous silicon alloy diode and TFT AMLCD display addressing*

In the 1960's, we pioneered amorphous thin-film active matrix switching, first in electroluminescent displays using OTS pixel addressing switches [51] and then, as liquid crystals became more stable and reliable, in liquid crystal displays, using a–Si alloy pixel addressing switches. The company originally set up by ECD, Ovonic Imaging Systems (OIS), was at one time the largest U.S. company in liquid crystal displays. It later became Optical Imaging

Systems (OIS), no longer associated with ECD. When still a subsidiary of ECD, we licensed Samsung and others in the technology. Prior to that, we had introduced Sharp in Japan to amorphous silicon alloy materials while working together in photovoltaics. Sharp has subsequently become a very major player in displays.

By far, the largest application of amorphous materials to date from the perspective of market size has been the use of amorphous silicon alloy thin-film devices for active matrix addressing of LCD displays. Within 10 years of the commercial introduction of the first active matrix LCD (AMLCD) displays in 1987, the market for these devices has grown to over $10 billion, with growth in the future expected to continue at over 10% per year.

AMLCD displays employ a microelectronic switching device, typically a thin-film transistor (TFT) or diode, in series with each liquid crystal pixel element in the display to improve pixel addressability. Without the active element present, the optoelectronic properties of the LCD material are not sufficiently non-linear to allow matrix addressing without significant loss of image contrast, especially in large-area and color displays. Amorphous silicon alloy is an excellent materials technology to achieve active matrix addressing because it can be used to make semiconducting devices at each pixel of the display by being deposited directly on large area glass substrates at low temperatures. Like a–Si alloy photovoltaic devices, a–Si alloy TFTs and diodes are fabricated at around 300°C, which is well below the softening temperature of the glass used for the LCD display. Two devices using amorphous silicon alloy have been successfully employed as addressing elements in AMLCD displays: diodes and thin-film transistors.

Amorphous silicon alloy TFTs behave similarly to their crystalline silicon counterparts with one notable exception: the field effect mobility of the semiconductor is around 0.5 cm^2/Vs instead of several hundred cm^2/Vs for crystalline FETs, leading to a corresponding drop in saturation current [52]. For low current applications, such as an active matrix element for an LCD display, however, a high quality a–Si alloy TFT switching device can be produced readily. The resistance of a typical TFT drain-source channel can be modulated between $\sim 10^{14}$ and 10^6 Ω, which is a sufficient range for operating an LCD display. To achieve this range, a high gate voltage is used, typically in excess of 12 V. By using both p and n contacts in an a–Si alloy TFT, the current in the forward-biased direction of the TFT can be increased by a factor of 20, but with a greatly increased off conductivity [53].

Refinements in a–Si alloy TFT manufacture have improved to a point where TFT performance characteristics no longer appear to be the limiting factor in display speed, size or density. Rather, the conductance of the metallization needed to carry the control signals to the transistor now limits display performance [54].

Due to the photoconductivity of amorphous silicon alloys, light shielding of the TFT is essential for operation in a display. This is usually accomplished by metallization on the top and bottom of the transistor channel. Also, just as in photovoltaic applications, Staebler–Wronski degradation of the performance of a–Si alloy TFTs occurs with use. However, this effect is not significant enough to adversely affect displays, which have a charge transfer of less than a coulomb per pixel during the entire lifetime of a display.

Amorphous silicon alloy's photosensitive nature, combined with the addressing capabilities of transistors and diodes, has also led to its use in optical detector arrays with amorphous silicon alloy transistors or diodes being used as charge-transfer devices. Such detector arrays have potential applications in optical, radiation and X-ray imaging.

Chalcogenides may play a role again in switching liquid crystal displays by means of recently designed Ovonic threshold switch circuits.

2.6. *Long-life amorphous silicon alloy copier drums*

The use of amorphous silicon (a–Si) alloys in electrophotographic photoreceptors was first demonstrated in the early 1980's [55, 56]. These devices were proposed as a replacement for photoreceptors based on the earlier chalcogenide amorphous semiconductor alloys, usually involving selenium or a selenium alloy such as a–Se, a–As$_2$Se$_3$, or a–Se$_{90}$Te$_{10}$. The a–Si alloy devices used non-toxic materials and exhibited excellent wear-life. However, because photoreceptors are typically 30 mm to 40 mm thick, deposition of these devices at 1.8 mm per hour using the typical RF excited plasma enhanced chemical vapor deposition process (PECVD) was very slow and, consequently, the a–Si alloy photoreceptors were expensive to manufacture.

In 1985, Hudgens and Johncock at ECD [57] described the use of a new low-pressure microwave PECVD process that would permit the deposition of a–Si alloys with acceptable electrophotographic performance at greatly enhanced rates — up to 36 mm per hour. This process employed 2.45 GHz microwave excitation of feedstock gas at a typical pressure of 1 m Torr. At these low pressures, the tendency to formation of gas phase polymers and powder particles was greatly suppressed even at the high levels of plasma excitation employed to achieve the high deposition rates. In addition, because of the long mean free path for collisions at these low pressures, deposition uniformity was high and the feedstock gas utilization efficiency approached 100%.

Measurements of the low pressure microwave plasmas [58] used to deposit the a–Si alloys indicated that the density of plasma ions was about an order of magnitude higher than seen in typical 13.56 MHz RF PECVD plasmas and the electron temperature was about a factor of five higher for the same applied power. Although the density of neutral free radicals in the low pressure microwave plasma was not measured directly, application of a substrate bias of up to 100 V had no effect on the deposition rate, suggesting that neutral species dominate the deposition process in the microwave excited system in the same way that they dominate a–Si alloy deposition in RF PECVD systems.

The low pressure microwave deposition process is being used at ECD [59, 60] to deposit dielectric thin-film materials at high speed on polymer substrates for use as visually-transparent barrier films in food packaging applications and as a means of depositing optical coatings on polymer sheets for display and solar control applications. Work continues toward applying this very high throughput process to solar cell manufacturing.

3. Energy Generation and Storage

There is an overwhelming societal need for clean energy. The use of fossil fuels has led to serious problems of pollution, global warming, national trade deficits which are connected to inflation and unemployment and, most serious of all, the growing strategic dependence by many nations on imported oil — a situation that has been historically a major cause of war. Furthermore, in various areas of the world, such as Europe, there is a serious problem of unemployment and degradation of the educational base, all of which could be addressed by the building of new industries to make the new products which are necessary to redress the problems caused by reliance on fossil fuels.

Development of clean energy technology entails the creation of a system that addressed both energy generation and storage. This section will describe how amorphous and disordered materials can be used to make photovoltaics (PV), the direct conversion of sunlight into electricity, practical and competitive with electricity generated from coal, gas, oil and, of course, nuclear energy. We will then discuss how hydrogen, reversibly stored in a solid (a hydride), has led to the development of a new rechargeable battery technology that has made electric and hybrid vehicles practical. We will then briefly discuss how metal hydrides can also be used to store hydrogen for fuel cells and, indeed, provide hydrogen as a fuel for internal combustion engines. Ovonic nickel-metal hydride (NiMH) batteries can also be used in combination with photovoltaics in homes and buildings to provide autonomous and distributed sources of electrical power — a field that will grow rapidly in the next century. Hydrogen is certainly the prime mover and the fuel of the future.

3.1. *Amorphous silicon alloy thin-film photovoltaics*

3.1.1. *Introduction*

Although the photovoltaic effect was discovered over 150 years ago [61], the role of photovoltaics as a viable terrestrial energy source has been evolving only over the last several decades. Originally developed as a source of energy in space, solar cells remain the major power source for most satellites, where they provide reliable service over extended periods of time.

Continual technological development over the years has resulted in lower costs for a variety of terrestrial applications where access to conventional electric power has been either impossible or impractical. As costs decreased further, new uses for PV have emerged. Today, PV is used to power a wide variety of applications from solar calculators to entire villages. The photovoltaic products now receiving the most attention are our PV shingles and other roofing products. Satellites for commercial rather than military communication require low cost, low weight, high energy density photovoltaics. Ovonic photovoltaics meeting these requirements are now on the MIR Space Station and are being designed into various commercial satellites.

There has been a steady growth of the PV industry (currently at 25% annually) and the 1998 world market size was 160 MW, approximately $1.5 billion. This growth is expected to continue as remote applications and niche markets expand and PV prices continue to decline.

These present uses, however, represent only a portion of the market potential for PV. Generating and distributing electricity is one of the world's largest businesses, with annual revenues estimated at more than $800 billion, roughly twice the size of the world auto industry. Moreover, energy forecasters are predicting enormous growth in the next few decades. The *new* electrical generating capacity need for U.S. utilities alone is 600 GW during the next ten years.

Enormous opportunities for PV also exist in Latin America, Asia and Africa. The U.S. Agency for International Development (USAID) forecasts that the market for power-generation systems between 1990 and 2010 in the developing world alone will be \$914 billion. Distributed power is the most desirable mode of electricity generation in many of these countries, allowing them to avoid billions of dollars of investment in central station plants and transmission and distribution networks. This makes PV systems very attractive.

Residential, commercial and industrial markets normally served by large-scale utilities are being addressed by our roofing products and by our building of larger machines which, for the first time, will make solar energy cost competitive to fossil fuels.

3.1.2. *Using science and technology to break the cost barrier*

The greatest challenge confronting the PV industry is, therefore, to continue to bring down costs while maintaining performance and reliability. In particular, the large utility markets will open up when PV module manufacturing costs are below \$1.00 per watt. Amorphous silicon (a–Si) alloy technology, coupled with large-scale roll-to-roll device manufacturing, is ideal for reaching this cost goal.

Crystalline silicon photovoltaic technology, originally patented in 1941 [62], has been the workhorse materials technology of the PV industry for many years. Crystalline silicon solar cells are fabricated from single crystal silicon semiconductor wafers. Producing these solar cells requires growing a solid crystal (a costly and energy intensive process), slicing it and then processing the cells into modules. The substantial advances made in the development of this technology over the years has resulted in a product which has proven itself to be reliable in a wide array of applications. However, the costs have reached a plateau, and it is generally conceded that the high material costs and many processing steps needed to manufacture the modules will prevent it from breaking the cost barrier.

The production of PV modules that can be economically viable for widespread application depends on three key factors: low material cost; high energy conversion efficiency with good stability; and a low-cost, high-yield production process. In addition, the modules must be environmentally benign. ECD began developing technology that can address these issues in the late 1970's and built production equipment beginning in 1981. ECD and United Solar

Systems Corp. (United Solar), a joint venture between ECD and Canon established in 1990, have continued commercialization of products and processes that address each of these factors to develop solar cell technology for large-scale, cost-effective production of solar electrical energy.

Low Material Cost

Because of the structural disorder in a–Si alloys, optical absorption can occur via the direct process — without the participation of a lattice phonon to conserve momentum as is required in crystalline silicon. Consequently, a–Si alloys absorb light more efficiently than their indirect bandgap crystalline counterpart; so, the a–Si alloy solar cell thickness can be 50 to 100 times smaller, thereby significantly reducing material cost. Thin-film deposition also requires significantly less energy than is used in the production of silicon crystals. It takes seven years for a crystalline silicon solar cell to produce the amount of energy that went into its production. By contrast, a–Si alloy solar cells have a 1.5 year energy payback. This reduced energy budget lowers costs further.

High Energy Conversion Efficiency with Good Stability

One of the most important measures of solar cell performance, and a major factor in its cost effectiveness, is the efficiency with which it converts sunlight to electricity. The higher the efficiency, the lower the cost. Commercial crystalline PV solar cells generally are 10%–12% efficient. A major problem for a–Si alloy technology in its early stages was low conversion efficiency. Another problem was that the conversion efficiency tended to degrade upon prolonged exposure to sunlight due to the Staebler–Wronski Effect [63].

ECD and United Solar addressed both of these problems by developing high quality materials and incorporating them in a multi-junction cell structure [64]. This structure is shown in Fig. 17. The total cell thickness is less than one micrometer. The individual cells are thin to improve stability, and efficient photon absorption is facilitated by using cells with different bandgaps to absorb efficiently the wide range of photon energies in the solar spectrum. Highly conductive doped layers with low optical losses are used as the window layer [65] as well as in the tunnel junctions between the sub-cells. The bandgap is changed by incorporating different amounts of germanium in the intrinsic layers [66]. ECD/United Solar holds the efficiency records for all relevant a–Si alloy solar cell structures (Table 2). In order to improve efficiency, new materials were developed and new cell structures were designed [67]. Progress in

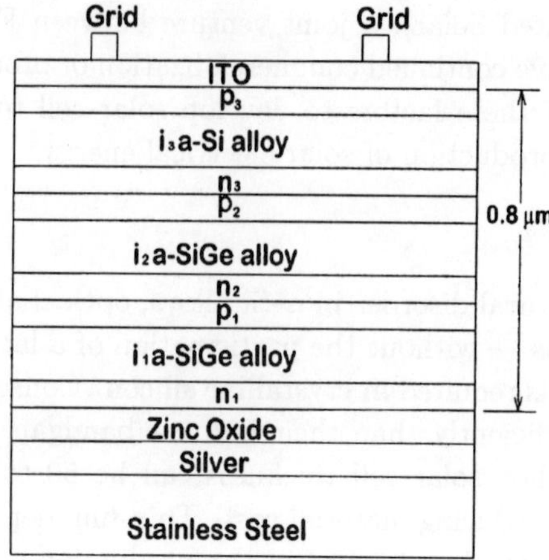

Fig. 17. Schematic diagram of ECD's multi-junction, three-cell, spectrum splitting thin-film *a*–Si alloy solar cell.

Table 2. Highest stable (> 1000 hours) cell efficiencies (Area ~ 0.25 cm^2) achieved by United Solar for different junction configurations.

Device		J_{SC} (mA/cm^2)	V_{OC} (V)	FF	Efficiency (%)
a–Si:H	Initial	14.65	0.992	0.730	10.6
	Stable	14.36	0.965	0.672	9.3
a–Si:H/*a*–Si:H	Initial	7.9	1.89	0.76	11.4
	Stable	7.9	1.83	0.70	10.1
a–Si:H/*a*–SiGe:H	Initial	11.04	1.762	0.738	14.4
	Stable	10.68	1.713	0.676	12.4
a–Si:H/*a*–SiGe:H/	Initial	8.57	2.357	0.723	14.6
a–SiGe:H	Stable	8.27	2.294	0.684	13.0

improving the stable energy conversion efficiency has been rapid and continued work at ECD and United Solar laboratories will lead to further improvement.

Low Cost Production with High Yield

Just as other low-cost, large surface-area products such as newsprint and photographic film are most economically produced in a roll-to-roll production process, ECD pioneered a proprietary continuous solar cell deposition

Fig. 18. ECD/United Solar new roll-to-roll solar cell manufacturing line has a 5 MW per year production capacity. Shown here is the nine-chamber PECVD machine that deposits the amorphous and microcrystalline semiconductors that comprise the multi-junction cell structure.

process. The novel process involves a proprietary gas gate design that effectively prevents dopants from diffusing from chambers that deposit doped semiconductor layers into the deposition chambers for undoped layers. A manufacturing line (Fig. 18) capable of producing the complex triple-cell structure has been designed and built by ECD for United Solar using the proprietary roll-to-roll approach. Products from this manufacturing plant, which has an annual capacity of 5 MW, are now being shipped worldwide for a wide range of applications.

The manufacturing process consists of the following steps. A roll of stainless steel, half-a-mile long, 14 inches wide and 0.005 inches thick, moves in a continuous manner at a speed of about two feet a minute through four separate machines that serve the purpose of (1) washing, (2) depositing the back reflector, (3) depositing a nine-layer stack consisting of a–Si alloy, a–SiGe alloy, and doped layers, and (4) depositing indium tin oxide (ITO) which serves as both a top contact to the cell and an antireflection coating. The coated web is next processed to make a variety of lightweight, flexible and rugged products. The processing steps involve (1) cutting of the web into 9.4 inch × 14 inch strips, (2) short and shunt passivation and etching of ITO to define strip-cell area, (3) attaching electrodes and grids, and (4) final assembly involving strip cutting, interconnection of the strips and lamination. The roll-to-roll production process offers significant economy-of-scale advantages compared to batch-process manufacturing used by conventional crystalline and amorphous solar cells and this results in dramatic cost savings as the volume of production increases.

Both flexible and rigid modules of different sizes and power ratings are now available from the 5 MW United Solar plant. The stable output from a nominal four-square-foot module is 32 W with open-circuit voltage of 23.8 V, short-circuit current of 2.4 A, and fill factor of 0.56. The power output is backed by a 20 year warranty and United Solar's standard and flexible modules are listed by Underwriter's Laboratories.

Thus, all three factors required to make PV economically viable are successfully addressed with the ECD/United Solar technology. Economies realized through more efficient utilization of materials, automation of the module assembly process and high-volume production of PV modules (75–100 MW annually) will bring the fully-loaded costs of PV module production down to less than $1.00 per watt.

Product Advantage

Many new products have to compete with more mature products which have been present in the market for a long time, and *a*–Si alloy products are no exception. During the last few years, however, several new and attractive features of *a*–Si alloy solar cells have been established resulting in their wider acceptance.

Since the semiconductor bandgap of *a*–Si alloys is much larger than that of crystalline or polycrystalline silicon, the power output from an *a*–Si alloy solar panel is remarkably independent of temperature. For example, at a typical ambient temperature of 30°C, the module temperature can be 60°C. The power output from an *a*–Si alloy panel barely changes [68] from room temperature up to this temperature range, whereas a crystalline silicon panel can have its output reduced by about 20%. For the same power rating, *a*–Si alloy PV products, therefore, perform much better than the crystalline counterparts under normal outdoor conditions. This is a significant advantage both for grid-connected and battery-charging applications. Independent studies [69] on ECD/United Solar *a*–Si alloy panels show that the battery-charging capacity is 30% higher than that of a polycrystalline panel with the same power rating. This advantage of *a*–Si alloy technology is a direct consequence of the solid state physics of its operation, yet it is still not very widely appreciated. Thin-film products can, of course, be made lightweight and flexible, and there is a tremendous opportunity for these products in the building-integrated PV market.

Ovshinsky invented PV shingles in 1980 [70]. ECD/United Solar is now aggressively pursuing the roofing market by introducing novel products such as PV shingles and PV metal roofs. The United Solar PV shingle was the recipient of both the *Popular Science* "Best of What's New Grand Award" in the Environmental Technology category in 1996 and the *Discover Magazine* "Technology Innovation Award" in the Environment category in 1997.

ECD in the early 1980's demonstrated ultralight solar cells with stainless steel and Kapton substrates with dramatic energy-to-weight ratios. This led to the use of very light weight *a*–Si alloy solar cells. The light weight of these cells now being manufactured by United Solar, coupled with their radiation tolerance and good performance at elevated temperature under the AM0 solar illumination typical of space applications, has opened up a range of new uses for this technology in satellite telecommunications. New cell designs, developed by United Solar for telecommunications applications, employ thin, 0.5 million stainless steel substrates or 1.0 million Kapton substrates with thin-film SiO_x encapsulation to achieve very high power density. The stainless steel based cells have achieved 10% efficiency and 450 W/kg, and the Kapton based cells at the same efficiency can achieve up to 3000 W/kg.

Prototype lightweight cells have also been subjected to electron and proton bombardment at various energies and dose levels. These experiments showed that prolonged operation (longer than 15 hours at 70°C) anneals out bombardment-induced defects and restores the original cell efficiency.

3.2. *Multi-phase disordered hydrogen storage materials*

A "hydrogen economy," based on the use of renewable, non-polluting hydrogen as a replacement fuel for hydrocarbons, has been proposed over the years as a solution to many of the problems mentioned in our introduction. Most of this discourse, however, has been in utopian terms, since the use of hydrogen as a fuel presently lacks the basic infrastructure necessary to be practical. This problem is particularly severe in the transportation sector. Hydrogen for vehicle propulsion, when stored as a gas or a liquid, presents well-known safety problems, and these and other technical and economic problems presently inhibit its widespread use. We have provided solutions to many of these problems by utilizing our principles of disorder which permit safe high-density storage and transport of hydrogen in a solid form — a solid metal hydride that can hold much more hydrogen by volume than can be stored in its liquid or gaseous form. By utilizing hydrogen in our batteries, we have taken a giant step towards the hydrogen economy since, in driving a car with Ovonic batteries, one is "riding on hydrogen."

Our metal hydrides have been proven very successful in establishing a new field in rechargeable batteries. The electric energy delivery infrastructure is already well established. Recharging of NiMH batteries is merely a matter of utilizing grid energy since the electric utilities are a hundred years old, electric power is ubiquitous and available for quick and efficient battery charging everywhere. Of course, we have utilized solar energy for charging as well.

In addition, hybrid vehicles capable of running on various fuels and using metal hydride rechargeable batteries have substantial advantages over conventional vehicles. The fuel efficiency of hybrid vehicles is greatly improved by allowing their fuel-burning engine (or fuel cell) to operate at peak efficiency while charging the battery pack, which is used either as the primary energy source for the electric drive or as an assist to the engine.

There were no commercial NiMH batteries prior to ours. By utilizing our principles of disorder and local order in the form of multi-elemental, multiphase d- and f-orbital materials, we made possible, for the first time, a battery which could be continuously improved in energy density and power.

3.2.1. *Ovonic nickel metal-hydride batteries*

The worldwide market for rechargeable batteries for consumer electronic applications is growing at a record pace due to increased consumer demand for portable devices such as cellular phones, laptop computers, camcorders, and other personal electronic devices. Nickel metal-hydride batteries today represent the fastest growing segment of this rechargeable battery market for consumer electronics due to their higher energy density and environmentally acceptable chemistry relative to NiCd. Current estimates indicate that the total portable rechargeable battery market will grow to over \$5 billion by 2000 [71].

Ovonic Battery Company (Ovonic Battery) is the recognized world leader in the development of NiMH battery technology and today essentially all of the worldwide major manufacturers of NiMH cells for consumer applications are producing under agreements with ECD and Ovonic Battery. Ovonic Battery manufactures and continues to develop advanced negative electrode and positive electrode material for its licensees. Other applications for NiMH technology include bicycles, scooters, remote and standby power for utilities, marine, agricultural and industrial use. General Motors and ECD have established a joint venture, GM Ovonic, which is in the production of batteries for electric vehicles and GM Ovonic has recently expanded its production capacity by opening another plant in Ohio.

Ovonic NiMH batteries today demonstrate specific energies of over 80 Wh/kg, energy densities in excess of 200 Wh/L in commercial cells and 95 Wh/kg in next generation production batteries to be introduced at the end of 1999. There is a continuous improvement process whose goal is 120 Wh/kg. We have already demonstrated over 1000 W/kg and are designing 3000 W/kg systems. Ovonic batteries are the only ones that have both high energy (range) and power. The standard lead acid starter, lighter and ignition (SLI) batteries can be replaced by cost-competitive, smaller, lighter, high-performance, long-life Ovonic batteries.

A four-passenger automobile designed specifically for electric propulsion and powered by Ovonic NiMH batteries went 373 miles on a single charge; a four-passenger Solectria EV went from Boston to New York on the equivalent BTU energy of less than one gallon of gas and had 15% energy remaining [72], hybrid electric vehicles (HEVs) with our batteries will offer at least 80 miles per gallon in the charge-sustaining mode, and over 100 miles per gallon in the charge-depletion mode.

3.2.1.1. NiMH cell chemistry

The basic cell reaction for NiMH can be written as:

$$MH + NiOOH \rightarrow M + Ni(OH)_2$$

where M represents a multi-elemental disordered intermetallic alloy capable of forming a metal hydride phase. It is interesting to note that the overall cell reaction consists of transfer of a hydrogen ion from one electrode to the other, in much the same manner as a Li ion battery functions through transfer of a Li^+ ion between two insertion electrodes. Therefore, one could equally well describe a NiMH battery as a "hydrogen ion" battery or, as some have called it, a "protonic" battery [6, 73].

An important feature of the NiMH chemistry is the cell's ability to tolerate both overcharge and overdischarge through gas recombination reactions that result in no net change in battery electrolyte and prevent a build up of pressure inside the sealed and totally maintenance-free cell. This ability to tolerate both overcharge and overdischarge is particularly advantageous for EV applications where system voltages over 300 volts are common. Under these conditions, with over 200 cells series connected, individual cells will be subjected to varying degrees of overcharge during charging and individual cells can experience overdischarge and cell reversal, particularly during acceleration or at very low states of battery pack charge. The ability of NiMH to accept overcharge and overdischarge eliminates the need for single-cell voltage monitoring, simplifying battery management in comparison to some of the high energy experimental systems such as NaS, Li ion, or Li polymer. The NiMH battery does not have the safety problems of sodium sulfur or lithium systems, is much lower in cost, and can provide much higher power and mileage.

3.2.1.2. Metal hydride alloy development

Early development work on hydride alloys for rechargeable batteries focused on conventional hydride materials of well-defined composition and crystal structure. The most widely studied of these materials were alloys of the $CaCu_5$ family, most notably $LaNi_5$ and its derivatives [74]. Their shortcomings prevented them from becoming commercial batteries due to the universally-attempted use of these simple, single-phase hydrides.

Working with a team of material scientists at ECD, I employed a fundamentally different approach using disordered materials to develop hydrides for various applications, specifically including rechargeable batteries [75]. The materials requirements for the use of metal hydrides in electrochemical applications were defined, and novel alloys were specifically engineered to meet the demanding and diverse materials requirements. Among the required properties for a negative electrode hydride material are:

- high hydrogen storage capacity
- proper metal to hydrogen bond strength
- oxidation and corrosion resistance
- fast gas recombination kinetics
- low raw materials and manufacturing cost.

The engineering of multi-component, multi-phase materials allows for alloys that satisfy these required properties by introducing compositional and structural disorder [76]. By adjusting the alloy composition, one can control the metal-to-hydrogen bond strength to achieve the desired value. Control of alloy microstructure provides for increased hydrogen storage sites as well as improved kinetics and corrosion resistance. Ovonic Battery has successfully employed this concept of compositionally complex, multi-phase materials to develop alloys that, in our commercial NiMH batteries, provide high energy, high power, and long cycle life [77].

Ovonic Battery has focused predominately on alloys of the V–Ti–Zr–Ni type due to their intrinsic ability to store very high amounts of hydrogen compared to rare-earth based systems. The specific role of each component in these new intermetallic compounds is well understood [78, 79], allowing Ovonic materials scientists to systematically vary the stoichiometry of the alloy to meet the application requirement. While typical rare earth based materials store up to 300 mAh/g specific capacity, commercial Ovonic alloys today store over 400 mAh/g Ovonic Battery is actively developing Mg-based materials such as MgNi-based alloys capable of storing 550 to 700 mAh/g hydrogen [77, 80]. The higher capacity of these Mg-based materials, and the lower alloy raw materials costs offer the potential for significant long-term performance and cost advances.

Ovonic Battery currently operates a manufacturing facility for the production of hydride alloys and negative electrode "belt" material for consumer battery and electric vehicle battery applications and for the assembly of complete batteries and battery packs.

3.2.1.3. Positive electrode development

By following the same principles of disorder that we utilize for the development of the negative electrode, we have been able to make a positive electrode which increases the energy storage capacity of the battery, lowers the cost and, at the same time, makes an important scientific advance. The nickel hydroxide electrode charge/discharge reaction is commonly written in a simple, one-electron transfer reaction. It is well known that this reaction is much more complex [81, 82]. In fact, values approaching 1.6 electrons transferred per Ni atom are theoretically possible for this reaction. Various researchers have demonstrated multi-electron transfer materials utilizing the (alpha) $Ni(OH)_2$ to (gamma) $NiOOH$ reaction, but they have not been able to maintain stability with electrode cycling [83]. ECD and Ovonic Battery researchers have been working on advanced $Ni(OH)_2$ materials incorporating the principles of compositional and structural disorder that have been demonstrated so successfully in metal-hydride materials. These proprietary $Ni(OH)_2$

Table 3. Ovonic NiMH battery performance characteristics.

	GMO 1 (Current Production EV)	GMO 2 (High Power EV)	GMO 3 (High Power EV)	HEV 60 (HEV Cell)	HEV 20 (HEV Cell)
Capacity (Ah)	90	100	100	60	20
Specific Energy (Wh/kg)	75	80	95	70	65
Energy Density (Wh/L)	170	200	230	170	160
Specific Power (W/kg) 50% DOD	225	300	350	630*	650*
Power Density (W/L)	600	750	850	1700**	1600**
Cycle Life (DST to 80% DOD)	600	600	1,000	Vehicle Life	Vehicle Life

*Next Generation 1000 W/kg

**Next Generation 2500 W/L

Table 4. Battery and vehicle performance characteristics.

	Current Production		Next Generation	
Characteristics	Battery	Vehicle	Battery	Vehicle
Specific Energy	75–80 Wh/kg	> 200 miles	95 Wh/kg	> 250 miles
Energy Density	220 Wh/L	—	275 Wh/L	—
Power Density 50% DOD 80% DOD	290 W/kg 220 W/kg	0–60 mph in 8 secs	355 W/kg 290 W/kg	0–60 mph in < 8 secs
Cycle Life	600–800	Life of Vehicle	> 1000	Life of vehicle
Charging	60% in 15 minutes	Quick recharge	60% in 15 minutes	Quick recharge
Battery Cost (High-volume production)	$235/kWh	$4,000–$5,000	$150/kWh	$3,000–$4,000

Ovonic Battery operated electric vehicles have achieved a fuel economy of over 130 miles per gallon of equivalent gasoline energy.

materials have demonstrated up to 1.5 electron transfer in thin-film electrodes and up to 1.3 electrons per Ni atom in battery electrodes containing bulk spherical powders produced through a proprietary process [80, 84]. Scale-up and the production of reversible phase change, high-gamma materials and their incorporation into cell designs yield energy densities of 95 Wh/kg and greater in commercial consumer and EV batteries, as has already been demonstrated. The performance of production EV Cells and high power EV and HEV cells is shown in Table 3. The relationships between various battery performance characteristics and vehicle range and performance are shown in Table 4.

3.2.2. *Hydrogen storage for fuel cell applications*

3.2.2.1. Fuel cells

A hydrogen fuel cell combines hydrogen and air directly to produce electricity. There are several different types of fuel cells that can be used this way, and they are identified according to the type of electrolyte and the temperature at which they operate. These are the polymer electrolyte membrane (PEM), alkaline, phosphoric acid, molten carbonate and solid oxide fuel cell systems. For transportation applications, the PEM fuel cell is being considered because of its low temperature operation and its potential for being compact and lightweight. While the average energy efficiency of a gasoline internal combustion engine is about 20 percent, hydrogen fuel cells can operate with efficiencies of approximately 50. Fuel cell vehicles also typically use a design similar to other hybrid vehicles with a rechargeable battery to provide the high power demanded by acceleration. This battery requirement is well served by nickel metal-hydride batteries being developed for other hybrid vehicles.

The success of fuel cells in transportation applications depends upon dramatically reducing their cost. For our part, we are able to solve the problem of low cost, reliable production and storage of hydrogen. We can also aid their development by our knowledge of how to make catalysts which are not easily poisoned and we have had unique approaches to solid state ionic conductors. Fuel cell vehicles are not considered near-term candidates and, even when they do appear, they will require our NiMH batteries as part of their system.

3.2.2.2. The fuel

While much effort has gone into the development of the "cell" part of the "fuel cell," until recently very little thought has been given to the "fuel" and the fuel infrastructure of the system. The fuel of choice for a fuel cell is hydrogen. However, much attention is being given to other gases containing hydrogen, such as methane, etc. that can be reformed on board to provide the hydrogen needed for fuel.

Direct hydrogen storage technologies include high-pressure gaseous hydrogen, cryogenic liquid hydrogen and metal hydrides. For large-scale stationary applications, the high-pressure and liquid hydrogen options may be practical choices; however, for mobile/transportation applications, solid-state storage as metal hydrides is the practical choice. Metal hydrides offer advantages of being safe, compact and low-pressure with no boil-off losses at ambient temperature. In fact, metal hydride storage is considered the best and safest means of transporting hydrogen to various locations.

Metal Hydrides — On-Board Vehicle Storage

Metal hydrides provide an ideal method for safe, reversible hydrogen storage, overcoming the problems of safety and shelf life. The volumetric hydrogen storage density of metal hydrides can exceed that of liquid hydrogen. Hydrogen storage by metal hydrides is based on the reversible reaction:

$$M + x/2\ H_2 \leftrightarrow MH_x$$

where M is a hydrogen-absorbing alloy. The equilibrium between the exothermic forward reaction and the endothermic reverse reaction is dependent on temperature and hydrogen pressure.

Hydride storage on board vehicles has been demonstrated in the U.S., Germany and Japan by Mercedes-Benz, Mazda and Toyota. However, vehicle range in these demonstrations has been limited because conventional low temperature hydrides have low storage capacity. In order to meet this objection, we have developed new, high-capacity disordered alloys.

In addition to low storage capacity, conventional metal hydrides also suffer from the problem of decrepitation, i.e. breakdown of the alloy particles into very fine powders on cycling. Decrepitation leads to denser packing of the alloy bed, which restricts the hydrogen gas flow. These problems can be overcome by novel approaches in engineering the materials to tailor the compositional structural properties of the alloys. This is the approach being followed by ECD.

According to the U.S. Department of Energy (US DOE), on-board storage of hydrogen providing a fuel cell vehicle range of 340 miles will require about

3.6 kg of hydrogen fuel. At four-weight percent storage capacity, this will require about 198 pounds of metal hydride.

Additional advantages of using metal hydrides for a PEM fuel cell include the following:

(1) Hydrides deliver very high purity hydrogen to the PEM fuel cell, which uses a platinum catalyst that can easily be poisoned by impurities.
(2) Hydrogen stored in metal hydrides is delivered over a narrow pressure range, which means less stringent pressure regulator requirements.
(3) The endothermic hydrogen desorption from a metal hydride can be used for cooling the fuel cell cooling water.
(4) Because metal hydrides are solid state materials, their storage containers can be modular and flexible in terms of size and shape.

4. Conclusion

There are, of course, many more applications for amorphous and disordered materials. We have attempted here to provide highlights of some of the more important areas to show the growth and significance of the field that has now gained worldwide commercial acceptance.

The applications of glasses and amorphous and disordered materials address fundamental problems of energy and information. In terms of energy, the answer to the problems of pollution, climate change, dependence on oil which causes wars, is to utilize triple-junction thin-film amorphous photovoltaics made in a continuous roll-to-roll manner for energy generation that can be competitive to conventional fuels. Equally as important is to as quickly as possible make hydrogen the preferred basis for transportation both as a fuel and as rechargeable batteries made of metal hydride materials. I believe that photovoltaics will become a ubiquitous form of energy generation and that the first giant step toward a hydrogen economy has already been taken since, as we have said, when one drives an electric or hybrid vehicle with Ovonic nickel metal-hydride batteries today, one is already participating in the hydrogen economy. One is "riding on hydrogen."

My vision of the future is very much the same as the one I had when I first started my work in amorphous materials. My first memory and switching devices were analog neurons which could be connected by synaptic nodes. I believe that intelligent machines using these devices will become practical in the not too distant future. Fortunately, now in the information field our optical phase change memories have become the enabling technology opening the doors for the near-term use of our Ovonic Unified Memory described here. The next step will be to combine our Ovonic threshold switch with the various Ovonic memory devices integrated in a single amorphous circuit which will include our amorphous semiconductor light sensors so that a three-dimensional learning machine all in thin-film form will become an actuality. Such a self-organizing machine would meet the criteria of cognitive scientists by having both evolved structured and learned functions.

It is certain that the field of amorphous and disordered materials will continue as an exciting area of materials science with new physics and new chemistry expressed in new materials, new products and new production technology.

Acknowledgments

I owe a special debt of gratitude to Steve Hudgens for his assistance in the preparation of this chapter and to my many other collaborators in the ECD/United Solar/Ovonic Battery team. I also want to acknowledge my collaboration with Hellmut Fritzsche and the late David Adler which has added so much value to the work discussed in this chapter. As always, Iris is the other half of our lone pair.

References

[1] S.R. Ovshinsky, *J. Non-Cryst. Solids* **42**, 335 (1980).
[2] S.R. Ovshinsky, presented at the 1978 Gordon Research Conference on Catalysis (unpublished).
[3] S.R. Ovshinsky, presented in May 1980 at Lake Angelus, Michigan (unpublished).
[4] K. Sapru, B. Reichman, A. Reger and S.R. Ovshinsky, U.S. Patent No. 4,633,597 (18 November 1986).
[5] S.R. Ovshinsky, M.A. Fetcenko and J. Ross, *Science* **260**, 176 (9 April 1993).
[6] S.R. Ovshinsky and R.C. Stempel, *Proc. Japanese Society of Electric Vehicles*, Tokyo, Japan (February 1997).
[7] S.R. Ovshinsky and D. Adler, *Contemporary Physics* **19**, 109 (1978).
[8] S.R. Ovshinsky, Physical Properties of Amorphous Materials, eds. D. Adler, B.B. Schwartz and M.C. Steele (Plenum Press, NY, 1985) p. 105.
[9] S.R. Ovshinsky, *J. Non-Cryst. Solids* **73**, 395 (1985).
[10] S.R. Ovshinsky, *Rev. Roum. Phys.* Tome **26**, 893 (1987).
[11] R. Tsu, J. Gonzales-Hernandez, J. Doehler and S.R. Ovshinsky, *Solid State Comm.* **46**, 79 (1983).

[12] S.R. Ovshinsky, Compositionally Varied Materials and Method for Synthesizing the Materials, U.S. Patent No. 4,520,039 (28 May 1985).

[13] S.R. Ovshinsky, *Memoires Optiques and Systemes*, No. 127, 65 (1994).

[14] M. Kastner, *Phys. Rev. Lett.* **28**, 355 (1972).

[15] S.R. Ovshinsky, *Proc. Int. Topical Conf. on Structure and Excitation of Amorphous Solids*, eds. G. Lucovsky and F.L. Galeener, AIP Conf. Proc. (Williamsburg, Virginia, 1976) p. 31.

[16] S.R. Ovshinsky and K. Sapru, *Proc. 5th Int. Conf. on Amorphous and Liquid Semiconductors*, eds. J. Stuke and W. Brenig, Garmisch–Partenkirchen (Taylor and Francis, UK, 1974) p. 447.

[17] J. Feinleib, S. Iwasa, S.C. Moss, J.P. deNeufville and S.R. Ovshinsky, *J. Non-Cryst. Solids* **8–10**, 909 (1972).

[18] S.R. Ovshinsky and H. Fritzsche, *Metallurgical Trans.* **2**, 641 (1971).

[19] S.R. Ovshinsky and H. Fritzsche, *Proc. 6th Int. Conf. on Amorphous and Liquid Semiconductors* (Leningrad, 1975) p. 426.

[20] S.R. Ovshinsky, E.J. Evans, D.L. Nelson and H. Fritzsche, *IEEE Trans. on Nuclear Science* **15**, 311 (1968).

[21] S.R. Ovshinsky, Disordered Materials, Science and Technology, Selected Papers, eds. D. Adler, B. Schwartz and M. Silver (Plenum Press, NY, 1991).

[22] S.R. Ovshinsky, The Nerve Impulse (1955) (unpublished).

[23] M.P. Southworth, *Control Engineering* **11**, 69 (1964).

[24] S.R. Ovshinsky, Electrical Activity of the Brain (1955) (unpublished).

[25] F. Morin, G. Lamarche and S.R. Ovshinsky, Functional Aspects of Celebellar Afferent Systems and of Cortico-Cerebellar Relationships, *Laval Medical* **26**, 633 (1958).

[26] S.R. Ovshinsky, *Phys. Rev. Lett.* **21**, 1450 (1968).

[27] This Week's Citation Classic, *Current Contents* (8 March 1982).

[28] S.R. Ovshinsky, *Proc. 5th Annual National Conf. on Industrial Research*, Applying Emerging Technologies, (Chicago, IL, 18–19 September 1969) p. 86.

[29] S.R. Ovshinsky, *J. Applied Photographic Engineering* **3**, 246 (1977).

[30] L. Lessing, The Printed Word Goes Electronic, *Fortune* (September 1969) p. 116.

[31] L. Lessing, Great Hopes from Ovshinsky's Little Switches Grow, *Fortune* (April 1970) p. 110.

[32] S.R. Ovshinsky, U.S. Patent No. 3,530,441 (22 September 1970).

[33] D. Strand, J. Gonzalez–Hernandez, B.S. Chao, S.R. Ovshinsky, P. Gasiorowski and D. Pawlik, *Spring Meeting Mat. Res. Soc. Symp.*, Anaheim, California, 1991.

[34] J. Gonzalez–Hernandez, B.S. Chao, D. Strand, S.R. Ovshinsky, D. Pawlik and P. Gasiorowski, *Appl. Phys. Commun.* **11(4)**, 557 (1992).

[35] M. Chen, K.A. Rubin and R.W. Barton, *Appl. Phys. Lett.* **49**, 502 (1986).

[36] F. Betts, A. Bienenstock and S.R. Ovshinsky, *J. Non-Cryst. Solids* **4**, 554 (1970).

[37] M.H. Cohen, H. Fritzsche and S.R. Ovshinsky, *Phys. Rev. Lett.* **22**, 1065 (1969).

[38] H.K. Henisch, E.A. Fagen and S.R. Ovshinsky, *J. Non-Cryst. Solids* **4**, 538 (1970).

[39] H. Fritzsche, *Ann. Revue Mat. Sci.* **2**, 697 (1972).

[40] K. Peterson and D. Adler, *Appl. Phys Lett.* **25**, 211 (1974).

[41] D. Adler, H.K. Henisch and N.F. Mott, *Rev. Mod. Phys.* **50**, 209 (1978).

[42] D. Adler, M.S. Shur, M. Silver and S.R. Ovshinsky, *J. Appl Phys.* **51**, 3289 (1980).

[43] H.K. Henisch, J.-C. Manifacier, R.C. Callarotti and E. Schmidt, Physics of Disordered Materials, eds. D. Adler, H. Fritzsche and S.R. Ovshinsky (Plenum Press, NY, 1985) p. 779.

[44] H.K. Henisch and R.W. Pryor, *Solid State Electronics* **14**, 765 (1971).

[45] H. Fritzsche, *March Meeting of the American Physics Society*, Philadelphia, 1969.

[46] S.R. Ovshinsky and H. Fritzsche, *IEEE Trans. on Electron Devices* **ED20**, 91 (1973).

[47] M.P. Shaw, S. Holmberg and S.A. Kostylev, *Phys. Rev. Lett.* **31**, 542 (1973).

[48] R. Shanks and C. Davis, *IEEE Int. Solid State Circuits Conf.* (1978).

[49] R.G. Neale, D.L. Nelson and G.E. Moore, *Electronics* **43**, 56 (1970).

[50] S.R. Ovshinsky, Amorphous and Disordered Materials — The Basis of New Industries, *Mat. Res. Soc.*, Boston, MA, 30 November–4 December 1998.

[51] J. Perschy, *Electronics* **40**, 74 (24 July 1967).

[52] W. Spear and P. LeComber, *J. Non-Cryst. Solids* **8–10**, 727 (1972).

[53] M. Hack, M. Shur and W. Czubatyj, *Appl. Phys. Lett.* **48**, 1386 (1986).

[54] W. Howard, *Journal SID*, **3(3)**, 127 (1995).

[55] I. Shimizu, T. Komatsu, K. Saito and E. Inoue, *J. Non-Cryst. Solids* **35 & 36**, 773 (1980).

[56] S.R. Ovshinsky and M. Izu, U.S. Patent No. 4,217,374 (12 August 1980).

[57] S.J. Hudgens and A. Johncock, *Mat. Res. Soc. Symp. Proc.* **49**, 403 (1985).

[58] S.J. Hudgens, A.G. Johncock and S.R. Ovshinsky, *J. Non-Cryst. Solids* **77 & 78**, 809 (1985).

[59] M. Izu, B. Dotter, S.R. Ovshinsky and W. Hasegawa, *36th Annual Technical Conf. Proc. SVC* (1993) p. 333.

[60] T. Ellison, B. Dotter, M. Izu and S.R. Ovshinsky, *40th Annual Technical Conf. Proc. SVC*, New Orleans (April 1997) p. 309.

[61] E. Becquerel, *Comptes Rendus de L'Academie des Sciences* **9**, 145 (1839).

[62] R.S. Ohl, U.S. Patent No. 2,402,622 (1941) and No. 2,443,542 (1941).

[63] D.L. Staebler and C.R. Wronski, *Appl. Phys. Lett.* **31**, 292 (1977).

[64] J. Yang, A. Banerjee and S. Guha, *Appl. Phys. Lett.* **70**, 2975 (1997).

[65] S. Guha, J. Yang, P. Nath and M. Hack, *Appl. Phys. Lett.* **49**, 218 (1986).

[66] S. Guha, J. Yang, A. Pawlikiewicz, T. Glatfelter, R. Ross and S.R. Ovshinsky, *Appl. Phys. Lett.* **54**, 2330 (1989).

[67] S.R. Ovshinsky, *Solar Energy Materials and Solar Cells* **32**, 443 (1994).

[68] B. Kroposki *et al.*, *IEEE PVSC-26*, Anaheim, CA (1997).

[69] Data obtained from tests done by Photocomm, Inc. at its Scottsdale, AZ facility.

[70] J. Glorioso, *Energy Management*, June/July 1980, p. 45. Shown in 1980 by Stan and Iris Ovshinsky to Domtar, a Canadian roofing company, and to Allside, an aluminum siding company of Akron, OH.

[71] Charging into Rechargeables, Business Week (7 October 1996) p. 142V.

[72] *IEEE Spectrum* **34**, 68 (December 1997).

[73] S.R. Ovshinsky, M.A. Fetcenko, S. Venkatesan, S.K. Dhar, A. Holland, R. Young, P.R. Gifford and D.A. Corrigan, presented at the *13th Int. Seminar on Primary and Secondary Battery Technology and Application*, Deerfield Beach, FL, March 1996.

[74] M.H.J. van Rijswick, *Hydrides for Energy Storage*, eds. A.F. Andresen and A.J. Maeland (Pergamon Press, Elmsford, NY, 1978) p. 261.

[75] S.R. Ovshinsky, K. Sapru, B. Reichman and A. Reger, U.S. Patent No. 4,623,597 (November 1986).

[76] S.R. Ovshinsky, M.A. Fetcenko and J. Ross, *Science* **260**, 176 (1993).

[77] S.R. Ovshinsky and M.A. Fetcenko, *Proc. 185th Meeting of the Electrochemical Society*, San Francisco, CA, May 1994.

[78] S.R. Ovshinsky, R.C. Stempel, S. Dhar, M.A. Fetcenko, P.R. Gifford, S. Venkatesan, D.A. Corrigan and R. Young, *Proc. 29th Int. Symp. on Automotive Technology and Automation*, Florence, Italy, June 1996.

[79] M.A. Fetcenko, S. Venkatesan and S.R. Ovshinsky, Selection of Metal Hydride Alloys for Electrochemical Applications, *Proc. 180th Meeting of the Electrochemical Society*, Phoenix, AZ, 15 October 1991.

[80] S.R. Ovshinsky, D. Corrigan, S. Venkatesan, R. Young, C. Fierro and M.A. Fetcenko, U.S. Patent No. 5,384,822 (September 1994).

[81] M. Oshitani, H. Yufu, K. Takashima, S. Tsuji and Y. Matsumaru, *J. Electrochem. Soc.* **136**, 1590 (1989).

[82] D.A. Corrigan and S.L. Knight, *J. Electrochem. Soc.* **136**, 613 (1989).

[83] C. Delmas, L Gautier and L. Guerlou, Meeting Abstracts, *190th Society Meeting*, Vol. 96–2, Abstract No. 883, The Electrochemical Society, Inc., Pennington, NJ, 1996.

[84] S.R. Ovshinsky, M.A. Fetcenko, C. Fierro, P.R. Gifford, D.A. Corrigan, P. Benson and F.J. Martin, U.S. Patent No. 5,523,182 (June 1996).

B.   AMORPHOUS CHALCOGENIDE PHOTOCONDUCTORS IN IMAGING TECHNOLOGIES

SAFA O. KASAP* and JOHN A. ROWLANDS†

*Department of Electrical Engineering, University of Saskatchewan,
Saskatoon, S7N 5A9, Canada
†Department of Medical Biophysics, University of Toronto,
Sunnybrook Health Science Centre, 2075 Bayview Avenue,
North York, M4N 3N5, Canada
safa-kasap@engr.usask.ca

Contents

1. Introduction

Amorphous selenium (a–Se) alloys are well known for their excellent photo-conductivity and large area imaging applications dating back to 1950's and 1960's when a–Se was successfully used as a xerographic photoreceptor (PR) in the first commercial Xerox photocopiers which literally revolutionized the office [1]. The first generation photoreceptors were essentially stabilized a–Se layers, typically 40–70 μm in thickness, vacuum coated onto Al substrate drums. Stabilized a–Se refers to a–Se alloyed with about 0.3–0.5% As to inhibit crystallization and doped with 10–40 ppm Cl to improve the charge carrier transport properties [2]. The relatively wide band gap ($E_g \approx 2.22$ eV) and the resulting high dark resistivity ($\rho > 10^{15}\Omega$ m) of a–Se means that the surface of an a–Se photoreceptor can be readily corona charged, that is electrostatically charged, to several hundreds of volts as illustrated in Fig. 1. When the a–Se photoreceptor is subsequently exposed to light reflected from a document, electron-hole pairs (EHPs) photogenerated in the photon absorbed areas discharge the surface voltage. The resulting electrostatic surface potential pattern on the photoreceptor then represents the black and white image on the document. This electrostatic image is subsequently used to transfer toner particles onto a plane paper in a xerographic process called the toner development which is not shown in Fig. 1. The toners are fused into paper to obtain a final "xeroxed"document. As much as the photoconductivity of a–Se lead to the development and commercialization of xerography in 1960's, its X-ray photoconductivity similarly lead to xeroradiography in 1960's and 1970's which was essentially photocopying the body-part using X-rays (Fig. 1). In its heyday, xeroradiography was highly successful and had its staunch followers based on the quality of the images [3, 4]. It suffered from the cumbersome

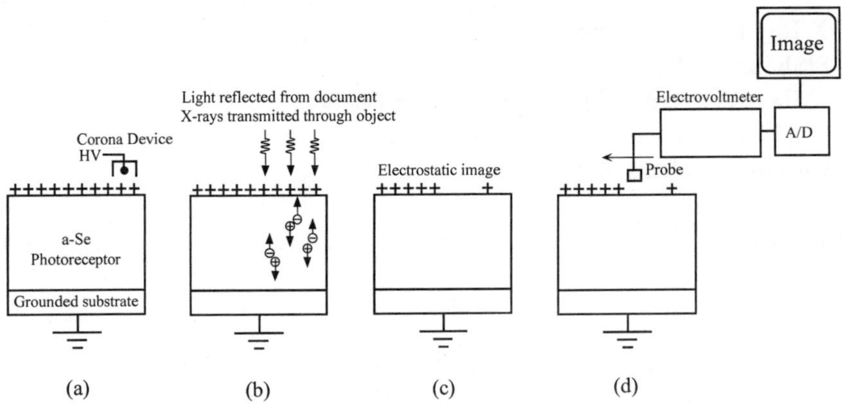

Fig. 1. (a) In xerography and xeroradiography, the a–Se photoreceptor PR surface is electrostatically charged to some positive voltage (e.g. 500 V in xerography and 2000 V in xeroradiography). (b) Light reflected from the white regions of the document becomes absorbed in the PR and thereby photogenerates EHPs that discharge the exposed region of the PR. In xeroradiography, X-ray photons transmitted through the object cause the generation of EHPs. (c) At the end of exposure, the image is in the form of an electrostatic potential distribution on the surface of the PR. In xerography, this electrostatic image provides the basis of the toner development method. (d) In Thoravision, the electrostatic image on the PR is readout using an array of electrostatic probes scanning the surface of the PR and an A/D (analog-to-digital) converter digitizes the image.

toner development method which limited its practical usefulness; xeroradiography is now obsolete.

The xerographic process has been extensively reviewed by a number of authors [1, 5–7]. The basic principles operating in today's copying machines remain unchanged. Although over the last decade organic photoconductors (OPCs) have, by and large, taken over the photoreceptor market, one nonetheless still finds amorphous inorganic semiconductor photoreceptors (primarily a–As_2Se_3 and a–Si:H) in a limited number of copying machines. Cost effectiveness of the OPC will probably see the eventual demise of amorphous inorganic semiconductors in xerography.

There are however new imaging technologies that have been emerging in which amorphous chalcogenide semiconductors cannot be simply replaced by OPCs due to a number of fundamental reasons. In X-ray imaging, the X-ray photoconductor has to absorb as much of the incident radiation as possible to convert the X-ray photons to collectable EHPs. The thickness L of an efficient photoconductive layer has to be several times the absorption depth δ ($= 1/\alpha$, α = absorption coefficient) which makes organic photoconductors consisting of low atomic number elements unsuitable for this application as they have very small absorption coefficients. Recently, Philips Medical Imaging Systems have commercialized a new digital chest X-ray imaging system, called Thoravision, that is based on the xeroradiographic process with the toner development process replaced by an electronic read-out technique, which enables the digitization of the X-ray image [8]. The basic principle is illustrated in Fig. 1(d) and is based on earlier similar work where an array of electrometers were used to read the electrostatic image [9, 10]. The photoconductor in the Thoravision system is a 500 μm thick stabilized a–Se layer coated on a large Al drum. One of most interesting properties of a–Se is that at sufficiently high electric fields, charge carriers in an a–Se layer exhibit avalanche multiplication which has enabled the development of super sensitive TV pick up tubes (the HARPICON vidicons [11, 12]) with gains as high as 800. There have been no reports of avalanche multiplication in OPCs.

Due to a number of unique features, a–Se alloy based photoconductors are presently used in a number of important imaging applications either in commercial products such as the HARPICON and the Philip's Thoravision, or in prototype units such as the flat panel X-ray image detectors developed at DuPont, Sterling Diagnostic Imaging and Direct Radiography Corp. [13, 14], Thermotrex [15] and the Sunnybrook Health Science Centre [16–20], and

Table 1. Amorphous selenium alloys as photoconductors in imaging applications. Typical range of compositions.

Major Application	Chalcogenide Photoconductor	Comment
Xerography [1, 2, 5]	a–Se:0.5% As + 20 ppm Cl	Stabilized a–Se no longer used in xerography.
Xerography [2, 5, 22]	a–Se$_{1-x}$Te$_x$ (+ y ppm Cl) Single and multi-layers	Panchromatic and red sensitive. Single and multi-layer photoreceptors. Still in use in old copying machines.
Xerography [2, 5]	a–As$_2$Se$_3$ (+ y ppm I)	Red sensitive PR. Thermally stable and excellent wear resistance. Long lifetime. Used in some copying machines.
Xeroradiography [3–5]	a–Se:0.5% As + 20 ppm Cl	Photocopying of body-part with X-rays. Now obsolete.
Electroradiography with laser scanning [23]	a–Se:0.5% As + 20 ppm Cl	Laser scanned read-out method replaces toner development in xeroradiography and leads to digital radiography.
Electroradiography with electrometer readout [8, 9, 24]	a–Se:0.5% As + 20 ppm Cl	Scanned electrometer readout replaces toner development in xeroradiography. Commercial Thoravision by Philips Medical Imaging.
SATICON [25, 26]	a–Se$_{1-x-y}$Te$_x$As$_y$	TV pick-up tube with no avalanche gain. Commercialized by Hitachi.
HARPICON [11, 12]	a–Se$_{1-x-y}$Te$_x$As$_y$ (+ ppm LiF)	TV pick-up tube with avalanche gain. Commercialized by NHK (Japan). Alloying a–Se with As inhibits crystallization and alloying with Te enhances red sensitivity.
Flat Panel X-Ray Image Detector for Digital Radiography [13–19, 27–29]	a–Se:0.3% As + 20 ppm Cl	Stabilized a–Se coated onto an active matrix array and electroded. X-ray image converted to charge stored in pixels and read out electronically. Poised for commercialization in 2000–2001.
Large area X-ray vidicon [21, 27]	a–Se:0.3% As + 20 ppm Cl	Large area X-ray vidicon using stabilized a–Se which could be charged negatively.

in a prototype large area X-ray vidicon [21]. In this review we first explain the fundamental principles of a–Se based flat panel X-ray imaging systems (Sec. 2) and discuss the necessary properties for an ideal X-ray photoconductor for use in such detectors (Sec. 3). We then highlight the properties of stabilized a–Se and how these unique properties have lead to the development of a successful flat panel X-ray photoconductor (Sec. 4). A key parameter that determines the efficiency of an X-ray photoconductor is the EHP creation energy which is defined and discussed in terms of our current understanding (Sec. 5). We review the principle of operation of the HARPICON and the avalanche multiplication that has enabled super sensitive TV pickup tubes for high definition TV (Sec. 6). The properties of a–Se will be examined in relation to its use in these imaging technologies. It will be apparent that although the technology has ploughed ahead with ingenuous and marvelous developments leading to commercial products, some of the fundamental physics still remains unresolved (Sec. 7). For example, to date, there has been no successful explanation of the avalanche effect in a–Se. Table 1 summarizes the use of a–Se based photoconductors in various imaging technologies from their historical use in xerography to their present uses in "new" technologies. The most significant development in a–Se photoconductor technology is undoubtedly the recent *flat panel digital X-ray image detectors* which promise to revolutionize X-ray imaging which is still mainly an analog film technology. One of the attractive advantages of the prototype flat panel X-ray image detectors

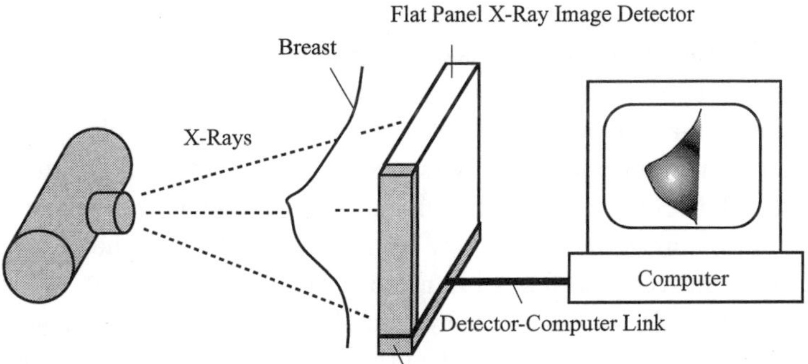

Fig. 2. Schematic illustration of a flat panel X-ray image detector for mammography. The connection from the detector to the computer can be a wireless link allowing a more versatile detector usage.

is their very nature of being flat as a convenient direct replacement of the present film cassette. Figure 2 illustrates an example of a flat panel X-ray image detector for use in digital mammography where the flat panel advantages are self-evident. (The electronic connection from the detector to the computer can be a wireless link which would make these flat panel detectors even more versatile as the data can be transmitted to a central computer where it can be analyzed/diagnosed and stored more efficiently.)

2. Flat Panel X-Ray Image Detectors for Digital Radiography

Very recent research carried out concurrently at DuPont [13, 14], and Sunnybrook Health Science Centre [16–18] has shown clearly that one of the most promising digital radiographic systems is based on using a large area thin film transistor (TFT) active matrix array (AMA, as used in flat panels displays, for example) with an electroded X-ray photoconductor. An important first step in the flat panel X-ray detector technology was the development of flat panel thin film transistor displays which matured as the fabrication and doping of large area hydrogenated amorphous silicon (a–Si:H) films became technologically possible in the early nineties [30]. Once large area flat panels with small pixel sizes became available at the component level it was only a matter of time before an X-ray photoconductor such as a–Se would be used to directly convert X-ray images to a charge distribution stored on the pixels of a flat panel. The combination of an active array matrix and an X-ray photoconductor then constitutes a *direct conversion X-ray image detector*; a term coined by DuPont [13] (the terms X-ray sensor and detector have been used interchangeably in the literature). Direct conversion here refers to the fact that the X-ray photons are directly converted to charges that are subsequently collected and detected viz-à-viz an intermediate conversion, via a phosphor, to photons (light) and then from photons to charges as in some other flat panel sensors [27, 31, 32]. The charge distribution residing on the panel's pixels are simply read out by scanning the arrays row by row using the peripheral electronics and multiplexing the parallel columns to a serial digital signal. This signal is then transmitted to a computer system (Fig. 2). The system is simple, inherently digital and has so many advantages that it has now become a major contending choice in digital radiography [17, 33].

An active matrix array, as shown in Fig. 3, consists of millions of individual pixel electrodes connected by transistors (one for each pixel) to electrodes passing over the whole array to subsidiary electronics on the periphery. The

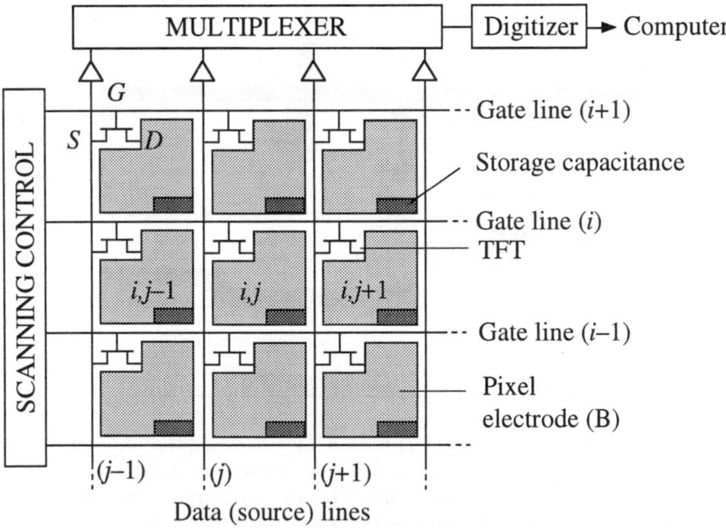

Fig. 3. Thin film transistor (TFT) active matrix array (AMA) for use in X-ray image detectors with self-scanned electronic readout. The AMA has of $M \times N$ pixels. The labels i and j are used to label rows and columns respectively. The amount of charge stored on each pixel is determined by the amount of incident radiation and the conversion efficiency of the X-ray photoconductor. The image is obtained by self-scanning. Rows are scanned sequentially from $i = 1$ to M and the parallel data ($j = 1$ to N) on the pixels of an activated row are multiplexed and converted to digital information. For example, when the scanning control addresses row i, all the pixels $(i, 1)$, ..., (i, j), ..., (i, N) are read as parallel data into a multiplexer. Next, row $(i + 1)$ is read so on until all the rows to M are scanned.

thin film transistors (TFTs) act as switches to control the clocking out of image charge a line at a time. Very large area (e.g. 30 cm $\times$ 30 cm) active matrix arrays are now becoming available and even larger ones will be possible in the future. The AMA consists of $M \times N$ (e.g. $1,280 \times 1,536$) TFT based pixels and each pixel (i, j) carries a charge collection electrode B connected to a signal storage capacitor C_{ij} whose charge can be read through properly addressing the TFT (i, j) via the gate (i) and source (j) lines. An external readout electronics and software, by proper self-scanning, converts the charges read on each C_{ij} to a digital image as explained below. The term self-scanning here refers to the fact that no external means, such as a scanning laser beam as in some other digital X-ray imaging systems, is used to scan the pixels and extract the information. The scanning operation is part of the flat panel detector electronics and its software.

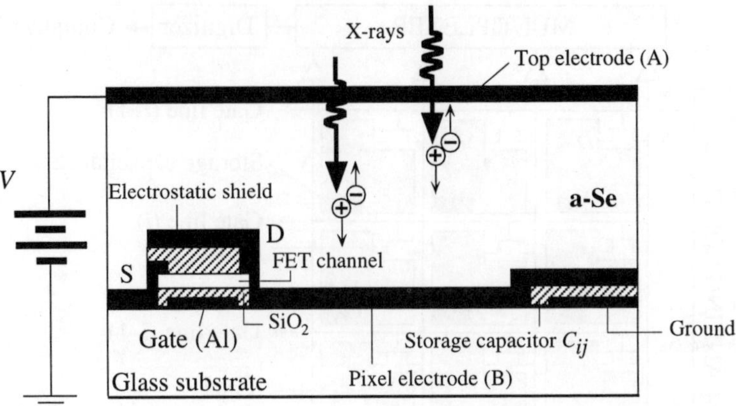

Fig. 4. Cross section of a single pixel (i, j) with a TFT showing the accumulation of X-ray generated charge on the pixel electrode and hence storage capacitor (exaggerated scale). The pixel is typically 150 μm $\times$150 μm in size. The photoconductive layer is vacuum coated onto a flat panel active matrix array and carries a top electrode for biasing and applying an electric field. Its thickness is typically 200–500 μm depending on the imaging application. Under X-ray irradiation, each pixel carries a charge proportional to the incident radiation because the a–Se photoconductor in that region release charge in proportion to the energy deposited in the photoconductor above each pixel.

The a–Se layer is coated onto the active matrix array to serve as an X-ray photoconductor as shown in Fig. 4. An electrode (labeled A) is subsequently deposited on the a–Se layer to enable the application of a biasing potential and hence an electric field in the a–Se layer. The electron hole pairs that are generated in the photoconductor by the absorption of an X-ray photon travel *along* the field lines. Electrons are collected by the positive bias electrode (A) and holes accumulate on the storage capacitor C_{ij} and thereby provide a charge-signal q_{ij} that can be read during self-scanning.

Each pixel electrode carries an amount of charge that is proportional to the amount of incident X-ray radiation by virtue of the X-ray photoconductivity of the photoconductor over that pixel. All TFTs in a row have their gates connected whereas all TFTs in a column have their sources connected. When gate line i is activated, all TFTs in that row are turned "on" and N data lines from $j = 1$ to N then read the charges on the pixel electrodes in row i. The parallel data are multiplexed into serial data, digitized, then fed into a computer for imaging. The scanning control then activates the next row, $i + 1$, and all the pixel charges in this row are then read and multiplexed so

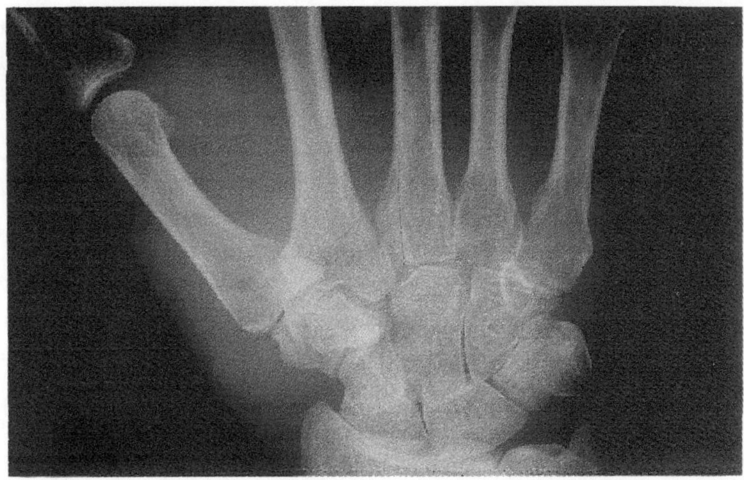

Fig. 5. An X-ray image obtained using a flat panel a–Se based detector (obtained by Rowlands and coworkers at Sunnybrook Health Science Centre, University of Toronto). The pixel size is ~ 160 μm.

on until the whole matrix has been read from the first to the last row (Mth row). It is apparent that the charge distribution residing on the panel's pixels are simply read out by self scanning the arrays row by row and multiplexing the parallel columns to a serial digital signal. This signal is then transmitted to a computer system. Figure 5 shows a typical X-ray image of a phantom wrist obtained by a flat panel X-ray image detector developed by Rowlands and coworkers at the Sunnybrook Health Science Center [33].

The high resolution and high sensitivity makes this system a leading contender in high resolution, low dosage digital radiography as recently analyzed and discussed in the literature [13, 14, 16–18, 33]. The resolution is determined by the pixel size which in present experimental image detectors is typically ~ 150 μm but is expected to be as small as 50 μm in future detectors to achieve the resolution necessary for mammography. Furthermore, the flat panel detector also has potential for use in real-time interactive X-ray imaging called *fluoroscopy*. An interesting feature of the a–Se based flat panel X-ray sensor is that this technology has been made possible by the use of two key elemental amorphous semiconductors: a–Si:H and a–Se. Although their properties are different, both can be readily prepared in large areas which is essential for an X-ray image detector. It will be impractically difficult and expensive to develop a large area detector using a single crystal technology.

3. X-Ray Photoconductors

Any candidate material for use as an X-ray photoconductor should possesses the following material properties:

(a) The photoconductor should have as high intrinsic X-ray sensitivity as possible, that is be able to generate as many EHPs as possible per unit of incident radiation. This means that the amount of radiation energy required, denoted as $W_\pm$, to create a single free electron hole pair (EHP) must be as low as possible because the charge ΔQ generated from an incident radiation of energy ΔE is simply $e\Delta E/W_\pm$. Experiments investigating the charge generated by high energy particles and X-rays have shown that this charge depends linearly on the incident radiation energy, i.e. $\Delta Q \propto \Delta E$. This has lead to the introduction of an EHP creation energy $W_\pm$ in semiconductors, i.e. $\Delta Q = \Delta E/W_\pm$. The EHP creation energy is a material property which increases with the bandgap energy E_g. High X-ray sensitivity implies a lower $W_\pm$ and hence a lower E_g.

(b) Nearly all the incident X-ray radiation should be absorbed within a practical photoconductor thickness to avoid unnecessary exposure of the patient. The absorption depth of the X-rays is determined by the X-ray absorption coefficient α which is highly energy dependent. It also depends on the atomic number Z of the material, α being larger for higher Z ($\alpha \propto Z^n$, where $n \approx 3$–5). The low Z is the primary reason for inexpensive organic semiconductors being excluded as X-ray photoconductor candidates. The minimization of dosage requires α to be such that the most of the radiation is absorbed within L, or $L > 1/\alpha$.

(c) There should be negligible dark current. This means the contacts to the photoconductor should be non-injecting and the rate of thermal generation of carriers from various defects or states in the bandgap should be small (i.e. dark conductivity is practically zero). We should note that typically low conductivity is associated with a wider band gap E_g which conflicts with the requirement in (a).

(d) There should be no bulk recombination of electrons and holes as they drift to the collection electrodes; EHPs are generated in the bulk of the photoconductor.

(e) There should be no deep trapping of EHPs. This means that both electrons and holes should have long *schubwegs* compared with the thickness (L) of the detector; otherwise charge trapping will prevent the carriers from reaching the electrodes. The schubweg is the mean distance traversed by a carrier before it is trapped and is given by $\mu\tau E$ where μ is the drift mobility, τ is the lifetime (trapping time) and E is the field. We need $\mu\tau E \gg L$. From (b), it is apparent that there is an important interplay and an inevitable compromise between α, L and the carrier schubwegs. For example, at an operating field of 5 V/μm, the schubwegs in device quality a–Se are typically in the range 1.2 mm–12 mm for holes and 300 μm–1500 μm for electrons. The absorption depth ($1/\alpha$) at 20 keV photon energy is about 50 μm but it is 1000 μm at 60 keV. It is clear that the operation in the high energy region requires a partial compromise between full absorption and schubweg limited sensitivity. It is also apparent that the quality control of the electron transport in a–Se is more critical than the hole transport.

(f) The longest carrier transit time, which depends on the smallest drift mobility (that of the electrons), must be shorter than the access time of the pixel and inter-frame time in fluoroscopy.

(g) The above should not change or deteriorate with time and as a consequence of repeated exposure to X-rays, i.e. *X-ray fatigue* and *X-ray damage* should be negligible.

(h) The photoconductor should be easily coated onto the AMA panel, for example, by conventional vacuum techniques. Special processes are generally more expensive.

The large area coating requirement in (h) over areas typically 30 cm × 30 cm, or greater, rules out the use of X-ray sensitive crystalline semiconductors which are difficult to grow in such large areas. Various polycrystalline semiconductors such as $Zn_x Cd_{1-x} Te$, PbI_2 have the feasibility to be prepared in large areas but their main drawback is the adverse affect of grain boundaries in limiting charge transport and, further, the high substrate and annealing temperatures required to optimize the semiconductor properties. Organic photoconductors that currently dominate the xerographic photoreceptor industry [7, 34] (in which they replaced a–Se, a–As$_2$Se$_3$), and can be cheaply prepared in large areas are useless because they do not satisfy (a) and (b). On the other hand, amorphous semiconductors such as a–Se, a–As$_2$Se$_3$ and a–Si:H are routinely prepared in large areas for such applications as xerographic photoreceptors and solar cells and are therefore well suited for flat panel X-ray detector applications. Amongst the three, a–Se is particularly well poised because it has a much greater X-ray absorption coefficient than a–Si:H, due to the greater Z (atomic number), and it possesses good charge transport properties for both holes and electrons compared with a–As$_2$Se$_3$ in which electrons become trapped and hole mobility is much smaller. Further, the dark current in a–Se is much smaller than that in a–As$_2$Se$_3$.

Due to its commercial use as a xerographic photoreceptor, a–Se is one of the most highly developed photoconductors [1, 2, 5]. It can be easily coated as thick films (e.g. 100–500 μm) onto suitable substrates by conventional vacuum deposition techniques and without the need to raise the substrate temperature beyond 60°–70°C. A large area detector (e.g. at least 24 cm × 18 cm for mammography) is essential in radiography since the lack of a practical means to focus X-rays necessitates a shadow X-ray image which is larger than the body part to be imaged.

4. Stabilized a–Se

Because of its xerographic importance in the sixties and seventies, amorphous form of selenium is probably one of the most widely studied elemental amorphous semiconductors as recently reviewed by one of the present authors [2]. The density of states (DOS) diagram for a–Se is the key to understanding its properties. This is shown in Fig. 6 and it is derived by combining results from a variety of experiments such as the time-of-flight drift mobility and xerographic cycled-up residual voltage measurements [2, 35]. Both holes and electrons are mobile in a–Se and both are *believed* to be shallow trap controlled as originally proposed by Spear [36, 37]. In other words, the effective electron drift mobility μ is the mobility μ_o in the extended states just above E_c reduced by the trapping and release events due to the presence of shallow traps at 0.35 eV below E_c, that is,

$$\mu = \frac{\tau_c}{\tau_c + \tau_r}\mu_o \tag{1}$$

where τ_c is the capture time and τ_r is the release (detrapping) time. Similarly, hole transport is controlled by shallow traps at 0.28 eV above E_v. Even though the exact nature of the shallow traps has not been established, both hole and electron drift mobilities are remarkably well defined and reproducible.

The deep localized states near the midgap have been associated with structural defects, such as the positively charged over coordinated center (Se$_3^+$) and

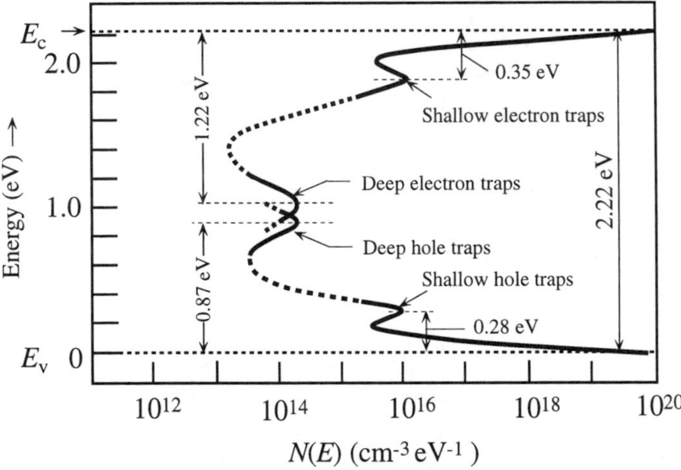

Fig. 6. Density of states for a–Se. The states between E_v and E_c are localized states. Data combined from transient photoconductivity, xerographic residual potential and other measurements (after Abkowitz in Ref. [35]).

the negatively charged undercoordinated center (Se_1^-), and are thought to be responsible for hole and electron trapping, that is the hole and electron lifetimes (deep trapping times). Experiments carried out at Xerox during the eighties have shown that these deep-states are thermodynamically derived, that is they are equilibrium defects [38, 39], and therefore cannot be simply eliminated from a–Se by various purification or preparation methods. They control the carrier lifetimes, or trapping times, and thus determine the carrier schubwegs and hence the photoconductor performance.

An important development in understanding the structure of chalcogenides towards explaining many of their optical and electronic properties was the absence of dangling bonds and the thermodynamic presence of what are called valence alternation pair (VAP) type defects [40–42]. In the case of a–Se, these defects correspond to a positively charged triply bonded center, Se_3^+, and a singly bonded, negatively charged center, Se_1^-, as illustrated in Fig. 7. The introduction of impurities into a–Se provides additional sources of structural VAP defects and thereby shifts the balance between Se_3^+ and Se_1^-. It has been therefore possible to control the charge transport parameters of a–Se by suitably alloying a–Se with other elements. After much research, again primarily at Xerox 5 [43–45], it was found that alloying a–Se with small amounts

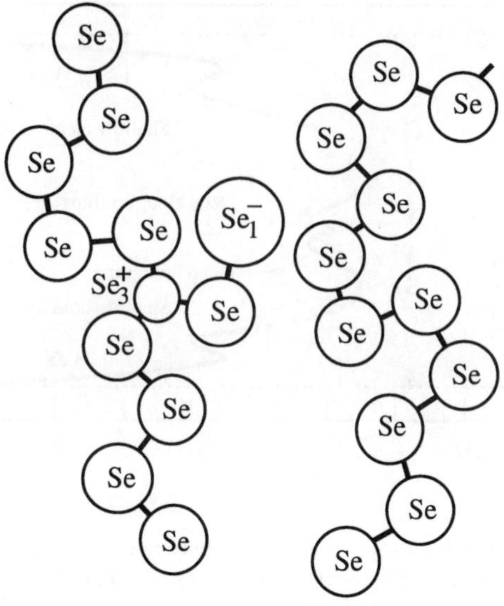

Fig. 7. The structure of a–Se has Se chains with under and overcoordinated charged defects, $Se_1{}^-$ and $Se_3{}^+$, called valence alternation pairs (VAPs). These VAPs are believed to be responsible for deep hole and electron traps in a–Se. A small amount of arsenic (0.2–1.0%) is added to a–Se to link some of the chains thereby increasing the viscosity and hindering crystallization.

of As (0.2%–0.5%As) and doping it with a few ppm (part per million) halogen (e.g. Cl) has two advantages. First, most As atoms (valency of III) in the structure are triply bonded and link Se chains. The resulting increase in the viscosity is sufficient to prevent the crystallization of a–Se which, left on its own, tends to crystallize over time (months to years depending on the ambient conditions and temperature). Secondly, the As and Cl in those amounts provide the right balance of VAP charged defects to result in good hole and electron transport. The compensation effects of As and Cl in the a–Se structure are still of topical interest and the fundamental issues have not been fully resolved [2, 46, 52]. The X-ray photoconductor that is used in the flat panel detectors is what is called *stabilized* a–Se, a–Se:(0.2–0.5)%As doped with 10–40 ppm Cl and quoted to have the nominal composition a–Se:0.3%As + 20 ppm Cl. Its optical and electrical properties are similar to pure a–Se. Table 2 summarizes the charge transport properties of stabilized a–Se. Charge carrier drift mobilities have been measured by conventional time-of-flight (TOF) transient photoconductivity experiments whereas the trapping times (lifetimes) have been obtained

Table 2. Stabilized a–Se (a–Se:0.2%–0.5%As + 10–40 ppm Cl) photoreceptor films.

Property	Typical range	Schubweg at $E = 5\,\mathrm{V}\,\mu\mathrm{m}^{-1}$	Comment
Hole mobility [2, 51–53]	0.12–0.14 cm^2 V^{-1} s^{-1}		Highly reproducible. Shallow trap controlled.
Electron mobility [2, 50–53]	0.003–0.006 cm^2 V^{-1} s^{-1}		Decreases rapidly with As addition [54]. Shallow trap controlled.
Hole lifetime [55]	20–200 μs	1.2–12 mm	Depends on the substrate temperature [56].
Electron lifetime [57, 58]	200–1000 μs	300–1500 μm	Sensitive to small quantity of impurities.

by the interrupted field time-of-flight (IFTOF) technique [47]. The charge carriers lifetimes measured from latter experiments have been found to agree with those obtained independently from the measurements of the first xerographic residual potential based on the Warter expression ($V_{\mathrm{residual}} = L^2/2\mu\tau$) for the residual potential [48]. The lifetimes reported in Table 2 represent the deep trapping of carriers into localized states located around the Fermi level in Fig. 6 from which the release times are hundreds of seconds for holes and hundreds of minutes for electrons.

Inasmuch as both electron and hole transport involve multiple trapping, the electron and hole schubwegs are relatively temperature insensitive even though both the drift mobility and the lifetime individually are thermally activated. For example, if τ_o is the trapping time in the absence and τ is the lifetime in the presence of shallow traps, then

$$\mu\tau = \mu_o\tau_o = \mu_o \frac{1}{S_t\, N_t} \tag{2}$$

where S_t is the capture coefficient and N_t the concentration of deep traps. For extended state transport as in a–Se, typically μ_o and S_t have weak algebraic temperature dependences so that $\mu\tau$ is not thermally activated [49, 50].

The wide band gap of 2.22 eV and the small concentration of localized gap states compared with other amorphous semiconductors means that the dark current in a–Se is relatively small. Under blocking contacts, it is the thermal generation of charge carriers from these midgap states that frequently controls the dark current in amorphous semiconductors. The electrical contacts to many chalcogenides semiconductors are generally ohmic and the current-voltage characteristics either represent bulk conduction (that is the material conductivity) or space charge limited conduction. Amorphous Se seems to be an exception in that the disagreement in the reported behavior of electrical contacts is as wide as there are papers on the subject. As a xerographic photoreceptor there was no technological need to understand contact effects to a–Se as the top surface was free for corona charging. Further, as a–Se was deposited on to oxidized Al substrates (copier drums) and the substrate contact was "reasonably blocking," that is, it did not allow electrons to enter the photoconductor and reach and discharge the surface potential. Understanding the physics of metal contact effects to a–Se remains one of the main current challenges in improving a–Se for flat panel detectors. The dark current and hence the detective quantum efficiency (DQE) depends on the metal/a–Se contacts.

5. X-Ray Sensitivity: The EHP Creation Energy

The creation of EHPs by an incident energetic particle or an X-ray photon first involves the generation of an energetic primary electron from an inner core shell, for example, the K-shell. As this energetic photoelectron travels in the solid it causes ionizations along its track and hence the creation of many EHPs. For many semiconductors the energy $W_\pm$ required to create and electron hole pair has been shown to depend on the energy bandgap E_g via the Klein rule [59, 60] $W_\pm \approx 2.8\,E_g + E_{\text{phonon}}$. The phonon energy term E_{phonon} is expected to be small (~ 0.5 eV) so that typically $W_\pm$ is close to $2.8\,E_g$. Further in many crystalline semiconductors, just like the optical quantum efficiency, $W_\pm$ is field independent and well-defined. This $W_\pm$ is so well-defined in crystalline semiconductors, such as high purity Ge for example, that they are used in spectrometers to measure the energy of the incident X-rays or high energy charged particles. Recently, Que and Rowlands [61] argued that if one relaxes the conservation of $\mathbf{k}$ rule in amorphous semiconductors than the EHP creation energy should be about $2.2\,E_g + E_{\text{phonon}}$. The field dependence of $W_\pm$ then arises from the recombination mechanism operating for the EHPs generated by the primary electron. The lowest or saturated $W_\pm$, denoted as $W_\pm^o$, at the highest fields should be $2.2\,E_g + E_{\text{phonon}}$. With $E_g \approx 2.2$ eV for a–Se, we would expect $W^o_\pm \approx 5$ eV. The situation for a–Se, as in the case for other low mobility solids, has proven to be difficult to understand. The measured $W_\pm$ shows a strong field dependence (just like the optical quantum efficiency) as shown in Fig. 8 as $W_\pm$ vs. $1/E$. The saturated $W_\pm$, i.e. $W^o_\pm$ (or the lowest $W_\pm$) has been only estimated by extrapolation to high fields [62] but seems to be in the range 5–6 eV as indicated in Fig. 8.

There are various reasons for the field dependence of the EHP creation energy. The primary electron generates many EHPs but only a certain fraction of these are collected because some disappear by recombination and some become trapped as they drift across the photoconductor. If we assume that there are

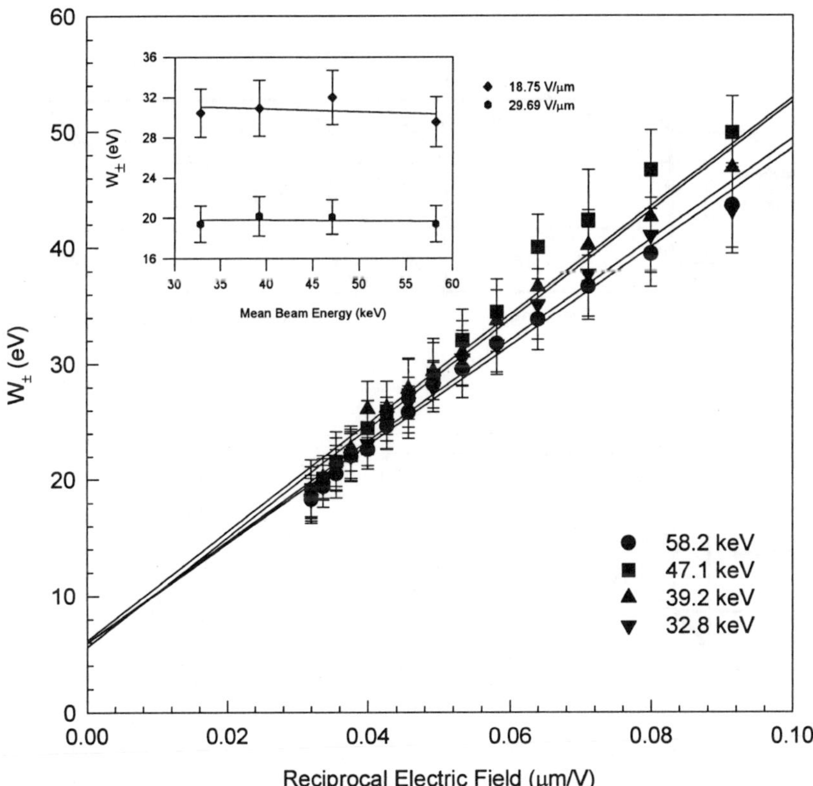

Fig. 8. EHP creation energy vs. reciprocal field at three different energies. The lines are extrapolated to $1/E = 0$ to estimate $W^o_\pm$ which is in the range 5–6 eV.

practically no carriers lost due to trapping, as will be the case for device-quality photoconductor material, then the recombination losses can be attributed to three sources as illustrated in Fig. 9. First, is simple bulk recombination, or bimolecular recombination, between drifting electrons and holes. The recombination rate is proportional to both the hole and electron concentrations so that the collected charge does not increase linearly with the intensity of the radiation. In fact, it depends on the square root of the intensity. Experiments, however, show that ΔQ increases linearly with intensity which rules out this type of recombination.

Second source is *geminate recombination* so named for gemini "the twins". In this case the electron and hole twin generated at the same time are attracted to each other by their mutual Coulombic force and may eventually

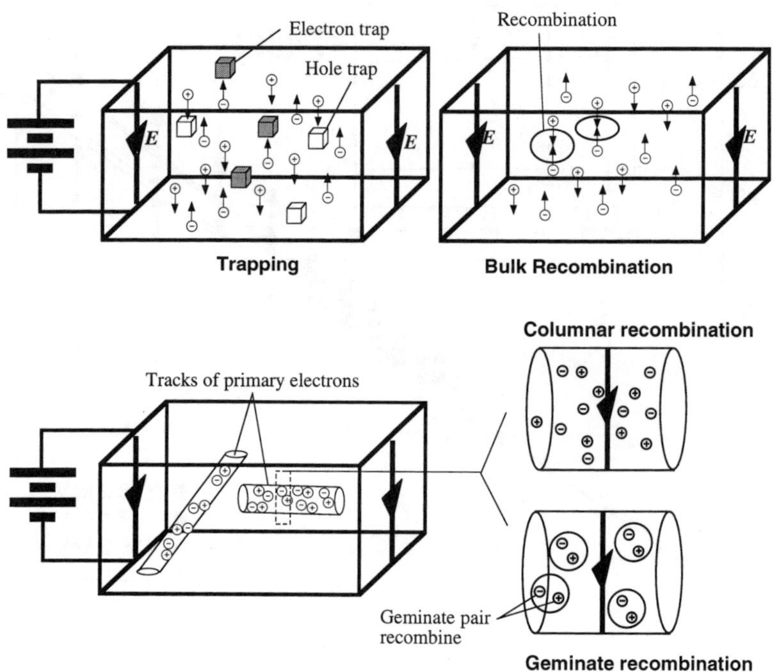

Fig. 9. Four possible mechanisms that reduce the number of collected (free) electron hole pairs and hence reduce the X-ray sensitivity. (a) Trapping of electrons and/or holes. (b) Bulk recombination. (c) Columnar recombination. This is bimolecular recombination within primary electron tracks. (d) Geminate recombination is the recombination of an EHP created at the same time and bound by their mutual Coulombic attraction.

recombine, hence the term geminate recombination. This is the basic model for the optical quantum efficiency where the number of EHPs collected, that is escape geminate recombination, is supposed to be governed by the Onsager model [63].

The third recombination loss mechanism is called *columnar recombination* which involves the recombination of non-geminate electrons and holes in the columnar track of a primary electron. As the intensity of the radiation is increased, the number of tracks is also proportionally increased, but the recombination within a track remains unaffected. This means that the collected charge increases linearly with the intensity in agreement with observations.

The question then comes down to whether it is the Onsager or columnar recombination that controls the field dependence and this has not been resolved. $W_\pm$ decreases with increasing field and at a high field of 30×10^6 V m^{-1}, $W_\pm$ is about 18 eV for a mean photon energy of 40 keV. The extrapolated value at infinite field is about 4–6 eV (Fig. 8) assuming that a reciprocal field extrapolation is palatable. If we take $E_g \approx 2.2$ eV, then 2.2 E_g and 2.8 E_g rules expect 5.4–6.7 eV. It is of interest to mention that the electron bombardment induced transient conductivity measurements by Hirsch and Jahankhani have indicated an EHP creation energy that is ~ 18 eV and an EHP yield that follows columnar recombination predictions [64]. Further experiments, particularly on the energy and temperature dependence of $W_\pm$, are needed to understand the origin of the X-ray sensitivity of a–Se and in similar amorphous semiconductors. The controversy between Onsager and columnar recombination interpretations for low mobility solids is not new [65, 66]. In the case of SiO$_2$, for example, early work was interpreted using the columnar recombination [67, 68] but later work was interpreted in terms of Onsager recombination [69]. Table 3 summarizes the bandgap E_g and the EHP creation energy for crystalline and non-crystalline semiconductors of interest and compares the values with 2.8 E_g and 2.2 E_g rules. The phonon energy E_{phonon} term has been taken as ~ 0.5 eV.

6. TV Pick-up Tubes and Avalanche Multiplication

A well-known and successful application of amorphous Se–As–Te alloys is in Hitachi's Saticon [25, 26], which is a commercially available TV pickup tube. Figure 10 displays the structure of a typical a–Se Saticon, which utilizes the high panchromatic photosensitivity of the Se–Te alloy and the relatively fast

Table 3. Bandgap (E_g) and EHP creation energy ($W_\pm$) in a–Se, a–Si:H and crystalline Si and comparisons with the theoretically expected values.

Material	E_g (eV)	2.8 E_g+ 0.5 (eV)	2.2 E_g+ 0.5 (eV)	$W_\pm$ (eV)	Comment
a–Se	2.22	6.7	5.4	5–6 (Fig. 8)	Extrapolated to $E = \infty$; 5–6 eV seems to agree with the 2.2 E_g rule.
				18 (Hirsch and Jahankhani [64])	
a–Si:H	~ 1.7	5.3	4.2	3.4–4.4 (Dubeau et al. [70, 71])	Agrees with the 2.2 E_g rule.
Crystalline Si	1.1	3.6	2.9	3.63	Agrees with the 2.8 E_g rule like many crystalline semiconductors [72].

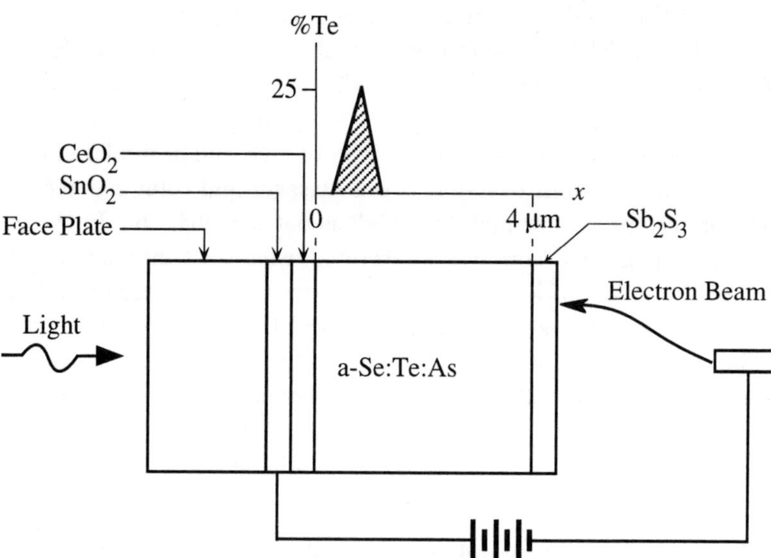

Fig. 10. Schematic diagram of a Saticon TV image pickup tube utilizing an a–Se:Te:As alloy. The Te composition is soncentrated in a narrow region, whereas the bulk is mainly a–Se:As. (after Maruyama in Ref. [26].)

hole drift mobility of a–Se. Layers of CeO_2 and Sb_2S_3 act as hole and electron blocking contacts, respectively. Electrons injected by the scanning electron beam become trapped in the Sb_2S_3 layer, forming a negative space charge in this layer. Photogenerated holes in a–Se, from exposure to the image, transit across toward Sb_2S_3 to recombine with the electrons trapped in Sb_2S_3. The photogeneration is contained in a region of high Te concentration. The crystallization of a–Se is inhibited by the addition of $\sim 1\%$ As to a–Se. Inasmuch as Saticon uses an amorphous material (i.e. grainless and uniform material), it exhibits high resolution.

Recently Tanioka and coworkers [11, 12] have developed a super-sensitive photoconductive target called the HARP (acronym for "high-gain avalanche

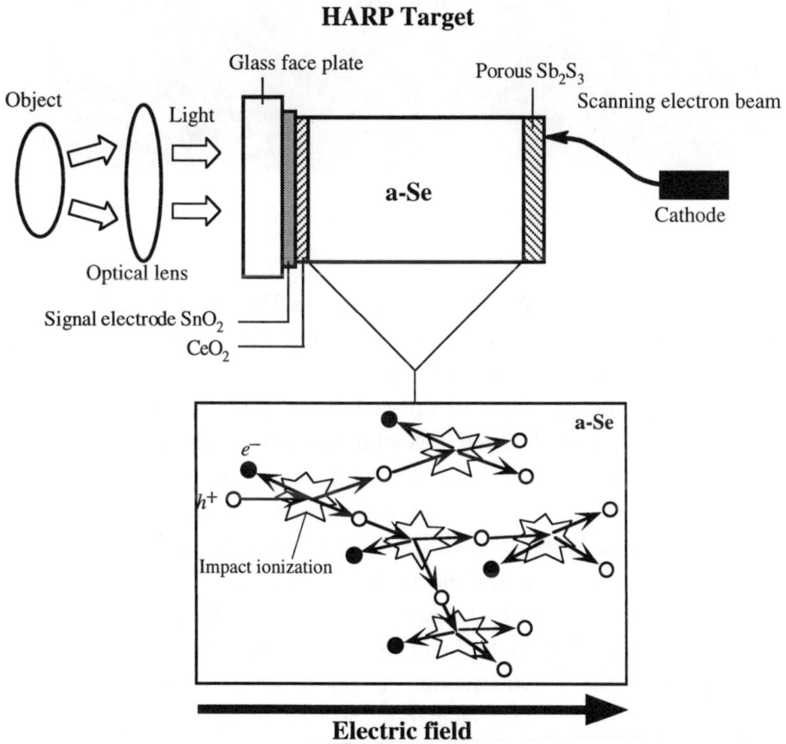

Fig. 11. A schematic illustration of a HARPICON. Avalanche breakdown by impact ionization in the a–Se layer is also illustrated. One photogenerated hole results in many electron hole pairs to be generated by impact ionization (avalanche breakdown) in the a–Se layer where the electric field is very large.

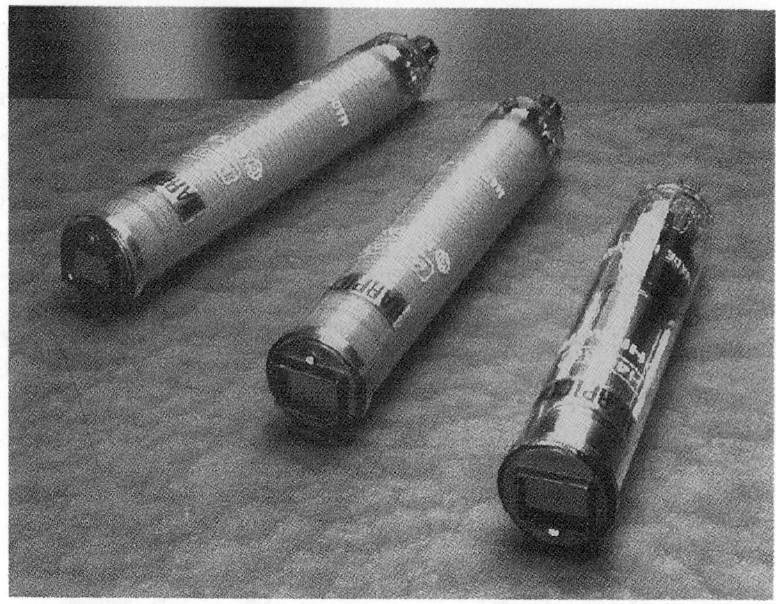

Fig. 12. HARPICON vidicon tubes (courtesy of Dr. Tanioka, NHK).

rushing amorphous photoconductor"), for use in HDTV (high definition television) camera pickup tubes. The vidicon using the HARP target has been called the HARPICON or a–Se avalanche vidicon. The basic structure of the HARP target and the principle of operation are schematically illustrated in Fig. 11. A photograph of HARPICON vidicon tubes is shown in Fig. 12. The entire target is typically about ~ 2 μm thick though it may be thicker in ultrasensitive targets. The transparent signal electrode (SnO_2) is biased positively with respect to the cathode. The CeO_2 and SbS_3 layers, as in the Saticon, act as blocking contacts for hole and electron injection respectively. The incident light from the object is absorbed mainly in the a–Se layer. The electron hole pairs photogenerated in the a–Se layer are then drifted by the applied electric field and constitute the signal current. As photogenerated holes drift through the a–Se layer towards the back electrode, as a result of the large applied electric field (greater than 8×10^5 V/cm or 80 V/μm), they experience avalanche multiplication and hence yield an effective quantum gain greater than unity (quantum efficiency is the number of EHPs generated per absorbed photon and the quantum gain is the number of EHPs collected per absorbed photon).

The effective quantum gain resulting from avalanche multiplication depends on the field as well as the photoconductor thickness. For example, in the 2 μm thick HARP target the quantum gain is about 10 at a field of 120 V/μm, whereas the gain is about $\sim$ 1000 in an a–Se target of thickness 24.8 μm at a field of 100 V/μm operating at wavelength of 400 nm. Figure 13 shows the dependence of the multiplication factor M on the electric field and also shows the rise in the dark current with the field. We should note that at $M = 1000$ (at $E = 100$ V/μm), the dark current is still only $\sim$ 5 nA. TV pickup tubes using such HARP targets clearly have far superior sensitivity than conventional TV pickup tubes and accordingly constitute ultra-high sensitive image pickup tubes (indeed, HAPICONS have been able to capture even star light images). The need for high gain vidicons in HDTV is due to the use of small aperture lenses with much greater depth of field than in conventional TV which allow realistic large screen effects.

It is apparent that such an improvement in gain could be useful in radiographic applications. Thus, if such avalanche gains could also be achieved in thick a–Se X-ray photoconductors, then the corresponding effective $W_\pm$ would only be a fraction of an electron volt and could translate to an improvement in

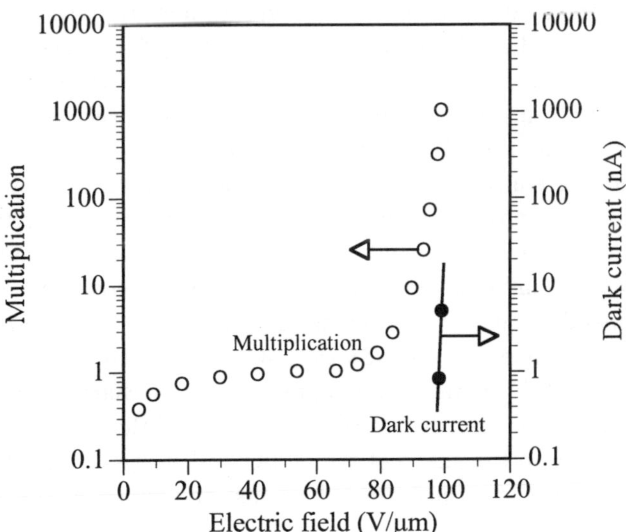

Fig. 13. Multiplication and dark current vs. electric field in a HARP target. a–Se layer is 24.8 μm thick and the illumination wavelength is 400 nm.

image quality especially in very low exposure techniques such as fluoroscopy. In fact, in the last few years, NHK has produced [73] an X-ray HARP tube that is claimed to have super-sensitivity, though the target area size so far is only 1 inch. The potential for a supersensitive large area flat panel detector that is usable both in projection radiology and fluoroscopy, however, seems feasible and achievable in the near future. Even though the avalanche effect in a–Se has enabled the super-sensitive HARP, its physics in amorphous semiconductors remains poorly understood and debated.

7. Current Problems and the Future

Although stabilized a–Se promises to be an excellent X-ray photoconductor, there are still a number of unresolved theoretical and practical issues. As noted above, there is very limited data on the temperature and energy dependence of the EHP creation energy $W_\pm$ given that this quantity directly determines the overall efficiency of the detector. Extended measurements will not only be able to answer the scientific curiosity of what limits $W_\pm$ (Onsager vs. columnar recombination) but also allow a better device design.

As apparent in Fig. 8, $W_\pm$ decreases with the electric field so that there is an incentive to operate the detector at high fields to improve the efficiency. Depending on the mode of X-ray application (mean energy), the photoconductor thickness has to be 100–500 μm to absorb most of the incident radiation. This requires large bias voltages e.g. greater than 1000 V which creates its own technological problems e.g. supplying these voltages, dielectric breakdown in the panel, protecting the TFTs in the flat panel etc.

The nature of contacts at A and B in Fig. 4 are extremely important as ideally the dark current should be zero (as for any ideal photodetector). Although during the sixties and seventies, many researchers have reported I–V measurements on a–Se films there has been no general conclusion on the behavior of metal/a–Se contacts and the resulting I–V characteristics. Most researchers have claimed to observe steady state dark currents and interpreted the observed dark I–V characteristics in terms of space charge limited currents (SCLC) and even obtained the energy distribution of the deep traps (or deep localized states) from the shape of the SCLC I–V curve. In nearly all cases, the models were not rigorously tested against the expected scaling law for SCLC i.e.

$$\frac{J}{L} = f\left(\frac{V}{L^2}\right) \qquad (3)$$

where J is the current density, V is the applied voltage, L is the sample thickness and f is any function that represents the SCLC I–V curve shape. Time-of-flight (TOF) transient photoconductivity experiments have been routinely and successfully performed by many researchers in the past on metal/ a–Se/metal structures which assume that the field inside the sample is uniform, $E = V/L$. This experimental fact is in direct contradiction with the observation of SCLCs inasmuch as the very assumption of SCLC means the field is non-uniform, e.g. E vanishes at the positive contact if the injected carriers are holes. More evidence against SCLCs comes from several sources. Pfister and Lakatos [74] found that SCLC could only be obtained by intensely illuminating one or both of the contacts with strongly absorbed ($\lambda = 3990$ Å) light. The illuminated contact is thereby made an injecting contact and resulting steady state I–V characteristics are then indeed SCL and follow Child's law ($J \sim V^2/L^3$ for one carrier injection and $J \sim V^3/L^5$ for two carrier injection). Müller and Müller [75] reported that the dark I–V characteristics of Au/a–Se/Cu devices are Schottky emission limited with experimental Schottky coefficients that are within 10%–20% of the theoretically expected value.

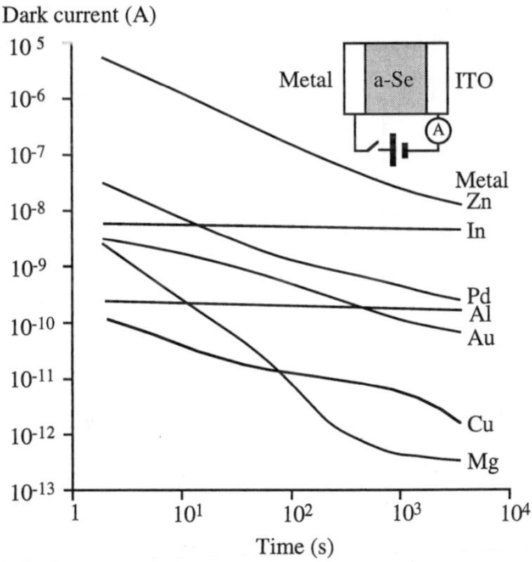

Fig. 14. Typical dark current vs. time characteristics immediately after the application of a voltage of 2000 V to a metal/stabilized a–Se/ITO devices. The metal electrode is positive with respect to ITO and the a–Se thickness is 500 μm. The electrode area is 0.5 cm², except for the Pd electrode which is 1 cm².

Given the present use of a–Se in various electroded imaging device applications, in contrast to its earlier applications as xerographic photoreceptors, there is now a technological need for systematic studies of I–V characteristics of various metal/a–Se/metal structures and the nature of metal contacts to a–Se. Most past papers on I–V characteristics do not specify the quality of the a–Se material used, i.e. there is no information on the charge carrier mobilities and lifetimes to gauge whether the results are representative of a device quality material.

Some of the ongoing recent work carried out on stabilized a–Se at the authors laboratories [76] indicates that the I–V characteristics depend on the nature of the metal/a–Se contact and do not follow well-established models (such as Schottky emission). The dark current immediately after the application of a bias voltage to a metal/a–Se/ITO (indium tin oxide coated glass substrate) device has been observed to decay with time (in a non-exponential manner), with decay characteristics that depend on the type of metal, as illustrated in Fig. 14. Further, as shown in Fig. 15, it has not been possible

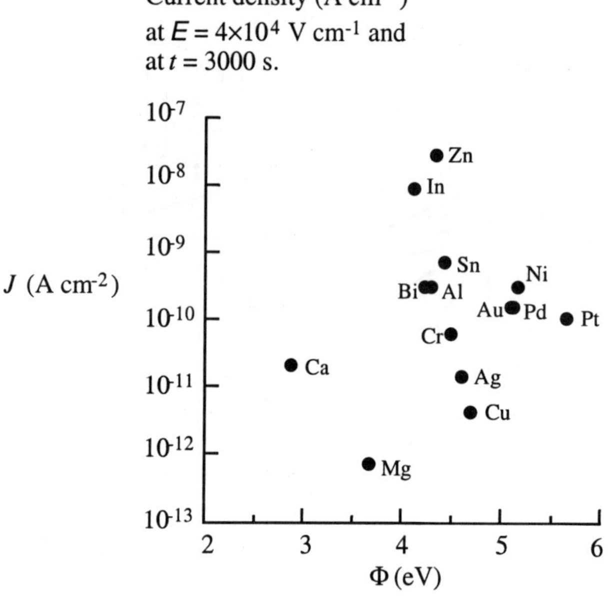

Fig. 15. The relationship between the dark current density at $E = 4 \times 10^4$ V cm^{-1} and at time 3000 seconds on the electrode metal work function Φ.

to simply correlate the dark current to the metal work function, which would be expected in the case of Schottky type of metal/a–Se contacts. Again, further work is needed to arrive at reproducible conclusions that are of scientific benefit to X-ray photoconductor design.

Despite the past and present commercialization of stabilized a–Se as an X-ray photoconductor, there is very little reported work on the observed effects of X-ray fatigue and X-ray damage. Over extended irradiation the alloys exhibit a temporary deterioration in the X-ray sensitivity [3]. Some initial limited work by Kasap and coworkers [77, 78] on stabilized a–Se photoreceptor films showed that X-ray irradiation induces defects in a–Se in the form of deep hole traps. The concentration of these defects depended on the X-ray exposure and that the recovery towards undamaged state was highly temperature dependent as shown in Fig. 16.

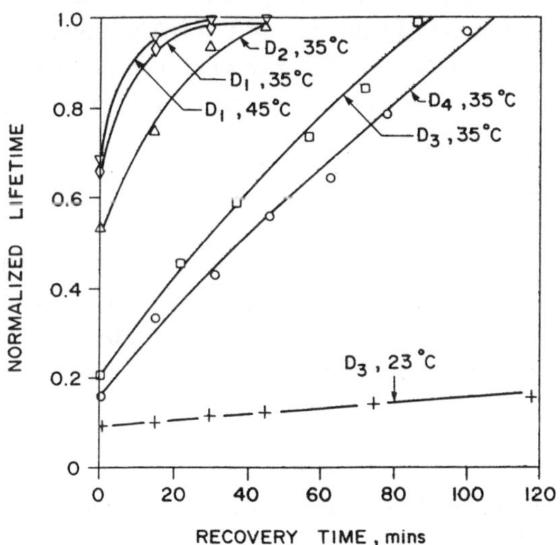

Fig. 16. Normalized hole lifetime vs. recovery time (measured immediately after exposure to X-rays) at different temperatures and under different amounts of X-ray pre-exposure. Normalized lifetime is the lifetime after exposure to lifetime before exposure to X-rays ("undamaged state"). The lifetime was obtained from xerographic first residual potential measurements. D represents the X-ray exposure where $D_1 = 0.1$ R, $D_2 = 0.47$ R, $D_2 = 0.95$ R and $D_4 = 2.44$ R. These exposures are much greater than typical exposures used in X-ray imaging (typically $10^{-3} - 10^{-2}$ R) and therefore likely to represent accumulated exposure where the photoconductor is exposed to a series of X-rays. Further, there is no applied electric field during exposure. (from Kasap *et al.* in Ref. [77].)

There is no doubt that the central reason for using a–Se as an X-ray photoconductor is that it can be readily coated on to a large substrate by simple vacuum deposition at a fast rate (e.g. 2–4 μm per min) with substrate temperatures which do not damage the electronics of the flat panel. The question remains whether a–Se can be replaced by another X-ray photoconductor. Some initial research towards new large area X-ray photoconductors has been encouraging but not conclusive. For example, PbI_2 photoconductive layers have shown good sensitivity but only in small areas of about 5 cm × 5 cm. Others are still in the laboratory research stage in small area detectors. Organic photoconductors obviously cannot be used on their own but it may be possible to use a composite material based on dispersing X-ray absorbing particles in an organic binder as recently reported in Science [79]. One of the best options at present seems to improve the design of a–Se photoconductor structures so that the devices can be operated at high fields with low dark currents. One possibility is to use various additional layers between the metal electrodes and the a–Se photoconductor as in the case of the HARP device.

Further reviews on a–Se based X-ray photoconductors may be found in Refs. [80, 81] which also include discussions on the proposed new density of states diagram for a–Se [82].

Acknowledgments

Authors are grateful to Natural Sciences and Engineering Research Council of Canada and Anrad (Noranda Advanced Materials, Montreal, Quebec) for supporting their X-ray photoconductor research program. We gracefully thank our past and present research fellows and graduate students for contributing to this project.

References

[1] J. Mort, *The Anatomy of Xerography* (McFairland and Co., London, UK, 1989), and references therein.

[2] S.O. Kasap, Chapter 8, Photoreceptors: The Chalcogenides, The Handbook of Imaging Materials, ed. A. Diamond (Marcel Dekker, New York, 1991), and references therein, p. 329.

[3] J.W. Boag, *Phys. Med. Biol.* **18**, 3 (1973), and references therein.

[4] A.G. Leiga, *Proc. 4th Int. Symp. on Uses of Selenium and Tellurium*, Selenium-Tellurium Development Association (STDA Inc., Grimbergen, Belgium, 1990) pp. 249–256 and references therein.

[5] R.M. Schaffert, Electrophotography (Focal Press, London, 1975), and references therein.

[6] D.M. Burland and L.B. Schein, *Physics Today* **39**, 46 (1986).

[7] D.M. Pai and B.E. Springett, *Rev. Mod. Phys.* **65**, 163 (1993), and references therein.

[8] U. Neitzel, I. Maack and S. Guenther-Kohlfahl, *Med. Phys.* **21**, 509 (1994).

[9] P.J. Papin and H.K. Huang, *Med. Phys.* **14**, 322 (1987).

[10] L.S. Jeromin, L.M. Klynn, *J. Appl. Photog. Eng.* **5**, 183 (1979).

[11] K. Tsuji, T. Ohshima, T. Hirai, N. Gotoh, K. Tanioka and K. Shidara, *Mater. Res. Soc. Symp. Proc.* **219**, 505 (1991).

[12] K. Tanioka, J. Yamazaki, K. Shidara, K. Taketoshi, T. Hirai and T.Y. Takasaki, *Adv. Electron. Electron Phys.* **74**, 379 (1988).

[13] D.L. Lee, L.K. Cheung and L. Jeromin, *SPIE Proc.* **2432**, 237 (1995).

[14] D.L. Lee, L.K. Cheung, E.F. Palecki and L. Jeromin, *SPIE Proc.* **2708**, 511 (1996).

[15] Personal communication; Continuing research work at Thermotrex, San Diego, California, USA.

[16] W. Zhao, J.A. Rowlands, S. Germann, D. Waechter and Z. Huang, *SPIE Proc.* **2432**, 250(1995).

[17] W. Zhao and J.A. Rowlands, *Med. Phys.* **22**, 1595 (1995).

[18] J. A. Rowlands, W. Zhao, I. Blevis, D. Hunt, S.O. Kasap, D.F. Waechter and Z. Huang, *SPIE Proc.* **3032**, 97 (1997).

[19] J.A. Rowlands and S.O. Kasap, *Physics Today* **50**, 24 (November 1997).

[20] W. Zhao, I.M. Blevis, S. Germann, J.A. Rowlands, D.F. Waechter and Z. Huang, *Med. Phys.* **24**, 1834 (1997).

[21] R. Luhta and J.A. Rowlands *Proc.Conf. Photoelectric Image Devices*, London, 1991.

[22] L. Cheung, G.M.T. Foley, P. Fournia and B.E. Springett, *Photogr. Sci. Eng.* **26**, 245 (1982).

[23] J.A. Rowlands, D.M. Hunter and N. Araj, *Med. Phys.* **18**, 421 (1991).

[24] W. Hillen, U. Schiebel and T. Zaengel, *Medical Imaging II*, eds. R.H. Schneider and S.J. Dwyer, *SPIE Proc.* **914**, 253 (1988).

[25] N. Goto, Y. Isozaki, K. Shidara, E. Maruyama, T. Hirai and T. Fujita, *IEEE Trans. Electron. Dev.* **ED2**, 662 (1974).

[26] E. Maruyama, E., *Japan J. Appl. Phys.* **21**, 213 (1982).

[27] M.J. Yaffe and J.A. Rowlands, *Phys. Med. Biol.* **42**, 1 (1997).

[28] US Philips Corporation, USA, 1995, U.S. Patent No. 5,396,072 (May 7, 1995), X-Ray Image Detector.

[29] DuPont de Nemours and Company, Wilmington. Del., USA., 1994, U.S. Patent No. 5,319, 206 (June 7, 1994), Method and Apparatus for Acquiring an X-Ray Image Using a Solid State Device.

[30] K. Suzuki, *Amorphous and Microcrystalline Semiconductor Devices: Optoelectronic Devices*, ed. J. Kanicki (Artech House, Boston, 1991) Chap. 3.

[31] L.E. Antonuk, J. Boudry, W. Wang, D. McShan, E.J. Morton, J. Yorkston and R.A. Street, *Med. Phys.* **19**, 1455 (1992).

[32] M. Hoheisel, M. Arques, J. Charbal, C. Chaussant, T. Ducourant, L. Fritsch, G. Hahm, H. Horbaschek, J. Michailos, R. Schulz, M. Soahn, V. Spinnier and G. Vieux, *Future Directions in Thin Film Science and Technology*, eds. J.M. Marshall, N. Kirov, A. Vavrek and J.M. Maud (World Scientific, Singapore, 1997) p. 112, and references therein.

[33] J.A. Rowlands, W. Zhao, I. Blevis, D. Waechter and Z. Huang, *RadioGraphics* **17**, 753 (1997).

[34] B.E. Springett, *Proc. 5th Int. Symp. on Uses of Selenium and Tellurium*, Selenium-Tellurium Development Association (STDA Inc., Grimbergen, Belgium, 1994) p. 187.

[35] M.A. Abkowitz, *Phil. Mag. Lett.* **58**, 53 (1988).

[36] W.E. Spear, *Proc. Phys. Soc. (London)* **B70**, 669 (1957).

[37] W.E. Spear, *Proc. Phys. Soc. (London)* **B76**, 826 (1960).

[38] M. A. Abkowitz, Disordered Semiconductors, eds. M.A. Kastner, G.A. Thomas and S.R. Ovshinsky (Plenum Publishing Co. New York, 1987) p. 205, and references therein.

[39] M. Abkowitz, *J. Non-Cryst. Solids* **66**, 315 (1984).

[40] D. Adler, *Sci. Am.* **236**, 36 (1977).

[41] S.R. Elliott, *Physics of Amorphous Solids* (Longman, London, UK, 1992).

[42] A Feltz, *Amorphous Inorganic Materials and Glasses* (VCH, Weinhein, Germany 1993) Chap. 2.

[43] J.C. Schottmiller, M. Tabak, G. Lucovsky and A. Ward, *J. Non-Cryst. Solids* **4**, 80 (1970).

[44] J.C. Schottmiller, *J. Vac. Sci. Technol.* **12**, 807 (1975).

[45] J.S. Berkes, J.S., *Electrophotography: 2nd Int. Conf.* ed. D. R. White, Society of Photographic Scientists and Engineers (Springfield, VA, 1974) p. 137.

[46] D.M. Pai, *J. Imag. Sci. Technol.* **41**, 135 (1997).

[47] S.O Kasap, B. Polischuk and D. Dodds, *Rev. Sci. Instrum.* **61**, 2081 (1990).

[48] S.O Kasap, V. Aiyah, B. Polischuk, A. Bhattarcharyya and Z. Liang, *Phys. Rev.* **B43**, 6691 (1991).

[49] D. Pai, *Amorphous and Liquid Semiconductors*, eds. J. Stuke and W. Brenig (Taylor and Francis, London, 1974) p. 355.

[50] S.O. Kasap and C. Juhasz, *Photogr. Sci. Eng.* **26**, 239 (1982).

[51] S.O. Kasap and C. Juhasz, *J. Phys. D: Appl. Phys.* **18**, 703 (1985).

[52] C. Juhasz and S.O. Kasap, *J. Phys. D: Appl. Phys.* **18**, 721 (1985).

[53] C. Juhasz, M. Vaezi-Nejad and S.O. Kasap, *J. Imag. Sci.* **29**, 144 (1985).

[54] J.M. Marshall, F.D. Fisher and E.A. Owen, *Phys. Stat. Solidi* **25**, 419 (1974).

[55] S.O. Kasap and B. Polischuk, *Canada J. Phys.* **73**, 96 (1995).

[56] S.O. Kasap and B. Polishuk, *Proc. 4th Int. Symp. on Uses of Selenium and Tellurium* (STDA Inc., Grimbergen, Belgium, 1989).

[57] S.O. Kasap, V. Aiyah, B. Polischuk, Z. Liang and A. Bekirov, *J. Non-Cryst. Solids* **137** & **138**, 1329 (1991).

[58] M.D. Tabak and W.J. Hillegas, *J. Vac. Sci. Technol.* **9**, 387 (1972).

[59] C.A. Klein, *J. Appl. Phys.* **39**, 3476 (1968).

[60] R.C. Alig and S. Bloom, *Phys. Rev. Letts.* **35**, 1522 (1975).

[61] S.O. Kasap, M. Nesdoly, C. Haugen and J.A. Rowlands, *J. Non-Cryst. Solids*, 2000 (in press).

[62] S.O. Kasap, V. Aiyah, B. Polischuk and A. Baillie, *J. Appl. Phys.* **83**, 2879 (1998).

[63] D.M. Pai and R.C. Enck, *Phys. Rev.* **B11**, 5163 (1975).

[64] J. Hirsch and H. Jahankhani, *J. Phys: Cond. Matter* **1**, 8789 (1989).

[65] R.C. Hughes, *J. Chem. Phys.* **55**, 5442 (1971).

[66] R.C. Hughes, *Conf. on Electrical Insulation and Dielectric Phenomena*, Annual Report (Washington Academy of Sciences, Washington DC, 1971) p. 8.

[67] G.A. Ausman Jr. and F.B. McLean, *Appl. Phys. Letts.* **26**, 173 (1975).

[68] D.M. Taylor and A.A. Al-Jassar, *J. Phys. D: Appl. Physics* **14**, 1531 (1981).

[69] D.M. Taylor and A.A. Al-Jassar, *J. Phys. D: Appl. Physics* **17**, 819 (1984).

[70] J. Dubeau, T. Pochet, L.A. Hamel, B. Equer, A. Karar, *Nuclear Instruments and Methods in Physics Research* **B54**, 458 (1991).

[71] L.A. Hamel, J. Dubeau, T. Pochet and B. Equer, *IEEE Trans. Nucl. Sci.* **38**, 251 (1991).

[72] G.F. Knoll, *Radiation Detection and Measurement* (John Wiley and Sons, New York, 1989).

[73] K. Tanioka, private communication.

[74] G. Pfister and A.I. Lakatos, *Phys. Rev.* **B6**, 3012 (1972).

[75] L. Müller and M. Müller, *J. Non-Cryst. Solids* **4**, 504 (1970).

[76] R.E. Johanson, S.O. Kasap, B. Polischuk and J.A. Rowlands, *J. Non-Cryst. Solids* **227**, 1359 (1998).

[77] S.O. Kasap, V. Aiyah, A. Baillie and A. Leiga, *J. Appl. Phys.* **69**, 7087 (1991).

[78] B. Polischuk, S.O. Kasap and V. Aiyah, *Int. J. Electronics* **76**, 1029 (1994).

[79] Y. Wang and N. Herron, *Science* **273**, 632 (1996).

[80] S.O. Kasap and J.A. Rowlands, *J. Mater. Sci. Mater. Elec.* (in press).

[81] S.O. Kasap and J.A. Rowlands, Selected Topics in Optoelectronics and Photonics: Principles and Practices, CDROM (Prentice Hall, Upper Saddle River, 2000) (in press).

[82] H.Z. Song, G.J. Adriaenssens, E.V. Emeliova and V.I. Arkhipov, *Phys. Rev.* **B59**, 10607 (1999).

C.   **REAL TIME OPTICAL RECORDING
ON THIN FILMS OF AMORPHOUS
SEMICONDUCTORS**

MARIA MITKOVA

*Central Laboratory of Electrochemical Power Sources,
Bulgarian Academy of Sciences, Acad. G. Bonchev str. Bl.10,
1113 Sofia, Bulgaria
selen@bas.bg*

Contents

1. Introduction

Visual impressions have always been very important to mankind. To have them
stored was the driving force for developing art (painting and sculpture) and

science (optical storage and information processing). Starting with the first paintings on the rocks, humans went through a long evolution till they discovered the present-day modern methods for image recording.

Conventional photography, based on the photochemical processes in the silver halides, was the first serious step in the optical recording process and many years it was the most important method for optical imaging as well as for data storage [1, 2]. The fact that one can obtain a very great magnification of the optical signal due to the chemical developing process, made it for a long time the best way for optical recording and caused its great recent progress concerning the resolution and the color processing. If we use for a quantitative description of the light induced reactions the quantum efficiency:

$$\eta_q = \frac{N}{N_0} \tag{1}$$

where N is the number of product particles of a given photoinduced reaction (e.g. the electrons in photoelectron emission) and N_0 is the number of the absorbed photons, the limit $\eta_q \leq 1$ usually holds only for the primary light induced process in silver halide photomaterials — the primary reaction of latent image formation, but in the subsequent chemical processing, the photoreaction (i.e. silver atom generation) is amplified by 10^6 to 10^9 times. The amplification is reciprocally dependent on the resolution of the emulsion and is proportional to the sensitivity of the material. However even with such a high efficiency the silver halide emulsions and the photographic process based on them are not able to obey all high issues that the modern optical computing and telecommunication require. Particularly they are not suitable for creating of real time optical recording.

In particular, the new era of optical recording started first with the discovery of the color centers and the photoinduced reactions in alkali halides, which gave rise to the great promotion of solid state photochemistry, and later with the discovery of the laser that offered the possibility to create a real time and three dimensional optical storage.

There are two main problems that have to be solved, when one wants to optimize the optical recording processes:

- to find materials that can result in recording with a higher speed, resolution and efficiency.
- to establish more advanced and highly efficient optical recording technologies.

Table 1. Principles of optical recording processes and related materials.

Recording Material Class or Process	Elementary Reaction	Example	Year, Reference
Inorganic photochromic materials	(1) Charge transfer (2) Orientation of dipols (3) Aggregation of defects	CaF_2 − Eu, Sm KCl-F_2 $F \rightarrow nF{=}X$ (NaCl, KCl etc.)	1965 [3] 1967 [4] 1974 [5]
Photorefractive electro-optic materials	Charge transfer with a following intrinsic electro-optic effect	$LiNbO_3$, $LiNbO_3$−Fe $LiTaO_3$ etc.	1968 [6]
Amorphous semiconductors	(1) Charge transfer and medium range order change (2) Phase transitions	As_2S_3 films SbSeIn, SbIn etc.	1971 [7] 80th [8]
Magneto-optic materials	Photoinduced thermal phase transitions in an external magnetic field	MnBi, CdTbFe etc.	1968 [9]
Organic materials (dyes, monomers etc.)	(1) Electron transfer (2) Proton transfer (3) Isomeric reactions (4) Polymerization	Bacteriorodopsin etc.	1985 [10, 11]
Hole burning recording (solids with zero phonon lines)	(1) Orientation of dipols (2) Charge transfer	H_2 − phtalocyanidin in n-octan matrix	1975 [12]
Time-domain holography	Laser pulse excitation during the lifetime of the excited state	2,3- dihidroporphirin in polyvinylbutyral matrix	1983 [13, 14]
Photoinduced luminescence	Charge transfer (recording) and recombination luminescence (readout)	CaS (doped)	1989 [15]
Laser assisted chemical vapor deposition (LCVD)	Computer controlled laser induced evaporation and deposition	3-D structure creation in Al, polymers etc.	90th [16, 17]

Table 1 gives an overview of how this was done during the second half of the 20th Century [1]. As one can see, an optimization of the optical recording process can be achieved by applying real time optical recording on thin films of amorphous semiconductors.

This paper will review some data about the real time optical recording on the most applied amorphous semiconductors. Their properties as optical storage medium are shown in Sec. 2. In Sec. 3, the basic principles of the real-time optical recording are discussed in terms of digital and holographic recording. The processes are illustrated with certain data of some particular systems. Some future trends are outlined together with the new trends in the application of these methods and materials in Sec. 4.

2. Amorphous Semiconductors as Optical Storage Medium

Developing of new materials is one of the key problems imposed to answer to the high requirements of the recording processes. Among the amorphous semiconductors most famous finding application in this area are chalcogenide glasses. They have been extensively studied in the recent decades partly because of their interesting basic properties and partly because of the great variety they promise on the application's side particularly in optical imaging, optical recording, infrared optics and in optical communications [18]. These glasses are named after the glass-former that they contain, namely, one or more of the elements from the 6 main group of the Periodic table — S, Se and Te. These elements can be combined with many others from the third to the seventh main groups of the Periodic table to form a wide variety of glass-forming systems [19], ranging from simple binary combinations such as the extensively studied As–S, As–Se and Ge–Se glasses to more complex multi-component systems, in which also metals e.g. Ag [20] or rare earths [21] can be included. Most of the examples in this review will be given just on such type of materials as at least seven effects are known to occur in these glasses [22, 23] ranging from relatively subtle effects involving minor atomic rearrangements [24], to more substantial atomic and molecular reconfigurations such as metal photodissolution [25], metal surface photodeposition [26, 27] and photo-crystallization [28, 29]. All that causes marked changes in the physical and chemical properties of the glasses.

It is generally assumed that the main reason for all these features in the chalcogenide glasses is the existence of lone-pair electrons in them forming different kinds of bonding defects. The first model to describe these defects was created by Mott, Davis and Street (the so called MDS model) [30]. It has been proved successfully in explaining externally induced phenomena in chalcogenides such as photoinduced processes, thermal processes and luminescence. According to MDS defects have three charged states — D^o, D^+ and D^-, depending on the local atomic configuration. The negatively charged defect D^- is a dangling bond associated with an under-coordinated atom, for example a chalcogen (say Se) bonded to *one* other atom or a pnictide (say As) bonded to *two* other atoms. When an electron is removed from the dangling bond (forming D^0), it is assumed that there is an attraction of the atom in question towards a fully coordinated neighboring chalcogen atom, one of the lone-pair electrons on the latter being used to form a bonding orbital and the other an antibonding orbital. When a second electron is removed, for then both lone-pair electrons from the neighboring chains are used in bonding and the former singly coordinated chalcogen becomes essentially three-fold coordinated — forming the D^+ center. D^o centers are unstable and react exothermically in the charge transfer process:

$$2D^o \rightarrow D^+ + D^- \tag{2}$$

A similar model was proposed by Kastner, Adler and Fritzsche (KAF) [31]. They introduced the term *valence alteration pairs* (VAP's) indicating some unusual bonding configurations occurring due to specific interactions between non-bonding orbitals. These configurations result for example when two chalcogenide atoms, each twofold coordinated in its ground state, forms instead one positively charged three-fold coordinated atom and one negatively charged singly coordinated atom. The creation of such a VAP requires a relatively small energy, so that the density of VAP's in most chalcogenide glasses is relatively large. This model is somewhat more flexible since it includes more possible combinations of the charge states and degrees of coordination. It is especially favorable in binary or more complex amorphous semiconductors as it allows one to describe atoms other than chalcogenides. In general, accordingly to it all atoms are designated by $C^z{}_i$ where z is the charge of the atom and i — the coordination number. Thus a normal Se atom i.e. neutral and doubly coordinated is designated as $C^0{}_2$ while the positively and negatively charged defects are designed respectively as $C_3{}^+$ and $C_1{}^-$. Of course one can easily transform between the MDS and KAF models. This problem is discussed in depth in the chapter by H. Fritzsche in this book.

The chalcogenide glasses can be specifically influenced by the heat, generated through the non-radiative recombination of photoexcited carriers. The medium can be heated to a temperature of phase transition (transformation temperature T_g, melting temperature T_m, or temperature of evaporation T_{ev}) due to the absorbed light energy and participate in this way in photoinduced thermal phase transition. The first step of this type photoinduced structural rearrangement occurs in a change of the short and medium range order. As shown for the As–S system [32, 33] some decoupling of pieces of the As_2S_3 is generated that may lead to increased steric freedom. The final step is the photocrystallization [34, 35] that brings the material to the equilibrium state. These processes can be reversible and the material can be brought by the influence of light back to the initial state.

In general, stable and homogeneous samples of these glasses are straightforward to prepare and so they can be obtained in bulk form or deposited by variety of techniques, including vacuum evaporation or CVD variations. They can also be drawn as fibers.

A particular part of the family of the amorphous semiconductors are the chalco-halide glasses, containing both chalcogen and halogen elements. These glasses combine the chemical durability of the chalcogenides with the low loss transmittance of the halide glasses in the infrared spectrum. The inclusion of halogens into chalcogenide glasses shifts their IR window towards longer wavelengths. Because the chalco-halide glasses are also photosensitive, there is a possibility to create diffraction gratings on them [36]. As these glasses find very often application in a fiber form, the optical recording technologies can be applied also for obtaining optical elements on the fibers, e.g. for sensing, or at their ends, for reducing coupling losses. The ability to reproducibly write both short and long period gratings in optical fibers and waveguides has allowed grating technology to move rapidly from research into development and on to commercialization. So, the chalco-halide glasses are a quite promising candidate for application in telecommunications. Many details concerning these glasses are given in the review by J. Lucas in this book and I will show only some of the possibilities to create optical image on thin films.

The typical semiconducting elements from the 4 group of the periodic table — the amorphous germanium and silicon, as well as the hydrogenated silicon, give some promising results as photosensitive materials and will be briefly listed in this review. As predicted by the atomistic model of Fedders *et al.* [37], the nature of the light induced effects in a–Si correlates with the massive reorganization that occurs following an electronic transition in it. Shimakawa *et al.* [38] discovered a great similarity in the photoinduced changes in the chalcogenide glasses and in the elementary amorphous semiconductors such as Ge and Si. There are some recent results showing that amorphous Ge can be used successfully as phase change recording material because of its fast memory and good resolution [39].

A great part of digital recording is due to the effects occurring in Te thin films, based mainly on the transition: pure $Te–TeO_x$. These effects are caused again primarily due to the photinduced changes in the chalcogenide matrix.

Almost all possible photoinduced effects and metastabilities in amorphous semiconductors and insulators are profoundly reviewed in the work of Shimakawa *et al.* [40].

3. Principles and Results on Real-time Optical Recording

In principle the interaction of semiconductors and insulators with light that has photon energy comparable with that of the band gap, results in the creation of electron — hole pairs or excitons. There are three ways for the electron-hole pairs to act — they can separate and contribute in this way to the photoconductivity of the material, they can recombine radiatively giving rise to the photoluminescence of the material or they can recombine non-radiatively. The last action is the basis for optical recording, as it brings the material to a condition different from the previous one (this could be manifested in photodarkening or photobleaching of the material), creating, for example, new defects using the energy released in the non-radiative recombination event via a transient exciton, helping overcome any activation barriers to defect formation. In the amorphous materials where constrains imposed by the lettice periodicity do not exist, there are many pathways in which this can be realized and this is the reason for the great variety of the effects that can occur in them. The flexibility associated with the atomic structure resulting from the relaxation of crystallographic constraints contributes also to the variety of the defects formation and variations in the local bonding configuration that can be different before and after the act of illumination [41]. All these possible differences in the condition of the material prior and after the illumination give rise to a great number of photoinduced effects whose contemporary understanding is given in an extremely clear and summarized form by Fritzsche [42].

The effects that occur in the amorphous semiconductors (usually for the optical recording thin films are applied) after illumination with light can result in general, in two types of real time optical recording — digital and holographic, depending upon the recording conditions:

3.1. *Digital optical recording*

The main characteristic of this type of optical recording is that the recorded area of the photosensitive material results in virgin areas that, in principle, reproduce the "0" state and illuminated (usually by a laser beam) spots representing the "1" state. In general, every light induced change in the property of a material can be used to make an optical recording. However, for practical applications, usually a change in the light induced complex index of refraction N given by

$$N = n - ik \tag{3}$$

where n is the index of refraction and k — the absorption index, or in the optical path length $d_{opt} = dn$, is acceptable. On the whole, a change in the complex refractive index always leads to changes in both Δn and Δk. However, often in particular materials change, either Δn or Δk is larger. These changes in the optical properties of the materials can be caused in the process of digital optical recording by creating holes or bubbles or by phase transition of the material in the illuminated spot.

The digital recording can be evaluated by: spatial resolution, time resolution, signal to noise ratio and storage capacity.

The *spatial resolution* (d_{rec}) of the recording medium depends on the physicochemical structure and the homogeneity of the material. Usually, optical recording occurs by the interaction of light with micro-defects of solids whose size is much smaller than the wavelength [43]. However, the distribution of defects, the cluster structure of glasses and amorphous solids, etc. are factors limiting the spatial resolution of the recording material, which is related to the minimum size that can be focused by the optical system (lenses, objectives etc.) This size is determined by diffraction and is a function of the wavelength of the incident light, which is the limitation factor. There is a special requirement that the resolution of the photosensitive material has to be greater than the one of the optical system in order to create good recording.

The *time resolution* is the shortest possible time for creating recording. For the visible spectra, where $\Delta E \sim 2$ eV, it is of the order of $\Delta t = 3.10^{-16}$ s. In optical memory discs the recording time is much longer (about 100 ns) than the resolution time. This is caused by the nature of the photothermal recording and limits the data rate in optical memory discs up to 100 Mbt/s. The main limitations in a real recording system are related both to the relaxation time of the phase transition (which is of the order of the lattice vibration period) and on the heat transfer time in the illuminated spot (which usually is much longer). This characteristic is very important for the developing of the optical computing where very high recording rates are required.

The signal to noise ratio in the case of the commercial digital memory discs is 40–60 dB. It can be reached by a strong primary optical recording and a selective electronic detection system [44].

The maximum *storage density* in a two-dimensional medium is $1/\lambda^2 \approx$ 5.108 bit/cm^2 and in a three-dimensional medium $- 1/\lambda^3 \approx 1013$ bit/cm^3 for the wave length 450 nm [1]. It is limited by the diffraction reasons by which the minimal size of the recorded spot is $d_{min} \geq \lambda$, where λ is the recording or read out light.

3.1.1. *Digital optical recording due to ablation of the films by illumination with light*

As shown on Fig. 1(a), this means to create some kind of holes in the optically active film. In order the holes to open, the films must be melted locally by the recording laser; therefore a sensitive ablative recording medium should have a low melting point. This is the reason why materials, such as tellurium,

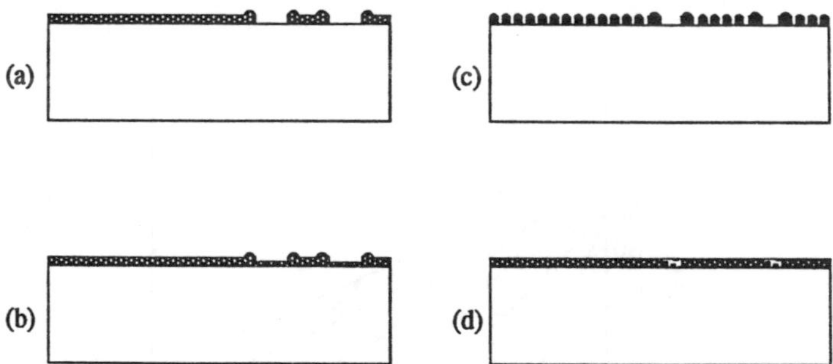

Fig. 1. Examples of digital real time optical recording in amorphous semiconductors and insulators [44]: (a) ablation of a thin tellurium film; (b) ablation in a bilayer system with a protective layer on the substrate; (c) coalescence of one island film; (d) crystallization of a metastable amorphous film.

selenium, bismuth and others received early attention [45] as write once digital recording materials. Recording requires much more than just the melted region because the continuous film is in a local energy minimum; a hole of some minimum size must be created by some other means before the surface tension forces can take over the dynamics and create a stable void. Merangoni forces (a surface-tension gradient due to the surface-tension-temperature gradient) could open the initial hole if the heating and cooling occur slowly. Chen *et al.* [46] confirmed these results showing that at lower laser power hole opening must be initiated by voids in the film, while at high laser power voids are not necessary for hole opening because of the kinetic reasons. They have received a contrast ratio up to 0.7 for 300 Å thick TeBi films on PMMA substrate at a laser power of 15 mW. For the high speed recording, actual vaporization of the film surface usually dominates the initial hole opening dynamics, hence the term ablative recording.

As mentioned above, the hole creation can be achieved due to evaporation or eventually sublimation of the illuminated part of the material as the chalcogen elements possess high vapor pressure and selenium, for example, tends to sublimate. This method has been used by Janai and Rudman [47], in order to create some image on thin films of amorphous As_2S_3. In this particular system they explain the effect of light enhanced vaporization of the material as a combination of photodecomposition and oxidation reactions caused by the light, and subsequent evaporation of the reaction products. Our investigations show that there is a good sublimation kinetics for some of the silver containing glasses — Fig. 2 and this makes them a suitable material for that kind of recording [48] as good contrast (up to 3.72 in reflection) is obtained.

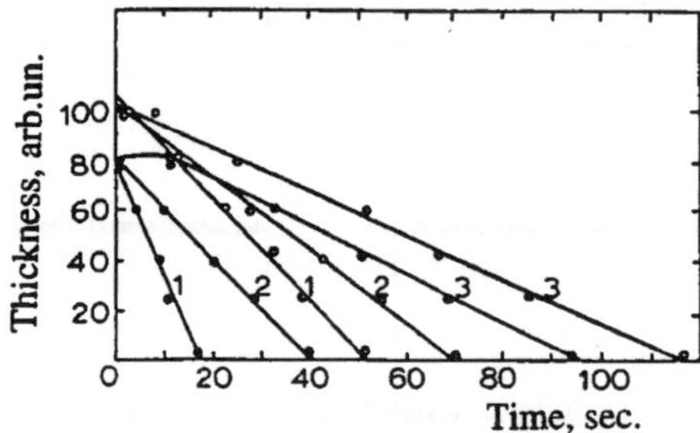

Fig. 2. Sublimation kinetics of thin films [48] $Se_{80}Te_{10}Ag_{10}$ (•) and $Se_{90}Te_5Ag_5$ (o) caused by illumination by a mercury lamp with different intensities starting with 85 mW/cm^2.

The image is visualized through higher transparency of the illuminated parts of the film (mainly due to material sublimation) and through darkening of the parts that have been covered by the non-transparent parts of the test pattern (as has been found mainly by crystallization of the active material caused by heating due to illumination with light).

The high signal to noise ratio (up to 60 dB) and the relatively low recording energy have made the ablative optical recording one of the most preferred for their wide application in everyday life.

3.1.2. *Digital optical recording due to coagulation of an island film*

The optical properties of a surface sometimes depend critically on its texture. This effect has been used for realization of digital optical recording on thin TeO_x films. As Te and TeO_2 do not mix and do not build up solid solutions [49], a film with composition TeO_x where $x = 1.1$ is actually built by a mixture of TeO_2 and crystalline or amorphous Te — Fig. 3(a). The optical recording in this system is in fact caused by the enlargement of the low dimensional colloid particles of Te. After the photoinduced reaction of recording (enlargement of the Te particles) relaxation in the dark occurs leading to a crystallization

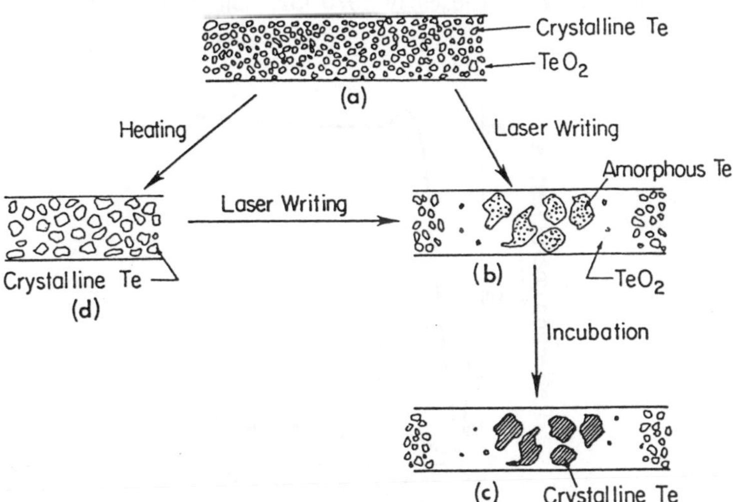

Fig. 3. Photo- and thermostimulated processes in the system $TeO_x = TeO_2 + Te$ [49] (a) fresh film with small dispersed particles; (b) larger particles after the action of light; (c) dark relaxation-transition of Te from amorphous to crystalline condition.

of the material. The electron microscopic data [50] confirm this model. This effect is successfully applied for optical recording by Mitsushita. However, the application of the TeO_x system is limited by the presence of slow dark reactions in the coagulated Te grains — Fig. 3(c). The coagulated amorphous Te particles are not stable and crystallize spontaneously with a relaxation time several seconds to several minutes.

3.1.3. *Digital optical recording due to phase transition*

Among the digital recording processes a special role has the one accompanied with a phase transition in the material caused by the light. A very impressive example is the phase change caused by light irradiation in SmS for which two stable phases — metallic (m) and semiconductor (s) exist [51]. The phase transition $s \Leftrightarrow m$ is reversible and correlates with large changes in the coefficient of reflection — Fig. 4. However, the use of SmS films as recording media is difficult because the transition reaction has very low rate even at high temperatures and the material has low light sensitivity (0.3 J/cm^2).

The reversible phase transition crystal $\Leftrightarrow$ amorphous phase gives reliable results, when one works in write — erase regime as the processes of crystallization $\Leftrightarrow$ amorphisation are well-defined. Research on laser-induced reversible phase transition started in the early 1970 [52, 53]. Figure 5 illustrates the

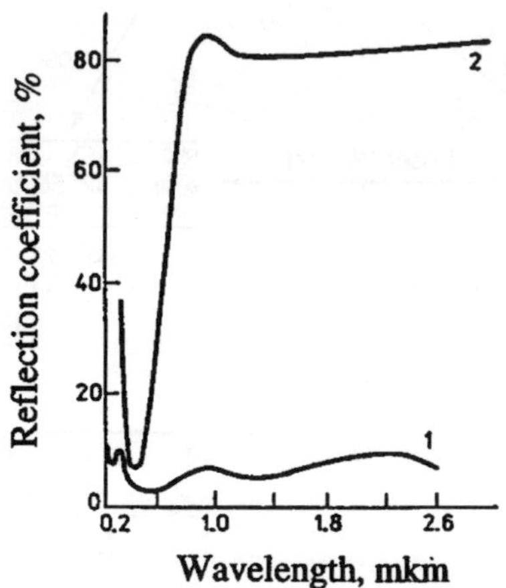

Fig. 4. Change in the reflectivity due to light induced phase transition from the semiconductor state — (1) to the metallic state (2) in SmS films [51].

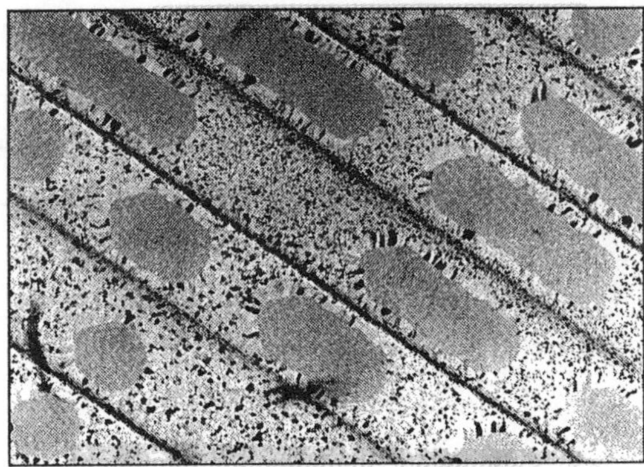

Fig.5. Transmission electron-microscope image of a Ge–Sb–Te alloy film [54]. Amorphous recording marks are formed in crystallized area along the tracks. The mark size is about 0.5 μm.

picture accompanying the reversible phase change optical storage. The main requirements towards these materials are: (1) Appropriate absorption edge — it should shift in the visible or near infrared wave length region with phase transitions; (2) Medium melting point — such that the most available lasers can cause crystallization of the material; (3) Rapid and stable phase transition process. These requirements suggest that most appropriate materials should be the semiconductors with melting points between 500°C and 1000°C with a large atomic mobility in the amorphous and supercooled states as well as a short atomic-diffusion distance from the atomic location in the amorphous state to the lattice sites of the crystalline state. In the last 20 years, the idea for the suitable material for these processes has evolved [54] as shown on Table 2.

Among the most popular chalcogenide glasses recently the photo-amorphization of crystalline $As_{50}Se_{50}$ films has been investigated [55] suggesting that the crystallization behavior of a–$As_{50}Se_{50}$ films is rather complex, depending on the annealing temperature, thickness of the film and the presence of a substrate. For the $GeSe_2$ films, it has been discovered that the electronic excitation by light is essential for the photo-induced crystallization process [56]. By more complicated chalcogenide systems like Te–Se–Br [57] the laser pulse irradiation causes almost complete evaporation of bromine. The Te–Se alloys formed after irradiation are amorphous or crystalline depending not on their composition but on the parameters of laser pulse length and power.

Table 2. Phase-change optical materials [54] — development in the material science.

Year	Compositions
1971	Te–Ge–Sb–S
1974	Te–Ge–As
1983	Te–Ge–Sn–O
1985	Te–Sn–Se, Te–Se–Ga
1986	Te–Ge–Sn–Au, Sb_2Se, In–Se, GeTe, Bi–Se–Sb, Pd–Te–Ge–Sn
1987	$GeTe–Sb_2Te_3$, $Ge_2Sb_2Te_5$, $GeSb_2Te_4$, In–Se–Tl–Co
1988	In–Sb–Te, In_3SbTe_2
1989	$GeTe–Sb_2Te_3$–Sb, Ge–Sb–Te–Pd, Ge–Sb–The–Co, $Sb_2Te_3–Bi_2Se_3$
1991	Ag–In–Sb–Te

In this case, the crystallization starts with surface filamentary growth exhibiting fractal network formation tending to coalescence at higher laser energy.

There are very encouraging results in short-pulse laser crystallization and structuring of a–Ge thin films [39] showing that the films crystallize at a laser pulse energy in the range 36–66 mJ/cm^2. Utilizing laser interference crystallization, selective crystallization can be caused by which Ge microstructures have been created (stripes and dots) indicating that explosive crystallization does not hinder the fabrication of μm-sized structure — Fig. 6.

Recently, the effect of anisotropic laser crystallization of a–Se [58] as well as polarization induced crystallization of chalco-halide glasses [59, 60] and polarization assisted crystallization in a–Se [61] have been reported. Polycrystalline films resulting after irradiation by polarized light beam display large linear dichroizm with the sign that depends on the direction of the electrical vector of light. At the same time irradiation by non-polarized light or by light with a periodically changing direction of the electrical vector leads to obtaining optically isotropic polycrystalline films. All these results suggest that there are great hopes for application as polarization depending crystallization offers one more degree of information density and can be of interest for the manufacture of binary phase gratings and optical Fourier processing.

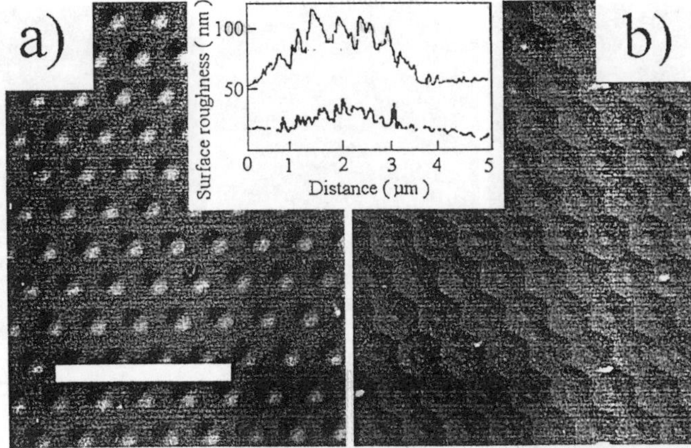

Fig. 6. Optical reflection micrograph of the laser interference crystallization dot array [39]: (a) before and (b) after thermal annealing. The white bar represents 20 µm. Insert: the top and bottom traces show the height profiles of the dots, measured by atomic force microscopy before and after thermal annealing, respectively.

The problem of phase-transition optical recording along with the production of commercial discs is profoundly considered in this book by S. Ovshinsky who did the pioneering work in discovering the great field of storage effects in the amorphous materials.

3.1.4. *Digital optical recording due to photoinduced surface deposition of metallic silver*

The effect of photoinduced surface photodeposition of silver has been profoundly investigated by Kavaguchi and Maruno [26, 27, 62–65]. The principles of this effect coincide with the fact that upon illumination with light the fine silver particles in silver containing glasses segregate and diffuse towards the surface of the films — Fig. 7. So, direct positive Ag patterning with high contrast has been attempted in different chalcogenide systems, such as Ge–Se–Ag, As–Se–Ag etc. which can be reversible written and erased, depending on the conditions of illumination [64]. The authors have investigated the nucleation processes in these systems as well as the influence of other dopants that can enhance the effect. They have found that analogous to the conventional photography, 0.037%–0.185% of gold increase the nuclear density by at least two orders of magnitude [26].

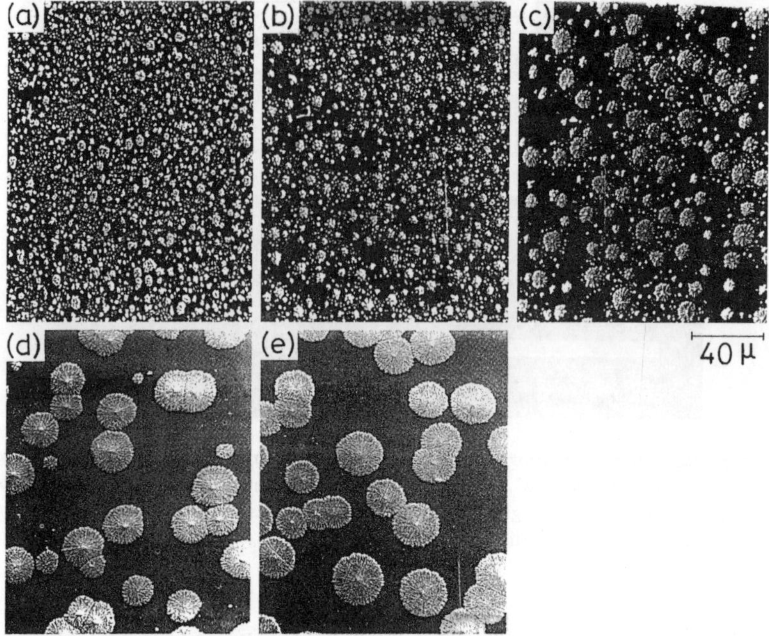

Fig. 7. Scanning electron microscopy photographs of $Ag_xAs_{60-x}S_{40}$ after illumination [27]: (a) $x = 35$; (b) 37.3; (c) 40; (d) 42.5; (e) 45. The illumination was provided with light from an ultrahigh-pressure Hg lamp through an IR-cut filter (100mW/cm^2) for 15 min at $20°$C.

3.2. *Principles of holographic recording*

Holographic imaging is a wave front reconstruction process, which allows one to obtain a real three-dimensional image of volume objects without lenses and objectives. For information processing, a multi-channel holographic recording and readout is possible which widely increases the field of application [66].

3.2.1. *Recording and readout*

Holography is a two step imaging process. The first step in producing a holographic image involves the illumination of the object with a coherent beam of light. The resulting scattered wave emanating from the object is allowed to interfere with the reference wave, whereby the resulting interference pattern is recorded on a photosensitive medium. In the case of the semiconducting amorphous films the material is real-time recording one and enables one to obtain the hologram immediately without any developing or fixing of the image.

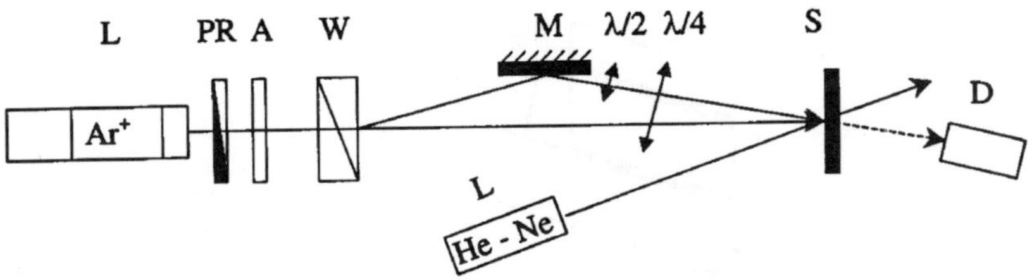

Fig. 8. Standard set up for recording of holographic gratings [67]. L_1 is an Ar$^+$ laser, PR — polarization rotator, A — gradual attenuator, W — Wolaston prism, M — mirror, L2 — He–Ne laser, $\lambda/2$ and $\lambda/4$ — phase plates, T — thermostat, D — photosensor.

The reading of the recorded image occurs by illumination of the hologram with the reference beam under the same angle of incidence as that used in the recording step. It may result applying light with different wavelength than the recording light. Normally this is preferred as that kind of light cannot produce any farther changes in the optical properties of the optical recording medium and thus, the image cannot be changed during the reading process.

Figure 8 shows a standard setup for recording a holographic gratings [67].

3.2.2. *Classification of the holograms*

If we follow the process of recording, the subject and the referred waves contact the photosensitive material, producing a diffraction pattern on it. Depending on the photosensitive material the grating could be produced transmitting the beams through it — Fig. 9(a) and (b) or reflecting from it — Fig. 9(c). Thus, one can produce *transmission or reflection holograms*. The period of the grating Λ is related to the angle of the incidence of the laser beam θ_S and θ_R as well as to the wave length of the laser beam light λ:

$$\Lambda = \lambda/(\sin\theta_S + \sin\theta_R) \tag{4}$$

The highest resolution that can be reached in a reflection hologram is

$$\Lambda = \lambda/2n \tag{5}$$

where n is the refractive index of the recording medium.

As mentioned in Sec. 3.1, any photoinduced effect is a result of change in the complex index of refraction N or the optical path length $d_{\mathrm{opt}} = dn$.

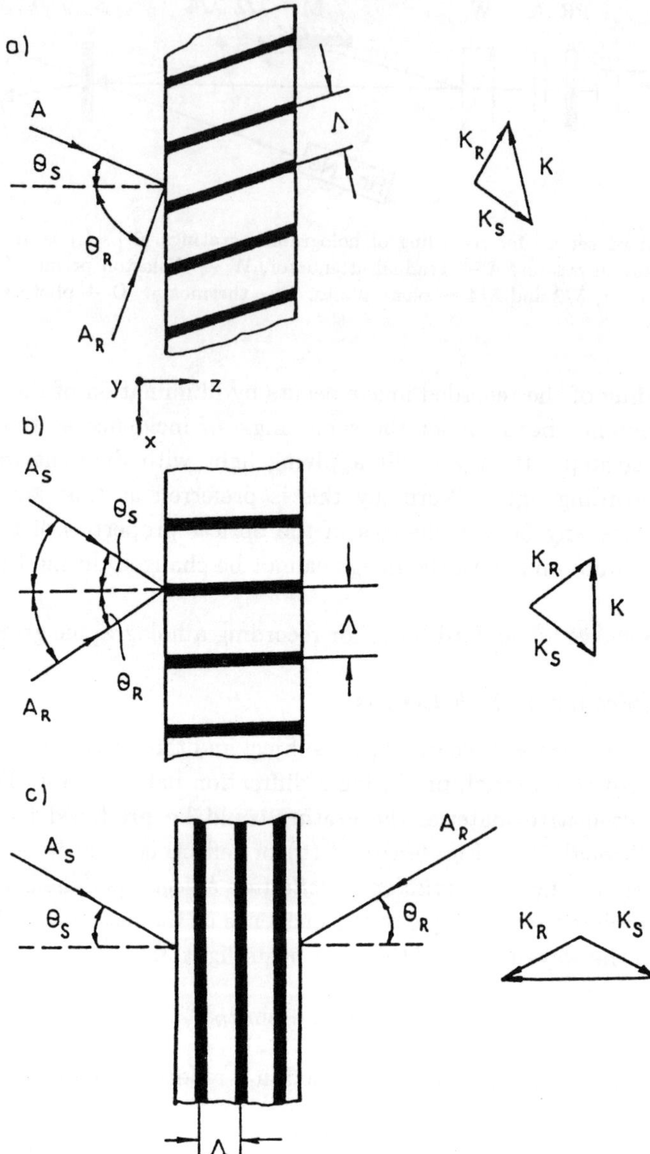

Fig. 9. Interference patterns in transmission (a, b) and reflection holograms (c) [1]: A_S, A_R subject and reference waves; θ_S, θ_R — angles of incidence for the subject and reference beams.

When $\Delta k \gg \Delta n$ one speaks about an *amplitude hologram*, since the change in the absorption index is felt most strongly by the amplitude of the transmitted wave. When the phase of the transmitted wave is more strongly modulated than the amplitude, then one speaks about a *phase hologram*. In this case $\Delta n \gg \Delta k$ or for the optical path length $\Delta(dn) = n\Delta d + d\Delta n \approx n\Delta d$ i.e. when exposure to light produces a change in the profile of the surface of the recording medium, the optical path length is also changed by variation of Δd, thus again a phase hologram is produced. Such recordings can only be made when the thickness of the medium is large enough to allow a change Δd without burning through the material. This type of holograms can be named also *thickness-modulated holograms*.

Depending on the thickness of the films of which the holograms are prepared, holograms are *thin* or $2D$ (two dimensional) and *volume* or $3D$ (three dimensional). The criterion to distinguish these two types of holograms is given by Klein [68]:

$$Q = 2\pi\lambda d/n\Lambda^2 \tag{6}$$

where λ is the wave length of the illuminating reference beam in a vacuum, n is the refractive index of the recording medium and Λ is the period of the holographic grating; d is the thickness of the film on which the hologram is recorded. Generally if $Q \geq 10$ the hologram is classified as a *volume*. For *thin* *holograms* $Q \leq 1$ and these holograms can be reconstructed only with coherent laser light.

A further classification concerns the recording technology and the kind of the incident light on the holograms. When the hologram is recorded with one unpolarized light beam, it is *scalar*. When the recording light is polarized, then one obtains a *polarization hologram*. Each of the latter two listed types of holograms can also be one of the previously described types depending upon their other characteristics. As the latest classification is mainly related to the recording technology that affects, to a great extend the properties of the holograms, it will be used for providing particular data on holographic recording.

3.2.3. *Diffraction efficiency*

The main particular characteristic of the holograms is their *diffraction efficiency η*. In the general case, it is determined by the relation of the power of the diffracted light beam ($P_{\text{diffr.}}$) to the incident power of the beam ($P_{\text{inc.}}$):

$$\eta = P_{\text{diffr.}}/P_{inc}. \qquad (7)$$

The power of the beam is determined by the integral of the light intensity (I) over the surface of the detector (S):

$$P = \int_S \int I \, dS \qquad (8)$$

If the light intensity of the subject and reference beams are I_S and I_R and the diffracted beam intensities are: I'_S and I'_R, then the diffraction efficiency is given by:

$$\eta = I'_{S(R)}/I'_{R(S)} \qquad (9)$$

The diffraction efficiency depends on the wavelength as well as on the thickness of the recorded material, on whether the hologram is 3-D or 2-D, phase or amplitude and on the type of the grating profile. The diffraction efficiency can be used to determine the type of the hologram (amplitude/phase or thick/thin) being scanned as it is much higher for the phase and thick gratings. It is a function of the amplitude of the induced holographic grating and its intensity.

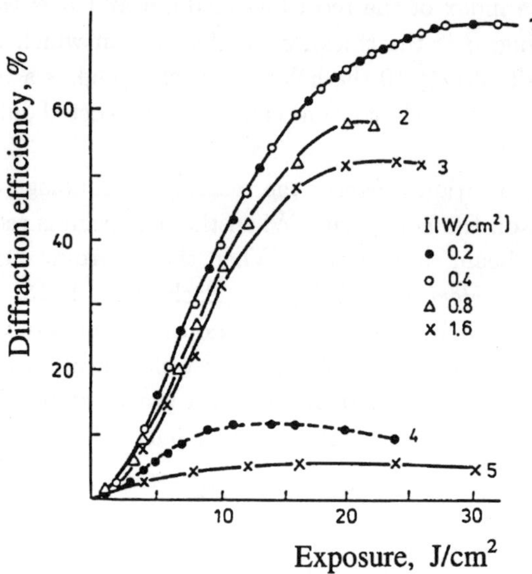

Fig. 10. Holographic recording kinetics [1], $\eta(I,t)$ in an As_2S_3 film (thickness $d = 5.2\ \mu m$, grating period $\Lambda = 0.7\ \mu m$) vs. light intensity for read out with $\lambda = 632.8$ nm (curves 1, 2, 3) and $\lambda = 514.5$ nm (curves 4 and 5).

3.2.4. *Results in holographic recording on chalcogenide glasses*

The holographic recording is mostly investigated for the As_2S_3 glasses and the production of holograms based on these materials is commercialized. A usual way to record *scalar holograms* on thin films of these materials is to illuminate them in a specific holographic set up with an Ar^+ laser with $\lambda = 514.5$ nm, which is in the vicinity of the band gap, in order to create photo-induced changes in the material. The readout process can be realized with light with the same λ or preferably with different one, for example with He–Ne laser ($\lambda = 632.8$ nm). The results on the diffraction efficiency are quite distinct depending on the different wavelengths — Fig. 10. This effect gives space for some optimization of the hologram's parameters. Besides, the recording kinetics is highly dependent on the laser beam intensity, the film's thickness and the grating period — Fig. 11. Recently, interesting results have been obtained by

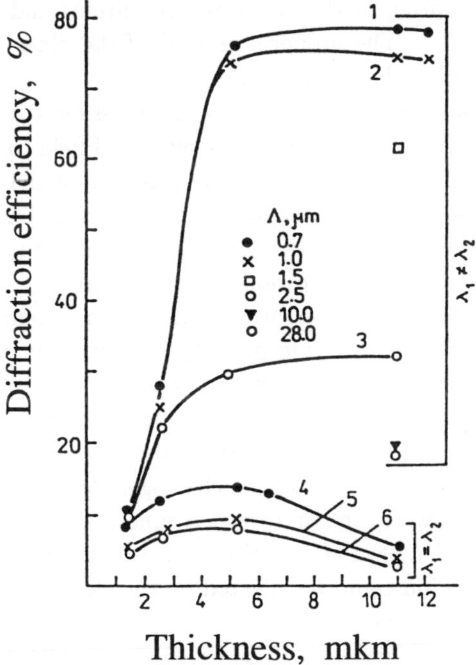

Fig. 11. The influence of the film thickness d and grating period Λ on the maximum diffraction efficiency of As_2S_3 films for readout with $\lambda = 632.8$ nm (curves 1,2,3) $\lambda = 514.5$ nm (curves 4, 5 and 6) [1]. Light intensity 0.2 W/cm^2.

a holographic recording with a subband-gap light [69] showing that the data for this recording considerably differ from that previously described, which occur in photoinduced refractive index increase, in weak photobleaching, absence of photoinduced thickness change, light polarization depending large exposures, holographic grating's shifts during the exposure and a peculiar two maxima spatial frequency response. A possible mechanism for these effects is photoinduced reorientation and generation of chalcogenide related D^- and D^+ centers counteracted by the photostimulated relaxational structural changes. The relaxation processes in these conditions give rise to a self-enhancement of the recorded image with a relaxation time decreasing exponentially as a function of the exciting laser beam intensity [70]. After a longer time storage of the films (of the order of years) a serious decrease of the diffraction efficiency has been observed [71]. An optimization of the recording processes can be achieved by modeling the grating fabrication [72], assuming that the grating will be lossless, unslanted and uniform in depth. The next step is then fabrication of the grating at the optimized conditions and also production of micro-lense arrays [73]. The proper storaging of the films can also contribute to better parameters of the holograms.

One way to receive better results concerning the sensitivity and the diffraction efficiency of holograms is to apply an electrical field to affect the medium simultaneously with the light. This idea was proposed in the seventies by Ovshinsky *et al* [74], Morikawa *et al* and Okuda *et al* [75, 76] and received presently again considerable interest [77, 78]. The construction of an element, that is working under these conditions, is shown on Fig. 12(a). The light is directed to the chalcogenide glass through the semitransparent SnO_2 electrode and *dc* electrical field with varying polarity is applied on the two electrodes. Especially promising results have been received for the Ge–Se–AgI glasses [78], where a drastic difference in the diffraction efficiency was found upon the electrical field polarity — Fig. 12(b). It is assumed that the reason for these results is the difference in the band gaps of the chalcogenide glass and the SnO_2 electrode building a structure similar to a *p–n* junction, while the nature of the contact with the golden electrode is ohmic, as well as the presence of the free silver ions whose motion could be affected by the electrical field and governed by the different types of electrodes.

The production of *polarization holograms* is of special interest because they allow us to obtain both image when the object and reference wave polarization vectors are perpendicular and they contain information about the polarization of the object wave giving in this way rise to the information capacity.

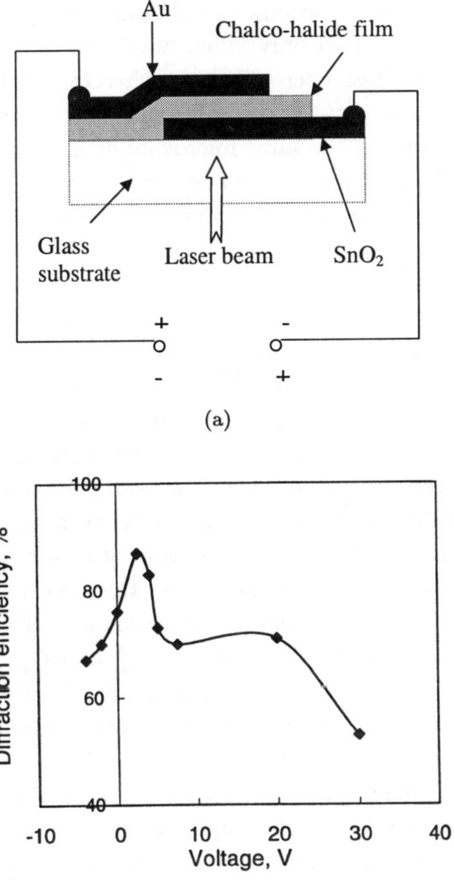

Fig. 12. (a) Construction of the elements for simultaneously application of light irradiation and electrical field for production of holograms; (b) diffraction efficiency depending on voltage polarization for the sample $(GeSe_4)_{90}AgI_{10}$.

They are recorded on the basis of both spatial-modulation parameters, which characterize the value of film's anisotropy and spatial modulation of the optical axis orientation of the material [79, 80]. The origin of the photoinduced optical anisotropy in chalcogenide glasses is due to non-radiative geminate recombination events that occur in the absorbing microvolumes and cause local changes in atomic bond configurations as it has been proposed by the model of Fritzsche [81–83]. He indeed has predicted that even unpolarized light will cause these materials to become optically anisotropic as a net anisotropy of the system can be realized only when hole-electron pairs recombine non-radiatively in their excited microvolumes, whereby some atomic rearrangements occur and rotate optical axes of some of the excited microvolumes. Occurrence of excitation in the same microvolume is essential to the existence of vector effects, whereas diffusion of excited carriers out of microvolumes results in scalar effects, because carriers lose the memory of preferentially excited microvolumes. All the data about the nature of this effect suggest that a disordered network of chalcogenide glasses is considerably dynamic under photoexcitation and possibly changes its middle range order or sometimes long-range order through the cooperative effect of local atomic events [84]. Some interesting results are obtained on "non-trivial" glasses from the system Se–Ag–I [85–87], where optimization of the recording process due to the composition and temperature of the films, as well as the writing laser beam intensity and wave length of the reading out laser has been performed — Fig. 13(a) and Fig. 13(b). The results obtained are discussed in terms of the occurrence of a Weigert effect (formation of chains, consisting of closely positioned spherical silver granules with a diameter much smaller than the wave length of the irradiation light) due to the presence of Ag atoms as well as the participation of the chalcogenide atoms and formation of C_3^+ and C_1^- defects with dipole character. The interaction of the polarized light with these structural units leads to dipole moment orientation predominantly in one direction and this results in the appearance of a corresponding macroscopic optical axis.

A very successful technology for creating diffraction gratings with high diffraction efficiency has been developed by using the process of metal-photodissolution. It was 33 years ago that Kostyushin *et al* [88] first reported the phenomenon, which is now generally known as metal-photodissolution in amorphous chalcogenides. It is based on the fact that the kinetics of the silver diffusion in chalcogenide glasses differs several orders of magnitude by illumination with light. The expertise accumulated by the group of Kostyushin in their 26 years of dealing with this problem is given in the book edited by Kurik [89], and the particular literature up to about 1990–1991 has been comprehensively reviewed by Kolobov and Elliott [25]. Illuminating bilayered structures formed by a layer of chalcogenide glass combined with a silver layer, one can obtain spots in the film with different refractive indexes and absorption coefficients depending upon the illumination. This can lead to obtaining gratings with quite high diffraction efficiency. Metal photodissolution effect occurs in many chalcogenide glasses and takes place in well-known films e.g. As_2S_3 [90]

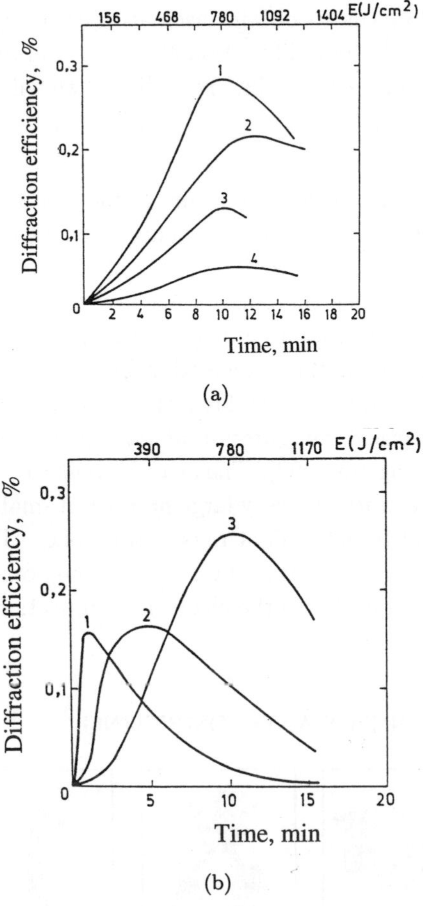

(a)

(b)

Fig. 13. (a) Dependence of the diffraction efficiency of polarization recording on composition for (1) $Se_{70}Ag_{15}I_{15}$; (2) $Se_{75}Ag_{12.5}I_{12.5}$; (3) $Se_{80}Ag_{10}I_{10}$; (4) $Se_{82.5}Ag_{7.5}I_{10}$; (b) dependence of the diffraction efficiency of polarization recording on the laser beam intensity — (1) $I = 1.3$ W/cm^2; (2) $I = 2.1$ W/cm^2; (3) $I = 2.5$ W/cm^2.

but also in metal containing systems [91]. A great disadvantage of the optical elements obtained in this manner is that they have very low reliability due to the continuos silver diffusion (even at dark), i.e. the elements are not stable with the time as an uniformly silver distribution spreads once the element has been created. This effect has been overcome by involving a process similar to the chemical developing of the silver halide emulsions based on the selective solubility of both (illuminated and non-illuminated) parts of the film. That particular process allows us to obtain fascinating relief optical elements and holograms with extremely high diffraction efficiency and resolution but as this technology is actually not a "real time recording", we will not discuss it here.

4. Future Trends

It is very tempting to think about the future of the real-time optical recording as this is a very fast developing area of science opening wide new horizons in the information technology. Production of optical components for the telecommunication window for which these materials have excellent optical properties and computing technology that requires application of these elements bring about the development of a close connection between optics and computing — Fig. 14. Perhaps we are not very far from the moment of the wide application of the digital optical computers that has been drawing the interest of the researchers for a long time [92]. That can be realized tiding together by a powerful connecting network of a very large number of small optical processors. The optical signals will travel as light rays in free space. The great advantage is that many independent rays may traverse the same portion of the $3D$-space without "crosstalk" in contrast to the electronic connections that are confined to material guides.

OPTICS AND COMPUTING

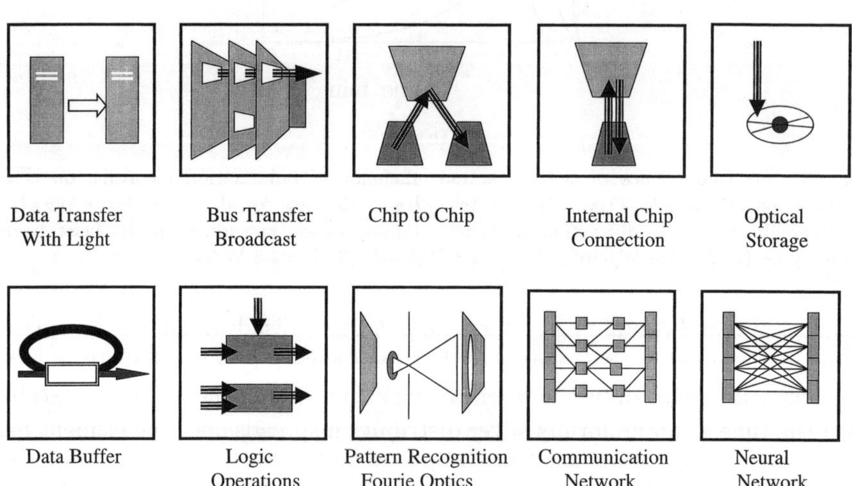

Fig. 14. Connections — optics and computing.

Expectations are good for improving the technologies and materials, so that commercial production of compact discs based on the amorphous semiconductors will surely grow up. The main competitor of the chalcogenide glasses in this respect are the organic polymers that show good features and find a wide application right now but of course this competition can be the reason for exciting results on both sides.

Real-time optical recording on the amorphous semiconductors is a very powerful method because of the absence of any chemical development and the direct storage, the possibility to realize phase recording and the high resolution of the material. These features can find a wide application in the future in the X-ray imaging as reported by S. Kasap *et al* in this book. At the same time, these glasses have quite a low sensitivity and this is the main limitation in their application. One of the best opportunities for application is the holography where usually there is a high level of illumination (so that the low sensitivity is not a problem) that allows producing holograms with high diffraction efficiency and resolution. The most fascinating application of these holograms that already has some commercialization and is in a process of fast progress is their

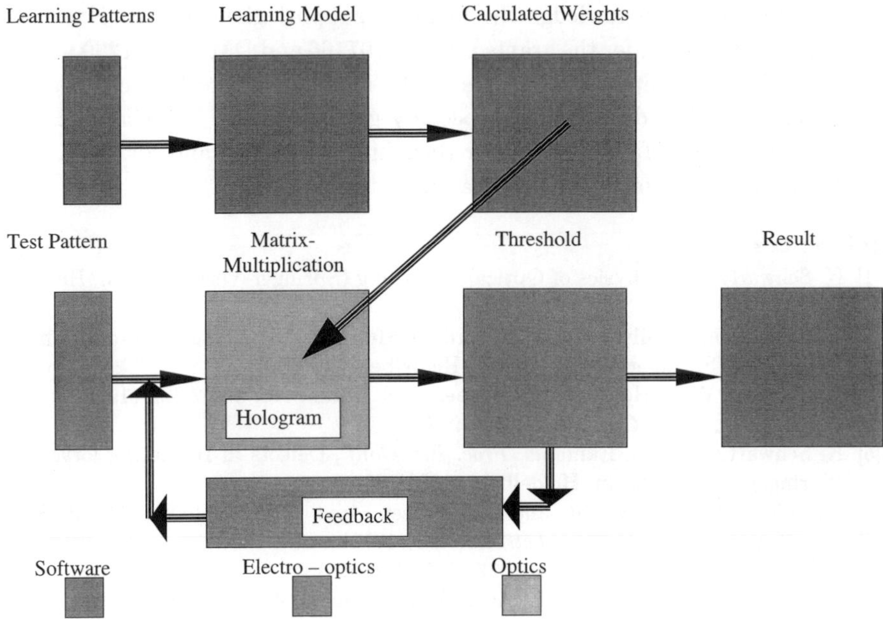

Fig. 15. Opto-electronic neural network.

inclusion in opto-electronic neural networks where they are actually the heart for the complex spreading the information in the whole system (Fig. 15).

One not enough explored area of application of the real-time optical recording that could be developed in the future is the holographic investigation of the photoinduced changes in thin films [93, 94]. The interest in the holographic technique as a tool for elucidating the mechanism of photoinduced changes is governed due to its high sensitivity and unique ability to follow the course of the photoinduced changes by measuring in real time the diffracted light from induced gratings. Owing to its high sensitivity, the holographic method is especially suitable for analysis of low sensitivity materials and thin films.

In 1995, Hisakuni and Tanaka [95] reported that light enhances the fluidity of the glasses, this process having photoelectronic nature. Shaping the glasses by stressing under light illumination can serve as a mode for producing micro-optic components, since it provides a way to bend plastically optical fibers on a small scale, a capability that also holds promise for the fabrication of integrated fiber circuits.

Acknowledgments

It is a pleasure to acknowledge the financial support given by the National Science Foundation under the grants DMR-92-07166 and DMR-97-01289 which made possible my visit to the University of Cincinnati where I have prepared the manuscript. My deep thanks especially go to Prof. P. Boolchand, Prof. K. Schwartz and Prof. V. Lyubin for their kind assistance and for the many fruitful discussions during the preparation of this work.

References

[1] K. Schwartz, The Physics of Optical Recording (Springer-Verlag, Berlin, Heidelberg, 1993).

[2] H.I. Bjelkhagen, Silver-Halide Recording Materials for Holography and Their Processing (Springer-Verlag, Berlin, Heidelberg, 1995).

[3] W.E. Bron, W.K. Heller and B. Weber, U.S. Patent No. 3.452.322 (1965).

[4] I. Schneider, *Appl. Optics* **6**, 2197 (1967).

[5] K. Schwartz and Yu. Ekmanis, *Proc. Int. Conf.*, Defects in Insulating Crystals (Springer-Verlag, Berlin, Heidelberg, 1981) 363.

[6] F.S. Chen, Y.T. La Machia and D.B. Fraser, *Appl. Phys. Lett.* **13**, 233 (1968).

[7] S.A. Keneman, *Appl. Phys. Lett.* **19**, 205 (1971).

[8] L.E. Ravich, *Laser Focus World* **25**, 115 (1989).

[9] D. Chen, J.E. Ready and G.E. Aagard, *Laser Focus World* **4**, 18 (1968).

[10] V.A. Poltoratskii and N.N. Vsevolodov, *Sov. Phys. Tech. Phys.* (USA) **30**, 1235 (1985).

[11] R.B. Gross, K.C. Izgi and R.R. Birge, *Proc. SPIE 1662*, Washington, 186 (1992).

[12] A. Szabo, U.S. Patent No. 3.893.442 (1975).

[13] H. Schwoerer, D. Erni, A. Rebane and U.P. Wild, *Adv. Mat.* **7**, 457 (1995).

[14] A.K. Rebane, R.K. Kaarli and P.M. Saari *JRTP Lett.* **38**, 383 (1983).

[15] J. Lindmayer, P. Goldsmith and Ch. Wrigley, *Laser Focus World* **25**, 122 (1989).

[16] F.T. Wallenberger, *Science* **267**, 1274 (1995).

[17] O. Lehmann and M. Stucke, *Mater. Lett.* **21**, 131 (1994).

[18] M. Asobe and K. Kubodera, *IEEE J. Quantum Electronics* **29**, 2325 (1993).

[19] Z.U. Borisova, Glassy Semiconductors (Plenum Press, New York and London, 1981).

[20] M. Mitkova and Z. Bontscheva-Mladenova, *Monatshefte fuer Chemie* **120**, 643 (1989); see also M. Mitkova, T. Petkova and A. Janakiev, *Mat. Chem. Phys.* **30**, 55 (1991), ibid **33**, 233 (1993).

[21] V. Krasteva, D. Machewirth, G. Sigel and Jr. Uspexi, *J. Non-Cryst. Solids* **213 & 214**, 304 (1997).

[22] P. Ewen and A.E. Owen, High Performance Glasses, M. Cable and S.C. Agarval (Blackie, 1992) p. 287.

[23] K. Tanaka, *Current Opinion in Solid State and Materials Science* **1**, 567 (1996).

[24] K. Tanaka, *Rev. Solid State Sci.* **4**, 641 (1990).

[25] A.V. Kolobov and S. R. Elliott, *Advances in Physics* **40**, 625 (1991).

[26] T. Kavaguchi, S. Maruno and S.R. Elliott, *J. Non-Cryst. Solids* **212**, 166 (1997).

[27] T. Kavaguchi, S. Maruno and S.R. Elliott, *J. Non-Cryst. Solids* **211**, 187 (1997).

[28] T. Ishida, M. Shoji, Y. Miyabata, Y. Shibata, E. Ohno and S. Ohara, *Proc. Soc. Photo-Opt. Instrum. Eng.* **2338**, 121 (1994).

[29] J. Gonzalez-Hernandez, B.S. Chao, D. Strand, S.R. Ovshinsky, D. Pavlik and P. Gasiorovski, *Appl. Phys. Comm.* **11**, 557 (1992).

[30] N.F. Mott, A.E. Davis and R.A. Street, *Phil. Mag.* **32**, 961 (1975).

[31] M. Kastner, D. Adler and H. Fritzsche, *Phys. Rev. Lett.* **37**, 1504 (1976).

[32] W. Zhou, M.A. Paesler and D.E. Sayers, *Phys. Rev.* **B46**, 3817 (1992).

[33] W. Zhou, D.E. Sayers, M.A. Paesler, B. Bouchet-Fabre, Q. Ma and D. Raoux, *Phys. Rev.* **B47**, 686 (1993).

[34] S. Sugai, *Phys. Rev. Lett.* **57**, 456 (1986).

[35] T. Ohta, N. Akahira, S. Ohara and I. Satoh, *Optoelectronics — Dev. Techn.* **10**, 361 (1995).

[36] V. Mateev, T. Petkova, M. Mitkova and Pl. Markovsky, *Thin Solid Films* **226**, 1203 (1993).

[37] P.A. Fedders, Y. Fu and D.A. Drabold, *Phy. Rev. Lett.* **68**, 1888 (1992).

[38] K. Shimakawa, A. Kondo, K. Hayashi, S. Akahori, T. Kato and S.R. Elliott, *J. Non-Cryst. Solids* **164–166**, 387 (1993).

[39] M. Mulato, D. Toet, G. Aichmayr, A. Spangenberg, P.V. Santos and I. Chambouleryonas, *J. Non-Cryst. Solids* **227**, 930 (1998).

[40] K. Shimakawa, A. Kolobov and S.R. Elliott, *Advances in Physics* **44**, 475 (1995).

[41] K. Tanaka, *J. Non-Cryst. Solids* **59 & 60**, 925 (1983); *Japan J. Appl. Phys.* **25**, 779 (1986).

[42] H. Fritzsche, Physics and Technology of Semiconductors **32**, 952 (1998) (1998).

[43] K. Schwartz, *J. Inf. Recording* **22**, 289 (1996).

[44] A.B. Marchand, Optical Recording a Technical Overview (Addison-Wesley, Reading Massachusetts, 1990), p. 41.

[45] P. Kivits, R. de Bont, B. Jacobs and P. Zalm, *Thin Solid Films* **87**, 21 (1982).

[46] M. Chen, V. Marrello and U.G. Gerber, *Appl. Phys. Lett.* **41**, 894 (1982).

[47] M. Janaj and P.S. Rudman, *Proc. 5th ICALS Garmisch-Partenkirchen*, eds. J. Stuke and H. Brenig, (Taylor and Francis, London, 1974), p. 425.

[48] M. Mitkova, E. Vateva and E. Skordeva, *J. Non-Cryst. Solids* **137 & 138**, 1013 (1991).

[49] S.B. Gurevich, N.N. Iliashenko and B.T. Kolomietz, *J. Techn. Phys.* **43**, 217 (1973).

[50] Y.S. Tyan, D.R. Preuss, F. Vazan and S.J. Marino, *J. Appl. Phys.* **59**, 716 (1986).

[51] V.V. Kaminskii, Yu. F. Solomonov, V.E. Yegorov, B.I. Smirnov and A.I. Smirnov, *Sov. Phys. Solid State* (USA) **17**, 1015 (1975).

[52] J. Feinleib, J. de Neufville, S.C. Moss and S.R. Ovshinsky, *Appl. Phys. Lett.* **18**, 254 (1971).

[53] A.W. Smith, *Appl. Opt.* **13**, 795 (1974).

[54] N. Yamada, *MRS Bulletin* **21**, 48 (1996).

[55] A.V. Klobov and S.R. Elliott, *J. Non-Cryst. Solids* **189**, 297 (1995).

[56] M. Nakamura, Y. Wang, O. Matsuda, K. Inoue and K. Murase, *J. Non-Cryst. Solids* **198–200**, 740 (1996).

[57] J.L. Adam, C. Ortiz, J.R. Salem and X.H. Zhang, *J. Mat. Sci.* **26**, 2900 (1991).

[58] V.K. Tikhomirov, P. Hertogen, G.J. Adriaessens, G. Glorieux and R. Ottenburgs, *J. Non-Cryst. Solids* **227**, 732 (1998).

[59] V.M. Lyubin, M. Klebanov, M. Mitkova and T. Petkova, *Appl. Phys. Lett.* **71**, 2118 (1997).

[60] V.M. Lyubin, M. Klebanov, M. Mitkova and T. Petkova, *J. Non-Cryst. Solids* **227**, 739 (1998).

[61] Ke Tanaka and K. Ishida, *Non-Cryst. Solids* **227**, 673 (1998).

[62] S. Maruno and T. Kawaguchi, *J. Appl. Phys.* **46**, 5312 (1975).

[63] T. Kawaguchi, S. Maruno and K. Tanaka, *J. Non-Cryst. Solids* **164–166**, 1231 (1993).

[64] T. Kawaguchi and S. Maruno, *Japan J. Appl. Phys.* **33**, 6470 (1994).

[65] T. Kawaguchi, S. Maruno and S.R. Elliott, *J. Non-Cryst. Solids* **202**, 107 (1996).

[66] G. Saxby, Practical Holography (Prentice Hall, New York, 1992).

[67] M. Mitkova, T. Petkova, V. Mateev and Pl. Markovsky, *J. Phys. Chem.* **96**, 8998 (1992).

[68] R.J. Colier, C.B. Burghardt and L. H. Lin, Optical Holography (Academic Press, New York 1971).

[69] A. Ozols, N. Nordman, O. Nordman and P. Riihola, *Phys. Rev.* **B55**, 14236 (1997).

[70] J. Teteris and O. Nordman, *Opt. Comm.* **138**, 279 (1997).

[71] O. Nordman, N. Nordman and A. Ozols, *Opt. Commun.* **145**, 38 (1998).

[72] A. Zakery, P.J.S. Ewen and A.E. Owen, *J. Non-Cryst. Solids* **198–200**, 769 (1996).

[73] N.P. Eisenberg, M. Manevich, M. Klebanov, V. Lyubin and S. Shtutina, *J. Non-Cryst. Solids* **198–200**, 766 (1996).

[74] S.R. Ovshinsky and H. Fritzsche, *IEEE Trans. Electron Devices* **20**, 91 (1973).

[75] T. Morikawa, T. Nakajima and K. Sakurai, *Appl. Phys. Lett.* **23**, 405 (1973).

[76] M. Okuda, T. Matsushita, T. Yamagami and K. Yamamoto, *Appl. Opt.* **13**, 799 (1974).

[77] M.S. Iovu and E.G. Khanchevskaya, *Phys. Stat. Solids (A)* **156**, 375 (1996).

[78] M. Mitkova, I. Iliev, V. Boev and T. Petkova, *J. Non-Cryst. Solids* **227**, 748 (1998).

[79] Sh.D. Kakichashvili and T.N. Kvinikhodze, *Kvantovaja Electron. (Moscow)* **2**, 1449 (1975) [*Sov. J. Quantum Electron*].

[80] T.D. Ebralidze and A.N. Mumladze, *Opt. Spectrosck.* **64**, 155 (1988).

[81] H. Fritzsche, *J. Non-Cryst. Solids* **164–166**, 1169 (1993).

[82] H. Fritzsche, *Phil. Mag.* **B68**, 561 (1993).

[83] H. Fritzsche, *Phys. Rev.* **B52**, 15854 (1995).

[84] K. Tanaka, *Science* **277**, 1786 (1997).

[85] V. Mateev, T. Petkova, M. Mitkova and Pl. Markovsky, *Thin Solid Films* **226**, 119 (1993).

[86] M. Mitkova, T. Petkova, Pl. Markovski and V. Mateev, *J. Non-Cryst. Solids* **164–166**, 1203 (1993).

[87] M. Mitkova, T. Petkova and Pl. Markovski, *Proc. ICDIM*, August, 16–20, 1992, eds. O. Kanert, J.-M. Spaet (World Scientific, Singapore, New Jersey, London, Hong Kong) p. 1145.

[88] M.T. Kostyushin, E.V. Mikhailovskaya and E.V. Romanenko, *Fizika tverd. tela* **9**, 571 (1966).

[89] M.V. Kurik, Photostimulated Interactions in Structures Semiconductor — Metal, Kiev, Naukova Dumka, 1992.

[90] E. Marquez, R. Jimenez-Garay, A. Zakery, P.J.S. Ewen and A.E. Owen, *Phil. Mag.*, **B63**, 1169 (1991).

[91] M. Mitkova and W. Fallmann, *Phys. Stat. Solids (a)* **K75**, 145 (1983).

[92] A.W. Lohmann, *Phys. Scr.* **T23**, 271 (1988).

[93] Pl. Markovski, V. Boev, M. Mitkova and K. Zlatanova, *Proc. 8th Int. School Cond. Mat. Phys.* (Res. Studies Press. England, 1995) 291.

[94] V. Boev, E. Sleekx, M. Mitkova, P. Markovsky, P. Nagels and K. Zlatanova, *Vacuum* **47**, 1211 (1996).

[95] H. Hisakuni and K. Tanaka, *Science* **270**, 974 (1995).

D.   DIAMOND BASED FIELD EMISSION DISPLAYS

JAMES E. JASKIE

*Advanced Display Technologies Motorola, Phoenix Corporate Research Laboratories,
2100 E. Elliot Rd, Tempe AZ 85284*
jim.jaskie@motorola.com

Contents

1. Introduction

Electronic display devices form part of the so-called "man-machine interface", transmitting information from a machine so that a human can visually recognize it. As electronic systems become more and more integrated, displays that can appropriately interact with the users will become even more important than they are today. Users will require high resolution displays that are bright, sharp, low power, lightweight and flat. Demand for such devices will continue to expand and diversify into the foreseeable future. In order to meet the ever expanding requirements of the information society, the development of new more effective displays is continuing all the time. The market for displays in the year 2000 is estimated to be \$21.8 billion [1]. The large size of this potential market has encouraged the funding of many efforts to explore new ways to make these devices.

Currently, the mass market for emissive displays consists of CRTs (Cathode Ray Tubes) and LCDs (Liquid Crystal Displays). They each have advantages and disadvantages, but both are old technology and do not produce the combination of bright, sharp images in a low power, lightweight, flat package that users want. These older technologies will be discussed and compared with the use of Diamond-based Field Emission to produce new generation displays.

2. Old Technology Emissive Displays

Displays convert various electrical signals into optical signals that can be recognized by a human as digits, characters, or graphics. If the optical signal is displayed by the emission of light, it is called an "emissive display". Displays that work by modulating incident light through reflection, scattering or other phenomena are called "non-emissive displays". Non-emissive displays are hard to read in dim light and have yet to demonstrate the sort of bright, vibrant colors that users want and expect. At this time, the only emissive display technologies on the market are CRTs and LCDs.

2.1. *CRT displays*

The most common high quality display is still the CRT (Cathode Ray Tube) which is a low-cost vacuum tube with a heated filament that emits electrons to excite the cathodoluminescent color phosphors. The CRT display, as used in the common television and computer monitor, has the longest history of any electronic display device. The CRT still produces the best display quality and leads the market economically. It is the display against which all other displays are measured for visual attractiveness. The efficient phosphors used in a CRT emit nearly pure colors. CRTs are perceived by humans as producing a "soft sharpness" that is considered to be very attractive. The sharpness comes from the CRTs ability to form an image at a higher resolution than the human eye can perceive, but there is also a small amount of bleed-over in both brightness and color which softens this sub-perceptive edge.

The packaging needs of these displays are not very attractive, however. Because of the CRT, a TV is typically as deep as it is tall, and the large glass package is very heavy. A one meter diagonal CRT can weigh over 200 kg. Even so, with the combination of the visual attractiveness of the color phosphor display and its low manufacturing cost, the CRT easily dominates the marketplace except where the mechanical package is totally unacceptable. The laptop computer market, for example, would grow even faster if it had a thin, flat, low-power, lighter weight display with the visual attractiveness of a color CRT.

2.2. *LCD displays*

The next leading product is the active matrix LCD (Liquid Crystal Display). LCDs have taken over much of the marketplace for portable color displays such as those on computer laptops and portable instruments because they are thin, flat and relatively lightweight compared to the CRT. Their first commercial introduction was for wristwatches and calculators in 1972 by Rockwell [2]. Unfortunately, LCDs suffer from several problems that limit their utilization. They are fundamentally inefficient. LCDs work by taking emitted light from a backlight, usually a fluorescent light, and then shuttering this light to produce an image. The shuttering is done through polarization where all the emitted light is first polarized (removing 50% of the light) and then the remaining light is modulated by the use of a liquid crystal, whose polarization is adjusted by an electric field. By shifting the polarization between parallel and perpendicular to the initially polarized light, the remaining light is modulated.

Color is also produced by an extraction method. That is, to produce a red pixel (pixel = "picture element") one must remove all the colors in the light, except red, through the use of optical filters. By the time a pixel of the right color, shade and intensity is produced, most of the light from the original source, which was the fluorescent tube behind the screen, has been removed. In fact, a good LCD display may only transmit about 3% of the original light at peak display brightness. That means, that for a typical display, the bulb must be about 33 times more intense than the brightest pixel of the brightest image that one ever intends to produce! This may not be a big deal for a computer monitor that is plugged into the electrical grid, but it is truly significant for a portable device such as a laptop computer. Typically, the backlit display is far and away the largest user of power in such a device. This excessive power usage leads to heavier devices which require larger batteries than would be the case if they were more power efficient.

The best LCD displays, called Active Matrix LCD (AMLCD), require a switching transistor at each pixel (the origin of the appellation "active matrix"). This gives the display a rapid response to changing input signals and better gray scaling. However, this transistor takes space away from the pixel on the display, producing dark ridges. This gives a granularity to the displayed image, besides further reducing the light transmission. This granularity contrasts vividly with the soft sharpness of a CRT display. Most people will only use an LCD based laptop computer when there are no other choices available. They will prefer a CRT based monitor at their desk, where the extra size and weight are not such an issue. In fact, most current high end laptops allow their users to plug-in a CRT in some manner for use at a permanent desk top location.

3. Field Emission Displays

A new kind of emissive display that is currently in the development phase is the Field Emission Display (FED). In terms of display quality, field emission displays are essentially flat CRTs. The thermally heated filament used for an electron emitter in a CRT, the grid structure used to modulate emitted current and the magnetic deflection system are all eliminated in the field emission display. These components are replaced with a flat structure, the cathode array, that is placed about a millimeter away from a fairly standard CRT faceplate, the anode. Each pixel on the anode is supplied with electrons produced by a corresponding set of electron emitters on the cathode array. This results in a display having the visual attractiveness and brightness of a CRT combined with the thin flatness of an LCD, but it is much lighter in weight than either one.

3.1. *History of field emission*

Originally, the theory of electron emission from a solid was developed by Richardson [3] based on the thermodynamic theory of vapor pressure. The electrons were treated as a substance that escapes from the solid state into a vacuum. Some of these electrons eventually get reabsorbed into the surface and an equilibrium is established. The equilibrium between these two rates changes with temperature. As the temperature of the system increases, electrons escape into the vacuum faster than they find their way back to the surface and become reabsorbed. The Richardson–Dushman equation expresses the current flux emission j, as

$$j = 120T^2 e^{\left(\frac{-e\phi}{kT}\right)} \tag{1}$$

where ϕ is the work function, k is Boltzmann's constant and T is the absolute temperature.

Only those electrons at the highest energy levels allowed by their probability distribution have the energy to escape. At room temperature, the average energy is about 0.025 eV. The energy necessary to overcome the barrier to emission is typically about 4.0 $\sim$ 4.5 eV for most materials of interest. At room temperature, a very small percentage of electrons have enough energy necessary to clear this hurdle.

3.1.1. *Increasing the emission by heating the emitter*

The simplest way to increase emission is to modify the statistics to increase the number of electrons that will have the energy to clear the hurdle and escape [4, 5]. This can readily be done by heating the emitter. Raising the temperature of the emitter by a factor of 5 also raises the average energy by a factor of 5 to 0.125 eV. The average electron is still not able to leave the surface. However, because of the sharp wings of the Fermi–Dirac energy distribution function, the percentage of electrons with more than 4.5 eV of energy goes up substantially, by a factor of 64 orders of magnitude! This can provide a usable amount of electron emission.

This technique works well and is used in electron guns in CRTs, cathodes in vacuum tube power amplifiers, etc. However, the burden associated with increasing the temperature of the emitter is high, because enough power to raise the emitter to a temperature that is so much higher than ambient must be supplied. The thermal radiation between a source at 1500 K and its environment at 300 K is about 29 watts/cm^2. This is more than wasted heat, as it must be supplied by the system power supply, at an efficiency of perhaps 50%, and then must be removed by a cooling system. The stress on packaging caused by these differences in temperature cause premature aging and failure. Worse are the startup and shutdown thermal transients that cause a shock effect and thereby weaken devices.

3.1.2. *Increasing the emission by tunneling*

There is another way to exploit the energy probability statistics to increase electron emission from a source. From quantum mechanics it is known that the position of an electron is best viewed in probabilistic terms. There is a finite probability that an electron will find itself outside of an energy barrier, in spite of whether it has enough energy to leap over the barrier. This effect, known as "*quantum mechanical tunneling* [6]", predicts that, under normal conditions, a disappearingly small percentage of electrons will tunnel through the energy barrier provided by the interface between an emitter and vacuum. However, these statistics can be improved by narrowing the width of the energy barrier. It is more likely for an electron to tunnel through a thin wall than through a thick wall of the same height. The width of the barrier can be varied by supplying a very large electric field at the surface of our 4.0 eV emitter. An electron that finds itself an infinitesimal distance outside of our emitter will be swept away and escape. An electric field of 3–6 $\times$ 10^7 eV/cm will create emission [7] of 10^3 amps/cm^2. Even on a microscopic scale, this amounts to several thousand volts per micron. Such a high electric field is both difficult and expensive to control. The governing equation for field emission, the Fowler–Nordheim equation [8] for the emitted electron flux j, is

$$j = 6.2 \bullet 10^6 \frac{(E_j/\phi)^{1/2}}{(E_j + \phi)} F^2 e^{(6.8 \bullet 10^7 \phi^{3/2}/F)} \tag{2}$$

where E_f is the Fermi energy, ϕ is the work function, and F is the electric field.

3.1.3. *Increasing the emission with Spindt Tips*

There are other ways to achieve these electric field strengths besides resorting to very large voltages. The electric field at the emitter is a function of the potential difference between the emitter and its anode and also of its geometry. By adjusting the geometry such that the radius of curvature of the emitter is very small, the local electric field will be very high. A controlling electrode, called a "gate" can be added close to the emitter to supply the potential needed to induce emission.

A common way to do this is by using a geometry called a *Spindt Tip* [9], named after the scientist at SRI who invented a relatively easy way to make these emitters in the early 1970's. Figure 1 shows a schematic view of one of these tips. This device is basically a sharp tip with a radius of curvature of a few hundreds of Angstroms and is about a micron tall. The tip sits in a hole that has a diameter of about a micron centered about the tip point. In this geometry, a potential difference of 50 volts between the emitter and the "gate" electrode can supply about 5×10^7 V/cm which is necessary for emission. The 50 V at the gate can be controlled with modern integrated electronic circuits.

The electrons are typically emitted over a half angle of about 30°. This produces an inherent constraint on the relationship between the electron spot focus, anode field anode spacing and gate potential. This constraint has led to several structures to add focusing control for displays with higher resolution

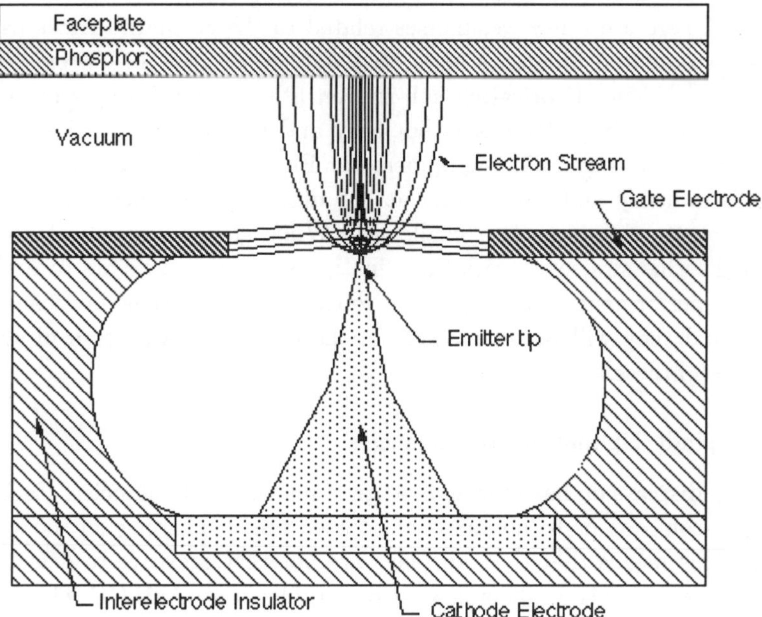

Fig. 1. Schematic view of a Spindt Tip.

than about 100μ per sub-pixel. While this is only of interest for a few monitors, it greatly increases cost and complexity.

3.1.4. *Using Spindt Tips in a field emission display*

Arrays of Spindt Tips are fabricated by the use of integrated circuit technology. Because of their small size, dozens of these micron size emitters can be used to supply electrons to each pixel in a display. This allows us to build a very thin flat field emission display (FED). This is the basis for a whole new industry being created to replace AMLCD displays in portable products. These displays will be brighter, prettier, and more power efficient than the backlit AMLCD displays that currently dominate the market. Candescent, Micron and Motorola have announced plans to manufacture these displays in the US. Pixtech in Europe and Futaba, Samsung and UMC in the Pacific Rim are also building production capability.

3.1.5. *Advantages and disadvantages of using Spindt Tips*

What are the disadvantages to this method of coaxing electrons out of an emitter? There are a few weaknesses related to the geometric tricks used to enhance a potential difference of $\sim$ 50 V/0.5 m, or 10×10^5 V/cm to look like 5×10^7 V/cm. Fabricating sub-micron lithographic features on a large (10" diagonal or larger) panel is not easy and is quite costly. The current generation of glass handling equipment for Flat Panel Displays is driven by the LCD industry and is currently at 20" diagonal. Submicron lithography on a substrate this large is a challenge. While steppers are available that can handle this size, even variations in tip geometries invisible to an SEM can affect performance. A stepper will repeat these variations in a consistent manner and they will become visible in the display as a pattern of slightly varying brightness.

3.1.5.1. Spindt Tip multiple redundancies

Spindt tip displays do have the advantage of multiple redundancies. Each sub-pixel has from dozens to hundreds of emitter tips, depending on the specific design. This means that these displays are not very sensitive to defects in a few emitters at a pixel. This is very preferable to an AMLCD where there are one, or possibly two transistors at each sub-pixel. Even a low defect rate, causing these one or two transistors at a pixel or two to fail will result in a visual flaw. The human eye is very sensitive to these kind of visual defects, and this fragility has been a prime driver in the cost of these displays. Scaling up the lithography for fabricating larger AMLCD displays (30" to 50") with individual transistors at each pixel will be very difficult.

3.1.5.2. The Spindt Tip dielectric layer

The Spindt structure has an effective requirement that the dielectric thickness needs to be about the same as the diameter of the gate hole. While this is not fixed in physical law, the standard fabrication method that guarantees self-alignment between cathode tip and gate hole requires this relationship. This means that for a one micron diameter gate hole (and designers would like to shrink this to 0.5 micron or better to lower the control voltage) the dielectric layer separating the emitter cathode from the gate is only about one micron thick. A dielectric this thin becomes sensitive to particulate contamination. Particles can act to create pinholes in the dielectric which then are filled with gate metal. This shorts the two electrodes, causing a defective pixel. This too can be resolved with redundancy, but each level of redundancy and repair adds to complexity and eventually to cost.

To minimize the risks of particulate damage, field emission displays are manufactured in clean rooms that mimic those of IC fabrication. This infrastructure, while well known, is very expensive. Any improvement that will reduce the sensitivity to particulates in such a way as to reduce the need for Class 10 clean rooms, massive air handlers, HEPA filters and bunny suits will be very welcome.

3.1.5.3. Control voltages for Spindt Tips

The drive voltage of Spindt Tip Field Emission Displays is higher than is desirable [10]. While fifty volts is within the range of current advanced IC technology, there is a strong motivation to lower this drive voltage because the drive chips typically cost as much as the actual display itself. Anything that can lower this cost will have a real effect on the cost of a display module. Lowering the voltage requirement of an IC allows the manufacturer to make the driver using chip design rules that allow closer placement of lines and devices thereby getting more chips from each wafer. This strongly effects cost.

The other burden associated with high drive voltage is that while the reactive power associated with the drive electrodes of a display structure varies linearly with the capacitance, it varies as the square of the voltage (CV^2). Hence lowering the voltage has a major effect on lowering the power requirements. Dropping the control voltage from 50 V to 40 V reduces this power term by over a third! Dropping it further to a more typical IC control voltage of 20 Volts reduces this by 80%! Since the primary market for flat panel displays is in wireless, cordless devices such as laptops, any reduction in drive voltage that reduces power consumption so effectively is highly desirable.

3.2. *Diamond-based field emission*

Two different ways to work around the statistical difficulty of extracting electrons from an emitter have been discussed thus far. The emitter can be heated to raise the energy level of the electrons until a reasonable number are able to readily leave. The effective barrier can also be penetrated by making use of the tunneling phenomena. A strong electric field, created by using geometric enhancement, can "thin" the barrier until a reasonable number of unexcited electrons can escape.

There remains yet another, more attractive method to making a good emitter. By choosing a material with a low effective barrier to electron emission, one could get the benefits of a field emission display while removing the constraints and cost of the emission tip. Previously, there had been no good candidate engineering materials that were both robust and readily emitted electrons, but practical modern techniques for producing diamond thin films has created a new opportunity.

3.2.1. *Properties of diamond and DLC*

Diamond is a crystalline form of Carbon, a group IV element in the Periodic Chart. Silicon and Germanium are also Group IV elements and also have the same crystalline lattice structure as diamond, so there has been theoretical interest in diamond's electronic properties since the beginning of the electronics age. In the 1920's, the use of diamonds was explored for photocathodes. However, the cost and poor crystalline quality of both natural and synthetic diamond had precluded any real industrial interest in diamond as an electronic material. Methods of low temperature and pressure diamond film deposition, developed initially by the Russians in the 1950's and 1960's (and hence by the Japanese, and eventually by others) have now made it possible to use this exotic material as an electronic substrate.

3.2.2. *Low work function materials and electron emission*

It was mentioned earlier that electron emission relies on electrons with enough energy to clear the hurdle presented by the surface, or on the use of an electric field to thin the barrier to the extent that comparatively unenergetic electrons can escape through tunneling. Another method to engineer an acceptable level of electron emission is simply to employ a material for the emitter that has a low barrier or effective work function. As can be seen from both the Richardson–Dushman equation and the Fowler–Nordheim equation, the electron emission is very sensitive to the energy barrier height, or work function. The work function of commonly employed electron emitters such as Tungsten or Molybdenum is about 4.5 eV. Finding a material that will lower this work function barrier to an effective 0.2 eV would increase electron emission many orders of magnitude.

The search for such materials has a long history, going back at least as far as Thomas Edison. Many different materials have been found that can emit electrons effectively. Some of the materials that have been used include Cesium and Cesium Oxide. There is typically however, a fundamental conflict between the willingness of a material to lose an electron and the stability of the material. If the material is structured atomically such that the outer electrons are loosely bound and easily lost, the general result is that the atoms are loosely bound to each other and do not form a highly stable material. This is the case with both Cesium and Cesium Oxide. The poor stability manifests itself in a comparatively high free energy and hence a high vapor pressure. This allows the material to be highly mobile, generally to the detriment of any electrical device geometry. Materials with loosely bound electrons adsorbed on insulating surfaces tend to lead to low voltage surface breakdown. Gates, grids and miscellaneous electrodes with monolayers of low work function gases start to emit electrons to more positively charged locations. High vapor pressure materials can also lead to gaseous conduction mechanisms between the electrodes at different potentials. This conduction can ruin the functioning of a device and even lead to an avalanche condition which develops into an arc which then destroys the device structure.

For example, when Cesium is put into a vacuum structure, it eventually migrates everywhere. It spreads to surfaces where electron emission is not desired, and it can make insulating surfaces undesirably conductive. It also spreads to interelectrode spaces where it will provide a relatively easy path for gaseous conduction and electrical arcs. Diamond, in contrast, behaves as a

Table 1. Physical Properties and Comparisons [11–15].

Lattice Spacing	1.78 Angstroms (200) plane	Silicon = 1.92 Angstroms (022) plane
Density	3.5 g/cm^3	Boron 5 g/cm^3
Thermal Expansion	0.81×10^{-6}K	Silicon = 2.6×10^{-6}K Alumina = 7×10^{-6}K
Thermal Conductivity	20 W/cm K	Copper = 4 W/cm K
Hardness	9000 kg/mm^2	Cubic Boron nitride = 4500 kg/mm^2 Silicon carbide = 2500 kg/mm^2
Coefficient of Friction (against steel)	0.1	Steel/steel = 0.58
Melting Point	3800°C	Silicon = 1412°C
Refractive Index (590 nm)	2.42	Fused quartz = 1.46
Electron Mobility	2.0×10^3 cm^2/V sec	Gallium arsenide = 8.5×10^3 cm^2/V sec
Hole Mobility	1.6×10^3 cm^2/V sec	Silicon = 600 cm^2/V sec
Band Gap	5.42 eV	Silicon = 1.1 eV
Resistivity	10^{16} ohm-cm	Silicon nitride = 10^{13} ohm-cm
Dielectric Constant	5.5	Silicon dioxide = 3.9
Electrical Breakdown	10^7 V/cm	Silicon dioxide = 6×10^6 V/cm
Mass Density	3.5 gm/cm^3	Silicon = 2.3 gm/cm^3
Bandgap	5.47 eV	Silicon = 1.12 eV
Intrinsic Resistivity	$> 10^{15}\Omega$-cm	Silicon = $10^5\Omega$-cm
Breakdown Electric Field	$1 - 20 \times 10^6$ V/cm	Silicon = 3×10^5 V/cm
Electron Saturated Carrier Velocity	$1.5 - 2 \times 10^7$ cm/s	Silicon = 1×10^7 cm/s
Hole Saturated Carrier Velocity	1.05×10^7 cm/s	Silicon = 9×10^6 cm/s

refractory material that does not melt but rather sublimes at 3550°C with an associated low vapor pressure ($\ll 10^{-11}$ torr at 100°C). This combination of a low barrier to electron emission in an otherwise stable, robust material, has attracted a great deal of attention to diamond from electronics researchers.

3.2.3. *The work function and electron affinity of diamond*

Diamond possesses very strong bonding as demonstrated by its high melting temperature, low vapor pressure, extraordinary hardness and high thermal conductivity (20×10^3 W/m-K). Many of its properties are extraordinary.

Diamond has been shown to have a Negative Electron Affinity (NEA) [16, 17], that is, the conduction band is above the vacuum level, and a low effective work function around $0.2 \sim 0.3$ eV [18, 19]. The diamond crystal is a face centered cubic (fcc) with two atoms per primitive unit cell. The measured cubic edge is 3.567 A. Figure 2 shows the band diagram for diamond. Note in this diagram that the band gap is about 5.5 eV and the conduction band (E_c) is above the vacuum level(E_{vac}). The difference between the conduction band and the vacuum level is the Electron Affinity, χ. For diamond, the affinity is negative. This relationship (strictly for the (111) crystalline direction) [20] is such that an electron in the conduction band near the surface would "prefer" to leave the crystal. By leaving the crystal it would lower its energy and become the escaped electron that is desired. Note that electron affinity, χ,

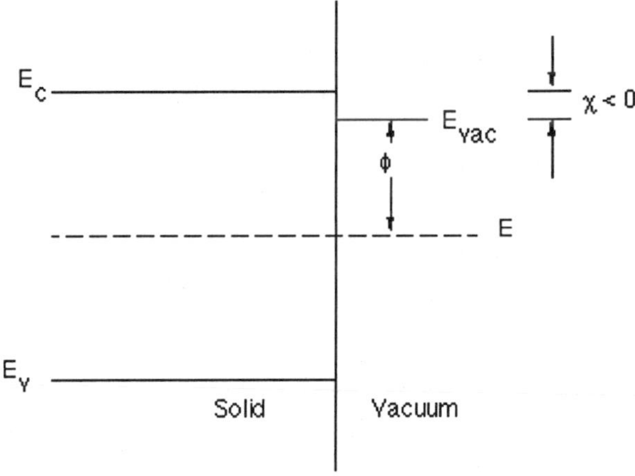

Fig. 2. The band diagram of a material with Negative Electron Affinity (NEA).

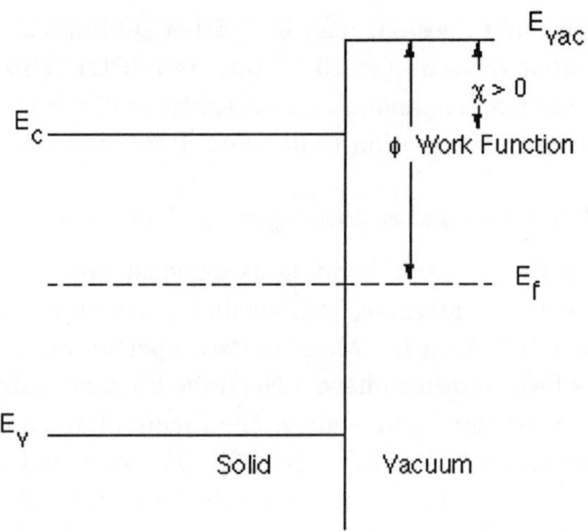

Fig. 3. Band diagram for a material with Positive Electron Affinity.

is not the work function, ϕ, which is the barrier presented to electrons at the Fermi energy level, E_f. This is a common misunderstanding. The work function must always be positive in order not to violate the Second Law of Thermodynamics. Figure 3 shows a band diagram for a more typical material with a positive electron affinity.

3.2.4. *The diamond structure*

Group IV elements such as carbon have half filled valence n-shells with the configuration ns^2p^2 where n is between 2 and 5. The "diamond structure" (in the same manner as Si and Ge) has the s and p states hybridizing to form tetrahedral sp^3 directed bonds. With this structure, diamond demonstrates strong bonding, extraordinary hardness, thermal conductivity, and a very high melting temperature. Figure 4 shows a drawing of the electron cloud charge density for carbon that illustrates the sp^3 hybridization that leads to the diamond lattice structure.

3.2.4.1. Diamond and graphitic phases of carbon

Carbon, mostly known as graphite or diamond, is a tetravalent atom that is polymorphic, existing in many metastable phases [21, 22]. Most of these phases still have no names. Several years ago, researchers identified giant spherical

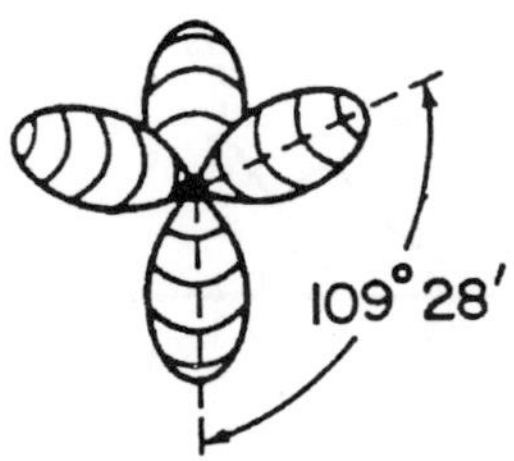

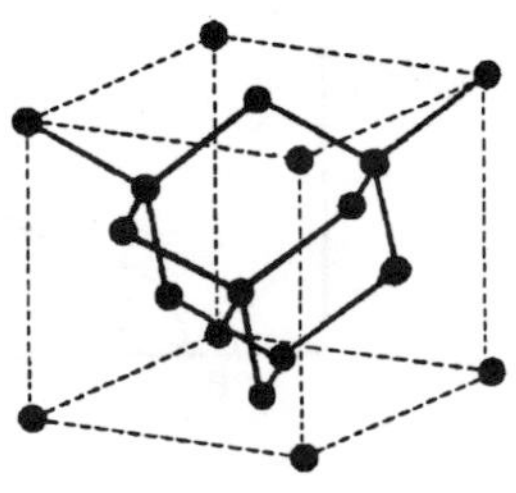

Fig. 4. Electron charge cloud density for carbon atom with sp^3 hybridization and crystal structure of diamond showing unit cell [41].

molecules containing 20, 28, 32, 60, 70, 900 or more carbon atoms (Buckminsterfullerenes) [23–25]. In the diamond phase, covalent bonds (sp^3) link all carbon atoms with four adjacent carbon atoms to form an isotropic, dense, face-centered cubic semiconducting crystal. In the graphitic phase the atoms bond (sp^2) strongly with three adjacent carbon atoms to form a two-dimensional hexagonal ring structure. Planes of atoms result, in an anisotropic semi-metallic structure. Figure 5 shows a drawing representing an electron charge cloud density for carbon with the sp^2 hybridization and the resulting highly anisotropic graphite crystal structure.

During the deposition process, as the graphitic sp^2 phase is increased at the expense of the otherwise sp^3 structure, preferred orientation can develop. The hexagonal graphitic axis will become associated with the triagonal diamond axis [26]. The spatial arrangements and interatomic distance of the carbon atom are very similar for the (111) diamond plane and the (0001) graphite

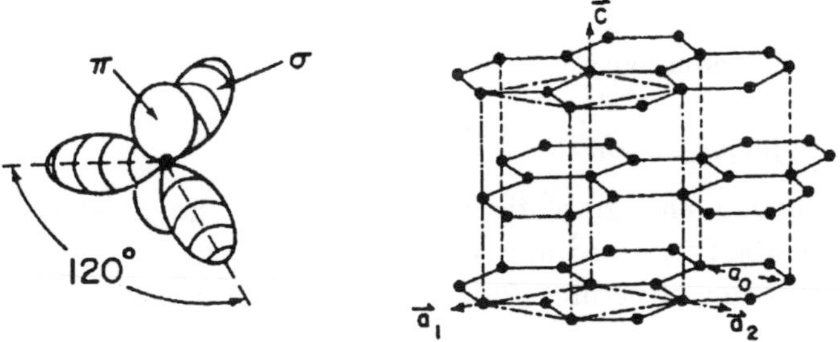

Fig. 5. Electron charge cloud density for carbon atom with sp^2 hybridization and crystal structure of graphite showing unit cell [42].

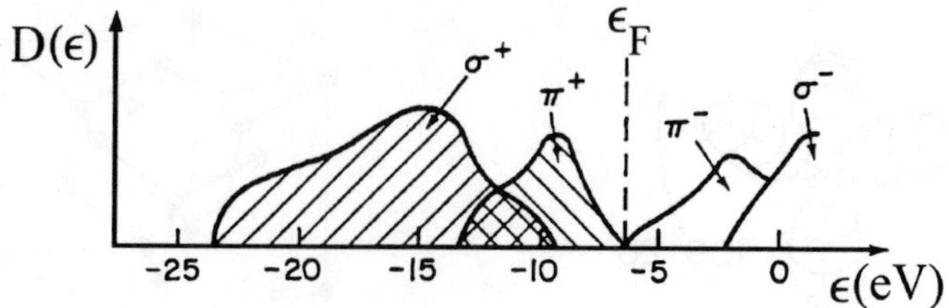

Fig. 6. Density of states for graphite, zero energy is taken at vacuum [43].

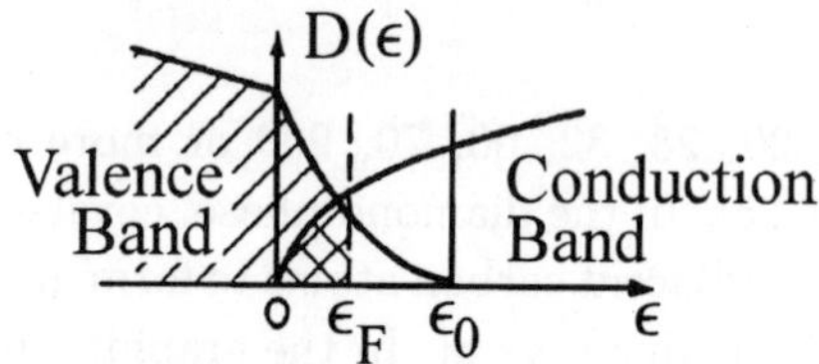

Fig. 7. Density of states for 3D graphite at region of band overlap, zero energy is taken at bottom of conduction (π^-) band [43].

plane. This introduction of hexagonal graphite basal plane intermediate to two diamond (111) planes causes a shift in position of the nearby carbon atoms and creates strain. This orientational effect is observed in diamond film and even in natural diamond [27]. Notice that in Fig. 6 the density of electronic states shows contact between the σ^+/π^+ and the σ^-/π^- bands. Figure 7 shows the density of states for graphite illustrating the conduction band and valence band structure. This is typical of a semimetal but unusual with such similar bonding to the dielectric phase of a single element. This gives us a element with two very similar phases that vary by 14 orders of magnitude in their electrical conductivity!

3.2.4.2. Amorphous carbon

Between diamond and graphitic phases lie many intermediate phases of amorphous carbon with sp^2 and sp^3 bonding resulting in 5, 6, and 7 (and possibly more) ring structures [28] intertwined with other structures. These materials are generally a cross-linked non-crystalline network of both sp^2 and sp^3 hybridized structure. The "phases" exhibit mixed properties which depend critically on the intermixed structure. Characterizing these materials is becoming increasingly complex as researchers seek to classify more structural forms. Figure 8 shows a ball and stick representation of a 64 atom amorphous tetrahedrally bonded carbon structure. This structure was calculated by energy considerations for the condensation of gaseous carbon atoms. These multi-phase films have also been shown to contain structures such as fullerenes [29] which have been linked through their very flexible band structure [30] to electron emission. These films can also be produced with a dispersion of crystallites of 100 Angstroms to several microns. These crystallites typically form in columnar grains that extend through the film normal to the surface. This structure [31], and the attendant grain boundaries can add greatly to the electrical performance and anisotropy. Figure 9 illustrates the density of states and the conduction and valence band for a single 2D sheet of graphite. This diagram ignores the quantum confinement that would be expected within this semimetal within a dielectric.

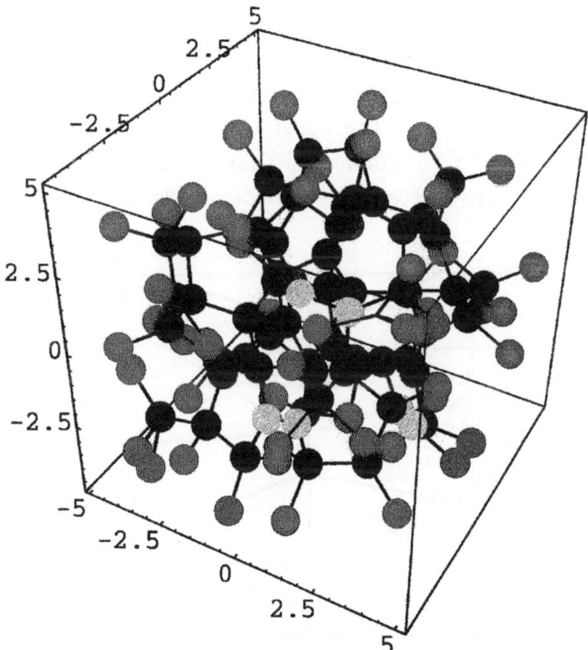

Fig. 8. Ball and stick representation of 64 atom α-tC network [44] (The light atoms depict 3-fold coordinated sp^2 carbon, the dark atoms are 4-fold coordinated sp^3 atoms and the medium gray are periodic replicas needed to complete coordination).

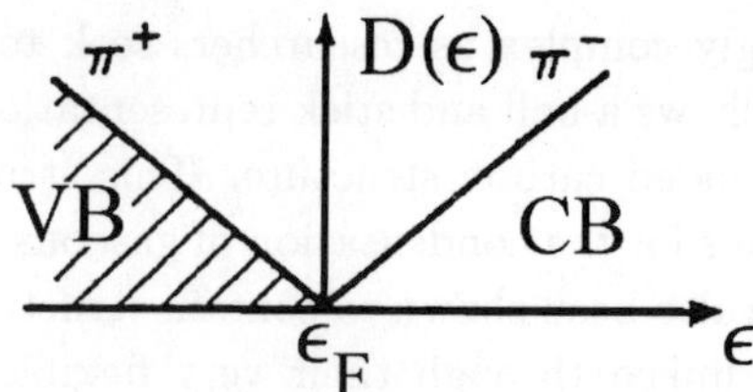

Fig. 9. Density of states for a single sheet of 2D graphite near the band touching region [43].

3.2.5. *Free carrier concentration and anomalous emission*

The widely varying electronic properties between the carbon phase of dielectric diamond and semimetallic graphite suggests great flexibility in tailoring band structures and properties. The Fig. 10 band diagram for diamond demonstrates some very valuable properties including the ability to emit electrons very easily. However, the current predicted by this band diagram is not very high. The free carrier concentration in a perfect diamond crystal is very low due to the high energy band gap. No one has yet demonstrated well-behaved n-type doping of single crystal diamond with a reasonably low dopant activation energy although groups around the world are working at it.

The primary source of free carriers in Diamond and DLC films comes from the graphitic phase. Many researchers have demonstrated what might be called

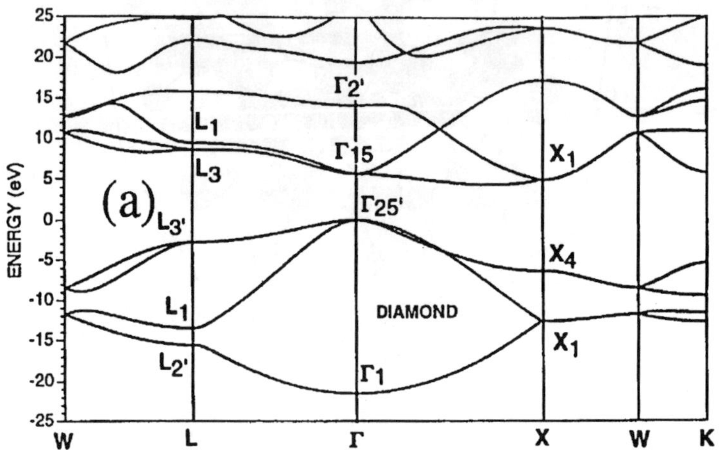

Fig. 10. The band diagram for diamond [45].

an "anomalous" emission current. This is an emission level much higher than is expected from mechanisms calculated from simple solid state models. There is much conjecture as to the cause of this anomalous emission. It has been proposed that this is related to hot carriers injected by ohmic contacts [32, 33], that it is related to surface states and surface conduction mechanisms [34], or that it comes about from a morphological state created by a propitious blending of the wide bandgap diamond structure with its cousin, the semimetallic graphitic structure.

This augmented emission is found in not only highly crystalline diamond but also in polycrystalline diamond and even diamond-like carbon (DLC). It is because this effect is found in material with poor crystalline quality that one can consider applying it over an entire flat panel display for electron emission. There are many techniques for depositing polycrystalline and DLC material over large areas. These include PECVD [35], RFCVD, laser ablation and even acetylene flame. Several of these methods are readily scalable to very large substrates.

3.2.6. *Emission barrier morphology*

The ease with which electrons can escape the diamond or DLC surface is readily observable in the laboratory [36, 37]. The emitted electrons can be used to study the material in trying to understand the emission mechanisms. Figure 11 shows an image taken from a DLC surface exposed to UV light. The UV photons provided enough energy to excite electrons into leaving the surface. Electron optics were used to collect these electrons and then focus them to create a highly magnified image. This shows the emission barrier morphology of the sample. Areas of low effective work function have high emission and appear as light regions in the image, while areas of high effective work function have low emission and appear as darker regions.

3.2.7. *Emitted current vs. applied fields*

Figure 12 shows a curve of emitted current vs. applied field for a DLC sample. Note that this is for a sample that does not have any apparent geometric field enhancement structure. It is always possible that there exists, within the film, an effective, but invisible enhancement geometry. The field shown in the drawing is a simple parallel plate field. The field required for emission turn-on is about three orders of magnitude lower than that required for more traditional emitter materials, such as the molybdenum which is typically used in a Spindt

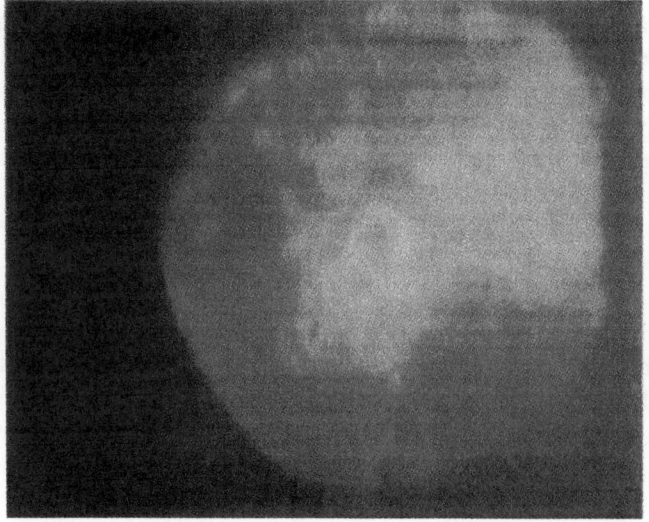

Fig. 11. Image from DLC surface exposed to UV light.

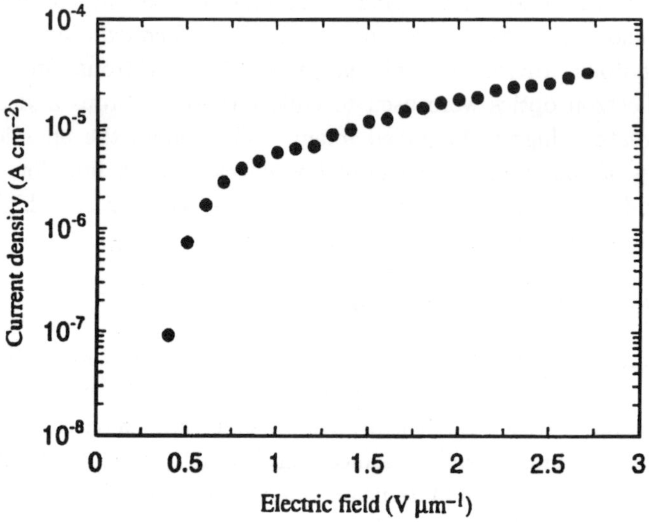

Fig. 12. Field emission current density vs. electric field [46].

tip field emitter. This emission at greatly lower field strength provides the display designer with the opportunity to make engineering trade-offs. In the Spindt structure described above, geometry was used to get a field enhancement of about 1.5 to 2 orders of magnitude. With an electrode that performs similarly at three orders of magnitude less, the geometric enhancement can be left out and it would still perform well at 1/10 the field, or possibly, for similar spacing, 1/10 the control voltage. The trade-off can be made between the low turn on voltage and producibility issues such as lithography requirements, power and of course, cost. These extra degrees of freedom can be used to much profit.

3.2.8. *Design advantages of a diamond-based field emission display*

Figure 13 shows a schematic view of a diamond based field emission display. Such structures, and similar ones, have been demonstrated in research labs around the world [38–40]. The advantages of such a structure are manifold. There is no lithography required beyond that necessary for the definition of the pixel. Thus, for example, in a display that is 20" on a side with a thousand pixels per line, the pixel size would be 0.020" or 0.5 mm. Even with color sub-pixels, this size is readily defined using printed circuit board techniques instead of sub micron steppers. The lithography requirements are loosened by a DLC emitter to the point that this is no longer a major cost issue. Sub-micron lithography systems are very sensitive to dust and particulates. With the minimization of lithography requirements goes a major particle sensitivity issue. The larger geometry that characterizes this type of device might even be fabricated with an ink-jet printer!

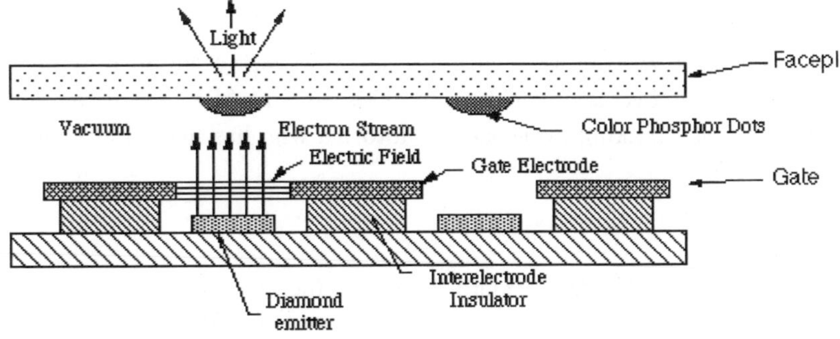

Fig. 13. Schematic view of diamond based FED.

The system designer can also trade lower voltage control for an increase in the interelectrode spacing. By increasing this distance, the sensitivity to particulate caused pinholes and visual defects is minimized. This increased spectrum of tradeoffs will allow display engineers to make the next generation of field emission displays with greater producibility and hence lower cost. Yield is an exceedingly critical component of cost. It affects direct cost, that is how many displays you make for every one you sell, and also affects indirect costs, such as how much floor space, level of clean rooms, and process equipment that you do not need to buy.

4. Conclusion

A planar emitter such as Diamond or DLC provides the next generation technology in electron emission after hot filaments and microscopically sharp tips. This type of emitter relieves many of the problems associated with the previous generations. They require neither the power nor the thermal burdens of a hot filament emitter, nor the lithographic and microstructural burdens of the sharp tip field emitter. The current ability to control the material's electronic properties is still relatively weak, although more than adequate electron emission is easily demonstrated. The industrial domestication of this property is still in a very early phase. This material will eventually allow engineers to develop flat panel displays that will be truly low in cost while providing the aesthetics of a high definition CRT at much lower power than a backlit AMLCD. This material technology is expected to accelerate the revolution in flat panel displays and eventually produce a true flat HDTV that hangs on the wall like a painting.

Acknowledgments

I could not have written this without the unwavering support of my wife Cynthia, and my two loving childeren, Kristen and Michael. I would also be remiss in not thanking everyone at Motorola who has helped at every step with this work.

References

[1] *Flat Information Display* Stanford Resources, Inc., 1997
[2] S. Matsumoto, *Electronic Display Devices* (John Wiley & Sons, 1984).
[3] P.W. Bridgman, *The Thermodynamics of Electrical Phenomena in Metal* (Macmillan, New York, 1934).

[4] K. Wandelt, The Local Work Function: Concept and Application, *Int. Vacuum Electron Sources Conf.* (Eindhoven, The Netherlands, 1996).

[5] N.D. Lang and W. Kohn, *Phy. Rev.* **B3**, 1215 (1971).

[6] That probabilistic wave mechanics predicts the possibility of tunneling was first recognized by Gamow who used it to explain alpha emission.

[7] R. Gomer, *Field Emission and Field Ionization* (Harvard University Press, Cambridge, Massachusetts, 1961).

[8] R.H. Fowler and L. Nordheim, *Proc. Roy. Soc.* **A119** (May 1928) p. 173.

[9] C.A. Spindt, I. Brodie, L. Humphrey and E.R. Westerberg, *J. of Applied Physics* **47**, 12 (1976) p. 5248.

[10] J.M. Kim, Field Emission Display Devices, *Proc. 3rd Int. Display Workshops* **2**, 123 (1996).

[11] D.L. Dreifus, Passive Diamond Electronic Devices, Diamond: Electronic Properties and Applications, eds. L.S. Pan and D.R. Kania (Kluwer Academic Publishers, 1995).

[12] I.M. Buckley–Golder and A.T. Collins, Active Electronic Applications For Diamond, *Dia. Rel. Mat.* **1**, 1083–1101 (1992).

[13] S.M. Sze, *Physics of Semiconductor Devices* (John Wiley and Sons, 1981).

[14] R.J. Trew, J.B. Yan and P.M. Mock, The Potential Of Diamond and SiC Electronic Devices for Microwave and Millimeter-Wave Power Applications, *Proc. IEEE* **79**, 5 (1991) p. 598.

[15] *Diamond Thin Films*, SRI International, 1995.

[16] B.B. Pate, W.E. Spicer, T. Ohta and I. Lindau, *J. Vac. Sci. Technol.* **17**, 1087 (1980).

[17] S.P. Bozeman *et al.*, Electron Emission Measurements from CVD Diamond Surfaces, Diamond Films 95, Barcelona Spain (1995).

[18] C. Xie, C.N. Potter, R.L. Fink, C. Hilbert, A. Krishnan, D. Eichman, N. Kumar, H.K. Schmidt, M.H. Clark, A. Ross, B. Lin, L. Fredin, B. Baker, D. Patterson and W. Brookover, *Int. Vacuum Microelectronics Conf. Proc.* **271**, 229 (1994).

[19] N. Eimori, K. Maeashi, A. Atta, T. Ito and A. Hiraki, *3rd Int. Symp. on Diamond Materials*, eds. J.P. Dismukes, K.V. Ravi, K.E. Spear, B. Lux and N. Setaka (The Electrochemical Society, Honolulu, Hawaii, USA, 1993) p. 934.

[20] S. Iarlori *et al.*, *Phys. Rev. Lett.* **69**, 20 (16 November 1992) p. 2947.

[21] D.A. Drabold, P. Stumm and P.A. Fedders, *Phys. Rev.* **B34**, 16415 (1994).

[22] J.S. Nelson *et al.*, *Phys. Rev.* **B52**, 9354 (1995).

[23] H.W. Kroto, J.R. Heath, S.C. O'Brien, R.F. Curl and R.E. Smalley, *Nature* **318**, 1744 (1985).

[24] T.W. Ebbesen, *Physics Today* (June 1996) p. 26.

[25] G. Galli and F. Mauri, *Phys. Rev. Lett.*, **73**, 25 (19 December 1994) p. 3471.

[26] P. Gonon, Y. Boiko, S. Prawer and D. Jamieson, *J. Appl. Phys.* **79**, 7 (1 April 1996) p. 3778.

[27] S. Jou, H. Doerr and R. Bunshah, *Thin Solid Films* **253**, 95 (1994).

[28] N.A. Marks *et al.*, *Phys. Rev. Lett.* **76**, 5 (29 January 1996) p. 768.

[29] V. Paillard *et al.*, *Phys. Rev. Lett.* **71**, 4170 (1993).

[30] Ph. Lambin, A. Fonseca, J.P. Vigneron, J.B. Nagy and A.A. Lucas, *Chemical Physics Letters* **245**, 85 (1995).

[31] A. DeVita, G. Galli, A. Canning and R. Car, *Nature* **379**, 523 (1996).

[32] J.C. Twichell, M.W. Geis, T.M. Lyszczaz, K.E. Krohn, N.N. Efremow and C.M. Marchi, Diamond Cold Cathodes, *6th European Conf. on Diamond, Diamond-like and Related Materials* (Barcelona, Spain, 1995) (in press).

[33] P. Lerner, P.H. Cutler and N. Miskovsky, Hot Electron and Quasiballistic Transport of Nonequilibrium Electrons in Diamond Thin Films, *9th Int. Vacuum Microelectronics Conf.* (St. Petersburg, Russia, 1996).

[34] P. Gonon, Y. Boiko, S. Prawer and D. Jamieson, *J. Appl. Phys.* **79**, 7 (1 April 1996) p. 3778.

[35] S. Jou, H. Doerr and R. Bunshah, *Thin Solid Films* **253**, 95 (1994).

[36] M.W. Geis, J.C. Twichell, T.M. Lyszczarz, K.E. Krohn, N.N. Efremow and C.M. Marchi, *Int. Vacuum Electron Sources Conf.* (Eindhoven, The Netherlands, 1996).

[37] E.I. Givargizov *et al.*, *Applied Surface Science* (September 1995) p. 1.

[38] A.S. Kupryashkin *et al.*, Graphite Field Electron Array for Flat-Panel Flourescent Display, *Technical Digest of Int. Vacuum Microelectronics Conf. 91* (Nagahama, Japan ,1991).

[39] A.Y. Tcherepanov, A.G. Chakhovskoi, V.B. Sharov, *J. Vac. Sci. Technol.* **B13**, 2 (March/April 1995) p. 482.

[40] N.S. Xu, Y. Tzeng and R.V. Latham, *J. Phys. D: Appl. Phys.* **27**, (1994) 1988.

[41] P.L. Walker Jr. and P.A. Thrower, *Chemistry and Physics of Carbon* **16** (Marcel Dekker Inc., New York, 1981) p. 126.

[42] P.L. Walker Jr. and P.A. Thrower, *Chemistry and Physics of Carbon* **16** (Marcel Dekker Inc., New York, 1981) p. 127.

[43] P.L. Walker Jr. and P.A. Thrower, *Chemistry and Physics of Carbon* **16** (Marcel Dekker Inc., New York, 1981) p. 128.

[44] J.S. Nelson *et al.*, *Phys. Rev.* **B52**, 9354 (1995).

[45] L.S. Pan and D.R. Kania, *Diamond: Electronic Properties and Applications* (Kluwer Academic Publishers, 1995) p. 6.

[46] Ken Okano *et al.*, *Nature* **381** (1996) p. 140.

Subject Index

Materials Index